T0297056

Advances in Intelligent Systems and Computing

Volume 427

Series editor

Janusz Kacprzyk, Polish Academy of Sciences, Warsaw, Poland
e-mail: kacprzyk@ibspan.waw.pl

About this Series

The series "Advances in Intelligent Systems and Computing" contains publications on theory, applications, and design methods of Intelligent Systems and Intelligent Computing. Virtually all disciplines such as engineering, natural sciences, computer and information science, ICT, economics, business, e-commerce, environment, healthcare, life science are covered. The list of topics spans all the areas of modern intelligent systems and computing.

The publications within "Advances in Intelligent Systems and Computing" are primarily textbooks and proceedings of important conferences, symposia and congresses. They cover significant recent developments in the field, both of a foundational and applicable character. An important characteristic feature of the series is the short publication time and world-wide distribution. This permits a rapid and broad dissemination of research results.

More information about this series at http://www.springer.com/series/11156

Ajith Abraham · Katarzyna Wegrzyn-Wolska
Aboul Ella Hassanien · Vaclav Snasel
Adel M. Alimi
Editors

Proceedings of the Second International Afro-European Conference for Industrial Advancement AECIA 2015

Editors
Ajith Abraham
Scientific Network for Innovation and Research Excellence
Machine Intelligence Research Labs (MIR Labs)
Auburn, WA
USA

Katarzyna Wegrzyn-Wolska
Membre du Groupe Efrei
ESIGETEL
Villejuif
France

Aboul Ella Hassanien
Faculty of Computers and Information
Cairo University
Giza
Egypt

Vaclav Snasel
VŠB-Technical University of Ostrava
Ostrava
Czech Republic

Adel M. Alimi
ENIS
University of Sfax
Sfax
Tunisia

ISSN 2194-5357 ISSN 2194-5365 (electronic)
Advances in Intelligent Systems and Computing
ISBN 978-3-319-29503-9 ISBN 978-3-319-29504-6 (eBook)
DOI 10.1007/978-3-319-29504-6

Library of Congress Control Number: 2015960822

Printed on acid-free paper

This Springer imprint is published by SpringerNature
The registered company is Springer International Publishing AG Switzerland

Preface

This volume of Advances in Intelligent Systems and Computing contains papers presented at the 2nd International Afro-European Conference for Industrial Advancement—AECIA 2015. The conference aimed at bringing together the foremost experts and excellent young researchers from Africa, Europe, and the rest of the world to disseminate the latest results from various fields of engineering, information, and communication technologies. The topics, discussed at the conference, covered a broad range of domains spanning from ICT and engineering to prediction, modeling, and analysis of complex systems. The 2015 edition of AECIA featured a distinguished special track on prediction, modeling, and analysis of complex systems—Nostradamus, and special sessions on advances in image processing and colorization and data processing, protocols, and applications in wireless sensor networks.

The second edition of AECIA was hosted by the AllianSTIC Laboratory at the Engineering School of Digital Sciences (ESIGETEL), Villejuif (Paris-sud), France, and took place between 9 and 11 of September. The conference was attended by participants from 25 countries and featured two plenary lectures, delivered by renown experts in their fields, Prof. Andries Engelbrecht, University of Pretoria, and Prof. Adel Alimi, University of Sfax. Two tutorials by Dr. Milos Kudelka, VŠB-Technical University of Ostrava, and Dr. Zeineb Chelly, High Institute of Management of Tunis, were held during the meeting.

The organizing committee expresses deep gratitude to the members of the International Program Committee, the hosting institution, ESIGETEL, local officials, the publisher of this volume and to all supporters and authors, for their impressive efforts and contribution to the success of the events.

September 2015

Ajith Abraham
Katarzyna Wegrzyn-Wolska
Aboul Ella Hassanien
Vaclav Snasel
Adel M. Alimi

Organization

Conference Chairs

Ajith Abraham, Machine Intelligence Research Labs, USA
Václav Snášel, VŠB-Technical University of Ostrava, Czech Republic
Adel M. Alimi, University of Sfax, Tunisia

Program Committee Chairs

Katarzyna Wegrzyn-Wolska, ESIGETEL, France
Ivan Zelinka, VŠB-Technical University of Ostrava, Czech Republic
Aboul Ella Hassanien, Cairo University, Egypt

Publicity Co-Chairs

Jan Platoš, VŠB-Technical University of Ostrava, Czech Republic
Habib M. Kammoun, University of Sfax, Tunisia
Pavel Krömer, VŠB-Technical University of Ostrava, Czech Republic
Khmaies Quahada, University of Johannesburg, South Africa

Conference Organizers

Hussein Soori, VŠB-Technical University of Ostrava, Czech Republic
Jakub Stolfa, VŠB-Technical University of Ostrava, Czech Republic
Svatopluk Stolfa, VŠB-Technical University of Ostrava, Czech Republic
Sylva Stefanisinova, VŠB-Technical University of Ostrava, Czech Republic
Elisabeth Colin, ESIGETEL, France
Catherine Marechal, ESIGETEL, France
Nicolas Sicard, ESIGETEL, France

International Program Committee

Heba Abdel-Atty, Port Said University, Egypt
Mohamed Abdel-Azim, Mansoura University, Egypt
Janos Abonyi, University of Pannonia, Hungary
Ajith Abraham, MIR Labs, USA
Ahmed Afifi, Menoufia University, Egypt
Muhanned Alfarras, Gulf University, Bahrajn
Ahmed Ali, Suez Canal University, Egypt
Carlos Henggeler Antunes, University of Coimbra, Portugal
Chaouki Aouiti, REGIM-Lab, Research Groups on Intelligent Machines, Tunisia
Eshetie Berhan, Addis Ababa Institute of Technology (AAiT), Ethiopia
Birhanu Beshah, Addis Ababa Institute of Technology (AAiT), Ethiopia
João P.S. Catalão, University of Beira Interior, Portugal
Hilmi Berk Celikoglu, Technical University of Istanbul, Turkey
Jiri Dvorsky, VŠB-Technical University of Ostrava, Czech Republic
Nashwa El-Bendary, SRGE (Scientific Research Group in Egypt), Egypt
Eid Emary, Cairo university, Egypt
Tarek Gaber, VŠB-Technical University of Ostrava, Czech Republic
Aboul Ella Hassanien, Cairo University, Egypt
Dusan Husek, Institute of Computer Science, Academy of Sciences of the Czech Republic
Konrad Jackowski, Wroclaw University of Technology, Poland
David Jezek, VŠB-Technical University of Ostrava, Czech Republic
Habib M. Kammoun, University of Sfax, Tunisia
Abeer Khalil, Qassim Private Colleges, Saudi Arabia
Daniel Kitaw, Addis Ababa Institute of Technology (AAiT), Ethiopia
Jan Kozusznik, VŠB-echnical University of Ostrava, Czech Republic
Bartosz Krawczyk, Wroclaw University of Technology, Poland
Pavel Kromer, VŠB-Technical University of Ostrava, Czech Republic
Mohamed Mostafa, VŠB-Technical University of Ostrava, Czech Republic
Santosh Nanda, Eastern Academy of Science and Technology, Odisha, India

Eliska Ochodkova, VŠB-Technical University of Ostrava, Czech Republic
Cyril Onwubiko, United Kingdom
Rasha Osman, Imperial College London, United Kingdom
Ahmed El Oualkadi, ENSA Tangier, Morocco
Nour Oweis, VŠB-Technical University of Ostrava, Czech Republic
Marcin Paprzycki, IBS PAN and WSM, Poland
Jan Platos, VŠB-Technical University of Ostrava, Czech Republic
Sg Ponnambalam, Monash University Sunway Campus, Malaysia
Petrica Pop, North University of Baia Mare, Romania
Radu-Emil Precup, Politehnica University of Timisoara, Romania
Khmaies Quahada, University of Johannesburg, South Africa
Nizar Rokbani, ISSAT, University of Sousse & Regim-lab, Tunisia
Mohammed Salem, SRGE (Scientific Research Group in Egypt), Egypt
Petr Saloun, VŠB-Technical University of Ostrava, Czech Republic
Noura Semary, SRGE (Scientific Research Group in Egypt), Egypt
Vaclav Snasel, VŠB-Technical University of Ostrava, Czech Republic
Tinus Stander, University of Pretoria, South Africa
Tammo Steenhuis, Cornell University, USA
Svatopluk Stolfa, VŠB-Technical University of Ostrava, Czech Republic
Thomas Stuetzle, IRIDIA, ULB, Belgium
Theo G. Swart, University of Johannesburg, Republic of South Africa
Alaa Tharwat, Suez Canal University, Egypt
Eiji Uchino, Yamaguchi University, Japan
Juan Velasquez, University of Chile, Chile
Michal Wozniak, Wroclaw University of Technology, Poland
Ivan Zelinka, VŠB-Technical University of Ostrava, Czech Republic
Ahmed Zobaa, Brunel University, UK

Sponsoring Institutions

AllianSTIC Laboratory, Engineering School of Digital Sciences (ESIGETEL), Paris, France
Department of Computer Science, VŠB-Technical University of Ostrava, Ostrava, Czech Republic
Research Groups on Intelligent Machines (REGIM Lab), University of Sfax, Sfax, Tunisia

Contents

Part I
Tutorials

Analysis of Significance and Evolution of Co-authorship Networks

Milos Kudelka

Abstract In co-authorship networks analysis, we may find many interesting tasks. The goals of networks analysis are: studying network structure, detecting communities and their evolution over time, identifying significant authors and relations, and many others. This tutorial is focused on analytical approaches applicable on analysis of co-authorship networks. Mainly three different directions are presented; weighting authors and relations over time, detection and evolution of small communities, and analysis of the evolution of research topics of individual authors.

Co-authorship networks belong to so-called social networks. Technically, we can represent a social network as a graph $G = (N, E)$ that defines a structure of nodes (actors) and edges (ties). Social networks are complex networks; it means they have non-trivial topological features that are neither purely regular nor purely random. The graph can be weighted, unweighted, directed and undirected. Moreover, there is plenty of semantics in the background of social networks; primarily some kinds of actors, interactions, relations, content and time aspects.

We can find some on-line solutions in the area of analytical tools focused on co-authorship network. *Arnetminer*[1] is designed to perform data mining and social network analysis to identify connections between researchers. *Microsoft Academic Search*[2] is a search engine for academic area developed by Microsoft Research. We present approaches used in our on-line tool *Forcoa.net*.[3] The system [1] measures and identify the significance of authors and evolution over time, small communities, and research topics of individual authors.

[1] http://aminer.org.
[2] http://academic.research.microsoft.com.
[3] http://www.forcoa.net.

M. Kudelka (✉)
IT4Innovations & Department of Computer Science, VŠB - Technical University of Ostrava, Ostrava, Czech Republic
e-mail: milos.kudelka@vsb.cz

A. Abraham et al. (eds.), *Proceedings of the Second International Afro-European Conference for Industrial Advancement AECIA 2015*, Advances in Intelligent Systems and Computing 427, DOI 10.1007/978-3-319-29504-6_1

Evolution and Weighting of Authors and Relations We proposed a simple version of the forgetting function inspired by Ebbinghauss *Forgetting Curve* [2]. The function is used for weighting nodes and edges in social networks. We can interpret the weight as a stability over time. The behavior of the function is simple and is based on the time of the repeated interaction and on the previous value of the stability [3]. As a result, the final value of stability includes the development of previous learning and forgetting. Following the stability, we can describe the publication trend of an author or his/her relationship (co-authorship) and the value of stability at a certain point in time then shows how stable (significant) the author or his/her relationship is.

Dependency and Community Detection Communities are a natural grouping element of social networks. We proposed a new local detection algorithm in [4]. The approach is similar to other locally-oriented methods that use so-called *Local Expansion*. The difference lies in the assessment of the membership of nodes in communities and also its applicability to weighted networks. Instead assessment of modularity, the membership is based on an analysis of dependency [5], which is a structural-weighted relation between two nodes.

Topic Evolution For each author or relationship between authors we are also able to identify topics which can describe this author or relation. We aggregate the topics from his/her papers and articles and apply the forgetting function. The topics of the relation between authors are detected as the topics of the joint papers of these authors. In [6], we presented a description of evolution of the research topic, including its visualization in the Forcoa.net system.

Acknowledgments This work was supported by the IT4Innovations Centre of Excellence project (CZ.1.05/1.1.00/02.0070), funded by the European Regional Development Fund and the national budget of the Czech Republic via the Research and Development for Innovations Operational Programme and by Project SP2015/146 "Parallel processing of Big data 2" of the Student Grant System, VSB - Technical University of Ostrava.

References

1. Horak, Z., Kudelka, M., Snasel, V., Abraham, A., Rezankova, H.: Forcoa.net: an interactive tool for exploring the significance of authorship networks in dblp data. In: Computational Aspects of Social Networks (CASoN), IEEE, pp. 261–266 (2011)
2. Ebbinghaus, H.: Memory: a contribution to experimental psychology. Ann. Neurosci. **20**(4), 155 (2013)
3. Kudelka, M., Horak, Z., Snasel, V., Abraham, A.: Social network reduction based on stability. In: Computational Aspects of Social Networks (CASoN), IEEE, pp. 509–514 (2010)
4. Zehnalova, S., Kudelka, M., Snasel, V.: Comparing two local methods for community detection in social networks. In: Computational Aspects of Social Networks (CASoN), IEEE, pp. 155–160 (2012)
5. Kudelka, M., Zehnalova, S., Horak, Z., Kromer, P., Snasel, V.: Local dependency in networks. Int. J. Appl. Math. Comput. Sci. **25**(2), 281–293 (2015)
6. Zehnalova, S., Horak, Z., Kudelka, M., Snasel, V.: Evolution of author's topic in authorship network. In: Advances in Social Networks Analysis and Mining (ASONAM), IEEE, pp. 1207–1210 (2012)

Data Pre-processing Based on Rough Sets and the Link to Other Theories

Zeineb Chelly

Abstract Data reduction is a main point of interest across a wide variety of fields. In fact, focusing on this step is crucial as it often presents a source of significant data loss. Many techniques were proposed in literature to achieve the task of data reduction. However, most of them tend to destroy the underlying semantics of the features after reduction or require additional information about the given data set for thresholding. Thus, this tutorial will be focused on presenting Rough Set Theory (RST) as a technique that can on the one hand reduce data dimensionality using information contained within the data set and on the other hand capable of preserving the meaning of the features. RST can be used as such tool to discover data dependencies and to reduce the number of attributes contained in a data set using the data alone, requiring no additional information. Basically, two main points will be discussed. First, presenting RST as a data pre-processing technique and, second, the link of RST to other theories; mainly to Fuzzy Set Theory.

Keywords Rough set theory · Feature selection · Pattern discovery

1 Content

Data reduction is a main point of interest across a wide variety of fields. In fact, focusing on this step is crucial as it often presents a source of significant data loss. Many techniques were proposed in literature to achieve the task of data reduction and they can be categorized into two main heads; techniques that transform the original meaning of the features, called the "transformation-based approaches", and the second category is a set of semantic-preserving techniques known as the "selection-based approaches" [3].

Z. Chelly (✉)
High Institute of Management of Tunis, LARODEC, University of Tunis, Tunis, Tunisia
e-mail: zeinebchelly@yahoo.fr

A. Abraham et al. (eds.), *Proceedings of the Second International Afro-European Conference for Industrial Advancement AECIA 2015*, Advances in Intelligent Systems and Computing 427, DOI 10.1007/978-3-319-29504-6_2

Transformation based approaches, also called "feature extraction approaches", involve simplifying the amount of resources required to accurately describe a large set of data. Feature extraction is a general term for methods that construct combinations of variables to represent the original set of features but with new variables while still describing the data with sufficient accuracy. The transformation based techniques are employed in situations where the semantics of the original database will not be needed by any future process. In contrast to the semantics-destroying dimensionality reduction techniques, the semantics-preserving techniques, also called "feature selection techniques", attempt to retain the meaning of the original feature set. The main aim of this kind of techniques is to determine a minimal feature subset from a problem domain while retaining a suitably high accuracy in representing the original features. In this talk, I will mainly focus on the use of a feature selection technique, instead of a feature extraction technique. Yet, it is important to mention that most feature selection techniques proposed in the literature suffer from some limitations. Most of these techniques involve the user for the task of the algorithms parameterization and this is actually seen as a significant drawback. Some feature selectors require noise levels to be specified by the user beforehand, some simply rank features leaving the user to choose their own subset. There are those that require the user to state how many features are to be chosen, or they must supply a threshold that determines when the algorithm should terminate. All of these require the user to make a decision based on its own (possibly faulty) judgment. To overcome the shortcomings of the existing methods, it would be interesting to look for a method that does not require any external or additional information to function appropriately. Rough Set Theory (RST) [1], which will be deeply explained in this tutorial, can be used as such tool.

RST was combined with other theories such as probability theory, fuzzy set theory and belief function theory. As each of these theories has inherent characteristics, works tended to explore the effectiveness of the mentioned theories to handle specific kinds of problems. For instance, both rough sets and the probability theory were hybridized in order to tackle problems dealing with incomplete data, and to solve the same kind of problems rough set theory was combined with the belief function theory. Another combination of rough sets was its hybridization with fuzzy sets which gave rise to the Fuzzy-Rough Set Theory (FRST) [2, 4]. The latter was, also, dedicated to handle incomplete data sets. Besides, it is applied to estimate the missing values in the learning process. FRST was first of all proposed as an extension to the rough set theory. This will be discussed in this tutorial as well.

References

1. Pawlak, Z., Polkowski, L., Skowron, A.: Rough set theory. Wiley Encyclopedia of Computer Science and Engineering (2008)
2. Jensen, R., Shen, Q.: Fuzzy-rough sets assisted attribute selection. IEEE Trans. Fuzzy Syst. **15**, 73–89 (2007)

3. Jensen, R., Shen, Q.: Semantics-preserving dimensionality reduction: rough and fuzzy rough-based approaches. IEEE Trans. Knowl. Data Eng. **17**(1), 1 (2005)
4. Dubois, D., Prade, H.: Putting rough sets and fuzzy sets together. In: Intelligent Decision Support: Handbook of Applications and Advances of the Rough Sets Theory, vol. 11, pp. 203–232. Springer (1992)

Part II
General Track

Checking the Compliance of Business Processes and Business Rules Using OWL 2 Ontology and SWRL

Tuan Anh Pham and Nhan Le Thanh

Abstract Business process modeling has become a popular method for improving the efficiency and quality of enterprise processes. Given the increasing complexity of the enterprise business processes and the richness of modeling languages, the ability to automatically validate a process model is a significant feature provided by modeling tools. This paper proposes an ontology-based approach for modeling and integrating business processes and business rules. It integrates a Colored Petri Nets-based ontology and also a business rule ontology. The integration is automated to facilitate the early detection of flaws and errors and to check the compliance of the business processes and the business rules.

Keywords Colored petri net · Business process validation · Business rules · OWL 2 · SWRL

1 Introduction

Business Process Management (BPM) has become more and more important with the methods, techniques and software developed to design, control, and analyze operational processes involving humans, organizations, applications, documents and other sources of information [1]. Organizations have understood the cost effectiveness of BPM that emerges from flaw detection and automation of business processes [2, 3].

To be able to validate a business process, one of the challenges is to check the compliance of business process with a set of business rules and to create the correspondence between them automatically. The other challenge is that the busi-

T.A. Pham (✉) · N. Le Thanh
WIMMICS - INRIA Sophia Antipolis, 2004 Route des Lucioles, 06902 Valbonne, France
e-mail: tuan-anh.pham@inria.fr

N. Le Thanh
e-mail: nhan.le-thanh@inria.fr

A. Abraham et al. (eds.), *Proceedings of the Second International Afro-European Conference for Industrial Advancement AECIA 2015*, Advances in Intelligent Systems and Computing 427, DOI 10.1007/978-3-319-29504-6_3

ness process and the business rules may be modified during the runtime because of changing in enterprise's policies. Therefore, the challenges for system designers are to build a flexible intelligent system which accepts and verifies the change on the business process and the business rules automatically. It can also work as a rule-based system which can reason and deduce a new knowledge or new decision based on a set of rules and facts.

In this paper, an ontology-based approach for business process modeling and checking is proposed, it gives full play to the excellence of ontology's semantic representation ability and automation reasoning. The business process model has been designed by using Coloured Petri Net (CPN) [4] and translated into a set of axioms in Business Process Ontology (BPO). The business rules have been also translated into a set of axioms in Business Rules Ontology (BRO). After finishing the translation, the consistency of the business process model and business rule has been validated by a reasoner. The system can detect flaws automatically either at design time or during runtime. If the user designs or modifies a business process and this action causes a conflict with the business rules, the system will notify automatically.

Our contributions in this paper are:

- An ontology-based method for modeling business process and business rule.
- Proposing an ontology-based method for integrating the business process and the set of business rules.
- Proposing a method for checking the compliance of business processes with business rules automatically during design time and runtime.

The rest of the paper is organized as follows. Section 2 analyses the differences with previous works. Section 3 presents the methodology of building BPO, BRO, and the technique for integrating BPO and BRO. Section 4 presents some discussions about the addressed topics, some conclusions and future research directions.

2 Related Works

Ensuring compliance of business processes is a challenging task: the number and complexity of business rules is increasing and the rules are subject to constant change. This calls for a holistic approach to manage compliance. Becker et al. define business process compliance management (BPCM) as "the steady modeling, refinement, and analysis of business processes regarding the fulfillment of regulatory compliance" [5]. Ramezani et al. outline the interdependencies between BPM and Compliance Management (CM) and describes CM as a "methodology to elicit, specify and formalize, implement, check and analyze, and optimize compliance requirements in organizations" [6].

Works on Business process compliance have focused on examining whether a given process model is compliant with a certain reference model/pattern. On the technical aspect, the business process pattern initiative has identified various patterns for the specification of control-flow [7], data-flow [8], and resources [8] in business process management systems. The work in [9] deals also with the planning layer by formalizing process patterns using UML concepts. These compliance works have focused on the structural level of process models, while another line of works focuses on the combination of data and structure [10–14]. The frameworks in [14–16], for example, provide general compliance criteria for assessing the compliance of processes with semantic constraints. In addition, some compliance works aimed at supporting specific purposes, for example: correcting process models at design time [15], verifying changes in existing models [15], identifying compliance in the context of process mining [16], and identifying violations of execution order compliance rules [10].

In our approach, we take advantage of CPN's color set for checking the compliance of processed data in data-flow with the constraints in Business Rule Ontology. A color set can be defined in many types: int, string or object. When firing a transition, the value of each color can be changed, the new value must respect to the constraint in BRO. This work will be explained in more detail in Sect. 3.

On the other hand, for representing the business rules, some works use an ontology approach [17–20]. They translate the Semantic of Business Vocabulary and Rule (SVBR) [21] vocabulary to OWL [22] and Semantic Web Rule Language (SWRL) [23], they provide the mapping or the rule in order to translate each property of SVBR (the definition of OMG) to a set of axioms in an ontology. In our approach, we classify the business rule into three main type of business rules, and we use Attempto Controlled English (ACE) [24] for defining the business rule, each ACE phrase will be translated into an axiom in BRO. We consider not only the business rule representation aspect but also the compliance of business process with a set of business rules. The advantage of our approach is to ensure that the business process is well-defined at design time and executed correctly during runtime. We must consider this aspect because a business process must always respect the rules that are defined inside the BRO. Therefore, the difference between our work and the previous ones, is that we allow the system to check not only the consistency of business process and business rule but also the consistency of the integration between them.

3 Modeling Business Process Ontology and Business Rule Ontology

In this section, we introduce two methods of building BPO and BRO, this method helps the user to represent a business process and a set of business rules by two ontologies. The advantage of this method is to allow the user to check the consistency of business rule and business process and also verify the compliance of business process with the business rules automatically by the reasoning.

3.1 Modeling Business Process Ontology

As mentioned above, we use Color Petri Nets for modeling business processes;

1. $CPN \sqsubseteq \geqq 1 hasPlace.Place \sqcap \geqq 1 hasTransition.Transition \sqcap \geqq 1\ hasInputArc.InputArc \sqcap \geqq 1 hasOutputArc.OutputArc$
2. $Place \sqsubseteq \geqq 0 hasToken.Token \sqcap \geqq 1 hasArc.(InputArc \sqcap OutputArc)$
3. $Transition \sqsubseteq \geqq 1 hasInputArc.InputArc \sqcap \geqq 1 hasOutptArc.OutputArc \sqcap \leqq 1 hasGuardFunction.Expression$
4. $InputArc \sqsubseteq 1 hasSourcePlace.Place \sqcap = 1 hasTargetTransition.Transition \sqcap \leqq 1 hasExpression.Expression$
5. $OutputArc \sqsubseteq = 1 hasTargetPlace.Place \sqcap = 1 hasSourceTransition.Transition \sqcap \leqq 1 hasExpression.Expression$

CPN is the concept for representing all CPN graphs. A CPN graph is well defined if and only if it has at least one place, one transition, one input arc and output arc. We define in BPO following classes: *CPN, Place, Transition, OutputArc, InputArc* and some properties which define the relations between them, *hasPlace, hasTransition, hasInputArc, hasOutputArc.* Place represents the properties of place, we define a concept *Place.* A place may have a token or not, it has also at least one *InputArc* or one *OutputArc.* The concept *Transition* is defined for all transitions. A transition must have at least one *InputArc* and one *OutputArc*. It's one of the minimum conditions for having a well-defined CPN graph. A transition may have only one guard function or not. The concept *InputArc* defined for all input arcs. An input arc has only one source place and one target transition. It may be marked by only one expression or not. An *OutputArc* has only one source transition and only one target place. It also may have only one expression or not.

3.2 Building Business Rule Ontology

Business rules are increasing parts of the system specifications used to design enterprise information systems. Business rules can be visualized as system requirement statements. A statement of a business rule falls into one of three categories:

1. Business terms and the relation between them
   It can be defined as a business rule that describes how people think and talk about things.
2. Activity Constraints

We define a set of actions (verb) in a domain and the relation between them. There are three main kinds of relations between two actions: dependence, parallel exclusion and sequential exclusion.

3. Derivations Business Rule
   Business rules defines that how knowledge in one form may be transformed into other knowledge, possibly in a different form. This kind of business can be expressed in the form "*IF* something *DO* something".

As mentioned in the introduction, Attemtpo Controlled English (ACE) [24] is proposed for representing the business rule on the user interface. The user can define and modify the rule by using ACE syntax; each ACE phrase will be translated to an OWL syntax in BRO. We introduce following table to describe the mapping between Attemtpo Controlled English and OWL functional syntax for building the business rule.

Business Terms and the Relation Between Them
In the Table 1, a set of mappings between OWL and ACE, we can use this table for defining a set of business rules (ontology domain) and building BRO. The user can use ACE syntax to define his own policy; the ACE phrase will be translated to a set of axiom in OWL language in BRO.

Table 1 Mapping between OWL and ACE

OWL properties and classes	Corresponding ACE verbs and noun phrases
Class(cat)	*Common noun,* e.g. cat
owl:Thing	Something; thing
ObjectComplementOf(*C*)	Something that is not a car; something that does not like a cat
ObjectIntersectionOf(*C*1 … *Cn*)	Something that is not a cat and that owns a car and that …
ObjectUnionOf(*C*1 … *Cn*)	Something that is a cat or that is a camel or that …
ObjectOneOf(*a*)	*Proper name,* e.g. John; something that is John
ObjectSomeValuesFrom(*R C*)	Something that likes a cat
ObjectExistsSelf(*R*)	Something that likes itself
ObjectMinCardinality(*n R C*)	Something that owns at least 2 cars
ObjectMaxCardinality(*n R C*)	Something that owns at most 2 cars
ObjectExactCardinality(*n R C*)	Something that owns exactly 2 cars
SubClassOf(*C D*)	Every man is a human
SubObjectPropertyOf (SubObjectPropertyChain(*R*1 … *Rn*) *S*)	Everything that owns something that contains something owns it
DisjointObjectProperties(*R*1 … *Rn*)	Nothing that is-child-of something is spouse-of it…
ClassAssertion(*a C*)	John is a man that owns at least 2 cars

Activity Constraint

In this section, we introduce the class *Action,* this class defined all activities in a domain. When a user defines an action, an individual of class *Action* is created. We also define the relation between actions. There a three type of relations that should be considered which are dependency relation, parallel exclusion relation and sequential execution relation. In order to represent the relations, we define three binary object properties of class *Action* as follow:

- *hasDependency*: this property defines an action that can be executed if another action was executed in a business process.
- *hasExclusionInParallelle:* this property defines two actions that cannot be executed in parallel in a business process.
- *hasExlusionInSequency:* this property defines two actions that cannot be executed in sequential in a business process.

We also define two object property chains for the system which are able to reason on a set of facts for instance if action *a* depends on action *b*, action *b* depends on action *c*, so the system can deduce that action *a* depends on action *c*. Additionally, if an action *a* depends on action *b* and action *b* exclude action *c*, so we can deduce that action *c* exclude action *a* (Table 2).

Table 2 Relation between two actions

OWL properties and classes	Corresponding ACE verbs and noun phrases
Class(Action)	Class *Action* defined all activities
ObjectProperty(:hasDependency) ObjectPropertyDomain(:hasDependency : Action) ObjectPropertyRange(:hasDependency : Action)	*Action* a depends on *Action* b
ObjectProperty(:hasExclusionInParallel) ObjectPropertyDomain(: hasExclusionInParallel :Action) ObjectPropertyRange(:hasExclusionInParallel: Action)	Action a and action b cannot be executed in parallel
ObjectProperty(:hasExclusionInSeqential) ObjectPropertyDomain(: hasExclusionInSequential:Action) ObjectPropertyRange(: hasExclusionInSequential:Action)	This property defines that action a and action b cannot be executed in sequential
SubObjectPropertyOf(ObjectPropertyChain(: hasDependency: hasDependency): hasDependency)	If action a depends on action b, action b depends c Then action a depends on action c
SubObjectPropertyOf(ObjectPropertyChain(: hasDependency :hasExlusionInSequential): hasExclusionInSequential)	If action a depends on action b, action b excludes action c Then action a exclude action c

Derivation Business Rule

This kind of business rules is represented in form "*IF* something *DO* something". For representing this form with OWL language, we use SWRL (Semantic Web Rule Language) [23]. The rule is represented in this form:

$$antecendent \Rightarrow consequent$$

where both antecedent and consequent are conjunctions of atoms written $a_1 \wedge \ldots \wedge a_{n.}$ Variables are indicated using the standard convention of prefixing them with a question mark (e.g., ?x). Using this syntax, a rule asserting that the composition of parent and brother properties implies the uncle property would be written:

$$parent(?x, ?y) \wedge brother(?y, ?z) \Rightarrow uncle(?x, ?z)$$

In this syntax, built-in relations that are functional can be written in functional notation, i.e., $op{:}\,numeric - add(?x, 3, ?z)$ can be written instead as $?x = op{:}\,numeric - add(3, ?z)$

3.3 Checking the Compliance of BPO with BRO

In Fig. 1, we introduce the sketch of our solution. Business processes (CPN graph) are designed by a graphic user interface. Each graphic item in a business process will be represented by an individual of the correspondence concept in BPO. Business rules are created and modified by an ACE editor. Each rule is represented by an ACE sentence.

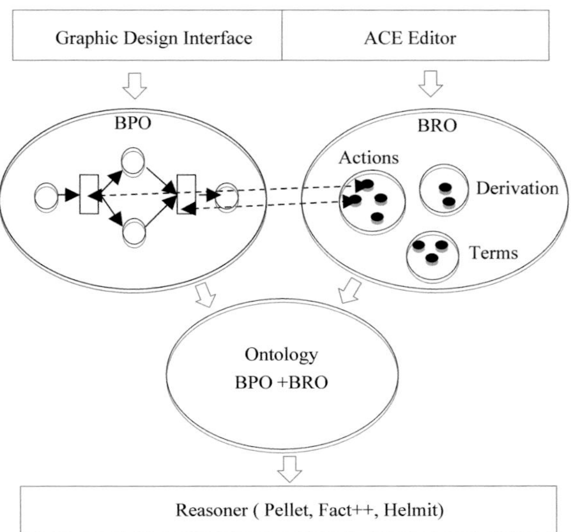

Fig. 1 Integration of BPO and BRO

In order to check the compliance of business process with business rules, we merge BRO and BPO into one ontology; two concepts *Transition* in BPO and *Action* in BRO are defined as two equivalence concepts. The business term individuals can be used as a color and a token in CPN graph (business process). During the execution of business process, the value of individual can be changed but it must respect the constraints inside BRO (TBox and Properties).

At design time, when a user defines a business process, the business term will be used to name an item. Each transition individual in BPO is mapped to an action individual in BRO. Depending on the user's given order, BPO editor will generate a set of negation rule inside BPO.

For example: there are two actions inside BRO, which is defined that b depends on a as follow:

$$ObjectPropertyAssertion(:hasDependency: b: a)$$

It means that a must be executed before b, but at design time a user define that a executes after b, therefore a negation rule is generated as follow:

$$NegativeObjectPropertyAssertion(:hasDependency: b: a)$$

Two rules above are opposite, so the merged ontology of BRO and BPO will be inconsistent. It can be checked by a reasoned (Pellet, Helmit).

At runtime, we use the same approach to verify the consistency of merged ontology. If a user modifies a business process, a negative rule is generated and insert into BPO; for each modification, the reasoner will check the consistency of merged ontology and notify to the user automatically.

4 Conclusion and Perspectives

In this paper, an ontology-based approach for checking the compliance of business process modeling and business rules is proposed. It takes important features of the ontology which are the reasoning capabilities, the possibility to express complex actions, and its declarative semantics to validate not only the consistency of business rules and business process but also the compliance of business process with a set of business rules. The advantage of this approach is to allow the system to detect the flaws of business process automatically at design time and run time by using the ontology's reasoning capabilities. Nevertheless, by using this approach, if BRO has many concepts and properties, the reasoning may take long time for checking the consistency of BPO and BRO ontology. According to that, future theoretic works involve three main issues. The first one is to focus on the distributed reasoning. The second one will be achieved by selecting the related rule of an action for the validation. And the last goal is to consider the business process execution and work with a data source.

References

1. Ko, R.K.L., Lee, S.S.G., Lee, E.W.: Business process management (BPM) standards: a survey. Business Process Manage. J. **15**, 744–791, 2009
2. Fasbinder, M.: Why Model Business Processes (2007)
3. Hommes, L.J.: The Evaluation of business process modeling techniques. Delft University of Technology, Ph.D. thesis 90-9017698-5 (2004)
4. Feng, L., Wei, Z.: Colored Petri net extended with price information and its applicaion. J. Comput. Appl. **201**, 2501–2503 (2007)
5. Becker, J., Delfmann, P., Eggert, M., Schwittay, S.: Generalizability and applicability of model-based business process compliance-checking approaches—A state-of-the-art analysis and research roadmap. BuR—Bus. Res. **5**, 221–247 (2012)
6. Ramezani, E., Fahland, D., van der Werf, J.M., Mattheis, P.: Separating compliance management and business process management. In: Daniel, F., Barkaoui, K., Dustdar, S. (eds.) BPM Workshops 2011, Part II. LNBIP, vol. 100, pp. 459–464. Springer, Heidelberg (2012)
7. Van Der Aalst, W.M.P., Ter Hofstede, A.H.M., Kiepuszewski, B., Barros, A.P.: Workflow patterns. Distrib. Parallel Databases **14**(1), 5–51 (2003)
8. Russell, N., ter Hofstede, A., Edmond, D., van der Aalst, W.: Workflow data patterns: Identification, representation and tool support. Conceptual Model.-ER **2005**, 353–368 (2005)
9. Tran, H.N., Coulette, B., Dong, B.T.: Broadening the use of process patterns for modeling processes. In: Proceedings of SEKE, Knowledge Systems Institute Graduate School, pp. 57–62 (2007)
10. Awad, A., Smirnov, S., Weske, M.: Towards resolving compliance violations in business process models. In: GRCIS. CEUR-WS. org (2009)
11. Weber, I., Homann, J., Mendling, J.: Semantic business process validation. In: Proceedings of International Workshop on Semantic Business Process Management (2008)
12. Governatori, G., Milosevic, Z., Sadiq, S.: Compliance checking between business processes and business contracts. In: Enterprise Distributed Object Computing Conference, 2006. EDOC'06. 10th IEEE International, pp. 221–232. IEEE (2006)
13. Thom, L., Reichert, M., Chiao, C., Iochpe, C., Hess, G.: Inventing less, reusing more, and adding intelligence to business process modeling. In: Database and Expert Systems Applications, pp. 837–850. Springer (2008)
14. Kharbili, M.E., Stein, S., Markovic, I., Pulverm, E.: Towards a framework for semantic business process compliance management. In: Proceedings of the Workshop on Governance, Risk and Compliance for Information Systems, pp. 1–15. Citeseer (2008)
15. Arbab, F., Kokash, N., Meng, S.: Towards using reo for compliance-aware business process modeling. In: Leveraging Applications of Formal Methods, Verification and Validation, pp. 108–123 (2009)
16. Gschwind, T., Koehler, J., Wong, J.: Applying patterns during business process modeling. In: BPM, vol. 5240, pp. 4–19. Springer (2008)
17. Reynares, E., Caliusco, M.L., Galli, M.R.: An automatable approach for SBVR to OWL 2 mappings (2013)
18. Bernotaityte, G., Nemuraite, L., Butkiene, R., Paradauskas, B.: Developing SBVR Vocabularies and Business Rules from OWL2 Ontologies. ICIST (2013)
19. Reynares, E., Caliusco, M.L., Galli, M.R.: SBVR to OWL 2 mappings: an automatable and structural-rooted approach. CLEI Electron. J. (2014)
20. Bernotaityte, Gintare, Nemuraite, Lina, Butkiene, Rita, Paradauskas, Bronius: Developing SBVR vocabularies and business rules from OWL2 ontologies. Inf. Soft. Technol. Commun. Comput. Inf. Sci. **403**, 134–145 (2013)
21. http://www.omg.org/spec/SBVR/
22. http://www.w3.org/TR/owl-features/

23. Horrocks, I., Patel-Schneider, P.F., Boley, H., Tabet, S., Grosof, B., Dean, M.: Swrl: A semantic web rule language combining owl and ruleml. W3C Member Submission May 21 2004. http://www.w3.org/Submission/SWRL/
24. Auer, S., Dietzold, S., Riechert, T.: OntoWiki—A tool for social, semantic collaboration. In: Proceedings of the 5th International Semantic Web Conference, vol. 4273, Lecture Notes in Computer Science, pp. 736–749. Springer (2006)

Introduction to Integration of the Process Mining to the Knowledge Framework for Software Processes

Jakub Stolfa, Svatopluk Stolfa, Michael Alexander Kosinar and Vaclav Snasel

Abstract Systems started to be more process oriented during last decades. The knowledge has moved from humans to systems and systems has more and more knowledge about the process. This paper presents an approach to the integration of the process mining to the process modeling supported by the formal layer. The process is modeled by framework that allows selection and formal description of the applicable meta-model parts, formally define process prescription and then model the process by graphical language. Process is executed and logs are stored. The presented process mining part of the framework then can discover the de facto process model, can check the differences between de facto and de jure model, can enhance de jure model according to the findings.

Keywords Process mining · Process modeling · Formal approach · OWL

1 Introduction

Process is ordered set of the steps that leads to some goal. In the reality there is a process that is executed by persons, machines, etc. Process that is followed in the information system is called workflow. Process oriented systems started to be worshiped as the only way to control the processes and activities that has to be

J. Stolfa (✉) · S. Stolfa · M.A. Kosinar · V. Snasel
Department of Computer Science, VŠB - Technical University of Ostrava,
17. Listopadu 15, Ostrava-Poruba, Czech Republic
e-mail: jakub.stolfa@vsb.cz

S. Stolfa
e-mail: svatopluk.stolfa@vsb.cz

M.A. Kosinar
e-mail: michal.kosinar@vsb.cz

V. Snasel
e-mail: vaclav.snasel@vsb.cz

A. Abraham et al. (eds.), *Proceedings of the Second International Afro-European Conference for Industrial Advancement AECIA 2015*, Advances in Intelligent Systems and Computing 427, DOI 10.1007/978-3-319-29504-6_4

enacted. The knowledge about the processes and their enactment was transferred to the systems by defining the workflow. The shift from the data oriented systems to the process oriented systems brought possibilities for further analysis of controlling and checking the enactment of the processes and resources that are involved.

Execution of the process in the information system generates data that is called event data and it can be examined by the process mining, that was introduced by Aalst in 2004 [10, 11] Process mining aims to discover, monitor and improve real processes by extracting knowledge from event logs readily available in today's information systems [9]. It is an approach that is used for the analysis of real enactment of the processes. Process mining uses logs of real process enactments to analyze the process itself. On the other hand stands the Business process modeling [1]. Business process modeling is the activity of creating a process model that should be followed then.

We have developed formal process modeling framework—Knowledge Support Framework for Software Processes. It includes modeling based on single process meta-model terminology defined in a formal way, utilization of explicit reusable knowledge profiles and intuitive visual modeling method based on diagrammatic language capable of software process modeling based on process patterns. In other words it offers complex framework for modeling software processes from the "top" and it is supported by the defined formal layer [5, 6].

The goal of this paper is to present our ideas and work that we have made in the area of connection between process mining that bring us the possibilities of the modeling from the "bottom" and the business process modeling that aim to model the process from the "top". The area of the process modeling is covered by our knowledge support framework that is so far applied in the software development [7] and healthcare processes [5]. In future work we would like to apply the framework to all business processes.

The paper is organized as follows: Sect. 2 describes Knowledge Support Framework for Software processes that is process modeling framework that we have created; Sect. 3 presents proposed approach of connection process mining and process modeling; Sect. 4 shows short model example of the usage of our approach. Finally, Sect. 5 presents the conclusion and intended future work.

2 Knowledge Support Framework for Software Processes

Like it is mentioned in the introduction, Knowledge Support Framework is a process modeling framework that bring us possibilities to create process model, set up formal knowledge base and all of that is supported by formal layer. The way of creating the framework, formalization of the process model, the integration of the process mining took some scientific research that we have done so far, we have published several papers [3–5, 7, 8].

The framework is able to model all aspects of software processes that are critical for process execution and further analysis [3, 7]—these aspects are usually referred to as aspects of functional perspective, behavioral perspective, and structural perspective—each concerns different scope (or viewpoint) of software houses. Framework's main parts that we use in this paper are: Layer 0 (meta-model), Layer 1 (Explicit Knowledge Profile) and Layer 2 (Visual Process modeling). These parts are described in the next sections.

3 Integration of Process Mining to Process Modeling

This section describes introduction how is the process mining integrated to the Knowledge support framework and how is modeling supported from the top and from the bottom. The structure and important parts of the connection between the framework for process modeling and process mining are depictured in Fig. 1. We can see that figure is divided into the three parts—Knowledge support framework for software processes, Execution of the process, and Integration of the process mining.

3.1 Process Modeling Part

First part Knowledge support framework for software Processes aims to create the process model of the software process in particular organization. From simplified point of view it has three layers:

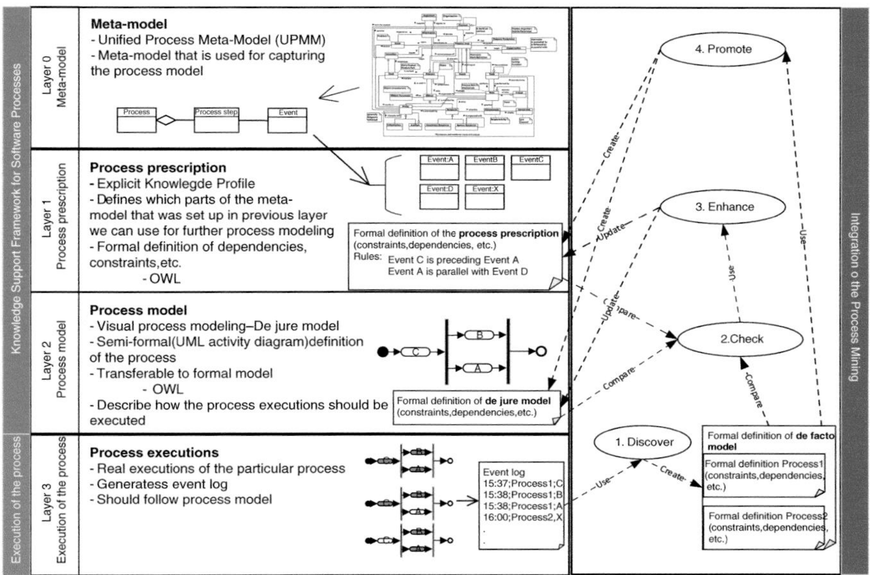

Fig. 1 Overview—process mining and process modeling

1. Layer 0—meta-model layer
This part is focused on adjusting of the meta-model named Unified Process Meta-Model according to organizational needs Adjusting means that we do not have to use all the possibilities that meta-model supports but we can use only part of that model.

2. Layer 1—Process prescription (Explicit Knowledge Profile)
Based on the selection of meta-model the process prescription is done here. The process prescription is developed and have e.g. defined constraints etc. Constraints are written down in formalized way in OWL language.

3. Layer 2—Process model (Visual Process modeling)
This layer is focused on visual process modeling, it means in our terminology creating the de jure model. A *de jure model* is normative, it means it specifies how things should be done. De jure model is assembled from the pieces that process prescription from the layer 1 provides to us. It has to follow the constraints of the prescription model and in case of need add new constraints.

We use UML Activity diagram as a tool for modeling—semiformal definition of the process. Thanks to the transferability of UML activity diagram to OWL we are able to write down process model in a formal definition too.

3.2 Process Mining Part

Second part of Fig. 1 is the Execution of the process. It represents execution of the process in the reality. Execution of the process is done by real people supported by information systems. During the execution lot of data are generated to event log.

Third part of figure represents integration of the process mining to the framework. There are four activities that are integrated for the usage in the framework:

1. Discover
This activity aims to discover the process from the event logs. Figure 1 depictures that in our case discover activity uses Event log of execution of the processes and create formal definition of de facto model. A de facto *model* is descriptive, it aims to capture reality not to steer or control reality. We use methods of the process mining for process model discovery [2, 7, 11] and we adjust the output of the methods to create formal definition that is written in the OWL language that we support in the framework

2. Check
This activity aims to compare the event log and de jure model. In our case it uses formal definition of the particular process cases and formal definition of de facto model and formal definition of the process prescription.

Method is comparing the formal definitions of the de jure model, process prescription and formal definitions of the particular process cases. We compare de facto, de jure and prescription model on the level of formal description:

- De facto and de jure level—this comparison aim to find the deviations between model of real process executions and the model that should be followed in the reality. If the models are identical it means that the process in the reality is executed in the way that we have modeled it. If not we can further analyze details of the deviation.
- De facto model and prescription model level—it compares de jure model with process prescription. It means if the execution of the process fits to main constraints in the prescription model at least. If we find out that the de jure model fits to prescription model from organizational point of view we can say that the process followed in the reality is not that what was desired (de jure model), but at least it do not break important constraints of the structure that were designed in the level of process prescription during the process modeling. If the de jure model breaks also the prescription model we have found remarkable deviation that we have to handle. How to handle it, depends on particular deviation but there exist some possibilities:

  - Enhance—The organization finds out that the deviation is not considered as a wrong deviation and update the process prescription, respective process model. More information in the enhance activity.
  - Analyze the deviation—The organization finds out that the deviation considered as a wrong deviation, so there can be used further analysis. For example we can find out which person executed the problematic event, using social networks, what are their connection, etc.

3. Enhance
Enhance activity uses information gathered from the activities check or diagnose and updates prescription model or process model. This activity we propose for future automatization according to the information that we are able to gather from the formal layer. For example we can have defined borders for the process that we cannot break. If the de facto process model do not break that rules we can adjust the process model. This can happen just for one instance of the process and then process model can be changed back. In reality it could be useful if we predict that some process instance if for example predicted that it will end late, so we can automatically skip some steps of the process model, respectively workflow in the system.

4. Promote
Promote activity uses de facto model and creates de jure model and particular process prescription, if there is no one. This case can happen when the framework is used in the company that has not used the framework before and their process model is not done properly.

4 Model Example

This section presents model example where proposed activities of the process mining integration—discover, check and enhance are used.

Let's imagine that we have an Organization O that has the process modeled by our framework. Now we apply integrated process mining methods to analyze how the process is executed in the reality, and according to that information, company can improve its processes.

The model example is following:

Process prescription
Process prescription is defined the following way (simplified view of the formal OWL description is presented):

- Events: A, B, C, D, E, X
- Constraints: A is followed by D; A is parallel with B;

Simplified formal definition of the process prescription:

```
<Ontology xmlns=http://www.w3.org/2002/07/owl
   <ClassAssertion><Class IRI="Task"/><NamedIndividual IRI="A"/></ClassAssertion>
   <ClassAssertion><Class IRI="Task"/><NamedIndividual IRI="B"/></ClassAssertion>
   <ClassAssertion><Class IRI="Task"/><NamedIndividual IRI="C"/></ClassAssertion>
   <ClassAssertion><Class IRI="Task"/><NamedIndividual IRI="D"/></ClassAssertion>
   <ClassAssertion><Class IRI="Task"/><NamedIndividual IRI="E"/></ClassAssertion>
   <ClassAssertion><Class IRI="Task"/><NamedIndividual IRI="X"/></ClassAssertion>
   <ObjectPropertyAssertion><ObjectProperty IRI="isFollowedBy"/><NamedIndividual IRI="A"/><NamedIndividual
IRI="D"/></ObjectPropertyAssertion>
   <ObjectPropertyAssertion><ObjectProperty IRI="isParallelWith"/><NamedIndividual IRI="A"/><NamedIndividual
IRI="B"/></ObjectPropertyAssertion>
</Ontology>
```

According to the prescription, one version of the process model could be e.g. modeled like is depictured in Fig. 2. First, A and B events are executed in parallel and followed by D event.

Process model
During the visual process modeling in the company they decided to put the event X before A and the process starts with event C. Process model is depictured in Fig. 3. It

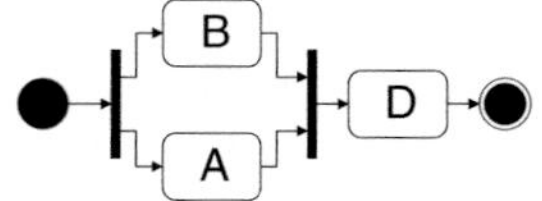

Fig. 2 Visualization of the process prescription possibility

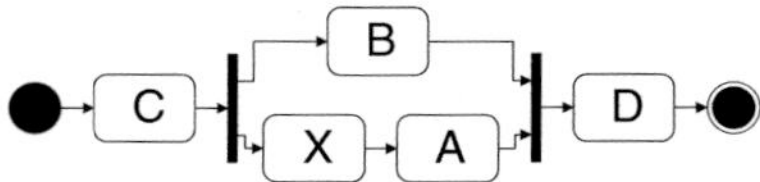

Fig. 3 Model example—de jure process model

do not breaks any constraints defined in the process prescription, so it is possible to do it. In our framework, we do not state that you cannot break the constraints defined in the process prescription, but you will be warned that you created the model that does not follow particular prescriptions. More information is in the paper [5, 6].

Formal definition of the process model:

```
<Ontology xmlns="http://www.w3.org/2002/07/owl"
    <ClassAssertion><Class IRI="Task"/><NamedIndividual IRI="A"/></ClassAssertion>
    <ClassAssertion><Class IRI="Task"/><NamedIndividual IRI="B"/></ClassAssertion>
    <ClassAssertion><Class IRI="Task"/><NamedIndividual IRI="C"/></ClassAssertion>
    <ClassAssertion><Class IRI="Task"/><NamedIndividual IRI="D"/></ClassAssertion>
    <ClassAssertion><Class IRI="Task"/><NamedIndividual IRI="X"/></ClassAssertion>
    <ObjectPropertyAssertion><ObjectProperty IRI="precedes"/><NamedIndividual IRI="A"/><NamedIndividual
IRI="X"/></ObjectPropertyAssertion>
    <ObjectPropertyAssertion><ObjectProperty IRI="precedes"/><NamedIndividual IRI="D"/><NamedIndividual
IRI="A"/></ObjectPropertyAssertion>
    <ObjectPropertyAssertion><ObjectProperty IRI="precedes"/><NamedIndividual IRI="D"/><NamedIndividual
IRI="B"/></ObjectPropertyAssertion>
    <ObjectPropertyAssertion><ObjectProperty IRI="precedes"/><NamedIndividual IRI="B"/><NamedIndividual
IRI="C"/></ObjectPropertyAssertion>
    <ObjectPropertyAssertion><ObjectProperty IRI="precedes"/><NamedIndividual IRI="X"/><NamedIndividual
IRI="C"/></ObjectPropertyAssertion>
    <ObjectPropertyAssertion><ObjectProperty IRI="isFollowedBy"/><NamedIndividual IRI="C"/><NamedIndividual
IRI="B"/></ObjectPropertyAssertion>
    <ObjectPropertyAssertion><ObjectProperty IRI="isFollowedBy"/><NamedIndividual IRI="B"/><NamedIndividual
IRI="D"/></ObjectPropertyAssertion>
    <ObjectPropertyAssertion><ObjectProperty IRI="isFollowedBy"/><NamedIndividual IRI="X"/><NamedIndividual
IRI="A"/></ObjectPropertyAssertion>
    <ObjectPropertyAssertion><ObjectProperty IRI="isFollowedBy"/><NamedIndividual IRI="A"/><NamedIndividual
IRI="D"/></ObjectPropertyAssertion>
    <ObjectPropertyAssertion><ObjectProperty IRI="isParallelWith"/><NamedIndividual IRI="A"/><NamedIndividual
IRI="B"/></ObjectPropertyAssertion>
    <ObjectPropertyAssertion><ObjectProperty IRI="isParallelWith"/><NamedIndividual IRI="X"/><NamedIndividual
IRI="B"/></ObjectPropertyAssertion>
    <ObjectPropertyAssertion><ObjectProperty IRI="isParallelWith"/><NamedIndividual IRI="B"/><NamedIndividual
IRI="A"/></ObjectPropertyAssertion>
</Ontology>
```

Like it was mentioned before the model is simplified so it miss information like by whom is the event performed, in which organization or in what context, which resources has been used, in what level of competence by some role, etc. All that possibilities are supported by our meta-model [5, 6].

Moreover there rises also a question: why the company put a new event during the modeling there and basically created new constraints—X is followed by A, X is predecessor of A? We allowed this variability to simplify the usage of our method. This came from the observation that in reality during the adjusting of the process prescription often happened that some constrain is forgotten or some relationship not mentioned. Then during the modeling of the particular process model this come to a mind of modeler and its modelled there. It is not necessary then to go back to the formal layer and define it immediately, but it can be done later by expert. Another reason may be that it is not always so important constraint so it gives us a possibility to specify it only during the process modeling.

4.1 Discover Activity

Performing of the activity discover creates a de facto process model that can be replied for all the cases of the process.

According to the discovered definition of de facto model we are able to create visualization that is depictured on the Fig. 4.

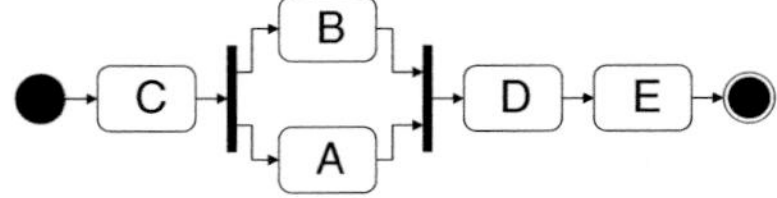

Fig. 4 Model example 1—de facto process model

Formal definition of discovered de facto model:

```
<Ontology xmlns="http://www.w3.org/2002/07/owl"
    <ClassAssertion><Class IRI="Task"/><NamedIndividual IRI="A"/></ClassAssertion>
    <ClassAssertion><Class IRI="Task"/><NamedIndividual IRI="B"/></ClassAssertion>
    <ClassAssertion><Class IRI="Task"/><NamedIndividual IRI="C"/></ClassAssertion>
    <ClassAssertion><Class IRI="Task"/><NamedIndividual IRI="D"/></ClassAssertion>
    <ClassAssertion><Class IRI="Task"/><NamedIndividual IRI="E"/></ClassAssertion>
    <ObjectPropertyAssertion><ObjectProperty IRI="precedes"/><NamedIndividual IRI="A"/><NamedIndividual
IRI="X"/></ObjectPropertyAssertion>
    <ObjectPropertyAssertion><ObjectProperty IRI="precedes"/><NamedIndividual IRI="D"/><NamedIndividual
IRI="A"/></ObjectPropertyAssertion>
    <ObjectPropertyAssertion><ObjectProperty IRI="precedes"/><NamedIndividual IRI="D"/><NamedIndividual
IRI="B"/></ObjectPropertyAssertion>
    <ObjectPropertyAssertion><ObjectProperty IRI="precedes"/><NamedIndividual IRI="B"/><NamedIndividual
IRI="C"/></ObjectPropertyAssertion>
    <ObjectPropertyAssertion><ObjectProperty IRI="precedes"/><NamedIndividual IRI="E"/><NamedIndividual
IRI="D"/></ObjectPropertyAssertion>
    <ObjectPropertyAssertion><ObjectProperty IRI="isFollowedBy"/><NamedIndividual IRI="C"/><NamedIndividual
IRI="B"/></ObjectPropertyAssertion>
    <ObjectPropertyAssertion><ObjectProperty IRI="isFollowedBy"/><NamedIndividual IRI="B"/><NamedIndividual
IRI="D"/></ObjectPropertyAssertion>
    <ObjectPropertyAssertion><ObjectProperty IRI="isFollowedBy"/><NamedIndividual IRI="A"/><NamedIndividual
IRI="D"/></ObjectPropertyAssertion>
    <ObjectPropertyAssertion><ObjectProperty IRI="isFollowedBy"/><NamedIndividual IRI="D"/><NamedIndividual
IRI="E"/></ObjectPropertyAssertion>
    <ObjectPropertyAssertion><ObjectProperty IRI="isParallelWith"/><NamedIndividual IRI="A"/><NamedIndividual
IRI="B"/></ObjectPropertyAssertion>
    <ObjectPropertyAssertion><ObjectProperty IRI="isParallelWith"/><NamedIndividual IRI="B"/><NamedIndividual
IRI="A"/></ObjectPropertyAssertion>
</Ontology>
```

4.2 Check Activity

Check activity focus on the conformance check of the discovered de facto model with de jure model and process prescription. First of all check activity compares the formalized process models to check out if these models are same or there are some deviations. Then we can check individual cases of the processes and see why in this case of the process the particular deviation happened.

First, conformance of de facto model and de jure model is checked. The result is:

- De facto and de jure models are not completely the same and some deviations exists:
  - Event X is missing and its causes that following definitions are missing

```
<ObjectPropertyAssertion><ObjectProperty IRI="precedes"/><NamedIndividual IRI="X"/><NamedIndividual
IRI="C"/></ObjectPropertyAssertion>
<ObjectPropertyAssertion><ObjectProperty IRI="isFollowedBy"/><NamedIndividual IRI="X"/><NamedIndividual
IRI="A"/></ObjectPropertyAssertion>
<ObjectPropertyAssertion><ObjectProperty IRI="isParallelWith"/><NamedIndividual
IRI="X"/><NamedIndividual IRI="B"/></ObjectPropertyAssertion>
```

– Event E is extra and it causes that the following definitions are extra

```
<ClassAssertion><Class IRI="Task"/><NamedIndividual IRI="E"/></ClassAssertion>
<ObjectPropertyAssertion><ObjectProperty IRI="precedes"/><NamedIndividual IRI="E"/><NamedIndividual
IRI="D"/></ObjectPropertyAssertion>
<ObjectPropertyAssertion><ObjectProperty IRI="isFollowedBy"/><NamedIndividual IRI="D"/><NamedIndividual
IRI="E"/></ObjectPropertyAssertion>
```

Since de facto and de jure models comparison showed deviations, conformance of the de facto and prescription models is checked:

- De facto model fulfils the constraints of the prescription.
  - De facto model contains known activities and do not contain unknown activities

    Activities: A, B, C, D, E

```
<ClassAssertion><Class IRI="Task"/><NamedIndividual IRI="A"/></ClassAssertion>
<ClassAssertion><Class IRI="Task"/><NamedIndividual IRI="B"/></ClassAssertion>
<ClassAssertion><Class IRI="Task"/><NamedIndividual IRI="C"/></ClassAssertion>
<ClassAssertion><Class IRI="Task"/><NamedIndividual IRI="D"/></ClassAssertion>
<ClassAssertion><Class IRI="Task"/><NamedIndividual IRI="E"/></ClassAssertion>
```

  - De facto model contains needed constraints from process prescription

    A is followed by D; A is parallel with B

```
<ObjectPropertyAssertion><ObjectProperty IRI="isFollowedBy"/><NamedIndividual IRI="A"/><NamedIndividual
IRI="D"/></ObjectPropertyAssertion>
<ObjectPropertyAssertion><ObjectProperty IRI="isParallelWith"/><NamedIndividual
IRI="A"/><NamedIndividual IRI="B"/></ObjectPropertyAssertion>
```

  - De facto model do not contains constraints that breaks needed constraints
  - De facto model contains new constraints that are not in the process prescription but do not break needed constraints

    E precedes D; D is followed by E

```
<ObjectPropertyAssertion><ObjectProperty IRI="precedes"/><NamedIndividual IRI="E"/><NamedIndividual
IRI="D"/></ObjectPropertyAssertion>
<ObjectPropertyAssertion><ObjectProperty IRI="isFollowedBy"/><NamedIndividual IRI="D"/><NamedIndividual
IRI="E"/></ObjectPropertyAssertion>
```

The result is that the de facto model does not fit the de jure model but at least fits the constraints in the prescription. Now the company may analyze the particular results and decide if they would like to enhance the process model, it means include extra constraints and delete not needed constraints or if the company wants to further analyze the cases that causes deviations.

4.3 Enhance Activity

Let's consider that company decided to enhance the process model by the data that were gathered by the check activity. The company also decided that the event X will stay in the de jure model. In reality, no case executed the activity X. So the company would like to force the people to follow the process model and execute the event X.

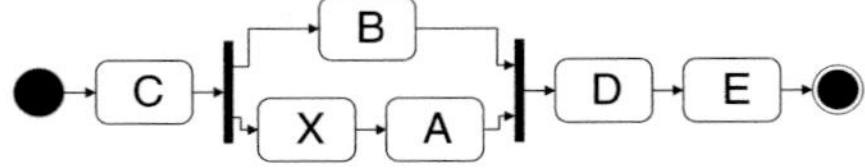

Fig. 5 Model example 1—final de jure process model

However, the process model will be enhanced with the event E. Information and constraints about the event E that we gathered during the execution of the activity check will be put to the formal definition of the process model. We have to also check if there is a need to enhance the process prescription. Anyway, the process prescription can be updated if the company wishes to. In this example we consider only enhancing of the process model.

Final process model is depictured on the figure Fig. 5.

Formal definition of enhanced process model:

```
<Ontology xmlns="http://www.w3.org/2002/07/owl"
    <ClassAssertion><Class IRI="Task"/><NamedIndividual IRI="A"/></ClassAssertion>
    <ClassAssertion><Class IRI="Task"/><NamedIndividual IRI="B"/></ClassAssertion>
    <ClassAssertion><Class IRI="Task"/><NamedIndividual IRI="C"/></ClassAssertion>
    <ClassAssertion><Class IRI="Task"/><NamedIndividual IRI="D"/></ClassAssertion>
    <ClassAssertion><Class IRI="Task"/><NamedIndividual IRI="E"/></ClassAssertion>
    <ClassAssertion><Class IRI="Task"/><NamedIndividual IRI="X"/></ClassAssertion>
    <ObjectPropertyAssertion><ObjectProperty IRI="precedes"/><NamedIndividual IRI="A"/><NamedIndividual
IRI="X"/></ObjectPropertyAssertion>
    <ObjectPropertyAssertion><ObjectProperty IRI="precedes"/><NamedIndividual IRI="D"/><NamedIndividual
IRI="A"/></ObjectPropertyAssertion>
    <ObjectPropertyAssertion><ObjectProperty IRI="precedes"/><NamedIndividual IRI="D"/><NamedIndividual
IRI="B"/></ObjectPropertyAssertion>
    <ObjectPropertyAssertion><ObjectProperty IRI="precedes"/><NamedIndividual IRI="B"/><NamedIndividual
IRI="C"/></ObjectPropertyAssertion>
    <ObjectPropertyAssertion><ObjectProperty IRI="precedes"/><NamedIndividual IRI="X"/><NamedIndividual
IRI="C"/></ObjectPropertyAssertion>
    <ObjectPropertyAssertion><ObjectProperty IRI="precedes"/><NamedIndividual IRI="E"/><NamedIndividual
IRI="D"/></ObjectPropertyAssertion>
    <ObjectPropertyAssertion><ObjectProperty IRI="isFollowedBy"/><NamedIndividual IRI="C"/><NamedIndividual
IRI="B"/></ObjectPropertyAssertion>
    <ObjectPropertyAssertion><ObjectProperty IRI="isFollowedBy"/><NamedIndividual IRI="B"/><NamedIndividual
IRI="D"/></ObjectPropertyAssertion>
    <ObjectPropertyAssertion><ObjectProperty IRI="isFollowedBy"/><NamedIndividual IRI="X"/><NamedIndividual
IRI="A"/></ObjectPropertyAssertion>
    <ObjectPropertyAssertion><ObjectProperty IRI="isFollowedBy"/><NamedIndividual IRI="A"/><NamedIndividual
IRI="D"/></ObjectPropertyAssertion>
    <ObjectPropertyAssertion><ObjectProperty IRI="isFollowedBy"/><NamedIndividual IRI="D"/><NamedIndividual
IRI="E"/></ObjectPropertyAssertion>
    <ObjectPropertyAssertion><ObjectProperty IRI="isParallelWith"/><NamedIndividual IRI="A"/><NamedIndividual
IRI="B"/></ObjectPropertyAssertion>
    <ObjectPropertyAssertion><ObjectProperty IRI="isParallelWith"/><NamedIndividual IRI="X"/><NamedIndividual
IRI="B"/></ObjectPropertyAssertion>
    <ObjectPropertyAssertion><ObjectProperty IRI="isParallelWith"/><NamedIndividual IRI="B"/><NamedIndividual
IRI="A"/></ObjectPropertyAssertion>
</Ontology>
```

5 Conclusion

This paper described introduction to the connection between process modeling that is presented by our framework and process mining that has possibilities to discover process model followed in the reality. All of that is supported by one formal layer that is common for process modeling and also process mining. We have set up approaches for several possible usages of this connection—discovery, check, enhance, and promote.

However we have lot of work to do, in future work, we would like to describe all possible usages of the approach, present case studies based on the real data, and also adjust the framework to support not just a software processes but also for all business processes. Connection of the process modeling and process mining that is proposed in this paper is the basic of the methodology that we are working on. It will include process modeling, process mining, one common formal layer, and automatized methods. In future work we would like to also present methods for process predictions that will be integrated in our methodology.

Acknowledgment The research was supported by the internal grant agency of VSB – Technical University of Ostrava, Czech Republic, project no. SP2015/85 'Knowledge modelling usage in software engineering'.

References

1. Aguilar-Saven, R.S.: Business process modelling: review and framework. Int. J. Prod. Econ. **90**(2), 129–149 (2004)
2. Carmona, J., Cortadella, J., Kishinevsky, M.: New region-based algorithms for deriving bounded Petri nets. IEEE Trans. Comput. **59**(3), 371–384 (2010)
3. Czopik, J., Košinár, M.A., Štolfa, J., Štolfa, S.: Addition of static aspects to the intuitive mapping of UML activity diagram to CPN. Adv. Intell. Syst. Comput. **334**, 77–86 (2015)
4. Czopik, J., Košinár, M.A., Štolfa, J., Štolfa, S.: Formalization of software process using intuitive mapping of UML activity diagram to CPN. Adv. Intell. Syst. Comput. **303**, 365–374 (2014)
5. Kosinar, M.A., Czopik, J., Štolfa, J., Penhaker, M.: Knowledge framework for clinical processes architecture and analysis. In: 13th International Symposium on Applied Machine Intelligence and Informatics (SAMI 2015), Herl'any, Slovakia (2015)
6. Kosinar, M.A., Czopik, J., Štolfa, J.: Formal knowledge framework for software processes architecture. Front. Artif. Intell. Appl. EJC 2015 (2015)
7. Kosinar, M.A., Stolfa, J., Stolfa, S.: Knowledge support for software processes. Front. Artif. Intell. Appl. **272**, 205–223 (2014)
8. Štolfa, J., Štolfa, S., Kožusznik, J., Moudrá, T.: Business process formal modeling in graphical ontology tool—Functional view. In: Proceedings of the International Conference on Management of Emergent Digital EcoSystems, MEDES, pp. 187–188 (2012)
9. Van Der Aalst, W.M.P.: Process mining in the large: a tutorial. Bus. Intell. 33–76 (2014)
10. Van Der Aalst, W.M.P., Weijters, A.J.M.M.: Process mining: a research agenda. Comput. Ind. **53**(3), 231–244 (2004)
11. Van Der Aalst, W., Weijters, T., Maruster, L.: Workflow mining: discovering process models from event logs. IEEE Trans. Knowl. Data Eng. **16**(9), 1128–1142 (2004)

Software Process Resource Utilization Simulation Using CPN

Jan Czopik, Jakub Stolfa, Svatopluk Stolfa, Michael Alexander Košinár and Ivo Vondrák

Abstract In software process engineering, the ability to simulate software process before it is deployed to some kind of workflow system for automated execution allows us to do a simulations and tune the process to maximum efficiency, stripping it off any defects it might have along the way. Our simulation of software process (the dynamic part) is extended by static aspects (mainly resources) and thus we could obtain even more accurate data on how software process behaves under scenarios possible only by adding resources into the equation. And of course, use those results to analyze, verify and improve the process itself. All of this using standard UML and CPN, based on our Unified Process Meta-model.

Keywords Software process · UML · CPN · OWL · Resource utilization · Unified process Meta-model

---

J. Czopik · J. Stolfa (✉) · S. Stolfa · M.A. Košinár · I. Vondrák
Department of Computer Science, VŠB - Technical University of Ostrava, 17. Listopadu 15, Ostrava-Poruba, Czech Republic
e-mail: jakub.stolfa@vsb.cz

J. Czopik
e-mail: jan.czopik@vsb.cz

S. Stolfa
e-mail: svatopluk.stolfa@vsb.cz

M.A. Košinár
e-mail: michal.kosinar@vsb.cz

I. Vondrák
e-mail: ivo.vondrak@vsb.cz

A. Abraham et al. (eds.), *Proceedings of the Second International Afro-European Conference for Industrial Advancement AECIA 2015*, Advances in Intelligent Systems and Computing 427, DOI 10.1007/978-3-319-29504-6_5

1 Introduction

Business process modeling basically means to capture the structural, behavioral and functional aspects of the modeled company, i.e. modeling from centralized perspective collecting information about activities, conditions, control paths, objects, roles and artifacts along with internal rules and constraints. Most of the time, semiformal specification was sufficient and fulfilled all the needs. Traditional way of specification became not sufficient when a complexity of processes started to grow. The need to have a stable reusable models that can be easily used for simulation has raised [1, 5, 8]. Many research groups have been investigating various methodologies, adaptive process techniques, etc. [7, 12]. Other solutions involve utilization of formal systems, knowledge bases and mathematical models (e.g. Discrete-Event Simulation, System Dynamics, etc.). By the implementation of formal rules and facts, the process model and its execution may be more adaptive to an unexpected situation and events.

This paper proposes a combined solution of UML's ease of use and understandability and CPN's strong semantics and verification, validation and simulation possibilities to allow process specialists to incorporate static aspects (mainly resources) to their models and thus enhance the simulation and analytical capabilities. The proposed approach benefits from both semiformal and formal approaches with the main aim to create an easily understandable and applicable method. Our approach is based on Unified Process Meta-model introduced in [9] and it is an outcome (and refinement) of our previous work presented in [2, 3].

The paper is organized as follows: Sect. 2 covers the state of the art in the field of our research; Sect. 3 introduces the approach and gives short introduction to the problematic of resource handling in process models; Sect. 4 introduces modeling details with UPMM and the transformation rule which is the very core of the approach; Sect. 5 briefly discusses methods of analysis and verification; Sect. 6 concludes benefits of this approach and outlines future work.

2 State of the Art

These days process modeling presents many modeling techniques as is mentioned in Vergidis's paper [16]. However according to paper [4] software process has been characterized as "the most complex endeavor humankind has ever attempted". It is caused by that the software process is quite specific [3].

In this paper we are focused on the converting software process models modeled in UML2 Activity diagrams to their Colored Petri Nets (CPN) counterparts to get advantages of well-formed formal language. Some of these advantages might include: verification of the model, performance evaluation, state space analysis and simulation.

Some of the researchers [4, 11, 13, 14] are focusing on using Petri nets for resource utilization modeling and simulation. They are mainly using colorless Petri nets meaning the expressivity of their models is limited and some of them are trying to overcome it by supplying the net with additional external software components which are not native to Petri nets [11] and therefore all the aspects of the Petri net itself are not present in the structure of the net and therefore the verification and analysis can't be sound. Our solution is based solely on CPN and their core features so everything is built-into the structure of the net and inscriptions providing sound analysis and verification. Others [4, 14] deal only with rules specification but there is no proper connection to an executable model.

For modeling, simulation and analysis we can use CPN tools that provide means for syntax checking, efficient simulation of either timed or untimed nets and their analysis.

The strongpoint of our approach is its simplicity with maintaining completeness and simple extensibility.

3 Our Approach and UPMM Metamodel

We have introduced the basics of our approach in previous papers [2, 3]. Now, we are focusing on definition of transformation rules mainly concerning resources. The whole approach is based on and benefits from our unified Process Meta-model (UMPMM). There are many process meta-models each focused on slightly different process aspects. Our UPMM is combining all the metamodels and adds some other aspects to form a complex generally usable metamodel and is a basement for all necessary models in our methodology for modeling [10].

Figure 1 shows basics of our modeling methodology. UPMM is a metamodel and is described in OWL. Based on UPMM, our modeling approach starts from formal definitions of modeling prescriptions e.g. some constraints, A has to follow B, etc. These prescriptions are written down in OWL. Then the semiformal model is done in UML. Semiformal model that follow formal process prescription, describes the desired process model, [9, 10]. Process models then can be performed in reality or simulated.

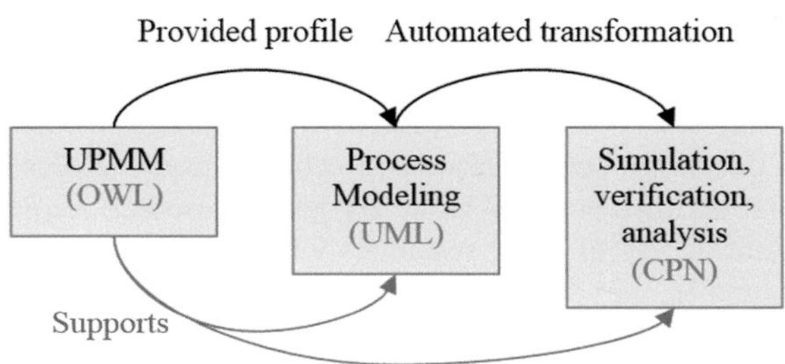

Fig. 1 Overview of the approach

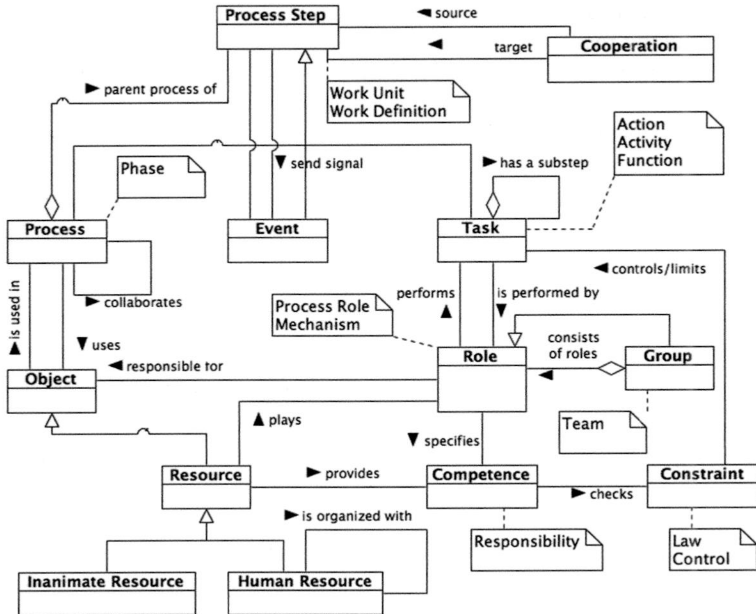

Fig. 2 Part of UPMM regarding resources and competences

Our UML to CPN transformation approach is based on a small part of the UPMM shown in Fig. 2. It shows the relationship between Process step (whether it is an Event, a Task or a sub Process) and some Role in which some Resource (Inanimate or Human) is then executing the Process step. It also shows a relationship between Resource and Role through Competence. The Competence can be described as some sort of skill a Resource might have to offer. In our previous work, we haven't taken competences into account because we wanted to make the approach as easy to understand as possible. But we designed the approach to be extensible.

To achieve even more precise definition of software process and get more precise simulation and analysis reports regarding resource utilization, one cannot settle for specifying resources just as role based. As in real world, resources (whether inanimate or human) can appear under roles. Of course, resource may serve under more than just one role. For example resource printer may serve the role printer but may also serve as coffee table. To give you one example that is more likely to appear in Software process we shall consider a resource employee which is a developer but can also be a technical leader. Roles are important from constraint and security point of view. By giving resource a role, we can then use the role to limit access of other resources which are not in that role to specific activity. For example, only a resource under role customer can create requirements (activity) in our Software process, by stating this, we are excluding developers, project managers and others from being able to execute this activity. Roles can be also useful for grouping resources and thus creating an alias which can be used, instead of stating individual resources in the model.

We stated that using role based resources is not enough. We need to introduce a more precise mechanism. In the real world, each of us has a skill set which defines what we can do. We call each individual skill a competence, which can be view as a pair of skill and level of the skill (how good we are in that skill). So competences define what we can do and how good we can do it. The skill level can represent how fast we can do certain task or how precise the result will be.

The resource is not defined just by its role, but have competence. The relation is depictured in Fig. 2.

4 Resource Modeling and Transformation

In previous chapter we have described the relationships between resources, roles and competences. Now let's have a look how are we incorporate it into Activity diagram and, of course, how are we transform it into CPN elements.

Main element of every Activity diagram is an activity (activities), it represents a task (or tasks in case when activity has actions), piece of work to be done. Activity is performed by a resource (preferably in a role) with certain competences. Activity can also consume (or utilize) passive resource (which is not performing the activity itself) and produce an artifact. The difference between consuming and utilizing is straightforward. When a resource is consumed, it is used and lost and can't be reused later in the process (e.g. printing paper). When a resource is utilized, it is used for certain amount of time but after that is free to be used again (e.g. printer).

A part of UML model in Fig. 3 shows how can we model an activity which is performed by a resource (represented by role) with some competence (on some level), consumes a passive resource a produces an artifact. This model follows the UPMM. We have decided to use object and object flow instead of swim lane for modeling an activity being performed by a resource (as we did in [2]) because of more versatility (more than one roles can perform certain activity). You can also notice that UPMM uses one more element between a Competence and Task called

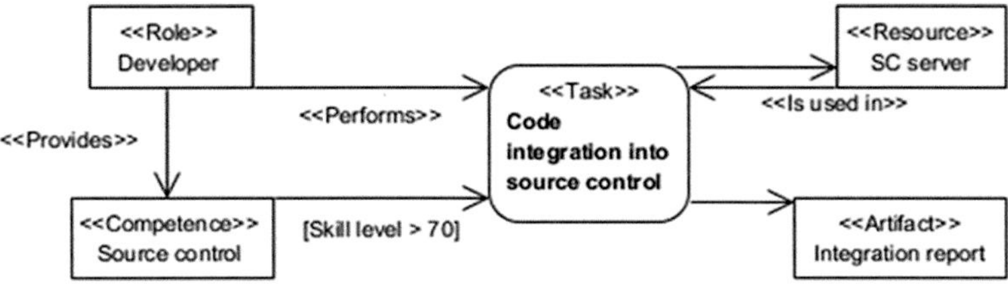

Fig. 3 UML resource modeling complaint to UPMM

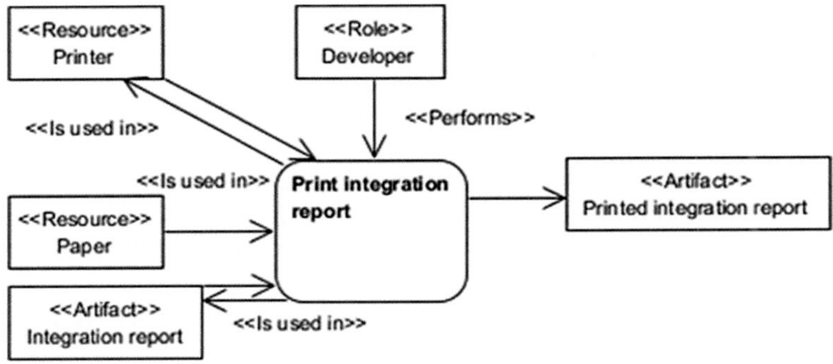

Fig. 4 Difference between resource consumption and utilization

Constraint. In example on Fig. 4 we are using an anonymous Constraint (not exactly specified) represented by object flow with a guard. The guard restricts level of the competence and it reflects the data property Skill level in UPMM. It means that activity *Code integration into source control* can be performed only by a *Developer* with competence *Source control* on level greater than 70.

Figure 4 shows the difference on an UML model between resource consumption and utilization. It goes without saying that an Artifact can be also utilized (as input) in an activity. You can clearly see the difference between consumption (only one output arc from object to activity) and utilization (two arcs between object and activity). You can also see that we can leave out the Competence if we feel it is not required.

4.1 CPN Insciptions

Based on newly added information from the UPMM we have extended [2] the generated colorsets definitions of two new colorsets, SkillLevel, Competences and Constraints.

All colorsets (except SkillLevel colorset, which is independent) are generated based on a concrete model (or set of models). Example of the colorsets regarding Fig. 3 would be generated like (example is written in CPN ML language and provided without core colorsets described in our previous paper [3]):

```
colset Resources = with SCServer;
colset Artifacts = with IntegrationReport;
colset Workers = with Developer;
colset SkillLevel = int with 0..100;
colset Competences = with SourceControl;
colset Constraints = product Competences * SkillLevel;
var sl : SkillLevel;
```

We create a Resources colorset which represents mostly inanimate resources in the model (mostly because all human resources are preferably represented by their roles). Next we have the Artifacts colorset, which declares all different artifacts in the model. The Workers colorset represents the human resources in their roles, i.e. the work force. It is preferred to use roles over specifying concrete resources. Newly added SkillLevel colorsets is just constrained integer colorset mirroring the Skill level property in the UPMM and it is not meant to be used as is (as oppose to Resources colorset for example). The Competences colorset represents the set of competences used in a model (models), it can't be used as is as it was with SkillLevel colorset. The two are meant to be used in conjunction forming a Constraints colorset which is a tuple of competence and Skill level value. The purpose of new variable sl is described in next section.

4.2 CPN Transformation Rule

Because the enhanced transformation rule can be a little bit overwhelming at first glance, let's break it down to two separate areas and describe them in more detail. But before we do that let's point out few generic differences from transformation rule introduced in [2]. At first, you can see that we are using object flow instead of swim lane to specify the role (roles) in which a resource needs to be to perform an activity. You can also see that arc inscriptions in the transformation rule are more generic including multiplicity part of arc inscription (Fig. 5).

In Fig. 6 is part of transformation rule that describes only roles and competences. We have added two new places, the Constraint_pool with tokes of type Constraints and an anonymous place with tokens of SkillLevel. The Constraint_pool represents the same idea as other pools, acting as a reservoir of competences (with different skill level) which resources might have. If an activity needs resource (in a role) with certain competence (possibly with certain skill level), the competence needs to be available in the pool, otherwise the activity has wait for its execution. Constraint_pool place is

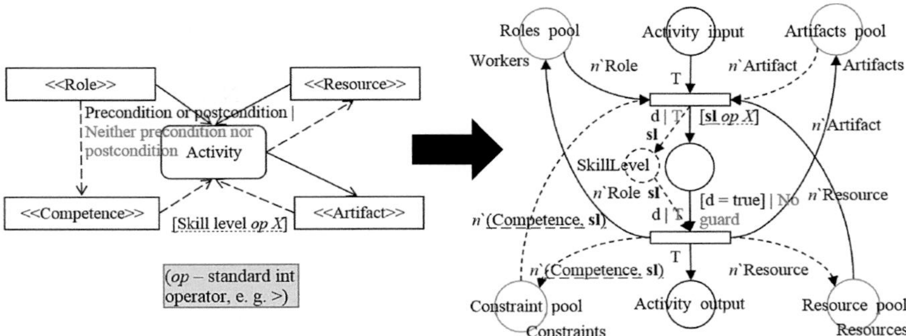

Fig. 5 Transformation rule

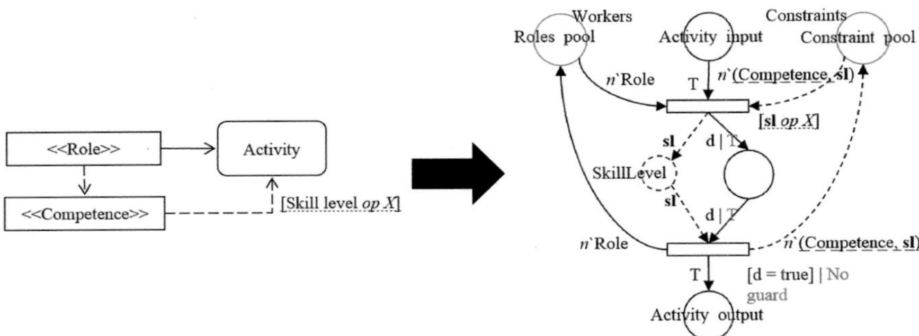

Fig. 6 First part of the transformation rule that concerns roles and competences

connected with outgoing arc with first internal transition of the activity and a n'(Competence, sl) is used as an arc expression. This expression evaluates to multiset tuple, where the tuple is of product colorset, in this case the Constraints colorset. The first part of the expression (in brackets) tells us that we require a specific Competence, the second is a variable which will be (if possible) bound to certain value (from multiset of the place), more in [6]. The variable represents concrete Skill level value of a Competence which might be required by the activity. We use variable for two reasons. The first is, the Skill level was not specified (so we can use any matching Competence without paying attention to concrete level value). The second is, we can use sign operators (like greater than, lesser than etc.) to constrain the level, not only equals or not equals operator. Speaking of constraining the level value, the transformation rule contains a guard expression [sl op X], on the first internal transition of the activity, which mirrors the guard in UML (of course the operator op is transformed from UML to CPN script language accordingly). As next, the anonymous place of type SkillLevel is then used for passing through the already binded sl variable so the incoming arc back to Constraint_pool place from second internal transition returns the same multiset of Constraints it took from Constraint_pool place when executing the first internal transition. Hence the same arc expression for the second arc between Constraint_pool place and an internal transition of the activity.

In Fig. 6 object flows are drawn in dashed line as well as certain matching arcs (and arc expression underlined with that very same dashed line) and the skill level guard is underlined in dotted underline. It is because we are trying to point out that the competence mechanism is optional (hence the dashed lines and underlines) and specifying the skill level is even more optional. If execution of an activity is not restrained by competence mechanism, the arcs from Constraint_pool to internal transitions are not generated as well as the anonymous place used for passing the sl variable. Nevertheless, the Constraint_pool is always generated, it just not connected in case were competences are not needed. If an activity requires a resource with some competence but it doesn't require specific skill level, the skill level guard is not generated into CPN structure.

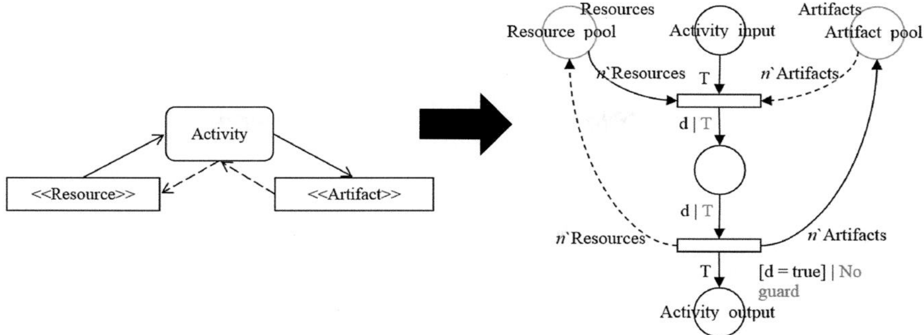

Fig. 7 Second part of the transformation rule that concerns resources and artifacts

On Fig. 7 is part of transformation rule that describes only resources and artifacts. We have decided to refactor artifact related arcs. We are not using the arc with n`ArtifactAsResource expression anymore to model situation where an artifact can be later used as resource. Instead we have replaced it with input arc to first internal transition from Artifact_pool which allows us to use artifacts also as resources. As with the dashed line in competence case, the dashed arc between first internal transition and Artifact_pool place suggest the arc is generated optionally based on whether any artifact is used as resource.

You can also notice the arc between Resource_pool and second internal transition is drawn as dashed. It means it is optional. The arc will be generated when specific resource is utilized. The arc won't be generated when resource is consumed, it means the concrete resource token won't return to Resource_pool place and therefore is lost, just as explained in Sect. 4.

5 Simulation and Analysis

When the models are ready, they are transformed into a CPN net using the information from both, UML and OWL (see on Fig. 1). We can then use two approaches to enhance our models via feedback from the results of the chosen approach.

1. The first approach to improve (not only improve, but also to verify) the process models is to use static analysis technique on the generated net. The technique is called a state space analysis. The basic idea of state spaces is to calculate all reachable states and state changes of the CPN model and to represent these in a directed graph where the nodes correspond to the set of reachable markings and the arcs correspond to occurring binding elements [6]. We can then use the reachability graph to examine properties of the CPN model and look for specific "quality" indicators. In that way we can expose the potential flaws (like a

dead-lock situation or unnecessary activities) and thus fix and improve the process model.
2. The second approach is to use the CPN's natural ability for simulation and run a simulation (or more simulation runs) on the CPN model. Before we actually run the simulation, we put markers on transition and places and mark the transitions and places which will be logged when state changes. We are logging the number of executions in case of transition and marking changes in case of places. From the number of execution we can find potential bottle necks in the process and marking changes give us excellent overview of the resource utilization, detail information can be found in [6].

6 Conclusion and Future Work

In this paper presents a refinement of our current research in terms of core modeling principles and the most important, the transformation rules. We have based our approach on our meta-model called Unified Process Meta-Model because we see great potential in this particular meta-model. The approach was enhanced by adding competences and few important changes were made mainly regarding modeling roles and artifacts. We have mentioned the ways you can analyze the CPN models and interpret data gained from simulation.

In the future, we would like to focus on enhancing the models even more with timed CPN which would allow us to incorporate duration information and make simulation and therefore simulation report (log) data more precise which would result in deeper process analysis and more precise error correction.

Acknowledgment The research was supported by the internal grant agency of VSB—Technical University of Ostrava, Czech Republic, project no. SP2015/85 'Knowledge modelling usage in software engineering'.

References

1. van der Aalst, W.M.P., van Hee, K.M., Houben, G.J.: Modeling workflow management systems with high-level petri-nets. In: De Michelis, G., Ellis, C., Memmi, G. (eds.) Proceedings of the Second Workshop on Computer-Supported Cooperative Work, Petri nets and Related Formalisms 1994, pp. 31–50 (1994)
2. Czopik, J., Kosinar, A.M., Stolfa, J., Stolfa, S.: Addition of static aspects to the intuitive mapping of UML activity diagram to CPN. In: Paper presented at the Proceedings of the 1st International Conference on Afro-European Conference for Industrial Advancement, Addis Ababa (2014)
3. Czopik, J., Kosinar, A.M., Stolfa, J., Stolfa, S.: Formalization of software process using intuitive mapping of UML activity diagram to CPN. In: Paper presented at the Proceedings of the 5th International Conference on Innovations in Bio-Inspired Computing and Applications, Ostrava (2014)

4. Gomez-Perez, A., Ramirez, J., Villazon-Terrazas, B.: An ontology for modelling human resources management based on standards. In: Presented on 11th International Conference on Knowledge-Based Intelligent Informational and Engineering Systems
5. Jennings, N.R., Faratin, P., Norman, T.J., O'Brien, P., Odgers, B.: Autonomous agents for business process management. Int. J. Appl. Artif. Intell. **14**(2), 145–189 (2000)
6. Jensen, K., Kristensen, L.M.: Coloured Petri nets: modelling and validation of concurrent systems. Springer, Dordrecht (2009). ISBN 978-3-642-00283-0
7. Kammer, P.J., Bolcer, G.A., Taylor, R.N., Hitomi, A.S., Bergman, M.: Techniques for supporting dynamic and adaptive workflow. Comput. Support. Coop. Work **9**(3–4), 269–292 (2000)
8. Klein, M., Dellarocas, C.: A knowledge-based approach to handling exceptions in workflow systems. Comput. Support. Coop. Work **9**(3–4), 399–412 (2000)
9. Kosinar, A.M., Czopik, J., Stolfa, J., Penhaker, M.: Knowledge framework for clinical processes architecture and analysis. In: Paper presented at the Proceedings of the 13th International Symposium on Applied Machine Intelligence and Informatics, Herlany (2015)
10. Košinár, M., Štolfa, J., Štolfa, S.: Knowledge support for software processes (revisited). In: Paper presented at the 24th European-Japanese Conference on Information Modeling and Knowledge Bases, Kiel, Germany (2014)
11. Kuchar, S., Martinovic, J.: Human resource allocation in process simulations based on competency vectors. In: Presented on 7th conference on Soft Computing Models in Industrial and Enviromental Applications (2013)
12. Narendra, N.C.: Flexible support and management of adaptive workflow processes. Inf. Syst. Front. **6**(3), 247–262 (2004)
13. Pesic, M., van der Aalst, W.M.P.: Modelling work distribution mechanisms using colored Petri nets (2007)
14. Radevski, V., Trichet, F.: Ontology-based systems dedicated to human resources management: an application in e-recruitment, France (2006)
15. Raffo, D.M.: Modeling software processes quantitatively and assessing the impact of potential process changes on process performance, Carnegie Mellon University (1996)
16. Vergidis, K., Tiwari, A., Majeed, B.: Business process analysis and optimization: beyond reengineering. IEEE Trans. Syst. Man Cybern. Part C Appl. Rev **38**(1), 69–82 (2008)

Scenario-Based Evolutionary Approach for Robust RCPSP

Hayet Mogaadi and Besma Fayech Chaar

Abstract The present paper deals with the resource-constrained project scheduling problem with uncertain activity durations. Based on scenarios, we investigate two robust models, the min-max model which focuses on the minimization of the absolute robustness objective and the min-max regret model having the object to minimize the absolute regret. We propose an adaptive robust genetic approach with a sophisticated initial population and a Forward-Backward Improvement heuristic. The proposed algorithm is applied for the PSPLIB J30 data set with modified activity durations. Obtained results show the performance of the genetic algorithm combined with the improvement heuristic compared with the basic version. Different perturbation levels were tested to determine the corresponding performance degradation.

Keywords RCPSP under uncertainty · Robustness · Scenario · Genetic algorithm · Forward-backward heuristic

1 Introduction

The Resource-Constrained Project Scheduling Problem (RCPSP) consists in scheduling a set of activities over scare resources subject to precedence and resource constraints while optimizing several objectives. The most common objective is to minimize the project duration (so called the makespan). The deterministic version of the RCPSP was classified as NP-hard [1]. However, in real-life, project scheduling process is subject to unexpected disruptions caused by different factors such that activity delays, new arrival activities to be inserted, resources broken down, etc.

Recently, the main researchers'reflects on project scheduling were concentrated in managing uncertainty to huge against schedule disruption. The robust optimization

H. Mogaadi (✉) · B.F. Chaar
National Engineering School of Tunis, Tunis El Manar University, Tunis, Tunisia
e-mail: Hayet.Mogaadi@insat.rnu.tn

B.F. Chaar
e-mail: Besma.Fayech@insat.rnu.tn

A. Abraham et al. (eds.), *Proceedings of the Second International Afro-European Conference for Industrial Advancement AECIA 2015*, Advances in Intelligent Systems and Computing 427, DOI 10.1007/978-3-319-29504-6_6

problem aims to find a solution having the best worst-case performance across a finite or infinite set of scenarios. A scenario represents a problem realization founded by matching fixed values to uncertain problem parameters. Consequently, a robust schedule is a schedule having the ability to absorb some level of unexpected events without rescheduling [2]. Typically the robustness objective optimizes the mean value or the worst case performance.

In this context, we are motivated to investigate the RCPSP with uncertain activity durations by applying heuristics and metaheuristics. Activity variability is modeled by scenarios. We tackled two robust models: the min-max and the min-max regret model. The first model aims to minimize the absolute robustness across all considered scenarios which corresponds to the worst case project makespan; the second model minimizes the absolute regret robustness among scenarios. A regret is defined by the deviation between a realized makespan and the optimal makespan for that project realization. We propose a Robust Genetic Algorithm (GA) combined with a Forward-Backward Improvement heuristic (FBI), and referred as (RGAFBI). The choice of the approximated approaches is justified by the high complexity of the studied scenario-based robust models.

This paper is organized as follows. Section 2 presents a literature review of the scheduling under uncertainty especially for the project scheduling problem. Section 3 gives the conceptual model for a static RCPSP and studied robust project scheduling models. Section 4 focuses on the proposed robust approach based on an improved genetic algorithm incorporating a robustness objective. Section 5 presents computational results. Finally, Sect. 6 concludes the paper.

2 Literature Review

Static RCPSP was widely studied in the literature. Heuristics and metaheuristics were successfully applied to the project scheduling such as Genetic Algorithm, Simulated Annealing, Local Search, and Sampling methods. Efficient surveys are proposed in the following references [3, 4].

However, in real-life, project scheduling process is subject to disruptions due to different causes: activity delays, unexpected events, resources broken down, etc. Many authors have investigated efficient efforts on scheduling under uncertainty yielding to two main classes of approaches: the reactive and proactive approaches. A complete survey was proposed by Herroelen and Leus [5] for the project scheduling under uncertainty. Next, we present, from the literature, the main works for robust scheduling problems.

In [6], Al Fawzen and Haouari have studied the RCPSP focusing on how to generate schedules having *good quality* which was defined in terms of performance and vulnerability facing to reworks and suppliers delays. To achieve this objective, they have proposed a bi-objective model with the minimization of the makespan and the maximization of the robustness. Robustness is defined by the *activities'slacks* and the problem is solved by a tabu search heuristic. A comparative work was proposed in [7] for the same bi-objective model for RCPSP. However, the robustness was evaluated in terms of project *floating time*.

Chtourou and al. [8] have investigated on robust RCPSP proposing a two-stage priority-rule-based heuristic. They studied various robustness measures based on *free activity slack.*

Based on scenarios, the work of Yamashita and al. [9] is focused on the project scheduling problem with resource availability cost when activity durations are uncertain. They performed the min-max regret and the mean-variance model. The problem was solved by a scatter search procedure compared with a multi-start heuristic.

The work presented in [10] deals with RCPSP under uncertainty. Artigues and al. have proposed a robust scenario-based bi-level model based on PLNE that minimizes the absolute and relative regret robustness. A solution was defined as a scheduling policy giving the rule to consider at decision times. They have developed, in first, an exact method that takes excessive computational time considering instances with medium size. So the authors have resort to heuristic procedures.

In [11], Kouvelis and Yu have proposed the application of scenario-based robustness to different operation research problems. The authors have reported that the min-max objective leads to conservative schedules when anticipating the worst case among all scenarios. However, less conservative schedules are generated by the min-max regret objective as it takes into account optimal solutions that should be scheduled. For other robust scheduling applications, in [12], the authors have applied metaheuristics to the robust one machine problem. The contribution of this work consists of the application of a modified genetic algorithm to the tackled problem with uncertain processing time.

3 Robust Project Scheduling

3.1 Standard Project Scheduling

A static version of RCPSP [3] is given by a set A of n activities to be processed on a set K of m resources. Every activity i has to be performed during a fixed processing time p_i and requires r_{ik} units of resource type k. Every resource k has a limited capacity R_k that must not be exceeded during the execution. Activities are not allowed to be interrupted during execution. Precedence constraints (3) perform that the start time of an activity i is permitted only when all its previous activities are finished. The resource constraints (4) specify that the use of every resource type, at every instant, does not exceed its capacity.

The objective of the standard RCPSP is to find a feasible schedule defined by a list of activities finish times F_i (or start times S_i) subject to precedence and resources constraints such that the total project duration (denoted by C_{max}) is minimized. Let $A(t)$ be the set of activities which are executed at time t, and P_j the set of predecessors of the activity i. A deterministic version of the standard RCPSP is formulated as follows:

$$\min F_{n+1} \tag{1}$$

$$\text{subject to} \quad F_h \leq F_j - p_j, j = 1, \ldots, n+1, h \in P_j, \tag{2}$$

$$\sum_{i \in A(t)} r_{ik} \leq R_k, k \in K, t >= 0, \tag{3}$$

$$F_i \geq 0. \tag{4}$$

The given problem can be represented by a graph $G = (V, E)$ where the set of nodes V is defined by project activities augmented with two dummy nodes 0 and $n + 1$. The set E represents arcs which are defined by precedence relations.

3.2 Robust Models Description

In this work, the activity uncertainty is modeled by a scenario approach. Let Σ be a set of l scenarios. We denote $f(x, \sigma_i)$ the performance of the solution x on scenario σ_i and f_i^* the best performance of that scenario where $\sigma_i \in \Sigma$. We investigate two scenario-based robust models: the min-max model (5) which minimizes the maximum of the makespan for all elements in Σ; and the min-max regret model (6) that optimizes the absolute regret robustness. Precedence and resource constraints for both robust models are defined by Eqs. (3) and (4), respectively. We suppose that a solution x for the deterministic optimization problem remains feasible in the uncertain environment in terms of precedence constraints.

The minmax robust model:

$$\min \max_{\sigma_i \in \Sigma} (f(x, \sigma_i)). \tag{5}$$

The minmax-regret robust model:

$$\min \max_{\sigma_i \in \Sigma} (f(x, \sigma_i) - f_i{}^*). \tag{6}$$

4 Robust Evolutionary Approach

4.1 Description

The proposed robust approach includes three steps. On the first step, we generate, according to the initial deterministic problem P_0, a set Σ of l scenarios which corresponds to the optimization set. For each scenario, we compute an approximated optimal solution by applying an iterative multi-pass sampling heuristic. The list of all approximated optimal solutions are then used on the next step which performs

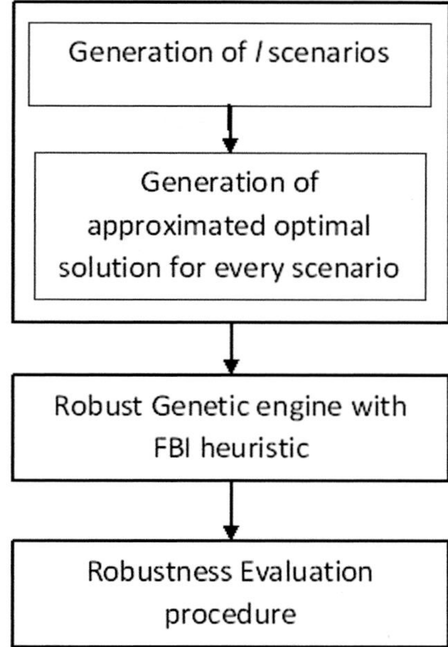

Fig. 1 The proposed robust scenario-based evolutionary approach

a GA combined with a FBI heuristic. Evaluation procedure is, then, running at step three using a set of new generated scenarios (Fig. 1).

To evaluate the min-max regret robustness, we need to compute approximated optimal solutions for each scenario of the optimization set which would automatically accentuate the problem complexity. To ensure a compromise between performance and computational time, we apply the Regret Based Biased Random Sampling method (RBRS) combined with the FBI heuristic [13].

4.2 Improved Genetic Algorithm

The Genetic Algorithm (GA) is a population-based metaheuristic that evolves simultaneously a set of individuals which are subject to genetic operators. This algorithm represents a sophisticated method for solving hard optimization problems. Next, we briefly describe the main operators and the applied improvement heuristic.

4.2.1 Representation

A chromosome is defined by an activity list which is composed by a sequence of $n + 2$ activities (including dummy activities 0 and $n + 1$) satisfying precedence constraints. Initial population is created using a *multi-pass sampling heuristic* in order to generate initial individuals with a good quality which could influence better on the global genetic performance.

4.2.2 Decoding Procedure

The Serial Schedule Generation Scheme (S-SGS) [3] represents the base of most scheduling algorithms for RCPSP. We choose to use the S-SGS as a decoding procedure which generates active (left-justified) schedules where any left shifting of an activity causes automatically a constraints violation.

4.2.3 Crossover and Mutation

The applied two-point crossover is the Linear Order Crossover (LOX). For each one of the selected parents P1 and P2, two crossover-points e_1 and e_2 are chosen randomly in the activity list. New offspring E1 is initially formed by copying the two partial of P1 delimited by indices (0, e_1) and (e_1 + 1, n + 1). The rest of activities is found by the scan of the parent P2 from index 0, whenever we have an activity that doesn't exist in E1 we copied it at the first place in the last offspring. The second child is constructed as E1. The mutation operator consists in performing all possible permutations between two activities in the current chromosome without violating precedence constraints.

4.2.4 Selection

The role of the selection operator is to choose individuals that will survive for the next generation. We adopt the *ranking selection*which consists in sorting parents and offspring on the same vector respect to their *fitness*. Best individuals having best performance values are then copied on the new population.

4.2.5 Robust Fitness

Each individual is decoded according to each scenario of the optimization set to get the corresponding project schedule. The set of project finish times for all scenarios are used to evaluate the robustness objective. For example, if we are interested by the min-max optimization model, the *robust fitness* will corresponds to the maximum makespan

4.3 The Forward-Backward Improvement Heuristic

As mentioned in the previous section, the FBI heuristic combined with the GA provides relevant results for project scheduling [14]. A left-justified schedule generated by the S-SGS procedure is right-justified in a forward recursion applied to the precedence-reverse graph. Considered priority list to generate schedule is obtained by sorting activities' finish times in a non-increasing order. In the backward scheduling, we construct, firstly, a priority list defined by the minimum of starting time which is computed as the total project completion time minus its finish time. Secondly, we re-apply the S-SGS according to the generated priority list.

5 Experiments

Based on the PSPLIB benchmark [15], we performed the J30 set that contains 480 projects with 30 activities and 4 cumulative resource types. The proposed algorithms were coded in java language, and run on a personnel computer with Intel Core i5, 2.50 GHz. For each instance, a fixed number of scenarios were generated independently for both optimization and evaluation phase. A project realization is obtained by selecting randomly a set of activities to be altered, for each of which, an activity delay δ is drawn from a uniform distribution $U(1, maxDealy)$, and added to the deterministic activity duration; where the parameter *maxDelay* denotes the maximum activity delay which is fixed to 10 time units.

5.1 Robust Performance Analysis

In order to evaluate the robust evolutionary approach, we define three performance measures calculated for every problem instance, and averaged by the number of tested instances:

- *dev_opt*: the deviation of the obtained schedule makespan C_{max} from the lower bound *LB* (7), if the problem is resolved to the optimality, we evaluate the deviation from the optimal solution (in percent);
- *gap_opt*: the gap between the simulated expected makespan $E(C_{max})$ and the lower bound (optimal makespan) of the corresponding deterministic problem, referred as *gap_opt* (in percent);
- *dev_std*: the standard deviation (9) of the obtained robust solution x_r simulated over l scenarios $\sigma_i \in \sigma$.

Table 1 Results of the J30 instances (simulation over 100 replications)

GA

♯ *schedules*	*dev_opt*	*nb_opt*	*gap_opt*	*dev_std*
1000	0.51	391	12.95	4.4489
5000	0.28	417	12.84	4.4507

Table 2 Results of the J30 instances (simulation over 100 replications)

GAFBI

♯ *schedules*	*dev_opt*	*nb_opt*	*gap_opt*	*dev_std*	*max dev_std*
1000	0.36	414	12.30	4.2796	10.1851
5000	0.17	441	12.19	4.2668	6.4358

$$dev\_opt = \frac{C_{max} - LB}{LB} * 100, \tag{7}$$

$$gap\_opt = \frac{E(C_{max}) - LB}{LB} * 100, \tag{8}$$

$$(dev\_std)^2 = \frac{1}{l} \cdot \sum (f(x_r, \sigma_i) - E(C_{max}))^2. \tag{9}$$

First experiments show the performance degradation of an optimal solution for the deterministic RCPSP when it is exposed to unexpected perturbations. So, we started by testing the two algorithms: the basic GA, and the improved version denoted by GAFBI with 1000 and 5000 generated schedules as stopping criteria. Results are given in Tables 1 and 2.

In both tables Tables 1 and 2, the first column shows the number of generated schedules, and the third column indicates the number of optimal solutions which are founded by the corresponding applied algorithm. In addition to the cited considered performance measures present in these tables, we calculated the maximum standard deviation given in column 6, Table 2.

According to the latter tables, we observe that obtained best schedules for the J30 set with a relative optimality deviation equal to 0.51 % have an optimality gap of 12.95 % on disturbed mode, with respect to 100 replications. As predicted, the improved GA outperforms the basic GA in terms of optimality gap and solution robustness. Nevertheless, the standard deviation has been slightly increased when increasing the number of generated schedules (5000). The main reflects are towards the effectiveness of the evolutionary algorithm as a robust approach.

The second set of experiments performs the Genetic Algorithms with the robustness objective (RGA) and the improved version denoted by RGAFBI. The stopping criterion is defined by 1000 generated schedules. Both Tables 3 and 4 show results with the same considered performance measures: *gap_opt*, and *dev_std* which are evaluated with the min-max objective, and the min-max regret robustness objective.

Table 3 Results of RGA for the J30 instances (simulation over 100 replications)

RGA	–1000	schedules		
robustness	*gap_opt*	*max dev_opt*	*dev_std*	*max dev_std*
MINMAX	13.46	27.05	4.2386	5.5664
MAX-REG	12.62	24.87	4.2069	6.3751

Table 4 Results of RGAFBI for the J30 instances (simulation over 1000 replications)

RGAFBI	–1000	schedules		
robustness	*gap_opt*	*max dev_opt*	*dev_std*	*max dev_std*
MINMAX	12.87	29.60	4.1502	5.5664
MAX-REG	12.31	24.28	4.1633	6.7931

The number of scenarios on the optimization phase is set to 10, while simulations were run for 100 test scenarios.

Compared with the deterministic mode, the deviation from optimality *gap_opt* has been increased with both RGA and RGAFBI algorithms. However, the standard deviation and the corresponding maximum value have been decreased. Results show that robust approaches guarantee solutions that are more controlled and do not diverge in disturbed mode unless the possible performance degradation. Another observed point in this experiment is that the min-max regret objective gives solutions with less optimality deviation than the min-max objective. As demonstrated for deterministic case, the effectiveness of the FBI heuristic remains valid for robust optimization.

5.2 *Varying Perturbations*

With stochastic approaches, activity durations are drawn from known stochastic distributions where deterministic activity durations are usually taken as expectation (mean values). However, with robust project optimization, distributions are not considered and the decision maker has to fix activity delays independently from deterministic durations. The aim of this experiment is to show the performance degradation, in terms of makespan, according to different activity delays. By this way, we give, in one hand, an idea about the robust algorithm behavior when varying uncertain parameters. On the other hand, we hope to get more opportunities comparing with others works in the literature especially in connection with the stochastic project scheduling.

The experiment consists in performing the RGA with the min-max objective for the first twenty instances of the J30 set with 1000 generated schedules. Obtained results are depicted in Fig. 2 drawing two graphs (a) and (b): the optimality gap

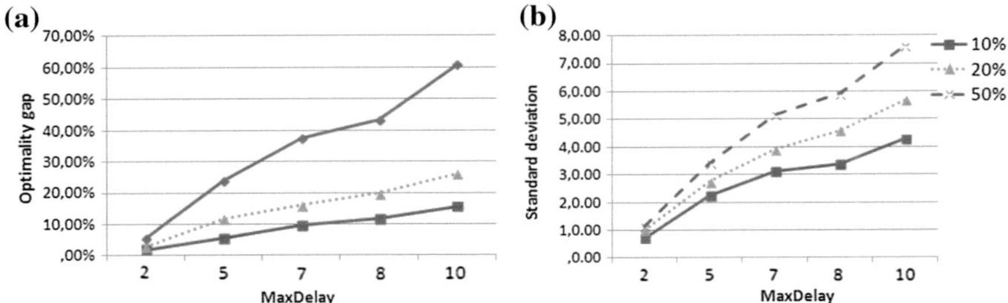

Fig. 2 Varying the maximum activity delay

variation and the standard deviation variation, respectively, when the maximum activity delay *maxDelay* takes values in $\{2, 5, 7, 8, 10\}$. For each graph, three curves are plotted respect to the percentage of the altered activities equal to 10, 20, and 50 %.

In [16], Ashtiani and al. have tackled the RCPSP problem with stochastic activity durations, three variability level were studied. A uniform distribution with a support $[0; 2 * p_i]$ and a variance $\frac{p_i^2}{3}$ was qualified as a medium variability. In our robustness case study, we consider that with a $maxDelay = 10$ should correspond to an important perturbation level in which the activity delay has a variance of $((MaxDelay - 1)^2/12)$. As shown in the Fig. 2, the optimality gap is increased when *maxDelay* is upper to 8 time units, especially with 20 and 50 % of altered activity.

6 Conclusion

This paper proposed a robust scenario-based approach for RCPSP with uncertain activity durations. We tackled two robust models: the min-max and the min-max regret model, that optimizes: the worst case scenario and the maximum makespan deviation between a scenario and the optimal solution of that scenario, respectively. We developed an evolutionary algorithm combined with FBI heuristic. Results on PSPLIB J30 set have shown the effectiveness of this hybridization for the robust optimization case. For further work, we aim to perform large sized problems and investigate more efficient perturbation levels for the decision making.

References

1. Blazewicz, J., Lenstra, J., Rinnooy Kan, A.: Classification and complexity. Discret. Appl. Math. **5**, 11–24 (1983)
2. Davenport, A.J., Beck, J.C.: A survey of techniques for scheduling with uncertainty (2000). http://tidel.mie.utoronto.ca/publications.php

3. Kolisch, R., Hartmann, S.: Heuristic algorithms for the resource-constrained project scheduling problem: classification and computational analysis. In: Weglarz, J. (ed.) Project Scheduling. International Series in Operations Research and Management Science, vol. 14, pp. 147–178. Springer, US (1999)
4. Kolisch, R., Hartmann, S.: Experimental investigation of heuristics for resource-constrained project scheduling: an update. Eur. J. Oper. Res. **174**, 23–37 (2006)
5. Herroelen, W., Leus, R.: Project scheduling under uncertainty: survey and research potentials. Eur. J. Oper. Res. **165**, 289–306 (2005)
6. Al-Fawzan, M., Haouari, M.: A bi-objective model for robust resource-constrained project scheduling. Int. J. Prod. Econ. **18**, 1–13 (2004)
7. Abbasi, B., Shadrokh, S., Arkat, J.: Bi-objective resource-constrained project scheduling with robustness and makespan criteria. Appl. Math. Comput. **180**(1), 146–152 (2006)
8. Chtourou, H., Haouari, M.: A two-stage-priority-rule-based algorithm for robust resource-constrained project scheduling. Comput. Ind. Eng. **55**, 183–194 (2008)
9. Yamashita, D.S., Armentano, V.A., Laguna, M.: Robust optimization models for project scheduling with resource availability cost. J. Sched. **10**, 67–76 (2007)
10. Artigues, C., Leus, R., Talla Nobibon, F.: Robust optimization for resource-constrained project scheduling with uncertain activity durations. Flex. Syst. Manag. J. **25**(1–2), 175–205 (2013)
11. Kouvelis, P., Yu, G.: Robust Discrete Optimiaztion and Its Applications. Klower Academic Publisher (1997)
12. Sevaux, M., Sorensen, K.: A genetic algorithm for robust schedules. 4OR 2, pp. 129–147 (2004)
13. Tormos, M.P., Lova, A.L.: A competitive heuristic solution technique for resource-constrained project scheduling. Annals OR, (1–4), 65–81 (2001)
14. Valls, V., Ballestin, F., Quintanilla, S.: Justification and rcpsp: a technique that pays. Eur. J. Oper. Res. **165**, 375–386 (2005)
15. Kolisch, R.: Psplib-a project scheduling problem library. Eur. J. Oper. Res. **103**, 483–496 (1997)
16. Ashtiani, B., Leus, R., Aryanezhad, M.B.: A novel class of scheduling policies for the stochastic resource-constrained project scheduling problem. Available at SSRN 1368714 (2008)

Internet of Things: Overview, Sources, Applications and Challenges

Nour E. Oweis, Claudio Aracenay, Waseem George, Mona Oweis, Hussein Soori and Vaclav Snasel

Abstract Nowadays, Internet of Things (IoT) is growing rapidly. Billions of devices are expected to be associated in the coming future. Smart and sensing devices have impacted the Big Data area through huge data generation and gathering during the communication between new physical object. The challenges in this communication process can be seen in the necessity to handle huge amount of data that can serve different purposes in various fields such as, medical, social, commercial, industrial and scientific fields. This study aims at presenting some of the most recent advances in IoT. We divided the study into four main parts: short history, Big Data concept, IoT sources (hardware and software), and finally some challenges and future expectations. The objective of the study is to give readers an updated description of the IoT and its impact on Big Data.

Keywords Internet of things · Big data · Smart devices · Internet of things applications

N.E. Oweis (✉) · W. George · H. Soori · V. Snasel
Faculty of Electrical Engineering and Computer Science,
Department of Computer Science and IT4Innovations,
VŠB-Technical University of Ostrava, 17.Listopadu 15/2172,
708 33 Ostrava, Poruba, Czech Republic
e-mail: nour.easa.oweis.st@vsb.cz

W. George
e-mail: waseem.george.oweis.st@vsb.cz

H. Soori
e-mail: sen.soori@vsb.cz

V. Snasel
e-mail: vaclav.snasel@vsb.cz

C. Aracenay
Department of Industrial Engineering, University of Chile Santiago, Santiago, Chile
e-mail: caracena@ing.uchile.cl

M. Oweis
Department of Computer Science, Middle East University, Amman, Jordan
e-mail: mona_owais2006@yahoo.com

A. Abraham et al. (eds.), *Proceedings of the Second International Afro-European Conference for Industrial Advancement AECIA 2015*, Advances in Intelligent Systems and Computing 427, DOI 10.1007/978-3-319-29504-6_7

1 Introduction

Currently there are more than two billion users of smart technologies including smart phones, home, city, business and entertainments applications [1, 2]. These smart object facilities allow machines to communicate with or without user intermediary means which leads to the term known as the IoT. However, all these modern techniques are creating a huge amount of structured, semi-structure and unstructured data which eventually leads to the increase in data capacity, variety of storage and processing systems. Haller et al. [3] define the term IoT as: "a world where physical objects are seamlessly integrated into the information network and where the physical objects can become active participants in business process".

We divide this study into four main sections: the first section is review and short history of IoT. The second section reviews the main concept of Big Data and its relation to IoT. The third section involves IoT sources (hardware and software). The fourth section review two main IoT challenges: data mining and security.

2 Short History of the IoT

IoT is not new to computer science, yet it has taken a new paradigm that combines the huge amount of smart objects—that are widely growing, their ability to be remotely connected—and data sharing from several resources.

In 1990 the first *things* was born and known as the *Internet Toaster* by John Romkey and Simon Hackett in their first connected Toaster device powered by the Internet. After that in 1991, Interop added a small robotic to pick up a slice of bread into the toaster to be a full automating machine [4].

Ten year later in 1999, the Internet Toaster was connected to the Internet. The name of the IoT has been widely known when Kevin Ashton coined the term *Internet of things* [5]. In the same year, Dr Andy Stanford-Clark of IBM, and Arlen Nipper of Arcom (now Eurotech) introduced the first machine-to-machine protocol for connected devices which is called (MQ Telemetry Transport (MQTT)) [6, 7].

One year later, LG Company announced plans for the first refrigerator connected to the Internet known as *LG Internet Refrigerator* [8]. Nowadays, after more than 40 years of Internet emergence, more than 20 years of the birth of *Internet Toaster*, and more than ten years after the coinage of the term *Internet of Things*, 13 billion devices are connected worldwide (2 devices per person) and by 2020, 50 billion devices are expected to be connected to the Internet (6 devices per person) [9]. In the next section Big Data concept and characteristics are presented with their relation to the IoT.

3 Big Data Concept

Massive set of data that is so large and complex to be managed by traditional processing applications lead to what is known as Big Data. The term Big Data includes a huge, complex, and abundant structured, semi-structure and unstructured, as well as, hidden data that are generated and gathered from several fields and resources. Gartner define the term Big Data as, "high-volume, high-velocity and high-variety information assets that demand cost-effective, innovative forms of information processing for enhanced insight and decision making" [10]. Big Data can be categorized by its 6Vs characteristics (Volume, Velocity, Variety, Veracity, Viability and Value) [11–14].

IoT is considered the most important source of Big Data, including the hardware devices, and software application. The following section presents the IoT hardware and software that cover the structure of the IoT.

4 IOT Sources (Hardware and Software)

Recently, there are many new IoT hardware devices and software applications that are already deployed and widely spared, some of them will be mentioned in the following subsections.

4.1 IOT Hardware

In today's digital world, there are huge amount of datacenters such as SAPs. This kind of huge and organized datacenter comes with a high tech building, security measures certificates, and cloud-solutions of Big Data [15].

Another example is WSN where there are multiple sensors interacting together for collecting data stream and sending these data to the distributed or centralized system for store, process, and analytics [16]. An example of WSN is a project in Chicago state called the Array of Things Project where they spread the WSN nodes all over the city parts. This makes the city healthier, more livable, and more efficient. Some additional eight nodes will be deployed along Michigan Avenue in Spring 2015 [17].

RFID Tags is a short range communication technology within one way transmission data with unpowered tags by using radio communication signal [18]. For example, Electronic Fuel Delivery System (EFDS) in Hung Gong, powered by Empress TM2.4 GHz Active RFID. The EFDS was developed to help oil companies enhance energy consumption efficiency. Another example of the RFID tags is the E-invoices which are issued instead of paper invoices annually. The Ministry of Finance of Taiwan set out to improve environmental conditions on the East Asia

Island. According to Taiwan Fiscal Information Agency, in 2013, the state was issuing more than 8 billion E-invoices in a year which is equivalent to 80,000 trees and 3200 tons of CO2 emissions. E-invoices issued by the Ministry of Finance of Taiwan was awarded the RFID Green Award by RFID Journal 2014 because it has demonstrated the best use of RFID technology to improve the environment and increase recycling or enhance sustainability. Electronic invoice applies RFID technology and multi devices to replace paper invoices, significantly reducing unnecessary paper waste and carbon dioxide emissions and revealing the key factor of winning the award [19].

NFC tags is a development technology of RFID Tags with a short range communication technology within two way transmission data in very short distances and unpowered tags by using radio communication that are widely commonly used in several applications and smart devices such as smart phone, smart travel, smart home, smart security, smart door, smart posters, electronic payment, smart fitness keychains, business and student cards, etc. [20].

Cloud computing is a highly feasible technology and attracts a large number of researchers to develop it and try to apply it on Big Data problems. At present, there are a huge number of cloud computing storage devices such as Google cloud and Drop box [21, 22].

Smart city is one of the most researched areas based on the application of the IoT data. The U.S. Miami-Dade County is one example. This project is a cooperation between Miami-Dade County in Florida and IBM company in which they closely connect 35 types of key county government departments and Miami City that reduce the water consumption by 20 % and intelligent policing helping police departments reduce the time taken for officers to identify leads, investigate crimes and remove barriers to information sharing with other law enforcement agencies [14, 23].

Smart phones are cell phone hybrid with digital computing capabilities to allow users to interact worldwide with latest technologies and techniques. Nowadays, mostly everyone has a touch screen smart phone/s that allow access to the Internet with all of its facilities. It is predicted that by 2018, each smart phone is expected to generate 2 GB of data monthly [24, 25].

After we described hardware and its extensive uses by smart and sensing device with some examples, in the next subsection, we review the different aspects related with software in the field of IoT.

4.2 IOT Software

Today, Internet technology allows the communication of almost every kind of devices, such as sensors or even appliances. For supporting all these structures, we, generally speaking, need operating systems, protocols, applications and platforms. This will be briefly described in this section to satisfy the requirements needed for each specific problem.

4.2.1 IoT Operating Systems

Windows Developer Program for IoT have been designed by OReilly Solid and Microsoft Build to allow users to run a version of Windows on the Intel Galileo board for both makers and Windows developers entering into the IoT space [26, 27]. Some OSs may be mentioned here as examples.

Contiki OS for the IoT is an open source operating system for memory-efficient networked embedded systems and wireless sensor networks that provides IP communication, both for IPv4 and IPv6 [28]. This OS is described in a study [29] presenting the low-power Constrained Application Protocol (CoAP) implementation for Contiki that leverages a generic radio duty cycling mechanism to achieve a high energy efficiency, such as, Distributed Real-time Control Systems (DRCS).

Representational State Transfer (REST) is a co-coordinated set of architectural constraints that attempts to minimize latency and network communication while at the same time maximizing the independence and scalability of component implementations. REST enables the caching and reuse of interactions, dynamic substitutability of components, and processing of actions by intermediaries, thereby meeting the needs of an Internet scale distributed hypermedia system [30]. The goal of REST is to achieve this in a more lightweight and simpler manner seamlessly integrated to the web.

Friendly OSs for IoT (RIOTS) use URIs for encapsulating and identifying services on the RIOT. RIOT is a standard C and C ++ programming OS that explicitly considers devices with minimal resources, provides multi-threading as well as real-time capabilities, and needs only a minimum of 1.5 kB of RAM [31].

Real Time Operating System for IoT (RTOS) is defined by [32] as a system in which the correctness of the system does not depend only on the logical results of computation but also on the time at which the results are produced. As mentioned in [33] the reinvented RTOS will add improved scalability, connectivity, security, safety, and an extended feature set to the solid real-time performance, low latency, and multi-core processor support of the RTOS of today. In addition to the above mentioned, there are many OSs designed especially for IoT such as, ARM OS, SPARK OS, and Tiny OS.

4.2.2 IoT Protocols

In this sub-section, we mention some of the main methods related to protocols for developing IoT technologies below.

Service Oriented Architecture (SOA) is a loosely-coupled architecture designed to meet the business needs of organizations. A SOA does not necessarily require the use of Web Services, but these are, for most organizations, the simplest approach for implementing a loosely coupled architecture [34], in particular for sensor and device functionality. SOA were primarily for connecting complex and static enterprise services because web service developing has not always been a straightforward task. Since sensors and devices have limited resources like computing such as communication and storage capabilities, this architecture require simplification, adaptation and optimization for a suitable performance [35]. In [36], authors propose an information-centric session mechanism to describe service behavior working upon distributed events called, event session. Another study [37] shows a SOA-based real-time service bus model which can be used to support constructing web.

Over the last decades, the Internet Protocol version 4 (IPv4) has emerged as the mainstream protocol for networking layer. However, this protocol was not designed for the IoT capabilities because it is inherently limited to about 4 Billion addresses. With the emergence of Internet Protocol version 6 (IPv6), it has scaled up the Internet to an almost unlimited number of globally reachable addresses. IPv6 provides 2^{128} unique Internet addresses, or 3.4×10^{38} addresses, which corresponds to over 6.67×10^{17} unique addresses per square millimeters of Earth surface [38]. In [39] the authors analyze the suitability of different IPv6 addressing strategies for nodes, gateways, and various network access deployment scenarios in the IoT.

4.2.3 IoT Applications

Clearly, the main contribution of IoT is its applications for facilitating and improving people's life and business processes. In this sub-section, we are citing some of the most recent applications in fields of search engine, social networks, health care, supply chains and retail.

In [40, 41] a search engine project is presented with focus on RFID devices based on Sphinx index module [42]. Ding et al. propose a Hybrid Real-time Search Engine Framework for the IoT based on Spatial-Temporal, Value-based, and Keyword-based Conditions due to data are dynamically changing so that the IoT search engine should support real-time retrieval. Additionally, the IoT search involves, not only keyword matches, but also spatial-temporal searches and value-based approximate searches, since IoT sampling data are generally from spatial-temporal scenario [43–45].

It has become evident that social networks are important platforms that allow people to share information and communicate for achieving different objectives. In that sense and in the IoT world Guo et al. have proposed a tool that enables information sharing and dissemination within/among opportunistic communities that are formed with the movement and opportunistic contact nature of human [46]. Atzori et al. propose to use this kind of social relationship to introduce a communication framework among devices and sensors in the IoT world called, SIoT paradigm (Social IoT) [47, 48].

Another area to be mentioned here is Health Care. Health care is a relevant area where IoT can help in a significant way due to several devices and sensors that allow monitoring and controlling diseases with low cost. In their study, Bui and Zorzi [49] explain about issues related with IoT protocols and requirement for being a suitable technology for health care. Ghose et al. in [50] explain the building of infrastructure for controlling and monitoring biological signals at home and small clinics through mobile health care devices and smart phones. In another study, they propose a method to improve accessibility to IoT data resources and IoT-based system for emergency medical services to demonstrate how to collect, integrate, and interoperate IoT data flexibly in order to provide support to emergency medical services [51].

In business, in the area of supply chains and retail, any improvement can be helpful since these businesses are multi- billionaire industries and a little positive change can impact in a drastic way in revenues and costs. In [52], the authors briefly discuss how IoT can have an effect on commercial process and supply chains and their future perspectives. There are also some proposals of new technologies for pedigree devices in food safety. The main objective is to track the processes in production, storage, transportation, sale, and even using phases of foods [53]. In this study [54], they presented several technologies that combat phony products in the global supply chain.

Some other applications such as Lee et al. [55] have designed an agricultural production system for monitoring and analyzing harvest statistics. The goal is to take smarter decision based on information from sensors in farms. In [56], they suggest a new schema for applying the IoT on intelligent traffic systems.

4.2.4 IoT Platforms

For the purpose of developing new applications and spreading fundamentals of IoT, IT companies have started to launch different platforms where it is possible to develop and learn more about latest applications and trends in IoT as is the case with IBM and Intel.

IBM has built a complete platform where it is possible to find different features and tools for developing and integrating to business. In IBM platform, there are several applications like Informix, storage environment, BlueMix, developer platform, MessageSight, and product dedicated to message communication among others [57].

In a lower level, Intel has been developing a platform that allows people to create new applications and discuss about new trends in IoT area. In addition to that, Intel has launched a developer kit that focuses on hobbyists, students and entrepreneurial developers, all this integrated with support, community and developer zone [58].

5 IoT Challenges

The IoT have changed the direction of the global Internet [59]. In this section, we will briefly discuss the two main challenges of IoT in two parts: IoT Big Data mining and IoT security.

IoT Data mining tools have become one of the most demanding necessities during the recent and future research challenges in terms of extracting valuable and useful data from the set of IoT and Big Data sources [59, 60].

Traditional data processing tools and its applications are not capable of managing such revolutionized amount of data. Nowadays, there are a few applications and platforms used to manage, extract, and execute IoT Big Data mining [61, 62]. Some of the latest software developed to manage both the IoT and Big Data are: NoSQL, which is used to manage Big Databases, MapReduce, which is used for data mining techniques, and Hadoop software used for processing, development, and execution of Big Data applications.

IoT security is one of the most critical areas needed for the development of different components and tasks. For example, device and process identification and tracking, sensing and actuation, communication, computational sensing, semantic knowledge processing, coordinated and distributed control, and user modeling. In addition, there are many constraints for IoT like cost, energy, lifetime and power. Nevertheless, one of the most important challenges is security, since the Internet is always a target of malicious attacks and intrusions. Usually these intrusions are fast and widely diffused [63].

Another aspect is trust and governance [64]. To create effective services, transmitted data have to be trustable. In other words, devices have to trust the data that other devices generated. Related with governance, some restricting policies can be applied. A classic example is limiting countries that can have access to the data.

6 Conclusion and Future Work

In this paper we presented some of the most recent advances in IoT field. We divided the study into four main parts: short history, Big Data concept, IoT sources (hardware and software), and finally some challenges and future expectations. The objective of the study is to give the reader an updated description of the IoT and its impact on Big Data. IoT is an area with a high potential of improving people's lives. However, this depends on how far we can reach to the challenge of the rapid growth in the day to day development in cell phones and WWW technologies. If predictions are true, it may ease human lives enabling us to enjoy the benefits of this technology by 2025. This is depending on the research and development that enterprises and governments will do in the coming years. We hope that this study will be helpful for researchers, students working in area of Big Data and IoT. Web of Thing (WoT) is another important part that we have not presented yet, but will be writing about in our future papers.

Acknowledgment This work was supported by the IT4Innovations Centre of Excellence project (CZ.1.05/1.1.00/02.0070), funded by the European Regional Development Fund and the national budget of the Czech Republic via the Research and Development for Innovations Operational Programme and by Project SP2015/146 "Parallel processing of Big data 2" of the Student Grand System, VŠB—Technical University of Ostrava.

References

1. Miorandi, D., Sicari, S., De Pellegrini, F., Chlamtac, I.: Internet of things: vision, applications and research challenges. Ad Hoc Netw. **10**(7), 1497–1516 (2012)
2. Rivera, J., van der Meulen, R.: Gartner's 2014 hype cycle for emerging technologies maps the journey to digital business. http://www.gartner.com/newsroom/id/2819918 (2014). Accessed 03 Dec 2014
3. Haller, S., Karnouskos, S., Schroth, C.: The internet of things in an enterprise context. Future Internet FIS **5468**, 14–28 (2009)
4. L. Internet: The internet toaster. http://www.livinginternet.com/i/iamythstoast.html (2000). Accessed 03 Dec 2014
5. Mattern, F., Floerkemeier, C.: From Active Data Management to Event-Based Systems and More. Springer (2010)
6. Collina, M., Corazza, G., Vanelli-Coralli, A.: Introducing the qest broker: scaling the iot by bridging mqtt and rest, pp. 36–41 (2012)
7. Mqtt organization, http://mqtt.org/faq (2015). Accessed 13 Jan 2015
8. Raduescu, C.: Towards an ethnographic study of information appliances: the case of LG internet refrigerator. Ann. Comput. Sci. Ser. J. **2** no. 1 (2004)
9. Evans, D.: The internet of things: how the next evolution of the internet is changing everything, CISCO white paper, vol. 1 (2011)
10. Gartner, Gartner it glossary. http://www.gartner.com/it-glossary/big-data/ (2014). Accessed 05 Jan 2015
11. Schroeck, M., Shockley, R., Smart, J., Romero-Morales, D., Tufano, P.: Analytics: the Real-world use of Big Data (2012)
12. Knilans, E.: The 5 vs of Big Data. http://blogging.avnet.com/ts/advantage/2014/07/the-5-vs-of-big-data/ (2014). Accessed 05 Jan 2015
13. Ding, G., Wu, Q., Wang, J., Yao, Y.-D.: Big spectrum data: the new resource for cognitive wireless networking (2014). arXiv preprint arXiv:1404.6508
14. Chen, M., Mao, S., Zhang, Y., Leung, V.C.: Big Data Applications, in Big Data, pp. 59–79. Springer (2014)
15. Center, S.D.: How a data center works, http://www.sapdatacenter.com/article/datacenterfunctionality/ (2014). Accessed 24 Nov 2014
16. Gubbi, J., Buyya, R., Marusic, S., Palaniswami, M.: Internet of things (iot): a vision, architectural elements, and future directions, vol. 29, no. 7, pp. 1645–1660, Sept 2013
17. The array of things project in Chicago state. https://arrayofthings.github.io/ (2014). Accessed 24 Nov 2014
18. Domingo, M.C.: Review: an overview of the internet of things for people with disabilities. J. Netw. Comput. Appl. **35**(2), 584–596 (2012)
19. RFID journal. https://www.rfidjournal.com (2014). Accessed 26 Nov 2014
20. Whitmore, A., Agarwal, A., Da Xu, L.: The internet of things a survey of topics and trends. Information Systems Frontiers, pp. 1–14 (2014)
21. Mell, P., Grance, T.: The NIST (National Institute of Standards and Technology) Definition of Cloud Computing, pp. 800–145 NIST Special Publication (2011)
22. Liu, L.: Computing infrastructure for big data processing. Front. Comput. Sci. **7**(2), 165–170 (2013)
23. Enbysk, L.: Big Data to the rescue in miami-dade county, florida. http://smartcitiescouncil.com/article/big-data-rescue-miami-dade-county-florida (2014), Accessed 05 Jan 2015
24. Ericsson, Ericsson mobility report. http://www.ericsson.com/res/docs/2013/ericsson-mobility-report-June-2013.pdf. Accessed 05 Jan 2015
25. Broll, G., Rukzio, E., Paolucci, M., Wagner, M., Schmidt, A., Huss-mann, H.: Perci: Pervasive service interaction with the internet of things. Internet Comput. IEEE **13**(6), 74–81 (2009)

26. Duncan, G.: Windows developer program for iot now rolling out. http://channel9.msdn.com/coding4fun/blog/Windows-Developer-Program-for-IoT-now-rolling-out (2014). Accessed 20 Dec 2014
27. Windows, Windows developer program for iot. http://dev.windows.com/en-us/featured/windows-developer-program-for-iot (2014). Accessed 20 Dec 2014
28. Contiki, Contiki: the open source os for the internet of things. http://www.contiki-os.org/ (2014). Accessed 20 Dec 2014
29. Kovatsch, M., Duquennoy, S., Dunkels, A.: A low-power coap for contiki. In: Mobile Adhoc and Sensor Systems (MASS), Conference on. IEEE 2011 IEEE 8th International pp. 855–860 (2011)
30. Fielding, R.T., Taylor, R.N.: Principled design of the modern web architecture. ACM Trans. Internet Technol. **2**(2), 115–150 (2002)
31. Baccelli, E., Hahm, O., Gunes, M., Wahlisch, M., Schmidt et al., T.: Riot os: towards an os for the internet of things (2013)
32. Hambarde, P., Varma, R., Jha, S.: The survey of real time operating system: Rtos, pp. 34–39 (2014)
33. Intel, The critical features and characteristics for an embedded oper-ating system. http://www.intel.com/content/www/us/en/internet-of-things/white-papers/real-time-operating-system-for-iot.html (2014). Accessed 20 Dec 2014
34. Library, M.: Chapter 1: Service oriented architecture (soa). http://msdn.microsoft.com/en-us/library/bb833022.aspx (2014), Accessed 01 Dec 2014
35. Guinard, D., Trifa, V., Karnouskos, S., Spiess, P., Savio, D.: Interact-ing with the soa-based internet of things: discovery, query, selection, and on-demand provisioning of web services. IEEE Trans. Serv. Comput. **3**(3), 223–235 (2010)
36. Zhang, Y., Duan, L., Chen, J.L.: Event-driven soa for iot services. In: 2014 IEEE International Conference on Services Computing (SCC), pp. 629–636, June 2014
37. Du, L., Duan, C., Liu, S., He, W.: Research on Service Bus for Distributed Real-time Control Systems, vol. 1, pp. 401–405, Aug 2011
38. Ziegler, S., Crettaz, C., Ladid, L., Krco, S., Pokric, B., Skarmeta, A., Jara, G., Kastner, W., Jung, M.: Iot6 moving to an ipv6-based future iot, vol. 7858, pp. 161–172 (2013)
39. Savolainen, T., Soininen, J., Silverajan, B.: Ipv6 addressing strategies for iot. Sensors J. IEEE **13**(10), 3511–3519 (2013)
40. Ostermaier, B., Romer, K., Mattern, F., Fahrmair, M., Kellerer, W.: A real-time search engine for the web of things. In: Internet of Things (IoT), 2010, pp. 1–8, Nov 2010
41. Jin, X., Zhang, D., Zou, Q., Ji, G., Qian, X.: Where searching will go in internet of things?. In: Wireless Days (WD), 2011 IFIP, pp. 1–3, Oct 2011
42. S.T. Inc.: Sphinx: Open source search server. http://sphinxsearch.com/ (2014). Accessed 02 Dec 2014
43. Ding, Z., Chen, Z., Yang, Q.: Iot-svksearch: a real-time multimodal search engine mechanism for the internet of things. Int. J. Commun Syst. **27**(6), 871–897 (2014)
44. Ding, Z., Dai, J., Gao, X., Yang, Q.: A hybrid search engine framework for the internet of things, pp. 57–60, Nov 2012
45. Ding, Z., Gao, X., Guo, L., Yang, Q.: A hybrid search engine framework for the internet of things based on spatial-temporal, value-based, and keyword-based conditions, pp. 17–25 (2012)
46. Guo, B., Yu, Z., Zhou, X., Zhang, D.: Opportunistic iot: exploring the social side of the internet of things, pp. 925–929 (2012)
47. Atzori, L., Iera, A., Morabito, G.: Siot: Giving a social structure to the internet of things. Commun. Lett. IEEE **15**(11), 1193–1195 (2011)
48. Atzori, L., Iera, A., Morabito, G., Nitti, M.: The social internet of things (siot) when social networks meet the internet of things: Concept, architecture and network characterization. Comput. Netw. **56**(16), 3594–3608 (2012)
49. Bui, N., Zorzi, M.: Health care applications: a solution based on the internet of things, pp. 131:1–131:5 (2011)

50. Ghose, A., Bhaumik, C., Das, D., Agrawal, A.K.: Mobile healthcare infrastructure for home and small clinic, pp. 15–20 (2012)
51. Xu, B., Xu, L.D., Cai, H., Xie, C., Hu, J., Bu, F.: Ubiquitous data ac-cessing method in iot-based information system for emergency medical services. IEEE Trans. Ind. Inf. **10**(2), 1578–1586 (2014)
52. Shen, G., Liu, B.: Research on application of internet of things in electronic commerce. In: Electronic Commerce and Security (ISECS), pp. 13–16 (2010)
53. Han, W., Gu, Y., Wang, W., Zhang, Y., Yin, Y., Wang, J., Zheng, L.-R.: The design of an electronic pedigree system for food safety. Inf. Syst. Front. pp. 1–13 (2012)
54. Li, L.: Technology designed to combat fakes in the global supply chain. Bus. Horiz. **56**(2), 167–177 (2013)
55. Lee, M., Hwang, J., Yoe, H.: Agricultural production system based on iot. In: Computational Science and Engineering (CSE), pp. 833–837 (2013)
56. Pyykonen, P., Laitinen, J., Viitanen, J., Eloranta, P., Korhonen, T.: Iot for intelligent traffic system, pp. 175–179 (2013)
57. IBM, The internet of things. http://www-01.ibm.com/software/info/internet-of-things/ (2014). Accessed 28 Nov 2014
58. Intel, The internet of things (iot). http://www.intel.com/content/www/us/en/internet-of-things/overview.html (2014). Accessed 28 Nov 2014
59. Bin, S., Yuan, L., Xiaoyi, W.: Research on data mining models for the internet of things, pp. 127–132 (2010)
60. Tsai, C.-W., Lai, C.-F., Chiang, M.-C., Yang, L.: Data mining for internet of things: A survey. Commun. Surv. Tutor. IEEE, **16**(1), 77–97, First 2014
61. Wu, X., Zhu, X., Wu, G.-Q., Ding, W.: Data mining with Big Data. IEEE Trans. Knowl. Data Eng. **26**(1), 97–107 (2014)
62. Barbierato, E., Gribaudo, M., Iacono, M.: Performance evaluation of NOSQL big-data applications using multi-formalism models,, vol. 37, no. 0, pp. 345–353 2014
63. Xu, T., Wendt, J.B., Potkonjak, M.: Security of iot Systems: Design Challenges and Opportunities, pp. 417–423. IEEE Press, USA (2014)
64. Roman, R., Najera, P., Lopez, J.: Securing the internet of things. Computer **44**(9), 51–58 (2011)

A Systematic Review of Security in Cloud Computing

Mohamed Amoud and Ounsa Roudiès

Abstract Security issues over cloud computing is definitely one of the major concerns that many companies are trying to recognize. However, some companies are also concerned about regulatory issues. Market observers say that a lot of the participants in the survey worry that they will be tied to one provider of cloud storage. Experts opine that cloud computing is at its nascent stage and providers will have to address issues related security, availability, and more to expand in the future. This paper presents a literature review of privacy and security approaches for cloud computing, and evaluates them in terms of how well they support critical security services and what level of adaptation they achieve. This work will be done following the Systematic Review approach. Our results concluded that the research field of security approaches for cloud is still poor of methods and metrics for evaluating and comparing different techniques.

Keywords Cloud computing · Security · Systematic review

1 Introduction

1.1 Cloud Computing

Cloud computing is the delivery of computing services over the Internet. These provide a shared pool of resources, including data storage space, networks, computer processing power, and specialized corporate and user applications. Cloud services allow individuals and businesses to use software and hardware that are managed by third parties at remote locations. The cloud model allows access to

M. Amoud (✉) · O. Roudiès
École Mohammedia d'Ingénieurs (EMI), Mohammed Vth University-Agdal, Rabat, Morocco
e-mail: amoudmohamed@gmail.com

O. Roudiès
e-mail: oroudies@gmail.com

A. Abraham et al. (eds.), *Proceedings of the Second International Afro-European Conference for Industrial Advancement AECIA 2015*, Advances in Intelligent Systems and Computing 427, DOI 10.1007/978-3-319-29504-6_8

information and computer resources from anywhere that a network connection is available.

Since cloud users do not have to invest in information technology infrastructure, purchase hardware, or buy software licences, the benefits are low up-front costs, rapid return on investment, rapid deployment, customization, flexible use, and solutions that can make use of new innovations. Many businesses are using cloud computing as it usually turns out to be cheaper, faster and easy to maintain. Now, not only businesses but regular Internet users are also using cloud computing services such as Google Docs, Drop box and more to access their files whenever and wherever they want.

In the Sect. 2 we describe our method for conducting the review. Results are presented in Sect. 3. Section 4 answers our questions. Finally, the Conclusion and future work Section close the paper.

1.2 Characteristics

The characteristics of cloud computing include:

1. *On-demand self-service*: consumers are able to provision cloud computing resources, such as server time and network storage, as needed automatically without requiring human interaction with each service provider.
2. *Broad network access*: allows services to be offered over the Internet or private networks. Cloud computing resources are accessible over the network through standard mechanisms that promote use by heterogeneous client platforms.
3. *Resource pooling*: the provider's computing resources are pooled to serve multiple consumers using a multi-tenant model, with different physical and virtual resources dynamically assigned and reassigned according to consumer demand.
4. *Rapid elasticity*: capabilities can be elastically provisioned and released to scale rapidly on demand. To the consumer, the capabilities available for provisioning often appear to be unlimited and can be appropriated in any quantity at any time.
5. *Measured service*: cloud systems automatically control and optimize resource use by leveraging a metering capability.

1.3 Service Models

The cloud computing service models are Software as a Service (SaaS), Platform as a Service (PaaS) and Infrastructure as a Service (IaaS) [12]:

In a *Software as a Service* model, a pre-made application, along with any required software, operating system, hardware, and network are provided. It provides automation in the lower layers (up to the virtualization layer).

Examples: Google Docs, Google Calendar, SAP Business by Design, Salesforce CRM.

In *PaaS*, an operating system, hardware, and network are provided, and the customer installs or develops its own software and applications. At this layer, the whole software stack needed by the application is managed by the PaaS provider.

Examples: Google App Engine, Jelastic, Force.com, Microsoft Windows Azure, Java.

The *IaaS* model provides just the hardware and network; the customer installs or develops its own operating systems, software and applications. For instance, Google Docs, for editing documents, or Gmail, a mail server, are provided as SaaS. Examples: Amazon S3, Windows Azure SQL.

1.4 Deployment Models

There exist four different deployment models:

1. *Private cloud*: the cloud infrastructure is provisioned for exclusive use by a single organization. It may be owned, managed, and operated by the organization, a third party, or a combination of them. For instance, building a private cloud can be done relying on technologies such as Eucalyptus or OpenStack.
2. *Community cloud*: the cloud infrastructure is provisioned for exclusive use by a specific community. Persons who are part of this community may be from different organizations, but sharing concerns: collaborative security requirements or policy.
3. *Public cloud*: is the most known one. The cloud infrastructure is provisioned for open use by anyone interested, and is managed by any academic, government or business organization. Examples: Amazon EC2, Windows Azure are public clouds.
4. *Hybrid cloud*: is the combination of two or more cloud infrastructures, either private, community or public. Such a model is useful when an organization requires an amount of resources to compute information about private data.

1.5 Security

The combination of continuous evolution in security threats and organizations' adoption of new information technologies has been met with a history of innovation in security. Beneficially in securing the cloud, security technologies and their foundational concepts that are already practiced in traditional environments can be adapted to the cloud. This does not mean that securing the cloud is a simple process of porting security technologies from traditional environments. Rather, the core precepts are present to be effective in securing cloud environments, but must be implemented with a thorough understanding of cloud's incremental risks and uniqueness [10].

The objective of this paper is to give an overview of the state of the art in the security issues for a Cloud Computer by doing a Systematic Literature Review (SLR) on simple and clear question in this regard. In particular, we identify and compare different cloud security approaches. Key security considerations and challenges which are currently faced in cloud computing will be highlighted.

2 Method

The aim of this study is identifying and comparing different security approaches for Cloud Computing. We used guidelines proposed by Barbara Kitchenham [13, 21] for performing our study. The main steps are explained in the next parts of this section.

2.1 *Systematic Literature Review (SLR)*

A Systematic Literature Review (SLR) is a rigorous method for assessing, reviewing and aggregating research results. Unlike an ordinary literature review consisting of an annotated bibliography, SLR analyzes existing literature with reference to specific research questions on a topic of interest. Furthermore, it can be considered as much more effort prone than an ordinary literature survey.

Guidelines on SLR have been defined and are quite stable in contexts such as medicine, social sciences, education and information sciences and used for analysing and synthesizing existing empirical results on a certain topic. Adaptations of these guidelines to software engineering have been made by Kitchenham [13, 21].

2.2 *Research Questions*

RQ1 What is the focus of research in security of cloud computing?

RQ2 What security vulnerabilities and threats are the most important in Cloud Computing which have to be studied in depth with the purpose of handling them?

RQ3 How can Cloud Computing autonomously evaluate changes and threats in their environment in order to adaptively reconfigure themselves?

RQ4 How confident are providers that customer security requirements of the cloud will be met?

With regards to RQ1,	looking for research and case studies needed to get insight in the research trends in security issues of Cloud, providing context for the study
With regards to RQ2,	the question focus was to identify the most relevant issues in Cloud Computing which consider vulnerabilities, threats, risks, requirements and solutions of security for Cloud Computing
With regards to RQ3,	we want to know how applications with stringent safety requirements require security mechanisms that reduce human intervention when Cloud Computing parameters change
The goal of RQ4	is to know customer concerns about the security of the cloud and to help deriving conclusions from the study

2.3 Research Process

Our search process for review was based on online searching in famous online databases which are addressed as Table 1. Since these databases cover almost all major journals and conference proceedings, manual review of journal was not required. Review has been carried on by mean of search facilities in these databases and using appropriate logical expressions. In the first stage, our focus was on title and abstract of articles found in search process and selecting appropriate and relevant studies. If there was any doubt, our decision was based on reviewing it at one glance.

2.4 Inclusion and Exclusion Criteria

Our primary goal is to understand the claims and supporting evidence of security in Cloud Computing, we excluded papers about theoretical aspects, as well as surveys and roadmap papers. We also excluded short papers of 1 or 2 pages.

There were some papers that were relevant to our study indirectly in our defined process. This will strengthen our review, because all relevant documents were included and our review covered sufficiently direct and indirect studies in this research.

Table 1 Studies resource

Source	Address
Scopus	www.scopus.com
IEEE Xplore	ieeexplore.ieee.org
ACM digital library	Portal.acm.org
Springer link	www.springerlink.com
Science direct	www.sciencedirect.com

All studies are assessed through a quality check, which is an inherent part of a thorough literature study. Checking the originality and quality of the studies is important for data synthesis and interpretation of results later on.

2.5 *Quality Assessment*

For assessing studies we defined the following questions

QA1 Does study agree with existence of the focus of research in security for Cloud Computing?
QA2 Does study report show how to face securely unexpected risks and activate appropriate countermeasures to respond to new threats?
QA3 Is it possible to avoid specifying all the behaviours in advance for Autonomous and Adaptive Security?
QA4 Does proper security become a priority for organizations contracting with a cloud computing provider?
QA5 Does study give confidence to the users of cloud computing security?

We scored questions as below:

QA1. Y (Yes) study explicitly agrees with existence of any objectives; P (Partially) study implicitly agrees and N (No) study disagrees with existence of any objectives
QA2. Y, study report shows how to react; P, so not effective and N, there is no reaction measurement
QA3. Y, study addresses possibility to avoid specifying all the behaviours in advance for Autonomous and Adaptive Security; P, study partly agrees (or implicitly) and N, there is no possibility
QA4. Y, the authors report provide sufficient arguments; P, so not enough and N, there is no argument
QA5. Y, study provides convincing arguments; P, so not enough and N, it remains unclear

We defined Y = 1, P = 0.5 and N = 0 or Unknown where information is not clearly specified. All authors assessed every article and if there is no agreement in scoring, we discussed enough to reach agreement.

2.6 *Data Collection*

These data were extracted from each article: the full source and references, the author(s) information and details, research issues, and main ideas.

All articles were reviewed and data was extracted and checked. This idea was chosen for better consistency in reviewing all papers and improving quality of review.

2.7 *Data Analysis*

Our collect data was organized to address:

- Whether study agrees with existence of the objectives of security for Cloud environment or not? (Addressing RQ1)
- Whether study provides sufficient arguments to face securely unexpected risks and activate appropriate countermeasures or no? (Addressing RQ2 and RQ3)
- Whether study agrees with existence of any needs for adaptation or no? (RQ3)
- Whether study shows the importance of security in the customer /provider relationship or no? (Addressing RQ3)
- Whether authors believe that cloud security is a goal that can be reached easily or no? (Addressing RQ4)

3 Results

In this section we explain results of our review.

3.1 *Search Results*

Table 2 shows the results of our selection procedure. In this table, results of searching in all databases are provided, but, some of the studies were repeated in more than one online database, so, final number of unique studies selected for our review was distinguished after elimination of repeated articles. Final selected studies are listed in Table 3.

Table 2 Results of study selection procedure

Source	Search results	Selected studies
Scopus	91	9
IEEE Xplore	65	9
ACM digital library	22	3
Springer link	31	9
Science direct	04	2
Total	**213**	**32**
Repeated articles		18
Finally selected articles		**14**

Table 3 Selected studies for conducting review

ID	Title	Author(s)	Main topic	Year
S1	Virtualized in-cloud security services for mobile devices	Oberheide et al. [5]	Propose malware detection system for mobile device	2008
S2	Securing elastic applications on mobile devices for cloud computing	Zhang et al. [6]	Presents security framework for elastic mobile application model	2009
S3	MobiCloud: building secure cloud framework for mobile computing and communication	Huan et al.[9]	Presents MobiCloud and cover security aspect in terms of risk management and secure routing	2010
S4	Energy-efficient incremental integrity for securing storage in mobile cloud computing	Itani et al. [14]	Proposes an energy efficient framework for mobile cloud	2010
S5	A secure data service mechanism in mobile cloud computing	Jia et al. [15]	Presents proxy re-encryption scheme to achieve secure service	2011
S6	Secure data processing framework for mobilecloud computing	Huang et al. [16]	Proposes secure data processing framework for MobiCloud	2011
S7	Security and privacy in mobile cloud computing	Hui Suo et al. [11]	Privacy issues from three aspects: mobile terminal, mobile network and cloud	2013
S8	Security and privacy issues in cloud computing innovation	Jaydip [7]	Cloud challenges: security, regulatory and privacy	2013
S9	An analysis of security issues for cloud computing	Hashizume1et al. [20]	Security vulnerabilities and threats cloud	2013
S10	Automated selection and configuration of cloud environments using software product lines principles	Clément Q.et al. [17]	How can cloud autonomously evaluate changes to adaptively reconfigure themselves	2014
S11	Comparative study on Cloud Computing (CC) and Mobile Cloud Computing (MCC)	Niranjanamurthy et al. [10]	Security risks and solution of Mobile Cloud Computing	2014
S12	Towards secure inter-cloud architectures	Oscar Encina et al. [2]	A security engineering to secure cloud	2014
S13	Building a security reference architecture for cloud systems	Fernandez et al. [22]	Activate appropriate countermeasures to respond to new threats	2015
S14	Cloud Access Security Broker (CASB): A pattern for secure access to cloud services	Fernandez et al. [23]	CASB controls access to application and protects data from malware	2015

Table 4 Quality evaluation

Source	QA1	QA2	QA3	QA4	QA5
1	Y	P	N	Y	Y
2	Y	Y	P	Y	Y
3	P	Y	Y	P	Y
4	P	Y	P	Y	P
5	P	Y	Y	P	Y
6	Y	Y	Y	Y	Y
7	N	P	P	Y	Y
8	P	Y	Y	N	Y
9	Y	P	P	Y	Y
10	P	N	P	Y	P
11	N	P	P	Y	P
12	P	Y	N	Y	P
13	N	P	P	Y	Y
14	P	N	Y	Y	Y
Average	**0.54**	**0.68**	**0.32**	**0.79**	**0.86**

3.2 *Quality Evaluation of Studies*

During this phase, we found that some of the selected articles discussing security in general or only the Cloud Computing, but, they do not provide any valuable information to our research, so, we decided to delete them from scope of our study. Assessment of each study was done by means of criteria explained in Sect. 2.4 and the scores for each of them are shown in Table 4.

3.3 *Quality Factors*

For assessing results of our quality questions, we use the average of total scores. This average is useful for some questions, but it is not useful for others. For instance, we cannot answer the question about possibility of integration with average of scores because of the nature of the question; instead, we use negative ideas for rejecting possibility.

4 Discussion

In this part, the answers to our study questions will be discussed.

4.1 What Is the Focus of Research in Security of Cloud Computing?

Most of the articles agree that there are the objectives of research in security for Cloud computing. By reviewing them, it seems that this research focus is derived from the following concerns: category of the study, subject, concrete focus and application domain. Overall, more than two thirds of the studies focus on one or more activities of security (monitoring, analyzing, planning, execution), runtime models, multiple control loops and on called SPI model (SaaS, PaaS and IaaS), identifying the main vulnerabilities in this kind of systems and the most important threats.

The majority of the articles agree that there is a need for secure cloud storage and cloud computing environments. For example, Molnar et al. [3] cite industry decision makers to emphasize the fact that security concerns are among the major factors that prevent businesses from deploying their data and computations into the cloud. Similarly, Chen et al. [4] cite opinions originating from academia, government and industry that point to security concerns as a barrier preventing a quicker adoption of cloud. Similar views are reported by other researchers within cloud computing security [18, 19].

4.2 What Security Vulnerabilities and Threats Are the Most Important in Cloud Computing Which Have to Be Studied in Depth with Purpose of Handling Them?

Most of the approaches list a number of vulnerabilities and threats focused on the Cloud, discussed how to identify, classify, analyze, and describe the relationship between these vulnerabilities and threats [20]; how these vulnerabilities can be exploited in order to perform an attack, and also present some countermeasures related to these threats which try to solve or improve the identified problems [22, 23], such as: Account or service Hijacking, Data scavenging, Customer-data Manipulation, Malicious VM creation, … The studies analyze the risks and provide a description of countermeasures: Dynamic credentials, Digital signatures, Homomorphic encryption, Fragmentation-redundancy-scattering, ….

4.3 How Can Cloud Computing Autonomously Evaluate Changes and Threats in Their Environment in Order to Adaptively Reconfigure Themselves?

Many researchers agree that it is possible to have an autonomous and adaptive security in Cloud Computing. Oberheide et al. [5] proposed Cloud AV platform, malware detection system for mobile device by moving detection capabilities to network service or cloud. Zhang et al. [6] present security framework for elastic mobile application model by dividing an application into easily configurable web lets. Wang and Wang [8] have proposed privacy preserving framework for mobile devices while using location based scheme by spatial cloaking. Huan et al. [9] presents framework—MobiCloud to enhance the functionality of MANET and cover security aspect in terms of risk management and secure routing. Itani et al. [14] proposed an energy efficient framework for mobile devices by using incremental message authentication code to ensure integrity of mobile users. Jia et al. [15] presents proxy re-encryption (PRE) scheme and identity based encryption (IDE) scheme to achieve secure data service. Huang et al. [16] proposed secure data processing framework for MobiCloud addressing issue of authentication on cloud.

4.4 How Confident Are We that User Security Requirements of Cloud Will Be Met?

As consumers transition their applications and data to use cloud computing, it is important that the level of security provided in the cloud environment be equal to or better than the security provided by their traditional IT environment. Failure to ensure appropriate security protection could ultimately result in higher costs and potential loss of business thus eliminating any of the potential benefits of cloud computing [22].

Regarding the sharing of security responsibility between Provider and customer, cloud security is a split responsibility between the cloud provider and its subscribers (customers); with the level of responsibility in the hands of the cloud provider growing in moving across the models from IaaS to SaaS. Similarly, identifying and mitigating security incidents and configuration errors attributed to the layers of the cloud infrastructure under the cloud provider's purview are also outside the line-of-sight of the customers who indirectly asked to trust without full verification. In cloud environments, the strength of security is partially dependent on the strength of the security operations and administration conducted by the cloud provider.

5 Conclusion and Future Works

The objective of this literature study was to summarize existing research on a rich set of cloud benefits and shed light on the claimed benefits of its security. Several security approaches exist. On one hand, approaches concentrate on a particular security mechanism or supporting a specific security attribute. On the other hand, some approaches are generic; they support different attributes and mechanisms.

Hence, it is difficult to select the most suitable adaptation approach for different usages. Moreover, it is difficult to know what research steps are needed in the future.

Future research is how to exploit adaptive security in Dynamic Software Product Lines (DSPL) to Develop Cloud Computing secure Applications. This solution is capable of providing more efficient mechanisms able to manage the dynamic and adaptive characteristics of security in Mobile Cloud Computing.

References

1. Suby, M.: Best Practice Security In a Cloud-Enabled WorldCloud. An Executive Brief Sponsored by IBM (2014)
2. Encina, O., et al.: Towards secure inter-cloud architectures. In: Proceedings of Nordic pattern conference on Pattern Languages of Programs, Sagadi Manor, Estonia (2014)
3. Cloud Computing Use Case Discussion Group. Cloud Computing UseCases V. 3.0 (2010)
4. ENISA: Cloud computing: benefits, risks and recommendations for information security (2010). http://www.enisa.europa.eu/act/rm/files/deliverables/cloud-computingrisk-assessment
5. Oberheide, J., Veeraraghavan, K., Cooke, E., Flinn, J., Jahanian, F.: Virtualized in-cloud security services for mobile devices. In: Proceedings of the 1st Workshop on Virtualization in Mobile Computing (MobiVirt), pp. 31–35, June 2008
6. Zhang, X., Schiffman, J., Gibbs, S., Kunjithapatham, A., Jeong, S.: Securing elastic applications on mobile devices for cloud computing. In: Proceeding ACM workshop on Cloud computing security, CCSW '09, Chicago, IL, USA, November. 2009
7. Sen, J.: Security and Privacy Issues in Cloud Computing Innovation. Labs, Tata Consultancy Services Ltd., Kolkata (2013)
8. Wang, S., Wang, X.S.: In-device spatial cloaking for mobile user privacy assisted by the cloud. In: Proceeding 11th International Conference on Mobile Data Management,MDM '10, Missouri, USA, May 2010
9. Huan, D., Zhang, X., Kang , M., Luo , J.: MobiCloud: building secure cloud framework for mobile computing and communication, In: Proceeding 5th IEEE International Symposium on Service Oriented System Engineering, SOSE '10, Nanjing, China, June 2010
10. Niranjanamurthy, M., et al.: Comparative Study on Cloud Computing (CC) and Mobile Cloud Computing (MCC). Int. J. Comput. Sci. Mob. Comput., **3**(10), 280–290 (2014)
11. Suo, H., et al.: Security and privacy in mobile cloud computing. In: Wireless Communications and Mobile Computing Conference (IWCMC). pp. 655–659. IEEE. Sardinia (2013)
12. Clément, Q : Cloud environment selection and configuration: a software product lines-based approach. Ph.D's thesis, University of Lille 1, France (2014)
13. Kitchenham, B., Charters, S.: Guidelines for performing Systematic Literature reviews in Software Engineering Version 2.3, Keele University (2007)
14. Itani, W., Kayssi, A., Chehab, A.: Energy-efficient incremental integrity for securing storage in mobile cloud computing. In: Proceeding International Conference on Energy Aware Computing, ICEAC '10, Cairo, Egypt, December 2010
15. Jia, W., Zhu, H., Cao, Z., Wei, L., Lin, X.,: SDSM: a secure data service mechanism in mobile cloud computing. In: Proceeding IEEE Conference on Computer Communications Workshops, INFOCOM WKSHPS, Shanghai, China, April 2011
16. Huang, D., Zhou, Z., Xu, L., Xing, T., Zhong, Y.: Secure data processing framework for mobilecloud computing, In: Proceeding IEEE INFOCOM Workshop on Cloud Computing, INFOCOM '11, Shanghai, China, June 2011

17. Clément Q., et al.:Automated selection and configuration of cloud environments using software product lines principles. In: Proceedings of the 7th IEEE International Conference on Cloud Computing, CLOUD'14.Anchorage, Alaska (USA) (2014)
18. Balachandra, R.K., Ramakrishna, P.V., Rakshit, A.: Cloud security issues. In: PROC'09 IEEE International Conference on Services Computing, 2009, pp 517–520 (2009)
19. Kuyoro, S.O., Ibikunle, F., Awodele, O.: Cloud computing security issues and challenges. Int. J. Comput. Netw. (IJCN), **3**(5) (2011)
20. Hashizume, K., et al.: An analysis of security issues for cloud computing. J. Internet Serv. Appl. **4**, 5 (2013)
21. Kitchenham, B., Pearl Brereton, O., Budgen, D., Turner, M.: Systematic literature reviews in software engineering–a Systematic literature reviews, Keele University (2008)
22. Fernandez, E.B., Monge, R., Hashizume, K.: Building a security reference architecture for cloud systems, Requir. Eng. (2015)
23. Fernandez, E.B., Yoshioka, N., Washizaki, H.: Cloud Access Security Broker (CASB): a pattern for secure access to cloud services. In: 4th Asian Conference on Pattern Languages of Programs, Asian PLoP'15, Tokyo, Japan (2015)

Towards a Model-Based Development Methodology for Evolvable Production Systems

A Domain-Specific Modeling Approach

Afifa Rahatulain, Tahir Naseer Qureshi and Mauro Onori

Abstract Evolvable production system (EPS) is one of the most promising emerging paradigms among the next generation of production systems dealing with challenges such as market unpredictability, high product variance and increasing automation costs. One of the major challenges faced by EPS for its wider industrial realization is the harmonization of its existing research activities such as the ontology and reference architecture with the agent-based control, dynamic skill-configuration methodology and self-organization algorithms while also considering the aspects of operation management and business models. In addition, the integration with existing industrial standards, targeting aspects like functional safety, system integrity, etc. is also required. This paper addresses the challenge by providing an extendible DSM (Domain Specific Modeling) based support for modeling an EPS. The work is a basis for a model-based and architecture-centric methodology applicable throughout the development life-cycle of an EPS.

1 Introduction

Typically an industrial environment comprises of large dedicated product-oriented systems, which are often designed for a particular product or a class of product. Any support for the product variance or market fluctuations is achieved only at the expense of high investment costs and extensive engineering efforts causing significant delays in the lead times. Several modular approaches have been proposed to deal

A. Rahatulain (✉) · M. Onori
KTH - The Royal Institute of Technology, Stockholm, Sweden
e-mail: afirah@kth.se

M. Onori
e-mail: onori@kth.se

T.N. Qureshi
HIAB AB, Hudiksvall, Sweden
e-mail: tahir.qureshi@hiab.com

A. Abraham et al. (eds.), *Proceedings of the Second International Afro-European Conference for Industrial Advancement AECIA 2015*, Advances in Intelligent Systems and Computing 427, DOI 10.1007/978-3-319-29504-6_9

with these challenges such as *Flexible Manufacturing Systems* (FMS) [30], *Reconfigurable Manufacturing Systems* (RMS) [16], *Holonic Manufacturing Systems* (HMS) [4] and *Evolvable Production Systems* (EPS) [23]. While RMS and FMS have certain limitations in their applicability to only pre-defined configurations, both HMS and EPS have wider coverage related to aspects such as flexibility, dynamics and adaptability.

As compared to HMS, EPS is a relatively new paradigm providing the concept of real-time *plug and produce* up to the level of fine granularity at the shop floor level [20]. Although promising, a wide scale industrial adoption of paradigms like EPS and HMS is limited by certain factors including, but not limited to, the concerns related to dependability [18], fulfillment of legislative requirements such as functional safety standards (e.g. IEC 61508) and tool support throughout the development life cycle. This in turn requires a well-defined methodology to enable development, operation as well as maintenance in a cost-effective manner.

The required methodology can ideally be achieved by a seamless model-based development platform similar to the one presented in [3]. The key characteristics of such a methodology include a centralized source of information and automated transformation of models between different abstraction levels as well as development tools. Domain-specific modeling, transformation engines and generators are two key technologies to realize such a methodology [29].

This paper presents an effort towards the envisioned methodology. The major contributions of the presented work are:

1. Development of EPS-DSL; a Domain Specific Language (DSL) and
2. Foundation for tool support for utilizing EPS-DSL

In addition to the above, the work also provides a basis for EPS requirements verification and support for automatic code generation for the equipment(s) under consideration.

The remaining paper is structured as follows: An introduction to the EPS concept along with the recent advancements and challenges is provided in Sect. 2. The main results, are presented in Sects. 3 and 4. Section 5 provides an overview of the related work. The paper is finally concluded with a discussion in Sect. 6.

2 Evolvable Production Systems

2.1 Introduction

Evolvable production system is one of the emerging paradigms which tend to revolutionize the manufacturing industry by incorporating intelligence, self-organization, adaptability and reconfigurability at the shop floor level [23]. A modular architecture with intelligent, agent-based, distributed control enables EPS to offer real-time *plug and produce* at the fine granularity level i.e. adaptability at the lowest level of system

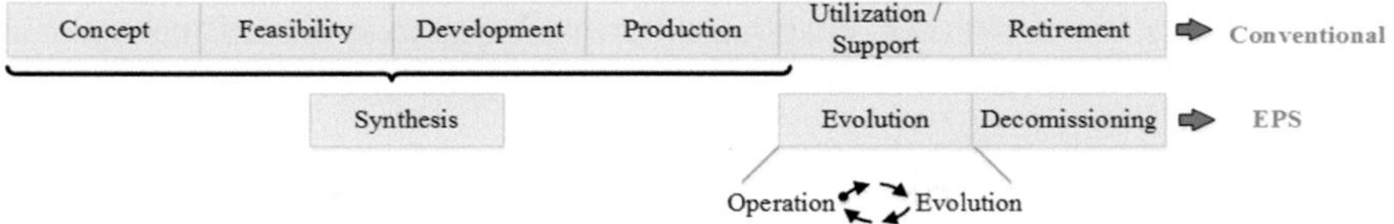

Fig. 1 EPS versus Conventional Life Cycle

hierarchy (e.g. sensors and actuators). The real-time coordination between the modules also enables the system to handle complex situations and respond efficiently to emergent behaviors [20, 23]. Unlike conventional systems, the EPS life cycle comprises of three stages (as shown in Fig. 1): Synthesis, Evolution, and Decommissioning [11, 22]. An iterative loop in the evolution stage, which caters for most of the life-cycle period, poses the need for an efficient development process to avoid any information inconsistencies resulting in unexpected system behavior.

2.2 Control Architecture

The core of EPS is its control architecture which is based on the concept of multi-agent systems [31]. Each physical module in an EPS is an intelligent, autonomous and skill-based entity called mechatronic agent [7, 27], which is an integration of the mechanical equipment, controller with a compatible communication interface and a software agent. As shown in Fig. 2 all the agents interact and communicate with each

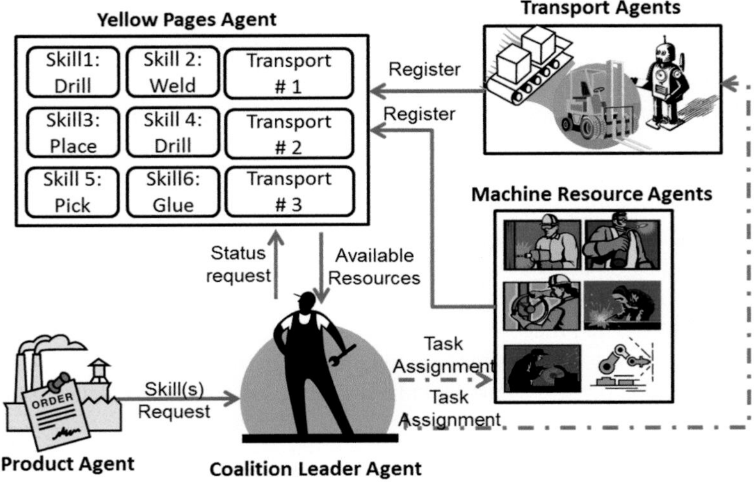

Fig. 2 EPS Control Architecture with Agent-Based Approach [6, 27]

other forming a social network, enabling the overall system to be self-manageable [10]. For more information the readers are referred to [6, 27].

2.3 Recent Advancements and Challenges

A few of the recent advancements related to the overall EPS development include; the concept of a reference architecture [23], an ontology to support evolvable assembly systems [26], utilization of JADE (Java Agent Development Environment) platform for implementation of agent-based control architecture [27], a visualization tool for retrieval of the information exchanged between different agents [7], a dynamic skill-configuration methodology [8], algorithms for self-organizing behavior [19], operational management through demand-responsive planning [2], and a innovative business model supporting the re-use of EPS modules [15], etc. One of the major challenges faced by EPS today is the harmonization of these efforts with each other as well as with the existing industrial standards targeting aspects like functional safety, system integrity, etc.

3 EPS-DSL

The EPS-DSL defined in this paper is mainly built upon the EPS ontology.[1] [12–14, 24]. The five major parts of the DSL are discussed as follows:

1. **Project View**: This part defines the overall scope of the project, such as business case, operational constraints, planning and scheduling, cost requirements, manufacturing environment, milestones, etc. The information regarding different assembly scenarios based on product variants is also defined in this module. Figure 3 shows the project view of the EPS meta model.
2. **Product View**: This view (left hand side artifacts in Fig. 4) models all the information related to the product and its variants, if any. For example, the product assembly requirements, volume, components, supporting materials, etc. The connection between the two components is represented by an assembly interface which is further identified as a *Male_Component_Port* or a *Female_Component_Port* depending on the liaison and connection type. All the interface types defined in the meta model are shown in Fig. 5.
3. **Process View**: The process view describes the sequence of operations and tasks needed according to the product assembly requirements defined in 'product view'. The product and process views from EPS-DSL are shown in Fig. 4 The processes are linked with each other via the process interfaces. The process interface, in general, consists of a *Control_Port* for determining operation sequence, a

[1]For the difference between meta model and ontology refer to [28]

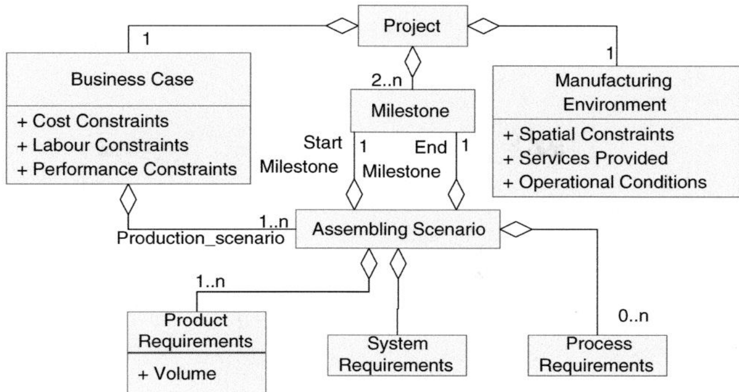

Fig. 3 Meta model for Project View: EPS-DSL

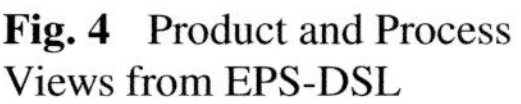

Fig. 4 Product and Process Views from EPS-DSL

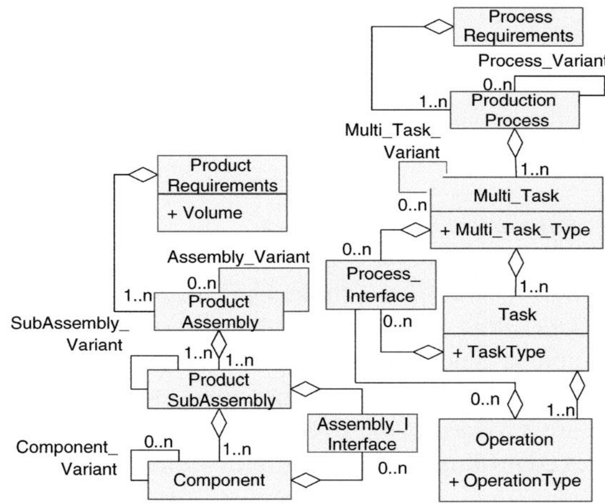

Fig. 5 Interface descriptions defined in meta model based on EPS ontology

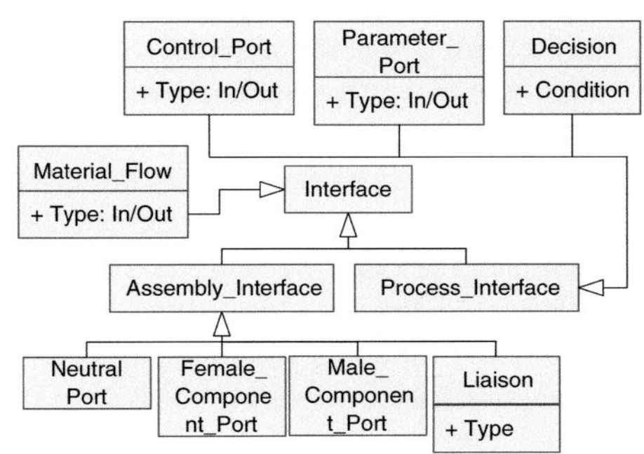

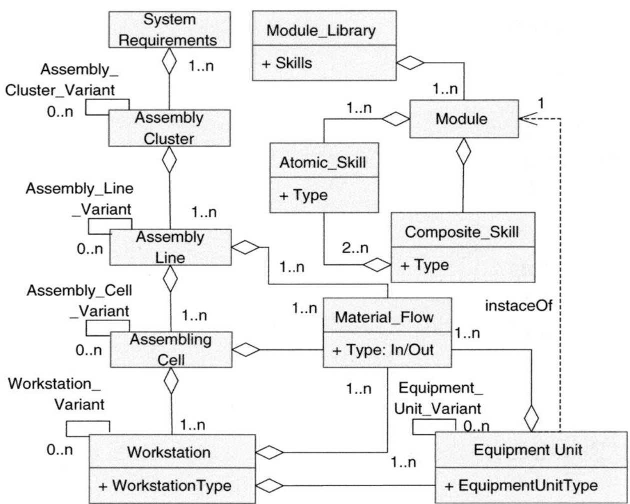

Fig. 6 Meta model for Assembling System Configuration: EPS-DSL

Parameter_Port for transfer of information regarding measured parameters between processes, and a *Decision* port to support the logical decision making during an operation/task. The processes are categorized as *Multi_Task*, *Task* and *Operation*. The type of each entity refers to the different types of processes available in general. For example, *OperationType* can be a handling operation', 'welding operation', 'loading operation', etc. Further types of processes and their details are provided in the EPS ontology [13].

4. **Assembling System Configuration**: Figure 6 provides an overview of the system configuration meta model. The process requirements serve as the input for modeling the assembling system configuration. The lowest hierarchical level in this view is the equipment unit which has one or more *Atomic Skills* required to perform a specific assembly process. The higher levels consisting of several equipments can be workstations, cells, lines or assembly clusters. The flow of material between each equipment is modeled via a *Material_Flow* port with each having one output flow and at least one input flow of material. An example of *EquipmentUnitType* is the 'PickAndPlaceUnit'. Other types and details of EquipmentUnitType and *WorkstationType* are provided in [12].
5. **Module to Process Mapping**: A *Module_Library* is created based on all the equipment modules and their respective skills available in the system. After the configuration of the assembly system based on the product and process requirements, the next step is the mapping of available modules to the required processes. It is to be ensured that each of the process activities is assigned a corresponding module providing the required skill. This step also helps in identifying the missing skills in the system.

4 Tool Support for EPS-DSL

The EPS-DSL has been implemented in MetaEdit+ (a DSM tool) [17]. The choice of MetaEdit+ is motivated by its features such as flexibility in terms of graphical symbols and reduced time for proof of concepts as compared to other tools based on technologies like EMF (Eclipse Modeling Framework). The following features are realized:

- **Graphical Modeling of EPS** providing a means for better understanding and interpretation between several stakeholders. Models are checked during modeling for basic constraints by utilizing live checking mechanism in MetaEdit+. For example, a product requirement should have at least one product assembly defined (see Fig. 4). A warning is generated in case of undefined assembly for a product requirement. This feature helps in guiding through the development process.
- **Automated Requirements Verification** is applied on the operation-skill/module mapping to verify the assembly process requirements. For this purpose a generator is developed which essentially navigates through the requirements and compares the mapping. A list of unmapped requirements are indicated which can be utilized for further action like their mapping to the system configuration, design modification or reasoning for not mapping a requirement.
- **Code Generation**: The code for a final system in place has one large part which is always static and another part of the code related to machine resource agents (Fig. 2) which varies from one machine to another. The tool support provides generation of code for the the skill related part of machine resource agents. The generated Java code is compatible with the industrial prototype from the IDEAS [21] and GASENS [1] project. The underlying generator extracts the information regarding the equipment modules and their associated skills from the developed system model and updates the code accordingly with the changes in the system design (i.e. addition/removal of modules).

Figure 7 shows the current status of the implementation as well as a part of the envisioned architecture-centric model-based framework. The green box depicts the part of meta-model derived directly from the EPS ontology. In addition to that,

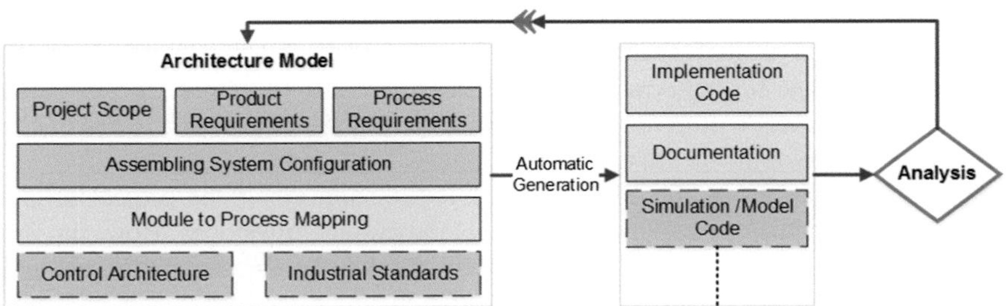

Fig. 7 Model-based development of EPS

the domain concepts have been extended with the system configuration and requirements mapping, shown by the yellow box. This enabled verification of the process, product and equipment requirements in relation to the overall system configuration.

The box shown in blue with slotted lines in the figure represent the additional viewpoints that are required as an extension of the meta model for a well-defined EPS methodology on a single unified platform. In terms of automatic generation, the implementation code for the equipment modules has already been generated, as discussed above. The generators for the simulation tools, testing models, etc., can be utilized in the extended model for a detailed system analysis. The results from various analysis tools can then be used as a feedback for updating the model and even the meta model itself, if required.

5 Related Work

A skill configuration methodology for modular assembly systems has been developed in [25] emphasizing the need for a unified platform. However, the methodology is based on a multi-agent approach, using industrial agents for system requirements and skill configuration which is a lower abstraction level as compared to the work presented in this paper. A virtual environment for EPS is proposed in [5]. The environment based on an integration of EPS ontological artifacts to EAST-ADL; a DSL for automotive embedded systems. Although relevant there exist several issues with the implementation of the virtual implementation. For example, a skill is mapped to a task which is contradictory to the EPS ontology. Another model-based approach towards run-time configuration for adaptable production systems is proposed in [9]. The focus is on ICT tools for reducing changeover time and efforts. In contrast this paper targets design phase of adaptable systems.

6 Discussion

The paper presented a DSM based approach as the first step towards a unified platform for an efficient and cost-effective development of EPS. The presented DSL and tool support covers most of the latest EPS developments. EPS is a relatively new field and hence subject to further developments in several areas. Therefore, the presented DSL and tool support will in turn be updated with new developments. A few of the possible new developments are as follows:

One of the features of the EPS ontology is the support for the project scope, description of manufacturing environment and the project milestones. However, the constraints (e.g. performance, cost, labor, etc.) need to be further elaborated and well-defined in the ontology so that it can be effectively utilized for system analysis. In addition, architectural elements (e.g. tasks, process, etc.) are also required to be complemented by properties such as timing constraints for analysis in the context of the overall project scope.

Moreover, the Industrial standards like IEC 61511, IEC 61508, ISO 13849 for functional safety, IEC 61131 for programmable controller and ISA 95 (ISO /IEC 26224) for enterprise and control system integration, etc. are currently addressed to a limited extent in the overall system design process. Especially the requirements for analysis, testing, verification and validation (V & V), hazard and risk identification, etc. needs to be incorporated in the overall methodology. This in turn implies addition of more views and viewpoints in the architecture modeling.

EPS and HMS both being based on the concept of industrial agents, share a lot of common concepts. Therefore, synergies can be exploited to develop a meta-model which is suitable for both the paradigms.

Acknowledgments The authors would like to thank SenseAir AB for financing and supporting this research.

References

1. UDI Project Information. http://www.vinnova.se/sv/Resultat/Projekt/Effekta/2011-01544/Hallbara-produktionsformer-for-hogteknologisk-tillverkning-av-MEMS-baserade-sensorsystem-i-Sverige/. Accessed May 2015
2. Akillioglu, H.: Evolvable Production Systems: Demand Responsive Planning. Technical report, Industrial Engineering and Management, KTH—The Royal Institute of Technology, Sweden, Nov 2011
3. Broy, M., Feilkas, M., Herrmannsdoerfer, M., Merenda, S., Ratiu, D.: Seamless model-based development: from isolated tools to integrated model engineering environments. Proc. IEEE **98**(4), 526–545 (2010)
4. Brussel, H.V., Wyns, J., Valckenaers, P., Bongaerts, L., Peeters, P.: Reference architecture for holonic manufactruring systems: PROSA. Comput. Ind. **37**, 255–274 (1998)
5. Chen, D.J., Maffei, A., Ferreira, J., Akillioglu, H., Khabazzi, M.R., Zhang, X.: A virtual environment for the management and development of cyber-physical manufacturing systems. In: 5th IFAC Workshop on Dependable Control of Discrete Systems. Cancun, Mexico (2015)
6. Colombo, A.W.: Industrial agent: towards collaborative production—automation—management- and organization. In: IEEE Industrial Electronics Society Newsletter, vol. 52, pp. 17–18. schneider Electric GmbH, Germany (2005)
7. Ferreira, J., Ribeiro, L., Neves, P., Akillioglu, H., Onori, M., Barata, J.: Visualization tool to support multi-agent mechatronic based systems. In: 38th Annual Conference on IEEE Industrial Electronics Society (IECON) (2012)
8. Ferreira, P., Lohse, N., Razgon, M., Larizza, P., Triggiani, G.: Skill based configuration methodology for evolvable mechatronic systems. In: IECON 2012–38th Annual Conference on IEEE Industrial Electronics Society, pp. 4366–4371, Oct 2012
9. Keddis, N., Kainz, G., Buckl, C., Knoll, A.: Towards adaptable manufactruring systems. In: IEEE International Conference on Industrial Technology (ICIT), CapeTown (2013)
10. Kephart, J.O., Chess, D.M.: The vision of autonomic computing. IEEE Comput. Soc. (2003)
11. Lindberg, B., Onori, M., Semere, D.T.: Evolvable production systems—a position paper. In: Swedish production symposium, SPS'07 (2007)
12. Lohse, N., Maraldo, T., Barata, J.: EUPASS: Assembling Equipment Ontology. Technical report (2008)

13. Lohse, N., Maraldo, T., Barata, J.: EUPASS: Assembling Process Ontology Specification. Technical report (2008)
14. Lohse, N., Maraldo, T., Barata, J.: EUPASS: Product Ontology Specification. Technical report, E (2008)
15. Maffei, A.: Characterisation of the business models for innovative, non-mature production automation technology. Ph.D. thesis, Department of Production Engineering, KTH—The Royal Institute of Technology, Sweden, Dec 2012
16. Mehrabi, M.G., Ulsoy, A.G., Koren, Y.: Reconfigurable manufacturing systems: key to future manufacturing. J. Intell. Manufact. **11**, 403–419 (2000)
17. MetaEdit+ Website: https://www.metacase.com/
18. Morel, G., Valckenaers, P., Faure, J.M., Pereira, C.E., Diedrich, C.: Manufacturing plant control challenges and issues. Control Engineering Practice **15**(11), 1321–1331 (2007), special Issue on Manufacturing Plant Control: Challenges and Issues INCOM 2004 11th IFAC INCOM'04 Symposium on Information Control Problems in Manufacturing
19. Neves, P.: System Evaluation and Learning in Evolvable Production Systems: Preliminary Considerations and Research Directions. Technical report, Industrial Engineering and Management, KTH—The Royal Institute of Technology, Sweden (2012)
20. Onori, M., Barata, J.: Mechatronic production equipment with process based distributed control. In: 9th IFAC Symposium on Robot Control, pp. 80–85 (2009)
21. Onori, M., Lohse, N., Barata, J., Hanisch, C.: The IDEAS project: plug & produce at shop-floor level. Assem. Autom. (2012)
22. Onori, M., Semere, D., Barata, J.: Evolvable assembly systems: from evaluation to application. In: Innovation in Manufacturing Networks, vol. 266, pp. 205–214. Springer (2008)
23. Onori, M., Semere, D., Lindberg, B.: Evolvable systems: an approach to self-X production. In: CIRP-sponsored International Conference on Digital Enterprise Technology, Advances in Intelligent and Soft Computing, vol. 66, pp. 789–802 (2010)
24. Onori, M., Semere, D.T., Barata, J.: EUPASS: Reference Architecture Specification. Technical report (2008)
25. Pedro: An Agent-Based Self-Configuration Methodology for Modular Assembly Systems. Ph.D. thesis, University of Nottingham (2011)
26. Ribeiro, L., Barata., J., Onori, M., Amado, A.: OWL ontology to support evolvable assembly systems. In: 9th IFAC Workshop on Intelligent Manufacturing Systems (2008)
27. Ribeiro, L., Rosa, R., Cavalcante, A., Barata, J.: IADE— IDEAS agent development environment: lessons learned and research directions. In: CIRP Conference on Assembly Technologies and systems (2012)
28. Saeki, M., Kaiya, H.: On relationships among models, meta models and ontologies. In: 6th OOPSLA Workshop on Domain—Specific Modeling (2006)
29. Schmidt, D.C.: Guest editor's introduction: model-driven engineering. Computer **39**(2), 25–31 (2006)
30. Tetzlaff, U.: Optimal design of flexiblemanufactruring systems. Contrib. Manag. Sci. (1990)
31. Wooldridge, M.: An Introduction to MultiAgent Systems. Wiley (2009)

The Using of Petri nets for Controlling of the Embedded Device

Jan Kožusznik and David Ježek

Abstract With the come of single board computers with sufficient performance and programmable components comes the possibility to program embedded devices using the techniques that are used in desktop or server solutions. This article explains how to use the formalism of Petri nets to control the functions of a device that we developed. Simulation based on The Petri nets runs on the device and controls visual effects that are realized by means of programmable LED strips. This paper in detail describes how the Petri nets based simulation process is interconnected with controlling of LED strips.

Keywords Petri net · LED strip · Enactment · Marking · Job · Task

1 Introduction

We had created a visual device that should serve as a promotional item for the IT4Innovations National Supercomputing Centre in Ostrava [4]. Later in this article we will call this device as the promotional item. It should demonstrate the principles of supercomputers using visual form together with the possibility of interaction. It consists of parts that may demonstrate nodes, a scheduler and terminals. Parts demonstrating nodes and the scheduler are decorated with LED strip that can indicate activity in a specific part by appropriate visual effect. Terminals are demonstrated by tablets that are connected to the promotional item. Tablets are used for a starting of "jobs" and a showing of basic descriptions about started jobs and a progress of their solution. We could say the tablets provide a capability of an interaction with

J. Kožusznik · D. Ježek (✉)
VŠB - Technical University Ostrava, 17. listopadu, Ostrava, Czech Republic
e-mail: david.jezek@vsb.cz
URL: http://www.vsb.cz

J. Kožusznik
e-mail: jan@kozusznik.cz

A. Abraham et al. (eds.), *Proceedings of the Second International Afro-European Conference for Industrial Advancement AECIA 2015*, Advances in Intelligent Systems and Computing 427, DOI 10.1007/978-3-319-29504-6_10

the promotional item. A detailed description of our work is available in [5]. In this article, we will focus on a description of a control of the device using Petri nets.

Embedded computers are often used to control many custom-made devices as our promotional item and many others. But in many cases, the functionality of a device is hard-coded in the source code of the program running on the embedded device. Our solution uses Petri nets as an abstraction of a performed process to increase reusability and changeability of our software. On the other hand, we do not describe the whole functionality by Petri nets. Rather, we have defined extension and enactment of Petri nets that allows to define a simpler and understandable net where some non-core functionality is not described as part of net but it is performed as sequence of commands in the used programming language.

Petri nets [2] were firstly described by Carl Adam Petri in 1962. It is a simple but robust mechanism for a modelling and a simulating of processes with the ability to describe a parallel processing in a straightforward way.

With the rise of cheap but powerful embedded devices a using of Petri nets as a direct model of a process management has also appeared. A device loads a definition of Petri nets and subsequently simulates the described process, individual steps of a simulation correspond to an implementation of partial activities of a system [1, 7].

The Sect. 2 briefly describes the promotional item and also contains a reference to the document that contains more particular information. The Sect. 3 contains an overview of the software architecture of the control program of the promotional item.

2 Basic Description

Promotional item (Fig. 1) is thoroughly described in [5]. This section contains only brief description of the promotional item with focus on parts that are closely related to the control mechanism implemented by Petri net.

User can manipulate the promotional item from one of three tablets that are physically attached to promotional item. Tablets hosts an application that serves to configure and start process that visually demonstrates processing of computational jobs by supercomputer. Another part is a scheduler, which contains one spiral LED strip. Other LED strips are stretched from tablets to scheduler. The last five elements that have a visual manifestation are the "computational" towers that consist of layers composed of LED strips. Each tower has a different number of layers. One of important part without visual effect is control part which is realized by microcomputer RaspberryPI and it is programmed using Java language (Fig. 2).

The user submits a request for the solving of his job in a supercomputer through one of the three tablets. Simulation of the computing runs in the control part after receiving a request from the tablet.

The device demonstrates sending a job to the job scheduler by shifting the illuminating LEDs through LED strip that is stretched between the tablet and the scheduler. The effect that demonstrates job scheduling begins after the previous effect. Time for

Fig. 1 Promotional item

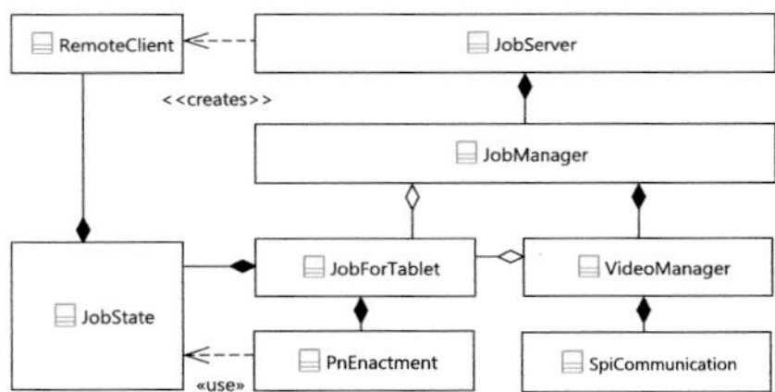

Fig. 2 Logical view of architecture

solving job is assigned according to origin of request. Our scheduling algorithm use the FIFO approach [8]. The arrival job is placed to end of the scheduler's queue. If a computational tower has no tasks assigned first job from scheduler's queue is divided to several tasks. The tasks are independently processed in computational towers. Processing of each task is demonstrated by single shining LED strip. It is required to wait for finishing of all tasks that are solved in computational towers. Completion the entire job (all tasks are finished) is demonstrated by similar visual effect as sending a job to the scheduler by sending information back to the tablet.

3 Architecture

A communication layer is responsible for receiving a request from a mobile device. A part of this layer is named as JobServer. Protocol TCP/IP is used for communication between it and remote clients. When a request is received a new object JobForTablet is created. It represents and encapsulates a state of a simulation of the job computing.

Object contains data that define resulting display in visual parts of promotional item (states of LEDs). JobServer also creates an object of RemoteClient. It is used for a communication from the job back to the client application in Android device (tablet). Also the state of simulation is sent to the tablet through that object.

A part named JobManager is responsible for controlling of running jobs. JobManager runs on a separate thread and sequentially removes jobs from queue. The manager assigns computational time to each job. Afterward, an every still running job is put at the end of the queue. Practically, JobManager works as a scheduler of jobs and use algorithm FCFS [8] for a scheduling.

Scenario of the visual simulation of a computational process is controlled by the mechanism of Petri nets [6] which is encapsulated in the part PnEnactment. Detail description of Petri net usage will be provided in the Sects. 4 and 5.

Module named VideoManager is responsible for visual demonstration of the status of a simulation. It includes self-running thread that periodically at an interval of 5 ms visits all running jobs to write the data into the output buffer (we use term "graphics buffer").

4 Petri net of Process

Initially, we wanted to describe scenario using finite automaton before we start with implementation of desired software. However, finite automaton was proved as unsuitable with regard to the fact that it is not able easily to capture parallelism. That is important because computational jobs runs in parallel on individual nodes. A very good alternative was a Petri nets. Originally, we planned the realization directly by means of the target technology—JAVA. We wanted to implement a sequence and a branching by program structures and a parallelism using threads.

Finally, we decided to create a mechanism that allows us to control and manage the sequence of the various visual effects based on defined Petri net.

There is a diagram of Petri net in Fig. 3 that captures modelled process. The fact that it is performed visual effect that illustrates an action, is represented by the presence of the token in the active place—symbol of a place having two arrows on the circuit.

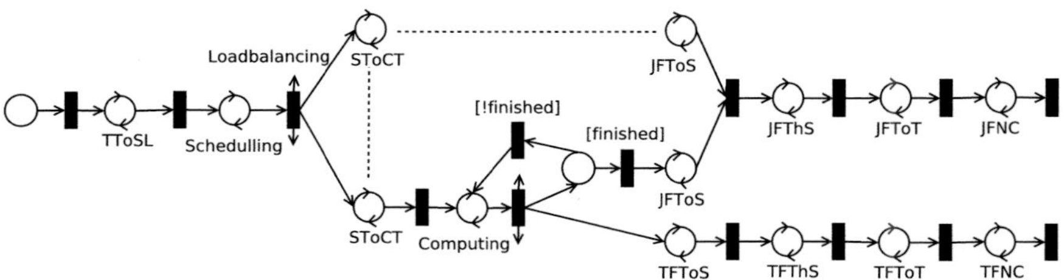

Fig. 3 PN-schema

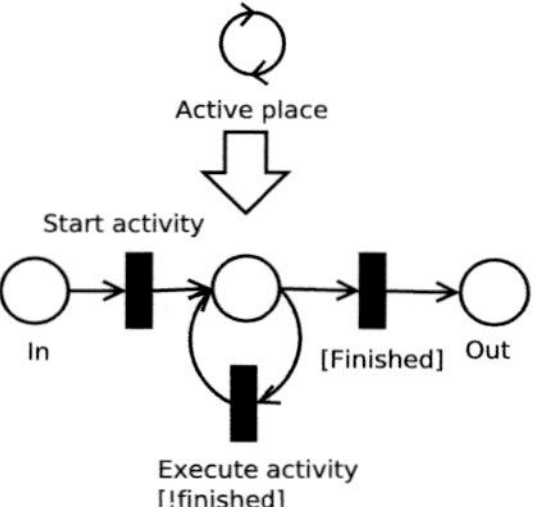

Fig. 4 Active place

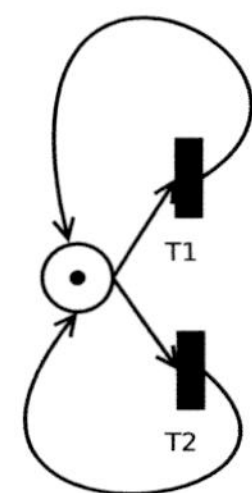

Fig. 5 Not mutually firable places

Definition 1 Let $N = (P, T, F)$ is the Petri net. Petri net with active places is a couple (N, P_{act}) where $P_{act} \subset P$. An item $p_{act} \in P_{act}$ is called active place. For purpose of enactment, a Petri net with active places is transformed into a Petri net so that each active place in the Petri net is replaced by a structure in the Fig. 4. Every arc that inputs into substituted active place is substituted with arc that inputs into place marked as "In" and every arc that outputs from substituted active place is substituted with arc that outputs from place marked as "Out".

Active places are finally implemented by structure which is shown in Fig. 4. When transition "Start activity" is "fired" the initialization code is called. Transition "Execute activity" is feasible only when the active place is not finished yet and it is "fired" again and again until the token leaves the place with a second alternative transition. How often is fired transition "Execute activity" it is explained in Sect. 5. We use extension of basic Petri nets with conditions defined for transitions which follow the same place. When a token is presented in the place it is "fired" transition that satisfy defined condition. During an execution of basic Petri net, it is "fired" one from such transitions and its selection is made randomly.

The reason why we used the active places for the representation of performing of some action, was that we wanted to represent action that last longer, and we do not wanted hide tokens during activity execution. Disappearance of tokens is undesirable because they contains information needed for the visual representation of the effect. Token from the network disappears only when it is assigned time to perform action, however due to the nature of these actions it is not too long. Our approach is thus possible only if the execution of the action code does not require 100 % of the duration of this action.

There are these active places in our Petri net:

- TToSL (Task to Scheduler)—visually demonstrates that the requirement to calculate the job is transmitted from the tablet to the scheduler.
- Scheduling—visually demonstrates the putting of jobs into the queue of the scheduler and subsequent extraction, if the job will be on first place in the queue.
- SToCT (Scheduler to Computational Tower)—visually should demonstrates that the requirement for calculation tasks is sent from the scheduler into the computational towers. The active place finally does not show anything, because LED strip between the scheduler and computational towers is not stretched in reality.
- Computing—the computational tower solves one task.
- TFToS respectively JFToS—visually should demonstrates that information about task completion and job transmission from the computational tower to the scheduler. Like in case SToCT, these places do not demonstrate anything.
- TFThS resp. JFThS—visually demonstrates that information about completed task respectively completed job goes through the entire scheduler.
- TFToT resp. JFToT—visually demonstrates the transfer of information about completion of the task respectively entire job from the scheduler to the tablet.
- TFNC resp. JFNC—via the TCP/IP connection the information about the solved task respectively the whole job is sent to the client application running on the tablet.

In Petri net, there are two transitions, followed by multiple places. It leads to the creation of new tokens. It will offer independent and therefore parallel visual effects. One transition is marked as "Loadbalancing". Each created token will match the progress of the solving of one task in certain computational tower. The transition in which these parallel branches are merged is located before the active place JFThS. This transition is "fired" when all tasks in all computational towers for given job are finished.

Each token carries peace of information about how many tasks will be computed. The initial token has this value set to N. Tokens, which were created in the transition "Loadbalancing" have these values set so that their sum gives value N. For these purposes, this transition is enhanced with additional semantics. Thus enhanced transitions are referred to as the active transitions. During their "firing", they are executed actions that are described programmatically. Such transitions are indicated by rectangles with vertical arrows that are directed out. Control code that is executed during "firing" the transition receives as input tokens these that enter into the transition, and pass as output tokens that leave the transition. Control code of the transition "Loadbalancing" realizes the separation of tasks for each computational tower.

After the active place "Computing", there is a second transition that causes parallelism. Into a tablet, it is sent information that was solved one task and independently is made a decision whether continue to solve another task or to send information that was solved whole job. The transition after "Computing" is also active transition. It creates a token for solving of another task with decreased number of tokens that should be still solved. It also creates token for notification about task solution and this token contains identifications of the solved task.

Decision making whether continue to solve another task or not is done with construction of a single place followed with two transitions. About which of the two transitions is "fired", determine conditions for these transitions.

Parallely running demonstrations of tasks solved in computational towers are merged again in the transition, which is located before the active place JFThS. This transition is fired only when the previous actions finish in all computational towers.

5 Enactment of Petri net

An instance of class "JobManager" calls a message "doOneStep" periodically on each active job. An active job is represented by an instance of "JobForTablet". "JobForTablet" delegates this call to the object of class "PnEnactment" and during method call it passes the states (the objects of class "JobState"), which includes marking of Petri net. The implementation of the method "doOneStep" in PnEnactment is described in the Algorithm 1. The algorithm is divided into 3 parts: the first part is processing of all enabled transitions and creation of set $\mathfrak{P}_{\mathfrak{sub}}$ that contains couples (transition, sub-marking). Before inserting into this set, every transition is tested whether is still enabled, since the creation of the sub-marking modifies the marking M and some transitions originally enabled may become not enabled. Second part of the algorithm fires transitions that are in the set $\mathfrak{P}_{\mathfrak{sub}}$. The last part of the algorithm merges all resulting sub-markings into a new marking M'. We divided the enactment this way because during one step are fired only those transitions that are actually "mutually fireable" not only mutually enabled but fireable sequentially. Transitions that are mutually enabled but not "mutually fireable" are transitions t_1 a t_2 in Fig. 5.

Parts of algorithm for "enactment" are protected by a "semaphore"[1] that is related to the object JobState. The reason is explained in Sect. 6.

[1] A "semaphore" is the special shared variable that is used by the classical mechanism for regulating access to shared resources [3].

Algorithm 1: Petri net enactment

$\mathfrak{P}_{\mathfrak{sub}} \leftarrow \emptyset$
$\mathfrak{T}_{\mathfrak{e}} \leftarrow \{t \mid t \in P \wedge IsEnabled(N, M, t)\}$
begin Capture semaphore
  forall the $t \in \mathfrak{T}_{\mathfrak{e}}$ **do**
    if *IsEnabled(N,M,t)* **then**
      $M_{sub}, M \leftarrow MakeSubmarking(N, M, t)$
      $\mathfrak{P}_{\mathfrak{sub}} \leftarrow \mathfrak{P}_{\mathfrak{sub}} \cup (M_{sub}, t)$
    end
  end
end
$\mathfrak{M}_{\mathfrak{sub}}{}' \leftarrow \emptyset$
forall the $(M_{sub}, t) \in \mathfrak{P}_{\mathfrak{sub}}$ **do**
  begin Capture semaphore
    $M'_{sub} \leftarrow FireTransition(N, M_{sub}, t)$
  end
  $\mathfrak{M}_{\mathfrak{sub}}{}' \leftarrow \mathfrak{M}_{\mathfrak{sub}}{}' \cup M'_{sub}$
end
$M' \leftarrow M$
begin Capture semaphore
  forall the $M'_{sub} \in \mathfrak{M}_{\mathfrak{sub}}{}'$ **do**
    $M' \leftarrow MergeMarking(M', M'_{sub})$
  end
end

In this algorithm, there are as yet undefined function *Submarking* and *MergeMarking*. Their definitions are in Eqs. 5.1 and 5.2. *Submarking* creates for a given Petri net, marking a transition two new markings. The first marking includes only the tokens in the places, which are necessary for "firing" specified transition. The second marking is the same as the original marking but it is missing tokens that are in the first marking. *MergeMarking* will create one marking from two markings such that the resulting markings has in every place only tokens from places which are simultaneously in both markings.

$$\begin{aligned} &Submarking = \\ &\quad (\lambda N, M, t)((\{(p, M(p)) \mid N = (P, T, F) \wedge p \in \bullet t_N\} \cup \\ &\qquad \{(p, 0) \mid N = (P, T, F) \wedge p \in P \wedge p \notin \bullet t_N\}, \\ &\qquad \{(p, 0) \mid N = (P, T, F) \wedge p \in \bullet t_N\} \cup \\ &\qquad \{(p, M(p)) \mid N = (P, T, F) \wedge p \in P \\ &\qquad\quad \wedge p \notin \bullet t_N\})) \end{aligned} \tag{5.1}$$

$$\begin{aligned} MergeMarkings = (\lambda M, M') \\ (\{(p, m + m') \mid (p, m) \in M \wedge \\ (p, m') \in M'\}) \end{aligned} \tag{5.2}$$

6 Connection of Active Place to Visual Effects

The execution of the active place represents a certain action with long duration. In our case visual effects control that this action demonstrates. The visual effects associated with the active places which demonstrate the transfer of information from one part to the other part (TToSL, TFToT, JFTOT) are implemented as movement of strip of a shining colored lights on the LED strip in the direction in which it is transmission of information demonstrated. If it is demonstrated the transfer of information from the scheduler to a tablet then the strip moves from the scheduler to the tablet. Similarly, looks the effects for active places TFThS and JFThS. If we demonstrate transfer of the entire job (active sites TToSL, JFThS, JFTOT) then the colour of the strip corresponds to the tablet from which was launched the request for a job solution. If the transfer relates to information on completing only one task, it is used one common colour (active places TFThS and TFToT).

Each type of visual effects correspond to a different implementation of an active place (see Fig. 6):

- ConnectionActivity—active places TToSL, TFToT, JFToT
- SchedulingActivity—active place Scheduling
- ComputingActivity—active place Computing
- NotificationThroughScheduler—active places TFThS and JFThS

Each visual effect is associated with an element which the promotional item is composed of (see Fig. 7). Elements in the model correspond to classes:

- Connection—connection between the tablet and the scheduler. Originally connection between the scheduler and computational towers should exist, which would be also implemented by instances of this class.
- SchedulerTower—scheduler.

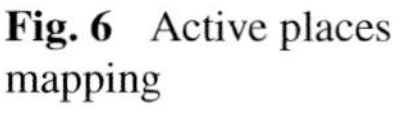

Fig. 6 Active places mapping

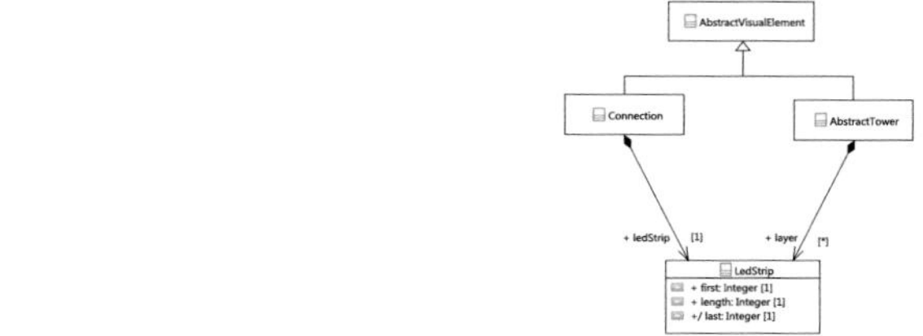

Fig. 7 PN-schema

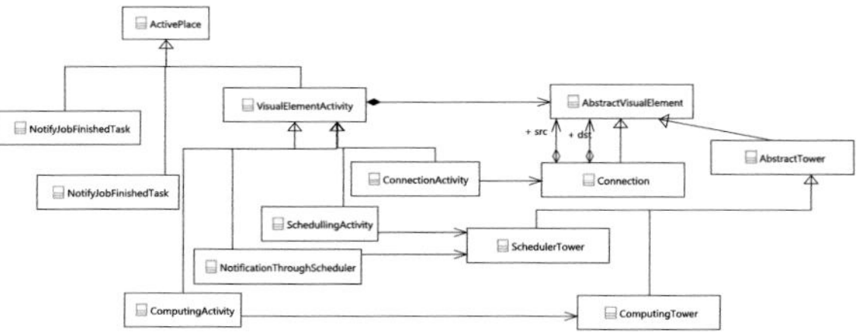

- ComputingTower—computational tower.

The individual elements consist of LED strips, which corresponds to a class LedStrip in the model (Fig. 6). Elements of towers contains more strips and then these strips forms tower's layers (in the scheduler, there is only one layer). All mappings of elements in the structure of LED strips that are installed on the promotional item are made by instance variables of objects LedStrip:

- first—Position of the first LED that is contained in the LED strip, which is mapped by this object LedStrip. The position is relativ to the beginning of the whole LED strip.
- length—Length of the LED strip.
- last—Position of the last LED, which is in the LED strip. This variable is derived from a value of variable "first" and "length".

If we want to change the mapping, we just change the initial values of these variables. This fact gives us much easier debugging of software of control parts, as the resulting promotional item was completed only at the end and it was a very short time for final deployment. So most of the things we debugged only on the continuous LED strip, where the mapping was done so that we could know whether the program does not have major defects and it runs smoothly. After composition of the promotional item, we only changed the resulting mapping and slightly tune the visual effects (speed, colour, and brightness), which took a several hours in total.

Now, we describe how they are handled visual effects in the active places. During the firing of internal transition within the active place (place marked as "Execute activity" in Fig. 4), one step of this effect is performed. New state (e.g. position) is derived base on the previous state (e.g. position in the case of effects that show the movement of information), the effect parameters (e.g. speed) and the time elapsed since the last execution of the visual effect. Or it is found that this effect ends and the active place is terminated. All status information is stored in the token for which the active place is performed.

During visual effects control, the state of resulting visual effect is also inserted into token (position, color and intensity of the LED in LED strip). When JobForTablet is asked by VideoManager to writes its visual data into graphical buffer, it is this action delegated to the object JobState and to all tokens in actual marking of Petri net or to all sub-markings that are created during enactment (as was demonstrated in Algorithm 1).

This is why we protected some parts of the code of enactment by a "semaphore". It must be ensured that during the writing of status of tokens into the graphic buffer is not "fired" any transition. Otherwise, the tokens that are on input to this transition, disappeared from Petri nets (from its marking) for a moment and visual effects running through these tokens would not be displayed. This was manifested by occasional flashing (Originally synchronization using a "semaphore" has not been used and flashing actually occurred.). Of course, the same "semaphore" must be suitably used for code protection where all the tokens write their data into a graphical visual buffer.

7 Conclusion

We showed the use of Petri nets for control device on which the user starts the actions, which can run in parallel. The actions consist of smaller actions. Petri nets allow us to define a unified approach for defining the behaviour of the device, which can be easily changed. Definition of structure of the Petri net was created during the development within the source code because it was not sufficiently advantageous to implement support for the definition of structure in more sophisticated manner. These methods include the definition in a text file or using a graphical editor.

Acknowledgments This research has been supported by the internal grant agency of VSB-TU of Ostrava - SP2015/85: "Knowledge modelling and its applications in software engineering".

References

1. Fernández, J.L., Sanz, R., Domonte, E.P., Alonso, C.: Using hierarchical binary Petri nets to build robust mobile robot applications: RoboGraph. In: ICRA, pp. 1372–1377. IEEE (2008)
2. Girault, C., Valk, R.: Petri nets for Systems Engineering. Springer, New York (2002)
3. Ii, J.D.D.: Computer Architecture: Fundamentals and Principles of Computer Design. CRC Press, Boca Raton, FL (2005)
4. IT4Innovations: What is IT4Innovations? http://www.it4i.cz/what-is-it4innovations/?lang=en. Acccessed 06 June 2014
5. Kozusznik, J., Seidl, D., Stolfa, S., Stolfa, J.: Construction of Supercomputer Demonstration Tool. http://www.kozusznik.cz/resources/docs/cisim_2014.pdf
6. Petri, C., Reisig, W.: Petri net. Scholarpedia **3**(4), 6477 (2008)
7. Richta, T., Janoušek, V., Kočí, R.: Petri nets-based development of dynamically reconfigurable embedded systems. CEUR Workshop Proceedings, vol. 989, pp. 203–217 (2013)
8. Tanenbaum, A.S.: Modern Operating Systems. Prentice Hall, Upper Saddle River, N.J (2007)

Cattle Identification Using Muzzle Images

Lukáš Zaorálek, Michal Prilepok and Václav Snášel

Abstract The quality of animal identification system plays an important role for producers to make management decisions about their herd or individual animals. The animal identification is also important to animal traceability systems to ensure the integrity of the food chain. Usually, recordings and readings of tags-based systems are used to identify an animal, but only effective in eradication programs of national disease. Recently, animal biometric-based solutions, e.g. muzzle imaging system, offer an effective and secure, and rapid method of addressing the requirements of animal identification and traceability systems. In this paper, we present an identification system based on muzzle images. The identification process is based on Support Vector Machine (SVM), Linear Discriminant Analysis (LDA) and Tucker Tensor Decomposition. This selected classifiers we compared on the same dataset of muzzle images with different experiment settings. The results we evaluated by F-score. The best F-score result gives us the Tucker Tensor Decomposition. It achieved the median of F-score 0.750.

Keywords Animal identification · Recognition · Identification · Tucker decomposition · Tensor · Support Vector Machine · Singular Value Decomposition · Linear Discriminant Analysis

L. Zaorálek (✉) · M. Prilepok · V. Snášel
Faculty of Electrical Engineering and Computer Science, Department of Computer Science and IT4 Innovations, VŠB-Technical University of Ostrava, 17. listopadu 15/2172, 708 33 Ostrava - Poruba, Czech Republic
e-mail: lukas.zaoralek@vsb.cz

M. Prilepok
e-mail: michal.prilepok@vsb.cz

V. Snášel
e-mail: vaclav.snasel@vsb.cz

A. Abraham et al. (eds.), *Proceedings of the Second International Afro-European Conference for Industrial Advancement AECIA 2015*, Advances in Intelligent Systems and Computing 427, DOI 10.1007/978-3-319-29504-6_11

1 Introduction

The quality of animal identification system plays an important role for producers to make management decisions about their herd or individual animals. The animal identification is also important to animal traceability systems to ensure the integrity of the food chain. Usually, recordings and readings of tags-based systems are used to identify an animal, but only effective in eradication programs of national disease. Recently, animal biometric-based solutions, e.g. muzzle imaging system, offer an effective and secure, and rapid method of addressing the requirements of animal identification and traceability systems.

In this paper, we are comparing three muzzle imaging identification system based on different classifiers. We chose following classifiers Support Vector Machine (SVM), Linear Discriminant Analysis (LDA) or Tucker Tensor Decomposition. The simulate real conditions by taking muzzle images, we added a different level of Gaussian white noise to the images. This additional noise simulates taking images in lower light conditions and other circumstances.

The organization of this article is as follows. The Sect. 2 describes proposal selected methods for classifying. The Sect. 3 contains the description of the muzzle images dataset. The Sect. 4 describes the experiments and it's set up. The Result of all selected methods are depict in Sect. 5. The last Sect. 6 contains the conclusion of this paper and future work plan.

2 Proposal Methods

Let us now introduce proposal methods of the experiment. There are used Linear Discriminant Analysis (LDA), Support Vector Machines (SVM) and Tucker Decomposition. Also the Singular Value Decomposition (SVD) and it's low-rank approximation we used as the important features selection method.

2.1 Singular Value Decomposition

Let A be a $m \times n$ matrix with $m \geq n$. The Singular Value Decomposition (SVD) of a matrix A we can be written as follow:

$$A = U \times \Sigma \times V^T \tag{1}$$

where $U^T U = V^T V = VV^T = I_n$ and $\Sigma = diag(\rho_1 \geq \cdots \rho_n)$. The matrix $U \in R^{m \times m}$ and the matrix $V \in R^{n \times n}$ are orthonormalized matrices. The matrix Σ express non-negative square roots of the eigenvalues of $A^T A$, they are called singular values. The singular values are positive numbers and we shall assume that [2, 9]:

$$\rho_1 \geq \cdots \rho_n > 0 \tag{2}$$

Let now assume that several last singular values of SVD are equal to zero:

$$\rho_{r+1} = \rho_{r+2} = \cdots \rho_n = 0 \tag{3}$$

then we can be define it as rank of the matrix A as follows:

$$rank(A) = r \tag{4}$$

If we choose rank r then this rank can be applied to the low rank approximation of SVD to the matrix A. Note that, the most biggest singular values $k = n - r$ represent the most significant features of a matrix A. The Eq. (5) defines matrix A_k in accordance with low rank approximation of SVD:

$$A_k = U * S_k * V^T, S = diag(\rho_1, \ldots, \rho_k, 0, \ldots, 0) \tag{5}$$

The Sect. 4 describes how we exactly are using the low rank approximation in our preformed experiments.

2.2 Linear Discriminant Analysis

Linear Discriminant Analysis (LDA) is a well-known algorithm which is used for features extraction and dimension reduction. The main goal of this algorithm is to achieve the best separation between between classes of the dataset and find linear combinations of the input features of the classes. The original method was published by Fisher in 1931. The original algorithm can classify object into only two classes. Let assume a matrix $A \in R^{m \times n}$ and x belongs to n-dimensional column vector. It is possible to construct a projection A onto x by the linear transformation Eq. (6) [4]:

$$\vec{y} = A\vec{x} \tag{6}$$

The Eq. (6) shows that y is an m-dimensional projected vector which has features of the matrix A. In order to find an optimal x it is introduced an objective function as shown the Eq. (7) [10].

$$J(x) = \frac{\left| x^T \hat{\Sigma}_b x \right|}{\left| x^T \hat{\Sigma}_w x \right|} \tag{7}$$

where Σ_b and Σ_w is defined as follow:

$$\hat{\Sigma}_b = \sum_{i=1}^{n} m_i \left(\bar{x}_i - \bar{x}\right) \left(\bar{x}_i - \bar{x}\right)' \tag{8}$$

$$\hat{\Sigma}_w = \sum_{i=1}^{n} \sum_{x \in c_i} \left(x - \hat{x}_i\right) \left(x - \hat{x}_i\right)' \tag{9}$$

The objective function Eq. (7) is called Fisher linear projection criterion. The optimal x can be computed by maximizing $J(x)$:

$$x_{opt} = \underset{x}{\operatorname{argmax}}\, J(x) \tag{10}$$

where x_{opt} is optimal x. The Eq. (10) is called Fisher optimal projection axis.

2.3 Support Vector Machine

The basic idea of Support Vector Machine (SVM) is to find a hyperplane which divide a dataset into two independent subsets. The method was originally developed by Vapnik in 1995. Since that time, we can find different modifications of the original method. If exist two independent subsets in a dataset, the margin between these subsets is defined by the distance of the hyperplane to the nearest of the positive and negative examples. The Eq. (11) shows the output of a linear SVM [6]:

$$u = \vec{w} \cdot \vec{x} - b \tag{11}$$

corresponding to decision functions Eq. (12):

$$f(x) = sign((\vec{w} \cdot \vec{x}) + b) \tag{12}$$

where $w \in R^N$ is the normal vector to the hyperplane, $x \in R^N$ is the input vector and finally $b \in R$ is a constant. In other words, the SVM is trying to find a gap between the two classes by maximizing the gap between the classes nearest points. The points lying on the boundaries are called *support vectors*. The dataset must meet the following necessary condition. Each sample has to belong to only one of the two classes of the dataset. The Eq. (13) shows the condition described above [3].

$$\{x_i, y_i\}, i = 1, \ldots N, x \in R^d, y \in \{-1, 1\} \tag{13}$$

2.4 Tucker Decomposition

A tensor is a multidimensional array. More precisely, Nth-order tensor $\underline{Y}$ is defined as follows: $\underline{Y} \in R^{I_1 \times I_2 \times \cdots \times I_N}$ where $I_1, I_2, \ldots, I_N \in N$ are index upper bounds. The Tensor is a general form of vectors and matrices. Therefore zero-order tensor is a scalar, first-order is a vector and second-order is a matrix [1]. However, a commonly used term tensor is the third-order tensor, which has three dimensions. This size of dimension we can find in the majority datasets originating from images recognition, or EEG/ECG time series. It is possible to choose subarrays from a tensor in accordance with their axis. In case of a second-order of a tensor, a matrix, these subarrays are rows and columns. On the other hand, subarrays, which are belonging to a third-order tensor, are called fibers and slices. Besides subarrays fibers and slices, a tensor has another no less important features and mathematic operations like unfolding and n-mode product [1, 5]. The following list summarizes definitions and notions of important features of a tensor which were described above:

1. *Fibers* are the higher analogue of matrix rows or columns. A fiber is defined by fixing every index but one. For example, $Y_{:jk}$, $Y_{i:k}$ or $Y_{ij:}$
2. *Slices* are two-dimensional sections (or fragments) of a tensor, defined by fixing all but two indices: $Y_{i::}$, $Y_{:j:}$ or $Y_{::k}$
3. *Unfolding* reshapes a tensor $\underline{X} \in R^{I_1 \times I_2 \times \cdots \times I_N}$ into a matrix, which is denoted by $Y_{(n)}$. It means the mode-n fibers are reshape into columns of a matrix $Y_{(n)}$.
4. *n-mode product* is described by Eq. (15)

Let introduce the Tucker decomposition which is usually called the Higher-Order Singular Value Decomposition (HOSVD). The Tucker decomposition of a tensor $\underline{Y} \in R^{n \times m \times p}$ is a product of four matrices N, M, P and C. The matrices N, M and P are called factors. The matrix C is called core (matrix). The decomposition can be written as:

$$A_{ijk} = \sum_{\alpha=1}^{nn} \sum_{\beta=1}^{mm} \sum_{\gamma=1}^{pp} C_{\alpha\beta\gamma} N_{i\alpha} M_{j\beta} P_{k\gamma} \tag{14}$$

$$(\underline{X} \times_n v)_{i_1,\ldots,i_{n-1} i_{n+1} \ldots i_N} = \sum_{i_n}^{I_n} x_{i_1 i_2 \ldots i_N} v_{i_n} \tag{15}$$

The relationship between (Higher-Order) Singular Value Decomposition and the Tucker decomposition can be described as following Eqs. 16, 17 and 18:

$$T_n = U^n * \Sigma^n * V^n \tag{16}$$

where $n = 1, 2, 3$ in case of three dimensional tensor. The Eq. (16) is used for the HOSVD as follows:

$$\underline{X} = C \times_1 U^1 \times_2 U^2 \times_3 U^3 \tag{17}$$

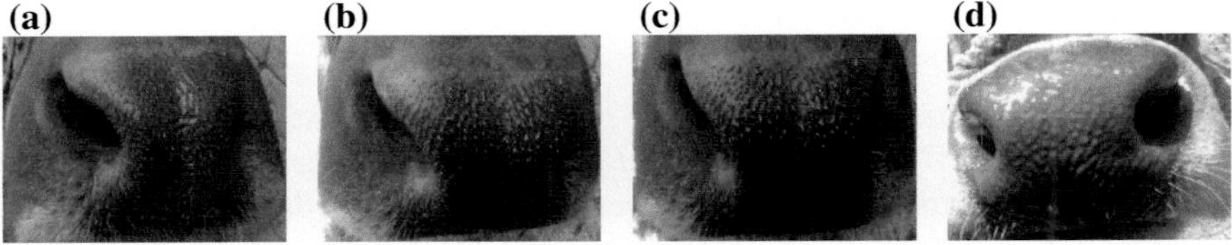

Fig. 1 Pictures of muzzles

and the core tensor C is computed by Eq. 18:

$$C = \underline{X} \times_1 U^1 \times_2 U^2 \times_3 U^3 \tag{18}$$

3 Dataset

To evaluate our proposed classifying methods we used our own dataset of muzzle images. This dataset contains 322 muzzle images of cows. It consists of 46 subjects. Each subject has 7 images taken from different perspective and light conditions [7]. Examples of the images are shown in Fig. 1 [8]. The first three images belong to the same object. The last image is from different object. It was also added a different level of Gaussian white noise to these images in order to simulate illumination condition.

This help us to explore how are the proposed methods are able to deal with noise in the muzzle images.

4 Experiment

The experiment used all three methods described in Sect. 2. Every experiment consist of following steps:

- filter important features of images,
- training the algorithm with a subset of images,
- and finally testing and measuring the performance of the proposal methods.

Because of needs to focus only to main features of images in the training phase, the SVD and it's low rank approximation we utilized in the first step. The SVD behaved as a filter in order to eliminate non-important features of the images by using the low rank approximation. More precisely, after applying SVD to an image we calculated the low rank approximation. It means that the following step of the experiment has worked with the reduced version of the image. According to the number of eigenvalues that we choose for low rank approximation. The number of most important features of an image was used to train of proposal methods. Following numbers we choose of eigenvalues: $2, 4, 8, 12, 18$.

Note that, this step was avoided in the experiment of Tucker decomposition due to nature character of the method. The Tucker decomposition uses SVD itself. The training phase of proposal methods was dedicated to the second step. Each training set was composed of images that were part of one of two subjects. The first class represents images belonged to investigated subject and second class represents images that were chosen from remaining subjects. The number of training sets was equal to the number of subjects in the dataset described in Sect. 3. The test set was created in the final step of the experiment and performance of the proposal methods was measured by computed confusion matrices for each combination of the number of eigenvalues and the noise thresholds. The Sect. 5 describes in more detail about the results of the experiment.

The way how the Tucker decomposition was employed in the context of classification describes Sect. 4.1. In case of SVM method it was choosed following parameters: *Kernel function*: Gaussian Radial Basis Function kernel (RBF) and *RBF sigma coefficient* equals 1.

4.1 Classification Based on Tucker Decomposition

First of all, we need to calculate the core matrix C according to the Eq. (14). The matrix C represents features that are obtained from the training set. The aim of this approach is comparing training and testing set by standard Euclidean distance. Before comparing training and testing set, it is necessary to transform the testing set to the same space where the matrix C is located. The transformation of testing set to the space of matrix C is achieved by n-mode product as shown the Eq. (15) by use factors matrices F of a tensor. Let now suppose a matrix T which is a result of the transformation described above. Now, it is possible to compare each row of the matrix T to each row of matrix C by euclidean distance. In other words, it is achieved comparison between each sample of the testing set and each sample belongs to training set by standard metric.

Algorithm 1 Classification by Tucker Decomposition

procedure TUCKER_CLASSIFICATION(X_{train}, Y_{test})
    $G_{tensor} \leftarrow calcCoreTensor(X_{train})$
    $F_{tensor} \leftarrow calcFactorsTensor(X_{train})$
    $Y_{tensor} \leftarrow (F_{tensor} \times_n Y_{test})_{i_1,\ldots,i_{n-1} i_{n+1} \ldots i_N}$
    for all i such that $0 \leq i < numImages(Y_{test})$ **do**
        $tres(i) \leftarrow euclideanDistance(G_{tensor}, Y_{tensor}, i)$
    $minTreshold \leftarrow percentile(tres(i))$
    for all i such that $0 \leq i < numImages(Y_{test})$ **do**
        $tres(i) \leftarrow minTreshold$
        if $tres < minTreshold$ **then**
            $class \leftarrow 1$
        else
            $class \leftarrow 0$

The measured distances are stored to a vector d. Before the last step, 95 % percentile is calculated from the similarity distances of the vector d. The computed percentile value is the minimum threshold for comparison training and testing set. If a value from the vector d was smaller than the threshold then estimated the class of current sample was set to 1, otherwise 0. The Algorithm 1 shows the algorithm in more detail.

5 Results

The experiment was evaluated by the F-score. The F-score is a statistics method for test's accuracy which is suitable with a binary classification. The Eq. (19) shows F-score formula:

$$F\text{-}score = \frac{2TP}{2TP + FP + FN} \tag{19}$$

where TP is a true positive, FP is a false positive and finally FN is a false negative. The values TP, FP and FN are possibly calculated by *confustionmatrix*.

In Sect. 3 was mentioned that the Gaussian noise was added to the images because of illumination conditions. It was used these following levels of the Gaussian noise: 0.1, 0.3, 0.6, 0.9, 1.2. In case of SVM and LDA they were choosed following values as a number of eigenvalues involved in low-rank approximation: 2, 4, 8, 12, 18. Let W denotes a number of factors in Tucker decomposition. The vector W has following settings for the experiment: $(-1, 2, 2)$, $(-1, 3, 3)$, $(-1, 4, 4)$, $(-1, 5, 5)$ and $(-1, 6, 6)$.

The Tables 1, 2 and 3 shows F-score results of proposal methods. It is clear evidence that the Tucker decomposition outperforms rest of methods. The minimum F-score for Tucker decomposition was 0.5 whereas F-score for SVM and LDA was 0.

Table 1 F-score values of Linear Discriminant Analysis

	0.1	0.3	0.6	0.9	1.2
Min	0.000	0.000	0.000	0.000	0.000
Max	0.500	0.000	0.000	0.000	0.000
Avg	0.022	0.000	0.000	0.000	0.000
Median	0.000	0.000	0.000	0.000	0.000

Table 2 F-score values of Support Vector Machine

	0.1	0.3	0.6	0.9	1.2
Min	0.000	0.000	0.000	0.000	0.000
Max	0.500	0.500	0.000	0.000	0.000
Avg	0.011	0.022	0.000	0.000	0.000
Median	0.000	0.000	0.000	0.000	0.000

Table 3 F-score values of Tucker decomposition

	0.1	0.3	0.6	0.9	1.2
Min	0.500	0.500	0.500	0.500	0.500
Max	0.750	0.750	0.750	0.750	0.750
Avg	0.739	0.674	0.576	0.576	0.560
Median	0.750	0.750	0.500	0.500	0.500

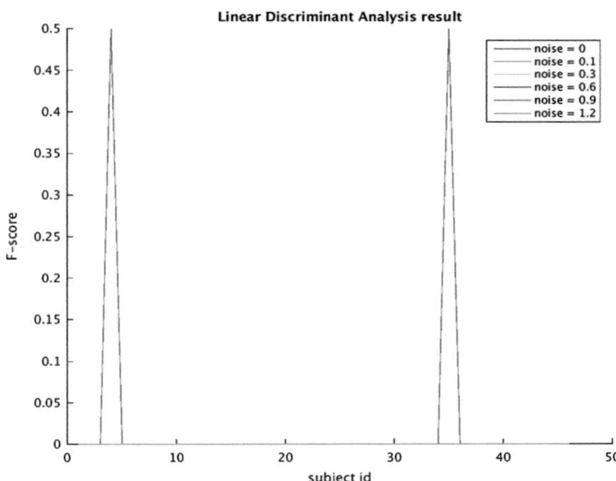

Fig. 2 F-score values of Linear Discriminant Analysis per experiment

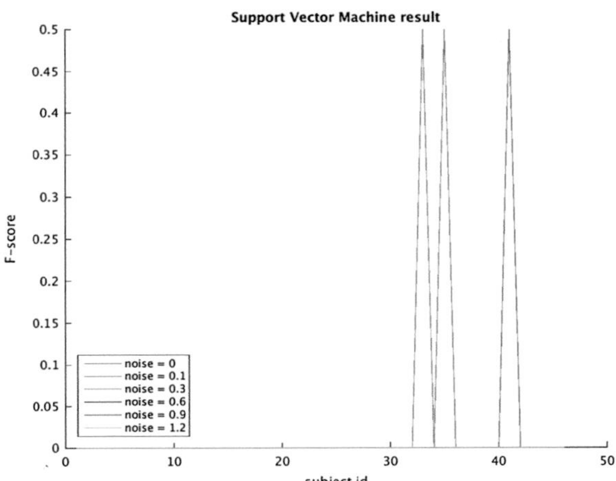

Fig. 3 F-score values of Support Vector Machine per experiment

Also, the highest average was achieved by Tucker decomposition. The Figs. 2, 3 and 4 show F-score accuracy depending on the subject. The Fig. 4 shows also that the Tucker decomposition has several drawbacks. These drawbacks are associated only with the highest value of Gaussian noise.

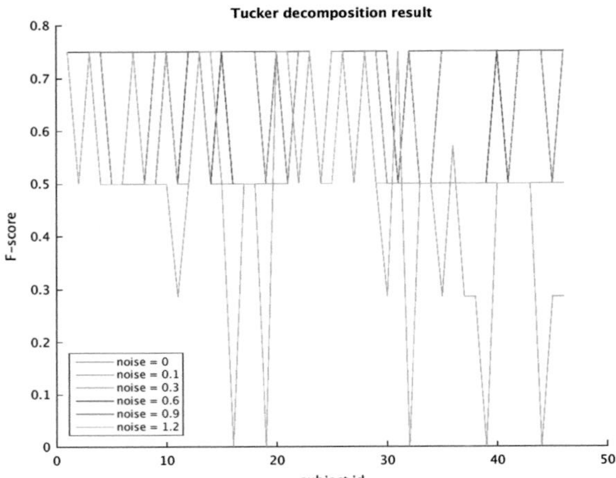

Fig. 4 F-score values of Tucker decomposition per experiment

6 Conclusion

The aim of the paper is compare three methods for classification in a task of cattle identification. Concretely, there we used Support Vector Machines (SVM), Linear Discriminant Analysis (LDA) and Tucker decomposition. Also, Gaussian noise was added to muzzle images in order to simulate realistic illumination conditions. Therefore, it was chosen SVD as a filter to obtain only the most important features from noisy images. The filter was omitted in the case of the classification by the Tucker decomposition. It seems that the classification by the Tucker decomposition outperforms the SVM and LDA in cases with almost all levels of Gaussian noise. In future work, we will focus on Parallel Factor analysis (PARAFAC) of muzzle recognition with used noisy images.

Acknowledgments This work was supported by the IT4Innovations Centre of Excellence project (CZ.1.05/1.1.00/02.0070), funded by the European Regional Development Fund and the national budget of the Czech Republic via the Research and Development for Innovations Operational Programme and by Project SP2015/105 "DPDM - Database of Performance and Dependability Models" of the Student Grand System, VŠB - Technical University of Ostrava and by Project SP2015/146 "Parallel processing of Big data 2" of the Student Grand System, VŠB - Technical University of Ostrava.

References

1. Cichocki, A.: Nonnegative matrix and tensor factorizations applications to exploratory multiway data analysis and blind source separation. Wiley, Chichester (2009)
2. De Lathauwer, L., De Moor, B., Vandewalle, J.: Higher-Order B.S.S.: singular value decomposition. In: Proceedings of EUSIPCO-94, Edinburgh, Scotland, UK, vol. 1, pp. 175–178 (1994)

3. Manners, D.N., Testa, C., Evangelisti, S., Gramegna, L.L., Bianchini, C., Cortelli, P., Tonon, C., Lodi, R.: Binary and multi-class Parkinsonian disorders classification using support vector machines. In: Pattern Recognition and Image Analysis: 7th Iberian Conference, IbPRIA 2015, Santiago de Compostela, Spain, 17–19 June 2015. Proceedings. vol. 9117, p. 379. Springer (2015)
4. Scholkopft, B., Mullert, K.R.: Fisher discriminant analysis with kernels. Neural Netw. Signal Process. **IX**(1), 1 (1999)
5. Sokolnikoff, I.S.: Tensor Analysis: Theory and Applications. Wiley (1951)
6. Steinwart, I., Christmann, A.: Support Vector Machines. Springer Science & Business Media (2008)
7. Tharwat, A., Gaber, T., Hassanien, A.: Cattle identification based on muzzle images using gabor features and SVM classifier. In: Hassanien, A., Tolba, M., Taher Azar, A. (eds.) Advanced Machine Learning Technologies and Applications, Communications in Computer and Information Science, vol. 488, pp. 236–247. Springer International Publishing (2014)
8. Tharwat, A., Gaber, T., Hassanien, A., Hassanien, H., Tolba, M.: Cattle identification using muzzle print images based on texture features approach. In: Krmer, P., Abraham, A., Snášel, V. (eds.) Proceedings of the Fifth International Conference on Innovations in Bio-Inspired Computing and Applications IBICA 2014, Advances in Intelligent Systems and Computing, vol. 303, pp. 217–227. Springer International Publishing (2014)
9. Thomasian, A.: Singular value decomposition, clustering, and indexing for similarity search for large data sets in high-dimensional spaces. In: Big Data: Algorithms, Analytics, and Applications, p. 39 (2015)
10. Ye, Q., Ye, N., Yin, T.: Fast orthogonal linear discriminant analysis with application to image classification. Neurocomputing **158**, 216–224 (2015)

Semantic and Similarity Measure Methods for Plagiarism Detection of Students' Assignments

Hussein Soori, Michal Prilepok, Jan Platos and Vaclav Snasel

Abstract This paper aims at detecting semantic plagiarism in Czech texts. The paper integrates a similarity measure technique previously used for text compression along with a synonyms structured thesaurus and a stemming algorithm to detect rewording and restructuring of texts in Czech language. Out of a 100 GB corpus, we extracted 884 files of B.A., M.A., and Ph.D. students' assignments, semester works and theses, from Computer Science major. The total size of the extracted testing data used was 1.98 GB of plain text for our initial experiment. The method is tested first on short texts. Then, the method is applied on longer texts of students' assignments. Our results on short texts showed more accurate results to detect paraphrased texts of semantic similarity, but lower accuracy was detected in case of identical texts with rearranged paragraphs. Our results experiment conducted on the long texts corpus of students' assignment and theses show a semantic plagiarism rate of 23.9 %. However, after manual scanning of documents, some noise results occur as a result of using the same technical terms and scientific definitions and references in bibliography lists in different documents. These results will be fine-tuned and optimized in the future by building a file—specific stop word list, additional exact match method and removing references and other standard text templates often used in certain parts of students' assignment works and theses.

H. Soori (✉) · M. Prilepok · J. Platos · V. Snasel
Faculty of Electrical Engineering and Computer Science, Department of Computer Science and IT4Innovations, VŠB-Technical University of Ostrava, 17. listopadu 15/2172, 708 33 Ostrava - Poruba, Czech Republic
e-mail: sen.soori@vsb.cz

M. Prilepok
e-mail: michal.prilepok@vsb.cz

J. Platos
e-mail: jan.platos@vsb.cz

V. Snasel
e-mail: vaclav.snasel@vsb.cz

A. Abraham et al. (eds.), *Proceedings of the Second International Afro-European Conference for Industrial Advancement AECIA 2015*, Advances in Intelligent Systems and Computing 427, DOI 10.1007/978-3-319-29504-6_12

Keywords Semantic plagiarism detection · Plagiarism detection methods · Similarity measures · Data compression · Czech thesaurus · Synonymy · Plagiarism detection techniques

1 Introduction

For more than a decade, research on plagiarism detection has been focusing on developing methods to detect plagiarism of exact matches in texts. However, the notion of plagiarism has expanded where paraphrasing may extend the notion of cutting and pasting a text into paraphrasing texts. According to Merriam Webster's Dictionary, plagiarism is 'the act of using another person's words or ideas without giving credit to that person' [1]. Since ideas may be reworded, this makes rewording without mentioning the source text—where the idea was taken from—an act of plagiarism, as well. Semantic plagiarism may take many forms including changing the structure of sentences (restructuring) and replacing words with their synonyms or rewriting (paraphrasing). In this paper we integrate a similarity measure technique previously used for text compression [2] along with a Czech synonyms thesaurus by Pala et al. [3] and a stemming algorithm to detect rewording and restructuring of Czech language texts.

2 Literature Review

Primarily, plagiarism detection methods were developed to detect plagiarism of programming codes in languages such as Java, C++, python, etc. Later, the idea was extended to include text plagiarism in English and other natural languages.

Mainly, research in this area uses different computational and linguistic techniques within wide range of areas such as data mining [4], information retrieval [5], similarity measures [6], cross lingual semantic detection techniques [7], etc.

Each of these techniques used approaches the problem from a different perspective. One of the most widely used methods is context based method that depends on similarity measurement between two documents. This method compares the fingerprints of documents where fingerprints are made by hashing subsets of documents. In this technique, the Winnowing algorithm [8] is used widely where small partial matches of a document are matched with other documents using k-grams where t represents the guarantee threshold and k represents the noise of the threshold. Stanford Copy Analysis Mechanism (SCAM) is another technique where a copy detection server is made of a repository and a chunker. At first, documents are broken up into small parts (sentences, words, ...etc.). After that, the documents are registered in the repository where each chunk is filtered and labeled. Eventually every new non-registered document is made into smaller parts and compared with the original documents already made in the repository.

Some other methods utilize text similarity measurements for data compression to detect text plagiarism. For example, Prilepok et al. [9] who use this measure to detect plagiarism of English texts and Soori et al. who use the technique to detect plagiarism of texts in Arabic and Czech [10, 11]. Other methods use a hybrid of computational and linguistic techniques. For example, Khan et al. [12] who use a method based on phrases taken from first sentence of paragraphs based on the cooccurrences of 5 tokens and eventually evaluated by precision, recall and f-measure.

Other approaches tend to deal with semantic plagiarism detection where the system is able to detect rewarding, restructuring and using synonyms and cross-lingual plagiarism [13–16].

Plagiarism detection methods extend the exact match and semantic detection into a third dimension where the text is examined intrinsically in terms of the consistency of style, vocabulary, usage of short or long sentences/ phrases, and the complexity of structure as in [17–20]. However, his method proves itself impractical if the writer has more than one writing style.

3 Similarity of Text

The similarity measure used in this paper is inspired by Chuda et al. [2]. The same technique was also used in [10, 11] for detection of exact matches in Arabic and Czech. However, it is used in this paper, with a synonyms thesaurus and stemming algorithm to detect semantic plagiarism. The main idea of text similarity lies in measuring the distance between two texts. In this case, the ideal scenario would be when the distance between the texts is metric. Formally, the distance is defined as a function over Cartesian product over a set S with nonnegative real value where the metric d is a distance that satisfies four conditions as follows:

1. $d(x, y) \geq 0$
2. $d(x, y) = 0$, if and only if $x = y$
3. $d(x, y) = d(y, x)$
4. $d(x, z) \leq d(x, y) + d(y, z)$

The conditions mentioned above as: (1), (2), (3) and (4) are called: axiom of non-negativity, axiom of identity, axiom of symmetry and the triangle inequality respectively. This kind of a definition is valid for any metric, e.g., Euclidean Distance. However, applying it on document or data similarity is a herculean task.

3.1 Comparison of Documents

Mainly, there are two properties in a dictionary where the comparison between two documents takes part: a list of word sequences and a word-sequences-count in the list. A dictionary is created for each of the compared files. The next phase involve comparison of these dictionaries where the main property for comparison lies in the number of common sequences in these dictionaries. The number, in this case, is represented by the parameter in the following formula which represents a similarity metric where the comparison is taking place between two documents.

$$SM = \frac{sc}{min(c_1, c_2)} \tag{1}$$

- sc—count of common word sequences in both dictionaries.
- c_1, c_2—count of word sequences in dictionary of the first or the second document.

Here the interval is the SM value which satisfies the condition, If $SM = 1$, then the documents are equal, but if the value is $SM = 0$, then the documents are totally different.

4 Rationale, Training Data and Thesaurus

The rationale behind the method we are proposing in this paper is, for someone to rewrite (paraphrase) a text, one needs to replace the original word either with a different word that carries—more or less—the same meaning, i.e.: synonym, or a variant of a synonym. For example, replacing a verb with a noun or a synonym, which involves restructuring of the sentence. One may also replace a word with a negated antonym by adding an article, e.g., replacing *important* with *far from insignificant, not inconsequential, not unessential, not irrelevant*, etc. However, antonyms detection is not included in this work.

For our experiment, we used a corpus of 100 GB of B.A., M.A., and Ph.D. students' assignments, semester works and theses, from different majors. After removal of metadata and sorting out the corpus into majors, semesters and academic years, the corpus size dropped to 77 GB with 13.671 files of plain text, out of which we used the Computer Science major students' works to apply our experiment on. The total size of the extracted testing data used for our initial experiment was 1.98 GB of plain text.

For our experiment, we integrated the above mentioned similarity measurement [2] with a Czech structured thesaurus called, Slovník českých synonym (Vocabulary of Czech synonyms) by Pala and Vsiansky [21] and a stemming algorithm for Czech language. Apache OpenOffice is currently using this thesaurus and stemming algorithm for their Czech language pack [22]. The thesaurus contains 386,891 tokens, 166,331 words and it is in UTF-8 format.

5 The Experiment

We started our experiment by a giving random short text (175 words), taken from a daily news paper, to three native Czech speakers. The three candidates were asked to rewrite this short text in their own words. At this point, we had four versions:

the original text (v1), and the three paraphrased versions by the three Czech native speakers, (v2, v3, v4). Another version called (v5) was added to the four versions. In (v5), we only changed the arrangement of paragraphs and sentences. In other words, (v1) and (v5) contained exactly the same words and word count. After that, we added another short text we called (v6). (v6) has the same word count as (v1) but the text contains totally different words from the other five versions.

For matching the words in the suspicious documents (v2, v3, v4, v5) with the source document (v1), we used only a first level—main entry—from the synonyms thesaurus. In other words, words in the source document were compared with their synonyms in the thesaurus, but not with the synonyms of their synonyms, i.e., subentries in the thesaurus. Before we searched the synonyms in the dictionary, we connected every word to its basic form of the word using the stemming algorithm. This helped us to find the synonyms in the dictionary where all words are in their basic forms.

After obtaining the results for the short texts, we applied this combined method on our corpus of students' assignments'. We measured the similarity between every document, in our short corpus and the corpus that consists of students' assignments, semester works and theses. Both corpora we processed in the same way. We applied the following steps for each document. In the first step we removed all punctuation marks such as comma, full stop, exclamation marks, etc. After removing the punctuation marks, the input text was converted into chains of words. In the next step we removed stop words from this series and all words were converted to their basic forms.

Every word in the series of words was looked up for the synonyms in the synonym dictionary for its several meanings. The synonyms were put in an entry of our internal dictionary of words. We built this entry of our dictionary every time whenever the current processed word is not found in our dictionary. In case, the current word occurs in more than one entry in our dictionary, we chose the entry with the higher count meanings.

After finding the synonyms to all words from the series of input words, we converted every text to a series of IDs in the dictionary entries. Each dictionary has its own ID. In the phase of measuring the similarity, our approach deals with these IDs. The similarity is measured by applying Chuda's method. Instead of words, this method works with the IDs of our dictionary entries. The threshold between the plagiarized and non-plagiarized documents was set up to 0.3. This value that we set is higher than other current plagiarism detection engines where the value is set to only 0.2. However, we set the value with a higher percentage because, in our experiment, we did not remove bibliography lists. In addition to that, in students' assignments, one may always find a lot of commonly used words such as technical terms, scientific definitions, etc. If the similarity of two documents is higher than the threshold set as 0.3, we considered these two documents as similar and, accordingly, suspicious. If the similarity is below or equal to 0.3 these two documents are not the same.

5.1 Results for Short Texts

In Table 1, we can see that the similarities of rewritten documents are above 0.5 and the similarities of two different documents with same word count are close to 0.1.

5.2 Pitfall of Using a Data Compression Similarity Measure for Plagiarism Detection

As depicted in Table 1, in our experiment on short texts, we noticed that the similarity rate between (v1) and (v5) was always more than 80 %, but never 100 %, in spite of the fact that both documents contain exactly the same word variants and word count. This illogical percentage is seen by us as a drawback to our combined method. This drawback maybe explained as follows. The traditional notion of data compression following Lempel-Ziv 78 [23] method mainly involves creating a dictionary as a major part of the encoding process. The input text is split into separate words. In case a word is not found in the dictionary, then it is added, but if the word is found, then the next word from the input is added. This creates a sequence of words. If this sequence is not found in the dictionary, then it is added, otherwise, it is extended with the next word from the input.

However, as a result of changing the arrangement of paragraphs and sentences in (v5), the sequence of words created yields totally different chains/sets of sequences. This affected the overall percentage of matches in our experiment because the sequences generated from (v1) were different from those generated from (v5). To overcome this drawback, in our future work, we will try to fine-tune the method by modifying the algorithm so to be able to detect rearranged sequences of sentences and paragraphs.

Table 1 Document similarities applied on short texts

	v1	v2	v3	v4	v5	v6
v1	1	0.556	0.632	0.673	0.827	0.086
v2	0.556	1	0.505	0.528	0.565	0.086
v3	0.632	0.505	1	0.621	0.621	0.054
v4	0.673	0.528	0.621	1	0.658	0.097
v5	0.827	0.565	0.621	0.658	1	0.108
v6	0.086	0.086	0.054	0.097	0.108	1

Table 2 Plagiarism detection results on long texts

	Plagiarized	Total count	Plagiarized in %
Results	211	883	23.9

5.3 Results for Long Texts (Students' Assignments)

In Table 2, we present the results for longs texts in a summarized way, despite the big number of files processes and the huge similarity matrix.

In this experiment the settings and preprocessing of text was the same as in the short text experiment. In our dataset, we found 211 similar documents with the rate of 23.9 % out of the total number of documents. This rate appears to be very high. However, when looking closer to the results, one may notice that this number is doubled. This can be explained as follows. In our experiments we compared all documents in the training data to all documents in our corpus. Hence, the tuples of documents were counted twice. For example, if we compared documents A and B and their similarity is above 0.3, then according to the metric definition, mentioned in Sect. 3, the two documents in the opposite order have the same similarity. This is the reason behind counting the documents twice.

6 Conclusion

In this paper we presented a plagiarism detection combined method based on stemming, synonyms and compression method. We ran two experiments on two corpora. The first corpus is small and contains one original text, three rewritten (paraphrased) text by Czech native speakers, one more version where paragraphs and some sentences in paragraphs are reordered, and one totally different text with the same word count as the original text. In the experiment on this short corpus, we can see how well the proposed method is able to distinguish between plagiarized and non-plagiarized texts. In the other long texts corpus, we can see the ability to detect suspicious documents in big corpora. In this corpus, we detected 23.5 % of suspicious tuples of documents. These documents may be considers as plagiarized. However, after scanning the documents manually, it appears that these documents contain a number of similar words—usually—technical terms, definitions, references in bibliography lists, etc. The results obtained were good in spite of the pitfall in case of (v5). However, we can say that this approach is usable to detect similar documents. Additional modifications are needed to fine-tune the obtained results by building a file—specific stop word list. In addition to that, an exact matched, but rearranged method is needed, as well as, removal of bibliography lists and other standard text templates often used in parts of students' assignment works and theses.

Acknowledgments This work was supported by the IT4Innovations Centre of Excellence project (CZ.1.05/1.1.00/02.0070), funded by the European Regional Development Fund and the national budget of the Czech Republic via the Research and Development for Innovations Operational Programme and by Project SP2015/105 "DPDM - Database of Performance and Dependability Models" of the Student Grand System, VŠB - Technical University of Ostrava and by Project SP2015/146 "Parallel processing of Big data 2" of the Student Grand System, VŠB - Technical University of Ostrava.

References

1. http://www.merriam-webster.com/dictionary/plagiarism
2. Chuda, D., Uhlik, M.: The plagiarism detection by compression method. In: Proceedings of the 12th International Conference on Computer Systems and Technologies, pp. 429–434. ACM June 2011
3. Pala, K., Vsiansky, J.: Slovník českých synonym. Lidové noviny (1996)
4. Tan, P.-N., Steinbach, M., Kumar, V.: Introduction to Data Mining. Addison Wesley (2005)
5. Konchady, M.: Building Search Applications: Lucene, LingPipe, and Gate 1st edn. Mustru Publishing (2008)
6. Lee, M.D., Pincombe, B., Welsh, M.: An empirical evaluation of models of text document similarity. In: Bara, B.G., Barsalou, L., Bucciarelli, M. (eds.) 27th Annual Meeting of the Cognitive Science Society, pp. 1254–1259 (2005)
7. Potthast, M., Barrón-Cedeño, A., Stein, B., Rosso, P.: Cross-language plagiarism detection. Lang. Resour. Eval. **45**(1), 45–62 (2011)
8. Schleimer, S., Wilkerson, D.S., Aiken, A.: Winnowing: local algorithms for document fingerprinting. In: Proceedings of the 2003 ACM SIGMOD International Conference on Management of Data, pp. 76–85. ACM June 2003
9. Prilepok, M., Platos, J., Snasel, V.: Similarity based on data compression. In: Advances in Soft Computing and Its Applications, pp. 267–278. Springer, Berlin, Heidelberg (2013)
10. Soori, H., Prilepok, M., Platos, J., Berhan, E., Snasel, V.: Text similarity based on data compression in Arabic. In: AETA 2013: Recent Advances in Electrical Engineering and Related Sciences, pp. 211–220
11. Soori, H., Prilepok, M., Platos, J., Snášel, V.: Utilizing text similarity measurement for data compression to detect plagiarism in Czech. In: Afro-European Conference for Industrial Advancement, pp. 163–172. Springer International Publishing Jan 2015
12. Khan, I.H., Siddiqui, M.A., Jambi, K.M., Imran, M., Bagais, A.A.: Query optimization in Arabic plagiarism detection: an empirical study. Int. J. Intell. Syst. Appl. (IJISA) **7**(1), 73 (2014)
13. Cosma, G., Joy, M.: An approach to source-code plagiarism detection and investigation using latent semantic analysis. IEEE Trans. Comput. **61**(3), 379–394 (2012)
14. Alzahrani, S., Salim, N.: Fuzzy semantic-based string similarity for extrinsic plagiarism detection. Braschler and Harman (2010)
15. Kent, C. K., Salim, N.: Web based cross language semantic plagiarism detection. In: IEEE Ninth International Conference on Dependable, Autonomic and Secure Computing (DASC), 2011, pp. 1096–1102. IEEE Dec 2011
16. Osman, A.H., Salim, N., Binwahlan, M.S., Alteeb, R., Abuobieda, A.: An improved plagiarism detection scheme based on semantic role labeling. Appl. Soft Comput. **12**(5), 1493–1502 (2012)
17. Stamatatos, E.: Intrinsic plagiarism detection using character n-gram profiles. Threshold **2**, 1–500
18. Seaward, L., Matwin, S.: Intrinsic plagiarism detection using complexity analysis. In: Proceedings of the SEPLN, pp. 56–61 Sept 2009

19. Stein, B., Lipka, N., Prettenhofer, P.: Intrinsic plagiarism analysis. Lang. Resour. Eval. **45**(1), 63–82 (2011)
20. Oberreuter, G., L'Huillier, G., Rıos, S.A., Velásquez, J.D.: Approaches for intrinsic and external plagiarism detection. In: Proceedings of the PAN (2011)
21. Pala, K., Vsiansky, J.: Slovník českých synonym. Lidové noviny (Vocabulary of Czech synonyms) (1996)
22. Pala, K., Všianský, J.: http://extensions.openoffice.org/en/project/czech-dictionary-pack-ceske-slovniky-cs-cz (1994–2008)
23. Ziv, J., Lempel, A.: Compression of individual sequences via variable-rate coding. IEEE Trans. Inf. Theory **24**(5), 530–536 (1978)

Neuro-Fuzzy Risk Prediction Model for Computational Grids

Sara Abdelwahab, Varun Kumar Ojha and Ajith Abraham

Abstract Prediction of risk assessment is demanding because it is one of the most important contributory factors towards grid computing. Hence, researchers were motivated for developing and deploying grids on diverse computers, which is responsible for spreading resources across administrative domains so that resource sharing becomes effective. Risk assessment in grid computing can analyses possible risks, that is, the risk of growing computational requirements of an organization. Thus, risk assessment helps in determining these risks. In this, we present an adaptive neuro-fuzzy inference system that can provide an insight of predicting the risk environment. The main goal of this paper is to obtain empirical results with an illustration of high performance and accurate results. We used data mining tools to determine the contributing attributes so that we can obtain the risk prediction accurately.

Keywords Risk assessment · Prediction · Adaptive neuro fuzzy system

---

S. Abdelwahab
Faculty of Computer Science & Information Technology, Sudan University of Science and Technology, Khartoum, Sudan
e-mail: saraabdelghani@gmail.com; saabdelghani@pnu.edu.sa

S. Abdelwahab
Computer Science & Information College, Princess Norah Bint Abddulrahman University, Riyadh, Saudi Arabia

V.K. Ojha (✉) · A. Abraham
IT4Innovations - Center of Excellence, VŠB - Technical University of Ostrava, Ostrava, Czech Republic
e-mail: varun.kumar.ojha@vsb.cz

A. Abraham
e-mail: ajith.abraham@ieee.org

A. Abraham
Machine Intelligence Research Labs (MIR Labs), Washington, USA

A. Abraham et al. (eds.), *Proceedings of the Second International Afro-European Conference for Industrial Advancement AECIA 2015*, Advances in Intelligent Systems and Computing 427, DOI 10.1007/978-3-319-29504-6_13

1 Introduction

Many risk factors are associated with grid computing that threaten security measures [1]. In this work, we applied a hybrid approach to model risk in computational grid. We used an adaptive neuro fuzzy inference system (ANFIS) for modeling risk prediction in computational grid environment. ANFIS is a fuzzy inference system learned using neural network type learning methods. By using a hybrid learning procedure, ANFIS can construct an input–output mapping based on both human-knowledge as fuzzy '*if-then*' rules and approximate membership functions from the stipulated input–output data pairs. ANFIS learning employs a hybrid method consisting of back-propagation for tuning the parameters associated with input membership and least-squares-estimation for tuning the parameters associated with the output parameters [2].

Researchers excessively used ANFIS in many significant research problems, such as industry, financial, weather-prediction, health, etc. [3, 4]. Beghdad et al. [5] used combination of ANFIS and clustering process applied on the CPU Load time series to predict values of CPU load. Their proposed model achieves significant improvement and outperforms the existing CPU load prediction models reported in literatures. In [6], ANFIS was used to predict average air temperature while authors in [2] used ANFIS to predict roughness surface in ball end milling aluminum.

The primary contributions of our research work are to conduct a pre-study based on ANFIS that can provide an insight of predicting the risk environment. We organize this paper as follows: Sect. 2 provides reviews on previous and related work; Sect. 3 presents the proposed methodology; the illustration of experimental results and discussion of the work presents in Sect. 4; Sect. 5 provides discusses followed by conclusions.

2 Related Research

Many hybrid approaches have been applied to predict the risk assessment of grid computing and achieved acceptable results. Risk assessment in grid computing was presented at two layers—resource provider (RP) and Broker by project of AssessGrid (AGP) [7]. In the beginning, the risk modeling of AGP project was conceded at RP considering the probabilistic as well as the possibility approaches. The risk assessment at the RP level in AGP was accomplished by Bayesian model and provided the values of risk at the node level. This approach followed the same context as that of node as the work proposed in [8]. Negoita [9] proposed the possibility modeling, which Broker level was intended to present a Broker. The Broker was introduced to facilitate the end-user to communicate as well as negotiate with RP. Also the level was designed so that it can make a selection of the relevant RP among many more existing [10, 11].

In this paper, we do not focus on Broker level work in the AGP. The risk modeling is accountable in AGP at node level rather than the component level. In addition, risk models of AGP do not reflect any insight of the grid failure data. In [12], authors used various reliability models to assess and evaluate on the basis of assuming Weibull distribution as the best-fit model. However, their work suffered a limitation of requirement of aggregation at both the levels of component and node level. Further, this concept was enhanced to improve reliability within the grids using stochastic model that extracted grid-trace-logs and thus enhanced the job resubmission strategy. Moreover, these works does not address the component-level risk assessment in grids, where the components could be either the disks, CPU, computer software, computer memory, etc. The types of grids also are not classified based on risk assessment, that is, whether the grids are replaceable or repairable. In [13], authors address the problem of risk assessment in computational grid considering security aspect, while the most of the earlier proposed model addressed risk assessment in grid considering resource failure aspect. In this work, we extend the work provided in [13] by using ANFIS method to predict risk in the computational grid environment.

3 Proposed Methodology

We divided our work into five phases. The descriptions of these phases are as follows:

3.1 Phase One—ANFIS Structure

Fuzzy inference system is a process of using fuzzy logic for formulating a nonlinear mapping from input to output, where this system has three parts. (1) A rule base containing fuzzy rules, which are selected. (2) Data base, which defines membership functions applied for the fuzzy rules. (3) A logical system performing the way of inference based on the rules and facts [2].

ANFIS comprises of five layers, where each layer contains several nodes described by the node function. In the first layer, every node is an adaptive node with a node function such as a trapezoidal membership function or a Gaussian membership function. In the second layer, each node multiplies incoming signals and the output is the product of all the incoming signals, where each node output represents the firing strength of a rule. In the third layer, each node calculates the ratio of the rules firing strength to the sum of all rules firing strengths. The normalized firing strengths are the output from this layer. In the fourth layer, each node calculates the contribution of the rule to the overall output. In the fifth (final) layer, the single node calculates the final output as the summation of all input signals [6].

3.2 Phase Two—Dataset Selection

In this phase, the dataset is obtained by simulating grid-computing environment and select the risk factors that threaten the grid-computing environment [1]. The dataset consisting of 20 risk factors and 1951 instances are used to predict the risk output for grid computing environment (Table 1).

3.3 Phase Three—Selection of the Best Input Model Variables

Feature selection is a preprocessing step that reduces dimensionality from a dataset to improve prediction performance. Feature selection can be viewed as a search problem, where searching of a subset from the search space in which each state represents a subset of the possible features. To avoid high computational cost and enhance the prediction accuracy, irrelevant input features are reduced from the dataset before constructing the prediction model. We used correlation feature selection (CFS) subset evaluator feature selection algorithm to search for the best-input model variables [14]. In this research, CFS was used along with the ANFIS to evaluate the merit of feature subsets.

3.4 Phase Four—Investigation of the Effectiveness of Data Splitting

The fourth phase is about partitioning data to investigate the effectiveness of data splitting. We randomly split our dataset into training and testing set as follows:

A: Split 60 % training, 40 % testing
B: Split 70 % training, 30 % testing
C: Split 80 % training, 20 % testing
D: Split 90 % training, 10 % testing

3.5 Phase Five—Development and Configuration of ANFIS

In this phase, the ANFIS model was constructed using grid partitioning and the membership function and consequent parameters were tuned using a hybrid learning process for 100 epochs. We used different membership functions to represent each input variable [23]. In this work, we used:

Table 1 Grid risk factors

Risk factor	Definition	References
Services Level Agreement Violation (SLAV)	SLA represents an agreement between a service user and a provider in the context of a particular service provision	[15]
Cross Domain Attack (CDA)	CDA in which the attacker compromises one site and can then spread his attack easily to the other federated sites	[16]
Job Starvation (JS)	In JS, stranger job scheduled on the host use local (host) resources	[17]
Resource Failure (RF)	It is a failure if: (i) resource stops because of resource crash; (ii) available resources does not meet the minimum levels of QoS	[18]
Resource Attacks (RA)	It is illegal use of host resources by attacker	[19]
Privilege Attack (PA)	User may gain excess privilege to accessing command shell. If grid computing allows access to command shell using a predefined scripts	[19]
Confidentiality Breaches (CB)	Unauthorized, unanticipated, or unintentional disclosure could result in loss of public confidence, or legal action against the organization	[20]
Integrity Violation (IV)	Integrity refers to the trustworthiness of data or resources, and it is usually phrased in terms of preventing improper or unauthorized change	[20]
Distributed Denial of Services (DDoS)	DoS attacks involve sending large number of packets to a destination to prevent legitimate users from accessing information or services	[21]
Data Attack (DA)	In grid security, DA is a scheme in which malicious code is embedded in innocuous-looking data which (when executed by a program) plays out the intended destructive results	[19]
Data Exposure (DE)	DE is other side of widespread connectivity in which (while improving productivity) makes it easier to obtain unauthorized to sensitive data	[19]
Credential Violation (CV)	Credentials are tickets or tokens used to identify, authorize, or authenticate a user. Comprise CV causes theft of user credentials	[20]
Man in the Middle Attack (MMA)	MMA is an attack, where the attacker secretly relays and possibly alters the communication between two parties	[17]
Privacy Violation (PV)	PV is the interference of a person's right to privacy by various means such as showing photos in public	[22]
Sybil Attack (SA)	In Sybil attacks, few entities fakes multiple identities. So it is concern for the systems that rely upon implicit certification	[17]
Hosting Illegal Content (HIC)	This can be done by exploiting the leased nodes	[19]

(continued)

Table 1 (continued)

Risk factor	Definition	References
Stealing Input or Output (SIO)	It is a way to steal the data received by the system or to steal data sent from it	[19]
Shared Use Threats (ShUTh)	Incompatibility between the attributes of grid usersand conventional users causes ShUTh. Hence, no strict separation between participants	[19]
Stealing or altering the Software (SS)	SS caused by unauthorized means entering altered data, false data, unauthorized data, or unauthorized instruction to a system	[19]
Policy Mapping (PM)	Multiple administrative domains with multiple policies causes difficulty to users to map different policies across the grid	[20]

Trapezoidal membership function (Trapmf): Trapezoidal curve is a function of a vector x and depends on four parameters *a*, *b*, *c* and *d*, as given by:

$$f(x,a,b,c,d) = \max\left(\min\left(\frac{\mathrm{x}-\mathrm{a}}{\mathrm{b}-\mathrm{a}}, 1, \frac{\mathrm{d}-\mathrm{x}}{\mathrm{d}-\mathrm{c}}\right), \mathrm{o}\right)$$

The parameter *a* and parameter *d* locate the "feet" of the trapezoid and the parameters *b* and *c* locate the "shoulder."

Triangular membership function (Trimf): The triangular curve is a function of a vector *x* and depends on three scalar parameters *a*, *b*, and *c*, given by:

$$f(x,a,b,c) = \max\left(\min\left(\frac{\mathrm{x}-\mathrm{a}}{\mathrm{b}-\mathrm{a}}, \frac{\mathrm{c}-\mathrm{x}}{\mathrm{c}-\mathrm{b}}\right), \mathrm{o}\right)$$

The parameter *a* and parameter *c* locates the "feet" of the triangle and the parameter *b* locates the peak.

Generalized bell function (Gbell): Depends on three scalar parameters *a, b* and *c*, given by:

$$f(x,a,b,c) = \frac{1}{1 + \left|\frac{x-c}{a}\right|^{2b}}$$

Where the parameter *b* is usually, positive and parameter *c* locates the center of the curve.

Gaussian membership function (Gaussmf): The symmetric Gaussian function depends on two parameters σ and *c* as given by:

$$f(x,\sigma,c) = e^{-\frac{(x-c)}{2\sigma^2}}$$

4 Experimental Dataset and Results

Using different feature selection techniques, eight different sub datasets were obtained [13]. These datasets were generated from our original dataset that has 20 risk factors attributes. Table 2 illustrates the number of attributes in each dataset and summarizes the search method.

In this paper, we extended the work reported in [13], which includes three attributes. The empirical result shows that, the prediction algorithm required the least number of attributes (3 attributes only out of 20 attributes) to achieve high performance. In this work, to verify the efficiency of the proposed method we used three features (CV, HIC, and SIO). To achieve the experimental result, different ANFIS parameters were tested as training parameters to maximize the prediction accuracy. Table 3 illustrates the ANFIS performance using different numbers of membership function (MF) shapes with different data splits. The lowest average testing error was obtained using Trimf membership functions with Dataset B.

Table 2 Different attribute selection method

Dataset	Evaluator	Search method	Selected attributes	Total
Original dataset			SLAV, CDA, JS, RF, RA, PA, CB, IV, DDoS, DA, DE, CV, MMA, PV, SA, HIC, SIO, ShUTh, SS, PM	20
1	Reliff attribute evaluation	Ranker	DDoS, PM, DE, SA, ShUTh, HIC, CV, RA, SIO, CDA, RF, SLAV, JS, MMA, SS, PA, PV, IV, CB, DA	20
2	Reliff attribute evaluation	Ranker	DDoS, PM, DE, SA, ShUTh, HIC, CV, RA, SIO, CDA, RF, SLAV, JS, MMA, SS, PA, PV, IV	18
3	Reliff attribute evaluation	Ranker	DDoS, PM, DE, SA, ShUTh, HIC, CV, RA, SIO, CDA, RF, SLAV, JS, MMA, SS	15
4	Reliff attribute evaluation	Ranker	DDoS, PM, DE, SA, ShUTh, HIC, CV, RA, SIO, CDA, RF, SLAV	12
5	Reliff attribute evaluation	Ranker	DDoS, PM, DE, SA, ShUTh, HIC, CV, RA, SIO	9
6	CFS subset evaluation	Evolutionary search	SLAV, JS, RA, CV, HIC, SIO	6
7	CFS subset evaluation	Best first search backward	CV, HIC, SIO	3
8	CFS subset evaluation	Exhaustive search	RA, CV, HIC	3

Table 3 ANFIS performance for different membership functions (MF)

	Data split	2 MF		3 MF		4 MF	
		RMSE					
		Train	Test	Train	Test	Train	Test
Trimf	A	0.0425	0.0433	0.0379	0.0381	**0.0139**	0.0146
	B	0.0424	0.0418	0.0325	0.0320	0.0143	**0.0137**
	C	0.0428	0.0424	0.0376	0.0374	0.0143	0.0141
	D	0.0428	0.0431	0.0372	0.0355	0.0144	0.0143
Gbellmf	A	0.0353	0.0357	0.0260	0.0271	0.0177	0.0195
	B	0.0355	0.0355	0.0255	0.0262	0.0189	0.0197
	C	0.0354	0.0353	0.0270	0.0263	0.0188	0.0198
	D	0.0352	0.0378	0.0222	0.0280	0.0184	0.0252
Guaussmf	A	0.0443	0.0444	0.0227	0.0250	0.0202	0.0217
	B	0.0409	0.0402	0.0229	0.0235	0.0216	0.0221
	C	0.0403	0.0409	0.0265	0.0282	0.0201	0.0214
	D	0.0401	0.0460	0.0235	0.0311	0.0188	0.0233
Trapmf	A	0.0398	0.0413	0.0381	0.0395	0.0290	0.0279
	B	0.0402	0.0406	0.0386	0.0389	0.0455	0.0453
	C	0.0396	0.0428	0.0379	0.0411	0.0451	0.0472
	D	0.0389	0.0507	0.0371	0.0497	0.0276	0.0255

A sample *if-then* rules obtained from the 64 rule based ANFIS using 4 input membership function for each input variables are illustrated below. One of the advantages of using ANFIS is its interpretability through simple *if-then* rules.

1. If (CV is VL) and (HIC is VL) and (SIO is VL) then (RO is VL) (1)
2. If (CV is VL) and (HIC is VL) and (SIO is Low) then (RO is VL) (1)
3. If (CV is VL) and (HIC is VL) and (SIO is M) then (RO is VL) (1)
4. If (CV is VL) and (HIC is VL) and (SIO is High) then (RO is VL) (1)
17. If (CV is Low) and (HIC is VL) and (SIO is VL) then (RO is Low) (1)
18. If (CV is Low) and (HIC is VL) and (SIO is Low) then (RO is Low) (1)
37. If (CV is Medium) and (HIC is Low) and (SIO is VL) then (RO is Medium) (1)
38. If (CV is Medium) and (HIC is Low) and (SIO is Low) then (RO is Medium) (1)
63. If (CV is High) and (HIC is High) and (SIO is Medium) then (RO is High) (1)
64. If (CV is High) and (HIC is High) and (SIO is High) then (RO is High) (1)

5 Conclusions

The problem of predicting risk assessment is a complex issue because there are many factors that affect the grid computing directly and indirectly. In this paper, we used 3 risk factors and the ANFIS model was selected based on the minimum value of root mean square error, which is constructed using four triangular-shaped membership function for each input variable and linear membership function for output. Hence we have developed a risk prediction model for computational grid environment using ANFIS.

Acknowledgments This work was supported by the IPROCOM Marie Curie Initial Training Network, funded through the People Programme (Marie Curie Actions) of the European Union's Seventh Framework Programme FP7/2007–2013/, under REA grant agreement number 316555.

References

1. Abdelwahab, S., Abraham, A.: A review of the risk factors in computational grid. J. Inf. Assur. Secur. **8** (2013)
2. Hossain, S.J., Ahmad, N.: Adaptive neuro-fuzzy inference system (ANFIS) based surface roughness prediction model for ball end milling operation. J. Mech. Eng. Res. **4**, 112–129 (2012)
3. Abraham, A.: Adaptation of fuzzy inference system using neural learning. In: Nedjah, N. et al. (eds.) Fuzzy Systems Engineering: Theory and Practice. Studies in Fuzziness and Soft Computing, pp. 53–83. Springer, Germany (2005)
4. Abraham, A.: Neuro fuzzy systems: state-of-the-art modeling techniques. In: Connectionist Models of Neurons, Learning Processes, and Artificial Intelligence, pp. 269–276. Springer (2001)
5. Bey, K.B., Benhammadi, F., Mokhtari, A., Guessoum, Z.: CPU load prediction model for distributed computing. In: Eighth International Symposium on Parallel and Distributed Computing, 2009. Ispdc'09, pp. 39–45 (2009)
6. Karthika, B., Deka, P.C.: Prediction of air temperature by hybridized model (wavelet-ANFIS) using wavelet decomposed data. Aquat. Procedia **4**, 1155–1161 (2015)
7. Djemame, K., Gourlay, I., Padgett, J., Birkenheuer, G., Hovestadt, M., Kao, O.V., Kerstin: Introducing risk management into the grid. In: Second IEEE International Conference on e-Science and Grid Computing, 2006. e-Science'06, pp. 28–28 (2006)
8. Sangrasi, A., Djemame, K.: Component level risk assessment in grids: a probablistic risk model and experimentation. In: Proceedings of the 5th IEEE International Conference on Digital Ecosystems and Technologies Conference (DEST), 2011, pp. 68–75 (2011)
9. Negoita, C., Zadeh, L., Zimmermann, H.: Fuzzy sets as a basis for a theory of possibility. Fuzzy Sets Syst. **1**, 3–28 (1978)
10. Carlsson, C., Fullér, R.: Risk assessment of SLAs in grid computing with predictive probabilistic and possibilistic models. In: Preferences and Decisions, pp. 11–29. Springer (2010)
11. Gourlay, I., Djemame, K., Padgett, J.: Reliability and risk in grid resource brokering. In: 2nd IEEE International Conference on Digital Ecosystems and Technologies, 2008. DEST 2008, pp. 437–443 (2008)

12. Raju, N., Gottumukkala, Y.L., Leangsuksun, C.B., Nassar, R., Scott, S.: Reliability analysis in HPC clusters. In: Proceedings of the High Availability and Performance Computing Workshop (2006)
13. Abdelwahab, S., Abraham, A.: Data mining approach for modeling risk assessment in computational grid. In: Computational Intelligence in Data Mining, vol. 3, pp. 673–684. Springer (2015)
14. Hall, M.A., Smith, L.A.: Feature subset selection: a correlation based filter approach. Presented at the Proceedings of the Fourth International Conference on Neural Information Processing and Intelligent Information Systems (ICONIP'97). New Zealand, Berlin (1997)
15. Rana, O.F., Warnier, M., Quillinan, T.B., Brazier, F., Cojocarasu, D.: Managing violations in service level agreements. In: Grid Middleware and Services, pp. 349–358. Springer (2008)
16. Syed, R.H., Syrame, M., Bourgeois, J.: Protecting grids from cross-domain attacks using security alert sharing mechanisms. Future Gener. Comput. Syst. **29**, 536–547 (2013)
17. Chakrabarti, A.: Taxonomy of grid security issues. In: Grid Computing Security, pp. 33–47. Springer (2007)
18. Lee, H.M., Chung, K.S., Jin, S.H., Lee, D.-W., Lee, W.G., Jung, S.Y.Y., Chang, H.: A fault tolerance service for QoS in grid computing. In: Computational Science—ICCS 2003, pp. 286–296. Springer (2003)
19. Smith, M., Friese, T., Engel, M., Freisleben, B.: Countering security threats in service-oriented on-demand grid computing using sandboxing and trusted computing techniques. J. Parallel Distrib. Comput. **66**, 1189–1204 (2006)
20. Chakrabarti, A., Damodaran, A., Sengupta, S.: Grid computing security: a taxonomy. IEEE Secur. Priv. 44–51 (2008)
21. Kar, S., Sahoo, B.: An anamaly detection system for DDOS attack in grid computing. Int. J. Comput. Appl. Eng. Technol. Sci. (IJ-CA-ETS), **1**, 553–557 (2009)
22. Naqvi, S., Riguidel, M.: Threat model for grid security services. In: Advances in Grid Computing-EGC 2005, pp. 1048–1055. Springer (2005)
23. Barua, A., Mudunuri, L.S., Kosheleva, O.: Why trapezoidal and triangular membership functions work so well: towards a theoretical explanation. J. Uncertain Syst. **8** (2013)

Neuro-Fuzzy Model for Assessing Risk in Cloud Computing Environment

Nada Ahmed and Ajith Abraham

Abstract Within a few years, cloud computing emerged as one of the hottest technologies in the IT field. It provides computational resources as general utilities that can be leased and released by users in an on-demand fashion. Cloud computing is rapidly growing interest in many companies around the globe, but adopting cloud computing comes with greater risks, which need to be assessed. In this research, an adaptive neuro-fuzzy inference system (ANFIS) has been applied to assess risk factors in cloud computing. Different membership functions were used for training the data. The model combined the modeling function of fuzzy inference with the learning ability of neural networks. Empirical results illustrate that ANFIS is very effective in modeling cloud-computing risks.

Keywords Adaptive neuro-fuzzy inference system · Feature selection · Cloud computing risk · Risk assessment

N. Ahmed (✉)
Faculty of Computer Science and Information Technology,
Sudan University of Science, Technology, Khartoum, Sudan
e-mail: naessa@pnu.edu.sa

N. Ahmed
College of Computer and Information Sciences,
Princess Nourah Bint Abdulrahaman University, Riyadh, South Arabia

A. Abraham
Department of Computer Science and IT4Innovations,
Faculty of Electrical Engineering and Computer Science, VŠB-Technical University of Ostrava, Ostrava, Czech Republic
e-mail: ajith.abraham@ieee.org

A. Abraham
Machine Intelligence Research Labs (MIR Labs), Scientific Network for Innovation and Research Excellence, Auburn, WA, USA

A. Abraham et al. (eds.), *Proceedings of the Second International Afro-European Conference for Industrial Advancement AECIA 2015*, Advances in Intelligent Systems and Computing 427, DOI 10.1007/978-3-319-29504-6_14

1 Introduction

The emergence of cloud computing represents a fundamental change in the way information technology service is invented, deployed, developed, maintained, scaled, updated, and paid [1]. Cloud computing provides on-demand service to its consumers, the consumers use these services as they need and pay only for what is used [2–4]. In recent years there is obvious migration to cloud computing with end users, quietly handling a growing number of personal data, such as photographs, music files, bookmarks, and much more, on remote servers accessible via a network [5]. The use of cloud computing services can cause greater risks to consumers. Before consumers start using cloud computing services they must confirm whether the product satisfies their needs and understand the risks involved in using this service [6].

Risk has been defined as "the chance that someone or something that is evaluated will be adversely affected by the hazard". Risk assessment is an essential tool for the safety policy of any organization. The essence of risk formulation is to estimate the level of the risk on the basis of likelihood of a risk scenario mapped against the estimated negative impact. The assessment of the level of risk is a complex issue covered in uncertainty and ambiguity. Risk level often assessed using linguistic terms, for this case fuzzy set analysis is an efficient tool, as it is able to work with problems where no sharp boundaries are possible. The imprecision may be dealt with using fuzzy rules in a fuzzy inference system. Adaptive Neuro-fuzzy inference system (ANFIS) is a fuzzy inference system implemented in the framework of adaptive networks [7]. The aim of this study is to create a reliable and an effective ANFIS model for risk assessment of cloud computing.

This paper is organized as follows: Sect. 2 gives a summary of literature related to risk assessment for adoption to cloud service followed by background and a brief overview on ANFIS in Sect. 3. The methodology used to develop ANFIS model is presented in Sect. 4. In Sect. 5 the result of the ANFIS model is presented and discussed followed by Conclusions in the last Section.

2 Related Works

ENISA identified 35 risks and classify then into four categories: policy and organizational, technical, legal, and other scenarios not specific to cloud computing. A group of experts determine the likelihood of each risk and their business impact [8]. Risk Management standard published by International Organization for Standardization (ISO) published a generic standard called ISO 31000 [9], for risk management. Later, ISO and International Electrotechnical Commission (IEC) complemented ISO 31000 by the joint publication ISO/IEC 31010 [10], about risk assessment techniques, also this standard is generic. COBIT is another generic framework for IT is introduced by Information Technology Governance

Institute and the Information Systems Audit and Control Association (ISACA) in 1996, which provides a common language to communicate the results, goals, and objectives of business. Enterprise risk management recommendations provided in the last versions of COBIT in 2013 [11]. Cloud adoption risk assessment model (CARAM) complement the various recommendation from European Network and Information Security Agency (ENISA), and Cloud Security Alliance (CSA) for a complete risk assessment framework. CARAM help in assessing various risks to business, privacy and security that cloud customers face when moving to use cloud computing [12].

3 Adaptive Neuro-Fuzzy Inference System (ANFIS)

System modeling based on mathematical and statistical methods is not well suited to represent data and knowledge, closer to human-like thinking. By contrast a Fuzzy Inference System (FIS) can be viewed as a real-time expert system used to model and utilize human experience, by employing fuzzy if-then rules [7, 13, 14].

*3.1 Fuzzy if-*then *Rules and FIS*

Fuzzy If-then rules are an expression of the form *If* ***A*** *Then* ***B***, where A and B are labeled of fuzzy sets [15] characterized by appropriate membership functions. Fuzzy if-then rules are employed to capture the imprecise modes of reasoning which represent a base role in human decision making [14]. Takagi and Sugeno [16, 17] proposed another form of fuzzy if-then rules, has fuzzy sets involved only in the premise part. The core part of Fuzzy Inference System was represented by fuzzy If-*Then* rules. Fuzzy Inference System is primarily applied to the cases that either if it is difficult to precisely model the system or it is ambiguous to describe the studying issues [18, 19]. The Fuzzy Inference System is the foundation of Adaptive Neuro-Fuzzy System (ANFIS).

Zadeh [7] proposed the fuzzy logic to describe complicated systems. The drawback of fuzzy logic is that there is no systematic procedure the design of a fuzzy controller. By contrast, a neural network has the ability to learn from the environment, self-organize its structure, and adapt to it in an interactive manner [20].

3.2 Adaptive Neuro-Fuzzy Inference System (ANFIS)

Adaptive Neuro-Fuzzy Inference System was first introduced by Jang [14, 21] ANFIS is a multilayer feed forward network, which uses neural network learning

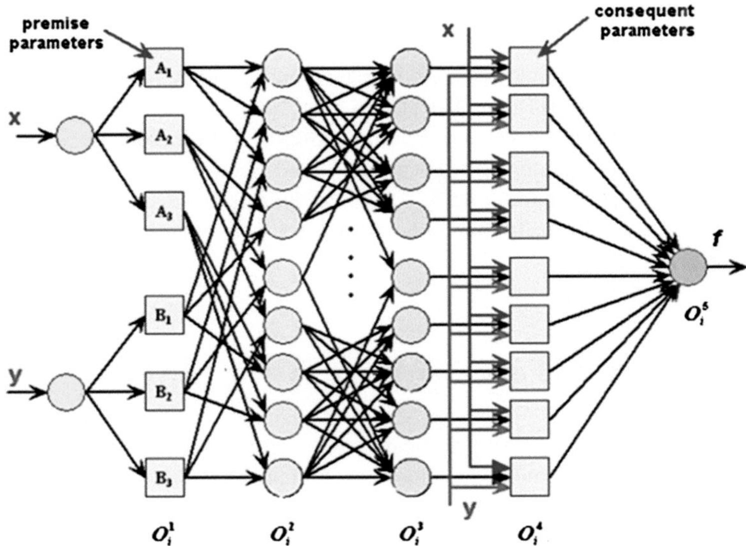

Fig. 1 The architecture of the ANFIS

algorithms and fuzzy reasoning to map input characteristics into input membership functions (MFs), input MFs to a set of if-then rules, rules to a set of output, characteristics, output characteristics to output MFs, and output MFs to a single-valued output [14, 18, 20, 22–24].

Figure 1 shows the architecture of ANFIS consisting of five layers used for training Sugeno-type FIS through learning and adaptation, it shows the structure for a system with (m) inputs (x_1, x_2, …, x_m) each with (n) membership functions (MFs), a fuzzy rule base of R rules, and one output y. The number of nodes N in layer 1 is the product of the number of inputs m and MFs n for each input, i.e., N = m*n. layer 2–4, the number of nodes is equal to the number of rules R in the fuzzy rule base. ANFIS implements a Takagi Sugeno Kang (TSK) fuzzy inference system. For a first order TSK model, a common rule set with two fuzzy *if-then* rules is represented as follows:

$$Rule\,1\text{: } If\; x\; is\; A1\; and\; y\; is\; B1,\; then\, f_1 = p_1 x + q_1 y + r_1 \quad (1)$$

$$Rule\,2\text{: } If\; x\; is\; A2\; and\; y\; is\; B2,\; then\, f_2 = p_2 x + q_2 y + r_2 \quad (2)$$

where x and y are linguistic variables and A1, A2, B1, B2 are corresponding fuzzy sets and p1, q1, r1 and p2, q2, r2 are linear parameters. ANFIS makes use of a mixture of back propagation to learn the premise parameters and least mean square estimation to determine the consequent parameters. ANFIS structure (Fig. 1) comprising of 5 layers are briefly described below:

- Layer 1: all nodes in this layer are adaptive nodes. Each node generates membership grades to which they belong to each of the appropriate fuzzy sets using membership functions.
- Layer 2: rule nodes, the nodes in this layer are fixed nodes. To obtain one output that represents the result of the antecedent for that rule the AND operator is applied.
- Layer 3: average nodes, the nodes in this layer are also fixed nodes. The output of this layer is called normalized firing strengths.
- Layer 4: consequent nodes, all nodes in this layer are adaptive node, whose output is simply the product of the normalized firing and a first-order polynomial.
- Layer 5: this layer includes fixed and single node, it computes the overall output by summing all the signals.

ANFIS applies a hybrid learning algorithm, the gradient descent method, and the least-squares method, to update the parameters. The least-squares method is used in the forward pass (offline learning) to identify consequent linear parameters, when attempting to minimize the error between the actual state and the desired state of the adaptive network, while the antecedent parameters (membership functions) are assumed to be fixed. Then, the gradient descent method is employed in backward pass to tune premise a non-linear parameters, by propagate the error rate from the output end towards the input end, while consequent parameters remain fixed [20, 25].

4 Experimental Methodology

4.1 Datasets

We conducted an online survey to determine and finalize the various risk factors. In the survey, we requested the participants to categorize risk factors to three levels (Important, Neutral, and Not important) according to their effect on cloud computing. 35 international experts responded to the survey from different countries and all of them agreed that the previously defined factors are important, which means that they have great effect over cloud computing. Next we mapped each risk factor to a qualitative scale, and finally we formulated expert rules and use some statistical methods to generate the data based on the rules. The Dataset contains 18 input attributes and comprise of 1940 instances. The 18 attributes were labeled as Data Transfer (DT), Insufficient due Diligence (IDD), Regulatory Compliance (RC), Business Continuity and Service Availability (BC&SA), Third Part Management (TPM), Interoperability and Portability (I&P), Data Loss (DL), Insecure Application Programming (IAP), Data location and Investigative Support (DL&IS), Recovery (R), Resource Exhaustion (RE), Service Level Agreement (SLA), Authentication and access control (A&AC), Shared Environment (ShE), Data

Table 1 Risk factors

Risk factor	Range value	Risk factor	Range value
DI	0–3	*R*	1–3
IDD	1–3	*RE*	0–2
RC	0–1	*SLA*	0–3
BC&SA	1–3	*A&AC*	0–3
TPM	0–2	*ShE*	1–3
I&P	0–1	*DB*	0–2
DL	0–3	*DS*	0–1
IAP	0–1	*VV*	1–3
DL&IS	0–3	*DI*	0–2

Breaches (DB), Data Segregation (DS), Virtualization Vulnerabilities (VV), Data Integrity (DI). Risk Factors and numerical ranges are illustrated in Table 1 with their range values. We use the following feature selection methods to extract subset instances from the original dataset.

Best-first: This algorithm search in the space of set as a graph called "feature selection lattice", then applies one of the standard graph searching algorithms [26].

Random search: First it randomly selects subset, and then continues in two different ways. One of them is to follow sequential search, the second is to continue randomly and generate the next subset randomly [27].

Feature selection reduced the number of features to first dataset with 3 attributes *(IDD, DL, and DL&IS),* and second dataset with 4 attributes (RC, DL, DL&IS, and VV). We also used random percentage split: 60–40 % (A); 70–30 % (B); 80–20 % (C); 90–10 % (D) to train and test the different algorithms [28]. ANFIS algorithm was developed to estimate the risk. There are 3 and 4 input variables for first and the second datasets respectively.

4.2 Development of ANFIS Model

ANFIS was developed in MATLAB. To generate the Sugeno fuzzy inference systems (FIS), 5 types of membership functions were used: Triangular, trapezoidal, Generalized bell, Gaussian, and Guassian2. For each type of the membership function two and three input membership functions were used for each variable. The following 5 types of MF were considered in our experiments:

- Tri: Triangular membership function.
- Trap: Trapezoidal MF.
- Gbell: Generalized bell curve MF given by $\mu_{Ai} = 1/(1 + (((x - c_i)/a_i)^2)^{b_i})$
- Gauss: Gaussian membership function given by $\mu_{Ai}(x) = \exp(-((x - c_i)^2/a_i))$
- Gauss2: A two-sided version of Gaussian membership function

4.3 Performance Evaluation of ANFIS Model

The performance of the model is assessed using some statistical parameters including Correlation Coefficient (CC), and Root Mean Square Error (RMSE).

5 Results and Discussions

The ANFIS models are compared based on their performance in test datasets. In our previous work the best RMSE obtained was 0.0006 [28]. It is observed that the ANFIS algorithm is accurate and can effectively predict the values of risk in cloud computing. Tables 2, 3, 4, 5 represent the results of first and second data with 2 and

Table 2 RMSE with 2 MF for the first dataset

MF shape	No of MF's	RMSE			
		60–40 %	70–30 %	80–20 %	90–10 %
Tri	2	**2.882e-06**	3.285e-06	4.022e-06	5.870e-06
Trap	2	**1.112e-05**	1.320e-05	1.937e-05	3.636e-05
Gbell	2	**1.874e-05**	2.294e-05	2.787e-05	3.669e-05
Gauss	2	**1.296e-05**	1.628e-05	2.440e-05	3.466e-05
Gauss2	2	**1.748e-05**	2.115e-05	2.739e-05	3.868e-05

Table 3 The RMSE with 3 MF for the first dataset

MF shape	No. of MF's	RMSE			
		60–40 %	70–30 %	80–20 %	90–10 %
Tri	3	**4.118e-06**	4.886e-06	5.789e-06	7.961e-06
Trap	3	**1.985e-05**	2.577e-05	3.320e-05	3.320e-05
Gbell	3	**1.871e-05**	2.115e-05	2.882e-05	3.991e-05
Gauss	3	**1.976e-05**	2.108e-05	2.840e-05	3.666e-05
Gauss2	3	**1.977e-05**	2.153e-05	2.826e-05	3.145e-05

Table 4 The RMSE with 2 MF for second dataset (4 attributes)

MF shape	No. of MF's	RMSE			
		60–40 %	70–30 %	80–20 %	90–10 %
Tri	**2**	**9.150e-06**	1.061e-05	1.407e-05	1.962e-05
Trap	2	**1.854e-05**	2.262e-05	3.092e-05	4.888e-05
Gbell	2	**1.801e-05**	2.032e-05	2.397e-05	3.209e-05
Gauss	2	**1.507e-05**	1.691e-05	2.328e-05	3.201e-05
Gauss2	2	**1.292e-05**	1.444e-05	1.636e-05	2.505e-05

Table 5 The RMSE with 3 MF for second dataset (4 attributes)

MF shape	No. of MF's	RMSE			
		60–40 %	70–30 %	80–20 %	90–10 %
Tri	3	**9.957e-06**	1.134e-05	1.537e-05	2.163e-05
Trap	3	**1.766e-05**	2.491e-05	3.454e-05	4.863e-05
Gbell	3	**1.819e-05**	1.94e-05	2.306e-05	3.392e-05
Gauss	3	**2.268e-05**	2.727e-05	3.101e-05	4.021e-05
Gauss2	3	**2.085e-05**	2.381e-05	2.899e-05	5.143e-05

3 MFs, where all the values of RMSE are smaller and correlation coefficients are close to 1. From the tables, one can conclude that the triangular membership function gives us the lowest error with the 60–40 % split of the first dataset (3 attributes) and 2 MFs.

6 Conclusions

In this research, we proposed the use of adaptive network-based fuzzy inference system (ANFIS) to construct a model to predict risk in a cloud-computing environment. ANFIS provides a method based on fuzzy if-then rule based system to learn information about the data set in a better way. The capability of ANFIS model is investigated through the use of two datasets (based on the number of features). In general, the ANFIS model provides accurate risk prediction, where the correlation coefficient is close to unity (0.999) and the RMSE is very small.

Acknowledgment This work was supported by the IT4Innovations Centre of Excellence Project (CZ.1.05/1.1.00/02.0070), funded by the European Regional Development Fund and the national budget of the Czech Republic via the Research and Development for Innovations Operational Programme and by Project SP2015/146 "Parallel Processing of Big Data 2" of the Student Grant System, VŠB-Technical University of Ostrava.

References

1. Avram, M.-G.: Advantages and challenges of adopting cloud computing from an enterprise perspective. Procedia Technol. **12**, 529–534 (2014)
2. Paquette, S.J.: Paul T Wilson, Susan C, Identifying the security risks associated with governmental use of cloud computing. Gov. Inf. Quart. **27**(3), 245–253 (2010)
3. Carroll, M., Van Der Merw, A., Kotze, P.: Secure cloud computing: benefits, risks and controls. In: 2011 IEEE Information Security South Africa (ISSA) (2011)
4. Sun, D., Guiran, C., Sun, L., Wang, X.: Surveying and analyzing security, privacy and trust issues in cloud computing environments. Procedia Eng. **15**, 2852–2856 (2011)
5. Zissis, D., Dimitrios, L.: Addressing cloud computing security issues. Future Gener. Comput. Syst. **28**(3), 583–592 (2012)

6. Chandran, S., Mridula, A: Cloud Computing: analysing the risks involved in cloud computing environments. In: Proceedings of Natural Sciences and Engineering, pp. 2–4 (2010)
7. Fragiadakis, N.G., Tsoukalas, V.D., Papazoglou, V.J.: An adaptive neuro-fuzzy inference system (anfis) model for assessing occupational risk in the shipbuilding industry. Saf. Sci. **63**, 226–235 (2014)
8. Catteddu, D.: Cloud Computing: benefits, risks and recommendations for information security. In: Web Application Security, p. 17. Springer, Heidelberg (2010)
9. Purdy, G.: ISO 31000: 2009—setting a new standard for risk management. Risk Anal. **30**(6), 881–886 (2010)
10. Commission, I.E., IEC/ISO 31010: 2009. Risk management-risk assessment techniques, 2009 http://www.iso.org/iso/catalogue_detail?csnumber=51073. Accessed on 16 Aug 2015
11. Association, I.S.A.C., COBIT 5: A Business Framework for the Governance and Management of Enterprise IT. 2012: ISACA, http://www.isaca.org/COBIT/Pages/default.aspx. Accessed on 16 Aug 2015
12. Cayirci, E., Alexandr, G., Santana, A., Roudier, Y.: A Cloud adoption risk assessment model. In: Proceedings of the 2014 IEEE/ACM 7th International Conference on Utility and Cloud Computing. IEEE Computer Society (2014)
13. Jang, J.-S.: ANFIS: adaptive-network-based fuzzy inference system. IEEE Trans. Syst. Man Cybern. **23**(3), 665–685 (1993)
14. Abraham, A.: Rule-Based expert systems. In: Handbook of Measuring System Design (2005)
15. Zadeh, L.A.: Fuzzy sets. Inf. Control **8**(3), 338–353 (1965)
16. Takagi, T., Sugeno, M.: Fuzzy identification of systems and its applications to modeling and control. IEEE Trans. Syst. Man Cybern. **1**, 116–132 (1985)
17. Takagi, T., Sugeno, M.: Derivation of fuzzy control rules from human operator's control actions. In: Proceedings of the IFAC Symposium on Fuzzy Information, Knowledge Representation and Decision Analysis (1983)
18. Khoshnevisan, B., Rafiee, S., Omid, M., Mousazadeh, H.: Development of an intelligent system based on ANFIS for predicting wheat grain yield on the basis of energy inputs. Inf. Process. Agric. **1**(1), 14–22 (2014)
19. Yang, Z., Yongqian, Liu, Chengrong, Li: Interpolation of missing wind data based on ANFIS. Renewable Energy **36**(3), 993–998 (2011)
20. Chang, F.-J., Ya-Ting, C.: Adaptive neuro-fuzzy inference system for prediction of water level in reservoir. Adv. Water Resour. **29**(1), 1–10 (2006)
21. Yüksel, S.B., Yarar, A.: Modelling uniform temperature effects of symmetric parabolic haunched beams using adaptive neuro fuzzy inference systems (ANFIS). In: Metaheuristics and Engineering, p. 83
22. Wang, Y.-M., Elhag, T.M.S.: An adaptive neuro-fuzzy inference system for bridge risk assessment. Expert Syst. Appl. **34**(4), 3099–3106 (2008)
23. Rezaei, E., Karami, A., Yousefi, T., Sajjad, M.: Modeling the free convection heat transfer in a partitioned cavity using ANFIS. Int. Commun. Heat Mass Transfer **39**(3), 470–475 (2012)
24. Hayati, M., Hayati, M.R.: Prediction of grain size of nanocrystalline nickel coatings using adaptive neuro-fuzzy inference system. Solid State Sci. **13**(1), 163–167 (2011)
25. Lima, C.A.M., Coelho, ALV., Von Zuben, F.J.: Fuzzy systems design via ensembles of ANFIS. In: Proceedings of the 2002 IEEE International Conference on Fuzzy Systems, FUZZ-IEEE'02. IEEE (2002)
26. Jain, A., Zongker, D.: Feature selection: evaluation, application, and small sample performance. IEEE Trans. Pattern Anal. Mach. Intell. **19**(2), 153–158 (1997)
27. Liu, H., Yu, L.: Toward integrating feature selection algorithms for classification and clustering. IEEE Trans. Knowl. Data Eng. **17**(4), 491–502 (2005)
28. Ahmed, N.A., Abraham, A.: Modeling cloud computing risk assessment using machine learning. In: Afro-European Conference for Industrial Advancement. Springer (2015)

The EC Sequences on Points of an Elliptic Curve Realization Using Neural Networks

Nikolay Ivanovich Chervyakov, Mikhail Grigorevich Babenko, Maxim Anatolievich Deryabin, Nikolay Nikolaevich Kucherov and Nataliya Nikolaevna Kuchukova

Abstract This paper shows that pseudorandom number generator based on EC-sequence doesn't satisfy the condition of Knuth k-distribution. A modified pseudorandom number generator on elliptic curve points built in neural network basis is proposed. The proposed generator allows to improve statistical properties of the sequence based on elliptic curve points so that it satisfies the condition of k-distribution i.e. the sequence is pseudorandom. Application of Neural network over a finite ring to arithmetic operations over finite field allows to increase the speed of pseudorandom number generator on elliptic curve points EC-256 by 1,73 times due to parallel structure.

Keywords EC sequences · Elliptic curve · Residue Number System · Neural network of a finite ring

1 Introduction

Everyday we use numbers that are chosen from some set in a random way, e.g. we throw a coin to make a decision. The use of random numbers in computer modeling helps us to make these models more similar to the real ones. They are also used in cryptography.

N.I. Chervyakov · M.G. Babenko (✉) · M.A. Deryabin · N.N. Kucherov · N.N. Kuchukova
North-Caucasus Federal University NCFU, Stavropol, Russia
e-mail: mgbabenko@ncfu.ru

N.I. Chervyakov
e-mail: k-fmf-primath@stavsu.ru

M.A. Deryabin
e-mail: maxim.deryabin@gmail.com

N.N. Kucherov
e-mail: nik.bekesh@mail.ru

N.N. Kuchukova
e-mail: knn.storage@yandex.ru

A. Abraham et al. (eds.), *Proceedings of the Second International Afro-European Conference for Industrial Advancement AECIA 2015*, Advances in Intelligent Systems and Computing 427, DOI 10.1007/978-3-319-29504-6_15

With the invention of computer scientists began to research effective methods of generating random numbers. The numbers that are generated by any algorithm only seem to be random. They follow some rule, so it is more correct to call them pseudorandom.

The main difficulty of generating pseudorandom numbers by computer is that computer can be in limited number of conditions only, although this number is big (the number of computer conditions is limited(finite)) therefore, any pseudorandom number generator (PRNG) definitely is periodical, but everything that is periodic is not random. So the aim of PRNG is to construct the sequence which is indistinguishable from the random one, has the period of reasonable length and corresponds to various criteria of random.

2 The EC Sequences on Points of an Elliptic Curve

In [1] it is shown that base-b sequence is k-distributed if $Pr(X_n, X_{n+1}, \ldots, X_{n+k-1} = x_1 x_2 \ldots x_n) = \frac{1}{b^k}$ for all base-b numbers $x_1 x_2 \ldots x_n$.

The concept of random sequence [1] is introduced on the basis of concept of k-distribution:

Definition 1 A base-b sequence of length N is random if it is k-distributed for all positive integers $k \leq \log_b N$

During the research we realized the PRNG on points of an elliptic curve. We have made such choice because it is convenient to calculate elements of such sequences and their statistical properties are good [2].

Choose a prime number $p > 3$. Then an elliptic curve in the form of Weierstrass [2] defined over a finite field F_p is the set of pairs (x, y), $x, y \in F_p$ satisfying NIST FIPS PUB 186-4.

$$y^2 = x^3 + ax + b \,(\text{mod} p) \tag{1}$$

where $a, b \in F_p$ and $4a^3 + 27b^2$ and 0 are inconrguent modulo p. Pairs (x, y), satisfying expression (1), are called points of an elliptic curve (Table 1).

Now define addition of two arbitrary points $P_1\left(x_1, y_1\right)$ and $P_2\left(x_2, y_2\right)$. The sum of $P_1\left(x_1, y_1\right)$ and $P_2\left(x_2, y_2\right)$ is a point $P_3\left(x_3, y_3\right)$ which coordinates are defined by congruence

$$\begin{cases} x_3 = \lambda^2 - x_1 - x_2 \,(\text{mod} p) \\ y_3 = \lambda\left(x_1 - x_3\right) - y_1 \,(\text{mod} p), \end{cases} \tag{2}$$

where the parameter λ is defined by congruences

$$\begin{cases} \lambda = \frac{y_2 - y_1}{x_2 - x_1} \,(\text{mod} p) \;\; \text{if } x_1 \neq x_2 \\ \lambda = \frac{3x_1^2 - a}{2y_1} \,(\text{mod} p) \;\; \text{if } x_1 = x_2 \text{ and } y_1 = y_2 \neq 0 \end{cases} \tag{3}$$

For division we will use the theorem [3].

Table 1 Parameters of studied elliptic curves

Value	
The curve I	
a	7
b	43308876546767276905765904595650931995942111794451039583252968842033849580414
p	57896044618658097711785492504343953926634992332820282019728792003956564821041
The curve II	
a	57797173260300591862887039728922683231990117698099726395553302688959615241517
b	7553699923012108603620040145431887057305706793702443677892317187823870284887
p	74071195152409579744075542254950126325253398248741892913923221064376398898449
The curve III	
a	61030749271824800554991559575506208894416059169666037759369625697772198293873
b	332904976132962158995036629330705489068186669667800999865470615866701901212386
p	67372339006199060004899298784615830852533077490386635296533159255966007376509

Theorem 1 *Fermat's little theorem. For any prime number p and an integer a if* $gcd\,(a,p) = 1$, *then the equality* $a^{p-1} = 1\,(\text{mod}p)$ *is true.*

On the basis of these curves we generated sequences of $100, 200, \ldots, 1000$ points. The statistical analysis of these sequences was made on the basis of Definition 1 and the concept of k-distribution.

In order for a sequence to be random k has to satisfy an inequality $k \neq \log_b N$ [2]. However testing 2^{14}–2^{17} subsequences is a resource-intensive task. Therefore for the analysis we chose the subsequences that consist only of 0 or only of 1. Let's denote them $A0$ and $A1$ respectively. As criterion of distribution we took an expression

$$\frac{1}{2^{k+1}} < P\left(x_n x_{n+1} \ldots x_{n+k+1} = a_1 a_2 \ldots a_k\right) \approx \frac{1}{2^k} < \frac{1}{2^{k-1}} \tag{4}$$

The values for which this condition is satisfied, are provided in Table 2.

Judging from these results, no curve satisfies the condition of random sequence by Definition 1. However it is visible from Table 2 that curves I and III generate pseudo random sequences with better statistical properties than curve II.

We apply a method from work [4] to improve traditional generators by means of neural network. Approach of creation of steady random number generators which can

Table 2 The values of k, satisfying (4) for $A0$ and $A1$

Quantity of points	Curve number	Sequence length	The values of k for $A0$	The values of k for $A1$
100	I	22,323	1–3	1–5
	II	22,464	1–3	1–6
	III	22,323	1–3	1–7
200	I	44,712	1–3	1–6, 15
	II	44,853	1–3	1–6
	III	44,755	1–3	1–8, 14–15
300	I	67,010	1–3	1–6, 14–16
	II	67,243	1–3	1–6
	III	67,131	1–3	1–7, 15–16
400	I	89,353	1–3	1–6, 15–16
	II	89,566	1–3	1–6
	III	89,705	1–3	1–7
500	I	111,670	1–3	1–6, 15–16
	II	111,930	1–3	1–6
	III	112,086	1–3	1–7
600	I	133,995	1–3	1–7, 15
	II	134,381	1–3	1–6
	III	134,590	1–3	1–7, 17
700	I	156,456	1–3	1–7
	II	156,744	1–3	1–6
	III	156,982	1–3	1–7, 17
800	I	178,727	1–3	1–7
	II	179,073	1–3	1–6
	III	179,474	1–3	1–7, 17
900	I	201,035	1–3	1–7
	II	201,388	1–3	1–6
	III	201,892	1–3	1–7, 17
1000	I	223,343	1–3	1–6
	II	223,823	1–3	1–6
	III	224,315	1–3	1–7, 17

be used in safety mechanisms for electronic trading systems is described below. It is based on feedforward neural networks (FNN). It is known that ANN are capable of approximation. This fact makes them a great instrument for many scientific disciplines. For example, FNN multilayer perceptron (MLP) is able to theoretically approximate the arbitrary non-linear functions as well as their differentials. It is supposed that such approximation with ANN is less resource-intensive than other competitive techniques (orthogonal polynomials, splines or a Fourier progression) because it has less parameters and arithmetics operations. Besides, as ANN are parallel and distributed processing devices, they can be carried out in parallel hardware. Therefore, they can be used in real-time computing.

ANN ability of generalization is the most important property which makes ANN useful for applications. It is possible due to ability to output true solution when data not submitted earlier is input [3].

Above mentioned abilities are gained by MLP in the course of learning on known input vectors. The condition that there was no retraining [5] has to be satisfied. On the one hand, in case of retraining network knows training selections very well. On the other hand, it is incapable to output the solution when unknown input data is given. It happens because the network draws the corresponding surface of training selections in the more complex way. It occurs because it is a surface of control and has higher degree, than it is necessary to display the actual selection. Then we do not receive an analytical formula to describe earlier mentioned complex surface. As a result the answer of MLP becomes unpredictable. Even for unknown data with small distances from training selections (similar input dates) output of a network will strongly differ from the obtained educational data. Though in a pattern recognition, management, the forecast such situation is very undesirable, it can be used in case of random number generation. In this work we show such retraining of MLP that the network can be used as the mechanism for generation steady pseudorandom sequences of bits as follows.

1. Train topology MLP, for example 4-6-6-1, a problem of frequency rate 4 so that to retrain, for example, on 2500 eras when 500 eras are enough. The purpose is that retraining has to happen in the course of MLP learning. Certainly, it is possible to use other similar binary estimates, as well as more complex topology of MLP.
2. Check trained MLP, using test input vectors. Their values are in an interval [0,1]. Components of vectors are obtained by a traditional (pseudo)random number generator such as ones mentioned in the experimental section of work.
3. Create a complex function of intercommunications of MLP. Calculate its value when the test input vector created in advance is input to the network.

Thus, we obtain the (pseudo) random numbers sequence which quality is estimated by means of statistical tests. The complex function of inner representations of MLP used in this work, has the following analytical formula

$$y = \mathrm{mod}f\left(1000 * \sum \frac{\left|\tanh\left(o_k\right) - \tanh\left(o_{k+1}\right)\right|}{\left|\tanh\left(o_{k+1}\right) - \tanh\left(o_{k+2}\right)\right|}\right),$$

where y—received random number and o_k, o_{k+1}, o_{k+2} values of activation of computing elements of layers of MLP. mod$f$—Unix-function which takes a fractional part of a real number.

Application of a neural network allows generating sequences on the basis of EC sequences with the improved statistical properties.

3 Neural Network of a Finite Ring

The hardware implementation of a neural network allows to carry out mass parallel execution of the elementary operations, and, the more the level of parallelism of computation is reached, the better is it [6, 7]. FPGA is ideal element base for realization of parallel structures such as neural network. Thus the data interchange between neurons is carried out in the same FPGA with a high speed. Besides, use of a neural network of a finite ring (NNFR) reduces number of the cascaded diagrams that allows to increase information processing rate. In case of restriction on arithmetical operations NNFR arithmetics provides their more effective implementation in comparison with binary arithmetics. With the advent of FPGA of the big area, creation of a single system of data handling on one crystal became possible, that opens new opportunities of NNFR application in processing systems and are hardwired large volumes of the graphic information.

The NN represents high-parallel dynamic system with topology of directed graph which can receive output information by means of response of its status to input influences. In NN processor elements and directed graphs are called as nodes [6, 7]. The structure of data processing algorithm, provided in residue number system, also as well as NN structure possess natural parallelism that allows to use NN as the formal device of the description of algorithm.

From this point of view algorithms of modular computation correspond to algorithms of computation by means of Basic processor Elements (artificial neurons). Artificial neural networks and the main modular structures represent the connective devices received by serial connection among themselves of Basic Elements. Neural and modular educations layerwise will be defined if the algorithm of connection of Basic Elements is set.

We will consider the general approach of application of NN to computation in finite rings and to formation of the NN model of a finite ring (NNFR). Thus neurons are arithmetical elements which have characteristics of the operator on the module, but not the ordinary non-linear functions of activation applied when training NN. The analysis of arithmetic of a finite ring showed that the computational model based on the iterative mechanism of reducing on the module is the main operation in case of modular data handling.

For effective implementation of the secret sharing scheme on matrix projections it is necessary to expedite addition and multiplication operations on the module. Computation in a finite ring (field) can be defined as follows.

NN architecture. The general interpretation of NN architecture is a mass parallel interdependent network of simple elements and its hierarchical arrangement. The NN structure somewhat models biological nervous system.

Power of NN is its ability of use of the initial knowledge base for the solution of the existing problem. All neurons work is competitive, and direct computation is influenced by the knowledge ciphered in communications of a network [7–10].

Interaction of neurons is considered in three-level hierarchy of the network consisting of layers [7, 11, 12]:

1. Display of parameter (this layer contains the residual connected to the weighed value of each computing discharge).
2. Display of bit computation (defines the function of a finite ring applied to each computing discharge).
3. Display of operation of a finite ring (defines the main operations which are used for implementation of arithmetics of a finite ring).

Well consider the structure of NNFR from work [13, 14]. Based on a computational model of a finite ring the main operator in which is the operator of extraction of separate discharges of binary representation of the converted number, multi-layer subnets can be constructed. Synaptic weights in the subnet are $w_i = \left|2^i\right|_p$, $i = 0, 1, \ldots, n - 1$.

1. Assembly (it is used for collection of the inputs belonging to one binary place of input sources). The result of binary to residue conversion operations, multiplication, addition calculated by means of NNFR is function of the amount of the weighed input discharges.
2. The computing. The result of computation is defined by the positive logic. The end result of NNFR will have the steady form.

4 Conclusion

This paper shows that pseudorandom number generator based on EC-sequence doesn't satisfy the condition of Knuth k-distribution. The modified pseudorandom number generator on elliptic curve points built in neural network basis is proposed. The proposed generator allows to improve statistical properties of the sequence based on elliptic curve points so that it satisfies the condition of k-distribution i.e. the sequence is pseudorandom. Application of Neural network over a finite ring to arithmetic operations over finite field allows to increase the speed of pseudorandom number generator on elliptic curve points EC-256 by 1,73 times due to parallel structure.

Acknowledgments Current work was performed as a part of the State Assignment of Ministry of Education and Science (Russia) No. 2563.

References

1. Knuth, D.: The Art of Computer Programming, vol. 2. Seminumerical Algorithms. Oscow, Publishing House Williams, p. 832 (2001)
2. Tarakanov, V.E.: Some remarks on arithmetic properties of the recursion sequences on elliptic curves over a terminating field. Math. Notes **82**(6), 836–842 (2007)
3. Bolotov, A.A., Gashkov, S.B., Frolov, A.B., Chasovskikh, A.A.: Algorithmic Bases of Elliptic Cryptography. RGSU Publishing House, Moscow, p. 499 (2004)

4. Chervyakov, N.I., Galushkin, A.I., Evdokimov, I.A., Lavrinenko, I.N., Lavrinenko, A.V.: Application of Simulated Neuronic Networks and System of Residual Classes in Cryptography. FIZMATLIT, Moscow, p. 280 (2012)
5. Patterson, D.W.: Artificial Neural Networks. Theory and Applications. Prentice Hall (1996)
6. Galushkin, A.I: The Theory of Neural Networks. INGNR, Moscow, p. 416 (2000)
7. Chervyakov, N.I., Sakhnyuk, P.A., Shaposhnikov, V.A., Makokha, A.N.: Neurocomputers in Residual Classes. Radiotechnique, Moscow, p. 272 (2003)
8. Zhang, C.N., Yun, D.Y.: Parallel designs for Chinese remainder conversion. In: IEEE 16-th International Conference on Parallel Processing—ICPP, pp. 557–559 (1987)
9. Zang, D., Jullien, G.A., Miller, W.C.: A neural-like approach to finite ring computation. IEEE Trans. Circuits Syst. **37**(8), 1048–1052 (1990)
10. Chervyakov, N.I.: The conveyor neural network of a finite ring. Patent RU 2317584 from (2008)
11. Zhang, D.: Parallel VLSI Neural System Designs. Springer, Berlin, Germany, p. 257 (1998)
12. Zhang, D., Jullien, G.A., Miller, W.C.: VLSI implementations of neural-like networks for finite ring computations. In: Proceedings of the 32nd Midwest Symposium on Circuits and Systems, vol. 1, pp. 485–488 (1989)
13. Chervyakov, N.I., Galkina, V.A., Strekalov, U.A., Lavrinenko, S.V.: Neural network of a finite ring. Patent RU 2279132 from (2003)
14. Chervyakov, N.I., Babenko, M.G., Kucherov, N.N., Garianina, A.I.: The Effective Neural Network Implementation of the Secret Sharing Scheme with the Use of Matrix Projections on FPGA. Advances in Swarm and Computational Intelligence, LNCS, vol. 9142, pp. 3–10 (2015)

Ensemble of Heterogeneous Flexible Neural Tree for the Approximation and Feature-Selection of Poly (Lactic-co-glycolic Acid) Micro- and Nanoparticle

Varun Kumar Ojha, Ajith Abraham and Vaclav Snasel

Abstract In this work, we used an adaptive feature-selection and function approximation model, called, flexible neural tree (FNT) for predicting Poly (lactic-co-glycolic acid) (PLGA) micro- and nanoparticle's dissolution-rates that bears significant role in the pharmaceutical, medical, and drug manufacturing industries. Several factor influences PLGA nanoparticles dissolution-rate prediction. FNT model enable us to deal with feature-selection and prediction simultaneously. However, a single FNT model may or may not offer a generalized solution. Hence, to build a generalized model, we used an ensemble of FNTs. In this work, we have provided a comprehensive study for examining the most significant (influencing) features that influences dissolution rate prediction.

Keywords Poly (lactic-co-glycolic acid) (PLGA) micro- and nanoparticle · Flexible neural tree · Function approximation · Feature selection

1 Introduction

Application domains, such as chemical, pharmaceutical, medical, biology, etc., yields a large volume of data with a large number of features. Researchers often use computational intelligence tools for data analysis, primarily, for discovering the relationship between independent and dependent variables. Traditional data-driven computational intelligence method are insufficient in dealing with such volume of

V.K. Ojha (✉) · A. Abraham · V. Snasel
IT4Innovations, VŠB Technical University of Ostrava, Ostrava, Czech Republic
e-mail: varun.kumar.ojha@vsb.cz; vkojha@ieee.org

A. Abraham
e-mail: ajith.abraham@vsb.cz

V. Snasel
e-mail: vaclav.snasel@vsb.cz

A. Abraham et al. (eds.), *Proceedings of the Second International Afro-European Conference for Industrial Advancement AECIA 2015*, Advances in Intelligent Systems and Computing 427, DOI 10.1007/978-3-319-29504-6_16

data with high accuracy. Mostly, a single function-approximation model may not be able to offer the most general solution to a problem. In this work, we solve a pharmaceutical industry problem that sought for a computational model that can predict the dissolution-rate profile of poly (lactic-co-glycolic acid) (PLGA) micro-and nanoparticles. PLGA dissolution profile prediction is a complex problem in terms of the number of feature that governs its dissolution-rate profile. A thorough examination of the academic literature provides us 300 potential factors that may influence the dissolution-rate of the PLGA protein particles [1, 2].

In this work, our approach was to find the most significant features that will govern the dissolution profile and to construct an approximation model with high accuracy without the loss of generality. The significance of the features selection is not limited only to finding most significant features, but it also reduces the computational complexity of the problem and significantly contributes in improving model's computational speed, predictability, and generalization ability. There are several techniques available for the pre-selection of the most probable features [3]. However, in this work, we presents a computational tool, namely, heterogeneous flexible neural tree-(HFNT) that offers automatic features selection and function approximation. Interestingly, HFNT also make sure diversity in the feature space that makes it an adaptive feature selector. In general, combined knowledge of many experts (predictors) offers better solution in terms of reliability and generality than that of a single expert. So, an ensemble of several HFNT would offer most general solution.

We provides a comprehensive discussion on the drug release problem and dataset collection mechanisms that tells the significance of macro-and nanoparticle dissolution-rate profile prediction (Sect. 2.1). In the following selection, we will offer a discussion on HFNT and ensemble methods. Finally, we discuss the experimental results followed by conclusion.

2 Methodology

2.1 The PLGA Dissolution-Rate Prediction Problem

PLGA micro- and nanoparticles plays a significant role in the medical application and toxicity evaluation of PLGA-based multi-particulate dosages [4]. PLGA micro-particles are important diluents used to produce drugs in their correct dosage form. Apart from playing the role as a filler, PLGA as an excipient, and alongside pharmaceutical APIs, plays other crucial roles in various ways. It helps in the dissolution of drugs, thus increasing the absorbability and solubility of drugs [5, 6]. It helps in pharmaceutical manufacturing processes by improving API powders' flow ability and non-stickiness.

The dataset collected from various academic literature contains 300 input features categorized into four groups, including protein descriptor, plasticizer, formulation characteristics, and emulsifier. Table 1 provides a detailed description of the dataset.

Table 1 The PLGA dataset description

#	Group name	# Features	Importance
1	Protein descriptors	85	Describes the type of molecules and proteins used
2	Formulation characteristics	17	Describe the molecular properties such as molecular weight, particle size, etc.
3	Plasticizer	98	Describe the properties such as fluidity of the material used
4	Emulsifier	99	Describe the properties of stabilizing/increase the pharmaceutical product life
5	Time in days	1	Time taken to dissolve
6	% of molecules dissolved	1	PLGA micro-nanoparticle dissolution-rate

2.2 Background Study

In the scope of this study, our focus was on PLGA dissolution properties and drug release rate. Szlęk et al. [2] and Fredenberg et al. [7] described two mechanisms: diffusion and degradation/erosion that is mainly governs the drug release from the PLGA matrix. Several factors influencing the diffusion and degradation rates of PLGAs described by Kang et al. [8, 9] Blanco and Alonso [10] and Mainardes and Evangelista [11] includes pore diameters, matrix active pharmaceutical ingredient (API) interactions, API-API interactions, and the formulation composition. Szlęk et al. [2] developed a predictive model to describe the underlying relationship between those influencing factors on the drug's release profile, they focused on feature selection, artificial neural network, and genetic programming approaches to come up with a suitable prediction model. In the past, several mathematical models, including the Monte Carlo and cellular automata microscopic models were proposed by Zygourakis and Markenscoff [12] and Gopferich [13]. A partial differential equation model was proposed by Siepmann et al. [14] to address the influence of underlying PLGA properties on the drug's release rate/protein dissolution. Ojha et al. [15] used a pool of several trained predictors and made an ensemble to get a model with high predictability. However, they had to use separate techniques for the feature-selection beforehand to make approximation models, in contrast to that, in this work, we proposed a tool for function-approximation and feature-selection simultaneously.

2.3 Flexible Neural Tree Approach

For a dataset with n many independent variables X and a dependent variable Y, an approximation model tries finds relationship between them. Moreover, it tries to find unknown parameter θ such that root mean square error (RMSE) between models' output $\hat{Y}$ and actual output Y be zero. Therefore, we may write RMSE as:

$$RMSE = \sqrt{\frac{1}{N} \sum_{i=1}^{N} (y_i - \hat{y}_i)^2}, \quad (1)$$

where N is number of examples.

A wide range application accepts artificial neural network (ANN) as most convenient tool for the approximation [16]. Thus, makes it a universal approximator. ANN performance heavily relies on its structure, parameters, and activation-functions [16] optimization. Researchers have investigated various ways in the past to optimize the individual components of ANN using evolutionary procedure [17, 18]. Chen et al. [19] proposed flexible neural tree (FNT) that addressed ANN optimization in all its components including structure and parameters. And, does an automatic feature selection. FNT was conceptualized around a multi-layered feed-forward neural network to build a tree based model, where network structure and parameters were optimized by using meta-heuristic optimization algorithms (the nature inspired stochastic algorithms for function optimization).

An FNT can be defined as a set of function-nodes and terminals, where the function-node indicates a computational node and terminals indicated a set of all input features. The function instruction set F and terminal instruction set T for generating an FNT model are described as follows:

$$S = F \cup T = \{+_2, +_3, +_4, \ldots, +_N\} \cup \{x_1, x_2, \ldots, x_n\}, \quad (2)$$

where $+_i$ $(i = 2, 3, \ldots, n)$ indicates that a function-node can take i arguments, whereas, the leaf node (terminal node) receives no arguments. Figure 1 illustrates a function-node/computational-node of an FNT.

In Fig. 1, the computational node $+_i$ receives i inputs through i connection weights (random real values) and two adjustable parameters/arguments a_i and b_i of the squashing (transfer) function, that limits the total output of the function-node within a certain range. A transfer-function used at the function-node is:

$$f(a_i, b_i, net_n) = e^{-\left(\frac{(net_n - a_i)}{b_i}\right)}, \quad (3)$$

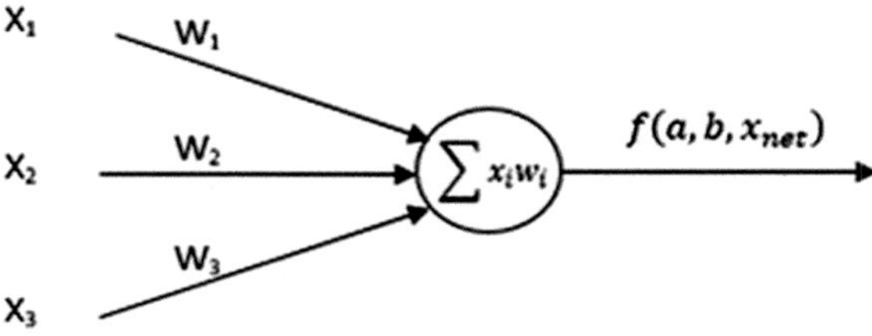

Fig. 1 A computational node of a flexible neural tree

where net_n is the net input to the ith function-node also known as excitation of the node computed as:

$$net_n = \sum_{j=1}^{n} w_j x_j, \tag{4}$$

where $j = 1, 2, 3 \ldots$ is the input to the ith node. Therefore, the output of the ith node is given as:

$$out_n = f(a_i, b_i, net_n) = e^{-\left(\frac{(net_n - a_i)}{b_i}\right)}, \tag{5}$$

Figure 2 illustrates an example of a typical FNT. The root node of the FNT given in Fig. 2 indicates the output of the entire tree-based model. The leaf nodes of the tree are indicating the selected input feature and the edges of the tree indicate the underlying parameters (or the weights) of the model. We can choose various kinds of activation-function at the function-nodes of an FNT and genetic evaluation give various structure at various instances. So, FNTs with heterogeneity in its function-nodes and structure is now on, called, heterogeneous flexible neural trees (HFNTs). Fitness of a HFNT is computed according to (1). The root node of a HFNT returns the output of the entire model.

Meta-heuristic framework for HFNT optimization: We can classify HFNT optimization in two different parts:

(1) The optimization of the HFNT structure, in other words, the first approach is to find a near optimum tree using genetic programming that works on a genetic population and uses operators like crossover and mutation to evolve new generation [20].
(2) The optimization of the parameters of the tree, i.e., the optimization of the edges and the arguments of the activation function used at the function-nodes. Swarm based meta-heuristic were used for optimizing parameters, where swarm-based meta-heuristics are the algorithms that are inspired from the foraging behavior of the swarm such as group of fish, birds, bees, etc. [21]

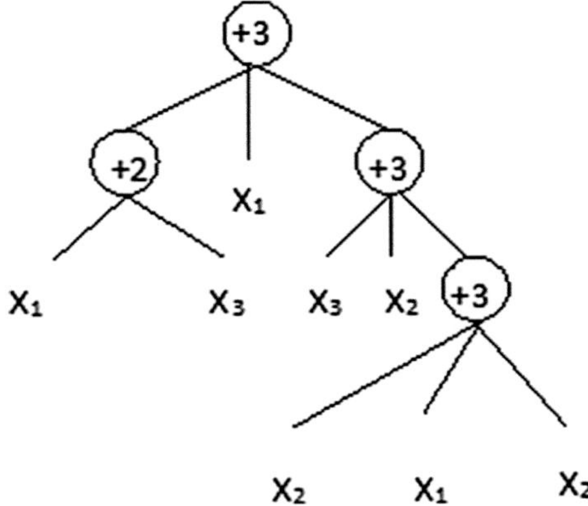

Fig. 2 A typical FNT with instruction set $F = \{+_2, +_3,\}$ and $T = \{x_1, x_2, x_3\}$

2.4 Ensemble of HFNT

A collective decision with consensus of many candidates is better than decision of an Individual. Hence, ensemble of many models (predictors) may offer the most general solution to a problem [22]. There are two components in ensemble system [23, 24] construction:

(1) Construction of as diverse and as accurate models as possible. To construct diverse and accurate models following techniques may be used: (a) training models with different sets of data, like bagging algorithm [25]; (b) training models with different set of input features, Random Sub-space algorithm; and (c) training models with different set of parameters.
(2) Combining the models using a combination rules. Once many models constructed with high diversity and accuracy, then we need to combine them for a collective decision. We used weighted mean combination method, where the weights for the models were computed by using meta-heuristic algorithm.

3 Experimental Set-up and Results

We conducted our experiment using a platform independent software tool that we developed for realizing FNT. We used the develop software tool for processing of PLGA dataset that has 300 features and 754 examples. We used training parameters setting as mentioned in Table 2.

We conducted several experiments with the parameter settings mentioned in Table 2. Since, the computation model mentioned is stochastic in nature, each instance of experiment offer distinct results in terms of accuracy and future selection. The Accuracy, in other words, fitness of an approximation model was measured in terms RMSE given in (1). However, the correlation that tells relationship between two variables (in this case, the actual output, and the models' output) reveals the quality of the constructed model.

We have selected four highly accurate and divers models for making ensemble. The obtained results is provided in Table 3.

We select model (second in Table 3) for comparing with the results available in literature. Our second model selects fifteen features and offer a cross validation RMSE 13.248. A comprehensive list of models available in literature for PLGA prediction is given in Table 4.

Main focus of the present experiment as promised at beginning was to find-out the significant features and the construction of accurate model. In previous part of results section, we have presented accurate model. However, we were resistant to demonstrate significant features. To obtain significant features, we repeated our

Table 2 The parameters setting of the HFNT software tool

#	Parameter name	Parameter utility	Values
1	Tree height	Maximum number of level of FNT	5
2	Tree arity	Maximum number of siblings a node	10
3	Tree node type	Indicates the type of activation at nodes	Gaussian
4	GP population	Number of candidates in genetic population	30
5	Mutation probability	The probability that a candidate will be mutated in a genetic programing	0.4
6	Crossover probability	The probability that candidates will take part in crossover operation	0.5
7	Elitism	The probability that a candidate will propagate to next generation as it is	0.1
8	Tournament size	It indicate the size of the pool used for the selection of the candidates	15
9	MH algorithm population	The initial size of the swarm (population)	50
10	MH algorithm node range	It defines the search space of the transfer-function arguments	[0,1]
11	MH algorithm edge range	It defines the search space for the edges (weights) of tree	[−1.0,1.0]
13	Maximum structure iteration	Maximum number of generation of genetic programing	100000
14	Maximum parameter iteration	Maximum number of evaluation of parameter optimization	10000

Table 3 RMSE and r of the constructed models and ensemble based on 10 cross validation

Exp.	Training		Test		Ensemble weights
	RMSE	r	RMSE	r	
1	12.885	0.908	12.741	0.909	0.355686
2	12.907	0.908	13.248	0.903	0.319278
3	13.855	0.897	13.776	0.897	0.259922
4	14.599	0.881	14.374	0.884	0.085882
Ensemble	–	–	**11.541**	**0.928**	–

experiment 20 times. We listed all the features selected by each instant of experiment according to their probability of selection. In Table 5, we have provided a list of 15 such features whose probability of selection were obtained higher than 0.2. Similarly, Table 6 was realized. In Table 6, we have presented a picture of overall group probability. The probability of feature selected from particular group was computed by averaging the probabilities of all the features those were fall under that particular group.

Table 4 Best models constructed for PLGA prediction problem (cross validation results)

#	RMSE	Features	Model	Literature
1	**13.248**	**15**	HFNT	Current work
2	13.34	15	REP tree [26]	Ojha et al. [15]
3	14.3	17	MLP [16]	Szlęk et al. [2]
4	14.88	15	GP regression [27]	Ojha et al. [15]
5	15.2	15	MLP [16]	Ojha et al. [15]
6	15.4	11	MLP [16]	Szlęk et al. [2]

Table 5 Most significant feature (according to probability of selection)

#	Feature	Index	Abbreviation	Probability
1	Time days	299	TD	0.94
2	Prod method	100	PM	0.83
3	PVA conc. inner phase	88	PVA	0.78
4	Ring atom count	110	RAC	0.61
5	Heteroaliphatic ring count	23	HIRC	0.50
6	Aliphatic bond count	104	ABC	0.44
7	Diss. add	98	DA	0.44
8	pH 11 msdon	195	PH11MD	0.39
9	pH 12 msacc	181	PH12MC	0.39
10	Ring count	23	RC	0.39
11	a(yy)	119	AYY	0.28
12	Chain bond count	213	CBC	0.28
13	Diss. add conc.	99	DAC	0.28
14	Fragment count	133	FC	0.28
15	Aromatic ring count	24	ARC	0.22

Individual assessment

Table 6 Most significant feature (Group wise assessment)

Sl. no.	Group name	# Features	Probability	Selected feature	# Most significant features	
					#	Name
1	Protein descriptors	85	0.15	17	3	HRC, RC, ARC
2	Formulation characteristics	17	0.28	10	4	PM, PVA, DA, DAC
3	Plasticizer	98	0.16	29	6	RAC, ABC, AYY, FC, PH11MD, PH12MC,
4	Emulsifier	99	0.1	38	1	CBC
5	Time in days	1	0.94	1	1	TD

4 Discussion

The aim of PLGA dissolution-rate prediction experiment was to find the significant variables that governs the prediction rate and to create a model for realizing the PLGA prediction profile. The results obtained from the experiment suggests significant improvement over the earlier models developed [2, 15]. Our current experiment provides an insight of the PLGA dissolution rate prediction. We have discovered a list of most significant features by computing their probability of selection using our model (higher the probability of selection, higher the significance on prediction). The probability was computed by examining the features those were repeatedly selected by the distinct experiments. The results of ensemble of distinct predictors (models) helped in achieving high accuracy, whereas, the individual models were using their own set of features. Hence, the ensemble model was using a union of all features sets by all models. However, the second model (see Table 3) gave us a better result than the earlier models reported in [2, 15]. The developed tool was hence able to address the problem related to the prediction modeling efficiently.

5 Conclusion

Computational Intelligent tools are widely used for data analysis in the industries such pharmaceutical, medical, chemistry, etc. In this article, a function approximation and feature-selection tool was used for the identification of the significant features that govern the prediction of PLGA micro-nanoparticles. The computational intelligence tool mentioned in this paper was a tree-based implementation of neural network that provides optimum network structure and parameters. The entire model was optimized using meta-heuristic algorithms. Hence, it helps in creation of diverse (heterogeneous) models in the ensemble system. The ensemble of heterogeneous models offered better result than that of all the earlier models. However, the size (complexity) of the tree was a concern. A multi-objective approach may be useful in dealing with the conflicting objectives such as complexity and accuracy.

Acknowledgment This work was supported by the IPROCOM Marie Curie Initial Training Network, funded through the People Programme (Marie Curie Actions) of the European Union's Seventh Framework Programme FP7/2007–2013/, under REA grant agreement number 316555.

References

1. Astete, C.E., Sabliov, C.M.: Synthesis and characterization of PLGA nanoparticles. J. Biomater. Sci. Polym. Ed. **17**(3), 247–289 (2006)
2. Szlęk, J., Paclawski, A., Lau, R., Jachowicz, R., Mendyk, A.: Heuristic modeling of macromolecule release from PLGA microspheres. Int. J. Nanomed. **8**(1), 4601–4611 (2013)

3. van der Maaten, L.J, Postma, E.O, van den Herik, H.J.: Dimensionality reduction: a comparative review. Technical report TiCC TR 2009-005
4. Langer, R., Tirrell, D.A.: Designing materials for biology and medicine. Nature **428**, 487–492 (2004)
5. Brodbeck, K.J., DesNoyer, J.R., McHugh, A.J.: Phase inversion dynamics of PLGA solutions related to drug delivery. Part II. The role of solution thermodynamics and bath-side mass transfer. J. Controlled Release **62**(3), 333–344 (1999)
6. Makadia, H.K., Siegel, S.J.: Poly lactic-co-glycolic acid (PLGA) as biodegradable controlled drug delivery carrier. Polymers (Basel) **3**(3), 1377–1397 (2011)
7. Fredenberg, S., Wahlgren, M., Reslow, M., Axelsson, A.: The mechanisms of drug release in poly (lactic-co-glycolic acid)-based drug delivery systems: a review. Int. J. Pharm. **415**(1–2), 34–52 (2011)
8. Kang, J., Schwendeman, S.P.: Pore closing and opening in biodegradable polymers and their effect on the controlled release of proteins. Mol. Pharm. **4**(1), 104–118 (2007)
9. Kang, J., Lambert, O., Ausborn, M., Schwendeman, S.P.: Stability of proteins encapsulated in injectable and biodegradable poly(lactide-co-glycolide)-glucose millicylinders. Int. J. Pharm. **357**(1), 235–243 (2008)
10. Blanco, M.D., Alonso, M.J.: Development and characterization of protein loaded poly (lactide-co-glycolide) nanospheres. Eur. J. Pharm. Biopharm. **43**(3), 287–294 (1997)
11. Mainardes, R.M., Evangelista, R.C.: PLGA nanoparticles containing praziquantel: effect of formulation variables on size distribution. Int. J. Pharm. **290**(1–2), 137–144 (2005)
12. Zygourakis, K., Markenscoff, P.A.: Computer-aided design of bioerodible devices with optimal release characteristics: a cellular automata approach. Biomaterials **17**(2), 125–135 (1996)
13. Gopferich, A.: Mechanisms of polymer degradation and erosion. Biomaterials **17**(2), 103–114 (1996)
14. Siepmann, J., Faisant, N., Benoit, J.P.: A new mathematical model quantifying drug release from bioerodible microparticles using Monte Carlo simulations. Pharm. Res. **19**(12), 1885–1893 (2002)
15. Ojha, V.K., Jackowski, K., Abraham, A., Snášel, V.: Dimensionality reduction, and function approximation of poly(lactic-co-glycolic acid) micro- and nanoparticle dissolution rate. Int. J. Nanomed. **10**, 1119–1129 (2015)
16. Haykin, S.: Neural Networks: A Comprehensive Foundation, 1st edn. Prentice Hall PRT, Upper Saddle River (1994)
17. Goldberg, D.E.: Genetic Algorithms in Search, Optimization, and Machine Learning. Addison-Wesley, Boston (1989)
18. Yao, X.: Evolving artificial neural networks. Proc. IEEE **87**(9), 1423–1447 (1999)
19. Chen, Y., Yang, B., Dong, J., Abraham, A.: Time-series forecasting using flexible neural tree model. Inform. Sci. **174**(3), 219–235 (2005)
20. Goldberg, D.E., Holland, J.H.: Genetic algorithms and machine learning. Mach. Learn. **3**(2), 95–99 (1988)
21. Kennedy, J.: Particle swarm optimization. In: Encyclopedia of Machine Learning, pp. 760–766. Springer, US (2010)
22. Dietterich, T.G.: Ensemble methods in machine learning. In: Multiple Classifier systems, pp. 1−15. Springer, Berlin (2000)
23. Mendes-Moreira, J., Soares, C., Jorge, A.M., et al.: Ensemble approaches for regression: a survey. ACM Comput. Surv. (CSUR) **45**(1), 10 (2012)
24. Polikar, R.: Ensemble based systems in decision making. IEEE Circuits Syst. Mag. **6**(3), 21–45 (2006)
25. Breiman, L.: Bagging predictors. Mach. Learn. **24**(2), 123–140 (1996)

26. Quinlan, J.R.: Simplifying decision trees. Int. J. Man Mach. Stud. **27**(3), 221–234 (1987)
27. Rasmussen, C.E., Williams, C.K.: Gaussian Processes for Machine Learning (Adaptive Computation and Machine Learning). MIT Press, Cambridge (2005)

Researches of Algorithm of PRNG on the Basis of Bilinear Pairing on Points of an Elliptic Curve with Use of a Neural Network

Nikolay Ivanovich Chervyakov, Mikhail Grigorevich Babenko, Nikolay Nikolaevich Kucherov, Viktor Andreevich Kuchukov and Maria Nikolaevna Shabalina

Abstract In this paper pseudorandom number generator based on elliptic curve bilinear pairing is developed. Residue number system and approximate method are used for effictive realization of modular operations over finite field that allows to increase the speed of pseudorandom number generator for −256 by 2,15 times compared to similar PRNG that uses positional notation. The developed pseudorandom number generator based on neural network has as good statistical properties as random sequences from site random.org and passes Diehard tests.

Keywords Pseudo random number generator · Elliptic curve · Bilinear pairing · Residue Number System · Neural network

1 Introduction

The modern information systems require special approaches to saving a secret. The generator of pseudo-random numbers (PRNG) is an important cryptographic primitive. Nowadays PRNG constructed on points of an elliptic curve are popular. This is because it meets two criteria—cryptographic and statistical safety. The first criterion is satisfied because the elliptic curve allows

N.I. Chervyakov · M.G. Babenko (✉) · N.N. Kucherov · V.A. Kuchukov · M.N. Shabalina
North-Caucasus Federal University NCFU, Stavropol, Russia
e-mail: mgbabenko@ncfu.ru

N.I. Chervyakov
e-mail: k-fmf-primath@stavsu.ru

N.N. Kucherov
e-mail: nik.bekesh@mail.ru

V.A. Kuchukov
e-mail: viktor-kuchukov@yandex.ru

M.N. Shabalina
e-mail: mari.qwertyqwerty@mail.ru

A. Abraham et al. (eds.), *Proceedings of the Second International Afro-European Conference for Industrial Advancement AECIA 2015*, Advances in Intelligent Systems and Computing 427, DOI 10.1007/978-3-319-29504-6_17

1. to build cryptography protocols with a key of smaller length
2. to keep the security level of the developed cipher.

So, the scientific task of development of a new algorithm of constructing PRNG combining high speed performance, good statistical and cryptographic properties of formed sequence is of high proirity.

2 Statistical Analysis

Random numbers are important for:

1. mechanisms of integrity (ISO 8731-2 [1])
2. cryptography mechanisms of key exchange (protocol of Diffie-Hellman [2])
3. digital signature schemes (ElGamal signature scheme)
4. digital signature scheme (DSS) [3]

Besides, random numbers are used for generating pseudonym, traffic and applications communication. It provides protection against traffic analysis attacks. Such approach allows to calculate steady and effective stream ciphers [3].

Use of random series of bits of high quality is desirable, it means, good behavior in statistical sense and sense of unpredictability. Otherwise, calculation of the subsequent bits would be possible for cryptanalysis, considering a segment of bit sequence with reasonable computer resources. In the last two decades considerable research in creation and analysis of pseudorandom number or bit sequences generators was carried out [4].

The level of sequence randomness can be defined by statistical tests. They emulate the computation met in practice. These tests allow to check if in case of research sequence properties can be predicted, if each bit (or number) satisfies uniform probability distribution [5].

Secure key generators must satisfy certain criteria, such as the long period, ideal distribution of a k-tuple, linear complexity, randomness, distribution and nonlinearity [5]. The majority of them are contained in the offered testing technique.

Theoretical statistics provides some quantitative measures of randomness. There is the infinite number of criteria to check randomness of a sequence [6].

The test on a random walk is one of the most powerful and sensitive test on existence of correlations in PRNG. In particular, the test with use of idea of a random walk was one of the tests which have revealed correlations in PRNG of type linear feedback shift register [7]. Other test for a random walk allowed to explain the nature of these correlations [8].

There are some variations of test for a random walk in different number of dimensions [9]. The one-dimensional model of a directional random walk was considered [8].

Algorithm 1. Test algorithm

1. Step 1.
   (a) The unidirectional walk begins in some node of a one-dimensional grid in case of the discrete values of time i with probability $1 - \mu$.
   (b) The unidirectional walk stops with probability *mu*. Transition to a step 2.

1. Step 2. In case of (b) transition to a Step 1. Start of a new walk.

The probability of a walk of the length n equals $P(n) = \mu^{n-1}(1 - \mu)$. The average length of a walk equals $\langle n \rangle = \frac{1}{1-\mu}$. It is necessary to mention, that the considered model represents Volfs effective method for one-dimensional Izings model [8]. It can be seen from the fact that the average cluster size in Volfs method is equal to average length of a walk for $\mu = \tanh\left(\frac{J}{k_B T}\right)$ where J is a constant of spins communication.

3 Elliptic Curve

In order to construct a PRNG we will introduce the operation of pairing on the elliptic curve, similar to the one considered in work [10]. For every point $P \in E\left(F_{p^k}\right)$ equality $(n + 1)P = O$ is true, where O is an infinite point. Let $E\left(F_{p^k}\right)$ denote a set of points of an elliptic curve which is defined by the equation $y^2 = x^3 + ax + b$, over fields F_{p^k}. It is important to know structure of elliptic curve points group for pairing construction. Following theorem describes structure of group when the elliptic curve is defined over a field F_{p^k}, where p is a prime number $k \geq 1$ and order of elliptic curve points group is expressed by a formula $E\left(F_{p^k}\right) = p^k + 1 - t$.

Theorem 1 *[11] If $t^2 = p^k, 2p^k, 3p^k$, group is cyclical. If $t^2 = 4p^k$, the group is isomorphic to $Z_{\sqrt{p^k}-1} \oplus Z_{\sqrt{p^k}-1} t = 2\sqrt{p^k}$, or is isomorphic to $Z_{\sqrt{p^k}-1} \oplus Z_{\sqrt{p^k}-1}$ if $t = 2\sqrt{p^k}$ or is isomorphic to $Z_{\sqrt{p^k}+1} \oplus Z_{\sqrt{p^k}+1}$ if $t = -2\sqrt{p^k}$.*

If $t = 0$, $p^k \neq 3\,(\mathrm{mod}4)$—group is cyclical. If $t = 0$, $p^k \equiv 3\,(\mathrm{mod}4)$, the group is cyclical or isomorphic to $Z_{\frac{p_i^k+1}{2}} \oplus Z_2$.

From the Theorem 1 it follows that if $t^2 = 4p^k$ and r is an even number, group of points $E\left(F_{p^k}\right)$ is represented in the form of the direct sum $Z_{\sqrt{p^k}-1} \oplus Z_{\sqrt{p^k}-1}$ or $Z_{\sqrt{p^k}+1} \oplus Z_{\sqrt{p^k}+1}$. Let $E[l]$ denote these groups, where $l = \sqrt{p^k} - 1$ or $l = \sqrt{p^k} + 1$.

As $E[l]$ is represented in the form of the direct sum of cyclic groups, it is possible to fix some generating pair of points G and H, so that it could be possible to express any point $E[l]$ by it. Let's consider points $P = a_1 G + b_1 H$ and $Q = a_2 G + b_2 H$ that are elements of $E[l]$, where $a_1, a_2, b_1, b_2 \in [0, l - 1]$. For some fixed whole $\alpha, \beta \in [0, l - 1]$ we will define a function:

$$L_{\alpha,\beta} : E[l] \times E[l] \to E[l] \text{ and } L_{\alpha,\beta}(P,Q) = \left(a_1, b_1 - a_2, b_2\right)(\alpha G + \beta H)$$

except a trivial case, when α and β are at the same time equal to zero.

Let G_1, G_2 and G_3 be three Abelians groups. Bilinear pairing is function $e: G_1 \times G_2 \to G_3$ among these three groups. This function must satisfy to property of bilinearity: for $\alpha, \beta \in G_1, \gamma, \delta \in G_2 \cdot e(\alpha+\beta,\gamma) = e(\alpha,\beta), e(\alpha,\gamma+\delta) = e(\alpha,\gamma) + e(\alpha,\gamma)$ is true.

The following theorem shows that function $L_{\alpha,\beta}$ defines bilinear pairing.

Theorem 2 *[10] Function has the following properties:*

1. *Identity.*

$$\textit{For all } P \in E[l],\ L_{\alpha,\beta}(P,P) = O$$

2. *Bilinearity.*

$$\textit{For all } P,Q,R \in E[l]\ L_{\alpha,\beta}(P+Q,R) = L_{\alpha,\beta}(P,R) + L_{\alpha,\beta}(Q,R) \textit{ and}$$
$$L_{\alpha,\beta}(P,Q+R) = L_{\alpha,\beta}(P,Q) + L_{\alpha,\beta}(P,R)$$

3. *Assimetry.*

$$\textit{For all } P,Q \in E[l]\ L_{\alpha,\beta}(P,Q) = -L_{\alpha,\beta}(P,Q)$$

4. *Nondegeneracy.*

$$\textit{For all } P \in E[l]\ L_{\alpha,\beta}(P,O) = O$$

$$\textit{Besides, if } L_{\alpha,\beta}(P,O) = O \textit{ for all } Q \in E[l],\ \textit{then } p = 0.$$

Function $L_{\alpha,\beta}$ is called pairing because $E[l] \times E[l] \to E[l]$ (similarly to Veylyas and Tates traditional pairings).

4 Algorithm of PRNG on the Basis of Bilinear Pairing on Points of an Elliptic Curve

So, we will propose PRNG algorithm.

Algorithm 1. PRNG algorithm.

Input. p – a prime number $a, b, \alpha, a_1, b_1 \in F_p$, where α – a quadratic non-residue in F_p^*.

Output. PRNG.

1. For $i = 1$ to $n-1$ execute: $\left(a_{i+1}, b_{i+1}\right) = \left(a_1, \alpha b_1\right)\left(a_i, b_i\right)$.

2. Output $a_nG + b_nH_n$.
3. END.

Condition for construction of sequence of the maximum period $p^2 - 1$ by this algorithm, is $ord\{a_1x + b_1\} = p^2 - 1$ in the field $F^*_{p^2} = \{ax + b \mid x^2 = \alpha, \alpha, a, b \in F^*_p\}$, where α—a quadratic non-residue in F^*_p. The PRNG algorithm which was considered above, was implemented in a programming environment Microsoft Visual Studio 2013 in a programming language C++. Then constructed PRNG was tested for existence of correlations "Criterion of a directional random walk". The testing algorithm was described above.

$20 \times \chi^2$—checks we made by criterion of a directional random walk with $\mu = \frac{1}{2}$. "The Algorithm P" was applied:

Algorithm 3 The Algorithm P

1. If $r < \mu$—the step to the right is made.
2. If $r \geq \mu$ a stop. (where r—a random number generated by the generator).

Thus, $\mu = \frac{1}{2}$ means, that in checks participates one bits of the generator. Results for the generator constructed on the basis of bilinear pairing of an elliptic curve points were received from test on correlation of a directional walk . Distribution of output values slightly differs from χ^2—distributions. Each of twenty tests was passed in the sense that acceptable results of the constructed pseudorandom number generator were received.

From all aforesaid it is possible to draw a conclusion that offered PRNG has long period. The carried-out tests showed that constructed PRNG has good statistical properties.

In work [12] the new techniques based on use of artificial neural network for improving traditional generators, such as ANSI X.9, basic for DES and IDEE. Also in this work the testing methodology with use of the neural networks, developed to estimate the property of unpredictability is offered. This test on unpredictability, along with traditionally used statistical tests and tests on nonlinearity, is offered as methodology to estimate stability of pseudorandom number generators. The classical and proposed generators were estimated with this methodology. The results shown that the proposed generators behave much better than traditional, particularly in terms of unpredictability [13].

5 Method of RNS to WNS Conversion

We will use the approximate method from work [14] for effective realization of encryption algorithms. The essence of the approximate method of modular numbers comparison is based on use of the relative value of the initial number to the full range of the Chinese Remainder Theorem(CRT) which connects positional number A to its representation in residuals $(\alpha_1, \alpha_2, \ldots, \alpha_n)$, where α_i is the smallest non-negative residue of number, concerning RNS modules $p_1, p_2, \ldots, p_n$ by the following expression

$$A = \left| \sum_{i=1}^{n} \frac{P}{p_i} \left| P_i^{-1} \right| \alpha_i \right|_P$$

where $P = \prod_{i=1}^{n} p_i$, p_i—RNS modules, $\left|P_i^{-1}\right|$—multiplicative inversion P_i relatively to p_i, and $P_i = \frac{P}{p_i} = p_1 p_2 \cdots p_{i-1} p_{i+1} \cdots p_n$.

If we divide (1) by constant, we will receive approximate value

$$\frac{A}{P} = \left| \sum_{i=1}^{n} \frac{\left|P_i^{-1}\right|_{p_i}}{p_i} \alpha_i \right|_1 = \left| \sum_{i=1}^{n} k_i \alpha_i \right|_1$$

where $k_i = \frac{\left|P_i^{-1}\right|_{p_i}}{p_i}$—constants of the chosen system, α_i—bits of number represented in RNS thus value of each amount will be in an interval [0, 1). The end result of the sum after addition and discarding of an integer part is defined of number only saving a fractional part of the sum. The fractional part can also be written as A mod 1, because $A = \lfloor A \rfloor + A$ mod 1. The quantity of bits of a number fractional part is defined by the greatest possible difference between adjacent numbers [14].

6 Conclusion

In this paper pseudorandom number generator based on elliptic curve bilinear pairing is developed. Residue number system and approximate method are used for effictive realization of modular operations over finite field that allows to increase the speed of pseudorandom number generator for −256 by 2,15 times compared to similar PRNG that uses positional notation. The developed pseudorandom number generator based on neural network has as good statistical properties as random sequences from site random.org and passes Diehard tests.

Acknowledgments Current work was performed as a part of the State Assignment of Ministry of Education and Science (Russia) No. 2563.

References

1. ISO 8731.: Banking—approved algorithms for message authentication, Part 1, DEA, IS 8731–1, Part 2, Message Authentication Algorithm (MAA), IS 8731–2 (1987)
2. Diffie, W., Hellman, M.E.: New directions in cryptography. IEEE Trans. Inf. Theory **22**(6), 644–654 (1976)
3. Schneier, B.: Applied Cryptography: Protocols, Algorithms, and Source Code in C. Willey (2007)

4. Zeng, K., Yang, C.-H., Wei, D.-Y., Rao, T.R.N.: Pseudo random bit generators in stream cipher cryptography. IEEE Comput. 8–17 (1991)
5. Simmons, G.J.: Contemporary Cryptology. The Science of Information Integrity. IEEE Press (1992)
6. Knuth, D.: The Art of Computer Programming. Seminumerical Algorithms, vol. 2, p. 832. Publishing House Williams, Moscow (2001)
7. Vattulainen, I., Ala-Nissila, T., Kankaala, K.: Physical tests for random numbers in simulations. Phys. Rev. Lett. **73**(19), 2513 (1997)
8. Shchur, L.N., Butera, P.: The RANLUX generator: resonances in a random walk test. Int. J. Modern Phys. C **9**(4), 607–624 (1998)
9. Binder, K., Heermann, D.W.: Monte Carlo Simulation in Statistical Physics. Springer, Berlin (1992)
10. Lee, H.-S.: A self-pairing map and its applications to cryptography. Appl. Math. Comput. **151**, 671–678 (2004)
11. Bolotov, A.A., Gashkov, S.B., Frolov, A.B., Chasovskikh, A.A.: Algorithmic Bases of Elliptic Cryptography, p. 499. RGSU Publishing House, Moscow (2004)
12. Chervyakov, N.I., Galushkin, A.I., Evdokimov, I.A., Lavrinenko, I.N., Lavrinenko, A.V.: Application of Simulated Neuronic Networks and System of Residual Classes in Cryptography, p. 280. FIZMATLIT, Moscow (2012)
13. Karras, D.A., Zorkadis, V.: Improving pseudorandom bit sequence generation and evaluation for secure internet communications using neural network techniques. In: Proceedings of the International Joint Conference on Neural Networks, 2003. IEEE, vol. 2, pp. 1367–1372 (2003)
14. Chervyakov, N.I., Babenko, M.G., Lyakhov, P.A., Lavrinenko, I.N.: An approximate method for comparing modular numbers and its application to the division of numbers in residue number systems. Cybern. Syst. Anal. **50**(6), 977–984 (2014)

A Decentralized Management Approach for On-Demand Transit Transportation System

Olfa Chebbi and Jouhaina Chaouachi

Abstract Challenges of sustainability require flexible and innovative approaches to manage transportation systems. On-demand transit transportation systems that provide an exclusive, on demand and flexible transportation service are becoming an important part of public transportation system. In this paper, we focus on Personal Rapid Transit (PRT) system. We propose a new decentralized approach for managing vehicles in PRT system. Our approach considers the PRT as a multi-agent system where various autonomous agents represent various elements of the system. We also propose a decentralized managing algorithm for the different PRT vehicles that allows to express interest of the vehicles for various transportation requests. Using the mechanism of our algorithm, we were able to manage efficiently the different actors in our system. The results of the multi-agent modeling approach have been presented and validated in a set of computational experiments.

Keywords Multi-agent systems · Intelligent transportation systems · Simulation · Vehicle routing

1 Introduction

Due to the several environmental problems observed in urban areas, more cities tend to invest in various activities in order to enhance the quality of their public transportation systems. The main goal is to attract users by increasing the quality of the transportation service and therefore the attractiveness of public transportation tools. We could note for example planning and control of transportation operation [16],

---

O. Chebbi (✉)
Institut Supérieur de Gestion de Tunis, Université de Tunis, 41,
Rue de la Liberté - Bouchoucha, 2000 Le Bardo, Tunisia
e-mail: olfaa.chebbi@gmail.com

J. Chaouachi
Institut des Hautes Etudes Commerciales de Carthage, Université de Carthage,
IHEC Carthage Présidence, 2016 Tunis, Tunisia

A. Abraham et al. (eds.), *Proceedings of the Second International Afro-European Conference for Industrial Advancement AECIA 2015*, Advances in Intelligent Systems and Computing 427, DOI 10.1007/978-3-319-29504-6_18

strategy evaluation [9] users perception of public transportation tools [15], and so on. However, it is very difficult to use direct implementation for testing and developing new transportation systems and solutions due to various reasons such as financial, legal, material and time constraints. We should note also that the size and complexity of developing new transportation systems make the implementation of a global and valid mathematical model very difficult. That is why simulation have been proposed as the best approach to analyze, implement and evaluate transportation systems and solutions.

Simulation is developed as just an abstraction of a more complex system in order to imitate specific behavior of a system. Therefore, a simulation program don't need to develop in details all the complex aspect of the system intended to be simulated. In the literature, many simulation approaches exists such as discrete event simulation, dynamic system and agent-based simulation approach(ABMS). A special focus could be put on the ABMS as it presents the advantage to study complex system from an individual point of view and is considered as an effective approach to model complex transportation systems.

Agents in ABMS represent the active entities which perform specific operations in a specific given space [4]. A simulation model would present the conceptual description of the agents as well as their actions and activity space. The simulation model therefore, presents a systematic and specific point of view of a much more complex and detailed system. On the other hand, simulation is the result of executing the simulation model in a computer. In ABMS, developing simulation models and execute them are generally identified as two distinctive stages of any ABMS. However the transition between these two stages is not clearly defined.

On the other hand, an increasing concern have been observed over the huge growth of number of private vehicles. In fact as the demand for transportation service continues to growth, current public transportation service have been unable to provide an efficient transportation service desired by passengers. That is why, more people tends to use their private automobile over the use of public transportation tools. Consequently, what is needed today is a new public transportation system that offers the advantages of the using private vehicles by offering a door-to-door and responsive service but without the congestion and pollution issues related with cars. It is within this context that the PRT was proposed as a new effective transportation tool being able to provide a fast point to point transportation service while being a public transportation tool.

In fact, PRT uses a set of small driverless electric vehicles that could take up to six passengers. PRT have been implemented in various contexts such as Morgan town West Virginia USA, the Heathrow airport London UK and Masdar City Abu Dhabi UAE. In the literature, PRT is considered as a clean and ecological transportation tools which have the ability to solve various problems related to the urban mobility.

Although the PRT systems ideas was first developed sine the 50s, PRT literature stills at its infancy. In fact, several problems related to PRT have been studied in the literature such as the waiting time for passengers [7], the fleet size [8] and energy consumption [12, 13]. For the simulation context, several PRT specific simulators have been implemented in order to simulate such a complex system [1]. However, one could note that the majority of these simulator don't take into consideration the battery constraints related to the PRT system. In fact, the limited electric battery for the PRT' vehicles is an important feature as it imposes a bounded maximum allowable traveling distance for the PRT' pods (i.e. vehicles). Therefore, the battery of PRT' vehicles needs to be charged periodically. As for simulator taken into account the battery issues for PRT vehicles one could note the one developed by Mueller and Sgouridis [14]. However, their simulation model uses a discrete event approach for modeling the PRT system.

According to us, ABMS requires a specific attention in the PRT literature due to its huge ability to simulate the PRT'vehicles behavior. In fact as the PRT vehicles are driverless, agent based simulation represents the perfect tool to model efficiently how the PRT vehicles are operated in this system. It is within this context that this paper tries to fill these gaps in the literature.

The contributions of this paper are several. We provide in this work a simulation study related to PRT. For simulating such a system, one could use discrete event simulation or ABMS. We provided in this paper an ABMS in order to simulate the PRT system. Also, a decentralized real time fleet management of PRT system is suggested in order to minimize the waiting time of passengers and the consumed electric energy of PRT system. Finally, we validate our proposed decentralized strategy based on a specific data set based on a PRT' real case study from the United Kingdom.

The structure of this paper is as follows. In Sect. 2, we explain the main concepts in the ABMS. In Sect. 3, we present our problem definition. Section 4 explains the proposed ABMS as well as our proposed decentralized management approach for PRT. In Sect. 5, computational results and analysis of our simulation model are provided. Finally, Sect. 6 concludes and gives directions for future research.

2 Multi Agent Simulation

Multi Agent simulations are often used in the literature to model traffic systems by simulating driver behavior. In fact in multi agent simulation, models for modeling transportation systems have a big impact on the different vehicles movements and speed. One big reason for the popularity of multi agent simulation models recently is related to the huge advance observed in many intelligent systems. In fact, it was observed that there is a tendency toward more distributed, decentralized large and open systems. The internet can be cited as one typical example of these systems. However, decentralized systems comes often with a large complexity in term of applications, management algorithms and hardware. That is why, there have been an increasing focus on multi agent systems as they are able to manage the different complex communications and interactions between autonomous entities. Multi agent systems aim mainly at applying sophisticated models derived from social interactions in human societies to intelligent complex systems in order to analyze them. By doing so, Multi agent simulation focus on solving the different conflicts in organizational structures of various types.

In particular, in the present paper we deal with agent-based techniques for the simulation and modeling as well as coordination and adaptation of the autonomous vehicles in PRT system. As an extensive definition of multi Agent system is outside the scope of this paper, we aim in this section to present general definition and related literature of Multi agent systems.

An agent can be defined as an autonomous computer program which behaves to reach a set of specific objectives. We should note that several definitions have been proposed in the literature [10]. However, one should note that one single definition of multi agent system have not been agreed by scientists over the world [11]. Nevertheless, multi agent systems have become recently a first choice for model real time decentralized distributed system due to the different specific features of multi agent technology [6].

Agents could be categorized into various types (reactive, proactive, interactive, adaptive, collaborative, autonomous, mobile, etc.). These categories are mainly based on the agent specific different behaviors exhibited by them.

A system where it exists multiple intelligent agents that coordinate with each other in order to solve a complex, hard problem is called multi-agent system (MAS). Mainly, Agents in a MAS could either compete or cooperate with each others to reach individuals or a global collective objective. Due to the huge potential of MASs, they have been used in the various domains. One could note for example supply chain management, transportation, manufacturing, health care, decision support, e-commerce, network management, banking and finance, intelligent systems, etc [5].

3 Problem Definition

In this section, we first define the problem for our PRT system. Consider a PRT system with k vehicles, M stations, and a PRT network of guideways that satisfy connectivity constraints by making it possible to move from any one station to another. We also assume there is a PRT depot at which all of the vehicles are initially located. Vehicles have to return to the depot whenever they need to recharge their battery. The batteries allow a vehicle to run for a predefined number of minutes B. We do not consider the cost matrix that defines the cost of moving between each pair of stations. Passengers are assumed to arrive at stations according to a Poisson process with rate λ. In a PRT system, a dispatching decision is needed when more than one PRT pod is waiting at a station and different passengers are waiting at different stations to be transported to their destination. As the PRT system is a private transportation tool, each vehicle can only handle one passenger(s) request at a time. This makes the development of a dispatch system for PRT a difficult task, as we aim to minimize passenger waiting time and the amount of energy wasted by empty vehicle movements.

As an innovative intelligent public transportation such as PRT must be evaluated from different levels such as waiting time, capacity of the system and energy consumption,etc, such an evaluation becomes very difficult.

For our problem and as a first step, we try to minimize delays of service against the tie that the system is informed. This is a very important objective for such an ondemand service, as the coming of passengers is supposed to be unknown in advance. In PRT it exists also the case of perfect demand responding service were each demand is already known in advance. Therefore the scheduling and routing for each vehicles could be computed at the starting of the transportation service In this case, the obtained evaluation function could be used as an index to evaluate dynamic strategies. This is made possible as the static problem would present lower bound for simulation. In the literature different studies were used to compare static and dynamic solution for vehicle routing problem [2].

4 Agent Based Simulation for Modeling Trips in PRT System

In the following subsections, we will first describe the used simulation software. Then, we describe the developed PRT simulation model.

4.1 The Used Simulation Software

Different simulation software was evaluated in order to develop our multi agent simulation model for the PRT system. Among evaluated software we could note Entreprise dynamics,[1] Jadex,[2] etc. Our decision was made in favor of the Anylogic software.[3] The Anylogic is a java based simulation software. It offers the advantages of combining three major simulation methodologies: Discrete event simulation; System Dynamic and Agent based simulation. The Anylogic offers modular and hierarchical modeling of complex models. The Anylogic provides also different tools such as diverse library which helps users to better build their models. This software offers also the possibility to work with action chart, state chart and a huge number of statistics tools. This makes possible for any developed simulation model to evaluate their output using the different built-in statistical tools.

4.2 Decentralized Management Strategy for PRT

PRT'agent based simulation model mainly imitates the behavior of PRT pods. However, it also contains depot and passengers of the PRT system. The top-layer model

---

[1] Source: http://www.incontrolsim.com/enterprise-dynamics/enterprise-dynamics.html.

[2] Source: http://www.activecomponents.org/bin/view/AC+Tool+Guide/09+Simulation+Control.

[3] Source: www.anylogic.com/.

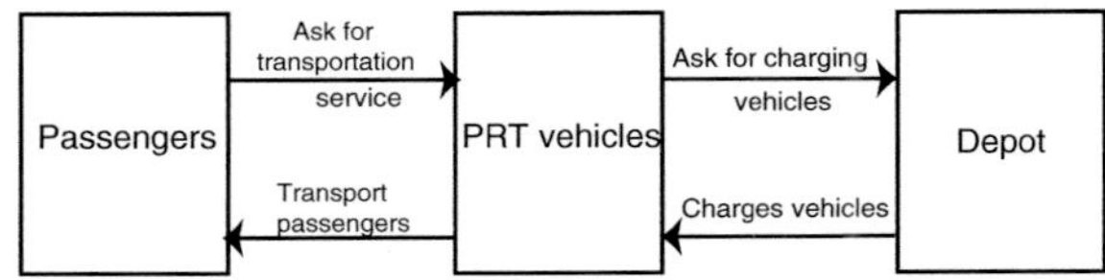

Fig. 1 Top-layer model of PRT system

of the PRT' agent based simulation model is shown in Fig. 1. In the model, there are three main classes: passengers, depot and PRT pods.

The PRT pods accepts passengers' demands, handles transportation requests checks their battery level, and so on. Passengers are the users of the system. They ask to be transported from one station to another in the PRT network. Passenger requests are accepted by the PRT system and send to the vehicles to be handled.

In the proposed simulation model, the PRT pods as well as the passengers are considered as an autonomous independent agents. They take their own decisions. These decisions will have an impact on the whole performance of the system.

The PRT vehicles achieves three main functions:

1. Passengers satisfaction by transporting them from their departure station to their arrival station.
2. Control of its level of energy by effectively minimizing its empty movements, a vehicle agent would perfectly manage its energy consumption.
3. Collection of statistics related to the whole PRT system.

The working mechanism of the PRT vehicle agent is shown in Fig. 2.

As for the passenger agent, its working mechanism is presented as follow.

First, a passengers arrive at its departure station. Passengers arrive at a station following a Poisson process with rate λ. The PRT system takes and records the

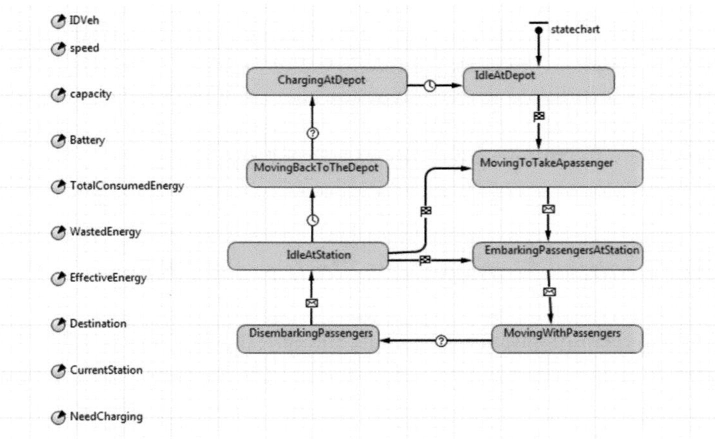

Fig. 2 Flow chart of a PRT pod

demand of the passengers and handle it according to a decentralized managing strategy.

In fact, the PRT system would handle passengers requests based on the principle of the **longest wait first**. This means that the vehicle that finishes serving its current transportation request would go to serve the passenger with the longest waiting time.

This strategy offers the advantage of reducing considerably the waiting time of passengers and thus offering a high quality transportation service.

However, based on our decentralized strategy a situation of conflict could exists where two vehicles would be able to serve the same longest waiting transportation' request. In this case, a set of simple rules is applied. First, the closest vehicle to the transportation request would be sent to serve it. Otherwise, the vehicle with the highest level of energy in its battery would be sent to serve the transportation request. Based on these simple rules, the PRT system would ensure a quick response to a passenger' request and thus an interesting transportation option for its users. After a vehicle would arrive to the departure station of a passengers it would take him directly to its final destination.

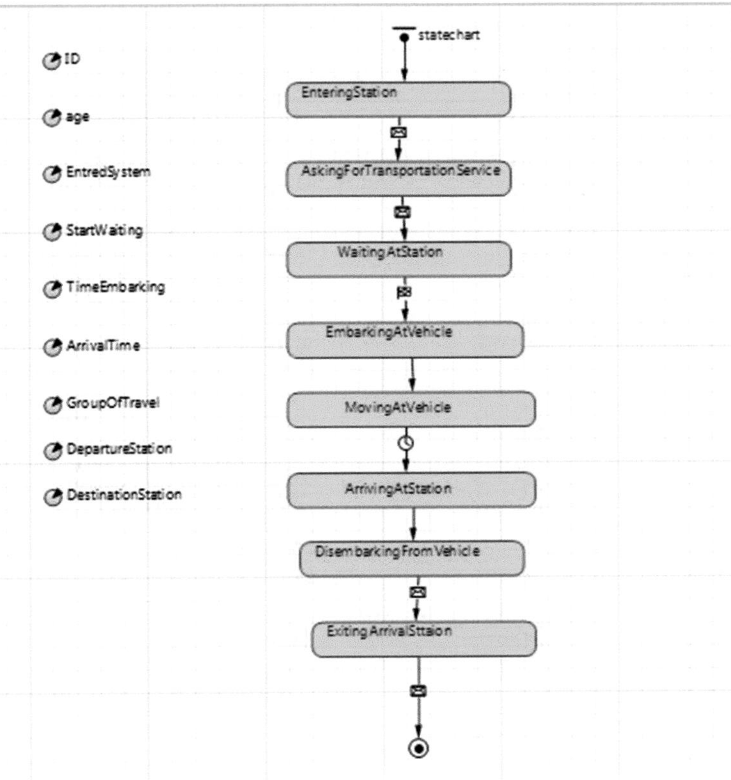

Fig. 3 Flow chart of passenger

A PRT vehicles after finishing serving a passenger would perform a check on it level of energy. If its battery level is below 10 % the vehicle would go immediately to the depot to recharge it battery.

The flow chart of passengers agent's is shown in Fig. 3.

5 Computational Results

The applicability of the proposed ABMS have been proved with the evaluation of its performance on the PRT Corby network [3]. The Corby PRT network presents an urban case study for testing various PRT management strategies. The proposed network has a multi-depots shapes. In our simulation study, we proposed a single depot shape and thus generated four distinctive networks based on one single depot at each time. The arrival rate of passengers is based on a Poisson process with rate λ. These rates were taken from the literature [7]. The results are shown in Table 1. Table 1 presents the results in term of average waiting time for passengers as well as

Table 1 Computational results

Scenario	Network layout	Effective energy (%)	Wasted energy (%)	Mean waiting time (s)
1	D1	56.517	43.482	408.484
2	D1	55.958	44.041	417.990
3	D1	58.369	41.630	483.351
4	D1	56.334	43.665	459.090
5	D1	57.753	42.246	481.8708
1	D2	60.999	39.0001	586.744
2	D2	59.470	40.529	538.557
3	D2	60.946	39.053	588.390
4	D2	62.688	37.311	516.721
5	D2	59.795	40.204	470.561
1	D3	61.964	38.035	513.046
2	D3	63.478	36.521	443.355
3	D3	63.581	36.418	556.867
4	D3	62.460	37.539	498.080
5	D3	63.125	36.874	454.282
1	D4	61.962	38.037	515.9762
2	D4	61.970	38.0291	482.943
3	D4	62.199	37.800	468.4740
4	D4	62.037	37.962	599.798
5	D4	61.699	38.300	503.134
Average		60.665	39.334	499.386

the percentage of energy used for the empty moves (i.e. wasted energy) as well as the percentage of energy used to move passengers (i.e. effective energy). We should note that the waiting time of a passenger is computed as the difference between the time where the transportation request was requested and the time of service. Results presented in Table 1 shows that the PRT system is an effective transportation system as it was able to offer a high quality transportation service. In fact, the mean waiting time was around 8 min which is considered as a good performance.

As a post analysis of our results, we wanted to estimate the impact of the different strategic decisions on the performance of the system. More specifically, we wanted to observe the effect on the depot location on the waiting time of passengers and the wasted energy of the vehicles. For that purpose, a correlation test was conducted to test the statistical relation between the network topology, the different scenarios used as well as the mean waiting time and the wasted energy. A Pearson correlation test is conducted for this purpose using the GraphPad software.[4] The results of our statistical analysis show that there is a strong correlation between the empty movements of vehicles and the network topology as well as the simulated scenarios. However as for the waiting time of passengers, results shows that there is not a strong statistical relation between the different analyzed variables. In fact, we found a P-value of 0.194. From our statistical analysis, we could conclude that the different strategic decisions related to the PRT systems(i.e. network design, station design) are of a high importance as they have a direct impact on the performance of the system. In fact as the distance between the depot and the different stations in the system increases, the empty movement from and to the depot of the vehicles would increase significantly. However and based on our preliminaries study, these factors have not a significance on the waiting time of passengers. Further studies are required to confirm these preliminaries results.

6 Conclusion and Future Research Directions

This paper has presented a new algorithm to model the real time decision making process of a PRT vehicles that can be used in a agent based simulations. The algorithm is easy to implement and mimic realistically what could actual happen in real PRT system. Priority in our proposed strategy was given to minimize the waiting time of passengers. This paper also presented multi agent simulation model which could mimic adequately the behavior of the different actors in the PRT system. The simulation model was developed using the Anylogic simulation software. Simulation experiments shown that the algorithm could take into account several parameters of the PRT system such as the depot location and the specific scenarios simulated. Also, our experiment using statistical tests have proven how the different parameters could interact within each other. Experiments presented promising computational results as we found a low mean waiting time of passengers as well as a reasonable

[4]more details about GraphPad software could be found in http://www.graphpad.com/.

empty vehicle movements. Extension to this work include the inclusion of operational research techniques in order to enhance the quality of the obtained solutions. Extensions includes also the inclusion of the multiple depot feature in our simulation model.

References

1. Anderson, J.E.: Some History of PRT Simulation Programs (2007)
2. Berbeglia, G., Cordeau, J.F., Laporte, G.: Dynamic pickup and delivery problems. Eur. J. Oper. Res. **202**(1), 8–15 (2010). http://www.sciencedirect.com/science/article/pii/S0377221709002999
3. Bly, P., Teychenne, P.: Three financial and socio-economic assessments of a personal rapid transit system. In: Proceedings of the tenth International Conference on Automated People Movers, p. 39 (2005)
4. Ghorbani, A., Dijkema, G., Bots, P., Alderwereld, H., Dignum, V.: Model-driven agent-based simulation: procedural semantics of a maia model. Simul. Model. Pract. Theory **49**, 27–40 (2014)
5. Guo, Q., Zhang, M.: A novel approach for multi-agent-based intelligent manufacturing system. Inf. Sci. **179**(18), 3079–3090 (2009)
6. Holmgren, J., Davidsson, P., Persson, J.A., Ramstedt, L.: Tapas: a multi-agent-based model for simulation of transport chains. Simul. Model. Pract. Theory **23**, 1–18 (2012)
7. Lees-Miller, J.D.: Minimising average passenger waiting time in personal rapid transit systems. Ann. Oper. Res. 1–20 (2013)
8. Li, J., Chen, Y.S., Li, H., Andreasson, I., van Zuylen, H.: Optimizing the fleet size of a personal rapid transit system: a case study in port of Rotterdam. In: International Conference on Intelligent Transportation, pp. 301–305 (2010)
9. Li, T., van Heck, E., Vervest, P., Voskuilen, J., Hofker, F., Jansma, F.: Passenger travel behavior model in railway network simulation. In: Proceedings of the 38th conference on Winter simulation, Winter Simulation Conference, pp. 1380–1387 (2006)
10. Manzoor, U., Nefti, S., Rezgui, Y.: Categorization of malicious behaviors using ontology-based cognitive agents. Data Knowl. Eng. **85**, 40–56 (2013)
11. Manzoor, U., Zafar, B.: Multi-agent modeling toolkit-mamt. Simul. Model. Pract. Theory **49**, 215–227 (2014)
12. Mrad, M., Chebbi, O., Labidi, M., Louly, M.: Synchronous routing for personal rapid transit pods. J. Appl. Math. 2014 (2014). http://dx.doi.org/10.1155/2014/623849
13. Mrad, M., Hidri, L.: Optimal consumed electric energy while sequencing vehicle trips in a personal rapid transit transportation system. Comput. Ind. Eng. **79**, 1–9 (2015). http://dx.doi.org/10.1016/j.cie.2014.09.002
14. Mueller, K., Sgouridis, S.P.: Simulation-based analysis of personal rapid transit systems: service and energy performance assessment of the masdar city prt case. J. Adv. Transp. **45**(4), 252–270 (2011). http://dx.doi.org/10.1002/atr.158
15. Stradling, S., Carreno, M., Rye, T., Noble, A.: Passenger perceptions and the ideal urban bus journey experience. Transp. Policy **14**(4), 283–292 (2007)
16. Toledo, T., Cats, O., Burghout, W., Koutsopoulos, H.N.: Mesoscopic simulation for transit operations. Transp. Res. Part C: Emerg. Technol. **18**(6), 896–908 (2010)

Hybrid Metaheuristic to Solve the Selective Multi-compartment Vehicle Routing Problem with Time Windows

Hadhami Kaabi

Abstract This paper presents a new variant of the multi-compartment Vehicle Routing Problem (MCVRP) with profits and time windows. This problem is called the selective MCVRP with time windows. In the proposed approach, a limited number of k vehicles with multiple compartment is available at the depot to serve a set of customers. Each vehicle has a limited capacity and each compartment contains one product. A customer is served only within a given time windows and when it is visited a profit is collected. Moreover, to the best of our knowledge, this problem is addressed for the first time. The goal of this research work is to determine a set of routes limited in length, such that a set of customers are visited within a given time windows, the total collected profit is maximized and the total routing cost is minimized. We present an hybrid metaheuristic to solve this problem.

Keywords Multi-compartment vehicle routing · Customer selection · Profit · Time windows · Hybrid metaheuristic

1 Introduction

Classical Vehicle Routing Problems (VRP) is introduced by Dantzig and Ramser (1959) and is considered as a combinatorial optimization problem which is crucial for a huge area of applications (e.g. logistics, production, distribution, ect). The VRP is an extension of the classical Traveling Salesman Problem in which a set of customers are served with a fleet of vehicles based at one depot. Classical VRP normally assume that the different customers requests (e.g. products) can be mixed and placed into one only compartment. However, in real world scenarios, the products cannot be transported together in one compartment due to homogeneity constraints. To deal with this situation, a multi-compartment Vehicle Routing Problem (MCVRP) is

H. Kaabi (✉)
Institut Supérieur de Gestion de Tunis, Université de Tunis,
41, Rue de la Liberté - Bouchoucha, 2000 Bardo, Tunisia
e-mail: hadhamikaabi@gmail.com

A. Abraham et al. (eds.), *Proceedings of the Second International Afro-European Conference for Industrial Advancement AECIA 2015*, Advances in Intelligent Systems and Computing 427, DOI 10.1007/978-3-319-29504-6_19

proposed. The MCVRP is introduced in 1979 and had practical significance [15]. The MCVRP consists in transporting different products that should be kept in separate compartments because of incompatibility constraints. Each compartment is dedicated to one product. The demand of each customer for a product is delivered by one vehicle. However, the entire delivery (e.g. demand) of a customer can be brought by several vehicles [7]. The MCVRP is NP-hard since it is a particular case of the VRP. Vehicle routing problems with compartments are well known in logistics but have little attention in the literature. The most published papers of the MCVRP concern these applications:

The distribution of various types of Fuel and oil [3–5, 12, 17], (Fagerholt et al. 2000) and some maritime applications [19], (Havattum et al. 2009). A second application involves the delivery of animal foods introduced by [8], and grocery products requiring different levels of refrigeration (e.g. dry, refrigerated and frozen) [2]. The pickup and delivery of livestock is also a known application of the MCVRP as developed in [14]. Animals should be kept separately to avoid contamination. Multi-compartment vehicles are also used in waste collection, where for example the general waste are loaded in a specified compartment and another compartment is dedicated to recyclables [13, 15].

The selective MCVRP with time windows (SMCVRPTW) is a variant of the MCVRP. First, each customer is served only within a given time windows, which is a common constraint in various routing problem (e.g. vehicle routing problem with time windows (VRPTW)) [18]. Second, the request of a customer is composed of different products with which profits are associated. Once a request is satisfied, a profit is collected. A customer is visited if and only if the total collected profit is maximized and the routing cost is minimized. This constraint implies that some customer requests may not be satisfied. This problem is addressed under different names in the literature. A traveling Salesman Problems with profits was provided by Feillet et al. (2011), [1, 20] have also proposed a selective vehicle routing problem. The two objectives functions (e.g. maximizing the collected profit and minimizing the total routing cost) are usually used in three different ways. The first prospect is to minimize the total routing cost under a minimum collected profit constraint. The second way is the opposite: maximize the total collected profit and impose a limit on the maximum route travel cost. In the third possible way, both objectives are present in the objective function. The problem is then to maximize the total profit minus the total traveled distance cost. In this paper, we propose an hybrid algorithm to solve the selective MCVRP with time windows. To the best of our knowledge, the MCVRP addressed in this paper is studied for the first time. The first assumption, is that a fleet of vehicles equipped with multiple compartments, serve customers demands and return to the depot. Each compartment is dedicated to one product. Indeed, a product may regroup different kinds of goods which share the same characteristics. This case is encountered in the real life application which consists in the delivery of groceries to convenience stores. The second assumption, is to serve a customer within a given time windows at minimum cost and a maximum collected profit. This implies that the vehicles can not visit all customers.

In this paper, we propose an hybrid approach based on the Genetic Algorithm (GA) and the Iterated Local Search (ILS), to solve the selective MCVRP with time windows. This approach exploits the usefulness of both the GA and ILS.

The remainder of this paper is organized as follows. In the next section, the problem formulation is addressed. In Sect. 3, the hybridization approach based on the GA and and ILS is presented. In section Sect. 4, preliminary results are established. Concluding remarks and future works are reported in section Sect. 5.

2 Problem Formulation

The selective multi-compartment vehicle routing problem with time windows (the selective MCVRPTW) proposed can be defined as an undirected complete graph G(N, E), in which N = $\{0, \ldots, n\}$ represents the set of nodes and E the set of edges. The depot is represented by the node 0, and a set of customers is represented by the set N'. In this definition, a non negative profit π_i is associated with each node. A non negative routing cost c_{ij} and a travel time t_{ij} are associated with each arc (i, j) $\in$ E. A fleet V = $\{1, \ldots, v\}$ of identical vehicles with m compartments, deliver a set P = $\{1, \ldots, m\}$ of m products. Each product p is loaded in compartment p which has a known capacity Q_p. Each customer i has a known request $d_{ip} \leq Q_p$ for each product p. The request of each product to each customer must be delivered by only one vehicle. However, the different products required by one customer can be brought by different vehicles. The customer is served only within a given time windows $\left[a_i, b_i\right]$ satisfying $a_i - b_i \geq s_i$, where s_i denotes the service time. A product is delivered to a customer if and only if the collected profit is maximized and the traveled distance cost is minimized.

$$max \sum_{i,j} \sum_{k \in V} \pi_i x_{ijk} \tag{1}$$

$$s.t \sum_{i \in N} x_{ijk} \leq 1 \quad j \in N', k \in V, \tag{2}$$

$$\sum_{i \in N} x_{ijk} = \sum_{i \in N} x_{jik} \quad j \in N', k \in V, \tag{3}$$

$$\sum_{i,j \in S} x_{ijk} \leq |S| - 1 \quad k \in V, S \subseteq N', |S| \geq 2, \tag{4}$$

$$y_{jkp} \leq \sum_{i \in N} x_{ijk} \quad j \in N', k \in V, p \in P, \tag{5}$$

$$\sum_{k \in V} y_{jkp} = 1 \quad j \in N', p \in P, \tag{6}$$

$$\sum_{j \in N'} y_{jkp}\, d_{jp} \leq Q_p \quad k \in V, p \in P, \tag{7}$$

$$s_{ik} + t_{ij} - K(1 - x_{ijk}) \leq s_{jk} \quad i,j \in N', k \in V, \tag{8}$$

$$a_i \leq s_{ik} \leq b_i, \quad i \in N, k \in V, \tag{9}$$

$$\sum_{i,j \in N'} \sum_{k \in V} c_{ij} x_{ijk} \leq C_{max} \quad i,j \in N', k \in V, \tag{10}$$

$$x_{ijk} \in \{0, 1\} \quad i,j \in N, i \neq j, k \in V, \tag{11}$$

$$y_{jkp} \in \{0, 1\} \quad j \in N', k \in V, p \in P, d_{jp} \neq 0. \tag{12}$$

The objective function (1) maximized the total collected profit. Constraints (2) ensure that each customer may be visited at most once by each route. Constraints (3) ensure the continuity of each route: if a vehicle visits node j it must leave it. Constraints (4) define the classical subtour elimination constraints. Constraints (5) set y_{jkp} to zero for each product p if vehicle k do not serve customer j. Due to constraints (6), each product required by a customer is brought by one single vehicle. Constraints (7) ensure the vehicle compartment capacities are respected. Constraints (8) states that a vehicle k cannot arrive at j before $s_{ik} + t_{ij}$ if it is traveling from i to j. Constraints (9) allow customer deliveries within a given time windows. Constraints (10) ensure the limited cost routing budget. Constraints (11) set x_{ijk} to 1 if and only if edge (i,j) is traversed by vehicle k. Constraint (12) set y_{jkp} to 1 if and only if customer j receives product p from vehicle k, 0 otherwise.

3 Hybrid Metaheuristic

In this section, the hybrid metaheuristic for solving the selective MCVRP with time windows is proposed. For this purpose, two meta-heuristics will be combined (e.g. The Genetic Algorithm (GA) and the Iterated Local Search (ILS)), to identify a good feasible solution to our problem.

3.1 Genetic Algorithm

Genetic algorithms are general-purpose search algorithms that use principles inspired by natural population genetics to evolve solutions to problems [10]. Genetic algorithms have been applied to a wide range of optimization problems. The basic idea of GA is to manipulate a population of chromosomes (a set of genes) that represents candidate solutions to the problem. The algorithm begins with the generation of the initial population. In each iteration, the chromosomes are evaluated by computing their fitness values, and then two chromosomes are selected according to their relative fitness. Then, the selected chromosome are altered using genetic operators. The first one is crossover which mixed two parents to create two new offspring. The second genetic operator is the mutation which alters one or more gene from the offspring to ensure population diversity. At the end, the new offsprings are evaluated and inserted back in the population. This process continues until a stop criterion is met.

3.2 Iterated Local Search

The Iterated Local Search (ILS) approach, is a metaheuristic used to deal with several combinatorial optimization problems. To diversify the search and to escape from local optima, the ILS alternates the local search phase around the current solution and the perturbation phase. The initial solution S_0 of the algorithm is obtained randomly or returned by a greedy heuristic. The local search procedure is construct to obtain a first local optimum. Iteratively, the perturbation is applied to the current local optimum in order to improve the solution. The algorithm is stopped when a termination condition is met. The pseudo code of the ILS approach [9] is presented in the following algorithm:

```
Begin
    s0= Generate initial solution;
    s*= Local Search (s0);
    repeat
    s'= Perturbation (s*,history);
    s*'= Local Search (s');
    s*= Acceptance criterion (s*,s*',history);
    until termination condition is met;
End.
```

3.3 Genetic Algorithm with Iterated Local Search

An hybrid metaheuristic which combines the GA and the ILS is proposed to solve the selective MCVRP. Since the GA may not converge to a global optimum and the local search may rapidly fall in a local optimum, our approach is addressed. The main idea behind the hybridization is to improve the GA solutions by using an ILS to intensify the search space. In what follows we will use the GA to obtain the best solution having the highest collected profit and which satisfy the temporal and capacity constraints, whereas the ILS is used to refine the GA solution and consider the total traveling cost constraint. The pseudo code of the hybrid approach is as follows:

```
Begin
  Initialize population
  Evaluate each candidate;
  repeat
  Select parents;
  Recombine pairs of parents: Crossover;
  Mutate the resulting offspring;
  Apply ILS to the offspring;
  Evaluate new candidates
  Replacement candidates for the next generation
  until termination condition is met;
End.
```

3.3.1 Solution Representation

In the proposed approach a solution x (i.e. chromosome, candidate) is represented by using a vector V(x). This vector is a permutation of nodes (customers), which tries to insert an order in the current route, while non violating the temporal and capacity constraints.

3.3.2 Selection and Crossover

The selection of chromosomes is an important step of the GA process to produce a new population. In the literature, several selection methods have been proposed, for instance the roulette wheel selection, binary tournament method, rank selection and some others. To randomly select a parent seems to be a non efficient method. Hence, we use the same selection method introduced by Reeves in 1995 [16]. In this method, a parent is selected according to the following probability distribution:

$$P(S) = \frac{2S}{M(M+1)} \tag{13}$$

where S is the Sth chromosome ranked in a descending order of its objective value (e.g. collected profit) and M is the population size. In addition, the chromosome which has the best objective value (e.g. the highest profit) has a higher probability to be selected. Then, the crossover operation swap parts of two parents in the population to generate new offsprings [6]. In our approach, one cutting point i such that $1 \leq i \leq \rho$ is randomly drawn where ρ is the number of requests. The subsequence of a chromosome is copied from the first parent till this point, the second parent is

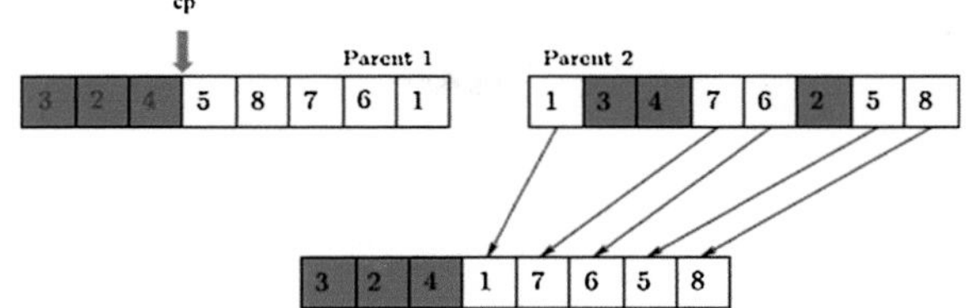

Fig. 1 Example of crossover operation

than scanned and the missed customers are added [6]. An example of the crossover operation with a crossover point cp = 3 is depicted by Fig. 1.

3.3.3 Mutation

This operator is a crucial step to escape from local optima since it contributes to the diversification in the population . The mutation process consists in removing a customer at random and then adding a random customer at a random position.

3.3.4 Local Search

The Local search (LS) is The first step of the Iterated Local Search (ILS) approach is the Local Search (LS). A random initial solution is generated. Due to the neighborhood exploration of the initial solution, a local optimum is produced. The choice of the neighborhood is one of the most important parameter when designing a LS. The neighborhood can be defined as a modification of the initial solution to reach a new better solution. The main purpose of the LS in our proposed approach is to find a solution which maximizes the total collected profit (the objective function) and minimizes the traveling cost. Moreover, we will search for solutions that satisfy the traveling cost constraint. We use three neighborhoods including insertion move, swap move and relocate. The neighborhoods are defined as follows:

Neighborhood *N1*: A request which is not satisfied in the current solution and which minimizes the total traveling cost is inserted. Given a route, the nodes which violate the traveling cost constraint are removed and replace by a new sequence of requests.

Neighborhood *N2*: the 2-opt move is applied. It consists of replacing two non-adjacent nodes belonging to two different routes, by two non visited nodes. The temporal, capacity and traveling cost constraints should be satisfied.

Neighborhood *N3*: the swap move is addressed inside the same route. More precisely, we consider one route and swap the position of two nodes belonging to this route.

After the definition of the LS neighborhoods, the LS proceeds follows: given an initial solution x (e.g. provided by the GA), we improve x according to *N1* until we found a solution x1 that can not be improved. x1 is the generated local optimum which is considered as the initial solution for the new LS using *N2*. We proceed in the same way with *N3*. This process is repeated until a local optimum is reached.

3.3.5 Perturbation

The ILS algorithm is based on the perturbation operator which is a large random move of the current solution. In fact, the process of the perturbation consists in modifying the current local optimum found by the LS, in order to obtain a new solution. The swap move is applied to modify a solution. We interchange two random selected customers (e.g.nodes) belonging to two different routes.

3.3.6 Evaluation Function

In any GA approach, the chromosomes are compared based on their evaluation functions. In this paper, this function is the total collected profit of a vehicle which leaves the depot, serves customers demands and return to the depot.

3.3.7 Replacement

The replacement operation consists in keeping a fixed population size. After the generation of the offsprings by using all the previous phases (i.e. GA operators and ILS phases), it is compared with the worst chromosome in the population and then the best one is kept.

4 Preliminary Results

The proposed approach will be tested on a set of instances issued from the well known Solomon's data set [18]. Since no MCVRP with time windows instances are available in the literature, a known instances have been transformed to deal with our proposed approach. In fact, the data set contains 56 problems divided into 6 instances: C1, R1, RC1,C2, R2 and RC2. Each instance contains 100 customers and a depot. Each customer i has an associated demand d_i and a time windows $[a_i, b_i]$. The proposed selective MCVRP with time windows instances are generated as follows. The number of compartments m for each vehicle is set to 2. The customer request is divided randomly into two parts [8]. For each customer i the request for product 1 is calculated as $d_{i1} = d_i/\lambda$, where λ is a random integer in [3, 5]. The request for product 2 is calculated as $d_{i2} = d_i - d_{i1}$. The compartment capacity is computed as $Q_p = (Q \times D_n)/(D_1 + D_2)$, where Q is the vehicle capacity in the original VRPTW [18] and D_p is the average demand for a product p. The profit π_{ip} is set to 1 for each request. The number of the vehicles is equal to 3 (m = 3).

Table 1 Average results on the selective MCVRP instances

Data set	m = 1			m = 2			m = 3		
	Profit	distance	Time(s)	Profit	Distance	Time(s)	Profit	Distance	Time(s)
C1	19	106.9	80.2	33	106.9	118.5	47	256.2	190.2
C2	25	220.4	86.6	37	312.7	147.2	55	465.3	207.1
R1	13	175.2	76.3	26	240.2	120.1	39	332.1	218.3
R2	17	101.3	83.7	27	320.3	132.4	38	319.3	209.7
RC1	12	140.7	71.2	17	278.8	112.7	34	411.9	211.2
RC2	16	168.5	75.1	23	286.1	117.2	37	389.2	198.3

The proposed approach will be coded in C and will be tested on an Intel Core I5 with 1.70 GHz. The experiments will be carried out and we will compare our proposed approach with some results from the literature, in order to assess the efficiency of the proposed approach.

Table 1 presents the average collected profit and the travel distance results obtained for each instance and the average CPU time for all the instances according to the number of vehicles. The first column contains the name of the instance for one, two and three vehicles. It can be seen from Table 1 that the total collected profit and the travel distances values increase when the number of vehicles increases and that the problem can be solved within a reasonable time.

Since the proposed approach is studied for the first time, there are nos solutions to compare with for instance. Then it is difficult to provide a detailed analysis at this moment. However, the average computing time revealed at least that the problem is solved in a reasonable time.

5 Conclusion

This paper proposes a research study dealing with the selective MCVRP with time windows. Despite its crucial practical applications, this problem is not yet studied in the literature. The proposed approach presents a first study, in terms of problem formulation (i.e. to consider time windows and traveling cost as constraints), and in terms of problem resolution. In fact, an hybrid metaheuristic was developed to address the problem. The mixture exploits the usefulness of ILS in diversifying the genetic search and improving the GA population. More precisely, The GA is used to generate a solution with a maximum total collected profit, while an ILS is used to optimize the latter solution, in which the temporal, capacity and traveling cost constraints are satisfied. The obtained results shows that the problem ca be solved in a reasonable time. An interesting avenue for further research would be to: (i) propose a MCVRP which has the minimum traveling cost as objective function and with profit and time windows as constraints, (ii) propose a multi objective MCVRP which combines the traveling cost and the collected profit in the objective function, (iii) propose a comparative study of the above MCVRP.

References

1. Aras, N., Aksen, D., Tekin, M.T.: Selective multi-depot vehicle routing problem with pricing. Trans. Res. Part C: Emerg. Technol. (2011)
2. Chajakis, E.D., Guignard, M.: Scheduling deliveries in vehicles with multiple compartments. J. Glob. Optim. (2003)
3. Cornillier, F., Laporte, G., Boctor, F.F., Renaud, J.: The petrol station replenishment problem with time windows. Comput. Oper. Res. (2009)
4. Cornillier, F., Boctor, F.F., Laporte, G., Renaud, J.: A heuristic for the multi-period petrol station replenishment problem. Eur. J. Oper. Res. (2008)
5. Cornillier, F., Boctor, F.F., Renaud, J.: Heuristics for the multi-depot petrol station replenishment problem with time windows. Eur. J. Oper. Res. (2012)
6. Derbali, H., Jarboui, B., Hanafi, S., Chabchoub, H.: Genetic algorithm with iterated local search for solving a location-routing problem. Expert Syst. Appl. (2012)
7. El Fallahi, A., Prins, C.H., Calvo, R.W.: A memetic algorithm and a tabu search for the multi-compartment vehicle routing problem. Comput. Oper. Res. (2008)
8. Feillet, D., Dejax, P., Gendreau, M.: Traveling salesman problems with profits: an overview. Trans. Sci. (2001)
9. Grosso, A., Jamali, A., Locatelli, M.: Finding maximin latin hypercube designs by iterated local search heuristics. Eur. J. Oper. Res. (2009)
10. Herrera, F., Lozano, M., Verdegay, J.L.: Tackling real-coded genetic algorithms: operators and tools for behavioral analysis. Artif. Intell. Rev. (1998)
11. Hvattum, L.M., Fagerholt, K., Armentano, V.A.: Tank allocation problems in maritime bulk-shipping. Comput. Oper. Res. (2009)
12. Lahyani, R., Coelho, L.C., Khemakhem, M., Laporte, G.: A multi-compartment vehicle problem arising in the collection of olive oil in Tunisia. Omega (2015)
13. Muyldermans, L., Pang, G.: On the benefits of co-collection: experiments with a multi-compartment vehicle routing algorithm. Eur. J. Oper. Res. (2010)
14. Oppen, J., Lokketangen, A.: A tabu search approach for the livestock collection problem. Comput. Oper. Res. (2008)
15. Reed, M., Yiannakou, A., Evering, R.: An ant colony algorithm for the multi-compartment vehicle routing problem. Appl. Soft Comput. (2014)
16. Reeves, C.: Modern Heuristic Techniques for Combinatorial Problems. McGraw- Hill Book Company Inc. (1995)
17. Relvas, S., Magatos, N.B., Neves, J.F.: Integrated scheduling and inventory management of an oil products distribution system. Omega (2014)
18. Solomon, M.M.: Algorithms for the vehicle routing and scheduling problems with time window and constraints. Oper. Res. (1987)
19. Stalhane, M., Rakke, J., Moe, C., Andersson, H., Christiansen, M., Fagerholt, K.: A construction and improvement heuristic for a liquefied natural gas inventory routing problem. Comput. Ind. Eng. (2012)
20. Valle, C., Martinez, L.C., Da Cunha, A.S., Mateus, G.R.: Heuristic and exact algorithms for a min-max selective vehicle routing problem. Comput. Oper. Res. (2011)

Solar Power Production Forecasting Based on Recurrent Neural Network

Tomas Burianek, Jindrich Stuchly and Stanislav Misak

Abstract The Active energy unit is a complex system that is independent of an external source of electric energy. It uses mainly renewable sources namely a wind turbine and a photovoltaic power plant. Because of a stochastic character of renewable sources it is important to implement the Active Demand Side Management to control energy flows in the system and manage plans of connected appliances to preserve safe working. Forecasting of solar power production from photovoltaic power plant is one of the most important parts in the system. This paper presents a solar power production forecasting model based on the Recurrent Neural Network.

Keywords Recurrent Neural Network · Time-series forecasting · Photovoltaic Power Plant

1 Introduction

Nowadays, the popularity of the Active energy units in the general public is rising. The Active energy unit is a system that is independent of an external energy source and it can be divided into two basic parts. The 1st one is a hardware part (power sources, part of electrical consumption, parts of energy transfer and energy conversion) and the 2nd one the software part. The software part includes the inputs such as weather forecast, energy production forecast, consumption plan (forecast), load configuration, Power Quality (PQ) forecast and implementation of a new protection

T. Burianek (✉) · J. Stuchly · S. Misak
FEECS, Department of Computer Science, Department of Electrical Power Engineering, VŠB - Technical University of Ostrava, 17. Listopadu 15, 708 33 Ostrava, Poruba, Czech Republic
e-mail: tomas.burianek.st1@vsb.cz

J. Stuchly
e-mail: jindrich.stuchly@vsb.cz

S. Misak
e-mail: stanislav.misak@vsb.cz

A. Abraham et al. (eds.), *Proceedings of the Second International Afro-European Conference for Industrial Advancement AECIA 2015*, Advances in Intelligent Systems and Computing 427, DOI 10.1007/978-3-319-29504-6_20

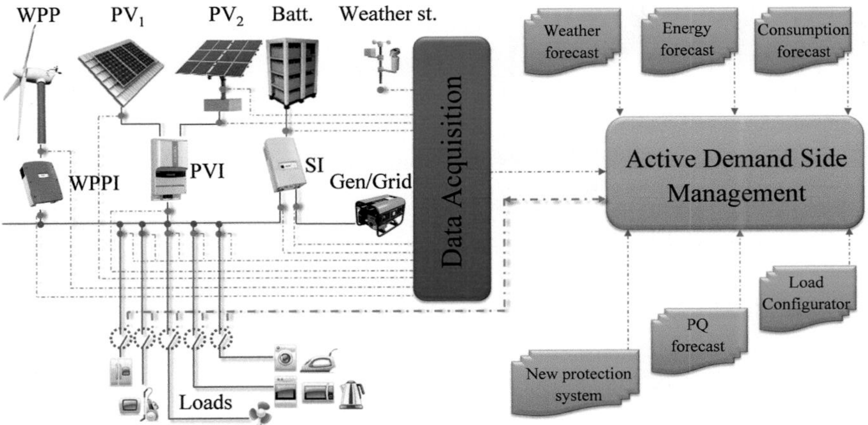

Fig. 1 Active energy unit

scheme. Basic block scheme of the Active energy unit is shown in Fig. 1. The parts of weather and energy forecast are one of the most important in the meaning of software inputs to the Active Demand Side Management (ADSM). These inputs present the forecasting of available amount of energy from renewable energy sources in the Active energy unit. This information is subsequently implemented into computing core of ADSM. Energy forecasting together with a consumption forecasting and a battery bank capacity present the basic inputs for ADSM. Based on comparison of these values the system adjust and optimize the consumption plan to keep a primary goal of the ADSM—to meet the basic requirements of house inhabitants, such as keeping food fresh, food reheating and cooking, thermal comfort, etc.

Solar power forecasting is a complex problem and many methods are used to solve it. One approach is to solve this problem by mathematical model and this solution was introduced by F. Bizzarri in [1]. Another approaches are based on machine learning algorithms for instance forecasting based on weather classification and Support Vector Machine(SVM) [6] or the hybrid model including ARIMA, SVM and Artificial Neural Network presented in [8].

Photovoltaic power plant production forecasting is mainly dependent on solar irradiance that is influenced by actual weather conditions. This paper introduces forecasting model based on a meteorological forecasting data. These data along with a real historical production from the reference solar power plant form time-series data. For time-series forecasting the Recurrent Neural Networks are suitable for this kind of task because it can capture sequence data relations in time in contrast to Feed-Forward Neural Network.

This paper is organized as follows: Sect. 2 introduces reference photovoltaic power plant. In Sect. 3, the data preparation for model is presented. Section 4 describes the Recurrent Neural Network in more details. Experiments and results are presented in Sect. 5 and the conclusion of results and the future work is given in Sect. 6.

2 Photovoltaic Power Plant

The Active energy unit consists of two strings with a total installwed power of 4–2 kWp each, see Table 1 for details. Figure 2 shows the power curves for each PV string, the blue curve (PV1) shows the dependency of global radiation of polycrystalline panels installed on the roof of the building and the red curve (PV2) monocrystalline panels installed on the tracker. Based on Fig. 2, it should be noticed that the power output from the 1st string of the PV panel almost linearly increases with solar irradiation whereas the power output from the 2nd string of the PV panel precisely follows the solar tracker throughout the day and maximum possible energy

Table 1 Table of settings

Parameters	Value	PV_1	PV_2
PV cell technology	–	Polycrystalline	Monocrystalline
PV area	m^2	19.5	19.2
Maximum power	W	2220	2160
Voltage at maximum power	V	29.8	36.0
Current at maximum power	A	6.2	5.0
Short circuit current	A	8.34	5.2
Open circuit voltage	V	36.8	45.0

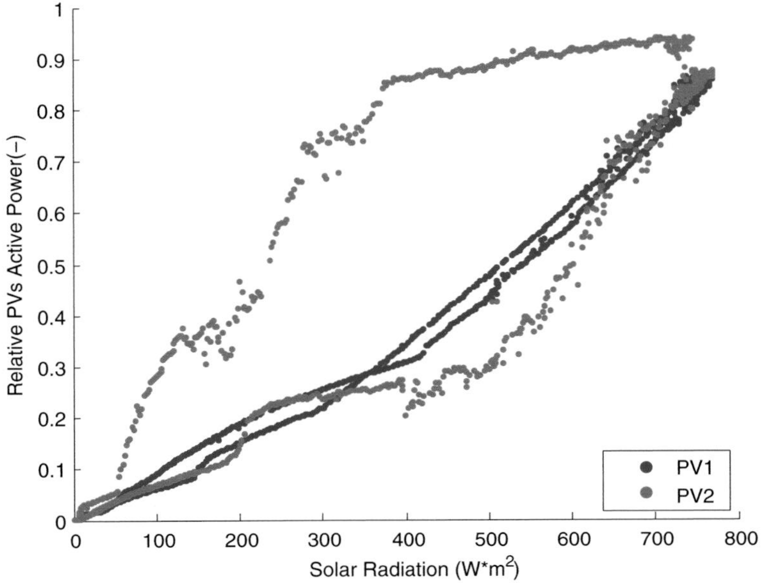

Fig. 2 PV's power curves

is extracted from it. PV1 is composed of 12 PV panels under angle about 15, where 6 PV panels have eastern and 6 PV panels have western orientation. PV2 consists of 12 panels that are installed on a two-axial pointing device, so-called tracker which uses the control electronics to achieve the PV panel's ideal angle and orientation to ensure maximum profits of electric energy throughout the day. The tracker is able to increase the power production up to 30 % in direct comparison with the permanent installation. The power consumption absorbed by the tracker for the self-consumption has been taken into account in this value. This argument is well-founded by using the database of measured values. The control unit allows maximum use of sunlight (even on cloudy days). Tracker uses to follow the sun linear motor and harmonic gearbox, which are controlled by control electronics.

3 Data Preparation

The proposed solar power production forecasting model is based on the data obtained from meteorological forecastings. Each meteorological forecasting is done every day four times a day with a six hour gap. Each forecasting contains 55 hours of predicted data ahead and it is comprised of many attributes namely *humidity*, *temperature*, *wind speed*, *atmospheric pressure* and *cloudiness*. The main effect on a solar irradiation has the *cloudiness* data. These data include four classes of cloudiness: *low*, *medium*, *high* and *total*. For a real production from the photovoltaic power plant affected by clouds is crucial to know unaffected clear production for every day of the year. The approximate clear production is possible to construct from a collected historical data of the reference solar power plant. Following approach is described in more details in [2] by Majer. This approach has three main steps. In the first step it is important to pick only sunny days. This should be easy if there were associated cloud data. But in the case of missing cloud data for whole year it is possible to pick sunny days by computing median and variance around a midday of each day and specify limits for a sunny day. In the second step each picked day is divided to 16 intervals and the concrete power is computed for each of 17 key points in the day. It is important to obtain time of sunrise and sunset for each day of the year. Then the first interval of the particular picked day starts at sunrise and the last interval ends at sunset. The other intervals are then equally distributed from sunrise to sunset. It is assumed that the irradiance progress of the day is symmetrical against midday. The final power for each key point is computed as a median around each point. In the third step the key points of picked sunny days are interpolated through the year. This results in the key points obtained for each day of the year. In the final step each day of the year is interpolated according to divided intervals by sunrise and sunset. Result of the clear production is shown in Fig. 3. The approximate clear production data and the cloudiness data are then used to construct dataset applied in supervised learning. The input part of the dataset contains one column of approximate clear production and four columns of cloudiness classes composed of forecasting data. The output part

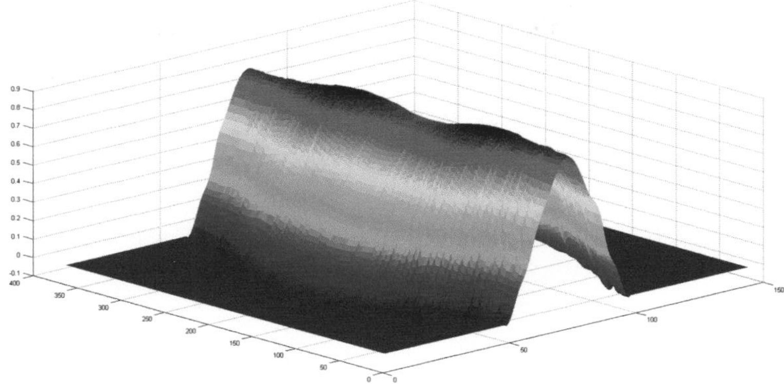

Fig. 3 Approximate clear production of photovoltaic power plant for every day of the year

is then one column of the real production from the photovoltaic power plant that is affected by presented inputs of cloudiness and the clear production at the same hour.

Proposed model for solar power production forecasting uses the Recurrent Neural Network as a supervised learning technique.

4 Recurrent Multilayer Neural Network

The Artificial Neural Networks (ANNs) are inspired in structure and functionality of a human brain [4]. The particular type of the ANNs is the Recurrent Neural Network(RNN). It allows feedback connections that is a main feature of this model. It is widely used in a signal processing of real-world problems with a complex relations in time or space [3]. Topology of the RNN consists of layers with neurons. Example of RNN structure is presented in Fig. 4. Network has at least 3 layers: input layer, one hidden layer and output layer. Number of hidden layers is varying and depends on design influenced by solving problem. The connectivity of neurons is different in comparison with the feed-forward neural network. The input layer and the hidden layer has an extra recurrent neuron for each regular neuron in the next layer. Inputs to these recurrent neurons are realized by an output signals from its regular neurons that are stored from the previously presented input pattern. Neurons between layers are connected by weighted connections. Each regular and recurrent neuron in the layer is connected to all regular neurons in the next layer.

Process of signal propagation through the RNN begins with presenting first input pattern ($t = 0$) to the input layer and these signals are without any change. Then output signals from this current layer are brought to the next layer. Each neuron of the next layer proceeds summation of the input signals x_i altered by connection weights w_i for each neuron i of n neurons from the current layer as a neuron excitation z:

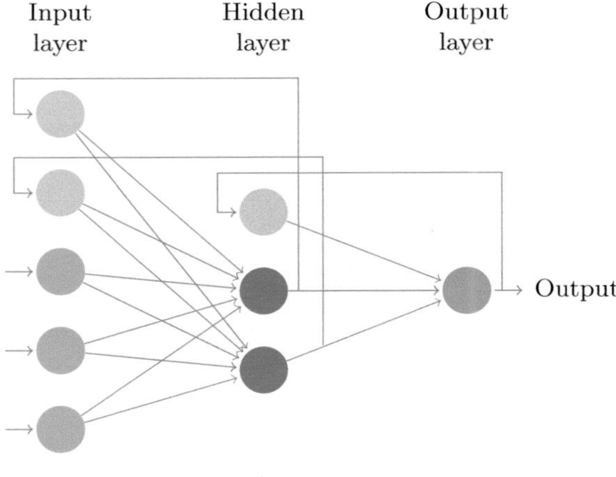

Fig. 4 Example of Neural Network design with one Hidden layer

$$z = \sum_{i}^{n} w_i x_i, \tag{1}$$

where input signal x_i for regular neuron is an output signal from the current layer and for recurrent neuron it is an output signal of associated regular neuron from the next layer that was computed for previous input pattern $(t-1)$. In case of the first input pattern when the previous output signals are not computed the summation for recurrent neurons is then skipped. Output from this neuron is computed by applying sigmoid activation function to excitation presented in equation:

$$y = \frac{1}{1 + e^{-\lambda z}}, \tag{2}$$

where λ is a slope of the sigmoid function. Then all output signals from this layer are presented to the next layer and also stored in the recurrent neurons in the previous layer for the next pattern and this is repeated until output layer is reached. Then output from the RNN for the current pattern $t = 0$ are excitations of output neurons. Computation for following patterns $(t = 1, \dots, T)$ is then performed. It is obvious that output for each pattern is influenced by its input and also by sequence of excitations stored in the recurrent neurons done by previous patterns. The RNN learning is performed by updating connection weights between neurons. This is done by training RNN using training set of knowledge. This process is called supervised learning and one of the algorithms is the Backpropagation Through Time learning algorithm.

4.1 Backpropagation Through Time Learning Algorithm

Backpropagation Through Time (BPTT) algorithm is a variant of the orignal Backpropagation algorithm for feed-forward neural networks [5, 7]. This algorithm is based on gradient descent method. An adaptation process starts with a forward signal propagation for all input patterns $t = 1 \ldots T$. Then errors of the neurons in the output layer are computed followed by computation of errors in previous layers back to the first layer. This is done for the last input pattern $t = T$ and then for previous patterns to the first pattern $t = 1$. Computed errors are then used for obtaining changes of weights that are accumulated for each connection and for each pattern. Weights of connections are then updated via changes of weights. The goal is to minimize final error function:

$$E_f(t) = E(t) + E(t+1), \tag{3}$$

where the errors for input pattern t and input pattern $t + 1$ are computed by equation:

$$E(t) = \frac{1}{2}\sum_{j=1}^{m}(y_j(t) - d_j(t))^2, \quad E(t+1) = \frac{1}{2}\sum_{j=1}^{M}(y_j(t+1) - d_j(t+1))^2 \tag{4}$$

where errors between real output from network y_j and desired output d_j are summed for all m output neurons in the output layer for the pattern t and pattern $t + 1$ respectively. Change of weight is done by equation:

$$\Delta w_i = -\eta \frac{\partial E_f}{\partial w_{ic}}, \tag{5}$$

where η is a learning coefficient. It is necessary to compute errors for each regular neuron for each pattern t. An output error for each current neuron c is obtained by expression:

$$\delta_c(t) = \sum_{j=1}^{m} \delta_j(t)\frac{dy_j(t)}{dz_j(t)} w_{cj}(t) + \sum_{k=1}^{n} \delta_k(t+1)\frac{dy_k(t+1)}{dz_k(t+1)} w^r_{ck}(t), \tag{6}$$

where the first sum are the errors from the next layer with m neurons influenced by weights w_{cj} connected from the current neuron c to each j neuron. Second sum are the errors from the current layer computed in the next step influenced by weights w^r_{ck} of the recurrent neuron in previous layer that has feed-back connection from the current neuron c and is connected to each neuron k in the current layer. In case of the last layer the first sum is simplified by difference between real and desired output $y_c(t) - d_c(t)$ and in case of the last pattern $t = T$ the second sum is omitted. The final expression for partial derivations of error E_f is obtained for regular neurons:

$$\frac{\partial E_f(t)}{\partial w_{ic}(t)} = \delta_c(t)\frac{dy_c(t)}{dz_c(t)} x_{ic}(t), \tag{7}$$

and for recurrent neurons:

$$\frac{\partial E_f(t)}{\partial w^r_{ic}(t)} = \delta_c(t)\frac{dy_c(t)}{dz_c(t)}y_c(t-1). \tag{8}$$

Derivation of the sigmoid function part in the all expressions are obtained by:

$$\frac{dy}{dz} = y(1-y)\lambda. \tag{9}$$

5 Experiments and Results

Experiments were performed on the constructed dataset with 2970 data rows representing hours with 5 input data columns and 1 output data column. This represents 124 days of the data. The dataset was divided to 66 % of the data for training and the rest for testing. Structure of the RNN was designed according to the dataset with 5 inputs, one output and one hidden layer that has 4 hidden neurons. The slope λ of each neuron is 0.6 and the learning coefficient η during training is 0.4. Training was performed for 450,000 iterations. This settings was selected experimentally. For measuring training and testing error the mean square error using difference between predicted $y(t)$ and desired $d(t)$ output from the network was used:

$$MSE = \frac{1}{T}\sum_{t=1}^{T}(y(t) - d(t))^2. \tag{10}$$

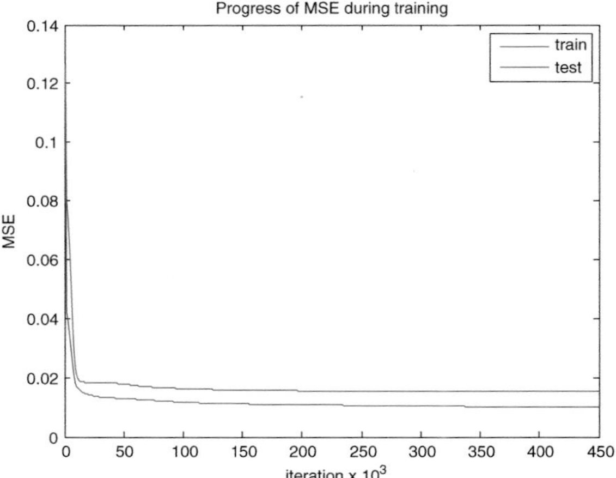

Fig. 5 MSE during training and testing

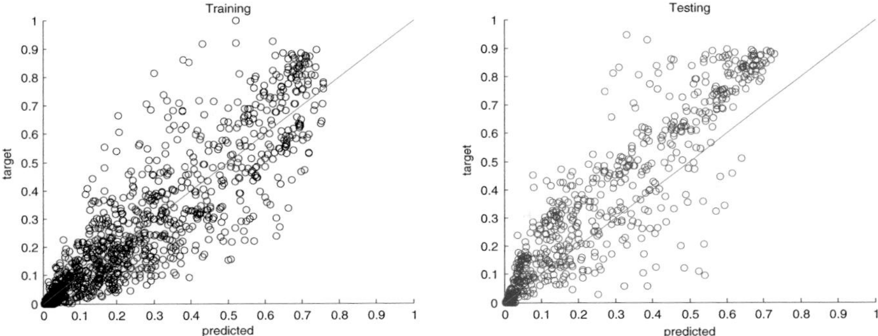

Fig. 6 Target and prediction comparison for training in *left* and testing in *right*

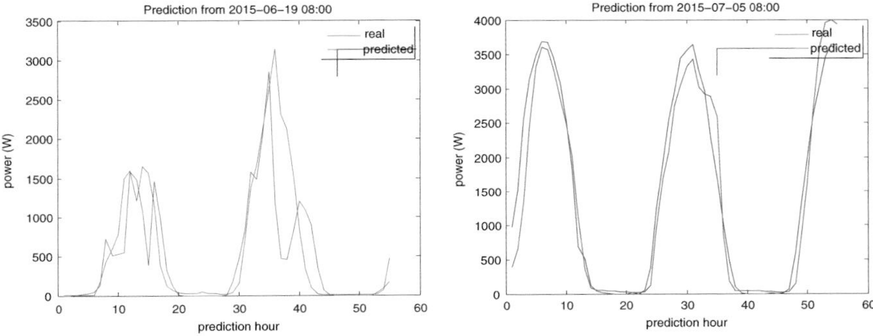

Fig. 7 Prediction for cloudy days in *left* and sunny days in *right*

Progress of MSE during training is shown in Fig. 5. Final MSE for training was $9.956e-3$ and for testing $15.329e-3$.

The comparison of results for training and testing is presented in Fig. 6. Example of predictions are shown in Fig. 7. For sunny days the model is more accurate because progress is near to clear production. It is obvious that accuracy of the model depends on the accuracy of the meteorological forecasting data.

Prediction for cloudy days are less accurate but the results are still close to the real power production and this information is valuable for optimizing plan of the appliances.

6 Conclusion and Future Work

This paper presents the solar power forecasting model based on the Recurrent Neural Network. The model was tested on the collected data and the obtained results are very promising. Final mean square error for training was $9.956e-3$ and $15.329e-3$

for testing. It is obvious that precision of the model is mainly affected by obtained meteorological forecasting data. Even though obtained results are very close to the real output production and the presented model is suitable for approximate solar power production forecasting. The whole model is now successfully implemented as a part of the Active Demand Side Management. In the future work the accuracy of the model will be improved by collecting bigger training dataset during longer time period.

Acknowledgments This paper was conducted within the framework of the project LO1404: Sustainable development of ENET Centre, Students Grant Competition project reg. no. SP2015/146, SP2015/170, SP2015/178, project LE13011 Creation of a PROGRES 3 Consortium Office to Support Cross-Border Cooperation (CZ.1.07/2.3.00/20.0075) and project TACR: TH01020426.

References

1. Bizzarri, F., Bongiorno, M., Brambilla, A., Gruosso, G., Gajani, G.: Model of photovoltaic power plants for performance analysis and production forecast. IEEE Trans. Sustain. Energy **4**(2), 278–285 (2013)
2. Majer, V.: Predikce vyroby elektriny z fotovoltaickych elektraren v libelarizovane energetice. Zapadoceska Univerzita v Plzni (2013)
3. Pearlmutter, B.A.: Dynamic recurrent neural networks. Technical report, Carnegie Mellon University (1990)
4. Rosenblatt, F.: Principles of neurodynamics; perceptrons and the theory of brain mechanisms. Spartan Books Washington (1962)
5. Rumelhart, D.E., Hinton, G.E., Williams, R.J.: Learning representations by back-propagating errors. In: Anderson, J.A., Rosenfeld, E. (eds.) Neurocomputing: Foundations of Research, pp. 696–699. MIT Press, Cambridge (1988)
6. Shi, J., Lee, W.J., Liu, Y., Yang, Y., Wang, P.: Forecasting power output of photovoltaic systems based on weather classification and support vector machines. IEEE Trans. Ind. Appl. **48**(3), 1064–1069 (2012)
7. Werbos, P.: Backpropagation through time: what it does and how to do it. Proc. IEEE **78**(10), 1550–1560 (1990)
8. Yuan-Kang, Wu: C.R.C., Rahman, H.A.: A novel hybrid model for short-term forecasting in pv power generation. Int. J. Photoenergy **2014**, 9 (2014)

An ICT Solution for Shared Mobility in Universities

Carlo Giglio and Roberto Palmieri

Abstract The aim of this study is that of depicting the novel architecture of a vehicle pooling platform—i.e. CityRide. Such a platform and, hence, this paper, adopt a broader amplitude compared to the traditional carpooling initiatives. In fact, this initiative is the first, to authors' knowledge, to include and to be dedicated to any kind of vehicle. The architecture of the platform is composed of a logistics optimization module (LOE) integrated with a social networking (SE) one. Such an integration is not available in other carpooling platforms so far, thus proving the innovative characteristic of CityRide. The platform has been designed by following a determinant-driven approach, which considers key social innovation user-related factors in order to identify and address the implementation of the main features of the SMOB project. The adoption of CityRide in a Southern Italian university acts as an application scenario. The study has been conducted with a qualitative research approach about carpooling associated with the SMOB project case study.

Keywords Carpooling · Vehicle pooling · Shared mobility · Social innovation · University application scenario · CityRide case study · SMOB project

1 Introduction Section

The emerging concept of Collaborative Consumption proves how consumers are eluding the behavioural patterns suggested by the current hyperconsumeristic society [1–5]. Such a novel resource consumption model originates and benefits from sharing, exchanging, negotiating or renting goods, thus avoiding to attribute any value to the concept of property. Although this philosophy was already put

C. Giglio (✉) · R. Palmieri
DIMEG Department, University of Calabria, Rende, Italy
e-mail: carlo.giglio@unical.it

R. Palmieri
e-mail: roberto.palmieri@unical.it

A. Abraham et al. (eds.), *Proceedings of the Second International Afro-European Conference for Industrial Advancement AECIA 2015*, Advances in Intelligent Systems and Computing 427, DOI 10.1007/978-3-319-29504-6_21

forward before the inception of the Consumerism, it has been strengthened by the major and emerging possibilities provided by the Social Technology, and by the current inescapable commitment linked to the global economic and environmental issues. In light of this, a consumer model based on trust and shared access has been developed and adapted to the present state of the economy and the society. This approach represents an alternative solution in order to re-design creatively the core of the society existing before the arrival of the Consumer Society. In this context, carpooling and comparable collaborative solutions are gaining increasing importance, since their number on the total of Collaborative Consumption initiatives is even more higher. Therefore, the general concepts and functionalities of carpooling platforms are worth to be analyzed. The study is realized by integrating a literature review about carpooling and the case study at hand. By a methodological point of view, the functionalities of the platform have been defined and implemented by analyzing the key social innovation factors concerning users needs. The application scenario of the University of Calabria in Southern Italy represents operating context of the platform.

2 Characteristics and Variants of Vehicle Pooling Services

By adopting the stakeholder-oriented value of innovation [6] slant, a set of widely used characteristics and functionalities has been identified. The outcomes of this first step of the design process are reported and discussed in the following.

Before starting a trip, carpoolers—i.e. drivers and riders—are informed about the people they will travel with and the time they will leave. The first step concerns with the driver's ride offer, which makes available his/her private vehicle, and the corresponding riders' demand, which is geared to allow riders to satisfy their need for a lift. Carpooling and, hence, vehicle pooling, are defined as sustainable mobility solutions, whose goal is to increase the occupancy rate of private vehicles by allowing more carpoolers to share the same trip. The transportation requirements of each participants must be the same, especially in terms of time and place of departure/arrival. The shared trip offers relevant advantages to both the environment and crew members as it is discussed below.

First, sharing trips means reducing the amount of vehicles circulating at the same time, thanks to the increase of the occupancy rate and, thus of the total number of carpoolers per vehicle. Vehicle pooling generates less pollution by decreasing the number of vehicles utilized and, hence, by limiting the emissions of dangerous substances for people and the environment. A further advantage of making a habit of carpooling is that private vehicles tend to be utilized with an increased occupancy rate in the long-term. Moreover, people tend to repeat such behavioural patterns, especially when they become aware that they could share trips, instead of owning a number of vehicles in the same. In the long-term, this may result in a decrease of the amount of cars, since people start thinking that it is not useful to maintain many of them. In turn, a lower number of vehicles would reduce the aforementioned negative environmental effects. Sharing trips means also lowering transportation costs, since fuel, tolls and parking fees would be shared among trip participants. This is an endogenous monetary incentive for carpoolers, since it is an intrinsic factor associated to the decision of taking part in the shared trip. Moreover, drivers and riders can change their role in each shared trip, thus making it possible to alternate the use of private cars. This results in less wear and tear issues for each vehicle. Again, it is a further endogenous monetary incentive, as it is directly linked to the participation in the ride. The reduction of such costs—e.g. amortization costs—tends to make cars more durable than in the past, thus avoiding a phenomenon of fast replacement and, hence, limiting the negative environmental impacts. Vehicle pooling also reduces the level of both physical and psychological stress, also due to the possibility to discuss for a certain mileage with travel companions having the same interests. The risk of accidents decreases because of the possibility for drivers to rest and alternate with each other, especially in long-distance trips. It represents a kind of endogenous non-monetary incentive, which ensures comfort and safety to carpoolers. This way, carpooling contributes to a relaxed trip with well-trained and non-stressed drivers. In turn, participants have a perception of increased safety and comfort, which results in a higher level of satisfaction. Moreover, the lower amount of cars circulating at the same time makes it easier to park, since more parking lots tend to be available on average. An improved availability of parking facilities, in turn, leads to a higher users satisfaction, both in urban and extra-urban contexts. In addition, the emerging attention to the environment may lead users to give importance to the green side of carpooling and to adopt a sustainable lifestyle, thus benefitting further the environment and the society as a whole. Vehicle pooling represents also a valuable alternative service for mobility of users in remote areas, which are usually not (sufficiently) covered by public or private transportation companies, due to the disadvantaged market scenario. Therefore, people living in remote areas could solve somehow their mobility problems. This way, carpooling systems prove to be a viable solution in order to limit social exclusion, while vehicle owners start utilizing more carefully and sustainably their private transportation means. Such a platform provides driver and riders with a more flexible way of moving compared to the traditional public system. Flexibility ensures users to make a decision together about the trip and its spatial and temporal constraints. This could never happen in the public transportation field, since timetables are planned in advance and routes are fixed. Flexibility affects positively also the pick-up/drop-off points during the trip. As a matter of fact, participants can agree to change the expected route in order to allow on-the-fly passengers to join the trip or to avoid traffic jam. An additional advantage of carpooling is related to the socialization among participants, which is another endogenous non-monetary incentive and adds value to their trip sharing experience. As a matter of fact, thanks to the social networking engine, carpoolers share their trips with people having the same interest, thus increasing the individual and overall user satisfaction.

However, it is important to mention also some issues inhibiting carpooling, as it is discussed in the following. Nonetheless, its flexibility compared to the public transportation, shared trips are somehow characterized by a rigid planning, which does not allow users to enjoy a totally flexible journey. Moreover, the online platform is exposed to risk of missing matching, It depends on whether the system is able to attract a high number of registered users and, hence, the required critical mass, or not. Based on this, the users have a higher or lower likelihood to find relevant trip offers/requests and to satisfy their needs, respectively. In the negative case, they may make the decision to not to use the platform again. People commuting to work or involved in long-distance trips may look for a higher level of flexibility while planning their trips. In fact, they may require flexible timetables and routes in order to interconnect with other transportation systems. Again, since carpooling is more flexible than public transportation, but still it requires some spatial and temporal constraints, users experience may prove not to be fully satisfying. Last, but not least, cultural and social differences may generate significant issues during shared trips. This may result in a negative user experience and, hence, tends to produce a word-of-mouth which may hinder the diffusion of the carpooling service.

Western countries like U.S.A., Canada, Germany, Austria, Portugal, Norway and Sweden. Locally-specific online carpooling systems developed distinctive functionalities, which determined a number of platform variants [1–5]. After the above discussion about general issues and features of carpooling systems worldwide, it is very problematic to find relevant contributions in literature concerning with specific application scenarios. Therefore, this study is based on the identification of the main clusters of such variants as it is described in the following. Corporate systems refer to carpooling in the private sector. The main goal of such systems is making it easier the commute of workers, which can be long and stressful, thus affecting negatively employees' performance. This way workers are more satisfied and also more productive, since they perceive a better human resource management and benefit from a value-added service. In addition, they have the opportunity to increase their socialization and team building skills by travelling with their colleagues. In the long-term, it proves to be beneficial also fro the company as a whole, since more satisfied individuals contribute to the achievement of its overall goal. University carpooling is a very peculiar variant, which is focused on providing students and academicians with an alternative solution for mobility. Therefore, such systems are geographically disperse, since they are available many continents, but their target is very restricted to primary university stakeholders. As in the case of corporate carpooling, students and academicians are provided with a value-added service, which increase their level of satisfaction and the overall appeal of the university, thus hiking up its talent attraction & retention rate. Moreover, university carpooling gives participants an additional possibility to expand their personal and professional network. Family carpooling is focused on those who want to utilize a lower number of vehicles within a family, thus reducing costs related to amortization and oil consumption. This is an endogenous non-monetary incentive to change families' behavioural patterns in favour of carpooling and to shift them towards a more sustainable approach. Random carpooling services include all the systems not listed above.

This cluster is characterized by the general pros and cons discussed in the initial part of this section.

Each carpooling variant cluster is characterized by some of the above mentioned functionalities and issues, as it is described in the following [1–5]. Family carpooling is not affected by the issues related to socialization and cultural differences, since trip members have already met each other and can share their interests also in other contexts, irrespective of the carpooling experience. Corporate and university systems, instead, tend to leverage such socialization opportunities in order to increase users satisfaction. Socialization is relevant also in random systems, where carpoolers do not know crew members and, hence, have a low level of trust towards their travel companions. Nonetheless, the role of the social networking engine is to face such issues thanks to the implementation of rating, reputation and recommendation sub-systems, which are geared to strengthen the level of trust among participants. The lack of trust led platform designers and managers to conceive some monetary incentive systems, which are geared to encourage users to utilize the platform. Monetary incentives are often utilized in university, corporate and random systems, since they are designed to compensate for the lack of social relationship among participants, which is not typical of family carpooling. An additional factor to be analysed is that of spatial flexibility, which is peculiar of random carpooling. In fact, it concerns with carpoolers with different needs and without pre-existing relationships. By contrast, family, corporate and university systems are targeted towards people having the same (or similar) interests and needs. Also socio-demographic issues are relevant for each cluster. As in the case of spatial flexibility, such an issue does not affect university and corporate platforms significantly, since their users are characterized by akin needs. Family-oriented platforms are targeted towards users with different socio-demographical configurations, thus such factors impact relevantly on the corresponding carpooling variant. Nonetheless, given the strong ties among family members, the above mentioned impacts are mediated by the existence of some interests and needs in common among family users. In random systems, differences tend to be more relevant, thus leading socio-demographical factors to play a key role in the corresponding platforms. Also the pool-size effects are relevant, especially in corporate- and university- oriented platforms, since such effects are related to the number of people participating in the trip. As a matter of fact, corporate and university systems target their services towards homogeneous sets of potential users, which are characterized by comparable transportation and socialization needs. In family and random carpooling contexts, which usually reach a lower occupancy rate, such effects are countered by the above mentioned hindering factors, especially in random systems. Also in family trips the occupancy rate tends to be below average because of the lower number of family members having the same interests and needs in order to share a trip. Also the temporal configuration is a relevant factor related to the shared interests and needs of participants. In particular, it is a key element for corporate- and university-related users, since time windows to commute to work or to attend university courses are widespread needs among such categories of carpoolers. In random carpooling services, the temporal configuration is fairly important, since travel companions make a shared decision about departure, pick-up/drop-off and arrival times. In family contexts, carpoolers are characterized by a more flexible time schedule about shared trips. An additional factor concerns with distance. In corporate, family and university contexts, users usually cover fairly short distances, since they live near their workplaces/universities. On the contrary, carpoolers utilizing random-oriented platforms cover above-average distances, thus making such a factor more relevant than in the other carpooling systems. Programming is more relevant for users of corporate and university platforms—mainly because of the compulsory deadlines—, likewise the temporal configuration. In fact, family members are quite flexible in terms of programming a trips altogether, while random systems target their services towards more heterogeneous segments of users, which still have to meet the deadlines about space- and time-related configuration of their rides. By considering the carpooling categories, corporate, random and university services are focused on more demanding users, which need to meet trip-related deadlines agreed with (usually) unknown members. Therefore, this users' characteristics affect significantly the functionalities of the resulting carpooling platform. On the contrary, family carpoolers tend to be more flexible and to leverage trust among members.

When it comes to analyze the market potential of a locally-specific vehicle pooling system, it is very important to conduct a research geared to determine the number of potential users, the distinctiveness of the geographic area, the available facilities and infrastructures, the behavioural, social and mobility patterns of potential users.

In light of this, it is essential to develop a close collaboration with public and private partners, like public administrations, public transportation companies, police, etc., in order to collect a lot of data and information about the local operating context. Therefore, such data and information can be used for a tailored design of platform architecture and functionalities.

The analysis depicted the urban and university area associated with the case study at hand as a residential zone characterized by a low occupancy rate of private vehicles and traffic flows such that specific destinations are clearly identified.

3 The CityRide Platform

The application scenario of this study is a urban context, which includes a university campus in Southern Italy. Its distinctive trait is the coexistence of a social networking engine and a logistics optimization module, which are geared to provide users with a comprehensive satisfaction of their mobility and socialization needs. In Fig. 1, the logical architecture of the online vehicle pooling system is represented. The core sub systems of the platform are associated with the above mentioned engines. In particular, the social networking one consists of a set of reputation management, social rating, recommendation and rewarding sub modules. Moreover, the social networking module is linked to widespread and well-known social

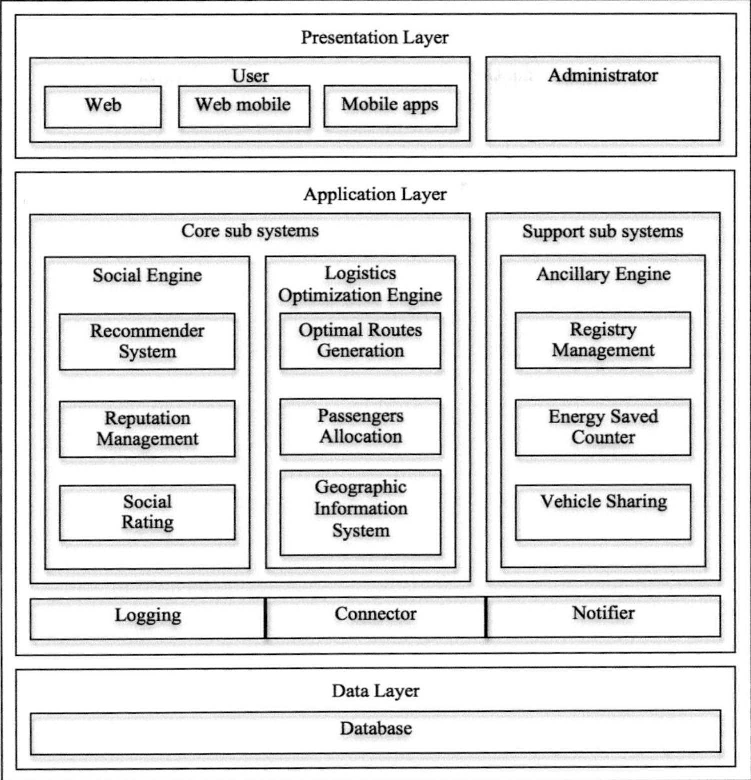

Fig. 1 Logical structure

networks such as Facebook and Twitter. The system is able to ensure users with a fast sign up thanks to the extraction of users' data from such social networks. This way it inherits the social relations from Facebook and Twitter and enhances a more rapid growth of the internal social graph. Such a sub system strictly cooperates with the logistics optimization one, which provides some additional services—like the calculation of the shortest path between two nodes, the visualization of georeferenced maps and routes, and the running of the implemented vehicle routing algorithms. The problem of the assignment is managed by generating routes from source to destination. An additional feature of the logistics sub system is related to the management of dial a ride services, especially from users with on-the-fly, urgent transportation needs, which requires a real-time response from the module. The software architecture model is multi-layered and can be accessed by desktop and mobile devices by means of the Internet. It adopts the inter-process communications widespread in many enterprise solutions. Internal communications among the different modules—e.g. the social networking and the logistics ones—are assured by a bus-based set of services of Switchyard JBoss 0.7.0—SY7 within the logistics optimization module.

Since the aim of this paper is to describe the novel architecture of the platform, but to analyze specific optimization techniques and algorithms implemented, thus this paper only focuses on the corresponding logical structure.

3.1 The Module of Social Networking

The social networking sub system consists of a set of reputation management, social rating, recommendation and rewarding sub modules, which are based on the architectural pattern MVC. The ESB and the MOM assures the communication interactions between the social networking component and the logistics one. A server side API model manages the interactions related to the Social Rewarding and Recommender Manager. The main scope of such a core sub system is to support and enhance the mechanisms managing the internal social graphs and, hence, to define the relationships associated with the trip sharing of carpoolers by means of the platform. The Social Rewarding component's goal is identifying and providing the virtuous users with monetary or non-monetary benefits. The main aim of the Recommender Manager is easing the decision-making process concerning with the problem of the assignment of users to vehicles.

3.2 The Module of Logistics Optimization

This sub section deals with the module geared to provide an optimal matching between passengers and vehicles, and to identify the cheapest routes for carpoolers. Such a sub system works totally separated from the other components of the platform and receives and sends data and information by means of a communication interface. It works thanks to the call of on-demand external services provided by open source frameworks. It is able to manage a set of constraints concerning with the logistics and social preferences of potential travel companions. As a matter of fact, some of such constraints come from the social networking component, namely from the recommendation and social rewarding sub module. Data and information considered by the logistics optimization engine are related to the following features: type of vehicle, ability to load passengers, ability to load luggage, time windows, vehicle for smoking or non-smoking users, minimum rating for passengers, favourite users, black list, etc. Based on such variables, the system is able to determine the most adequate vehicles, and the corresponding crews, to carpoolers. Users' requests are processed by means of advanced heuristic procedures and choose the most fitting trips among those satisfying the above mentioned constraints. The logistics optimization module is geared to ensure a high perceived quality to users, which provides a reliable assessment of the travel experience of each crew member. Such an evaluation of the perceived quality depends on different factors and metrics, as it is described in the following. The overall cost met by users is a key element, since carpoolers try to minimize it, but often this implies disadvantages in terms of trip duration or total mileage. The opportunity for drivers to share the cost of a trip is linked to the payment of an overall quota defined in advance and divided by the riders. This way it is possible to reduce the cost for drivers and to provide riders with a more flexible, comfortable and convenient transportation service than the public one. When it comes to meet deadlines, the duration of the trip plays a key role. In fact, users want to be dropped-off at the desired destination on time, while several factors may intervene and cause possible delays. An additional factor is the mileage of the trip. Covering the minimum distance from source to destination often is the choice of carpoolers. Nonetheless, this does not ensure the minimum travel time. Travelling with companions sharing the same interests and needs is an important factor in order to provide the highest possible users satisfaction. Such a metric allows to perform an overall evaluation of the reliability of crew members and is divided by three sub metrics, namely reliability and compliance with commitments, punctuality, and compliance schedules. In addition, it is very important to deal with some constraints, which affect the algorithmic procedures of the logistics optimization engine. In particular, temporal flexibility is a value-added feature for carpoolers, especially for those wanting to modify the duration of the trip. Nonetheless, meeting the deadlines is equally important for users, especially when it comes to respect-t work and university-related commitments. Therefore, it is important to define a limit for changes in the duration of the trip. This way adding or removing passengers, hence, modifying the route and the travel timetable, results in changes of the individual rides and the schedule of daily commitments of each carpooler. Moreover, changes in terms of the distance covered by a trip affect significantly the quality perceived by users, likewise the above mentioned temporal modifications. Therefore, a limit on the change of the total distance is beneficial for the majority of carpoolers. The outcomes provided by the heuristic procedures of the logistics optimization module are sent to an external sub module, which is in charge of managing trip requests/offers, their status updates and the calculations associated with the performed trips. In detail, it considers the set of deals, the set of travel requests assigned to a vehicle and confirmed, the set of travel requests assigned to a vehicle, but not confirmed, the set of travel requests unassigned. In summary, the advanced services and algorithmic solutions provided by the logistics sub system are geared to optimize the routes chosen by carpoolers taking part in trips arranged privately, to manage unconfirmed trip demands/offers by satisfying the corresponding constraints, provide on-the-fly solutions in order to satisfy urgent transportation needs of users, visualize routes by means of interactive maps, quantify CO_2 saved for each trip and segment of trip, sum up CO_2 grouped by geographic and administrative regions.

4 Conclusions

This paper provides field scholars and mobility managers with an advanced architecture and the corresponding features of a vehicle pooling platform. To authors' knowledge, it includes the most advanced functionalities and discusses the hottest topics in terms of concept development, system requirements and industrial research issues. The description of the free and open source-based design approach for alternative shared mobility solutions highlights a more comprehensive perspective adopted in this study compared to the traditional one in this field. One of the most relevant innovations is the integration of the two core sub systems, that is the social networking and the logistics optimization engines. By means of such modules it is possible to ensure a higher users satisfaction, since carpoolers perceive a value-added service provided by both the optimization of every aspect of their trip and the improved socialization experience. As a matter of fact, the social networking sub system suggests the most adequate travel companions depending on the user preferences stored in the platform database. On the other hand, the logistics optimization module is in charge of improving the users experience thanks to the algorithmic procedures geared to reduce the overall travel time and/or cost. Moreover, the platform is endowed with an advanced virtual credits mechanism. It is intended to provide a further incentive towards the adoption of novel behavioural patterns, which favour the diffusion of the shared-transportation service of the platform. It is also an incentive towards a more environmentally friendly mobility approach. Also the adoption of a sub module for the management of real-time requests and the geo-referencing and geo-locating functionalities are among the most advanced features of such a platform. As for the data coming from the locally-specific analysis, the current occupancy rate is about 1.3 users per private vehicle [2]. Therefore, the diffusion of such an alternative transportation service provided by the platform is very promising and represents an interesting field test for mobility managers. The innovative architecture and features of the platform at hand address many and relevant research topics for the scientific community in this field. In addition, this work provides also a novel design process in order to develop advanced online vehicle pooling systems. In fact, the design approach consists of the analysis of the general and locally-specific determinants, factors and constraints pertaining to the planning, the development and the application in operating contexts of vehicle pooling initiatives. Moreover, such an approach emphasizes the key social innovation user-related factors and it is among the small number of cases worldwide to develop carpooling platforms by adopting a social innovation and non-profit slant. The relevance of locally-specific characteristics related to the application scenario at hand impacts on the possible generalization of the outcomes of this study, since some specific features of this platform can not be considered for the implementation in initiatives operating in different contexts. By a research perspective, there also some limitations due to the lack of contributions in literature pertaining to comparable application scenarios. Therefore, this paper considers mainly some benchmarking studies and competitor analyses, which come out form the industrial research efforts of the SMOB project. In summary, the relevance of the locally-specific scenario at hand and the lack of adequate contributions in this niche literature lead not to consider the very general contributions in literature, which do not deal with the most relevant factors pertaining to the subject at hand. Moreover, such general works currently available in literature are not too many, since the majority of online carpooling platforms are very recent. Although carpooling is a well-consolidated mobility solution, it is still difficult to find a large number of academic researches in this field. Field scholars started working regularly on it when the Sharing Economy and the corresponding consumption patterns spread in many Western societies. Further limitations are related to the significant architectural differences of the system at hand from other existing carpooling platforms and to the hither-to unexplored design approach, which is based on the key social innovation user-related factors. In light of this, the authors did not consider the majority of the few existing studies in this field. Finally, an additional topic concerns with project sustainability. In detail, both administrative and private partners should contributing in order to make the virtual credits mechanism viable in the long-term and to help going beyond possible cultural barriers.

Acknowledgments SMOB project (PON04a3_00164) and CityRide platform have been funded by the European Union and the Italian Ministry of Education, University and Research.

References

1. Ferreira, J., Trigo, P., Filipe, P.: Collaborative Car pooling system. World academy of science. Eng. Technol., Int. Sci. Index 30, **3**(6), 365–369 (2009)
2. Giglio, C., Carrozzino, G., Ceravolo, R., Cosma, A.M.I.: Social MOBility—SMOB. Italian Ministry of Education, University and Research—MIUR (2012)
3. Giglio, C.: Profiling the Innovation Management discipline: a comprehensive perspective focused on creativity-based vs. knowledge-based strategies. Ph.D. Thesis, University of Calabria (2014)
4. McKinsey&Company: The social economy: Unlocking value and productivity through social technologies. McKinsey Global Institute (2012)
5. Tafaro, S.: Modelli di e-business e analisi di servizi per sistemi di carpooling. MScEng Thesis, University of Calabria (2012)
6. Palmieri, R., Giglio, C.: Seeking the stakeholder-oriented value of innovation: a CKI perspective. Measuring Business Excellence (ISSN 1368–3047), vol. 18, no. 1, pp. 35–44 (2014)

Simulated Annealing Approach for Solving the Fleet Sizing Problem in On-Demand Transit System

Olfa Chebbi and Jouhaina Chaouachi

Abstract Over the last years, operating expenses for on demand transit system have been increased as the demand for this type of transportation service has expanded. The on-demand transit system that we studied consists on moving a set of driverless electric taxi with bounded battery capacity. Many management algorithms have been proposed to improve the efficiency of such a system. In this paper, we propose to deal with the problem of determining the optimal fleet sizing of driverless electric taxis under a known transportation demand. We present a Simulated Annealing to solve the proposed problem. Evidence for the efficiency of our algorithm is proposed where computational results prove that our algorithm provide good quality solutions.

Keywords Routing · Simulated annealing · Fleet sizing · On-demand transit system · Transportation

1 Introduction

1.1 Background of the Paper

Automated Transit Networks (ATN) is a system of small, driverless electric vehicles that could carry small groups of people up to six persons. Transportation service in ATN is done non-stop between pairs of stations in a network of exclusive guideways [1]. The guideways could be elevated or at grade level. On the other hand, the stations are all "off-line". This feature allows vehicles to travel non-stop to their final destination.

O. Chebbi (✉)
Institut Supérieur de Gestion de Tunis, Université de Tunis,
41, Rue de la Liberté - Bouchoucha, 2000 Le Bardo, Tunisia
e-mail: olfaa.chebbi@gmail.com

J. Chaouachi
Institut des Hautes Etudes Commerciales de Carthage, Université de Carthage,
IHEC Carthage Présidence, 2016 Tunis, Tunisia

A. Abraham et al. (eds.), *Proceedings of the Second International Afro-European Conference for Industrial Advancement AECIA 2015*, Advances in Intelligent Systems and Computing 427, DOI 10.1007/978-3-319-29504-6_22

ATN provides personalized transportation service (like a taxi) [2]. Therefore, it is expected to be highly attractive to its users. This concept of transportation service is effective from both economic and environmental perspective and can be used as a complement to other transportation service. Economic advantages includes a reduced operational costs, efficient use of land as well as a high level of on-demand transportation service. Environmental advantages includes a reduced noise and environmental pollution due to the use of small electric cars.

ATN vehicles run on exclusive guideways. The guideways are designed in order to eliminate any interference with other transportation modes. Thus, ATN would reduce congestion on urban roads and contribute on increasing the sustainability of urban areas. ATN have been implemented in many real case applications such as in Korea [3], Sweden [4] and United Arab Emirates [5].

1.2 Motivation of the Proposed Problem

This paper focus on the fleet sizing of an ATN system. In fact, we deal with a specific routing problem related to the strategic level of decisions. Our aim is to find the best size of ATN'fleet in order to satisfy a set of transportation requests in a static deterministic setting. Earlier work on fleet sizing for ATN was proposed by Li Jie et al. [6] without considering battery issues of ATN'vehicles. Battery constraints involve that a small set of the ATN' fleet of vehicles would not be available for service as they are charging their battery. That is why, optimizing of the fleet size of an ATN system is of a high importance for such an intelligent system.

1.3 Contribution of the Paper

In this paper we:

1. Proposes a routing problem related to ATN considering several practical constraints simultaneously, i.e., battery capacity, time windows, waiting time of transportation requests while aiming to reduce the total fleet size of vehicles.
2. Design an effective metaheuristic algorithm to solve the problem. Although such a combinatorial optimization problem can be solved by Cplex[1] in some cases under small size context, it is not possible to obtain good quality solution by Cplex for large instances size in a small computational time. That is why, an heuristic approach based on simulated annealing metaheuristic(SA) is developed to obtain good quality solution **in a reasonable computational time**.
3. Provide tests to our method by applying it to benchmarks instances from the literature.

[1]Source: http://www-01.ibm.com/software/commerce/optimization/cplex-optimizer/.

1.4 Outline of the Paper

The remainder of this paper is organized into four sections. Section 2 describes our problem definition based on the treated several constraints. The subsequent section presents the proposed solution method based on the SA algorithm. Section 4 presents the computational experiments of our algorithm and discusses the obtained results in the context of the hypotheses. The final section summarizes our findings.

2 Problem Definition

The ATN problem studied in this paper is a public transportation problem that aims to serve a set of ATN'users transportation request under battery and time windows constraints. In this context, we extend the problem definition presented in [7], [8].

Let suppose to have a ATN system with a specified ATN network of a fully connected guideways. Let us suppose to have M station and one depot. In what follows, we suppose that we have a list of ATN passengers request T. Each passenger request i is characterized by departure station Ds_i, departure time Dt_i, arrival time At_i and arrival station As_i. Our ATN routing problem focus on minimizing the fleet size to satisfy all the available trips without any delay with respect to the battery and time windows constraints. In addition, we suppose that ATN pods(vehicles) need to visit the depot whenever it is necessary to load their batteries. The ATN routing problem can be described briefly as follows. Let $G = (C, E)$ be a directed graph where $C = \{c_0, c_1, \dots, c_n\}$ is a vertex set, and $E = \{(c_i, c_j) : i \neq j\}$ is an arc set. Vertex c_0 denotes a depot at which an unlimited number of identical ATN pods are based. The number of used ATN vehicles will be minimized and used as our objective function. The remaining vertices of C represent ATN' user travel requests. Each arc (c_i, c_j) has an associated nonnegative electric consumed energy c_ij and a nonnegative travel time t_ij. The ATN fleet sizing problem consists of designing a set of least cost vehicle routes such that:

- **a** Every route starts and ends at the depot.
- **b** Every trip is visited exactly once by exactly one vehicle.
- **c** The total demand of any vehicle route does not exceed the battery capacity of the vehicle B.
- **d** Each trip has a specific departure time, arrival time and departure station. If the ATN vehicles arrive at a station to satisfy a trip x before the departure time of x, the ATN vehicle should wait at its current location until the trip x is trigged.

$V^* = V \setminus c_0$ Additionally, the total consumed electric energy of any route can not exceed the battery capacity B. The set of edges E will be defined by the following rules:

- for each nodes i and j in V^* we add an arc $(i; j)$ only if: $j > i$ and $At_i + Sp(As_i; Ds_j) < Dtj$ (the cost of this arc is noted by c_{ij} and it represents the energy consumed from arrival station of trip i (As_i) to arrival station of trip j (As_j)).
- for each node i we add an arc $(0; i)$ (the cost of this arc is c_{0i} and it represent the energy used from the depot to the arrival station of trip i).
- for each node i we add an arc $(i; 0)$ (the cost of this arc is c_{i0} and it represents the energy used from the arrival station of trip i to the depot).

$E^* = \{(c_i, c_j) : i \neq j \text{ and } i, j \in V^*\}$.

From this problem definition, we can see that our problem is closely related to the asymmetric distance-constrained vehicle routing problem (ADCVRP) [9]. The distance constraint is imposed by the battery capacity in this case. Our problem is NP-hard, and is asymmetric because the distance from node i to node j is different to the distance from node j to node i.

3 Simulated Annealing as a Solution Approach

Simulated annealing is an iterative stochastic computational technique that finds non-optimal solutions to combinatorial problems. The simulated annealing method is inspired by the process of annealing metals in order to get the lowest free energy state. Annealing is widely in use in metallurgy. In this process, the material is heated so that its molecule gain energy and start to move freely. Then, the temperature is decreased gradually in steps until the material reach its equilibrium. In each step, the temperature and the total material energy are decreased [10].

The mathematical simulation of this process is done in a fashion where the molecule are considered to be decision variable of our problem. Each solution is conceived as a special arrangement of the molecule (aka decision variables). The energy level corresponding to a specific molecule arrangement is considered to be the objective function [11]. The simulated annealing is simple and powerful heuristic method that has proven to be efficient in solving different types of optimization problems (discrete, continuous, integer, etc.). The method has the main advantage of accepting some degrading solutions in order to escape from local minimal solutions which is considered as the main drawbacks of many local search algorithms [12].

In a more formal way, this overall annealing process is used and mimicked by the simulated annealing search algorithm to solve optimization and combinatorial problems. The main feature of this algorithm is its ability to accept solutions that could be worse than the actual solution in order to escape from local minima, and therefore locate and reach the global optimal solution. The simulated annealing algorithm accepts worse solutions following the Boltzmann probability [11]:

$$probability(P) = exp(\frac{-\Delta E}{K_B T}), \tag{1}$$

Algorithm 1 SA Procedure

$r = 0$
$X_{best} = \oslash$
Generate X_0 Randomly
$X_{best} = X_0$
while $(R \leq MT) AND (T_r \geq 0)$ **do**
    $n = 0$
    while $n \leq EL$ **do**
        Choose a Neighborhood Operator
        $X_n \longrightarrow X_{new}$
        $\Delta = C(X_{new}) - C(X_{best})$
        if $\Delta < 0$ **then**
            $X_{best} = X_{new}$
            $X_n = X_{new}$
        else
            y=RAND(0, 1)
            $Z = e^{-\Delta/T}$
            if $y < Z$ **then**
                n=n+1
                $X_n = X_{new}$
            end if
        end if
    end while
    r=r+1
    $T_r = T_{r-1} - \alpha T_{r-1}$
end while

where ΔE is the difference between the fitness values of the current solution and the candidate solution, T is a control parameter that refers to a temperature that regulates the search process, and K_B is the Boltzmann constant. In general, at the start of the simulated annealing algorithm, any move within the search space is acceptable. This helps the algorithm to discover the search space in a more extensive way. The control parameter T is then decreased, making the algorithm more selective in accepting new solutions. This is done iteratively until we reach a state where only moves that improve the best solution are accepted.

According to this formal description of our algorithm (see Algorithm 1), three main components have to be defined in order to use SA for solving our problem: the objective function, neighborhood operators, temperature control.

In the next we present the main components of our SA algorithm adapted to solve the presented problem.

3.1 The Objective Function

As an objective function, we use and adapt the route first cluster second principle used for classic vehicle routing problem (VRP) [13] in order to solve our proposed fleet sizing problem. For that purpose, the Split function of Prins [13] is applied to our context. This method builds feasible subtours in an auxiliary graph in order to cover all the transportation requests. Then, a shortest path algorithm is used in order to get the best feasible option that determine the evaluation of a given solution. More details could be found in [13]

3.2 Temperature Control

The SA escape from the premature convergence that generally characterize local search algorithms by using a specific temperature control process. This process represents one of the unique and essential feature of the SA algorithms. It determines the rate of decreasing the temperature parameter in our algorithm. The temperature T is decreased as follows:

$$T_i = Ti - 1 * 0.95 \tag{2}$$

Therefore, T is decreased by 5 % in each step of the algorithm.

3.3 Neighborhoods Operator

The neighborhood operator used in our algorithms are the exchange mutation (see Fig. 1), the displacement mutation (see Fig. 2), and the insertion mutation (see Fig. 3).

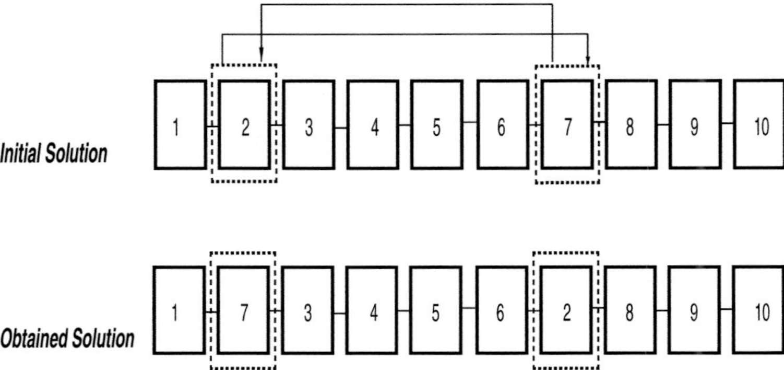

Fig. 1 Exchange mutation

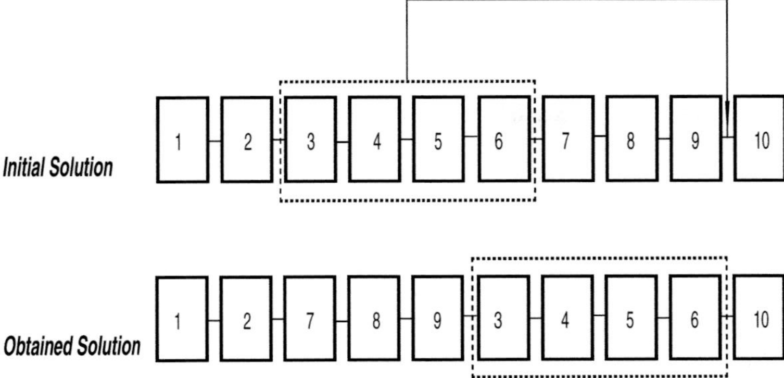

Fig. 2 Displacement mutation

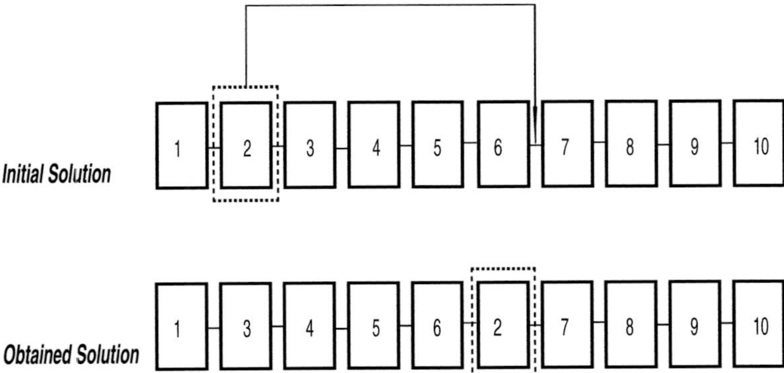

Fig. 3 Insertion mutation

3.4 Problem Specific Initial Solution

In this section, we present an heuristic for the problem considered above.

We first solve the following relaxed linear program:

$$\textbf{ATN} : \text{Minimize} \sum_{i \in V^*} x_{0i}$$

$$\sum_{j \in \delta^+(i)} x_{ij} = 1 \forall i \in V^*$$

$$\sum_{j \in \delta^-(i)} x_{ji} = 1 \forall i \in V^*$$

$$x_{ij} \in \{0, 1\} \forall \ (i,j) \in E$$

However by solving this linear program, we can get infeasible tour which consume more energy than what the Battery allows. To fix this dilemma, we propose a construct a permutation of trips using the different obtained roads. Then, we use the evaluation function to get feasible solution and the perfect evaluation starting from this obtained permutation. By doing so, we ensure to get a good quality solution obtained by linear programming techniques that could enhance the quality of the SA algorithm.

4 Computational Results

In this section, we describe the experiments made to apply our algorithms to a set of problems generated randomly. For each instance, tests are made using a program coded in C++ and simulations are performed on a computer with a 3.2 GHZ CPU and 8 GB of RAM. All the mathematical models were solved using Cplex 12.2 commercial solver.

4.1 Test Problems

We test our algorithm on 19 different classes for each size $n \in \{10; 15; 20; 25; 30; 35; 40; 45; 50; 55; 60; 65; 70; 75; 80; 85; 90; 95; 100\}$ 40 instances were generated.

We base our instance generator on the one proposed in the literature for solving a closely similar problem to reduce the energy'cost [7, 8].

So, in general we did tested the proposed heuristics on 760 different instances.

For the evaluation of results we used the GAP witch is defined as follow:

$$GAP = (\frac{S_{Heuristic} - LB}{LB}) * 100$$

- $S_{Heuristic}$ is the solution of the heuristic.
- LB is the minimum between our SA and the results of a valid mathematical formulation to solve this problem [14].

The SA parameters were sets as follows:

- $EL = 50$
- $MTT = 200$
- $\alpha = 0.95$
- $T_0 = 5.0$

Table 1 Result of the SA

Statistics	GAP (%)	Time (s)
Minimum	0	1
25 % percentile	17.21	1.965
Median	26.32	2.665
75 % percentile	33.33	3.703
Maximum	50	8.26
Mean	24.82	3.114
Std. deviation	12.15	1.58
Std. error of mean	0.8815	0.1146
Lower 95 % CI of mean	23.08	2.888
Upper 95 % CI of mean	26.56	3.34

4.2 *Result of the Heuristic*

Table 1 presents the results of our heuristic. Table 1 presents descriptive statistics for the GAP in % and computational time in seconds. We should note from our results the goods quality of the obtained solutions. Also, our method is fast as it founds its solution in an average time equal to 3.11 s. Especially for large instances our method found feasible solution in less than 9 s. Therefore, we should note that our approach is successful as our primary objective was to find feasible solution in a small computational time.

5 Conclusion

This paper address a problem routing related to ATN. Our problem aim to minimize the fleet size of ATN while respecting the battery capacity of the vehicles. We developed a metaheuristic based on SA which was proved to provide good results for various sizes of ATN instances 'size. Our metaheuristic include specific linear programming based initial solution and specific evaluation function. The SA were tested on the randomly generated instance with specific structure and include passenger request up to 100 trips. These tests shows that the initial linear programming solution included in our algorithm brings a significant improvement in term of deviation of lower bound. We should note also that on all instances, our algorithm have very reasonable running time. To extend this approach, future directions of this research could consist on the development of tight lower bounds to better evaluate our metaheuristics. We could also develop a more sophisticated metaheuristic that found good upper bound for very large instance'size.

References

1. Anderson, J.E.: A review of the state of the art of personal rapid transit. J. Adv. Transp. **34**(1), 3–29 (2000)
2. Zheng, H., Peeta, S.: Network design for personal rapid transit under transit-oriented development. Transp. Res. Part C: Emerg. Technol. (0) (2015)
3. Suh, S.D.: Korean apm projects: status and prospects. In: Proceedings of the 8th International Conference on Automated People Movers, San Francisco, CA. (2001)
4. Tegnér, G., et al.: Prt in sweden: from feasibility studies to public awareness (2007)
5. Mueller, K., Sgouridis, S.P.: Simulation-based analysis of personal rapid transit systems: service and energy performance assessment of the masdar city prt case. J. Adv. Transp. **45**(4), 252–270 (2011)
6. Li, J., Chen, Y.S., Li, H., Andreasson, I., van Zuylen, H.: Optimizing the fleet size of a Personal rapid transit system: a case study in port of rotterdam. In: International Conference on Intelligent Transportation, pp. 301–305 (2010)
7. Mrad, M., Hidri, L.: Optimal consumed electric energy while sequencing vehicle trips in a personal rapid transit transportation system. Comput. Ind. Eng. **79**, 1–9 (2015)
8. Mrad, M., Chebbi, O., Labidi, M., Louly, M.A.: Synchronous routing for personal rapid transit pods. J. Appl. Math. **2014**(1), 1–8 (2014)
9. Toth, P., Vigo, D.: The vehicle routing problem. SIAM monographs on discrete mathematics and applications. Society for Industrial and Applied Mathematics (2002)
10. Lin, S.W., Yu, V.F., Lu, C.C.: A simulated annealing heuristic for the truck and trailer routing problem with time windows. Expert Syst. Appl. **38**(12), 15244–15252 (2011)
11. Kirkpatrick, S., Vecchi, M., et al.: Optimization by simmulated annealing. Science **220**(4598), 671–680 (1983)
12. Humberto César, B.d.O., Germano Crispim, V.: A hybrid search method for the vehicle routing problem with time windows. Ann. Oper. Res. **180**(1), 125–144 (2010)
13. Prins, C.: A simple and effective evolutionary algorithm for the vehicle routing problem. Comput. Oper. Res. **31**(12), 1985–2002 (2004)
14. Kara, I.: Two indexed polonomyal size formulationsfor vehicle routing problems. Technical report. Baskent University, Ankara/Turkey (2008)

Fuzzy Logic Based Human Activity Recognition in Video Surveillance Applications

Slim Abdelhedi, Ali Wali and Adel M. Alimi

Abstract Automatic fall detection using computer vision is a particular case for real time video analysis, efficient in kindergartens. This paper is focused on the design and implementation of a Human activity analysis system. The multiple cameras sends captured frames to the monitoring system via the local network. Through the use of human silhouette, acquired from a smart camera, a shape representation of the human beings was built in real-time and a fuzzy logic inference system was developed for fall detection. The system also allows tracking and localizing children within an authorized area. The alarm is triggered in case of transgression. Experimental results prove that the fuzzy inference system is efficient.

Keywords Video surveillance · Fall detection · Human behaviour analysis · Fuzzy logic

1 Introduction

Children in kindergartens are prone to the risk of falls and the outcoming injuries might be serious. Therefore, it is essential to monitor their movement in a real time in order to prevent injuries and intervene to provide the necessary help as soon as possible. In this work, we introduce an intelligent fall detection system based on video surveillance and background subtraction technique.

The proposed system allows a good detection, activity analysis and monitoring of children in a kindergarten, and triggers alarms when a child leaves the permitted

S. Abdelhedi (✉) · A. Wali · A.M. Alimi
REGIM: REsearch Groups in Intelligent Machines, National Engineering School of Sfax (ENIS), University of Sfax, BP 1173, 3038 Sfax, Tunisia
e-mail: slim.abdelhedi.tn@ieee.org

A. Wali
e-mail: ali.wali@ieee.org

A.M. Alimi
e-mail: adel.alimi@ieee.org

A. Abraham et al. (eds.), *Proceedings of the Second International Afro-European Conference for Industrial Advancement AECIA 2015*, Advances in Intelligent Systems and Computing 427, DOI 10.1007/978-3-319-29504-6_23

zone whether indoors or outdoors. Instead of using a physical fence, the system uses virtual fences positioned within each camera image. For the surveillance of a large area, one or more cameras are installed and a kindergarten video dataset [1] is used.

Our work deals with the design and implementation of an intelligent security system using the multi-camera and OpenCV library.[1] This approach allows to prevent children falls and their possible adverse results. In fact, it allows taking every precaution to inform the staff of the real-time children status. A video surveillance cameras were placed in different zones so as to cover the whole institution venue and video sequences were continuously logged. The falls and non-falls were respectively detected using automated algorithms to process camera board video frame in real kindergarten environments.

The system has the capability of detecting the motion of a human and their localization by the background subtraction algorithm [1]. When the moving human is detected, the system can classify them as a fall or no-fall [2]. If there is fall detection, the system sends out an alerting signal to the staff of a kindergartens via the Local Area Network (LAN).

Many studies have been done on human activities recognition, especially on fall detections based on computer Vision. Foroughi et al. [2] developed an approach for fall detection using a combination of the Eigen Space method an integrated time motion images. In [3] an approximated ellipse was used around the human body for shape change. In [4], an Omni-Camera called MapCam was employed to capture images and detect falls using the Background Subtraction method. Miaou et al. [5] developed a fall detection system based on foreground extraction and height over width parameters.

This paper is arranged in 2 sections. Section 2 provides a proposed system architecture using multi-camera. Section 3 presents the image processing techniques and explains the basics of the fuzzy inference engine used to classify human action and the obtained experimental results.

2 Proposed System for Kindergarten Video Surveillance

For kindergarten safety, the area is covered by multiple cameras. Our system is capable of tracking persons and determining their location and activity. It detects breaches to secure the zones, unauthorized activity, falling in specific areas and perimeter intrusion then generates an alarm. All videos frames or control signals are transferred over the network in real time for control and visualization.

The human detection system is based on a IP cameras attached to the switch port at FastEthernet speeds and monitoring system [6]. In a LAN network, the IP camera is attached to the switch port and records video sequences continuously and detects human motion in the kindergarten rooms. When the security system detects an abnormal behaviour, the monitoring system sends an alarm indicating the breach location.

[1]http://sourceforge.net/projects/opencvlibrary/.

The Emergency Alert Notification System (EANS) is used to alert staff by sending warnings via text messages and e-mail messages including the video and image of the children in case of emergency. The system allows to access all information of children in kindergartens from anywhere at anytime using the web server, and perform real-time acquisition.

3 Image Processing

Many techniques are used to extract the background model depending on the video scene. Extracting moving objects is difficult due to many factors, such as motion changes and frame moving. The fall detection system process can be described in three steps: first the background is modeled by the Type-2 Fuzzy Gaussian Mixture Models (T2 FGMMs) algorithm presented by Abdelhedi et al. [1] for each input frame, then the object is segmented and extracted, and finally the human activity is analyzed by a fuzzy inference engine. The implemented Fuzzy logic controls the parameters of the output and sets fuzzy rules for human activity analysis. The image processing block is illustrated by the flowchart in Fig. 1.

3.1 Background Subtraction for Motion Detection

To extract an efficient silhouette segmentation, the Type-2 Fuzzy Gaussian Mixture Models (T2 FGMMs) algorithm [1] was applied to extract foreground images. Our goal was to create a robust activity analysis and an adaptive tracking system [7] for the video surveillance domain [7–10]. In many security surveillance systems [11, 12], several problems give false foreground detections. For this reason, we used morphological operators and filtering methods to obtain more accurate results. The application of a median filter [1] consists on eliminating small size foreground regions. In this context, the essential morphological operations are used to reduce the noise and remove undesirable elements in a video frame.

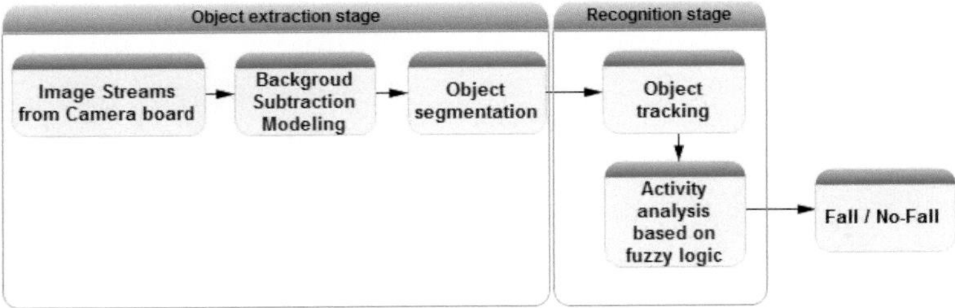

Fig. 1 Block diagram of the proposed surveillance system

3.2 Human Detection and Tracking

Our human behaviour analysis system is based mainly on measuring the information of the human silhouette. The principal idea is the shape extraction of the person and the orientation to distinguish their state. We used the bounding box dimensions (Width (W) and Height (H)) corresponding respectively to the width and height of human silhouette to detect falls. Three conditions are calculated as follows:

- If human is in a vertical position, the height of the human silhouette is bigger than the width ($H > W$).
- If human is in an in-between position, the height of the human silhouette is equal to the width ($H \approx W$).
- In case of an horizontal position, the height of the human silhouette changes to a value smaller than the width ($H < W$).

Combined with this method, we determined an approximated ellipse in the bounding box, and its orientation. Falls are detected by analyzing the orientation of the ellipse as well as the ratio of the major and minor axes of the ellipse.

- If human is in a vertical position, the orientation $0 < \Theta < 50$.
- If human is in an in-between position, the orientation $40 < \Theta < 60$.
- In case of an horizontal position, the orientation $50 < \Theta < 90$.

Figures 2 and 3 illustrate the proposed method, using dataset in [13].

Fig. 2 Extraction of fall features: Width (W), Height (H) and ellipse Orientation. **a** Behaviour class = Walk. **b** Behaviour class = Fall. **c** Behaviour class = Fall

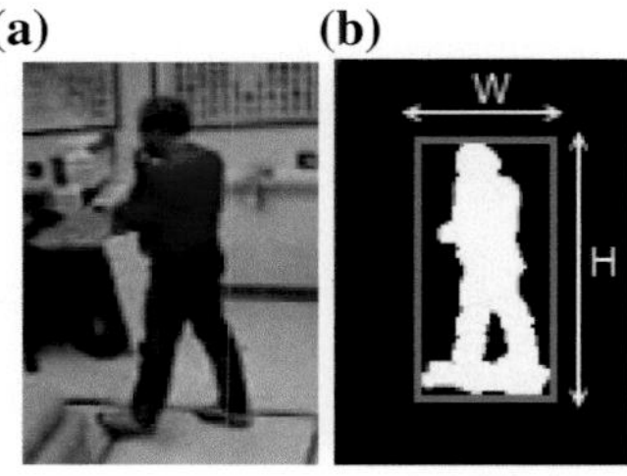

Fig. 3 Width (*W*) and height (*H*) measurements of the human silhouette. **a** Original frame. **b** Silhouette

3.3 Design of a Fuzzy Logic Controler

Fuzzy logic is typically used to ease the modeling and processing of complex tasks by allowing uncertain states. A fuzzy system [14] can be used to determine whether a fall has occurred or not.

Falls are detected by analyzing the orientation of the ellipse as well as the ratio of the major and minor axes of the ellipse.

To determine the membership functions, 2 input variables are used (Human Orientation (O) and bounding box dimensions (Width (W) and Height (H))) and 1 membership function for the output stage for defuzzification (Human behaviour (HB)). After a series of essays, the triangular membership functions were employed because it led to the most accurate results as compared to other alternative membership functions. In our work, we applied Mamdanis fuzzy inference method [15].

In our fuzzy rule system, we used as an input the measurements of the human silhouette calculated as shown in Fig. 3.

In this work, the Mamdani fuzzy rule system uses the four rules below:

- If Orientation is Vertical and Dimensions is Bigger then Human behaviour is No-fall,
- If Orientation is Horizontal and Dimensions is Middle then Human behaviour is Fall,
- If Orientation is In-between and Dimensions is Middle then Human behaviour is Fall,
- If Orientation is Vertical and Dimensions is Middle then Human behaviour is Fall.

Figure 4 illustrates the membership functions:

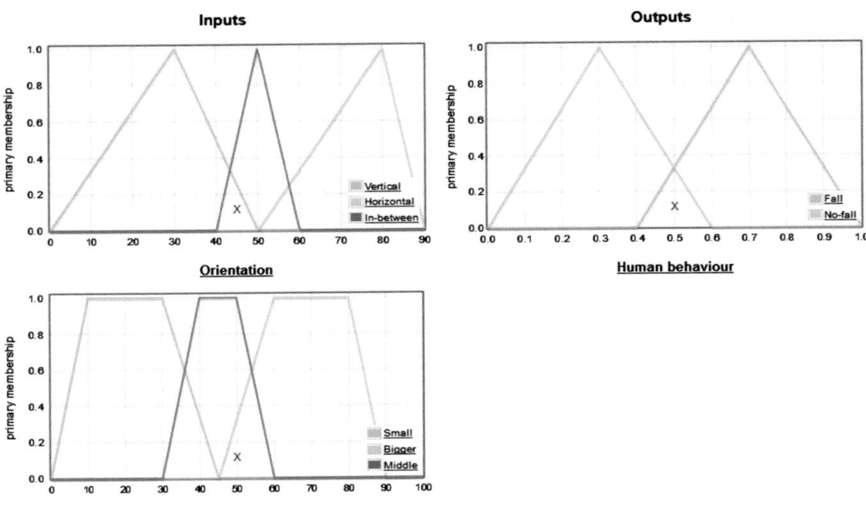

Fig. 4 Input 1: Dimensions (%)—The bounding box dimensions (Width (W)/Height (H)), Input 2: The orientation (Degrees), Output: Human behaviour (Fall or No-fall)

4 Experimental Results

The dataset was designed to be more realistic, natural and challenging for a video surveillance domain. Data was collected in natural scenes showing humans performing normal actions in standard contexts, with controlled and cluttered backgrounds [1]. The video sequences were made of foreground objects (children in kindergarten) moving over a real background. To access the efficiency of the designed system, the used dataset [13] contained 24 scenarios recorded with 8 IP video cameras. The first 22 first scenarios contained fall and confounding events, the last 2 ones contain only confounding events.

Figures 5 and 6 illustrate the experiment results of silhouette extraction using, respectively, dataset [13] and [1]:

For measuring the accuracy of the our methodologies [1], three different metrics have been adopted as an evaluation method for fall detection algorithms. These two values are computed as follows:

$$Recall = TP/(TP + FN) \tag{1}$$

$$Precision = TP/(TP + FP) \tag{2}$$

$$Accuracy = TP + TN/(TP + FP + TN + FN) \tag{3}$$

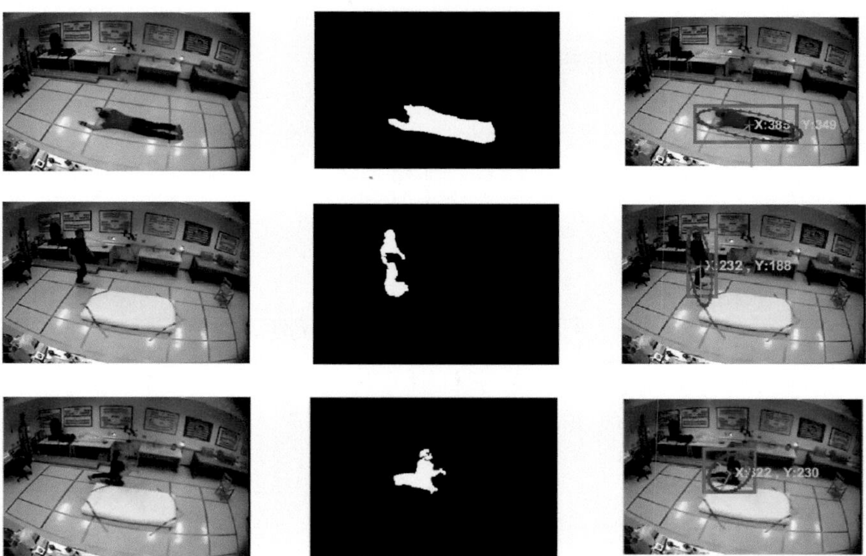

Fig. 5 *First column* Video frame. *Second column* Human silhouette. *Third column* Experiment results of fall detection

Fig. 6 *First column* Video frame. *Second column* Human silhouette. *Third column* Experiment results of fall detection

where, TP (True Positive) is the number of correct fall detections, FP (False Positive) is the number of wrong fall detections and FN (False Negative) is the number of missed fall detections. The system evaluation results are shown in the following Table 1.

The system evaluation results are shown in the following Table 1 using dataset in [13].

We used precision, recall and accuracy to evaluate the performance of our system using dataset [1]. The results obtained for human localisation are shown in Table 2.

Table 1 True and False alarm table for fall detection

	Precision (%)	Recall (%)	Accuracy (%)
System [16]	95,65	98,36	96,92
Proposed system	95,72	97,54	98,34

Table 2 Human localization

	Precision (%)	Recall (%)	Accuracy (%)
KVSS system [1]	90,33	93,67	77,22
Proposed system	91,82	92,25	91,80

5 Conclusion

The present work aims at designing a security system that issues preventive alarms that inform the staff of a kindergartens of children's fall in real-time by means of an intelligent system containing a multiple cameras and monitoring system that uses a Type-2 Fuzzy Gaussian Mixture Models (T2 FGMMs) background subtraction and a fuzzy logic system to analyse the children's activity and send data through the Local Area Network.

Our system shows efficacy in the detection of children falls in kindergarten environments despite the noisy environment and the need for real-time alert. It can be improved through better analysis of the falls provided by false children movements. Future work may also include a discriminative identification of the people in the scene, teacher, child or intruder using descriptors and classification systems. The use of an intelligent system for monitoring, control and identifying falls shall continue to progress to become the standard fall prevention intervention used in educational institutions. Precautions taken by teachers and headmasters are important to reduce the risk of deficiency and injury of children in kindergartens. Using the intelligent control system inside this institution offers promising information in real-time to detect the abnormal behavior of children and capture the events leading up to a fall.

Acknowledgments The authors would like to acknowledge the financial support of this work by grants from General Direction of Scientific Research (DGRST), Tunisia, under the ARUB program. The authors are grateful to the owner and staff of the 'King Kid' kindergarten in Sfax, Tunisia for allowing access to the institution, and for their help in the implementation and test of the designed surveillance system.

References

1. Abdelhedi, S., Wali, A., Alimi, A.M.: Toward a kindergarten video surveillance system (KVSS) using background subtraction based Type-2 FGMM model. In: Proceedings of (SoCPaR), Tunisia, pp. 440–446, Aug 2014
2. Foroughi, H.: A eigen space-based approach for human fall detection using integrated time motion image and neural network. In: Proceedings of (ICSP), pp. 1499–1503 (2008)
3. Muhammad, M.: A survey on fall detection: principles and approaches. Neurocomputing **100**, 144–152 (2013)

4. Liu, H., Pi, W., Zha, H.: Motion detection for multiple moving targets by using an omnidirectional camera. In: IEEE International Conference on Robotics, Intelligent Systems and Signal Processing, vol. 1, pp. 422–426, Oct 2003
5. Miaou, S., Sung, P., Huang, C.Y.: A customized human fall detection system using omni-camera images and personal information, pp. 39–42 (2006)
6. Abdelhedi, S., Taouil, K., Hadjkacem, B.: Design of automatic vision-based inspection system for monitoring in an olive oil bottling line. Int. J. Comput. Appl. **51**, 39–46 (2012)
7. Wali, A., Alimi, A.M.: Event detection from video surveillance data based on optical flow histogram and high-level feature extraction. In: Proceedings of DEXA'09, pp. 221–225, IEEE (2009)
8. Wali, A., Adel, A.: Multimodal approach for video surveillance indexing and retrieval. J. Intell. Comput. **4**(1), 165–175 (2010)
9. Bouwmans, T., Porikli, F., Hferlin, B., Vacavant, A. (eds.): Background Modeling and Foreground Detection for Video Surveillance, No. 978-1-4822-0538-1. Chapman and Hall/CRC Press (2014)
10. Wali, A., Aoun, N.B., Karray, H., Amar, C.B., Alimi, A.M.: A new system for event detection from video surveillance sequences. In: Advanced Concepts for Intelligent Vision Systems, pp. 110–120. Springer, Berlin, Heidelberg (2010)
11. Hu, W., Tan, T., Wang, L., Maybank, S.: A survey on visual surveillance of object motion and behaviors. IEEE Trans. Syst. Man Cybern. Part C: Appl. Rev. **34**, 334–352 (2004)
12. Wali, A., Alimi, A.M.: Incremental learning approach for events detection from large video dataset. In: Proceedings of AVSS, pp. 555–560. IEEE (2010)
13. Auvinet, E., Rougier, C., Meunier, J., St-Arnaud, A., Rousseau, J.: Multiple cameras fall dataset, Technical report 1350, DIRO - Universit de Montral, July 2010
14. Khawandi, S., Daya, B., Chauvet, P.: Real time recognition of elderly daily activity using fuzzy logic through fusion of motion and location data. Int. J. Comput. Appl. **54**(3), 0975 8887 (2012)
15. Mamdani, E.H., Assilian, S.: An experiment in linguistic synthesis with a fuzzy logic controller. Int. J. Man-Mach. Stud. **7**(1), 1–13 (1975)
16. Mirmahboub, B., Samavi, S., Karimi, N., Shirani, S.: View-invariant fall detection system based on silhouette area and orientation. In: Proceedings of ICME, pp. 176–181, July 2012

Application of Bio-inspired Methods Within Cluster Forest Algorithm

Jan Janoušek, Petr Gajdoš, Michal Radecký and Václav Snášel

Abstract Cluster Forest (CF) is relatively new ensemble clustering method inspired by Random Forest algorithm. The main idea behind of the existing algorithm consists in a construction of a larger number of partial clusterings for feature subsets using K-means algorithm. At the end, these clusterings are aggregated using a method of spectral clustering. This article describes a new application of bio-inspired methods that replaces the K-means algorithm in the computation pipeline. Several bio-inspired methods were tested on eight different datasets and compared with the original CF and others well known clustering methods.

Keywords Cluster Forest · Clustering · Ensemble clustering · Optimization · Soft computing

1 Introduction

Clustering represents one of the main machine learning research area. It aims to partitioning data points into groups based on their similarity. Points in the same group should be as similar as possible in comparison to the points in another group. Concrete number of clusters is mostly unknown. Clustering is often called unsupervised learning, because there are no information on classes. A large number of clustering algorithms has been developed in past decades. One of the best known clustering algorithms is k-means [1]. This method assign each point into the cluster with the

J. Janoušek (✉) · P. Gajdoš · M. Radecký · V. Snášel
Department of Computer Science, FEECS, VŠB - Technical Univesity of Ostrava, Ostrava, Czech Republic
e-mail: jan.janousek@vsb.cz

P. Gajdoš
e-mail: petr.gajdos@vsb.cz

J. Janoušek · P. Gajdoš · V. Snášel
IT4Innovations, Centre of Excellence, VŠB - Technical Univesity of Ostrava, 17. listopadu 15, 708 33 Ostrava, Czech Republic

A. Abraham et al. (eds.), *Proceedings of the Second International Afro-European Conference for Industrial Advancement AECIA 2015*, Advances in Intelligent Systems and Computing 427, DOI 10.1007/978-3-319-29504-6_24

nearest mean. k-means is very popular method because it is fast and provide relatively good results. Other examples of well known clustering algorithms are hierarchical clustering [2, 3], DBSCAN [4], OPTICS [5], Fuzzy c-means [6], Gaussian mixture models [7], and many others [8, 9]. The mentioned methods have a wide application area, e.g. recommendation systems [10], image segmentation [11], gene expression [12], pattern recognition [13], and many others [14, 15]. The ensemble methods became popular in the past years. One of the best known ensemble method is Random Forests (RF) [16], that is commonly used for classification and regression. The same idea, that is behind Random Forests, was introduced as "Cluster Forests" [17] that proposed a new clustering method. Authors made the comparison of CF with other ensemble methods like Bagged clustering, Evidence accumulation, and Random Projection as well. Finally, they shown that CF can outperform these algorithms.

A new modification of CF is introduces in this article. It utilizes a selected bio-inspired methods that replaces k-means clustering algorithm within CF.

2 Cluster Forest Algorithm

As is was already mentioned, Cluster Forest was inspired by Random Forests algorithm. Both algorithms can be divided into two main phases. The first phase covers an application of multiple weak learners (base clustering/classification algorithms). Particular application results in assignment of input points to clusters. Then, all the assignments are aggregated in the second phase of algorithm. The original CF uses k-means in the second phase. There exists a hidden constrain that weak learners must not produce the same assignments. Otherwise, the aggregation does not bring any additional information. This is the main disadvantage of k-means algorithm whose frequent application produces almost the same results. In RF, the mentioned problem is solved using bagging and random selection of features. CF solves this problem using the *clustering vector growing method* (CVG) shown in Algorithm 1. Here, the Cluster vector is a subset of features used for clustering, hence it will be called the feature vector. In the first step of CVG, the feature vector is initialized with n randomly selected features. Then, k-means algorithm is applied and quality of obtained clustering is evaluated using Formula 1, where c is the clustering, SS_w is the within cluster sum [18] of squares and SS_b is the between cluster sum of squares [18].

$$\rho(c) = \frac{SS_w(c)}{SS_b(c)} \tag{1}$$

After this, other n features are selected randomly and merged with the previous feature vector. k-means algorithm is applied to this temporary created vector and if the quality of obtained clustering is higher, then the previous feature vector is replaced by the new one.

All features are selected randomly in each step. This can lead to poor selection if the data contains only small number of significant features. To solve this problem, feature competition is used for feature vector initialization. In this procedure, base clustering algorithm is applied to multiple randomly selected feature subsets. And subset with the highest quality is selected as initial vector.

The output of the CVG described above is represented by the final feature vector and related clustering. This clustering is equivalent to a single tree in the RF algorithm. The algorithm must be executed multiple times to build the set of clusterings.

Algorithm 1 Cluster vector growing method

Require:
f: list of all features
$sampleFeatures$: randomly sample n features
$KMeans$: k-means clustering

$iter = 0$
v = featureCompetition(f)
$clustering_{old}$=KMeans(v)
while $iter < iter_{max}$ **do**
    $iter = iter + 1$
    v_{tmp} = sampleFeatures(f)
    $clustering_{new}$ = KMeans(v_{tmp})
    if $\rho(clustering_{old}) > \rho(clustering_{new})$ **then**
        $v = v_{tmp}$
        $clustering_{old} = clustering_{new}$
        $iter = 0$
    end if
end while

In the second phase, all results of the weak learners are aggregated. Spectral clustering [19] was selected as an optimal aggregation method, that is based on the spectral graph theory. The spectral clustering is applied on the matrix M whose elements a_{ij} are give by Formula 2. Matrix M is regularized using thresholding, when element a_{ij} is set to zero, if it is less then configuration parameter β with value in interval (0, 1]. After this, all elements of matrix D are scaled using $a_{ij} = exp(\gamma a_{ij})$, where γ is a configuration parameter with the value in interval [0, 1].

$$a_{ij} = exp\left(\frac{1}{T}\sum_{t=1}^{T}\left\{\begin{matrix} 1 & if\ c_t(i) = c_t(j) \\ 0 & otherwise \end{matrix}\right\}\right) \tag{2}$$

where $i, j \in \{1, \ldots, n\}$, and T is the total number of iterations, and c_t returns a cluster index to which a point j was assigned to in the iteration t.

3 Bio-inspired Methods

Bio-inspired methods make a subset of algorithms that are inspired from nature. A large proportion of these methods is represented by optimization algorithms and can be applied to a wide range of machine learning methods focused on finding global optima, solutions of real problems. In our case, selected Differential Evolution was implemented within the clustering algorithm. Its implementation leads to function minimization problem where the clustering error is computed as an output of fitness function, e.g. *within-cluster sum of squares* [18], *Davies-Bouldin index* [18], or *Dunn index* [18]. Davies-Bouldin index was used in our experiments that is described by Formula 3.

$$fit = \frac{1}{K}\sum_{i=1}^{K} \max_{j\in K, j\neq i}\left(\frac{\frac{1}{N_i}\sum_{\vec{x}\in C_i} d(\vec{x}, c_i) + \frac{1}{N_j}\sum_{\vec{x}\in C_j} d(\vec{x}, \vec{c}_j)}{d(\vec{c}_i, \vec{c}_j)}\right) \tag{3}$$

where K is the number of clusters, N_i is the number of points in cluster C_i for $i \in \{1, \dots, K\}$, N_j is the number of points in cluster C_j for $j \in \{1, \dots, K\}$, c_i is the center of cluster i, and d is the distance function.

All clusters have to be represented as single vector $v = (c_{1,1}, \dots, c_{1,n}, \dots, c_{K,1} \dots, c_{K,n})$ to be able to apply bio-inspired methods, where K is number of clusters as n is the number of features.

Finally, five different bio-inspired methods were implemented to compare their influence on quality of clustering. Two of them were evolutionary algorithms and three were swarm intelligence algorithms.

3.1 Particle Swarm Optimization (PSO)

PSO [20] is one of the best known swarm intelligence algorithms. PSO consists of multiple candidate solutions (particles), that moves around search-space using Formula 4. Where x_i is the current particle position and $\vec{v}_i$ is a velocity vector computed using Formula 5. Variables ω, φ_p and φ_g represent configuration parameters that affect particle movement. Vector p_i is the best known particle position and g is the best known global position (of all particles). Initial positions and velocity vectors are set randomly. In each step of the algorithm, every particle is moved.

$$\vec{x}_i^{new} = \vec{x}_i + \vec{v}_i \tag{4}$$

$$\vec{v}_i^{new} = \omega \times \vec{v}_i + \varphi_p \times rand(0,1) \times (\vec{p}_i - \vec{x}_i) + \varphi_g \times rand(0,1) \times (\vec{g} - \vec{x}_i) \tag{5}$$

3.2 *Differential Evolution (DE)*

DE [21] is a representative of evolutionary algorithms. In this method, candidate solutions are called "agents". These agents are initialized randomly. In each step of the algorithm, every agent picks N (N is set to three by default) different agents, e.g. "a", "b" and "c", and generates random number R between 0 and F, which is the number of features of the agent. After this, a new agent is created using Formula 6, that is applied to each feature independently, if $rand(0, 1) < CR$. If $rand(0, 1) >= CR$ then $x_i^{new} = x_i$. Variables CR and W are configuration parameters called "crossover probability" with value in interval $[0, 1]$ and "differential weight" with value in interval $[0, 2]$. If the new created agent is better then the original one, then the original agent is replaced.

$$x_i^{new} = a_i + W \times (b_i - c_i) \tag{6}$$

3.3 *Artificial Bee Colony (ABC)*

ABC [22] is a swarm intelligence method inspired by behavior of honey swarm during searching for food. All candidate solutions are called "food sources" in this method, and these solutions are initialized randomly and the computation is divided into three phases. In the first phase called "employed", a food source is assigned to each of bee in the swarm. Next, the second food source is selected randomly for every bee and a new solution is created using Formula 7, where the vector $\vec{\phi}_i$ contains random numbers in interval $[-1, 1]$. The new obtained solution is evaluated and the present candidate solution is replaced, if it is worse.

$$\vec{x}_i^{new} = \vec{x}_i + \vec{\phi}_i \times (\vec{x}_i - \vec{x}_j) \tag{7}$$

The second phase called "onlookers" behaves like the first one, but the second food source is selected base on its quality (better sources are preferred). In the third phase called "scout", the number of unsuccessful attempts to improve every food source is counted. If the number reaches threshold t, the food source is regenerated randomly and the number is set to zero.

3.4 *Grey Wolf Optimizer (GWO)*

GWO [23] represents a relatively new swarm intelligence method inspired of behavior of grey wolves. All candidate solutions are represented by wolf positions. These positions are initialized randomly and their are updated in each iteration using Formula 8, where α, β and δ are indexes of the three best wolf positions. Parameter $a_{p,i}$ in Formula 9 is changing based on the current iteration using Formula 10.

$$\vec{x}_i^{new} = \frac{\vec{P}_i(\alpha, i) + \vec{P}_i(\beta, i) + \vec{P}_i(\delta, i)}{3} \tag{8}$$

$$\vec{P}_i(p, i) = \vec{x}_p - a_{p,i} \times (2 \times rand(0, 1) \times \vec{x}_p - \vec{x}_i) \tag{9}$$

$$a_{p,i} = 2 \times (2 - \frac{2 \times iter}{maxIter}) \times rand(0, 1) - (2 - \frac{2 \times iter}{maxIter}) \tag{10}$$

The positions are updated even they are worse than the previous.

3.5 Harmony Search (HS)

HS [24] is the second representative of evolutionary algorithms used in this article. The algorithm starts with randomly generated candidate solutions. The set of these solutions is called "harmony memory". Only one new solution is created in every iteration. All solution features are affected by the configuration parameter called harmony memory considering rate (*hmcr*), which is a value in interval [0, 1].

A new solution is constructed such that for every feature the following steps must be evaluated. First, the feature is generated randomly with the probability *hmcr*, otherwise a second solution S_2 vector is selected randomly and the feature is copied from this solution S_2 and modified with probability *par* (pitch adjusting rate) using Formula 11, where *fw* is called "fret width". The value of "fret width" can be constant (i.e. configuration parameter), or can change based on the current iteration.

If the new solution is better then the worst solution in the harmony memory, the worst solution is replaced the new one.

$$\vec{x}_i^{new} = \vec{x}_i + rand(-1, 1) \times fw \tag{11}$$

4 Data Collections

For our experiments we used 8 freely available dataset from UC Irvine Machine Learning Repository [25] (Table 1).

These collections were chosen for purposes of comparison with the original version algorithm, published in the paper [17].

Table 1 Data collections

Dataset	Instances	Features	Classes
Statlog (Heart)	270	13	2
Image segmentation	2100	19	7
Madelon	2000	500	2
Robot execution failures	164	90	5
Soybean	47	35	4
SPECT heart	267	22	2
Wine	178	13	3
Breast Cancer Wisconsin (Diagnostic)	569	30	2

5 Experiments

Two main experiments were performed. The quality of clustering was compared using already described bio-inspired methods in the first experiment. In the second experiment, the results achieved by the best bio-inspired methods were compared with the results made by the original CF and others clustering methods.

There exists a lot of metrics that can be used for clustering quality comparison. Two metrics based on classification performance were used. The first metric is ρ_r:

$$\rho_r = \frac{\sum_{i=1}^{N}\sum_{j=1}^{N}\sigma(O_i, O_j)\sigma(A_i, A_j)}{N(N-1)} \qquad i \neq j \tag{12}$$

where σ is the function that returns 1 if the values of both attributes are the same, 0 otherwise. O is the vector of the original (expected) classes and A is the vector of assigned classes. N is the number of classified vectors. This metric is based on an assumption that all possible pairs of vectors that have the same class in the original class vector, should have the same class in the assigned class vector.

The second metric is ρ_c:

$$\rho_c = \max_{\tau \subset T}\left\{\frac{1}{N}\sum_{i=0}^{N}\sigma(\tau(O_i), A_i)\right\} \tag{13}$$

where T is the set of all class permutations. σ, O_i and A_i have the same meaning like in Formula 12. This metric is equivalent to the standard classification accuracy. For both metrics, the higher value the better result.

All published results in both experiments are means of 20 subsequent executions of the algorithms.

The maximum threshold of the fitness function was set to 300 for all methods, just to be able to compare selected bio-inspired methods. This number was chosen empirically. Quite similar results were obtained for most of used methods in case of higher threshold. Moreover, the key point of this experiment consists in the selection

Table 2 Comparison of bio-inspired methods results for metric ρ_r

Dataset	Bio-inspired method				
	DE	PSO	GWO	HS	ABC
Statlog (Heart)	62.39	61.22	**64.76**	54.59	58.39
Image segmentation	80.25	79.98	75.39	**81.94**	75.73
Madelon	51.21	51.1	50.01	50.03	**51.4**
Robot execution failures	**68.53**	56.53	66.38	64.01	51.8
Soybean	95.46	**97.89**	97.07	94.01	76.79
SPECT heart	59.28	51.65	54.5	52.45	**60.82**
Wine	**91.26**	77.17	64.68	76.26	79.21
Breast Cancer Wisconsin	**83.92**	52.11	78.95	52.11	61.91

Table 3 Comparison of bio-inspired methods results for metric ρ_c

Dataset	Bio-inspired method				
	DE	PSO	GWO	HS	ABC
Statlog (Heart)	74.94	73.16	**77.25**	65.07	69.05
Image segmentation	**42.52**	41.19	31.52	38.42	31.21
Madelon	55.78	55.47	51.35	51.75	**57.42**
Robot execution failures	**46.58**	41.73	41.95	42.61	40.23
Soybean	91.7	94.46	**95.00**	93.29	58.19
SPECT heart	71.59	59.08	65.21	61.32	**73.33**
Wine	**93.11**	77.5	57.05	76.34	80.08
Breast Cancer Wisconsin	**91.18**	60.52	88.04	60.52	74.51

of the best method that can product the best results with the smallest number of fitness function evaluations, which is computationally the most demanding part of the algorithm.

The results of bio-inspired methods are presented in Tables 2 and 3. It can be seen that there is no single method that would provide the best results for all data collections. Based on the results, differential evolution was chosen as the winner because it provides the best results for most of the collections. Moreover, all values obtained from both metrics are relatively height.

There are results of comparisons between cluster forest based on DE (CF-DE) with the original cluster forest and standard clustering methods like *k*-means and three others ensemble methods called "bagged clustering" (bC2) [26], "evidence accumulation" (EA) [27] and "random projection" (RP) [28] in Tables 4 and 5. All results for bC2, EA and RP are taken from the paper [17].

From Table 4 can be seen that CF-DE outperforms the other clustering methods, including the original version of CF in all but one cases for metric ρ_r. In Table 5, there are results for metric ρ_c, where CF-DE outperforms other algorithms in 7 of 8 cases.

Table 4 Comparison of results for metric ρ_r

Dataset	Cluster forest variants					
	CF-DE	CF	*k*-means	bC2	EA	RP
Statlog (Heart)	**62.39**	56.8	47.3	51.50	53.20	52.41
Image segmentation	80.25	79.71	23.61	82.19	85.75	**85.88**
Madelon	**51.21**	50.76	49.98	49.98	49.98	50.82
Robot execution failures	**68.53**	63.42	53.26	39.76	58.31	41.52
Soybean	**95.46**	92.36	85.14	83.16	86.48	87.04
SPECT heart	**59.28**	56.78	49.22	50.61	51.04	49.89
Wine	**91.26**	79.7	32.05	71.97	71.86	71.94
Breast Cancer Wisconsin	**83.26**	79.66	82.14	74.87	75.04	74.89

Table 5 Comparison of results for metric ρ_c

Dataset	Cluster forest variants					
	CF-DE	CF	*k*-means	bC2	EA	RP
Statlog (Heart)	**71.94**	68.26	57.3	59.10	59.26	60.54
Image segmentation	47.82	48.24	43.42	49.91	**51.30**	47.71
Madelon	**55.78**	55.12	51.73	50.20	50.30	55.19
Robot execution failures	**46.58**	41.2	42.05	35.37	37.19	35.50
Soybean	**91.7**	84.43	82.142	72.34	76.59	71.83
SPECT heart	**71.59**	68.02	55.9	56.28	56.55	61.11
Wine	**93.11**	79.19	46.53	70.22	70.22	70.79
Breast Cancer Wisconsin	**91.18**	88.7	82.142	85.38	85.41	70.79

6 Conclusion

In this article, different approaches and improvements of the original Cluster Forest (CF) algorithm were presented. The main goal was substitute the standard clustering method, that was used within CF, by selected bio-inspired method. Several bio-inspired methods were tested and finally, Differential Evolution was selected as it brought the highest improvements. Eight datasets that differ in the number of instances, features and classes were used in performed experiments. As it can be seen in the result tables, the Cluster-Forest algorithm with Differential Evolution (CF-DE) does not bring the better result just in one case that will be the subject of further research and observations.

Acknowledgments This work was supported by the IT4Innovations Centre of Excellence project (CZ.1.05/1.1.00/02.0070), funded by the European Regional Development Fund and the national budget of the Czech Republic via the Research and Development for Innovations Operational Programme and by Project SP2015/105 "DPDM - Database of Performance and Dependability Models" of the Student Grand System, VŠB Technical University of Ostrava and by Project SP2015/146 "Parallel processing of Big data 2" of the Student Grand System, VŠB Technical University of Ostrava.

References

1. Hartigan, J.A., Wong, M.A.: Algorithm AS 136: A k-means clustering algorithm. Appl. Stat. **28**(1), 100–108 (1979). doi:10.2307/2346830
2. Hastie, T., Tibshirani, R., Friedman, J.: The Elements of Statistical Learning: Data Mining, Inference, and Prediction, 2nd edn. (Springer Series in Statistics). Springer (2009)
3. Cinar, G., Principe, J.: Clustering of time series using a hierarchical linear dynamical system. In: 2014 IEEE International Conference on Acoustics, Speech and Signal Processing (ICASSP), pp. 6741–6745 (2014). doi:10.1109/ICASSP.2014.6854905
4. Ester, M., Kriegel, H.P., Sander, J., Xu, X.: A Density-based Algorithm for Discovering Clusters in Large Spatial Databases with Noise, pp. 226–231. AAAI Press (1996)
5. Ankerst, M., Breunig, M.M., Kriegel, H.P., Sander, J.: Optics: Ordering Points to Identify the Clustering Structure, pp. 49–60. ACM Press (1999)
6. Bezdek, J.C.: Pattern Recognition with Fuzzy Objective Function Algorithms. Kluwer Academic Publishers, Norwell, MA (1981)
7. Yang, M.-S., Lai, C.-Y., Lin, C.-Y.: A robust {EM} clustering algorithm for gaussian mixture models. Pattern Recogn. **45**(11), 3950–3961 (2012). doi:10.1016/j.patcog.2012.04.031
8. Yang, P., Huang, B.: A spectral clustering algorithm based on normalized cuts. In: 2008 International Conference on Computer Science and Software Engineering, vol. 4, pp. 329–331 (2008). doi:10.1109/CSSE.2008.910
9. Madhulatha, T.S.: An overview on clustering methods. arXiv:1205.1117
10. Lu, M., Qin, Z., Cao, Y., Liu, Z., Wang, M.: Scalable news recommendation using multi-dimensional similarity and Jaccard-Kmeans clustering. J. Syst. Softw. **95**, 242–251 (2014). doi:10.1016/j.jss.2014.04.046
11. Zeng, S., Huang, R., Kang, Z., Sang, N.: Image segmentation using spectral clustering of gaussian mixture models. Neurocomputing **144**, 346–356 (2014). doi:10.1016/j.neucom.2014.04.037
12. Sirbu, A., Czibula, G., Bocicor, M.: Dynamic clustering of gene expression data using a fuzzy approach. In: 16th International Symposium on Symbolic and Numeric Algorithms for Scientific Computing, SYNASC 2014, Timisoara, Romania, 22–25 Sept 2014, pp. 220–227 (2014). doi:10.1109/SYNASC.2014.37
13. Schwenker, F., Trentin, E.: Pattern classification and clustering: a review of partially supervised learning approaches. Pattern Recogn. Lett. **37**, 4–14 (2014). doi:10.1016/j.patrec.2013.10.017
14. Bose, I., Chen, X.: Detecting the migration of mobile service customers using fuzzy clustering. Inf. Manage. **52**(2), 227–238 (2015). doi:10.1016/j.im.2014.11.001
15. Modzelewski, M., Dojer, N.: MSARC: multiple sequence alignment by residue clustering. Algorithms Mol. Biol. **9**, 12 (2014). doi:10.1186/1748-7188-9-12
16. Breiman, L.: Random forests. Mach. Learn. **45**(1), 5–32 (2001). doi:10.1023/A:1010933404324
17. Yan, D., Chen, A., Jordan, M.I.: Cluster Forests. arXiv:1104.2930
18. Halkidi, M., Batistakis, Y., Vazirgiannis, M.: Clustering validity checking methods: part ii. ACM Sigmod Record **31**(3), 19–27 (2002)
19. Shi, J., Malik, J.: Normalized cuts and image segmentation. In: 1997 IEEE Computer Society Conference on Computer Vision and Pattern Recognition, 1997. Proceedings, pp. 731–737 (1997). doi:10.1109/CVPR.1997.609407
20. Kennedy, J., Eberhart, R.: Particle swarm optimization. In: IEEE International Conference on Neural Networks, 1995. Proceedings, vol. 4, pp. 1942–1948 (1995). doi:10.1109/ICNN.1995.488968
21. Storn, R., Price, K.: Differential evolution a simple and efficient heuristic for global optimization over continuous spaces. J. Glob. Optim. **11**(4), 341–359 (1997). doi:10.1023/A:1008202821328
22. Janousek, J., Platos, J., Snasel, V.: Clustering using artificial bee colony on cuda. In: 2014 IEEE International Conference on Systems, Man and Cybernetics (SMC), pp. 3803–3807 (2014). doi:10.1109/SMC.2014.6974523

23. Mirjalili, S., Mirjalili, S.M., Lewis, A.: Grey wolf optimizer. Adv. Eng. Softw. **69**, 46–61 (2014). doi:10.1016/j.advengsoft.2013.12.007
24. Geem, Z.W., Kim, J.H., Loganathan, G.: A new heuristic optimization algorithm: harmony search. Simulation **76**(2), 60–68 (2001)
25. Lichman, M.: UCI machine learning repository (2015). http://archive.ics.uci.edu/ml
26. Dudoit, S., Fridlyand, J.: Bagging to improve the accuracy of a clustering procedure. Bioinformatics **19**(9), 1090–1099 (2003). doi:10.1093/bioinformatics/btg038
27. Fred, A.L., Jain, A.K.: Data clustering using evidence accumulation. In: 16th International Conference on Pattern Recognition, 2002. Proceedings, vol. 4, pp. 276–280. IEEE (2002)
28. Fern, X.Z., Brodley, C.E.: Random projection for high dimensional data clustering: a cluster ensemble approach. In: ICML, vol. 3, pp. 186–193 (2003)

Modelling of Fetal Hypoxic Conditions Based on Virtual Instrumentation

Radek Martinek, Adam Sincl, Jan Vanus, Michal Kelnar, Petr Bilik, Zdenek Machacek and Jan Zidek

Abstract This article is dedicated to the design and implementation of a software simulator of the fetal ECG signal (fECG). Primarily, it aims at modelling the influence of physiological and pathological changes. Fetal electrocardiogram is a valuable carrier of information and its potential can be utilized in the diagnosis of hypoxic conditions. Analysis of real records and study of methods intended for obtaining the ECG signal in clinical practice resulted in creating a simulator of fECG waveforms based on virtual instrumentation. In a suitable user environment, the signal formed in this manner can be combined with physiological and pathological parameters that modify the signal. The respective changes can be observed and analysed with the help of graphical output elements of the user interface. The generator allows the user to model the transvaginal monitoring method (invasive), transabdominal monitoring method (non-invasive) and cardiotocograph (CTG) with ST analyser (STAN). Due to their morphology, the simulation results correlate with the established theoretical assumptions and approximate to real records.

R. Martinek (✉) · A. Sincl · J. Vanus · M. Kelnar · P. Bilik · Z. Machacek · J. Zidek
VŠB – Technical University of Ostrava, 17. listopadu 15, Ostrava, Czech Republic
e-mail: radek.martinek@vsb.cz
URL: http://www.vsb.cz

A. Sincl
e-mail: adam.sincl@vsb.cz

J. Vanus
e-mail: jan.vanus@vsb.cz

M. Kelnar
e-mail: michal.kelnar@vsb.cz

P. Bilik
e-mail: petr.bilik@vsb.cz

Z. Machacek
e-mail: zdenek.machace@vsb.cz

J. Zidek
e-mail: jan.zidek@vsb.cz

A. Abraham et al. (eds.), *Proceedings of the Second International Afro-European Conference for Industrial Advancement AECIA 2015*, Advances in Intelligent Systems and Computing 427, DOI 10.1007/978-3-319-29504-6_25

Keywords ECG · fECG · mECG · CTG · STAN · Virtual instrumentation · Disorders of cardiac function · Transabdominal monitoring · Transvaginal monitoring

1 Introduction

Development of fetal electrocardiography is driven forward by new advances in microprocessor technology, analysis and fECG processing. Increasing computational power allows faster and more complex analysis which basically leads to more accurate results or diagnostics.

Efficiency of fECG sensing is degraded by technical problems—see details in [6, 7]. These include the low signal-to-noise ratio (SNR) of fECG compared with the maternal ECG (mECG), low sensitivity of devices, imperfect extraction techniques and methods of processing fECG signals, etc. [3, 4, 6, 7, 10]

fECG signal itself is a carrier of valuable information (fetal heart rate—FHR, shape and length of the individual waves, dynamic behaviour, etc.)—see details in [3, 4, 10]. Due to this information, it is possible to diagnose hypoxic conditions, the maturity of the fetus or its congenital heart defect. Unfortunately, the potential of fECG is currently used only partially. Only CTG is used in the prenatal period and STAN (analysis of the ST segment) in the intrapartal period. Today, the latter represents a complementary function to the CTG. At present, a dominant challenge for modern obstetrics is the effort to maximally refine diagnostic methods for the determination of fetal hypoxic conditions as supported in [1, 9, 12].

Nowadays, the morphological analysis of fECG still remains a great challenge for experts. A big factor, that hinders the research, is the constantly missing database of physiological and pathological fECG records [4]. Pathological records are undoubtedly a key source for the development of systems automatically classifying fetal heart vitality.

Obtaining records from clinical practice is currently still very difficult, especially in cases of conditions immediately threatening the fetus (e.g. asphyxia) when the pregnancy is interrupted and operatively terminated.

The realized simulator of fetal hypoxic conditions can greatly contribute to the future research in the area of automatic classification of CTG records and ST analysis.

2 Design and Implementation of the Simulator of Fetal Hypoxic Conditions

2.1 Modelling of Fetal Cardiac Function

This article utilizes a mathematical modelling of fECG which is based on [2, 11]. The modelling is implemented using three below-defined Hermit functions [8]:

$$h(t) = K(b; t) \cdot \varrho(t), \tag{1}$$

$$h_0(t) = \frac{1}{\sqrt{b \cdot \sqrt{\pi}}} \cdot \exp\left(-\frac{1}{2} \cdot b^2 \cdot t^2\right) \tag{2}$$

$$h_1(t) = \sqrt{\frac{2}{b \cdot \sqrt{\pi}}} \cdot \frac{t}{b} \cdot \exp\left(-\frac{1}{2} \cdot b^2 \cdot t^2\right) \tag{3}$$

$$h_2(t) = \sqrt{\frac{2}{b \cdot \sqrt{\pi}}} \cdot \left[\frac{t}{b}^2 - 1\right] \cdot \exp\left(-\frac{1}{2} \cdot b^2 \cdot t^2\right) \tag{4}$$

In Eqs. 1 up to 4, it determines the width of the modelled curve in ms and time t also in ms. Based on multiplication and mutual superposition (e.g. Eq. 5) of the aforementioned functions, we can easily and efficiently model fECG signals with sufficient shape variability:

$$H_k(n) = \begin{cases} h_0[n] \cdot \frac{10-k}{10} + h_1[n] \cdot \frac{k}{10}, \\ 1 \leq k \leq 10, \\ h_0[n] \cdot \frac{(k-10)}{10} + h_1[n] \cdot \frac{(20-k)}{10} + h_2[n] \cdot \frac{(k-10)}{20}, \\ 11 \leq k \leq 20, \\ \\ h_0[n] \equiv h_0(n \cdot T_s), n \in \mathbb{Z}, \\ h_1[n] \equiv h_1(n \cdot T_s), n \in \mathbb{Z}, \\ h_2[n] \equiv h_2(n \cdot T_s), n \in \mathbb{Z}, \end{cases} \tag{5}$$

where k determines the shape of QRS complex.

2.2 Implementation of the Simulator of Fetal Hypoxic Conditions Based on Virtual Instrumentation

Mathematical models were implemented into LabVIEW development environment [5]. Using an intuitive user interface, the so-called front panel, the created application enables us to adjust a variety of parameters—see Fig. 1.

The front panel is divided into 4 main functional parts. The first tab is used for parameter setting. Using this tab, it is possible to set the parameters of the pulse, amplitudes, CTG and ST segment. It also contains a preview of the final waveform and amplitude spectrum of the signal. The second tab is used for transvaginal view. In clinical practice, this view is recorded by a special scalp electrode. This tab contains representation in time and amplitude spectrum. An option is to export records to EDF format. The third tab is called trans-abdominal view and refers to a non-invasive method. In practice, this sensing is implemented using abdominal electrodes. This tab allows the user to set parameters of maternal and fetal heartbeat, representation in time and spectrum, and select the view of leads and data export. The fourth tab is used to display the CTG and ST segment and approximates to the classic CTG/ST monitor. The output is a display of pulse value and T/QRS ratio in time with a detailed illustration of how this measurement takes place in clinical practice. Furthermore,

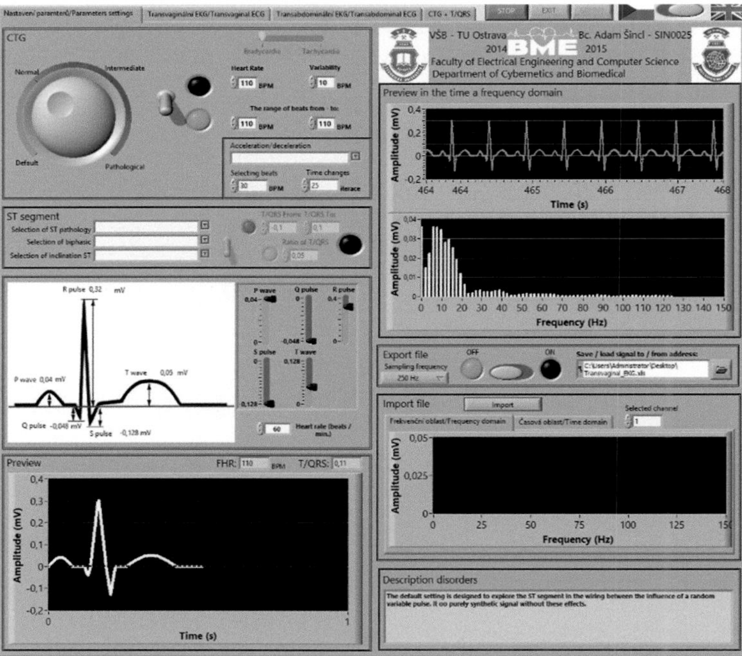

Fig. 1 *Front panel* of the simulator of fetal hypoxic conditions

the system displays one ECG period calculated from 30 samples as it is in case of actual ST monitor.

3 Results of Conducted Experiments

Complexity of the simulator consists in its ability to present the currently best-known imaging techniques for the diagnosis of fetal heart vitality, starting with the basic CTG monitor evaluating heart rate, i.e. the R–R interval, up to the added ST monitor capable of evaluating the whole fECG curve.

As a significant marker for the diagnosis, the simulator uses a numerical value of T/QRS ratio and signal period averaged from 30 samples. It also includes the most common view of fECG curve—the transvaginal method—which, however, still can not be called a conventional method. The scale ends with the unconventional transabdominal method which, despite its advantage of noninvasiveness, faces the problems of separating the ECG fetal signal from ambient noise—see details in [4, 6, 7]. All imaging methods of the simulator characteristically display all the elements and parameters together with pathological disorders that can be set.

3.1 CTG and ST Monitor

Currently, the most common imaging method is CTG with the monitor of ST interval. This part of the simulator was modelled in order to include all relevant information related to the fetal heart. It tries to copy the displayed data available to a real CTG with ST analyser. Therefore, it contains information about FHR and T/QRS ratio. Both figures are displayed numerically and graphically. Equally important information is the increase in the T/QRS ratio above a specified baseline. This figure indicates a change in the ratio and is a guide for doctors. This increase is defined according to Tab. X. Samples of waveforms and a detailed view of CTG and STAN simulations are shown in Fig. 2.

The figure above demonstrates a simulation of unstable CTG. At the second minute of the simulation, the curve of pulse frequency shows greater variability. Then we see a brief deceleration which is compensated by fetal overreaction. At 15:56, the fetus gets into tachycardia. Fetal defence reacts sharply by reducing the heart rate to 110 beats. This value persists for more than 3 min.

It is again interrupted by a jump acceleration which returns to normal after 1 min and 30 s. As a basal line, we can mark the value of 110 BPM (beats per minute). The curve of T/QRS ratio is stable. Its drop is recorded between the 6th and 7th minute. This decline can be due to a physiological or pathological reason. As a pathological problem, we can designate the possibility that the decline in the T/QRS ratio was the result of depletion of energy stores. Compensation was an increase in pulse rate until the ratio settled at the basal line.

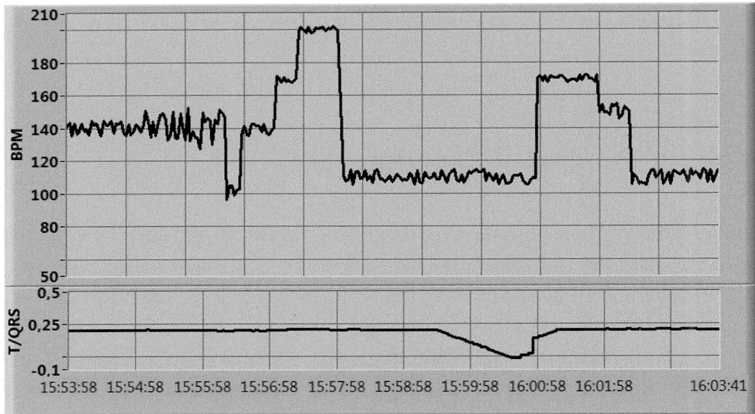

Fig. 2 Ten-minute simulation of CTG+STAN

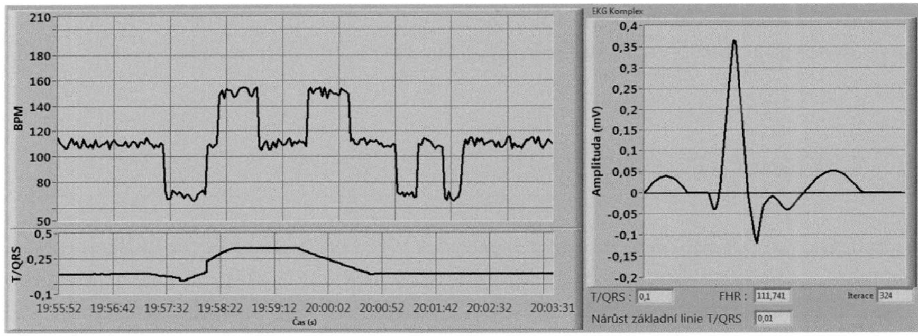

Fig. 3 Results of modelling the CTG+STAN record with the following parameters: FHR: 111 BPM, biphasicity of degree 3, T/QRS: 0.1, growth: 0.20. Episodic accelerations and decelerations

Figures 3 and 4 demonstrate the simulator variability in the case of CTG method and STAN monitor. The main aim of the article was to create an fECG simulator that can represent hypoxic conditions.

3.2 Biphasicities and ST Segment Elevations/Depressions in the fECG Signal

Figure 5 demonstrates the simulator functionality to model elevations and depressions of the ST segment and T wave. The magnitude of the ratio between the T wave amplitude and QRS complex can be specified within the limits of a clinical

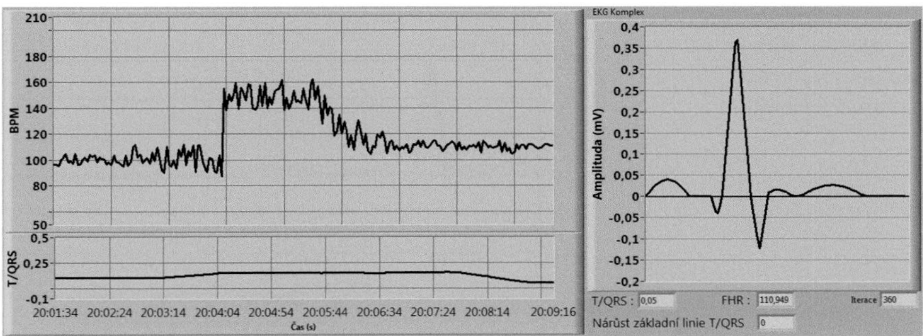

Fig. 4 Results of modelling the CTG+STAN record with the following parameters: FHR: 100 BPM, biphasicity of degree 1, T/QRS: 0.1. Episodic decelerations

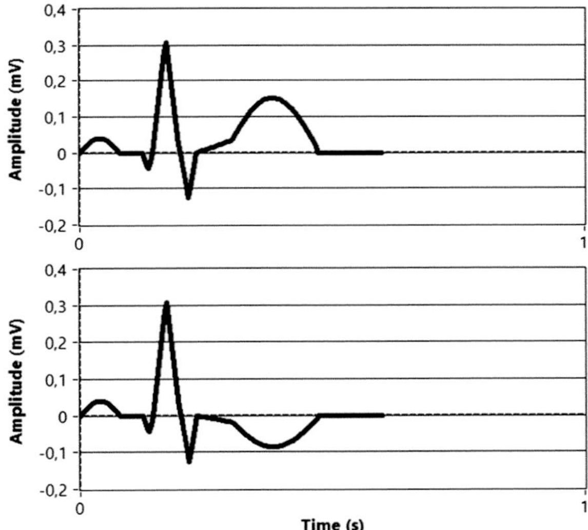

Fig. 5 Results of modelling the ST segment elevation and depression using the proposed simulator

recommendation. Literature [4] points to different possibilities of the ST segment shape. Horizontal (Fig. 5), falling (Fig. 5) or rising ST segments can be included among standard situations. In hypoxic conditions, a special group is biphasicity of the ST segment—see Fig. 6.

The results show that the simulator can mimic the basic theoretical assumptions—see Fig. 6. In the case of real records, the most common method for monitoring hypoxia is CTG+STAN. The comparison of hypoxic conditions would therefore appertain to the respective method which is also contained in the simulator. Unfortunately, quality records of fECG complex, that would be a source for such a

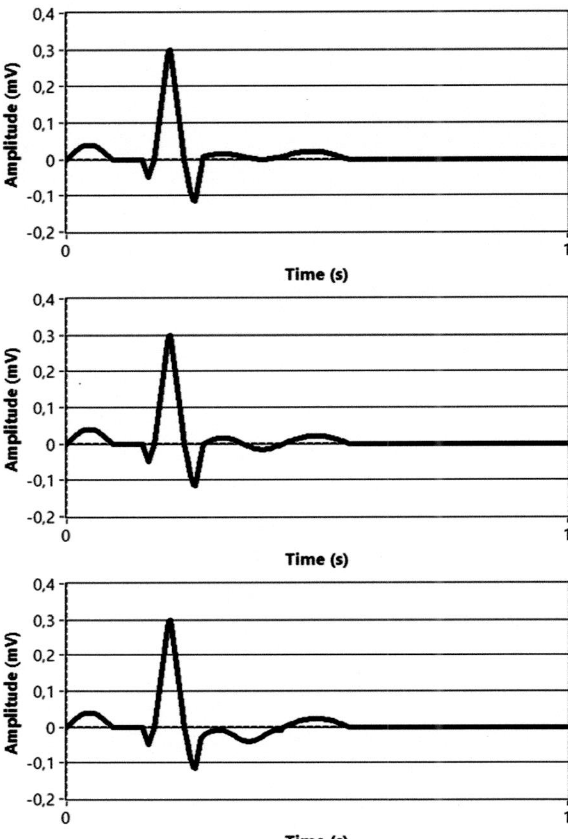

Fig. 6 Results of modelling BP1, BP2 and BP3 biphasicity using the proposed simulator

comparison, are not available due to the almost non-existent database. Database Stancases (see [9, 12]) includes only CTG+STAN records where CTG records contain graphically designated values of T/QRS ratio. Nevertheless, they do not retain the source fECG complexes from which the ratio is calculated. For this reason, it can be said that the fECG complexes of hypoxic conditions cannot be objectively compared with real records at the present time.

In clinical practice, the diversity of shapes of individual oscillations of the fECG period is extensive and depends on many factors such as the effect of noise, apparatus sensitivity or the method used for sensing. The simulator tries to approximate to the basic nuances such as the slope of the ST segment curve, negative values of the T wave amplitude or uncertainty in the biphasic averaging.

4 Discussion of the Achieved Results

The issue of identifying hypoxic conditions using fECG can be considered a very actual topic, both by technicians and physicians. The technological development brings new methods into this issue in the field of improved quality of monitoring. Firstly, non-invasive methods will provide a unique way of obtaining useful information hidden in the curve of abdominal fECG. Already today, it is possible to get records of transabdominal fECG, thanks to devices such as Monica AN24 [4]. However, such devices are currently used only for determining FHR (CTG). They not allow a deeper fECG analysis such as determination of the T/QRS ratio (ST analysis). This is possible, for example, with the device STAN S21 [9]. However, its disadvantage is the invasiveness in the form of the scalp electrode which must be introduced to the head or buttocks of the fetus.

In this article, we used theoretical knowledge based primarily on the literature [3, 4, 6, 7, 10] dedicated to hypoxic conditions and the method or technique of monitoring. We used valid clinical recommendations which objectively and parametrically define decision rules in the diagnosis of hypoxic conditions. Based on the clinical recommendations for the CTG and ST analysis, the doctors decide on the termination of pregnancy.

The Stanceases [12] database was used for comparison with real records. Thanks to this database, it was possible to get a real idea of how CTG and STAN records look like. The CTG simulator, together with the ST monitor, draws the fECG curve in time and displays the fECG complex which is a moving average calculated from the previous 30 samples (in clinical practice, the STAN monitors use arithmetic mean). Nevertheless, there is a second project which includes uterine contractions and other outside interference. It is expected that both projects will be combined in the future due to easy compatibility of the NI LabVIEW and similar program architecture. Furthermore, the CTG+STAN simulator allows continuous measurement of the T/QRS ratio. Today's devices are not able to continuously measure this ratio. They require a high-quality signal to make the necessary calculations. In today's practice, we are unable to sense high-quality signals which would allow the method to measure continuously.

The CTG+STAN simulator enables us to model arbitrarily long records which corresponds with findings from clinical practice. So it is possible to simulate data reflecting real records obtained in clinical practice.

The transvaginal method was compared with the theoretical presumptions and CTG+STAN records. Unfortunately, the quality of real records of the fECG curve does not allow objective comparison. The signal is facing high interference due to fluctuations of the isoelectric line. Because of the filtration, we observe deformations of the ST segment and reduced oscillations of the QRS complex. Direct comparison with practice is also impeded by the absence of databases that would retain the records. Extending the signal representation in the form of amplitude-frequency spectrum gives the user a different perspective on how hypoxic conditions change the fECG complex. It focused mainly on dominant changes in hypoxic conditions.

Based on providing real records from clinical practice, the transabdominal method benefited exactly from this kind of information and analysed parameters. It combined the method of eight electrodes circularly placed in the abdominal area of the pregnant woman according to the study [4] and analysed data from real records. The condition for fECG amplitudes took into account the results of the [4]. The simulation again provides an ideal signal which would be obtained by this method in practice. The transabdominal signal also included the effect of hypoxic conditions which is observable. To avoid the situation in which the generated signals are available only within the NI LabVIEW environment, the signals can be exported into EDF format. The generated signal can be further analysed in more sophisticated environments (e.g. ECG Feature Extractor) or used as a signal source when creating algorithms for the detection of hypoxic conditions, FHR detectors, STAN analysers, etc. The simulator allows the user to import signals from other or similar projects.

5 Conclusion

The main aim of this article was a simulation of hypoxic conditions endangering proper fetal development and reducing the chance for a smooth childbirth without complications. The key point was the classification of hypoxic conditions within the available clinical recommendations that try to characterize significant signs of coming changes in vitality of the fetus. The work also draws from scientific articles and specialized literature dealing with the changes in the fetal ECG curve.

The core of the article was the utilization of current knowledge about the classification of hypoxic conditions. The result was the design and implementation of a virtual simulator in the NI LabVIEW development environment. Properties of the simulator enable the user to set heart rates and amplitudes of the individual oscillations and waves within amplitude limits stipulated for the fetal ECG. Furthermore, they allow the user to select states corresponding to the diagnosis of hypoxic conditions. This relates to the possibility to choose pulse rates reflecting the normal, intermediate or pathological condition of the fetus in terms of CTG clinical recommendations. The simulator also allows the user to adjust the ST segment through a numerical selection of the T wave—QRS complex ratio. Within the ST segment changes, it is again possible to choose between normal, intermediate or pathological conditions. In addition to these options, the user can choose biphasicity of the ST segment in three degrees. All these features are complemented by the ability to display waveforms in the time as well as frequency domains.

The realized simulator is able to accurately demonstrate the fECG curve. This has been confirmed by comparisons of its courses with real records. The simulator is a practical tool for demonstrating the effects which are signs of hypoxic conditions. With the help of export functions, the simulation results can be further presented or used as a source for other analyses of FHR, ST analysers, for doctors, engineers, next research, etc.

This source of synthetic fECG records has the potential for further utilization in the development of non-invasive methods for monitoring fECG signals due to clearly defined parameters. Thanks to this fact, the signal can be used as a reference in the development of CTG or ST analysers, in the case of invasive as well as non-invasive methods of monitoring.

References

1. Clifford, G.D., Silva, I., Behar, J., Moody, G.B.: Non-invasive fetal ECG analysis. Physiol. Meas. **35**(8), 1521 (2014)
2. Estes, S., Utsey, K., Kalobwe, E.: Mathematically modeling fetal electrocardiograms. Pursuit-J. Undergrad. Res. Univ. Tennessee **5**(1), 10 (2014)
3. Jagannath, D., Selvakumar, A.I.: Issues and research on foetal electrocardiogram signal elicitation. Biomed. Signal Process. Control **10**, 224–244 (2014)
4. Martinek, R.: The use of complex adaptive methods of signal processing for refining the diagnostic quality of the abdominal fetal electrocardiogram. Ph.D. dissertation, VSB—Technical University of Ostrava, Department of Cybernetics and Biomedical Engineering, Czech Republic (2014)
5. Martinek, R., Al-Wohaishi, M., Zidek, J.: Software based flexible measuring systems for analysis of digitally modu-lated systems. In: 2010 9th Roedunet International Conference (RoEduNet), pp. 397–402. IEEE (2010)
6. Martinek, R., Zidek, J.: Refining the diagnostic quality of the abdominal fetal electrocardiogram using the techniques of artificial intelligence. J.: Przeglad Elektrotchniczny 155–160 (2012)
7. Martinek, R., Zidek, J.: A system for improving the diagnostic quality of fetal electrocardiogram. J.: Przeglad Elektrotchniczny (Electr. Rev.) **R 88**, 164–173 (2012)
8. Matonia, A., Jezewski, J., Kupka, T., Wrobel, J., Horoba, K.: Modelling of non-invasively recorded maternal and fetal electrocardio-graphic signals. Biocybern. Biomed. Eng. **25**(2), 27 (2005)
9. Noventa Medical AB: Stan s31 (2012). http://www.neoventa.com/products/stan
10. Sameni, R., Clifford, G.D.: A review of fetal ecg signal processing; issues and promising directions. The Open Pacing Electrophysiol. Therapy J. **3**, 4 (2010)
11. Sameni, R., Clifford, G.D., Jutten, C., Shamsol-lahi, M.B.: Multichannel ECG and noise modeling: application to maternal and fetal ECG signals. EURASIP J. Appl. Signal Process. **2007**(1), 94–94 (2007)
12. Stancases: Stancases—cases tagged fECG (2015). http://www.stancases.com/tags/fecg

A New Hybrid Weighted Optimization Model for Multi Criteria ABC Inventory Classification

Hadhami Kaabi and Khaled Jabeur

Abstract ABC analysis is a commonly used inventory classification technique which consists in dividing a set of inventory items into three categories: category A contains the most important items, category B includes the moderately important items and category C contains the least important ones. The purpose of this classification is to manage inventory items in an efficient way by relaxing controls on low valued items and applying more meticulous controls on high valued items. In this paper, we propose a new hybrid weighted optimization model which combines the usefulness of two well-known inventory classification models (ZF-model [13] and H-model [10]). To measure the performance of the proposed model with respect to some existing classification models, a comparative study—based on a service-cost analysis—is conducted.

Keywords ABC analysis • Multi criteria classification • Weighted optimization models • Service-cost inventory

1 Introduction

ABC analysis is a well-known inventory classification technique used to categorize a huge number of inventory items, since, in practice items cannot be managed with equal attention (e.g. specialized surgical equipment versus latex gloves in a hospital). This technique consists in classifying inventory items into three predefined

---

H. Kaabi (✉)
Institut Supérieur de Gestion de Tunis, Université de Tunis,
41, Rue de la Liberté – Bouchoucha, 2000 Bardo, Tunisie
e-mail: hadhamikaabi@gmail.com

K. Jabeur
Institut Supérieur de Commerce et de Comptabilité de Bizerte,
Université de Carthage, rue Sadok el Jaouani - Zarzouna, 7021 Bizerte, Tunisie

A. Abraham et al. (eds.), *Proceedings of the Second International Afro-European Conference for Industrial Advancement AECIA 2015*, Advances in Intelligent Systems and Computing 427, DOI 10.1007/978-3-319-29504-6_26

and ordered categories: category A contains the most valuable items, category B includes the moderately valuable items and category C contains the least valuables ones. In fact, ABC analysis is based on the Pareto principle which assumes that a small percentage of items (A-items) accounts for a high percentage of value. This value may be profits, sales, or any other measure of importance. The main aim of ABC analysis is to identifying different categories of inventory items, also known as Stock-Keeping Units (SKUs), which will require different management and control policies. Traditional ABC analysis consists in classifying a set of inventory items based on just one criterion which is the Annual Dollar Usage (ADU). However, recent literature on the ABC classification techniques has identified many other criteria such as, Lead Time (LT), Average Unit Cost (AUC), Inventory Holding Cost (IHC), Durability (D) and Demand Distribution (DD) may be significant to classify items into different categories. Hence, ABC analysis is inherently a Multi Criteria Inventory Classification (MCIC) problem. In any ABC MCIC model, the classification of items into one of the three above categories is based on their weighted scores. The score of each item is the result of an aggregation function that combines the item evaluations on the different criteria and the criteria weights.

In the ABC analysis literature, several classification models issued from different methodologies—such as Mathematical Programming (MP), Artificial Intelligence (AI) and Multi Criteria Decision Making (MCDM)—have been proposed to solve the ABC inventory classification problem. This paper specially focuses on classification models based on techniques issued from Mathematical Programming (MP) methodology. For this purpose, we propose a new hybrid weighted optimization model which combines the usefulness of two well-known inventory classification models: the ZF-model [13] and the H-model [10]. The ZF-model, which is a linear programming model, has the advantage of computing a global score for each item based on two opposite scores: a pessimistic score (also called good score) and an optimistic score (also called bad score). We believe that this way of doing produce a more realistic global score for each item in which the decision maker may express his own values system. The H-model, which is a non-linear programming model, has the advantage of maintaining the effects of weights in the computation of the global score of each item. To measure the performance of our hybrid model with respect to some other existing classification models, a comparative study based on a service-costs analysis is proposed.

The rest of the paper is organized as follows. In Sect. 2, related literature of the ABC inventory classification models is reviewed with a focus on the weighted optimization models (MP methodology). The proposed hybrid weighted optimization model is described in Sect. 3. Section 4 presents a comparative study of the proposed hybrid model with respect to some existing weighted optimization models by using a data set of items consumed in a Hospital Respiratory Therapy Unit (HRTU). Finally, concluding remarks and suggestions for further research are reported.

2 Related Research

In any ABC MCIC models, two main topics are often addressed: (i) The specification of the aggregation model used in the computation of the item score, and (ii) The estimation of the aggregation model parameters (e.g. criteria weights, discrimination thresholds, etc.). Many techniques issued from different methodologies such as Mathematical Programming (MP), Artificial Intelligence (AI) and Multi Criteria Decision Making (MCDM) are proposed to establish a classification of inventory items that optimize an objective function (e.g. minimizing an inventory cost function, maximizing the items scores, etc.).

Classification models based on MCDM methodology proceed in two steps to classify an inventory item into one of the three ABC categories. In the first, an MCDM method—essentially the Analytic Hierarchy Process (AHP) [7]—is applied once to determine the criteria weights. In the second step, an aggregation function is used to compute the global score of each inventory item. The models proposed by Flores et al. [3], Partovi and Burton [7], Bhattacharya et al. [2] and Vencheh and Mohamadghasemi [11] are considered as typical MCDM based classification models.

Classification models issued from Artificial Intelligence (AI) methodology propose a learning process that estimates the criteria weights in order to produce a classification of items that minimize some objective functions (e.g. inventory cost function, misclassified items rate, correct classified items rate, etc.). The assessment of the criteria weights is carried out by some well-known metaheuristics such as Genetic Algorithm (GA) [4], Particle Swarm Optimization (PSO) [9] and Simulated Annealing (SA) [5]. Once the criteria weights are determined, a score is computed for each item by using some aggregation rules (e.g. weighted sum. Yu [12] proposed a comparative study in which three AI based classification models (Support Vector Machines (SVMs), Back Propagation Networks (BPNs), and the K-Nearest Neighbor (KNN)) and the traditional Multiple Discriminant Analysis (MDA) are tested.

Classification models based on Mathematical Programming MP methodology propose linear and nonlinear programming to generate a weight vector (when these weights are established for each criterion) or matrix (when these weights are established for each (criterion, item) pair) that maximize or minimize the weighted score of each item. Ramanathan [8] proposed a linear optimization model—called R-model—to obtain a set of optimal weights which maximizes the score of each item expressed by a weighted additive function.

The main drawback of the R-model is that an inventory item with a high evaluation on an irrelevant criterion may be considered as a critical item, i.e. belonging to category A.

To address this drawback, Zhou and Fan [13] proposed an extended version of the R-model (called ZF-model). A detailed description of the ZF-model will be presented in the next section. Ng [6] proposed a linear optimization model (called Ng-model) in which the decision maker may integrate—for each item—a ranking of the criteria weight set. The main weakness of the Ng-model is that the optimal score of items are independent of the criteria weights. Vencheh [10] provides an improved version of the Ng-model by proposing a nonlinear optimization model, called H-model, which maintains the effects of weights in the optimal score of items. A detailed description of the H-model will be presented in the next section.

3 A New Hybrid Optimization Model for ABC MCIC

In this section, the different steps of our new hybrid optimization model, hereafter called ZF-H-model, will be detailed. As mentioned earlier, this model combines the usefulness of two well-known inventory classification models: a linear optimization model (the ZF-model) and a nonlinear optimization model (H-model). Before presenting the details of these three optimization models (ZF-model, H-model and ZF-H-model), some mathematical notions should be introduced: i is the item index, $i = 1, \ldots, M$ where M is the number of items; j is the criteria index, $j = 1, \ldots, N$ where N is the number of criteria; x_{ij} is the evaluation of the item i according to the criterion j; w_{ij} is the decision variable which represents a non-negative weight assigned to each evaluation $x_{ij;}$ s_i is the optimal score of item i.

3.1 The ZF-Model

According to the ZF-model, the determination of the optimal score s_i of each item i is based on the determination of two scores: the good score gs_i and the bad score bs_i. These scores are obtained by solving—for each item i—the two following linear programming models:

$$(LP_i^g): \begin{cases} gs_i = Max \sum_{j=1}^{N} w_{ij}x_{ij} \\ subject\ to \\ \sum_{j=1}^{N} w_{ij}x_{kj} \leq 1 \qquad k = 1, 2, 3, \ldots M \\ w_{ij} \geq 0 \qquad j = 1, 2, 3, \ldots N \end{cases} \quad and\ (LP_i^b): \begin{cases} bs_i = Min \sum_{j=1}^{N} w_{ij}x_{ij} \\ subject\ to \\ \sum_{j=1}^{N} w_{ij}x_{kj} \geq 1 \qquad k = 1, 2, 3, \ldots M \\ w_{ij} \geq 0 \qquad j = 1, 2, 3, \ldots N \end{cases}$$

The final optimal score s_i of each item i is computed by combining the good score gs_i and the bad score bs_i as follows:

$$s_i = \lambda\left(\frac{gs_i - gs^-}{gs^+ - gs^-}\right) + (1 - \lambda)\left(\frac{bs_i - bs^-}{bs^+ - bs^-}\right)$$

where $gs^+ = \max\{gs_i, i = 1, \ldots M\}$, $gs^- = \min\{gs_i, i = 1, \ldots M\}$, $bs^+ = \max\{bs_i, i = 1, \ldots M\}$, $bs^- = \min\{bs_i, i = 1, \ldots M\}$ and $0 \le \lambda \le 1$ is a control parameter which may reflect the optimism degree (or pessimism degree) of the decision-maker in the computation of optimal score s_i.

3.2 The H-Model

The H-model determines a set of optimal criteria weights $w_j, j = 1, \ldots N$[1] that maximizes the weighted score s_i of each item i. This optimization model is presented as follows:

$$(LP_i^H): \begin{cases} s_i = Max \sum_{j=1}^{N} w_j x_{ij} \\ subject\ to \\ \sum_{j=1}^{N} w_j^2 = 1 \\ w_j - w_{j+1} \ge 0 \qquad j = 1, 2, \ldots N-1 \\ w_j \ge 0 \qquad j = 1, 2, \ldots N \end{cases}$$

3.3 The Proposed Hybrid ZF-H-Model

In the proposed hybrid ZF-H-model the determination of the optimal score s_i of each item i is based on the determination of two scores: the good score gs_i and the bad score. The different steps of our proposed model are as follows:

- **Step 1. Normalization of the evaluations** x_{ij}
  Convert all evaluations x_{ij} in a 0–1 scale using the following linear transformation:

$$x_{ij}^N = \frac{x_{ij} - \min\limits_{j=1..N} x_{ij}}{\max\limits_{j=1..N} x_{ij} - \min\limits_{j=1..N} x_{ij}}$$

[1]In this model, a weight is assigned to each criterion j and not—as the other optimization models—to each evaluation x_{ij}.

- **Step 2. Generate for each item** i **the optimal good score** gs_i
  To obtain for each item i the optimal good score gs_i, the following non-linear programming model should be solved:

$$\left(LP_i^{ZF-H-g}\right): \begin{cases} gs_i = Max \sum_{j=1}^{N} w_{ij} x_{ij}^N \\ subject\ to \\ \sum_{j=1}^{N} w_{kj}^2 \leq 1 \qquad k = 1, 2, \ldots M \\ w_{ij} - w_{i(j+1)} \geq 0 \qquad j = 1, 2, \ldots N-1 \\ w_{ij} \geq 0 \qquad j = 1, 2, \ldots N \end{cases}$$

- **Step 3. Generate for each item** i **the optimal bad score** bs_i
  To obtain for each item i the optimal bad score bs_i, the following non-linear programming model should be solved:

$$\left(LP_i^{ZF-H-b}\right): \begin{cases} bs_i = Min \sum_{j=1}^{N} w_{ij} x_{ij}^N \\ subject\ to \\ \sum_{j=1}^{N} w_{kj}^2 \geq 1 \qquad k = 1, 2, \ldots M \\ w_{ij} - w_{i(j+1)} \geq 0 \qquad j = 1, 2, \ldots N-1 \\ w_{ij} \geq 0 \qquad j = 1, 2, \ldots N \end{cases}$$

- **Step 4. Generate for each item** i **the final optimal score** s_i
  The final optimal score s_i of each item i is computed by using a convex combination of the good score gs_i and the bad score bs_i as follows:

$$s_i = \lambda \left(\frac{gs_i - gs^-}{gs^+ - gs^-} \right) + (1 - \lambda) \left(\frac{bs_i - bs^-}{bs^+ - bs^-} \right)$$

  where $gs^- = \min\{gs_i, i = 1, \ldots M\}$, $bs^+ = \max\{bs_i, i = 1, \ldots M\}$, $bs^- = \min\{bs_i, i = 1, \ldots M\}$ and $0 \leq \lambda \leq 1$ is a control parameter which may reflect the optimism degree (or pessimism degree) of the decision-maker in the computation of optimal score s_i. In this paper we assume that $\lambda = 0.5$ (neutral attitude).

4 Computational Results

To measure the performance of the proposed weighted nonlinear optimization model, i.e. the ZF-H-model, with respect some other optimization models (R-model [8], Ng-model [6], ZF-model [13] and H-model [11]), a comparative study based on a service-costs analysis is proposed in this section. For this purpose, a benchmark data set of 47 items used in a Hospital Respiratory Therapy Unit is employed. Each item in this data set is evaluated according to three criteria: Annual Dollar Usage (ADU), Average Unit Cost (AUC) and Lead Time (LT) (see Table 1). The final classification of the 47 items is obtained by splitting items according to the commonly used distribution in the ABC inventory classification literature: the 10 first ranked items are classified in category A, the next 14 items in the ranking are classified in category B and the remaining 23 items are classified in category C. The final classifications of inventory items according to the five tested optimization models are reported in Table 1. These results show that 20 of 47 items (42.55 %) are classified in the same categories by all tested models: 2 items belonging to the category A, 3 items belonging to the category B and 15 items belonging to the category C. When our model is compared to the ZF-model (respectively to the H-model), 31 of 47 items (65.95 %) (respectively 30 of 47 (63.82 %)) are classified in the same categories by both models. The above result is quit quite understandable since the proposed model is an hybridization of the ZF-model and the H-model.

To measure objectively the performance of all tested optimization models, their final item classifications are evaluated according to the inventory cost and the fill rate service level as suggested in Babai et al. [1]. Before presenting these two measures of performance, the following notations should be introduced (we assume that the inventory system is controlled with a record point, record quantity, i.e. (s, Q)): N is the number of items in the inventory systems (47); n is the number of categories (3); D_i is the demand of item i; σ_i is the standard deviation of the demand of item $i(1 \times D_i)$; L_i is the Lead Time of item i; Q_i is the order quantity of item i $\left(\sqrt{\frac{2W_iD_i}{h_i}}\right)$; CSL_i is the cycle service level of item i (99 % $i \in$ category A, 95 % $i \in$ category B, 90 % $i \in$ category C); $\Phi(.)$ is the standard normal probability distribution function; C_T is the total safety stock inventory cost $\left(\sum_{j=1}^{N} h_i k_i \sigma_i \sqrt{L_i}\right)$; k_i is the safety factor for each item $i\left(\phi^{-1}(CSL_i)\right)$; FR_i is the fill rate of each item $i\left(1 - \frac{\sigma_i\sqrt{L_i}}{Q_i} G(k_i)\right)$; G($k_i$) is the loss function of the standard normal distribution $\left(\frac{1}{\sqrt{2\pi}}e^{-\frac{k_i^2}{2}} - k_i[1-\phi(k_i)]\right)$; FR_T is the overall fill rate of the inventory system $\left(\frac{\sum_{j=1}^{N} FR_i D_i}{\sum_{j=1}^{N} D_i}\right)$, h_i: Inventory holding cost of item i (20 % of the average unit cost); W_i is the unit ordering cost of item i.

The results presented in the bottom of Table 1 show that the proposed model provides an inventory cost value (1947.208) close to those obtained by the ZF-model (1890.714) and the H-model (1999.784). In fact, the proposed ZF-model outperforms the H-model in terms of safety stock cost. This is an expected result since the proposed model is an hybridization of the ZF-model and the H-model. In the other hand, the Ng-model obtained the highest fill rate service level, which is the fraction of demands that are satisfied directly from stock on hand, and therefore the highest inventory cost.

Table 1 Comparison of ABC classification using the weighted optimization models

Item no	ADU	AUC	LT	ZF-model	R-model	Ng-model	H-model	ZF-H-model
1	5840.64	49.92	2	A	A	A	A	C
2	5670	210	5	A	A	A	A	A
3	5037.12	23.76	4	A	A	A	A	A
4	4769.56	27.73	1	C	B	A	A	B
5	3478.8	57.98	3	B	B	A	A	A
6	2936.67	31.24	3	C	C	A	B	B
7	2820	28.2	3	C	C	B	B	B
8	2640	55	4	B	B	B	B	A
9	2423.52	73.44	6	A	A	A	C	A
10	2407.5	160.5	4	A	B	A	C	A
11	1075.2	5.12	2	C	C	C	C	A
12	1043.5	20.87	5	B	B	B	B	B
13	1038	86.5	7	A	A	A	C	A
14	883.2	110.4	5	A	B	B	C	A
15	854.4	71.2	3	C	C	C	C	B
16	810	45	3	C	C	C	C	B
17	703.68	14.66	6	C	C	C	C	C
18	594	49.5	6	A	A	B	B	B
19	570	47.5	5	B	B	B	B	B
20	467.6	58.45	4	B	C	C	C	B
21	463.6	24.4	4	C	C	C	C	C
22	455	65	4	B	C	C	C	B
23	432.5	86.5	4	B	C	B	B	B
24	398.4	33.2	3	C	C	C	C	C
25	370.5	37.01	1	C	C	C	C	C
26	338.4	33.84	3	C	C	C	C	C
27	336.12	84.03	1	C	C	C	C	C
28	313.6	78.4	6	A	A	B	B	B
29	268.68	134.34	7	A	A	A	C	A
30	224	56	1	C	C	C	C	C
31	216	72	5	B	B	B	B	B
32	212.08	53.02	2	C	C	C	C	C
33	197.92	49.48	5	B	B	B	B	C
34	190.89	7.07	7	B	A	B	B	C
35	181.8	60.6	3	C	C	C	C	C
36	163.28	40.82	3	C	C	C	C	C
37	150	30	5	B	B	C	C	C
38	134.8	67.4	3	C	C	C	C	C
39	119.2	59.6	5	B	B	B	B	C
40	103.36	51.68	6	B	B	B	B	C

(continued)

Table 1 (continued)

Item no	ADU	AUC	LT	ZF-model	R-model	Ng-model	H-model	ZF-H-model
41	79.2	19.8	2	C	C	C	C	C
42	75.4	37.7	2	C	C	C	C	C
43	59.78	29.89	5	C	B	C	C	C
44	48.3	48.3	3	C	C	C	C	C
45	34.4	34.4	7	B	A	B	B	B
46	28.8	28.8	3	C	C	C	C	C
47	25.38	8.46	5	C	B	C	C	C
Safety stock inventory cost				1890.714	1855.034	2022.014	1999.784	1947.208
Fill rate				0.984	0.986	0.991	0.99	0.972

5 Conclusion

In this paper we have proposed a new hybrid weighted optimization model which combines the usefulness of two well-known inventory classification models: the ZF-model [13] and the H-model [11]. To measure the performance of our hybrid model with respect some other existing classification models, a comparative study based on a service-costs analysis is proposed. This analysis has shown that our proposed model has obtained very promising results since it outperforms the H-model and Ng-model in term of minimizing the inventory cost. Some interesting directions for further researches consist in: (i) considering quantitative and qualitative criteria in the inventory classification models and (ii) combining the weighted optimization models with classification models issued from other methodologies such as Multi Criteria Decision Making and Artificial Intelligence.

References

1. Babai, M.Z., Ladhari, T., Lajili, I.: On the inventory performance of multi-criteria classification methods: empirical investigation. Int. J. Prod. Res. (2014)
2. Bhattacharya, A., Sakar, B., Mukherjee, S.K.: Distance based concensus method for ABC analysis. Int. J. Prod. Res. **45**, 3405–3420 (2007)
3. Flores, B.E., Olson, D.L., Dorai, V.K.: Management of multi criteria inventory classification. Math. Comput. Model. **16**(12), 71–82 (1992)
4. Guvenir, H.A., Erel, E.: Multicriteria inventory classification using a genetic algorithm. Eur. J. Oper. Res. **105**(1), 29–37 (1998)
5. Mohammaditabar, D., Ghodsypour, S.H., O'Brien, C.: Inventory control system design by integrating inventory classification and policy selection. Int. J. Prod. Econ. **140**, 655–659 (2011)
6. Ng, W.L.: A simple classifier for multiple criteria ABC analysis. Eur. J. Oper. Res. **177**(1), 344–353 (2007)
7. Partovi, F.Y., Burton, J.: Using the analytic hierarchy process for ABC analysis. Int. J. Oper. Manage. **13**, 29–44 (1993)

8. Ramanathan, R.: ABC inventory classification with multiple-criteria using weighted linear optimization. Comput. Oper. Res. **33**(3), 695–700 (2006)
9. Tsai, T.H, Yeh, S.W: A multiple objective particle swarm optimization approach for inventory classification. Int. J. Prod. Econ. **114**(2), 656–666 (2008)
10. Vencheh, A.H: An improvement to multiple criteria ABC inventory classification. Eur. J. Oper. Res. **201**(3), 962–695 (2010)
11. Vencheh, A.H., Mohamadghasemi, A.: A fuzzy AHP-DEA approach for multiple criteria ABC inventory classification. Expert Syst. Appl. **38**, 3346–3352 (2011)
12. Yu, M.C.: Multi-criteria ABC analysis using artificial-intelligence-based classification techniques. Expert Syst. Appl. **38**(4), 3416–3421 (2011)
13. Zhou, P., Fan, L.: A note on multi-criteria ABC inventory classification using weighted linear optimization. Eur. J. Oper. Res. **182**(3), 1488–1491 (2007)

Proposition of a Protocol for E-voting Systems that Breaks the Link Between the Voter and His Vote

Dunja Majstorović

Abstract One of the most challenging aspects in computer-supported voting is to combine the apparently conflicting requirements of privacy and verifiability. In this paper, some of the definitions related to the electoral procedure are introduced, different mechanisms for identification and authentication of voters are discussed, and then a protocol is proposed which breaks the link between identity of the voter and the vote itself very early in the electoral process, enabling both privacy and verifiability. The protocol is best suited for a system where votes are casted in polling stations, but recommendations for use in a remote voting scheme are also given.

Keywords Electronic voting · Anonymity · Verifiability

1 Introduction

Freedom to vote and correctness of the election process are fundamental requirements for democratic elections. Even though the right to vote might seem rather a political issue than a technical one, it can be assured by means of anonymity: the system ensures that the voter's preference is not revealed to anyone. Correctness on the other hand, is assured by means of verifiability: a voter can verify that his vote affects the election result as he intended [1]. This paper introduces a new protocol that is resistant to some of the well-known attacks and problems in today's e-voting systems, while giving enough freedom to election organizers in terms of customizing their system for their own needs. For example, we don't presume any specific encryption protocols, but leave it to the election committee to decide which one they want to use. The offered solutions are simple and logical, which is a very important characteristic, as these systems have to be easily understandable to the general public in order for voters to trust them.

D. Majstorović (✉)
School of Computers, Union University, Belgrade, Serbia
e-mail: majstorovic.dunja.rs@ieee.org

A. Abraham et al. (eds.), *Proceedings of the Second International Afro-European Conference for Industrial Advancement AECIA 2015*, Advances in Intelligent Systems and Computing 427, DOI 10.1007/978-3-319-29504-6_27

2 Fundamentals

When discussing a system as complicated as electronic voting, it is necessary to set up a definition. The Council of Europe adopted this one: "Voting can be considered electronic if electronic means are used at least for casting of the vote" [2]. Both in traditional (pen and paper) and electronic voting schemes, the entire voting procedure can be defined as a cycle which consists of three stages:

1. Pre-election period: This is the time from calling an election until the official start of polling.
2. Election period: This is the actual Election Day where the vote casting takes place.
3. Post-election period: This is the time during which the results are announced and a new election is called [3].

Whichever voting style is used, the roles of identification and authentication are the same: (1) ensuring that only eligible voters may cast a vote—those who have the right to vote, (2) ensuring that eligible voters cast their vote at most once—they can either not vote at all or vote only once, and (3) ensuring that eligible voters are not forcefully prevented from voting—meaning, if they want to vote, they should be able to do so. Unlike for the roles of identification and authentication, the literature is divided on the function of anonymity and verifiability. In this paper, the notion from [1] is adopted:

1. Verifiability is split into two parts:

   (a) Individual verifiability: The voter can verify that his vote affects the result correctly.
   (b) Universal verifiability: Anyone can verify that the announced result is a correct accumulation of individual votes.

2. The notion of anonymity is further refined into three following forms:

   (a) Privacy: Ensures that no observer learns how a voter voted.
   (b) Receipt-freeness: Ensures that the voter is forced to keep his vote private, even if he would like to share it. It makes sure that the voter cannot prove how he voted after the elections, thus preventing vote buying.
   (c) Coercion-resistance: Ensures receipt-freeness and resistance to attacks such as randomization (the voter is forced to vote for a random candidate), simulation (the voter is forced to give his voting credentials to the adversary, who then votes instead), forced abstention (the voter is forced not to vote at all).

3 Mechanisms for Identification, Authentication, and Privacy Preservation

The most commonly used techniques for identification and authentication in e-voting, both in remote and local (non-remote) systems are:

1. Username and password: Identification relies on voter knowing a secret.
2. Transaction number (TAN): The voter possesses something that identifies him.
3. Biometrics: The voter with his individual biometric properties identifies himself. A reader for the biometric feature is needed.
4. Smart Cards: The voter knows a secret that in combination with possession of the card identifies him, or a property pattern of the voter is stored on a smart card that is checked against the voter's property when casting a ballot—either way, a reader for the smart card is needed [4].
5. In person—makes sense only in the non-remote voting scenario: Just like in the traditional elections, voter identifies himself with an identity document on the spot, and then proceeds to vote electronically.

As seen above, there is a variety of mechanisms for identification and authentication to choose from, but that would only be half of the job done. The second half is made of efforts to preserve privacy of the voter. So far there's been a high number of techniques proposed for this purpose, and most of them were cryptographic solutions where the vote is crypted. These ideas won't be discussed in too much detail, as it isn't necessary to study them thoroughly to be able to understand why they are not appropriate for this kind of systems. Primary reason for marking them as inappropriate is that cryptographic schemes may turn out to be broken, and messages encrypted with such schemes may (eventually) be decrypted without using the decryption key—a particular risk for e-voting, as often, sensitive data is published to provide verifiability [1]. Although the timeframe in which votes must remain secret is a subject that should be discussed within social sciences, it is something that has to be considered when designing algorithms for these systems. Most algorithms that include cryptosystems rely on the assumption that, with today's technology, it is impossible to decipher encrypted text without knowing the key. Unfortunately, "today" is not enough for electoral systems. From the legal point of view, votes should remain anonymous forever.

4 The Protocol

In this chapter, a protocol for electronic voting is presented. All of the steps are distributed in stages (periods) of the electoral process.

4.1 Pre-election Period

At some point before Election Day, pre-election period begins when actions described below are taken.

1. Initialization of the server: Electoral Commission configures the server in several steps.

   (a) Loading of the candidates: Since the number of candidates is usually small, this can be done manually. At this point, the order of the candidates is defined.
   (b) Loading of the voters: It is assumed that a database of all eligible voters exists and is accurate—if a person is on that list, they have the right to vote.
   (c) Creation of credentials for the voters: This is where the anonymity of the voter is established. Since authentication is implemented in the preceding step, in this step we can simply take the number of eligible voters from the list and create the same number of username and password pairings. In this scenario, the right to vote can be viewed simply as a token chip—they are all identical, and the only thing that matters is that everyone can use theirs (at most once), and that everybody gets exactly one. For example, a random generated username-password pair for voter 'John Smith' would be 'q9b4pyy-x111fg3gws'.
   It is desirable that usernames and passwords differ from each other as much as possible, which is achieved by using more characters in credentials. It is up to election organizers to choose characters that will be used, but since they'll most likely be printed, it would be a good idea to avoid these: '0' (zero), 'O' (letter 'o'—both capital and small), '1' (number one), 'l' (small letter 'L') and 'I' (capitalized letter 'i'). This is because these symbols, particularly printed, look very similar in different fonts, and therefore can create confusion.
   In this step, the committee should also define the number of attempts given to the user to enter his credentials.
   (d) A pair of keys (a public and a private one) is generated. The private key is used to decrypt the votes once the election day is over, and should be kept in secret until it is used. It is left to the committee to decide how they will guard the key. Optionally, the private key can be subdivided into m parts, where each part is given to a trusted person (a member of the election committee for example) and then recollected at the end of the election day. The public key (hereinafter referred to as 'voting public key') is used by voters to crypt their votes (chosen candidate + username/password) and will be sent to each voter (client application) as they open their session with the server. At the beginning of the election period, another set of keys will be generated to be used for encryption of

voters' credentials during the login client-server message exchange. The election committee can choose to generate a pair of keys for this purpose per elections, per polling station (every machine in the station gets the same public key) or per voting machine (every machine in the station gets a different public key).
Although this protocol leaves room for the election commission to choose the cryptosystem they want to use (most likely the one they trust the most), it is recommended that an asymmetric key encryption scheme be employed. If the organizers decide to go with a well known system such as RSA, they can use the guidelines from official documentation, such as RFC 2437 (RSA Cryptography Specifications Version 2.0) while preparing the election softvare.

2. Delivery of credentials: Depending on the importance of elections, credentials can be delivered to voters in different ways. Assuming that considered elections are of high priority, a scenario where voters cast their votes in dedicated polling stations is described.

   (a) Step 1. If there is n eligible voters, there should be n username/password combinations.
   (b) Step 2. Each combination is printed and then covered with a scratch-off sticker. This paper is inserted into an envelope.
   (c) Step 3. Closed envelopes are then mixed, after which they can be distributed to local polling stations.

If this protocol is implemented in a remote voting scheme, credentials could be sent via mail in which case voters' names and addresses are to be put on envelopes after they've been mixed. One may wonder whether this process could have been shortened—the credentials will serve as a receipt after the vote has been casted so why not simply let the machine print the credentials? There are two scenarios to contemplate regarding this question: (1) the machine prints the receipt after the vote is casted and (2) the machine prints the receipt before the vote is casted. In both cases, we assume that the voter was identified prior to accessing the machine, and that the server does not ask for login credentials. If the machine prints the receipt (the credentials) after voting, than the system would be vulnerable to a "clash attack", described in [5]. The main idea behind this attack is that voting machines manage to provide different voters with the same receipt, which in this protocol, would mean that two different voters, after they've voted the same way, each get a copy of the same receipt (username/password pair). Although this attack is simple, [5] shows it can be carried out on prominent voting systems ThreeBallot and VAV [6], the Wombat voting system [7] and one variant of Helios [8]. Obviously, this attack can't be executed if the voter gets his receipt before he chooses his candidate (scenario 2), because the attacker (the malicious software) doesn't know how his victim is going to vote. If two persons get the same credentials but vote differently, one of them will notice that his vote was modified and therefore the attack would be

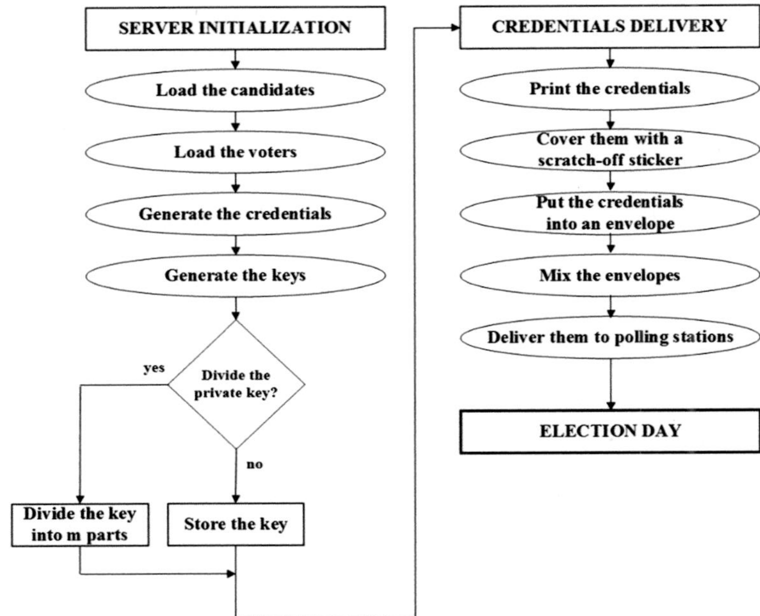

Fig. 1 Pre-election period

unsuccessful. Although it might seem that printing the receipt before voting solves the problem, if the rest of the protocol remains unchanged, a new issue emerges: there's no guarantee that the voter won't vote multiple times, once he is given access to the machine. This issue would actually be present in both scenarios, so for that reason, the idea of having the voting machine print the credentials is dismissed.

The entire pre-election period (for a non-remote scheme) is concisely depicted in Fig. 1.

4.2 Election Period

At the beginning of the Election Day, the server side of the e-voting application and voting machines in polling stations should be activated.

As soon as connection is established, server generates and sends the 'credentials public key' to every voting machine. This is when the server is ready to take on clients and start collecting votes. A client is a program that runs on voting machines located in polling stations. Prior to activation, two lists should be created: one for votes and the other one for credentials that had already been used by other voters. A scenario that demonstrates how the protocol works in this stage (period) is given bellow.

1. Voter arrives to the polling station and identifies himself with the local electoral commission with an identity document (ID). The electoral commission either hands the voter (exactly) one closed envelope or he chooses one himself.
2. Voter approaches the voting machine, which creates a session with the server. Voter then opens his envelope and scratches the sticker to find his credentials.
3. The server asks for credentials. If they are entered correctly, it offers a list of candidates for the voter to choose from. There are two requirements for credentials to be accepted as correct: (1) they match one of those in the database generated in pre-election period and (2) they haven't been used before.
4. In case the user enters the wrong credentials, the server will let him try again only if it's been configured to do so (in pre-election period).
5. After the choosing of the candidate, the client program merges voter's credentials to his chosen candidate, crypts that message with the voting public key and sends it to the server. The reason for crypting the vote in this step is simply to make sure that it doesn't get sniffed and/or altered on it's way to the server. Since this protocol enables verifiability regardless of which cryptosystem is deployed (or whether any is used at all), the encryption part can completely be left out—as long as voters remember their usernames, they'll know if their votes have been counted correctly once they see the results. Nevertheless, it is not recommended to send votes in plaintext, because an observer who somehow eavesdrops on network traffic, could relate votes to voters. The worst possible scenario in this protocol regarding privacy and anonymity would be a situation where the channel is tapped and votes are transferred in plaintext. In this case, a malicious observer could potentially figure out the links between voters and their votes by simply memorizing the times at which they used the voting machines.
6. As soon as the server receives the message, it can confirm it's arrival and close the connection with the client. It stores the crypted message and continues to serve clients until the Election Day ends.
7. Server inscribes the vote in its database after a random period of time, thus preventing an observer to link the time at which the voter used the machine with database's entry timestamp.

Assuming that the voter enters his credentials correctly at the first attempt, an overview of client-server communication is given in Fig. 2.

Before the beginning of the exchange illustrated in Fig. 2, voter was identified by the local electoral commission with his ID and given a randomly chosen envelope with credentials. Also, the voting machine had previously obtained the 'credentials public key'.

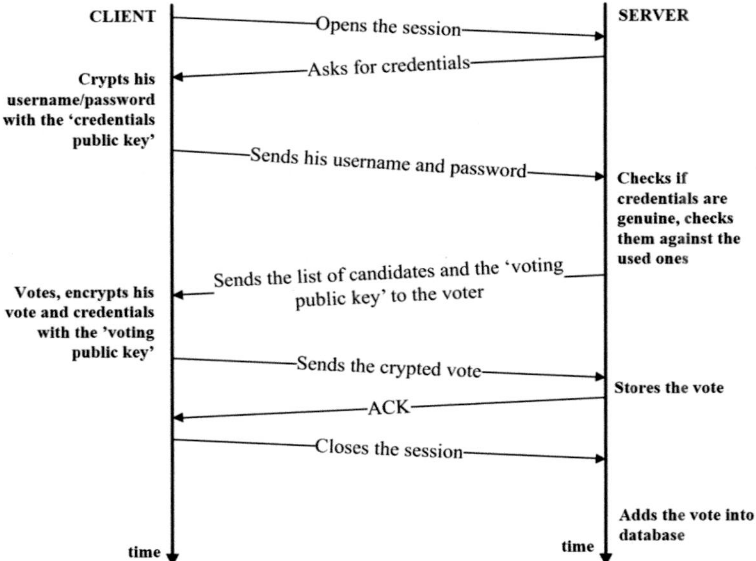

Fig. 2 Client/Server exchange during voting

4.3 Post-election Period

This period begins the moment Election Day ends. Events that occur at this stage are listed below.

1. The server side of the election application should be stopped, which means it will not take requests from clients anymore. From this moment on, the unused credentials do not pose a danger as clients (voting machines) can't connect with the server anymore, but they should still be destroyed. Prior to publishing of the results, all used usernames should be made public, in order to prevent any corrupt behavior with the unused ones after the official end of the Election Day. At the end of the Election Day, the list of all used usernames is already decrypted, which can be seen in Fig. 2, and therefore can be published immediately.
2. If the private key was divided into m parts, all m trusted persons then come together with their pieces and decrypt the votes. If not, the server uses the key to decrypt the votes. This is when the counting begins.
3. As soon as the server (or more of them) finishes with counting, results can be published. Beside the announcement of the winner, percentages and other statistics, all votes (username + chosen candidate) can also be published.
4. At this point, voters can verify that their votes have been counted as intended, by simply finding their own username in the list and checking if the candidate

they've chosen is linked to it. Since every voter is able to verify his vote, universal verifiability is provided.
As for anonymity, only the voter checking the results knows his own credentials. Therefore, nobody else can relate usernames to candidates from looking at the election results.

Figure 3 outlines the steps taken in this period, assuming that it is up to the specific election committee to find a convenient way to publish the results and make them easily accessible to voters.

5 Slight Changes for Coercion-Resistance

Maybe the biggest and the most obvious flaw of this protocol is that it isn't receipt-free, and therefore, isn't coercion-resistant. The envelope every voter gets when he identifies himself at the polling station is a receipt that he could use to prove to somebody else how he voted once the results are published. Also, he could be forced to vote in a certain way. Regarding the implementation of the protocol in a remote e-voting system, vote selling becomes even easier, as credentials can be sold and then easily used by another person. One way to avoid this problem (in the non-remote system) is to slightly alter the protocol. We explore the idea of letting coerced voters create fake votes (presented in [9]) through proposing upgrades that would give similar abilities to voters in this protocol.

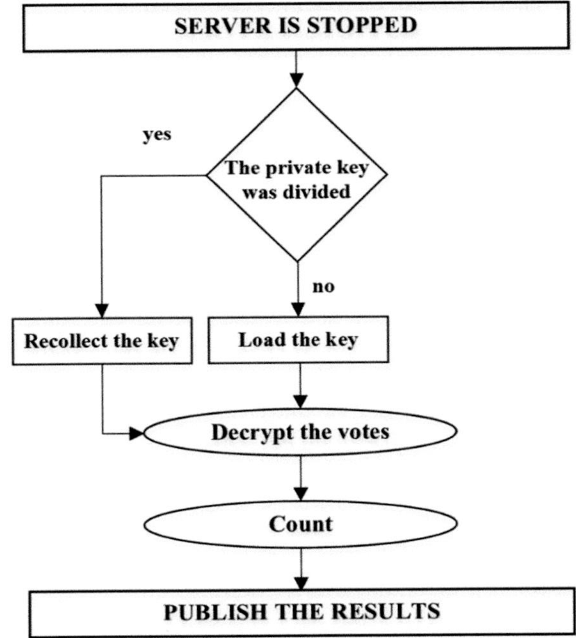

Fig. 3 Post-election period

1. Pre-election period. In addition to creating n username/password pairs, where n is the number of voters, server generates another m pairs, where m is the number of candidates. We now have a pair of credentials for each candidate. Right before the Election Day begins, each one of those m pairs are linked to one candidate: m initial votes are created. These are added to the votes database and will not be counted afterwards. Also, the private 'credentials key' is not shared but is stored on the server, and will be used to decrypt the votes as soon as they arrive.
2. Election period. Voter V_1 is identified by the polling station commission and is given his random credentials. He approaches the machine and votes for candidate C_1. His vote is instantly decrypted by the server and memorized. At this moment, the voter asks for an 'coercion receipt' for candidate C_2. The server checks its votes database, randomly chooses one username/password pair that is linked to C_2 and sends it back to him. If no one has previously voted for C_2, he'll get the initial credentials linked to candidate C_2, created in step 1. The machine prints out the credentials and covers them with a scratch-off sticker. The voter can now safely throw away one of the receipts he has.
   This lowers the level of trust between those who sell their votes and those who buy them.

If these recommendations are obeyed, not only will the system be resistant to coercion and vote buying, but also to 'trash attacks' [10]. As the name implies, a trash attack is an attack most likely carried out by election authorities, who have somebody pick up papers with credentials from the trash and then modify the votes linked with those credentials. Obviously, the voters who have thrown away their receipts will not verify their votes so the authorities take advantage of that.

These upgrades are shown as an additional option and not implemented by default in the protocol for two reasons: firstly, their implementation would require additional investment, and secondly, depending on the importance of the election, the organizers might predict low interest in manipulation and therefore no need for extra safety.

6 Conclusion

This paper proposes a new mechanism for electronic voting that promises to assure anonymity and verifiability. The goal of this work was originally to solve some of the well-known issues of electronic voting systems in an easily understandable way. For that reason, the language used through the entire paper is simple, and the mechanisms used to mitigate some of the problems seem almost trivial. This is because both organizers of the elections and voters must well understand the system in order to trust it, and trust is a very important part of every electoral process.

Acknowledgments The author would like to acknowledge the valuable comments and suggestions on the form of this paper from Professor Desimir Vucic.

References

1. Langer, L., Jonker, H., Pieters,W.: Anonymity and verifiability in voting: understanding (un) linkability. In: Proceedings of the 12th Conference on Information and Communications Security, pp. 1–3 (2010)
2. Council of Europe: Operational and technical standards for e-voting, Recommendation and explanatory memorandum. Strassbourg: Council of Europe, (2004)
3. Suksi, M.: The Electoral Cycle: On the Right to Participate in the Electoral Process. In: Hinz, V.U., Suksi, M., (eds.) Election Elements: On the International Standards of Electoral Participation, pp. 1–42. Turku/Abo: Institute for Human Rights, Abo Akademi University (2003)
4. Krimmer, R., Triessnigm, S., Volkamer, M.: The development of remote e-voting around the world: a review of roads and directions, E-Voting and Identity (VOTE-ID 2007), p. 5 (2007)
5. Kusters, R., Truderung, T., Vogt, A.: Clash attacks on the verifiability of e-voting systems. In: IEEE Symposium on Security and Privacy (2012)
6. Rivest, R., Smith, W.: Three voting protocols: ThreeBallot, VAV and Twin. In: USENIX/ACCURATE Electronic Voting Technology (EVT 2007) (2007)
7. Wombat voting system (2011). http://www.wombat-voting.com/
8. Adida, B.: Helios: web-based open-audit voting. In: SS'08: Proceedings of the 17th conference on Security symposium, Berkeley, USENIX Association, pp. 335–348 (2008)
9. Juels, A., Catalano, D., Jakobsson, M.: Coercion-resistant electronic elections. In: 4th ACM Workshop on Privacy in the Electronic Society, pp. 61–70. Alexandria, USA (2005)
10. Benaloh, J., Lazarus, E.: The Trash Attack: An Attack on Verifiable Voting Systems and a Simple Mitigation. Technical Report MSR-TR-2011-115, Microsoft (2011)

Geometric Properties on the Perfect Decoupling Disturbance Control in Manufacturing Systems

Paolo Mercorelli and Manuel Schimmack

Abstract This paper proposes a control to localize disturbances in manufacturing systems based on a geometric approach. The investigated system is represented by a Multi Input Single Output (MISO) model and a controller is proposed such that a disturbance localisation is obtained. Such kind of localisation problems are very important to minimize errors in manufacturing systems. The main purpose of this work is, starting from a linear model, to show some geometric structural properties to realise a perfect localisation of the vibrations during the milling of a manufacturing phase. The localisation of the vibrations represents a key point in the motion control problems. To realise the proposed controlled strategy an observer is required for estimation of the unmeasurable state variables. Additionally, Computer simulation results are shown.

Keywords Geometric approach · Disturbance localisation · Manufacturing systems

1 Introduction

The main purpose of this work is to localise disturbances using a control strategy based on a geometric approach in manufacturing systems. In the past three decades, research on the geometric approach to dynamic systems theory and control has allowed this approach to become a powerful and a thorough tool for the analysis and synthesis of dynamic systems [1, 2, 8]. Over the same time period, mechanical systems used in industry and developed in research labs have also evolved rapidly.

P. Mercorelli (✉) · M. Schimmack
Institute of Product and Process Innovation, Leuphana University of Lueneburg, Volgershall 1, 21339 Lueneburg, Germany
e-mail: mercorelli@uni.leuphana.de

M. Schimmack
e-mail: schimmack@uni.leuphana.de

A. Abraham et al. (eds.), *Proceedings of the Second International Afro-European Conference for Industrial Advancement AECIA 2015*, Advances in Intelligent Systems and Computing 427, DOI 10.1007/978-3-319-29504-6_28

Mobile robotics is a notable case of such evolution. The robotics community has developed sophisticated analysis and control techniques to meet increasing requirements on the control of motions of mechanical systems. These increasing requirements are motivated by higher performance specifications and an increasing number of degrees-of-freedom. Mercorelli and Prattichizzo [6], Prattichizzo and Mercorelli [7] and Marro and Barbagli [3] mark progress in the analysis and synthesis of geometric controllers for mechanical systems, and particularly in Mercorelli [4, 5], a non-interacting force-motion control in robotic manipulation is proposed. In Mercorelli and Prattichizzo [6], a robust decoupling controller using an algebraic state input feedback is presented. The main purpose of this work is, starting from a linear model, to show some geometric structural properties to realise a perfect localisation of the vibrations during the milling of a manufacturing phase. The localisation of the vibrations represents a key point in the motion control problems. The paper is organised in the following way. Section 2 presents a possible linear model of a manufacturing system. Section 3 shows some structural properties for the localisation strategy using the geometric approach. Section 4 is devoted to the simulation and discusses the obtained results. The conclusions section closes the paper. At the end, numerical computer simulations considering real data of a manufacturing system are shown.

The main nomenclature
m_r: mass of the arm of the robots
M: mass of the object not adjacent to the arm of the robot
m: mass of the object to be milled
$x_{rl}(t)$: position of the left arm of the robot
$x_{rr}(t)$: position of the right arm of the robot
$x_m(t)$: position of the mass of the object to be milled
$x_M(t)$: position of the mass of the object
M_s: mass of the object plus mass of the arm of the robot
$\mathbf{A}$: state matrix of the electrical model
$\mathbf{B}$: input matrix of the electrical model
$\mathscr{B} = \mathsf{im}\mathbf{B}$: image of matrix $\mathbf{B}$
(subspace spanned by the columns of matrix $\mathbf{B}$)
$\mathsf{min}\mathscr{I}(\mathbf{A}, \mathscr{B}) = \sum_{i=0}^{n-1} \mathbf{A}^i \mathsf{im}\mathbf{B}$: minimum $\mathbf{A}$
(invariant subspace containing $\mathsf{im}(\mathbf{B})$)
$\mathbf{F}$: decoupling feedback matrix field
$\mathbf{F}_{rl}(t)$: left arm robot force
$\mathbf{F}_{rr}(t)$: right arm robot force
$\mathscr{C} = \mathsf{ker}\mathbf{C}$: kernel of matrix $\mathbf{C}$
(subspace spanned by the columns of matrix $\mathbf{C}$)
$\mathscr{E} = \mathsf{ker}\mathbf{E}$: kernel of matrix $\mathbf{E}$
(subspace spanned by the columns of matrix $\mathbf{E}$)

2 Model Description

Figures 1 and 2 show two possible scenarios of a manufacturing system using a unidirectional robot to hold the object to be milled. The case depicted in Fig. 1 is a particular one of that represented in Fig. 2. If the rigid linear mathematical model is considered, then the following equations can be written for the system depicted in Fig. 2

$$m_r\ddot{x}_{rl}(t) = F_{rl}(t) - k_1\left(x_{rl}(t) - x_m(t)\right) - b_1\left(\dot{x}_{rl}(t) - \dot{x}_m(t)\right) \tag{1}$$

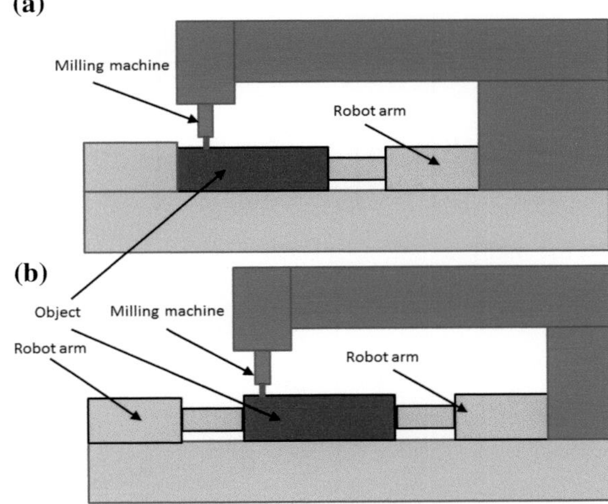

Fig. 1 Scheme of an unidirectional manufacturing system using one (**a**) and two arm robot (**b**)

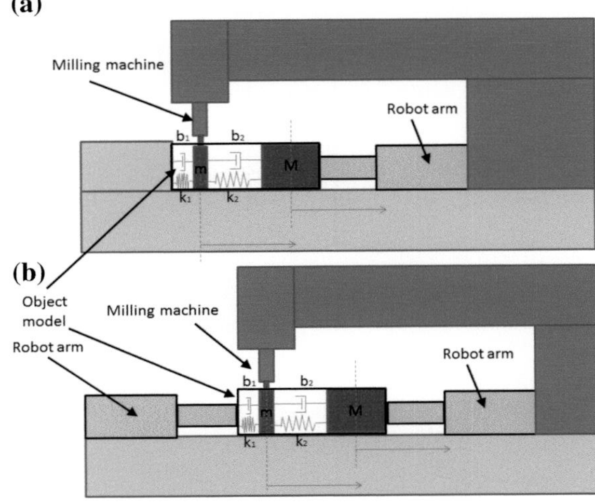

Fig. 2 Model of an unidirectional manufacturing system using one (**a**) and two arm robot (**b**)

$$m\ddot{x}_m(t) = d(t) - k_1\left(x_m(t) - x_{rl}(t)\right) - b_1\left(\dot{x}_m(t) - \dot{x}_{rl}(t)\right) \\ - k_2\left(x_m(t) - x_{rr}(t)\right) - b_2\left(\dot{x}_m(t) - \dot{x}_{rr}(t)\right) \tag{2}$$

$$\left(M + m_r\right)\ddot{x}_{rr}(t) = F_{rr}(t) - k_2\left(x_M(t) - x_m(t)\right) - b_2\left(\dot{x}_M(t) - \dot{x}_m(t)\right). \tag{3}$$

Converting (1), (2) and (3) in matrices, the following structure is obtained

$$\begin{cases} \dot{\mathbf{X}}(t) = \mathbf{A}\mathbf{X}(t) + \mathbf{B}\mathbf{F}(t) + \mathbf{D}d(t) \\ \mathbf{Y}(t) = \mathbf{E}\mathbf{X}(t), \end{cases} \tag{4}$$

with

$$\mathbf{X}(t) = \begin{bmatrix} x_{rl}(t) \\ \dot{x}_{rl}(t) \\ x_m(t) \\ \dot{x}_m(t) \\ x_{rr}(t) \\ \dot{x}_{rr}(t) \end{bmatrix}, \qquad \dot{\mathbf{X}}(t) = \begin{bmatrix} \dot{x}_{rl}(t) \\ \ddot{x}_{rl}(t) \\ \dot{x}_m(t) \\ \ddot{x}_m(t) \\ \dot{x}_{rr}(t) \\ \ddot{x}_{rr}(t) \end{bmatrix}, \qquad \mathbf{F}(t) = \begin{bmatrix} F_{rl}(t) \\ F_{rr}(t) \end{bmatrix}, \tag{5}$$

and the matrix

$$\mathbf{A} = \begin{bmatrix} 0 & 1 & 0 & 0 & 0 & 0 \\ -\frac{k_1}{m_r} & -\frac{b_1}{m_r} & \frac{k_1}{m_r} & \frac{b_1}{m_r} & 0 & 0 \\ 0 & 0 & 0 & 1 & 0 & 0 \\ \frac{k_1}{m} & \frac{b_1}{m} & -\frac{k_1+k_2}{m} & -\frac{b_1+b_2}{m} & \frac{k_1}{m} & \frac{k_2}{m} \\ 0 & 0 & 0 & 0 & 0 & 1 \\ 0 & 0 & \frac{k_2}{M_s} & \frac{b_2}{M_s} & -\frac{k_2}{M_s} & -\frac{b_2}{M_s} \end{bmatrix}, \tag{6}$$

matrices **B** and **D** result as follows

$$\mathbf{B} = \begin{bmatrix} 0 & 0 \\ \frac{1}{M_s} & 0 \\ 0 & 0 \\ 0 & 0 \\ 0 & 0 \\ 0 & \frac{1}{M_s} \end{bmatrix}, \qquad \mathbf{D} = \begin{bmatrix} 0 \\ 0 \\ 0 \\ \frac{1}{m} \\ 0 \\ 0 \end{bmatrix}. \tag{7}$$

$d(t)$ represents the disturbance due to the milling process and matrix **E** represents the output matrix to select the position of mass M and can be written in the following way

$$\mathbf{E} = \begin{bmatrix} 0 & 0 & 0 & 0 & 1 & 0 \end{bmatrix}. \tag{8}$$

3 Analysis of Geometric Structural Properties to Localise Disturbances

Let us consider a system described by a five-map system $(\mathbf{A}, \mathbf{B}, \mathbf{C}, \mathbf{D}, \mathbf{E})$, modelled by

$$\dot{\mathbf{x}}(t) = \mathbf{A}\,\mathbf{x}(t) + \mathbf{B}\,\mathbf{u}(t) + \mathbf{D}\,\mathbf{d}(t)\,, \quad \mathbf{x}(0) = x_0, \tag{9}$$

$$\mathbf{y}(t) = \mathbf{C}\,\mathbf{x}(t), \tag{10}$$

$$\mathbf{e}(t) = \mathbf{E}\,\mathbf{x}(t), \tag{11}$$

where $\mathbf{x} \in \mathscr{X}\ (= \mathbb{R}^n)$, $\mathbf{u} \in \mathscr{U}\ (= \mathbb{R}^p)$, $\mathbf{d} \in \mathscr{D}\ (= \mathbb{R}^d)$, $\mathbf{e} \in \mathscr{H}\ (= \mathbb{R}^m)$ and $\mathbf{y} \in \mathscr{Y}\ (= \mathbb{R}^q)$ respectively denote the state, the manipulable input, the disturbance, the regulated output and the informative output (measured output). In the following the short notations $\mathscr{B} := \mathrm{im}\mathbf{B}$ for the *image* of the matrix $\mathbf{B}$, $\mathscr{C} := \ker\mathbf{C}$ for the *kernel* of the matrix $\mathbf{C}$, $\mathscr{D} := \mathrm{im}\mathbf{D}$ for the *image* of the matrix $\mathbf{D}$ and $\mathscr{E} := \ker\mathbf{E}$ for the *kernel* of the matrix E will be used.

Problem 1 Given the system described in Eqs. (4), find a stabilising state feedback control law such that the unaccessible disturbance $d(t)$ is perfectly localized into the kernel of matrix $\mathbf{E}$ ($\mathscr{E}$).

According to Basile and Marro [1], Problem 1 can be solved without considering the requirement of stability if the following condition holds

$$\mathscr{D} \subseteq \mathscr{J}, \tag{12}$$

where $\mathscr{J}$ represents the maximum controlled invariant subspace containing in $\mathscr{E}$. According to the usual notation, the maximal invariant controller containing in $\mathscr{E}$ is expressed as follows

$$\mathscr{J} = max(\mathrm{A}, \mathscr{B}, \mathscr{E}). \tag{13}$$

It is worth while recalling that subspace $\mathscr{J}$ is said to be a controlled invariant if

$$\mathbf{A}\,\mathscr{J} \subseteq \mathscr{J} + \mathscr{B}, \tag{14}$$

and that if a subspace is a controlled invariant one, then there exists a state feedback matrix $\mathbf{F}$ such that

$$(\mathbf{A} + \mathbf{BF})\,\mathscr{J} \subseteq \mathscr{J}. \tag{15}$$

Concerning the meaning of (15), it is possible to see that through a state feedback matrix **F** it is always possible to obtain an invariant subspace with respect to the state input controlled matrix $\mathbf{A} + \mathbf{BF}$. The problem of the perfect localisation of the disturbance with stability is more complex. To achieve the stability (allocability of the internal eigenvalues to subspace $\mathscr{J}$) it is necessary that the linear model of the object-manipulator can satisfy the following condition

$$\mathscr{J} \cap \mathscr{B} \neq 0 \tag{16}$$

Condition (16) states the no-left invertibility of the linear model which describes the manipulator (b) of Fig. 2. It is straightforward to show that the linear model of the object-manipulator (a) of Fig. 2 is left invertible and thus in this case the free allocability of the eigenvalues internal in subspace $\mathscr{J}$ is structurally not possible. In fact, for manipulator (a) $F_{rl} = 0$ and matrix **B** consists of just the second column. Moreover, condition (16) is not a sufficient condition to guarantee the internal allocability of the eigenvalues in $\mathscr{D}$. According to Basile and Marro [1], the perfect localisation with internal allocability of the eigenvalues can be stated with two conditions. Condition one deals with

$$\mathscr{D} \subseteq \mathscr{J}. \tag{17}$$

The second condition is an internally stabilizable $\mathscr{V}_m$, where

$$\mathscr{V}_m = \mathscr{J} \cap \mathscr{K}_1 \tag{18}$$

with

$$\mathscr{K}_1 = min\mathscr{S}(\mathrm{A}, \mathscr{J}, \mathscr{B} + \mathscr{D}). \tag{19}$$

Expression (19) indicates the minimal conditioned invariant subspace containing $\mathrm{im}(\mathscr{B} + \mathscr{D})$. Expression $\mathscr{V}_m$ internally stabilizing means that all not allocable internal eigenvalues (transmission zero of the transfer function) are stable. In general, if

$$\mathscr{V}_m \cap \mathscr{B} \neq 0, \tag{20}$$

then $\mathscr{V}_m$ results internally stabilizable, if not, we have to hope that its internal eigenvalues are stable.

Remark 1 A possible interpretation of condition (25) is that, if this intersection is not zero, then all the internal eigenvalues in $\mathscr{J}$ are allocable and no zero transmissions are present. In fact, $\mathscr{V}_m$ represents the minimal self-bounded controlled invariant included in $\mathscr{E}$ which is the reachable subspace included in $\mathscr{E}$, see Basile and Marro [1]. An application can be found in Prattichizzo and Mercorelli [7].

4 Simulations

In Fig. 3 the conceptual and simulated control scheme is shown. Considering the data $M = 1Kg$, $m_r = 0.5Kg$, $m = 0.2Kg$, $k_1 = 2N/m$ and $b_1 = 1000Ns/m$, the simulated results shown in Figs. 4 and 5 are obtained. In particular, Fig. 4 shows the simulated object position in presence of a force disturbance. Figure 5 shows how the compensated object position remains constant due to the compensating control law presented above. It is easy to calculate the following subspace

$$\mathscr{D} \subseteq \mathscr{E} = \mathrm{im} \begin{bmatrix} 1 & 0 & 0 & 0 & 0 \\ 0 & 1 & 0 & 0 & 0 \\ 0 & 0 & 1 & 0 & 0 \\ 0 & 0 & 0 & 1 & 0 \\ 0 & 0 & 0 & 0 & 0 \\ 0 & 0 & 0 & 0 & 1 \end{bmatrix}. \tag{21}$$

Moreover,

$$\mathbf{A}\mathscr{D} \subseteq \mathscr{D} + \mathscr{B}. \tag{22}$$

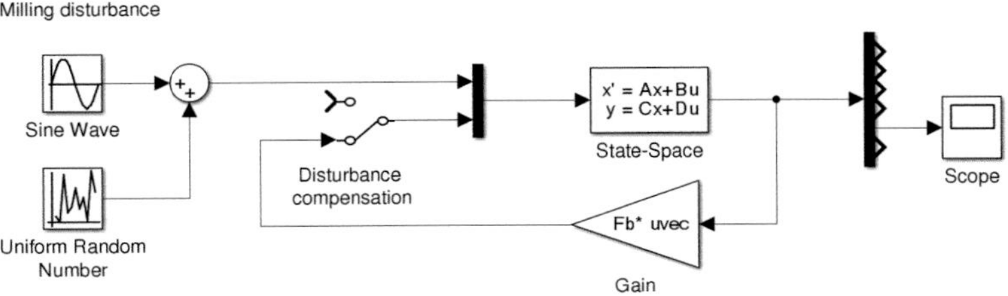

Fig. 3 Simulink scheme for the simulations

Fig. 4 Simulation results of the uncompensated milling disturbances

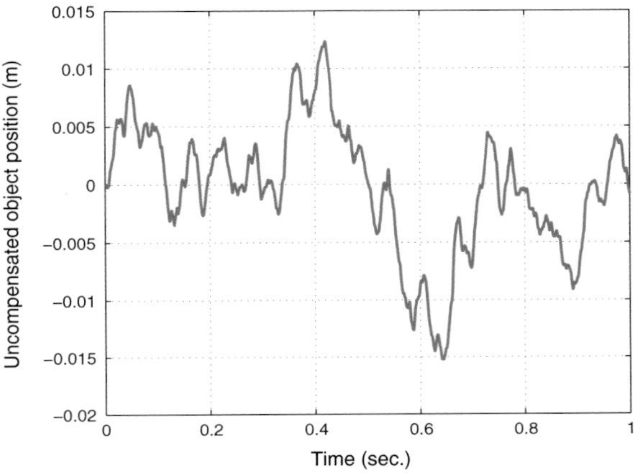

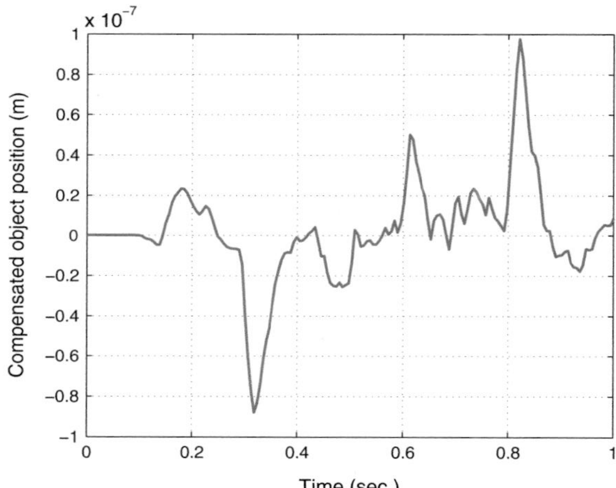

Fig. 5 Simulation results of the compensated milling disturbances

For manipulator (b) subspace it results

$$\mathscr{J} = \operatorname{im}\begin{bmatrix} 1\,0\,0\,0 \\ 0\,1\,0\,0 \\ 0\,0\,1\,0 \\ 0\,0\,0\,1 \\ 0\,0\,0\,0 \\ 0\,0\,0\,0 \end{bmatrix} \text{and } \mathscr{K}_1 = \mathscr{D} =\subseteq \mathscr{E} = \operatorname{im}\begin{bmatrix} 1\,0\,0\,0\,0 \\ 0\,1\,0\,0\,0 \\ 0\,0\,1\,0\,0 \\ 0\,0\,0\,1\,0 \\ 0\,0\,0\,0\,0 \\ 0\,0\,0\,0\,1 \end{bmatrix}. \tag{23}$$

Subspace $\mathscr{V}_m = \mathscr{J} \cap \mathscr{K}_1$ results to be

$$\mathscr{V}_m = \mathscr{J} = \operatorname{im}\begin{bmatrix} 1\,0\,0\,0 \\ 0\,1\,0\,0 \\ 0\,0\,1\,0 \\ 0\,0\,0\,1 \\ 0\,0\,0\,0 \\ 0\,0\,0\,0 \end{bmatrix}. \tag{24}$$

So, as already explained all the internal eigenvalues are allocable. In case manipulator (b) is considered, then a matrix $\mathbf{F}$ exists which localises the disturbance in $\mathscr{D}$ and locates all eigenvalues. In case manipulator (a) is considered, then matrix $\mathbf{F}_a$ exists which localises the disturbance in $\mathscr{D}$, but it cannot allocate the internal eigenvalues, so the hidden dynamics cannot be arbitrarily chosen. In fact, in case of manipulator (a) subspace $\mathscr{J}$ and $\mathscr{K}_1$ are equal like in case (b) and thus also $\mathscr{V}_m$ results to be the same. Because of the restricted input subspace it follows that

$$\mathscr{J} \cap \mathscr{B} = 0, \tag{25}$$

the system results left invertible and no internal eigenvalues are allocable.

After the control feedback for the case (b) the controlled system consists of the following structure

$$\mathbf{A}_f = \mathbf{A} + \mathbf{B}\mathbf{F}_b, \tag{26}$$

and

$$\mathbf{A}_{fb} = \begin{bmatrix} \mathbf{A}'_{11} & \mathbf{0}_{2\times 2} \\ \mathbf{0}_{2\times 4} & \mathbf{A}'_{22} \end{bmatrix}, \tag{27}$$

where $\mathbf{A}'_{11}$ is a 4×4 matrix and $\mathbf{A}'_{22}$ is a 2×2 matrix. Furthermore, minor $\mathbf{A}'_{11}$ represents the invariant dynamics which results localised in $\mathscr{E}$ and $\mathbf{A}'_{22}$ represents the external dynamics. In case of using manipulator (b) of Fig. 2, defined by the triple $(\mathbf{A}, \mathbf{B}, \mathbf{E})$, both dynamics are arbitrarily allocable. Minor $\mathbf{0}_{2\times 4}$ comes from the effect of the localisation. It is possible to remark how using the manipulator (a) of Fig. 2 a localisation of disturbance is obtainable but the internal dynamics which are included in $\mathscr{E}$, not visible from the output, are not allocable. In the specific case that these dynamics are stable, the no-allocability of a suitable zero dynamics cannot generate unsuitable constraints. In fact, the following controlled system

$$\mathbf{A}_{fa} = \mathbf{A} + \mathbf{B}_a\mathbf{F}_a, \tag{28}$$

using $\mathbf{B}_a$ which consists of just the second column of input matrix $\mathbf{B}$ results to be

$$\mathbf{A}_{fa} = \begin{bmatrix} \mathbf{A}_{11} & \mathbf{0}_{2\times 2} \\ \mathbf{0}_{2\times 4} & \mathbf{A}'_{22} \end{bmatrix}, \tag{29}$$

where minor

$$\mathbf{A}_{11} = \begin{bmatrix} 0 & 1 & 0 & 0 \\ -\frac{k_1}{m_r} & -\frac{b_1}{m_r} & \frac{k_1}{m_r} & \frac{b_1}{m_r} \\ 0 & 0 & 0 & 1 \\ \frac{k_1}{m} & \frac{b_1}{m} & -\frac{k_1+k_2}{m} & -\frac{b_1+b_2}{m} \end{bmatrix}, \tag{30}$$

is equal to the original matrix $\mathbf{A}$ defined by (30). The eigenvalues of this minor represent the transmission zeros of the triple $(\mathbf{A}, \mathbf{B}_a, \mathbf{E})$, where $\mathbf{B}_a$ consists of just the second column of matrix $\mathbf{B}$, which is not allocable by $\mathbf{B}_a$.

5 Conclusions

The paper deals with some structural properties concerning a geometric approach to a milling manufacturing system. An object is considered together with two Cartesian manipulators which should be milled. An analysis of structural conditions on the manipulators is investigated to find the control structure and to localise the milling disturbance. To realise the proposed controlled strategy an observer is required for estimation of the unmeasurable state variables.

References

1. Basile, G., Marro, G.: Controlled and conditioned invariants in linear system theory. Prentice Hall, New Jersey (1992)
2. Isidori, A.: Nonlinear control Systems: an introduction. Springer, Berlin (1989)
3. Marro, G., Barbagli, F.: The algebraic output feedback in the light of dual-lattice structures. Kybernetika **35**(6), 693–706 (1999)
4. Mercorelli, P.: A geometric algorithm for the output functional controllability in general manipulation systems and mechanisms. Kybernetika **48**(6), 1266–1288 (2012a)
5. Mercorelli, P.: Invariant subspace for grasping internal forces and non-interacting force-motion control in robotic manipulation. Kybernetika **48**(6), 1229–1249 (2012b)
6. Mercorelli, P., Prattichizzo, D.: A geometric procedure for robust decoupling control of contact forces in robotic manipulation. Kybernetika **39**(4), 433–455 (2003)
7. Prattichizzo, D., Mercorelli, P.: On some geometric control properties of active suspension systems. Kybernetika **36**(5), 549–570 (2000)
8. Wonham, W.: Linear multivariable control: a geometric approach. Springer, New York (1979)

Multi-level Significant Bit (MLSB) Embedding Based on Weighted Container Model and Weighted F5 Concept

Sergey Bezzateev, Natalia Voloshina and Konstantin Zhidanov

Abstract In this paper the weighted container concept of cover image jointly with error correcting codes perfect in weighted Hamming metric (ECCPWH) is consider. Such approach can improve the efficient of the ECC embedding schemes based on syndrome coding. To show the effectiveness of ECCPWH based scheme the example of multi-level significant bit (MLSB)embedding scheme applied to digital image is provided. ECCPWH based method is compared with the linear error-block codes (LEBC) steganography method that is based on linear error-block codes perfect in π- metric. The examples show that proposed MLSB embedding scheme that is based on ECCPWH approach is efficient for syndrome coding embedding algorithms.

Keywords Digital watermarking · Least significant bit embedding · Multi-level significant bit (mlsb) embedding · Weighted container model · Error correcting codes perfect in weighted Hamming metric (ECCPWH) · Weighted f5 (wf5) concept · Linear error-block codes (lebc)

1 Introduction

One of the most popular digital right management method for author right protection of multimedia data is a digital watermarking. Steganographic methods of information embedding are often used for invisible digital watermarking. The development of modern embedding methods moves toward increasing the embedded

S. Bezzateev · N. Voloshina (✉) · K. Zhidanov
Saint-Petersburg State University of Aerospace Instrumentation,
Saint-petersburg, Russia
e-mail: natali@vu.spb.ru

S. Bezzateev
e-mail: bsv@aanet.ru

K. Zhidanov
e-mail: konstantin.zhidanov@gmail.com

A. Abraham et al. (eds.), *Proceedings of the Second International Afro-European Conference for Industrial Advancement AECIA 2015*, Advances in Intelligent Systems and Computing 427, DOI 10.1007/978-3-319-29504-6_29

information volume and minimizing the possibility of embedding detection, which is often looked at as a number of appeared distortions/embedded errors [1].
It was shown that the optimal embedding methods are based on error correcting codes approach [2]. Recently there were described several steganographic methods that are based on error correcting codes approach [3–6]. For example, the F5 method [6] demonstrates high efficiency by embedding information in the JPEG image files using Hamming code. This F5 approach that is also called syndrome coding was expanded for the other types of ECC codes [1, 4, 5] such as BCH, Reed-Solomon, Goppa codes, etc. Also it was suggested to use more complicated model of the embedding area by extended LSB model to two LSB and weighted significant bits schemes.

For example in [7] it was proposed to use two least significant bits embedding (2LSB) scheme for digital watermarking. In this approach it was shown how to use higher than first bit plain that has more distortion value if some errors occur on it while embedding process realization.

In [8] it was shown that it is possible to use more than two bit planes if taking into account the significance of the errors in the corresponding bit planes and positions of corresponding pixels in the image. It means that the higher bit plane is the more significance it has and since less errors should occur in it during embedding process. Also the more heterogeneity of the corresponding pixel and its neighbor pixels is the higher bit plane could be used for embedding. To realize this concept it was suggested to use linear error-block codes [9]. It was shown that the efficiency of embedding could be increased if the difference between bit plains significance is used. Different significance of the bits was used for embedding also in [5].

In [10] the difference of significance of different parts of the image was represented as a weighted container model (WCM). This means that for different significance zones of the container it is possible to assign their weight. For such weighted container model it is possible to use syndrome coding embedding approach based on error correcting codes that are perfect in weighted Hamming metric [10–12]. To measure the effectiveness of embedding for weighted container model the penalty function estimation was proposed [10]. It was shown that it is possible to increase the effectiveness of embedding by using weighted embedding approach. In this paper comparison of two approaches that take into account the weighted significance structure of initial image is made.

2 LEBC in Steganography

In paper [8] it is shown that by using more than two bit planes of the digital image to embed information it is possible to obtain high effectiveness of steganographic process. It becomes possible when taking into account the significance of the errors in the corresponding bit planes. That means that the higher the bit plane of image is the more significance it has and the less errors should occur in it while embedding the message. To increase the efficiency it is also suggested to take into account the heterogeneous structure of image to guaranty perceptual distortions. To realize this concept it was suggested to use linear error-block codes [9]. The specific of LEBC codes that are used for embedding is briefly described below.

2.1 LEBC Codes Description

In [8, 13] it was shown how to use linear error-block codes for steganography. The main feature of these codes is that there defines the special partition π of the codeword length. This partition is defined as a composition of blocks $v_1, v_2, \ldots, v_s$ with lengths $\ell_1, \ell_2, \ldots, \ell_s$, $(\ell_1 \geq \ell_2 \geq \cdots \geq \ell_s)$. The main difference between linear error-block codes and classical linear error correcting codes is the usage of so-called $\pi-$ distance instead of classical Hamming distance. The $\pi-$ distance $d_\pi(\mathbf{u}, \mathbf{v})$ between two vectors $\mathbf{u} = (u_1, u_2, \ldots, u_s)$ and $\mathbf{v} = (v_1, v_2, \ldots, v_s)$ is defined as

$$d_\pi(\mathbf{u}, \mathbf{v}) = \sharp\{i : 1 \leq i \leq s, u_i \neq v_i\}. \tag{1}$$

In [9, 13] authors show that there exist a perfect LEBC codes of minimal distance 3, 4 and 5. In general case the existence of perfect LEBC codes and the method of their construction is still an open problem.

2.2 LEBC Based Embedding Approach

To embed a message in LEBC steganography method it is necessary to determine a partition π according to the partition of the initial image. In [13] the image partition scheme is as on the Fig. 1 below.

That means that the higher bit plane of the image is used the more significance of its bits for invisibility of distortions, the less errors should occur in corresponding bit plane. Also the useful part of corresponding bit plane (black marked parts on the figure) is defined according to the homogeneity of its pixels. For information embedding it is necessary to define a partition as $[v_1][v_2] \ldots [v_s]$ and create a

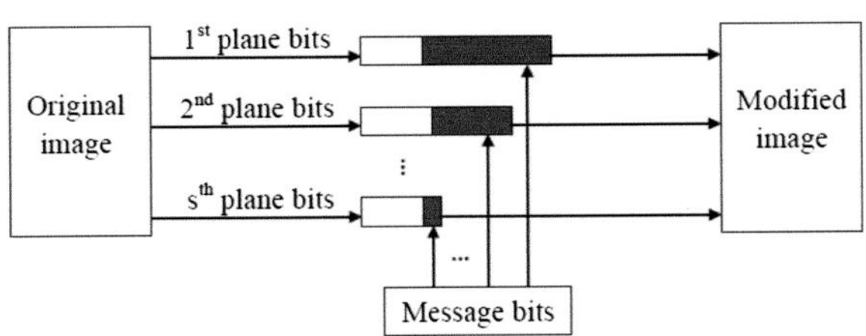

Fig. 1 Image partition scheme

bit stream according to the image partition scheme (pixel levels matrix [14]). Then the parity check matrix H should be created. The embedding process is performed according to the $F5$ concept [6] i.e. by realizing syndrome coding process.

3 MLSB Embedding

Similar approach is suggested in [5]. In this paper it is proposed to take into account different significance of errors that occur if different parts of the image while information embedding process. It is shown that it is possible to use this concept consequently with ECCPWH for message embedding in time domain of color images. It is shown that this concept (so-called weighted embedding) could be effectively used for embedding methods that use syndrome coding (F5 [6]). As a generalization the weighted container model was proposed [10] and several applications of this model was shown [7, 12].

4 Weighted Container Model

It is known that image contains several zones with variable perceptibility of distortions. The perceptibility for some zones is much more than for the others. Taking into account this property, in [10] it is shown that container could be represented as weighted structure. This structure was named as weighted container model. In this model the container is represented as a set of s zones $\mathbf{I} = \{I_1, I_2, \ldots, I_s\}$ with consequently assigned significance $\mathbf{Y} = \{y_1, y_2, \ldots, y_s\}$ according to contribution of the errors in particular zone to resulting distortion. Model of weighted container is shown on the Fig. 2.

According to the significance of zones the weights of the errors that occur while embedding process could be defined. For the simplest case weights set could be defined as equal to $\mathbf{Y}$. To compare the efficiency of different codes that are used for embedding, it is necessary to take into account the weighted structure of the

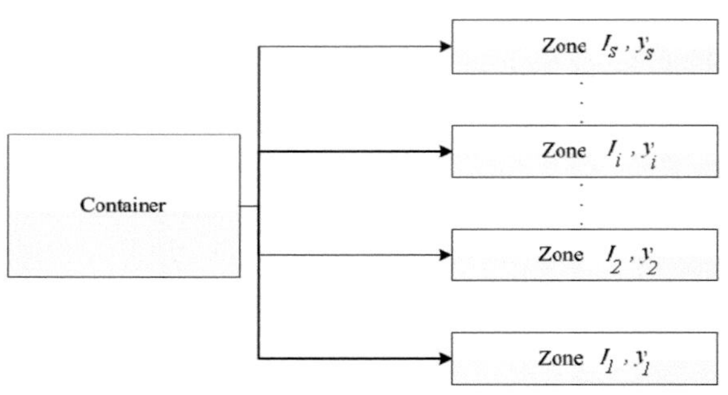

Fig. 2 Weighted container model

container. For this purpose we define the concept of penalty function. Penalty function is defined as the function of the number of added errors and specified distortion weights [10]. Simplest function could be defined as weighted sum of average number of errors in corresponding zones of significance:

$$P = \sum_{i=1}^{s} t_i p_i, \tag{2}$$

where t_i– average number of errors in zone I_i, p_i– penalty for bit inversion in significance zone I_i.

The set of penalty values $\mathbf{P} = \{p_1, p_2, \dots, p_s\}$ should be predefined for each type of container. For example for digital images it is possible to define penalty values p_i by using human vision system features. In simplest case it could be defined that $p_i = y_i$.

4.1 ECC that Are Perfect in Weighted Hamming Metric

For weighted container model it is natural to use error correcting codes in the weighted Hamming metric [11]. To define such codes let's use the same approach as for LEBC codes. At first the length of codeword is divided into s blocks $v_1, v_2, \dots, v_s$ with corresponding lengths $l_1, l_2, \dots, l_s$. Let's associate with each block v_i its unique weight $w_i \in \mathbf{Z}, w_1 < w_2 < \dots < w_s$. Now we have the following definition for distance $d_{WH}(\mathbf{u}, \mathbf{v})$ between two vectors $\mathbf{u} = (u_1, u_2, \dots, u_s)$ and $\mathbf{v} = (v_1, v_2, \dots, v_s)$ in weighted Hamming metric:

$$d_{WH}(\mathbf{u}, \mathbf{v}) = \sum_{i=1}^{s} d(u_i, v_i) w_i, \tag{3}$$

where $d(\mathbf{u}, \mathbf{v})$—the distance between two vectors $\mathbf{u}$ and $\mathbf{v}$ in classical Hamming metric. The Goppa codes perfect in such weighted Hamming metric is introduced and described more precisely in [11].

5 Examples

In this section two main examples are represented. First one for the case when it is possible to create a pixel levels matrix [14] and send it to the receiver side. The second example shows how to use and compare MLSB and LEBC approaches when the matrix could not be created or sent and as a result is not used.

5.1 Weighted Embedding with Usage of Pixel Levels Matrix

In example that is introduced for the LEBC embedding [8] four least significant bit planes are taken for embedding. The partition is defined as [3][2][1][1] so that the length of the codeword $n = 7$ and the syndrome length is equal to 4. The maximal error weight $wt_\pi = 2$. In this case the first block v_1 consists of the least significant bits and allows up to 3 errors while embedding process, second block v_2 allows up to 2 errors and consists of bits from second bit plane, next blocks v_3 and v_4 allow 1 error on its length and consist of the bits from third and fourth bit plane respectively according to the predefined pixel levels matrix [14] that should be known both to sender and receiver of the message. The embedding/extracting process is performed by syndrome coding/decoding technique as it is described for $F5$ steganographic method. The parity check matrix $H_{\pi[3211]}$ is defined in [8] and it is represented below.

$$H_{\pi[3211]} = \begin{bmatrix} 1\,1\,1\,0\,0\,0\,1 \\ 0\,1\,1\,1\,1\,0\,0 \\ 1\,0\,0\,1\,0\,1\,0 \\ 1\,0\,1\,0\,0\,1\,0 \end{bmatrix}$$

To estimate the distortions that could occur while information embedding process (syndrome coding) we use penalty function P (2). Penalty for errors that occur in i- th bit plane is defined as p_i. In simplest case penalties could be defined as $p_i = i$ where i is the number of bit plain when $i = 1$ for the least significant bit plain. For the partition [3][2][1][1] that was defined in [8] the error vectors was found. The parameters of this code are represented in Table 1.

The penalty function that is defined according to (2) for this partition and defined coset leaders [14] is equal to

$$P_{\pi[3211]} = \frac{20p_1 + 7p_2 + p_3 + p_4}{16}. \tag{4}$$

This penalty function $P_{\pi[3211]}$ shows the aggregate penalty for all 16 messages that are defined for the syndrome with length equal to 4.

For our multi-level significance bit approach the weighted container model [10] could be defined in accordance to the structure [3][2][1][1]. To construct similar structure for weighted information embedding it is necessary to define weights of corresponding blocks [11]. For the first block with length n_1 we define weight $w_1 = 2$, for the second block with length n_2 we define weight $w_2 = 3$ and for the third block with length n_3 we define weight $w_3 = 5$. The total length of the codeword $n = n_1 + n_2 + n_3 = 3 + 2 + 1 = 6$. The syndrome length is equal to 4. For all error vectors the following inequality is valid.

$$\tau_1 w_1 + \tau_2 w_2 + \tau_3 w_3 \leq 5$$

Table 1 Coset leaders, syndromes and corresponding penalties defined for partition [3][2][1][1] for LEBC embedding

Coset leader	Syndrome	Penalty
000 00 0 0	0000	0
100 00 0 0	1011	p_1
010 00 0 0	1100	p_1
001 00 0 0	1101	p_1
101 00 0 0	0110	$2p_1$
110 00 0 0	0111	$2p_1$
011 00 0 0	0001	$2p_1$
111 00 0 0	1010	$3p_1$
000 01 0 0	0100	p_2
000 11 0 0	0010	$2p_2$
000 00 1 0	0011	p_3
000 00 0 1	1000	p_4
100 01 0 0	1111	$p_2 + p_1$
001 01 0 0	1001	$p_2 + 2p_1$
011 01 0 0	0101	$p_2 + 2p_1$
111 01 0 0	1110	$p_2 + 3p_1$

the parity check matrix $H_{WH[321]}$ is

$$H_{WH[321]} = \begin{bmatrix} 0\,0\,0\,1\,1\,0 \\ 0\,0\,1\,0\,1\,1 \\ 0\,1\,0\,0\,1\,1 \\ 1\,0\,0\,0\,1\,1 \end{bmatrix}$$

The coset leaders and their syndromes for this code are represented in Table 2.

The penalty function that is defined according to (2) for this weighted codeword structure and defined coset leaders is equal to

$$P_{WH[321]} = \frac{15p_1 + 8p_2 + p_3}{16}. \tag{5}$$

This penalty function $P_{WH[321]}$ shows the average penalty for all 16 messages that are defined for the syndrome with length equal to 4. Thus the difference between penalties $P_{\pi[3211]}$ and $P_{WH[321]}$ is equal to

$$P_{\pi[3211]} - P_{WH[321]} = \frac{5p_1 - p_2 + p_4}{16}.$$

Table 2 Coset leaders, their syndromes and weights in the weighted Hamming metric and corresponding penalties defined for partition [3][2][1]

Coset leader	Syndrome	wt_{WH}	Penalty
000 00 0	0000	0	0
100 00 0	0001	2	p_1
010 00 0	0010	2	p_1
001 00 0	0100	2	p_1
101 00 0	0101	4	$2p_1$
110 00 0	0011	4	$2p_1$
011 00 0	0110	4	$2p_1$
000 01 0	1111	3	p_2
100 01 0	1110	5	$p_2 + p_1$
001 01 0	1011	5	$p_2 + p_1$
010 01 0	1101	5	$p_2 + p_1$
000 10 0	1000	3	p_2
100 10 0	1001	5	$p_2 + p_1$
001 10 0	1100	5	$p_2 + p_1$
010 10 0	1010	5	$p_2 + p_1$
000 00 1	0111	5	p_3

That means that the MLSB weighted embedding method is more efficient if

$$5p_1 + p_4 > p_2.$$

As $p_1 < p_2 < p_3 < p_4$ proposed MLSB method is more efficient then LEBC method.

5.2 *Weighted Embedding Without Usage of Pixel Levels Matrix*

This example shows how to use MLSB weighted embedding in case when it is impossible to get or send pixel levels matrix of the image. In this case it is possible to use such weighted structure:

- for the first block with length $n_1 = 3$ we define weight $w_1 = 1$,
- for the second block $n_2 = 3$ we define weight $w_2 = 4$ and
- for the third block $n_3 = 3$ we define weight $w_3 = 6$.

The total length of the codeword is $n = n_1 + n_2 + n_3 = 3 + 3 + 3 = 9$. For this partition [3][3][3] it is possible to construct ECC in weighted Hamming metric so that syndrome length is equal to 5. The parity check matrix $H_{WH[333]}$ for such ECCPWH code is constructed

Table 3 Coset leaders and their weights in the weighted Hamming metric and corresponding penalties defined for partition [3][3][3]

Coset leader	wt_{WH}	Penalty	Coset leader	wt_{WH}	Penalty
000 000 000	0	0	001 010 000	5	p_1+p_2
100 000 000	1	p_1	100 001 000	5	p_1+p_2
010 000 000	1	p_1	010 001 000	5	p_1+p_2
001 000 000	1	p_1	001 001 000	5	p_1+p_2
110 000 000	2	$2p_1$	110 100 000	6	$2p_1+p_2$
101 000 000	2	$2p_1$	011 100 000	6	$2p_1+p_2$
011 000 000	2	$2p_1$	101 100 000	6	$2p_1+p_2$
111 000 000	3	$3p_1$	110 010 000	6	$2p_1+p_2$
000 100 000	4	p_2	011 010 000	6	$2p_1+p_2$
000 010 000	4	p_2	101 010 000	6	$2p_1+p_2$
000 001 000	4	p_2	110 001 000	6	$2p_1+p_2$
100 100 000	5	p_1+p_2	011 001 000	6	$2p_1+p_2$
010 100 000	5	p_1+p_2	101 001 000	6	$2p_1+p_2$
001 100 000	5	p_1+p_2	000 000 100	6	p_3
100 010 000	5	p_1+p_2	000 000 010	6	p_3
010 010 000	5	p_1+p_2	000 000 001	6	p_3

$$H_{WH[333]} = \begin{bmatrix} 1&0&0&0&0&0&1&1&1 \\ 0&1&0&0&0&0&1&1&1 \\ 0&0&1&0&0&1&0&1&1 \\ 0&0&0&1&0&1&1&0&1 \\ 0&0&0&0&1&1&1&1&0 \end{bmatrix}.$$

The coset leaders and their weights for this code are represented in Table 3.

The penalty function is represented as

$$P_{WH[333]} = \frac{39p_1 + 21p_2 + 3p_3}{32}. \tag{6}$$

This penalty function shows the average penalty for all 32 messages that are defined for syndrome with length equal to 5. Thus the difference between penalty $2P_{\pi[3211]}$ for two syndromes of the length 4 in LEBC method and penalty $P_{WH[333]}$ for one syndrome of the length 5 is equal to

$$2P_{\pi[3211]} - P_{WH[333]} = \frac{p_1 - 7p_2 - p_3 + 2p_4}{32}.$$

That means that MLSB weighted embedding method is more efficient for [3][3][3] partition if

$$p_1 + 2p_4 > 7p_2 + p_3$$

wherein the number of messages in MLSB approach is twice more than in LEBC method and there is no need to use pixel level matrix. The third advantage is that in MLSB weighted embedding the fourth bit plane should not be changed at all.

6 Conclusion

The effectiveness comparison of two syndrome coding methods that are based on using of error correcting codes perfect in weighted Hamming metric [12] and on using of linear error-block codes [15] is given. The relations of the significance p_i of the areas I_i of weighted container model that is used for embedding under which one method has the advantage over another one is derived.

Acknowledgments We are grateful to the reviewer for his/her careful reading of the paper and his/her thoughtful comments. His/her comments and advices has helped us improve and clarify the paper.
This work was partly financially supported by the Russian Ministry of Education and Science within a framework of the basic task to the university in 2015 (project number 2452).

References

1. Berger, T., Medeni, M.: Maximum likelihood decoding algorithm for some Goppa and BCH Codes: Application to the matrix encoding method for steganography. In: INTERNET 2012: The Fourth International Conference on Evolving Internet, pp. 85–89 (2012)
2. Galand, F., Kabatiansky, G.: Information hiding by coverings, In: Proceedings of IEEE Information Theory Workshop, pp. 151–154. Paris (2003)
3. Munuera, C.: Steganography and error-correcting codes. Signal Process. **87**, 1528–1533 (2007)
4. Medeni, M.B., Souidi, E.M.: A novel steganographic protocol from error-correcting codes. J. Inf. Hiding Multimed. Signal Process. **1**(4), 339–343 (2010)
5. Bezzateev, S., Voloshina, N., Minchenkov V.: Special class of (L, G) codes for watermark protection in DRM. In: Proceedings of Eighth International Conference on Computer Science and Information Technologies, pp. 225–228. Yerevan, Armenia (2011)
6. Westfeld, A.: A steganographic algorithm. Proceedings of the 4th International Workshop on Information Hiding, pp. 289–302 (2001)
7. Liao, X., Wen Q.: Embedding in two least significant bits with wet paper coding. CSSE'08: Proceedings of the 2008 International Conference on Computer Science and Software Engineering, pp. 555–558 (2008)
8. Darit, R., Souidi, E.M.: An application of linear error-block codes in steganography. Int. J. Digit. Inf. Wirel. Commun. **1**(2), 426–433 (2011)
9. Feng, K., Xu, L., Hickernell, F.: Linear error-block codes. Finite Fields Appl. **6**, 638–652 (2006)
10. Bezzateev, S., Voloshina, N., Zhidanov, K.: Steganographic method on weighted container. In: Proceedings of XIII Int. Symposium on Problems of Redundancy in Information and Control System, pp. 10–12. Saint-Petersburg, Russia (2012)
11. Bezzateev, S., Shekhunova, N.: Class of generalized Goppa codes perfect in weighted Hamming metric. Des. Codes Cryptogr. **66**(1–3), 391–399 (2013)

12. Voloshina, N., Zhidanov, K., Bezzateev, S.: Optimal weighted watermarking for still images. In: Proceedings of XIV International Symposium on Problems of Redundancy in Information and Control System, pp. 98–102 .Saint-Petersburg, Russia (2014)
13. Darit, R., Souidi, E.M.: New families of perfect linear error-block codes. Int. J. Inf. Coding Theory (IJICOT) **2**(2/3), 84–95 (2013)
14. Darit, R.: These de doctorat: Linear Error-Block Codes and Applications. Universite Mohhamed V Agdal, Faculte des sciences, Rabat (2012)
15. Darit, R., Souidi, E.M.: A steganographic protocol based on linear error-block codes. In: Proceedings of the 11th International Conference on Security and Cryptography, pp. 178–183. Vienna, Austria, 28–30, August 2014

An Extended Kalman Filter as an Observer in a Sliding Mode Controller for a Metal-Polymer Composite Actuator

Manuel Schimmack and Paolo Mercorelli

Abstract This paper presents adaptive Sliding Mode Control combined with an Extended Kalman Filter used as an observer to control metal-polymer composite fibers as an actuator. The mechanismus based on the characteristic of the thermoplastic polymer which is coated with silver particles. Usual the interface between the polymer and the silver surface connected by physical or chemical methods to promote strong interactions between metal and polymer. To control this actuator a sliding mode control (SMC) is combined with an Extended Kalman Filter (EKF) and used as an observer. Despite of the particular simplified model of the considered actuator the EKF presents a nonlinear Jacobian Matrix. The parameter setting of system and the measurement covariance matrix together with their initial values are done heuristically. The sliding mode control scheme is designed using the well known Lyapunov approach. The simulation results indicate that the proposed algorithm is effective and robust.

Keywords Nonlinear actuators · Sliding Mode Control · Lyapunov approach · Observer · Extended Kalman Filter

1 Introduction

Sliding mode control (SMC) is well-known as a robust approach and it is useful for controller design in systems with parameter uncertainties. This approach thanks to the Lyapunov approach having such properties as stability, rate of convergence of tracking problems and also working point of the system to be stabilized is one

M. Schimmack (✉) · P. Mercorelli
Manuel Schimmack and Paolo Mercorelli, Institute of Product and Process Innovation, Leuphana University of Lueneburg, Volgershall 1, 21339 Lueneburg, Germany
e-mail: schimmack@uni.leuphana.de

P. Mercorelli
e-mail: mercorelli@uni.leuphana.de

A. Abraham et al. (eds.), *Proceedings of the Second International Afro-European Conference for Industrial Advancement AECIA 2015*, Advances in Intelligent Systems and Computing 427, DOI 10.1007/978-3-319-29504-6_30

of the most widely used in industrial application and also in theoretical field such as chaotic system control. Recently very innovative contributions were presented in this field. Recently, there has been a remarkable interest in applications of sliding mode control for actuators. For example, in [1, 2] position controls using a sliding mode technique are proposed for various different actuator structures. The robustness of this approach against parameter uncertainties is demonstrated. In particular, to achieve a soft landing, a high switching frequency is required near the landing point. A detailed analysis of this issue was presented in [3] and also in [4], along with a control strategy for the proposed design. Notably, in that paper an alternative approach to dealing with the switching mode was suggested. The second factor is that if there is high inductance in the electrical circuit in which the switching signals are involved, it is known to be difficult to switch the current quickly. In particular, in sliding mode control the phenomena associated with a high switching frequency are referred to as chattering. Chattering properties of various control approaches must be considered. These phenomena are classified into three types. The first one is harmless and cannot be avoided. On the other hand, the second and third types of chattering phenomena are dangerous but it was proven in [5] that they can be eliminated by the proper use of high-order sliding modes (HOSM). Moreover, in [6] it is shown that, for some sliding magnitudes in higher orders of the homogeneous sliding modes, the chattering effect results are not amplified in the presence of actuators.

2 Model of Metal-Polymer Composite Actuator

Conductive metal-polymer composite fibers, which are depicted in Fig. 1 are the fundamental core for wearable technology and an integral part of smart textiles. There is a wide field of applications both in use and in development, which is analysed in [7]. As traditional utilization, the yarn is used as a textile electrode as well as a transmissionline for monitoring of physiological parameters as exhibited in [8]. Combined with an analysing algorithm the fibers has been presented in context of a wearable system in [9] for monitoring during the predictive rehabilitation. A single fiber of the conductive multiple filaments consists of a polyamide kernel and is surrounded by a thin metallic covering. This Example focusses on the application as an metal-polymer composite actuator. The mechanismus is based on the characteristic of the thermoplastic polymers polyamide which is coated with silver particel. Usual the interface between the polymer and the silver surface connected by physical or chemical methods to promote strong interactions between metal and polymer. The multifilament actuator can be modelled using a linear system of the second order which represents a standard RLC circuit as follows

$$\begin{bmatrix} \frac{\mathrm{d}u_c(t)}{\mathrm{d}t} \\ \frac{\mathrm{d}i(t)}{\mathrm{d}t} \end{bmatrix} = \begin{bmatrix} 0 & \frac{1}{C} \\ -\frac{1}{L} & -\frac{R}{L} \end{bmatrix} \begin{bmatrix} u_c(t) \\ i(t) \end{bmatrix} + \begin{bmatrix} 0 \\ \frac{1}{L} \end{bmatrix} u_{in}(t), \quad (1)$$

in which the informative state is represented by the current $i(t)$ and the controlled state is the capacitor voltage $u_c(t)$. The input parameters are

$$i(t) = \begin{bmatrix} 0 & 1 \end{bmatrix} \begin{bmatrix} u_c(t) \\ i(t) \end{bmatrix}, \tag{2}$$

and

$$u_c(t) = \begin{bmatrix} 1 & 0 \end{bmatrix} \begin{bmatrix} u_c(t) \\ i(t) \end{bmatrix}. \tag{3}$$

The metal-polymer composite fiber is used as an actuator in the linear domain of the system. So the model can be reduced to variable resistance as follows

$$R\big(x_1(t)\big) = a_0 x_1(t), \tag{4}$$

in which $x_1(t)$ represents the position of the mass posted on the fiber as depicted in Fig. 1 and a_0 is an experimental parameter which depends on the temperature and on the orthogonal surface with respect to the current of the fiber.

It is possible to write $x_1(t) = l/2 \tan(\Theta)$, where l is the length of the fiber in the horizontal position, see Fig. 1. Considering the mechanical model of the system, together with the electrical one which we reduced to a resistance depending on the mass position, then the following equations can be written

$$\begin{bmatrix} \frac{dx_1(t)}{dt} \\ \frac{dx_2(t)}{dt} \\ \frac{R(t)}{dt} \end{bmatrix} = \begin{bmatrix} x_2(t) \\ -\frac{K}{m}x_1(t) - \frac{b}{m}x_2(t) + g \\ a_0 x_2(t) \end{bmatrix} + \begin{bmatrix} 0 \\ \frac{b_0}{\big(R_0 + a_0 x_1(t)\big)m} \\ 0 \end{bmatrix} u_{in}(t) + \begin{bmatrix} 0 \\ \frac{1}{m} \\ 0 \end{bmatrix} d(t), \tag{5}$$

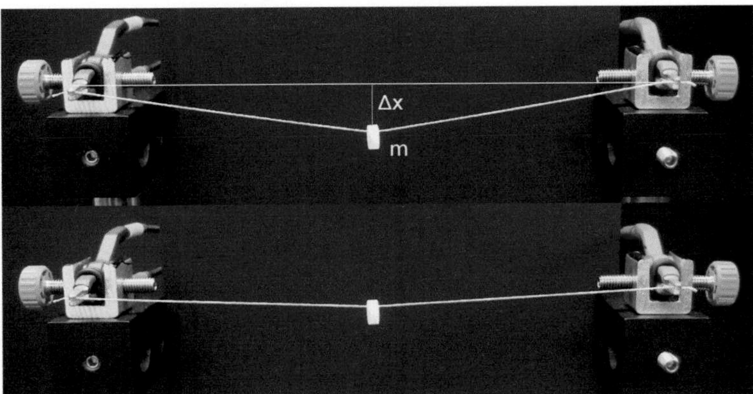

Fig. 1 Depiction of the mechanical actor model and its components

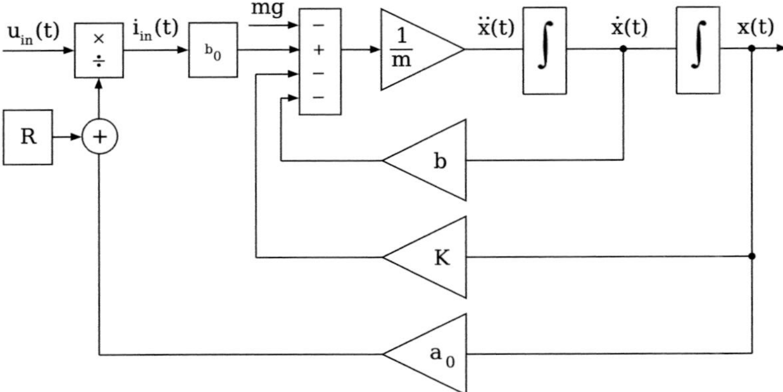

Fig. 2 General block diagram of the system

Fig. 3 General block diagram of the system and its components

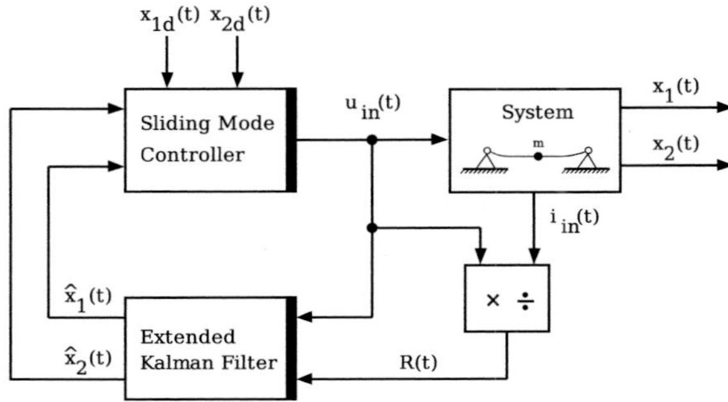

where K represents the elasticity of the fiber, b is the friction constant of the fiber, m is the mass of the posted corp on the fiber, g represents the gravitational acceleration, b_0 is an experimental parameter and $d(t)$ is a possible disturbance acting on the fiber. Figures 2 and 3 show the general block diagram of the system and its components.

3 Design of a Sliding Mode Controller

Considering the following controlled error as

$$x_e(t) = x_{1d}(t) - x_1(t), \tag{6}$$

where $x_{1d}(t)$ represent the desired position of the fiber and $x_1(t)$ represents its actual position. If the following sliding mode surface is defined

$$s(t) = k_e x_e(t) + \dot{x}_e(t), \tag{7}$$

where $k_e \in \mathbb{R}$ with $k_e > 0$. Then

$$\dot{s}(t) = k_e \dot{x}_e(t) + \ddot{x}_e(t) = k_e\left(\dot{x}_{1d}(t) - \dot{x}_1(t)\right) + \ddot{x}_{1d}(t) - \ddot{x}_1(t), \tag{8}$$

where in (8) Eq. (6) is used. Considering now (1), the first system model in which $\dot{x}_1(t) = x_2(t)$ is set, its following that

$$\dot{s}(t) = k_e\left(\dot{x}_{1d}(t) - x_2(t)\right) + \ddot{x}_{1d}(t) - \ddot{x}_1(t). \tag{9}$$

Using (5) its following that

$$\begin{aligned}\dot{s}(t) = k_e\left(\dot{x}_{1d}(t) - x_2(t)\right) + \ddot{x}_{1d}(t) + \frac{K}{m}x_1(t) + \frac{b}{m}x_2(t)\\ - g - \frac{u_{in}(t)}{m} - \frac{d(t)}{m}.\end{aligned} \tag{10}$$

At $s(t) = 0$ we obtain that

$$k_e x_e(t) + \dot{x}_e(t) = 0, \tag{11}$$

which admits the following solution

$$x_e(t) = \eta \exp^{-k_e t}, \tag{12}$$

where η is a constat which is determinated by the initial conditions. k_e states the rate of convergency. To realise an region of attraction around this sliding surface a Lyapunov function can be considered. The system model described in (5) results to be totally controllable and observable so, as typically, the following Lyapunov function can be chosen to design the sliding mode controller

$$V(t) = \frac{1}{2}s^2(t). \tag{13}$$

It follows that

$$\begin{aligned}\dot{V}(t) = s(t)\dot{s}(t) = s(t)\Big(k_e\left(\dot{x}_{1d}(t) - x_2(t)\right) + \ddot{x}_{1d}(t) + \frac{K}{m}x_1(t)\\ + \frac{b}{m}x_2(t) - g - \frac{u_{in}}{m} - \frac{d(t)}{m}\Big)\end{aligned} \tag{14}$$

The control loop can be designed as follows

$$\begin{aligned}u_{in}(t) = m\Big(k_e\left(\dot{x}_{1d}(t) - x_2(t)\right) + \ddot{x}_{1d}(t) + \frac{K}{m}x_1(t) + \frac{b}{m}x_2(t)\\ - g + \lambda s(t) + \beta \text{sign}\left(s(t)\right)\Big),\end{aligned} \tag{15}$$

where $k_e \in \mathbb{R}$ with $k_e > 0$, $\lambda \in \mathbb{R}$ with $\lambda > 0$ and $\beta \in \mathbb{R}$ with $\beta > 0$.

If (15) is inserted into (14), then

$$\dot{V}(t) = s(t)\dot{s}(t) = s(t)\Big(-\frac{d(t)}{m} - \lambda s(t) - \beta\ \mathrm{sign}(s(t))\Big)$$
$$< s(t)\Big(-\frac{d(t)}{m} - \beta\ \mathrm{sign}(s(t))\Big) \tag{16}$$

and it follows

$$\dot{V}(t) = -\frac{d(t)}{m} s(t) - \beta |s(t)|. \tag{17}$$

The sufficient condition

$$\dot{V}(t) < 0 \tag{18}$$

is guaranteed if

$$\beta > \max |\frac{d(t)}{m}|. \tag{19}$$

Remark 1 In case of no correct estimation of the parameters of the model the following expression can be considered for the asymptotical stability

$$\beta > \max |\frac{d(t)}{m}| + \max |\Delta(R, m, K, b)|, \tag{20}$$

where $\Delta(R, m, K, b)$ represents a maximal margin of uncertainty due to the parameters.

3.1 An Extended Kalman Filter as an Observer

The model described in (5) can be rewritten as follows

$$\dot{\mathbf{x}}_{\mathbf{w}}(t) = \mathrm{f}_{\mathbf{w}}\left(\mathbf{x}_w(t), u(t)\right) + \mathbf{w}_w(t), \tag{21}$$

$$\text{where} \quad \mathbf{x}_w(t) = \begin{bmatrix} x_1(t) \\ x_2(t) \\ R(t) \\ d(t) \end{bmatrix}. \tag{22}$$

Here, $\mathbf{f}_w$ represents the functional vector field and according to system in (5) can be expressed by

$$\mathbf{f}_w(\mathbf{x}_w(t)) = \begin{bmatrix} x_2(t) \\ +\frac{K}{m}x_1(t) - \frac{b}{m}x_2(t) + g + \frac{b_0 u_{in}(t)}{\left(R_0 + a_0 x_1(t)\right)m} \\ a_0 x_2(t) \\ 0 \end{bmatrix},$$

where $\mathbf{w}_w(t)$ is the vector model uncertainty and disturbance $d(t)$ is considered to be a constant. The measured output can be written as follows

$$\mathbf{y}_w(t) = \mathbf{H}_w \mathbf{x}_w(t) + \zeta_w(t), \tag{23}$$

where $\mathbf{H}_w = [0 \quad 0 \quad 1 \quad 0]$ and $\zeta_w(t)$ is the measurement noise covariance. The following new notation is introduced: $\mathbf{x}_w(t) \in \mathbb{R}^{n_w}$, $\mathbf{u}_w(t) \in \mathbb{R}^{m_w}$, and $\mathbf{y}_w(t) \in \mathbb{R}^{p_w}$, with $n_w, m_w, p_w \in \mathbb{N}$. An EKF is based on linearising the dynamics and output functions at the current estimate, and on propagation an approximation of the conditional expectation and covariance. Considering the field in (23), if the standard Euler discretisation is used, then the following expression is obtained

$$\mathbf{f}_w(\mathbf{x}_w(k/k)) = \begin{bmatrix} x_1(k/k) + T_s x_2(k/k) \\ x_2(k/k) + T_s g - \frac{T_s K}{m} x_1(k/k) - \frac{b}{m} T_s x_2(k/k) + \frac{T_s b_0 u_{in}(k/k)}{\left(R_0 + a_0 x_1(k/k)\right)m} \\ R(k/k) + T_s a_0 x_2(k/k) \\ d(k/k) \end{bmatrix}. \tag{24}$$

The predicted state is

$$\mathbf{x}_w(k+1/k) = \mathbf{x}_w(k/k) + T_s\left(\mathbf{f}_w(\mathbf{x}_w(k/k)) + \mathbf{B}_w \mathbf{u}_w(k)\right), \tag{25}$$

where T_s is the sampling time and $k \in \mathbb{Z}$. The predicted error covariance matrix is

$$\begin{aligned} \mathbf{P}_w(k+1/k) = \mathbf{P}_w(k/k) + T_s\big(&\mathbf{F}_w(x_w(k/k))\mathbf{P}_w(k/k) \\ &+ \mathbf{P}_w(k/k)\mathbf{F}_w(x_w(k/k))\big) + \mathbf{R}_{f_w}, \end{aligned} \tag{26}$$

where $\mathbf{R}_{f_w}$ is the process noise covariance matrix, $\mathbf{P}_w(k/k)$ is the error covariance matrix, $\mathbf{F}_w(x_w(k/k))$ is the Jacobian matrix of the system calculated for the estimated state. In particular, the Jacobian matrix $\mathbf{F}_w(\mathbf{x}_w(k/k))$ can be defined as follows

$$\mathbf{F}_w(\mathbf{x}_w(k/k)) = \begin{bmatrix} 1 & T_s & 0 & 0 \\ -\frac{T_s}{m}K - \frac{a_0 T_s b_0 u_{in}(k/k)}{\left(R_0 + a_0 x_1(k/k)\right)^2 m} & (1 - \frac{b}{m}T_s) & 0 & 0 \\ 0 & a_0 T_s & 1 & 0 \\ 0 & 0 & 0 & 1 \end{bmatrix}. \tag{27}$$

In (27) we assumed that the force disturbance $d(t)$ is a constant. The following equation states the correction of the Kalman Filter

$$\begin{aligned}\mathbf{K}_w(k) &= \mathbf{P}_w(k/k-1)\mathbf{H}_w^T\left(\mathbf{H}_w\mathbf{P}_w(k/k-1)\mathbf{H}_w^T + \mathbf{R}_{f_\zeta}\right)^{-1}\\ \mathbf{x}_w(k/k) &= \mathbf{x}_w(k/k-1) + \mathbf{K}_w(k)\left(\mathbf{y}_w(k) - \mathbf{H}_w\mathbf{x}_w(k/k-1)\right)\\ \mathbf{P}_w(k/k) &= \mathbf{P}_w(k/k-1) - \mathbf{K}_w(k)\mathbf{H}_w\mathbf{P}_w(k/k-1),\end{aligned} \tag{28}$$

where $\mathbf{K}_w(k)$ is the Kalman gain, and $\mathbf{R}_{f_\zeta}$ is the measurement noise matrix. Concerning matrix $\mathbf{H}_w$, it is known that this represents the output measurement matrix and in the presented case it is represented by the resistance of the fibers. In this sense this matrix is as follows

$$\mathbf{H}_w = \begin{bmatrix}0 & 0 & 1 & 0\end{bmatrix}. \tag{29}$$

4 Validation and Analysis

Figure 4 shows the graphical representation of the desired resistance with $R = 6\ \Omega$ and the observed values. Figure 5 shows the graphical representation of the desired and observed velocity. Figure 6 shows the desired position of the fibers with $\Delta x = 30$ mm and the observed values. The filtered switching function operating in the sliding controller is shown in Fig. 7.

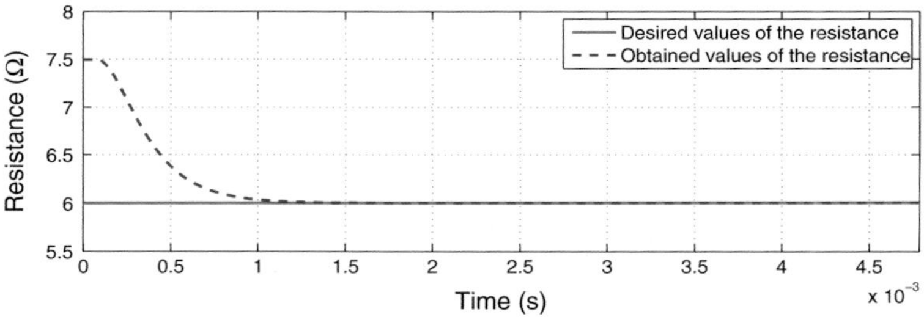

Fig. 4 Graphical representation of the desired and observed resistance

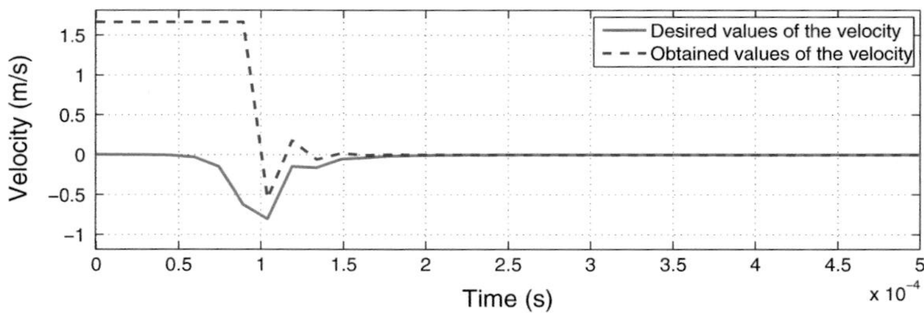

Fig. 5 Graphical representation of the desired and observed velocity

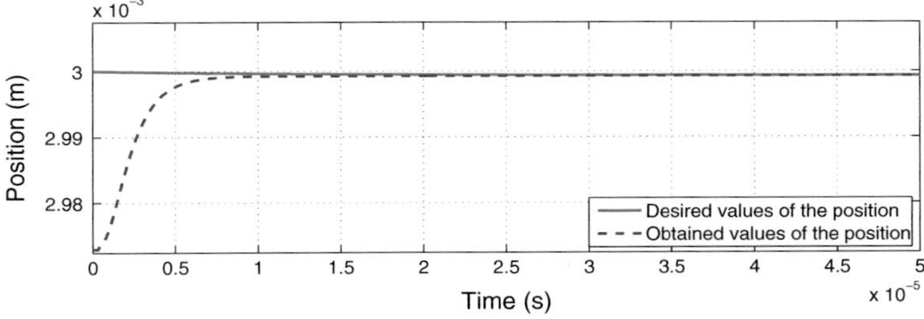

Fig. 6 Graphical representation of the desired and observed position

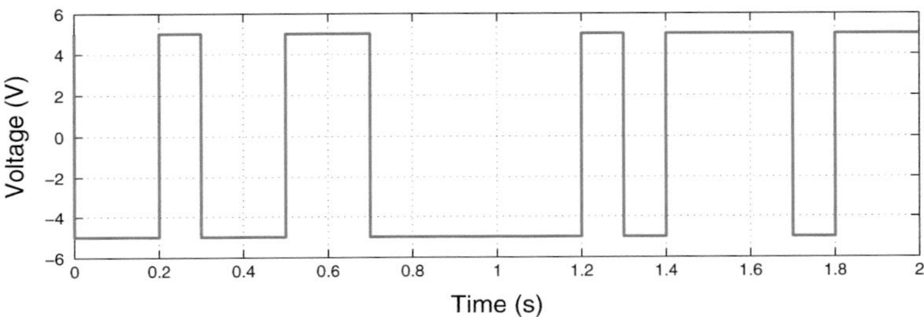

Fig. 7 Graphical representation filtered switching variable of the sliding controller

5 Conclusions

A single-input and single-output (SISO) adaptive Sliding Mode Control combined with an Extended Kalman Filter used as an observer to control metal-polymer composite fibers as an actuator was presented. The simulation results indicate that the proposed algorithm is effective and robust. Simulink®/Matlab® were used for the simulation and experimental setup. However, the performance of the Kalman Filter is strongly dependents on the a priori information of the process and measurement noise. Furthermore, a Kalman Filter often suffers from the problem of dropping-off and loses then the ability to match abrupt parameter changes. For future work in order to consider a more complex fiber model which can include its inductance and capacitive effect an Extended Kalman Filter in cascade configuration mode can be used.

References

1. Jian-Xin, X., Abidi, K.: Discrete-time output integral sliding-mode control for a piezomotor-driven linear motion stage. IEEE Trans. Ind. Electron. **55**(11), 3917–3926 (2008)
2. Pan, Y., Ozgiiner, O., Dagci, O.H.: Variable-structure control of electronic throttle valve. IEEE Trans. Ind. Electron. **55**(11), 3899–3907 (2008)

3. Mercorelli, P.: A switching Kalman Filter for sensorless control of a hybrid hydraulic piezo actuator using mpc for camless internal combustion engines. In: Proceedings of the IEEE International Conference on Control Applications, pp. 980–985, October 2012
4. Mercorelli, P.: An anti-saturating adaptive preaction and a slide surface to achieve soft landing control for electromagnetic actuators. IEEE/ASME Trans. Mechatron. **17**(1), 76–85 (2012)
5. Levant, A.: Chattering analysis. IEEE Trans. Autom. Control **55**(6), 13801389 (2010)
6. Levant, A., Fridman, L.M.: Accuracy of homogeneous sliding modes in the presence of fast actuators. IEEE Trans. Autom. Control **55**(3), 810814 (2010)
7. Pantelopoulos, A., Bourbakis, N.G.: A survey on wearable sensor-based systems for health monitoring and prognosis. IEEE Trans. Syst. Man Cybern. Part C Appl. Rev. **40**(1), 1–12 (2010)
8. Noury, N., Dittmar, A., Corroy, C., Baghai, R., Weber, J.L., Blanc, D., Klefstat, F., Blinovska, A., Vaysse, S., Comet, B.: A smart cloth for ambulatory telemonitoring of physiological parameters and activity: the vtamn project. In: IEEE HEALTHCOM—Proceedings of the 6th International Workshop on Enterprise Networking and Computing in Healthcare Industry, pp. 155–160. Japan (2004)
9. Schimmack, M., Hand, A., Mercorelli, P., Georgiadis, A.: Using a seminorm for wavelet denoising of sEMG signals for monitoring during rehabilitation with embedded orthosis system. In: IEEE MeMeA—Proceedings of the 10th International Symposium on Medical Measurements and Applications, pp. 277–282, Italy (2015)

Algebraic Identification Approach of Multiple Unknown Time-Delays of Continuous-Time Linear Systems

Asma Karoui, Kaouther Ibn Taarit and Moufida Ksouri

Abstract In this paper, a fast online identification algorithm is proposed for linear time invariant multiple time-delays systems (LTI-MTDS) in presence of both unknown initial conditions and static disturbance. It is based on an algebraic, non-asymptotic estimation technique initiated in the framework of systems without delay by Fliess and Sira-Ramirez in [5]. Such technique, developed here in the distributional framework, leads to simple realization schemes involving integrators, multipliers by exponential time functions. We consider the identification problem of multiple-input single-output (MISO) system with two unknown time-delays. Only estimating multiple time-delays from structured input signals will be performed. Simulation studies with noisy data show the performance of the proposed approach.

Keywords Multiple time-delays systems · Online identification · Algebraic method

1 Introduction

The linear time invariant time-delay systems (LTI-TDS) appear in many practical applications such as thermal processes, chemical processes, biological processes, metallurgical processes, etc. These processes are modelled by differential equations with one or multiple delays. The presence of time-delays is a classical source of complexity in the control problems. Its presence degrades the control system performances and may even lead to destabilization.

A. Karoui (✉) · K.I. Taarit · M. Ksouri
National Engineering School of Tunis, Analysis, Conception and Control of Systems Laboratory, University of Tunis El Manar, BP 37, Le Belvédère, 1002 Tunis, Tunisia
e-mail: asma.karoui@gmail.com

A. Abraham et al. (eds.), *Proceedings of the Second International Afro-European Conference for Industrial Advancement AECIA 2015*, Advances in Intelligent Systems and Computing 427, DOI 10.1007/978-3-319-29504-6_31

Correct time-delays identification is relevant for the design of efficient controllers. Many advances in this field have been reported and numerous online estimation methods for the identification of the constant delay have been suggested (see, e.g., [6, 9, 10] for adaptive techniques or [11] for modified least-squares techniques, [8, 20] for nonlinear least-squares and instrumental variables methods or [2–4, 14, 16–18] in the framework of finite dimensional models).

For multiple time-delays system, the problem is much more difficult and has attracted considerable interest (see, e.g., [1] for modulating functions technique or [7] for nonlinear least-squares method and [19] for nonlinear least-squares and instrumental variables methods). These approaches have considered only multiple-input single-output (MISO) systems with multiple unknown time-delays and have generally a poor convergence speed. They are also, essentially asymptotic, recursive or complex, what that leads to unrealistic implementations.

Among the various existing methods in the literature, our interest is focused on the algebraic identification procedure initiated by [5] in the framework of finite dimensional models and extended by [2–4, 14, 16–18]. This approach is a fast estimation technique for simultaneous parameters and delay identification of continuous dynamical systems. It is not asymptotic and does not need any statistical knowledge of the noises corrupting the data. This approach presents a new standpoint based on a differential algebra, a module theory and an operational calculus, which are most classical tools in the control system theory.

Nevertheless, this identification method has not been developed within the multiple delays system framework. Only estimating transmission delays has been treated for Networked Control Systems [18].

This paper considers the identification problem for linear time invariant multiple time-delays systems (LTI-MTDS). The proposed estimation technique derives from the fast algebraic identification technique proposed by Fliess [5] and joins in the continuity of the researches of Belkoura [2–4] and Ibn Taarit [16–18]. In these works, the authors have considered the online identification of the systems with single delay.

In our work, we only consider the identification of time-delays. All the other parameters are assumed to be known. Strictly for purposes of easier conveyance of the methodology, we treat in this paper the case of systems with two delays. The process can have nonzero initial conditions and a static disturbance, which are structured unknown and large noise measurements in the output. To prove the effectiveness of the proposed approach, a MISO system with two delayed structured inputs is considered.

The paper is organized as follows. Section 2 gives the main theoretical tools used for our identification problem. Section 3 exposes the algebraic identification procedure based on a distributional formalism. The proposed delay identification approach for the multiple time-delays systems is developed in Sect. 4 with illustrative example. Finally, numerical simulations of a well-stirred tank system are presented to show the effectiveness of the proposed approach.

2 Mathematical Framework

2.1 Notations

Let recall here some language and standard results from the distribution theory [12] and fix the notations to be considered throughout this paper. In a general context, the space of distributions with left bounded support is an algebra with respect to the convolution product with identity δ, the Dirac distribution. The familiar convolution product is defined as $(u * v)(t) = \int_0^\infty u(\theta)v(t-\theta)d\theta$ in case of a locally bounded function u and v. Derivation, integration and translation can be defined from the convolution products by the respective forms $\dot{u} = \delta^{(1)} * u$, $\int u = H * u$, $u(t-\tau) = \delta_\tau * u$ where H and $\delta^{(1)}$ are respectively the Heaviside step function and the derivative of the Dirac distribution. With a slight abuse of notations, we shall write $H^k u$, with k the order iterated integration of u and, broadly, T^k the iterated convolution product of order k. We also recall the known property of the convolution product $u(t) * v(t-\tau) = u(t-\tau) * v(t) = \delta_\tau * u * v$. The following property allows local considerations, supp $u * v \subset$ supp $u +$ supp v, where the sum on the right hand side is defined by: $\{x + y;\ x \in \text{supp } u,\ y \in \text{supp } v\}$. Finally, and without any confusion, we denote $u(s)$, $s \in \mathbb{C}$, the Laplace transformation of u.

2.2 Background

The specific need for the distributional framework lies in its ability to cancel the singular terms derived from the initial conditions by means of multiplication by some appropriate functions. The multiplication of two distributions (noted α and T) always makes sense when at least one of the two terms (noted α) is a smooth function. Several properties can be derived from such a product, particulary, the cancellation procedure, which is derived from the following theorem:

Theorem 1 *If T has a compact support K and is of a finite order m, $\alpha T = 0$ whenever α and all its derivatives of order $\leq m$ vanish on K [12].*

In this paper, we shall be mainly concerned with the annihilation of singular distributions by exponential functions, and as an illustration of the previous theorem:

$$(1 - e^{-\gamma t})\delta = 0,\ (1 - e^{-\gamma(t-\tau)})\delta_\tau = 0. \tag{1}$$

In forming the product αT, the delay τ involved in the argument $T(t-\tau)$ appears as a coefficient. In the distribution framework, straightforward computations can be carried out using the Leibniz formula related to the product of a Dirac derivative with the smooth function α:

$$\alpha\delta_a^{(n)} = \sum_{k=0}^{n} (-1)^{(n-k)} C_n^k \alpha^{(n-k)}(a)\,\delta_a^{(k)}. \tag{2}$$

where C_n^k are the familiar binomial coefficients.

The combination of multiplication by the exponential function with the convolution product $S * T$, if one of the S or T distributions has a compact support, it leads to the following two very useful properties in practice:

$$e^{-\gamma t}(S * T) = e^{-\gamma t}S * e^{-\gamma t}T, \tag{3}$$

Combining the Leibniz rule (2) respectively with (3), with $S = \delta^{(p)}$ and $T = y$ allows us to transform terms respectively of the form $e^{-\gamma t}y^{(q)}$ into linear derivative sums of the product: $z = e^{-\gamma t}y$ as shown as follows:

$$e^{-\gamma t}y^{(2)} = \gamma^2 z + 2\gamma z^{(1)} + z^{(2)}, \tag{4}$$

where, $z^{(1)}$ and $z^{(2)}$ are respectively the first and the second derivative of $z = z = e^{-\gamma t}y$. Note that integrating these expressions twice results in the by-part integration formula.

3 Algebraic Method

Structured entries have been introduced (see e.g. [3, 5, 13]) to refer to entities (mainly perturbations) that can be annihilated by means of simple multiplications and derivations. The identification procedure mainly consists in the following three steps: Differentiations, Multiplication with an appropriate C^∞ function and Integrations.

Derivation: This operation allows to reduce the dirac impulses, possibly derived and delayed, the perturbations as well as the specific entries appearing in the equation of the system.

Multiplication: In the general case, the multiplication of two distributions is not always defined. However, this operation always make sense when one of the two terms is a smooth function. Several properties can be derived from such product and the most important for our developments are derived from the Schwartz' Theorem 1. This theorem is the key result from which most of the parameters (including the delays) can be identified from step input responses.

Integration: This last step makes it possible to be freed from any derivation and led to the final system, from which the parameters are estimated.

4 Multiple Time-Delays Identification

Consider the following strictly stable multiple-input single-output (MISO) continuous time system with unknown time-delays under unknown nonzero initial conditions and static disturbance:

$$\sum a_i y^{(i)} = \sum K_j u_j(t - \tau_j) + \psi_0 + \gamma_0 H, \tag{5}$$

where γ_0 is the constant perturbation, K_j, a_j and τ_j are constant parameters, $u_j(t)$ is the structured delayed input. $\psi_0 = \sum_{i<n} y^{(i)}(0)\,\delta^{(i)}$ is the initial condition term and is an infinite dimensional.

In this work, we only consider the single delays identification. All the other parameters are assumed to be known. The class of systems with two unknown delays is considered.

In order to better understand its aims let us consider a first-order system with two delays governed by the following general equation:

$$y^{(1)} + a_0 y = \psi_0 + \gamma_0 H + \beta_1 u_1(t - \tau_1) + \beta_2 u_2(t - \tau_2), \tag{6}$$

where a_0, β_1, β_2, τ_1 and τ_2 are constant parameters. Consider also the two step inputs $u_j = u_{0j}H, j = 1, 2$.

The Eq. (6) is written as follows:

$$y^{(1)} + a_0 y = \psi_0 + \gamma_0 H + \beta_1 u_{01} H(t - \tau_1) + \beta_2 u_{02} H(t - \tau_2). \tag{7}$$

Suppose that at $t = 0$, two structured input test are applied to the process. The task is to estimate the two time-delays τ_j from the inputs $u_j(t)\, j = 1, 2$ and the output measurement $y(t)$ in the presence of the unknown static disturbance γ_0 and an initial condition ψ_0 which could be nonzero. The time constant $\frac{1}{a_0}$ is assumed to be known.

Following the steps of algebraic approach, a first order derivation yields:

$$y^{(2)} + a_0 y^{(1)} = \psi_0^{(1)} + \gamma_0 \delta + \beta_1 u_{01} \delta_{\tau_1} + \beta_2 u_{02} \delta_{\tau_2}, \tag{8}$$

where $\psi_0^{(1)} = (y^{(1)}(0) + a_0 y(0))\delta + y(0)\delta^{(1)}$, of order 1 and support $\{0\}$, contains the contributions of the initial conditions.

By virtue of Theorem 1, the multiplication of the Eq. (8) by a function such that $\alpha(0) = \alpha^{(1)}(0) = 0$ and $\alpha(\tau_j) = 0, j = 1, 2$ cancels the structured terms of the right side.

The candidate function is chosen as:

$$\alpha(t) = (1 - e)^2 (1 - \lambda_1 e)(1 - \lambda_2 e), \tag{9}$$

where $e = e^{-\gamma t}$, $\lambda_1 = e^{\gamma \tau_1}$, $\lambda_2 = e^{\gamma \tau_2}$ and γ is a tunable positive parameter.

Multiplying by $\alpha(t)$, we get the following equation:

$$(1-e)^2(1-\lambda_1 e)(1-\lambda_2 e)(y^{(2)}+a_0 y^{(1)})=0. \tag{10}$$

A polynomial in $\lambda_j, j=1,2$ is then obtained and expressed as follows:

$$\begin{aligned}&(\lambda_1+\lambda_2)\left[ey^{(2)}+a_0 ey^{(1)}-2e^2y^{(2)}-2a_0e^2y^{(1)}+e^3y^{(2)}+a_0e^3y^{(1)}\right]\\&-(\lambda_1\lambda_2)\left[e^2y^{(2)}+a_0e^2y^{(1)}-2e^3y^{(2)}-2a_0e^3y^{(1)}+e^4y^{(2)}+a_0e^4y^{(1)}\right]\\&=\left[y^{(2)}+a_0y^{(1)}-2ey^{(2)}-2a_0ey^{(1)}+e^2y^{(2)}+a_0e^2y^{(1)}\right].\end{aligned} \tag{11}$$

This equation shows that $k \geq 2$ successive integrations avoid the derivation in the delay identification algorithm. A linear system with unknown parameters $\Lambda_1 = \lambda_1+\lambda_2$ and $\Lambda_2 = \lambda_1\lambda_2$ is obtained by using different integration orders:

$$\begin{pmatrix} H^2\omega_1 & H^2\omega_2 \\ H^3\omega_1 & H^3\omega_2 \end{pmatrix}\begin{pmatrix}\Lambda_1\\ \Lambda_2\end{pmatrix}=\begin{pmatrix}H^2\omega_0\\ H^3\omega_0\end{pmatrix}, \tag{12}$$

where:

$$\begin{aligned}\omega_0 &= \left[y^{(2)}+a_0y^{(1)}-2ey^{(2)}-2a_0ey^{(1)}+e^2y^{(2)}+a_0e^2y^{(1)}\right],\\ \omega_1 &= \left[ey^{(2)}+a_0ey^{(1)}-2e^2y^{(2)}-2a_0e^2y^{(1)}+e^3y^{(2)}+a_0e^3y^{(1)}\right],\\ \omega_2 &= -\left[e^2y^{(2)}+a_0e^2y^{(1)}-2e^3y^{(2)}-2a_0e^3y^{(1)}+e^4y^{(2)}+a_0e^4y^{(1)}\right].\end{aligned} \tag{13}$$

Each of the terms, $H^k\omega_i, (i=0,\ldots,2)$ involved in (12) is achieved using the rules described in (3) and the same calculation developed in (4).

The vector of parameters $\Lambda_j, j=1,2$ to be estimated is obtained by the resolution of this last linear system (12). Then, λ_j, $j=1,2$ are deduced from the estimated $\Lambda_j, j=1,2$. The estimated delay is then obtained from the relation $\tau_j = \log(\lambda_j)/\gamma$, $j=1,2$. Note that since supp $H_k\delta_{\tau_j} \subset (\tau_j,\infty)$, the delay is not identifiable on $(0,\tau_j)$, $j=1,2$.

Numerical simulation results from the Eq. (12) are illustrated in the Fig. 1 for the parameters $y(0)=0.3$, $\gamma_0=0.3$, $a_0=2$, $\beta_1=4$, $\beta_2=0.5$, $u_{01}=1$, $u_{02}=3$, $\tau_1=1.2$ s, $\tau_2=0.5$ s, $\gamma=0.6$ and $k=2,3$ successive integrations. The simulation step size has been fixed to 0.005 s.

We note that the values of the parameters Λ_1 and Λ_2 converge to the desired value $\hat{\Lambda}_1=2.9$ and $\hat{\Lambda}_2=2.18$ Fig. 1a. Solving the following system of equations (14) leads to two solutions for each $\lambda_j, j=1,2$:

$$\begin{cases}\Lambda_1=\lambda_1+\lambda_2\\ \Lambda_2=\lambda_1\lambda_2\end{cases} \tag{14}$$

The correct solution of $\hat{\Lambda}_j$ is based on the prediction error minimization criterion. The estimated delays are then deduced from the relation $\hat{\tau}_j=\log(\hat{\lambda}_j)/\gamma, j=1,2$. The

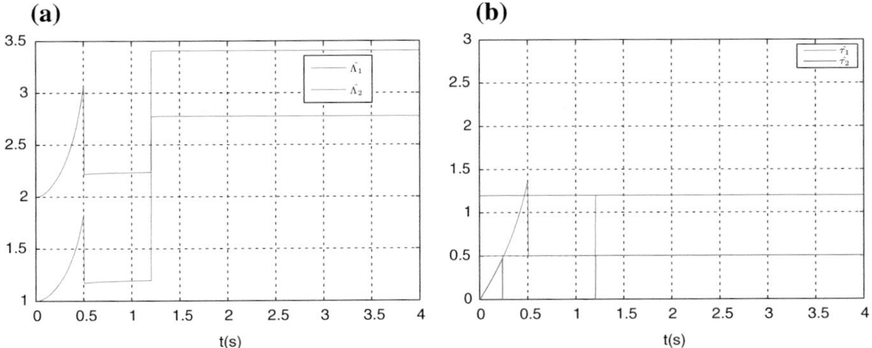

Fig. 1 Estimated time-delays τ_1 and τ_2. **a** Estimated parameters of the problem (12). **b** Estimated delays τ_j, j = 1, 2

Fig. 1b shows the trajectories of the two estimated time-delays $\hat{\tau_1}$ and $\hat{\tau_2}$. Note that the estimated delays $\hat{\tau_j}$ vanish on $(0, \tau_j), j = 1, 2$ and they are, therefore, not identifiable on this interval. For $t \geq \tau_j$, $j = 1, 2$ Fig. 1b shows good convergence speed of the proposed identification algorithms.

In order to test the robustness of the proposed method, a simulated noise obtained using the additive white Gaussian noise is added to the output signal. The signal-to-noise ratio, SNR, is fixed to 35 dB. The output signal is presented in the Fig. 2a. The resulting numerical simulations are shown in Fig. 2b for $\gamma = 0.6$, $k = 5, 6$ successive integrations. A good estimate of the two delays in a noisy context is obtained.

The selection criteria based on the prediction error minimization which permit to select the right solution of $\lambda_j, j = 1, 2$, of the linear equation system (12) is sensitive

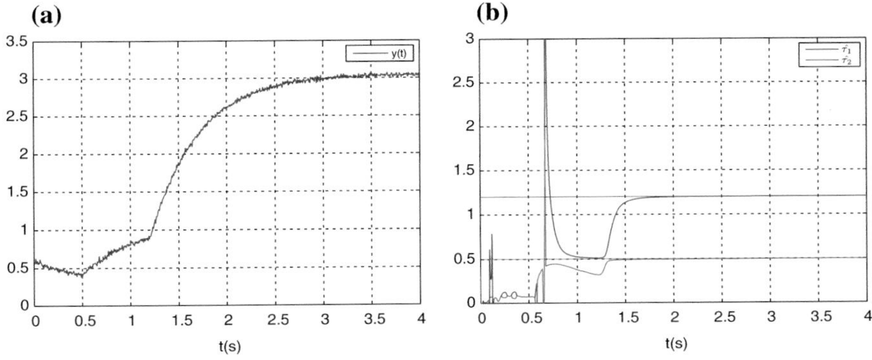

Fig. 2 Estimated time-delays τ_1 and τ_2 in the noisy case. **a** Output signal. **b** Estimated delays τ_j, j = 1, 2

in the presence of a high noise level. The main idea is to filter the real output signal by low pass filter such as:

$$F(s) = \frac{1}{\eta s + 1}. \tag{15}$$

The following simulations in Fig. 2b show, for $\eta = 0.05$, favorable robustness properties of the proposed delays identifier with respect to noise corrupted data.

5 Application to a Stirred Tank Reactor

The process considered is a well-stirred tank in which the liquid is assumed to be well mixed and the tank is well insulated in the Fig. 3. The outlet temperature, $T(t)$, is sensitive to changes in inlet temperature, $T_i(t)$. The following equation describes the dynamics of this process:

$$f\rho_i C_{pi} T_i(t) - f\rho C_p T(t) = \frac{\mathrm{d}\left[V\rho C_v T(t)\right]}{\mathrm{d}t}, \tag{16}$$

where f is a volumetric flow, ρ_i and ρ are the inlet and the outlet liquid densities, respectively, V is the volume of liquid in tank, C_{pi} and C_p are the inlet and outlet heat capacities at constant pressure, respectively and C_v is the liquid heat capacity at constant pressure [15].

By considering the deviation variables $\Delta T_i(t) = T_i(t) - T_i^0(t)$ and $\Delta T(t) = T(t) - T^0(t)$, we obtain the following equation:

$$f\rho C_p \Delta T_i(t) - f\rho C_p \Delta T(t) = V\rho C_v \frac{\mathrm{d}\left[\Delta T(t)\right]}{\mathrm{d}t}. \tag{17}$$

By removing the assumption that the tank is well-insulated, the outlet temperature is than sensitive to the surrounding temperature $T_s(t)$. In this case, the process can be defined by the following equation:

Fig. 3 Thermal process [15]

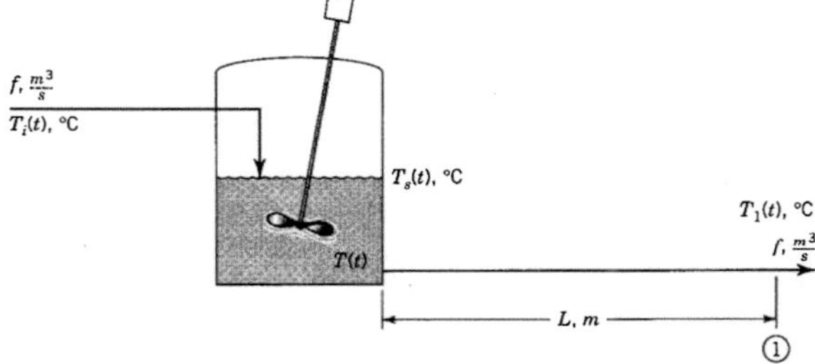

$$f\rho C_p T_i(t) - UA\left(T(t) - T_s(t)\right) - f\rho C_p T(t) = V\rho C_v \frac{\mathrm{d}\,[T(t)]}{\mathrm{d}t}, \tag{18}$$

where U is an overall heat transfer coefficient and A is a heat transfer area. Taking the Laplace transform of the Eq. (18), we obtain the following transfer function:

$$\Delta T(s) = \frac{K_i}{as+1}\Delta T_i(s) + \frac{K_s}{as+1}\Delta T_s(s), \tag{19}$$

where $\Delta T_s(t) = T_s(t) - T_s{}^0(t)$, $a = \dfrac{V\rho C_v}{f\rho C_p + UA}$, $K_i = \frac{f\rho C_p}{f\rho C_p + UA}$ and $K_s = \frac{UA}{f\rho C_p + UA}$.

Let now consider the process output as the outlet temperature $T_1(t)$ which is obtained at the end of the pipe assumed to be well-insulated. The measure of $T_1(t)$ is the same as $T(t)$ except that it will be delayed by some amount of time. In this case, the Eq. (19) becomes:

$$\Delta T(s) = \frac{K_i e^{-\tau_1 s}}{as+1}\Delta T_i(s) + \frac{K_s e^{-\tau_2 s}}{as+1}\Delta T_s(s). \tag{20}$$

Written in the time domain and taking into account the initial conditions, the system is governed by the following differential equation:

$$a\Delta T^{(1)} + \Delta T = K_i \Delta T_i(t-\tau_1) + K_s \Delta T_s(t-\tau_2), \tag{21}$$

After the derivation of the Eq. (21), one proceeds to the multiplication by the same exponential function used in the previous illustrative example (9). This candidate function satisfies the criteria of Schwartz' Theorem 1 and leads to the cancellation of the right-hand side of (21). This operation and after $k \geq 2$ successive integrations, leads to the linear equation system (12) involving $\Lambda_1 = \lambda_1 + \lambda_2$ and $\Lambda_2 = \lambda_1 \lambda_2$:

The main objective is namely to identify the two time-delays. All the other parameters are assumed to be known. The estimated delays are deduced from the relation $\tau_j = \log(\lambda_j)/\gamma, j = 1, 2$.

Numerical simulation results are illustrated in the Fig. 4 for $k = 2, 3$ integrations and the parameters $a = 398.9542$ s, $K_s = 0.4103$, $K_i = 0.5897$, $\tau_1 = 25$ s, $\tau_2 = 30$ s [15]. The simulation step size has been fixed to 1 s. The results of the estimated delays given by the Fig. 4a and enlarged in the Fig. 4a show its identifiability for $t \geq \tau_j, j = 1, 2$ with a fast convergence and thus the effectiveness of the proposed approach is enlightened.

6 Conclusion

In this paper, an online algebraic identification approach of multiple unknown time-delays for linear systems taking into account the initial conditions and static disturbances is proposed. The simulation results show that the estimated delays converge

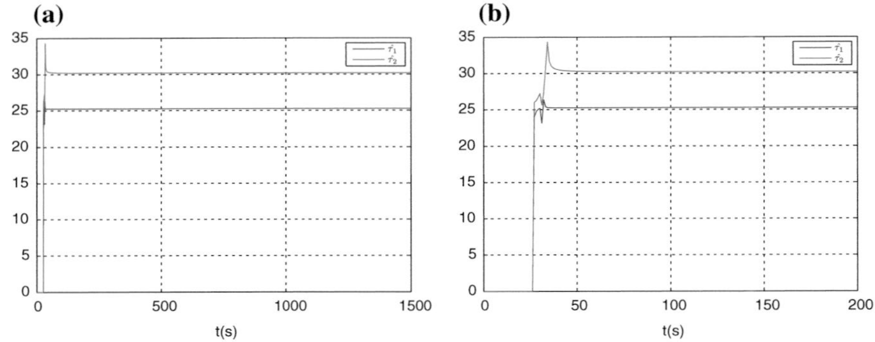

Fig. 4 Estimated time-delays of the well-stirred tank system. **a** Estimated delays τ_j, j = 1, 2. **b** Zoom on the estimated delays

to desired solutions with good performances using the proposed identification algorithm. Consistent estimated delays are also obtained in the presence of high noisy output measurement. The proposed method can be easily extended to linear-time systems with n unknown multiple delays.

References

1. Balestrino, A., Landi, A., Sani, L.: Parameter identification of continuous systems with multiple-input time delays via modulating functions. In: Control Theory and Applications, IEE Proceedings (2000)
2. Belkoura, L.: Identifiability and algebric identification of time delay systems, vol. 423, pp. 103–117. Springer, Berlin (2012)
3. Belkoura, L., Fliess, M., Richard, J.P.: Parameters estimation of systems with delayed and structured entries. Automatica **45**, 1117–1125 (2009)
4. Belkoura, L., Richard, J.P., Fliess, M.: A convolution approach for delay systems identification. In: 17th IFAC World Congress (2008)
5. Fliess, M., Sira-Ramirez, H.: An algebraic framework for linear identification. ESAIM Control Optim. **9**, 151–168 (2003)
6. Gomez, O., Orlov, Y., Kolmanovsky, I.: On-line identification of siso linear time-delay systems from output measurements. Automatica **43**, 2060–2069 (2007)
7. Iemura, H., Yang, Z., Kanae, S., Wada, K.: Identification of continuous-time systems with multiple unknown time delays by global nonlinear least-squares method. In: IFAC Workshop on Adaptation and Learning in Control and Signal Processing, and IFAC Workshop on Periodic Control Systems. Yokohama, Japan, 30 Aug, 1 Sept 2004
8. Kaur, D., Dewan, L.: Identification of delayed system using instrumental variable method. J. Control Theory Appl. **10**, 380–384 (2012)
9. Naa, J., Ren, X., Xia, Y.: Adaptive parameter identification of linear siso systems with unknown time-delay. Syst. Control Lett. **66**, 43–50 (2014)
10. Orlov, Y., Kolmanovsky, I., Gomez, O.: On-line identification of siso linear time-delay systems from output measurements: theory and applications to engine transient fuel identification. In: American Control Conference. IEEE, Minneapolis, Minnesota, USA, June 2006

11. Ren, X., Rad, A., Chan, P., Lo, W.: Online identification of continuous-time systems with unknown time delay **50**, 1418–1422 (2005)
12. Schwartz, L.: Théorie des Distributions, 2nd edn. Hermann, Paris (1966)
13. Sira-Ramirez, H.: Some applications of differential algebra in systems identification. In: 3rd IFAC Symposium on Power System, Structure and Control. Mabu Thermas Convention Center, Brazil (2007)
14. Sira-Ramirez, H., Rodriguez, C.G., Romero, J.C., Juarez, A.L.: Algebraic Identification and Estimation Methods in Feedback Control Systems. Wiley, United Kingdom (2014)
15. Smith, C., Corripio, A.: Principles and Practice of Automatic Process Control, 2nd edn. Wiley (1997)
16. Taarit, K.I.: Contribution l'identification des systèmes à retards et d'une classe de systèmes hybrides. Thèse de doctorat (2010)
17. Taarit, K.I., Belkoura, L., Ksouri, M.: Estimation en ligne par approche algébrique : application au cas des frottements secs. In: 6ème Conférence Internationale Francophone d'Automatique, CIFA 2010. Nancy, France, June 2010
18. Taarit, K.I., Belkoura, L., Ksouri, M., Richard, J.P.: A fast identification algorithm for systems with delayed inputs. Int. J. Syst. Sci. **42**(3), 449–456 (2011)
19. Yang, Z.: Iterative Methods for Identification of Multiple-input Continuous-time Systems with Unknown Time Delays, pp. 339–362. Springer, London (2008)
20. Yang, Z., Iemura, H., Kanae, S., Wada, K.: A global nonlinear instrumental variable method for identification of continuous-time systems with unknown time delays. In: 16th IFAC World Congress. Prague, Czech Republic, July 2005

Fetal Risk Classification Based on Cardiotocography Data: A Kernel-Based Approach

Khaoula Keddachi and Foued Theljani

Abstract The use of classification-based approaches for health state diagnosis is still an ongoing issue in medicinal applications. In this paper, we address the problem of fetal risk anticipation using cardiotocography (CTG) measurements. We adopt, for this purpose, a classification approach based on Support Vector Domain Description (SVDD). Indeed, the SVDD is an efficient kernel method employed to solve one-class problems known also as novelty detection problems. The fundamental goal of one-class learning is to generate a rule that distinguishes between a set of typical objects called target class and aberrant objects designated as outliers. Based on CTG data, the abnormalities anticipation/detection can be basically perceived as multi-classification problems since it consists in recognizing CTG patterns and classifying them into several classes according to the risk category. To do so, we have used a modified version of SVDD algorithm endowed with useful tools to manage the multi-classification problems. The aim is to generate, from a small number of CGT samples, a classification model that can be generalized and applied on a wider number of unknown samples. The proposed approach is assessed on real CTG database to prove its effectiveness.

1 Introduction

Diagnosis of fetal risks based on cardiotocography (CTG) classification becomes a growing tendency to generate decision support systems in medicinal field [1–3]. Cardiotocography (CTG) is a typical medical signal that records two types of information during pregnancy; Fetal Heart Rate (FHR), and Uterine Activities (UA). After treatment, the information extracted from the CTG measurements are useful early for making decision about the fetus state. Well, the FHR can be serving in finding

K. Keddachi · F. Theljani (✉)
Esprit: School of Engineering, Tunis, Tunisia
e-mail: foued.theljani@esprit.tn; foued_theljani@yahoo.fr

K. Keddachi
e-mail: khaoula.kaddachi@esprit.tn

A. Abraham et al. (eds.), *Proceedings of the Second International Afro-European Conference for Industrial Advancement AECIA 2015*, Advances in Intelligent Systems and Computing 427, DOI 10.1007/978-3-319-29504-6_32

out the amount of oxygen that a fetus is acquiring during pregnancy. We can classify, thus, the fetus state as Normal, Suspicious or Pathological.

Several works have addressed this kind of problem and used pattern recognition approaches for fetal monitoring based on CTG data. Various algorithms have been used for this purpose, among which we mention Artificial Neural Networks (ANN) [4] and Support Vector Machines (SVM) [5, 6]. However, there is still a method that is underused and can be solve efficiently the drawbacks discussed above. This method is called Support Vector Domain Description, well known as SVDD. It is developed at first by [7, 8]. It is an efficient technique employed to solve one-class problems (known also novelty detection problems). The fundamental goal of one-class learning is to extract a model that allows descerning between a set of representative objects called targets and unseen-novel objects designated as outliers. The optimization process aims to find the sphere or domain with minimum volume that encloses the whole target data or most of the target data. It consists, thus, in resolving an optimization problem with constraints using a quadratic programming (QP). Most of SVDD-based classifiers are designed to deal with one-classification problems. This may be not adequate for various real-life applications which usually need training data for at least two classes, as the case of CTG-based cardiac diagnosis. In view of that, we have used an adapted SVDD to deal with this dilemma [9, 10].

Taking into account these topics, the paper is organized into four sections. In Sect. 2, we give a global overview about the SVDD technique and its theoretical foundation. As for Sect. 3, it is reserved to describe the way according to which we have adapted the SVDD to manage the multi-class data. The electronic fetal monitoring methodology, based on CTG data classification, is explained in Sect. 4. Finally, we show some experimental tests performed on real CTG database taken from 2126 patients. As conclusion, we summarize the presented work and we discuss the interest that yields.

2 SVDD: Theoretical Foundation

Let $\chi = \{x_1, \ldots, x_i, \ldots, x_N\}$ be a target training set, with $\chi \subseteq R^d$. SVDD is a data description kernel machine, which aims to find a smallest hyper-sphere containing most of instances of the target class with some relaxation defined by slack variables. Its original form is formulated as a constrained optimization problem as follows:

$$\min R^2 + C \sum_{i=1}^{N} \xi_i \tag{1}$$

$$s.t \quad (x_i - a)^T (x_i - a) \leq R^2 + \xi_i, \quad \xi_i \geq 0 \;\; \forall i = 1, \ldots, N, \tag{2}$$

where R and a denote the radius and the center of the hyper-sphere and C is the regularization parameter which gives the trade-off between the volume of the sphere and the misclassification errors. The variable ξ_i is a slack variable designating the distance of ith data point from the sphere boundary. This is a quadratic optimization problem and can be solved efficiently by introducing Lagrange multipliers for constraints. The Lagrangian formulation of the problem is given thus by the following formula:

$$L = R^2 + C\sum_i \xi_i - \sum_i \alpha_i \left(R^2 + \xi_i - (x_i - a)^T (x_i - a)\right) - \sum_i \gamma_i \xi_i \tag{3}$$

where α_i and γ_i are Lagrange multipliers, $\alpha_i \geq 0,\ \lambda_i \geq 0$. Note that for each training data point x_i, a corresponding α_i and γ_i are defined. L has to be minimized with respect to R, a, and ξ_i and maximized with respect to α_i and γ_i. Taking the derivatives of (3) by setting $\partial L/\partial R = 0,\ \partial L/\partial a = 0$ and $\partial L/\partial \xi_i = 0$, we obtain the Karush-Kuhn-Tucker (KKT) conditions given by the following relations:

$$\begin{aligned} \frac{\partial L}{\partial R} &= 0 \quad \Rightarrow \quad \sum_{i=1}^{N} \alpha_i = 1, \\ \frac{\partial L}{\partial a} &= 0 \quad \Rightarrow \quad \sum_{i=1}^{N} \alpha_i x_i = a, \\ \frac{\partial L}{\partial \xi_i} &= 0 \quad \Rightarrow \quad 0 < \alpha_i < C. \end{aligned} \tag{4}$$

The QP equations are obtained by substituting the above KKT conditions in (3). We obtain a dual problem expressed by:

$$\max \ \tfrac{1}{2}\sum_{i=1}^{N} \alpha_i (x_i{}^T x_j) - \tfrac{1}{2}\sum_{i=1}^{N}\sum_{j=1}^{N} \alpha_i \alpha_j (x_i{}^T x_j), \tag{5}$$

$$s.t \ \sum_{i=1}^{N} \alpha_i = 1\ , \quad 0 \leq \alpha_i \leq C \ \ \forall i = 1, \ldots, N. \tag{6}$$

After solving such a standard Quadratic Programming (QP), we obtain the solution $\alpha_i = \alpha^*$, whose corresponding training instances can be classified as Boundary Support Vectors (*BSVs*) outside the hyper-sphere, if $\alpha_i = C$, Non-Support Vectors (*NSVs*) inside the hyper-sphere, if $\alpha_i = 0$ and Non-Boundary Support Vectors (*NBSVs*) just at the hyper-sphere, if $0 < \alpha_i < C$. Any data satisfying one of the above KKT conditions is designated as target class. Otherwise, it is an outlier. Furthermore, all points with $\alpha_i > 0$ are called Support Vectors (*SVs*) which restrict the data domain and can fully describe the one-class boundary. We can write thus $SVs = BSVs \cup NBSVs$. According to (4), the center a can be easily calculated as:

$$a = \sum_{i=1}^{N} \alpha_i x_i. \tag{7}$$

To make prediction on an unknown instance z, the squared distance to the center of the sphere must be calculated using the following formula:

$$d^2(z, a) = (z^T z) - 2 \sum_{i=1}^{N} \alpha_i (z^T x_i) + \sum_{i=1}^{N} \sum_{j=1}^{N} \alpha_i \alpha_j ({x_i}^T x_j). \tag{8}$$

Afterward, we define the decision function as:

$$G(z) = R^2 - d^2(z, a) \tag{9}$$

Now we say that some test instance belongs to the target class or lies inside the hyper-sphere if $G(z) \geq 0$. Otherwise, it is an outlier lying outside the hyper-sphere. Similarly, in traditional support vector machines and other kernel machines, the inner product between two vectors in (7) and (8) can be replaced by various kernels satisfying the Mercer theorem [11]. By introducing a kernel function, the formula (8) becomes as follow:

$$F(z, a) = K(z^T z) - 2 \sum_{i=1}^{N} \alpha_i K(z^T x_i) + \sum_{i=1}^{N} \sum_{j=1}^{N} \alpha_i \alpha_j K({x_i}^T x_j) \tag{10}$$

As we see, the methodology above gives efficiently an optimized data domain. Nevertheless, the challenge becomes valuable when dealing with datasets containing several classes. In the next section, we address this paradigm.

3 Multi-classification with SVDD

The SVDD algorithm, as presented in its initial version, is designed to separate only one target class of similar patterns from others called outliers. Therefore, we need to extend the scope of the application and adapt the algorithm in order to support multi-class data. A variety of techniques for decomposition of a multi-class problem into several 2-class problems have been proposed.

3.1 Conventional Methods

3.1.1 One-Against-All (OAA)

For a k-classes problem, the One-Against-All method constructs k SVDD classifiers. The ith SVDD classifier is trained by attributing all patterns in the ith class positive labels and all remaining classes negative labels. The final output of OAA method is the class that corresponds to the SVDD. Usually, classification of an unknown pattern is done according to the maximum output among all SVDD classifier.

3.1.2 One-Against-One (OAO)

The One-Against-One (OAO) approach, also known as pair wise coupling, consists in constructing one SVDD classifier per each pair of classes. Thereby, for a k-classes problem, we need to train $k(k-1)/2$ SVDD classifiers to separate patterns of one class from patterns of another class. The membership of a novel unseen pattern is done according to the maximum output provided by each SVDD classifier, where each classifier promotes one class.

3.1.3 Binary Tree (BT)

This practice is known as the binary tree or classification multi levels. We consider an initial class set that regroups all clusters. Thereafter, the set is partitioned into two sub-classes, S^{left} and S^{right}. We obtain, as a result, two subsets derived from the same node. At the internal node, a SVDD classifier makes the discrimination between the S^{left} and S^{right} subsets. This process is applied to each of the left and the right subsets recursively and so on. When the class set has only one class, the recursive process terminates. The classification of a new sample is done by browsing the tree from top to bottom.

3.2 Multi-SVDD Based on MKD

Several alternatives may be used to solve multi-classification problems. A manner to proceed consists in decomposing a multi-class problem into a set of two-class sub-problems and then use a discriminant function to make the final decision. These include methods OAA and OAO. Despite their generalization ability, these methods are characterized by a huge computational complexity. To overcome this drawback, we introduce an efficient architecture, particularly convenient when the information about labeled classes might potentially increase. Assume we have at first a dataset partitioned into K-disjoint subsets $\left\{\omega_k\right\}_{k=1}^{K}$ according to their output classes. For example, the kth class data set ω_k contains N_k samples, so that:

$$\omega_k = \left\{x_{k,1}, \ldots, x_{k,i}, \ldots, x_{k,N_k}\right\} \tag{11}$$

Here, $x_{k,i}$ is the ith element of the kth class data set. The main goal is to classify a new instance z_t in one of the possible K classes. Further, we assume that we can train a discriminant function $F(z_t, \omega_k)$ giving the dissimilarity degree between the pattern z_t and a given class ω_k. Instead, we only need to check the outputs of K classifiers and decide which class ω_k an arbitrary input vector z_t belongs to. Each input vector should be recognized only by one trained classifier. If the input vector belongs to an unknown class, then it should be rejected by all classifiers. However, we may fall in the case where several classifiers accept the input vector simultaneously. To manage this situation, the input vector is assigned to the most likely class defined by the following criterion:

$$\omega_{win} = \arg\left(\min_{k=1\ldots K} F(z_t, \omega_k)\right) \tag{12}$$

where ω_{win} designates the winning class. $F(z_t, \omega_k)$ is a predefined discriminant function. In the case of SVDD classifiers, the discriminant function provides as output some distance measurement. This distance gives the neighborhood between an unknown pattern and a class model. The discriminant function can be defined by means of two formulations. The first one is expressed as:

$$F(z_t, \omega_k) = K(z_t, z_t) - 2\sum_{i=1}^{N_k} \alpha_i K(z_t, x_{k,i}) + \sum_{i=1}^{N_k}\sum_{j=1}^{N_k} \alpha_i \alpha_j K(x_{k,i}, x_{k,j}) \tag{13}$$

The above function estimates the neighborhood between a new instance z_t and the center of the class model. Therefore, the formula (12) assigns a testing pattern to the class having the closest center in terms of distance. This method is called "Closest Center" (CC) [12]. This way to proceed may be practical when dealing with Gaussian mixtures, but it is not suitable for random distributions. It gives rather improper estimation of neighborhood leading to erroneous classification. To deal with this, we can introduce the second formulation of the discriminant function proposed by [12]. It aims to estimate the distance between a testing pattern z_t and the Closest Support Vector (CSV) of a given class ω_k, so that $F(z_t, \omega_k)$ becomes as follow:

$$F\left(z_t, \omega_k\right) = \min_{m=1,\ldots,N_{SV_k}} K(z_t, z_t) - 2K(z_t, SV_{k,m}) + K(SV_{k,m}, SV_{k,m}) \tag{14}$$

where $SV_{k,m} \in SV_k$, SV_k contains the whole support vectors of the class ω_k and N_{SV_k} is the total number of support vectors belonging to SV_k. The approach chooses the closest class to the newly coming instance. According to (12), when using the CC method, we obtain:

$$\omega_{win} = \arg\min_k K(z_t, z_t) - 2\sum_{i=1}^{N_k} \alpha_i K(z_t, x_{k,i}) + \sum_{i=1}^{N_k}\sum_{j=1}^{N_k} \alpha_i \alpha_j K(x_{k,i}, x_{k,j}), \tag{15}$$

Otherwise, when using the CSV method, we proceed as follow:

$$\omega_{win} = \arg\min_k \min_m K(z_t, z_t) - 2K(z_t, SV_{k,m}) + K(SV_{k,m}, SV_{k,m}), \tag{16}$$

The term K denotes all along the paper the used Kernel function. In the following, we present an exhaustive description of the methodology adopted to perform a fetal state monitoring based on information extracted from CTG data. We will employ, for this aim, the Multi-SVDD presented above.

4 Fetal Risks Anticipation Using CTG Data

Along this section, we focus on prenatal diagnosis of fetal risks based on CTG data. We employ for this purpose the Multi-SVDD algorithm previously presented. To better drive simulations, we even carry out a comparison of performances with some conventional methods from literature. We choose, mainly, the Multi-Layer Perceptron (MLP) and the Radial Basis Function network (RBF).

4.1 Data Description and Methodology

For assessment, simulations will be conducted on real dataset containing 2126 fetal cardiotocograms and 21 features. We give in Table 1 a summary of all CTG features used for classification. The CTGs were classified by three expert obstetricians and a consensus classification label assigned to each of them. Classification is done with respect to the Fetal State Class (FSC); Normal (N), Suspect (S) or Pathologic (P). We can, also, categorize the FSC as Normal (N) or Tainted (T). Therefore the dataset can be used either for 2-class or 3-class experiments.

In methodological terms, data are divided into two subsets (Fig. 1). The first one, devoted to train the classification models, comprises roughly the one-third (1/3) of the entire samples. The second subset, gathering the remaining samples (2/3), is dedicated to test and validate the models trained prior. Note that the schema is applied on the Multi-SVDD algorithm by making a comparison with two other conventional methods (MLP and RBF) in terms of their performances.

4.2 Simulations and Results

We have chosen to operate particularly on this CTG data because its taxonomy is extremely hard to distinguish and it represents an ideal case of overlapping data as shown by Fig. 2. So, samples are inseparable and not easy to classify. Furthermore, we consider two types of data divided as training dataset and testing dataset. Each sample is a vector consisting of 21 attributes. As well, the experiments have been conducted by considering both 2-class and 3-class data. For assessment, on the basis of the confusion matrix, we compute the classification accuracy for each classifier.

Table 1 List of attributes and their signification

Symbol	Attribute description
LB	FHR baseline (beats per minute)
AC	Number of accelerations per second
FM	Number of fetal movements per second
UC	Number of uterine contractions per second
DL	Number of light decelerations per second
DS	Number of severe decelerations per second
DP	Number of prolonged decelerations per second
ASTV	Percentage of time with abnormal short-term variability
MSTV	Mean value of short-term variability
ALTV	Percentage of time with abnormal long-term variability
MLTV	Mean value of long-term variability
Width	Width of FHR histogram
Min	Minimum of FHR histogram
Max	Maximum of FHR histogram
Nmax	Number of histogram peaks
Nzeros	Number of histogram zeros
Mode	Histogram mode
Mean	Histogram mean
Median	Histogram median
Variance	Histogram variance
Tendency	Histogram tendency
FSC	Fetal State Class: N = Normal; S = Suspect; P = Pathologic (3-class) N = Normal; T = Tainted (2-class)

This last one is a performance index signifying the percentage of cases correctly classified. The Table 2 summarizes the obtained results during experiments.

In an objective analysis of these results, we can simply extract the following interpretations; with respect to the classification accuracy, the adapted algorithm Multi-SVDD achieves the best results comparatively to those provided by MLP and RBF, and this is in the cases of 2-class and 3-class. On the other hand, the algorithms MLP and RBF present a lowest accuracy percentage when passing from 2-class to 3-class experiments. It infers from this that MLP and RBF methods lose their generalization ability when increasing the number of classes. This drawback does not arise in the case of Muli-SVDD algorithm. This, thereby, seems proper and practical since, in real applications, often we are faced with multi-classification problems.

To better illustrate the principle of Minimal Kernelized Distance (MKD) on which the Multi-SVDD algorithm is founded, we propose to perform some tests to explain nearly the detection process based on MKD computation. On the basis of limited number of samples, Fig. 3 shows the MKD variation, respectively in the space of 2-class and 3-class. Indeed, Fig. 3a shows a 2-class case where samples are partitioned

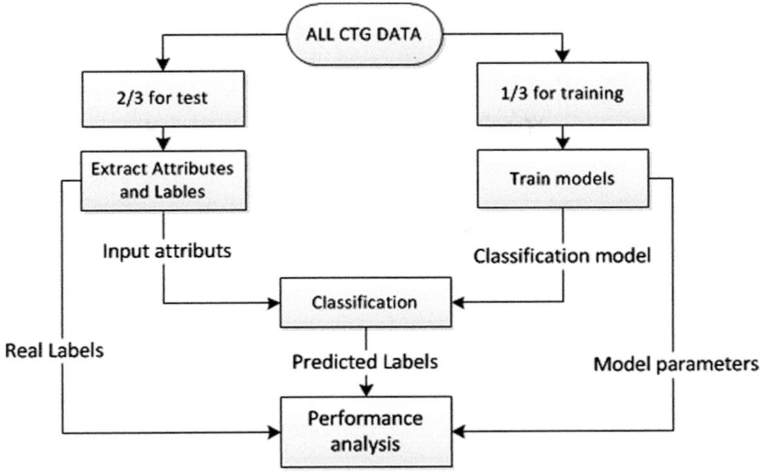

Fig. 1 Adopted methodology for CTG data classification

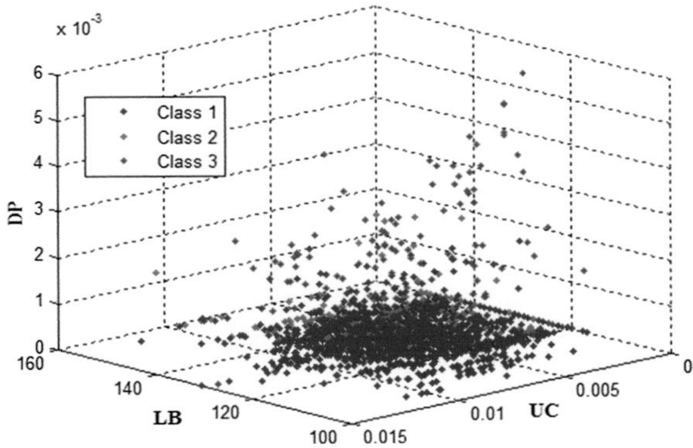

Fig. 2 3D graphical projection in the space of variables {LB, UC, DP} of 3-class data

Table 2 Classification accuracy of Mutli-SVDD compared to MLP and RBF methods

	Classification accuracy (%)	
	2-class data	3-class data
MLP	77.84	55.11
RBF	77.84	77.91
Mutli-SVDD	**80.16**	**82.20**

into "Normal" and "Tainted". By evaluating the Kernelized Distance, we can interpret easily that the first five samples belong to class 2 insofar as their Kernelized Distance from the center of class 2 is minimal. Similarly, we can classify the last five samples as class 1 since, with respect to the Kernelized Distance, they are closer

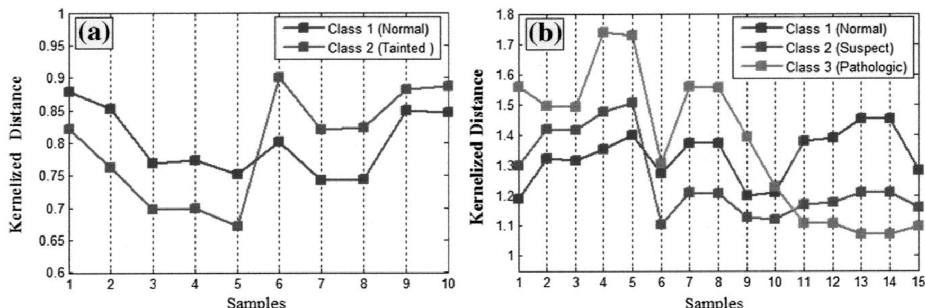

Fig. 3 Graphical illustration of Minimal Kernelized Distance

to this class. According to the same principle, in the case of 3-class classification problem as shown by Fig. 3b, we can draw the following conclusions:

- Samples {1, 2, 3, 4 and 5} belong to class 1 (Normal).
- Samples {6, 7, 8, 9 and 10} belong to class 2 (Suspect).
- Samples {11, 12, 13, 14 and 15} belong to class 3 (Pathologic).

5 Conclusion

The application of decision support systems in the medicinal field to provide a risk diagnosis, even a prediction, becomes an ever-increasing trend in view of their approved performance. In this paper, we have chosen to focus on the problem of fetal risks diagnosis using cardiotocography measurements (CTG). The main goal is to generate a classification model having the ability to learn from a limited number of samples the requisite knowledge that can be applied later on huge number of unseen samples. This is generally referred to as "generalization ability". To do so, we have employed a modified version of the well known SVDD algorithm. Based on Minimal Kernelized Distance (MKD), this last one is adapted in order to support learning data consisting of more than two classes. By performing a comparison with some conventional method from literature, the Multi-SVDD has proved a considerable accuracy and generalization ability, mainly when dealing with high dimensional data. As forthcoming works, the algorithm may be improved to be applied on other medicinal data base characterized by huge number of classes and attributes.

References

1. Jezewski, M., Czabanski, R., Leski, J.: The influence of cardiotocogram signal feature selection method on fetal state assessment efficacy **32**, 51–58 (2014)
2. Ravindran, S., Jambek, A.B., Muthusamy, H., Neoh, S.-C.: A Novel Clinical Decision Support System Using Improved Adaptive Genetic Algorithm for the Assessment of Fetal Well-Being, vol. 2015, Article ID 283532, p. 11 (2015)

3. Setiawan, N.A., Venkatachalam, P.A., Hani, A.F.M.: Diagnosis of coronary artery disease using artificial intelligence based decision support system. In: Proceedings of the International Conference on Man-Machine Systems (ICoMMS), pp. 11–13. Batu Ferringhi, Penang, MALAYSIA, Oct 2009
4. Sundar, C., Chitradevi, M., Geetharamani, G.: Classification of cardiotocogram data using neural network based machine learning technique. Int. J. Comput. Appl. **4**(14) June 2012
5. Yilmaz, E., Kilikcier, C.: Determination of fetal state from cardiotocogram using LS-SVM with particle swarm optimization and binary decision tree. Comput. Math. Methods Med. **2013**, 8 (2013)
6. Sahin, H., Subasi, A.: Classification of the cardiotocogram data for anticipation of fetal risks using machine learning techniques. Appl. Soft Comput. **33**, 231–238 (2015)
7. Tax, D., Duin, R.: Support vector domain description. Pattern Recognit. Lett. **20**, 1191–1199 (1999)
8. Tax, D., Duin, R.: Support vector data description. Mach. Learn. **54**, 45–66 (2004)
9. Theljani, F., Laabidi, K., Zidi, S., Ksouri, M.: Tennessee eastman process diagnosis based on dynamic classification with SVDD, ASME. J. Dyn. Syst. Meas. Control, **137**(9), 10, June 2015
10. Theljani, F., Laabidi, K., Zidi, S., Ksouri, M.: A new kernel-based classification algorithm for systems monitoring: comparison with statistical process control methods. Arab. J. Sci. Eng. **40**(2), 645–658 (2015)
11. Vapnik, V.N.: Statistical Learning Theory. Wiley, New York (1998)
12. Sachs, A., Thiel, C., Schwenker, F.: One-class support-vector machines for the classification of bioacoustic time series. Int. J. Artif. Intell. Mach. Learn. **6**(4), 29–34 (2006)

A Bag of Words Model for Improving Automatic Stress Classification

Aurelia Ciupe, Camelia Florea, Bogdan Orza, Aurel Vlaicu and Bogdan Petrovan

Abstract Most of the existing stress assessment frameworks rely on physiological signals measurements (EEG, ECG, GSR, ST, etc.), which involve direct physical contact with the patient in a medical setup. Present technologies rely on capturing moods and emotions through remote devices (cameras), further processed by computer vision and machine learning techniques. The proposed work describes a method of automatic stress classification where stress information is modeled based on pupil diameter non-intrusive measurements, recorded by an eye tracking remote system. The signal extracted from the pupil Dataset has been processed using the Bag-Of-Words model, with a SVM classification and results have been compared to similar experiments in order to validate the applicability and consistency of the Bag-Of-Words model on stress assessment and classification.

Keywords Automatic stress classification · Bag-Of-Words model · Pupil diameter · Biomedical time series · SVM

A. Ciupe (✉) · C. Florea · B. Orza · A. Vlaicu · B. Petrovan
Multimedia Technologies and Telecommunications Research Centre, Technical University of Cluj-Napoca, C. Daicoviciu 15, 400020 Cluj-Napoca, Romania
e-mail: Aurelia.Ciupe@com.utcluj.ro

C. Florea
e-mail: Camelia.Florea@com.utcluj.ro

B. Orza
e-mail: Bogdan.Orza@com.utcluj.ro

A. Vlaicu
e-mail: Aurel.Vlaicu@com.utcluj.ro

B. Petrovan
e-mail: Bogdan.Petrovan@com.utcluj.ro

A. Abraham et al. (eds.), *Proceedings of the Second International Afro-European Conference for Industrial Advancement AECIA 2015*, Advances in Intelligent Systems and Computing 427, DOI 10.1007/978-3-319-29504-6_33

1 Introduction

1.1 Problem Statement and Motivation

Stress load influences human responses in task-oriented environments with operational demands.

Stress recognition and assessment has become a research theme of interest in the Human Factors and Ergonomic Fields, as controlled or uncontrolled stressors influence social behavior, work performance and tasks executions. Coping solutions are to be developed based on stress recognition, measurement and classification. In real-world driving tasks, stress assessment may prove to be potentially valuable in terms of safety implications such as awareness increase to external disruptive factors. Therefore, stress becomes a strong factor in critical situations, where control is required.

Technological evolution provides new means of automatic stress detection and measurement through non-intrusive methods.

Intrusive methods for stress measurement have mainly relied on physiological signals (Electroencephalogram—EEG, Electrocardiogram-ECG, Heart Rate Variability—HRV, Galvanic Skin Response—GSR, Skin Temperature—ST). Although reliable and accurate, they lack in comfort in real-world applications, as measurement devices need to be attached to participants. Such environment diminishes the natural context of reactions due to an external-precedence acknowledgement. In the current computer vision and Human-Computer Interaction revolution, portable electronic devices and systems make use of computational imaging for face expression detection with high resolutions on eye movements and signal extraction, under lower hardware costs.

Bag-of-words model provides an accurate representation and classification of physiological time series.

Specifically used in human behavior analysis, human action recognition, face and expression recognition or visual categorization, the bag-of words model has proven a higher robustness to noise, providing enhanced results on wider time-windows for aperiodic waveforms, while other representations have been considered effective for short time series or time series with periodic waveforms.

1.2 Proposed Approach

Due to the success of the BoW model in text and object categorization, we have tested its consistency on pupil diameter time series, where stress presence has been related to amplitude fluctuations of the pupil diameter signal. Experiments have been performed on the existing PUPILLARY dataset [1] identified as relevant for the purpose of the study. A classic BoW model generation has been computed on the Pupil Diameter (PD) signal extracted from the given dataset. Support Vector

Machine (SVM) method has been chosen to classify stress levels and to provide results that further to be analyzed in comparison to other classification methods used in biomedical/pupil diameter time series, with or without stress appraisal.

An important issue that challenges stress assessment and particularly the proposed approach, is to identify/develop a relevant, complete dataset to be used in the experiment. Mainly datasets contain recordings of face expressions, human actions, and even physiological signals, but their experimental setups do not have any stress-related stimuli. For our study, we have used the PUPILLARY dataset, collected in a stress context, under various stimuli.

Given the selected approach and the existing challenges, the proposed work is structured as follows: Sect. 2 provides a brief introduction on the stress framework, assessment methods and possible implications of the BoW model. In Sect. 3 we describe the standard BoW model for pupillary data assessment, computed on time series (un-periodic waveforms). Section 4 describes the implementation of the proposed framework, the outcomes for each step of the BoW model generation and the classification results, in comparison to existing solutions.

2 Background and Related Work

2.1 Objective Measurements for Stress Classification

Involving multidimensional facets, stress studies have been grounded in several disciplines (i.e. Computer Science, Psychology, Physiology, Neurology) [2], with difficulties in term definition. In order to define stress and workload, measurement tools models needed to be defined [3]. Early models [2] have involved subjective, behavioral and psychophysiological measurements. Pupillary response has been evidenced [1, 4] to better perform than GSR, close to EEG and HRV, but poorly than ECG and heat flux, although from task performance point of view a dominance of cognitive load over emotion in eye features resulted. Same measurement metrics have been lately considered for workload measurement and classification in computational models [2, 5]. As physical contact is involved in physiological signal acquisition, such measures have been considered having an obtrusive effect on individuals. Therefore, a new direction of workload measurement has started moving towards non-intrusive methods using facial imaging or video recording [5]. Classical experiment set-ups involve controlled and predefined stimuli (demanding tasks, attention-focused simulation, high pressure situations) [2, 3, 5, 6]. In image-video eye-tracking experimental set-ups for, a challenge in extracting accurate data is related to calibration procedure [7] and eye-blink periods, in which loss of information occurs.

In order to classify the stress measurement results, several classification algorithms and models have been applied. SVM methods, originally associated to facial expression and recognition, have provided successful results in stress recognition, when stress patterns have been captured with histograms rendered from dynamic thermal patterns [2]. Eye-tracking logs providing real data sequences have been transformed to analytic signal through linear operators (Hilbert Transformation) [8]. Machine learning algorithms (simple logistic, k-nearest neighbor, linear support vector machine and decision trees) were used single and combined to provide cognitive load classification on 2 different levels (low, high), in an unobtrusive environment, where feature selection has been compiled using Weka data mining package [9]. Better accuracy in cognitive load classification for three levels has been obtained by using multiple regression analysis for pupil features with Gaussian mixture model classifier [4]. Real-time stress recognition and classification has been achieved through a full recognition system (face detection, eyes and mouth detection, eye centers localization, feature vector extraction, feature vector normalization and preprocessing and classification) [10]. Combination of Neural Networks and Wavelet have been applied for feature extraction and classification in pupil diameter monitoring with rmANOVA analysis [1].

2.2 *Psychological Time Series Processing in Stress Assessment*

In real world scenarios, manual inspection and monitoring of physiological signals lacks in accuracy and precision, affecting the diagnosis process [11], therefore, an automated system to analyze long-term biomedical signals becomes a real value in building applications [12]. Biomedical signals are mainly characterized through their statistical nature, therefore automatic analysis with machine learning techniques has been applied in several real-world ergonomic and health-care scenarios: epileptic seizure detection, brain computer interaction and mental fatigue detection [12]. Such representations extract local temporal or frequency information to characterize time series, which are very effective for short time series or time series with periodic waveforms, but yet, ineffective for repetitive and aperiodic waveforms, due to a low level of similarity interpretation.

2.3 *Implications of the Bag-Of-Words Framework in Stress Assessment*

In Neurotechnologies, physiological measures assessed for improving task execution performance in complex and demanding environment, has been challenged by machine learning techniques [13], where the Bag-Of-Words (BoW) representation model with machine learning techniques, has become a method used in computer vision, for psychological signals analysis and classification [12]. Fields of applications are related to Human-Computer Interfaces, applied in Human Behavior Analysis [14] with a particular reference to Brain-Computer Interfaces [13]. Several approaches have been alternatively considered: bag-of-patterns representations (BoP) with Symbolic Aggregate approXimation (SAX), regarded as a method which decreases computational complexity and reduces the data's dimensionality, guaranteeing a lower bound on Euclidean distances between time series [11]. Due to its drawback of high dimensions when computing large datasets, a refactoring of the method consisted of adopting BoW representations, where a time series is treated as a text document and local segments are extracted from the time series as words [12]. Also specific in applying data mining to time series, alternatives of time series representations include: Discrete Fourier Transformation, Discrete Wavelet Transformation, and Piecewise Linear Approximation [11]. In BoW representations, DWT, DFT or DCT, etc. can be applied to extract local features from local segments contained in windows of the given time series.

3 The BoW Model Generation and Classification Methods

The flowchart of the BoW model generation applied on physiological time series is described in Fig. 1. The first stage of the model construction is to build a description of the further-needed "words". "Interesting points" are to be identified in the signal waveform, local patches to be extracted and computed in a set of features, further used in processing. A set of local patches is extracted from the set of the given time series (TS1, TS2, …, TSN), based on a sliding-window mechanism applied to each TS [12], with a predefined window size. Local patches are represented as vectors in high dimensional spaces [15].

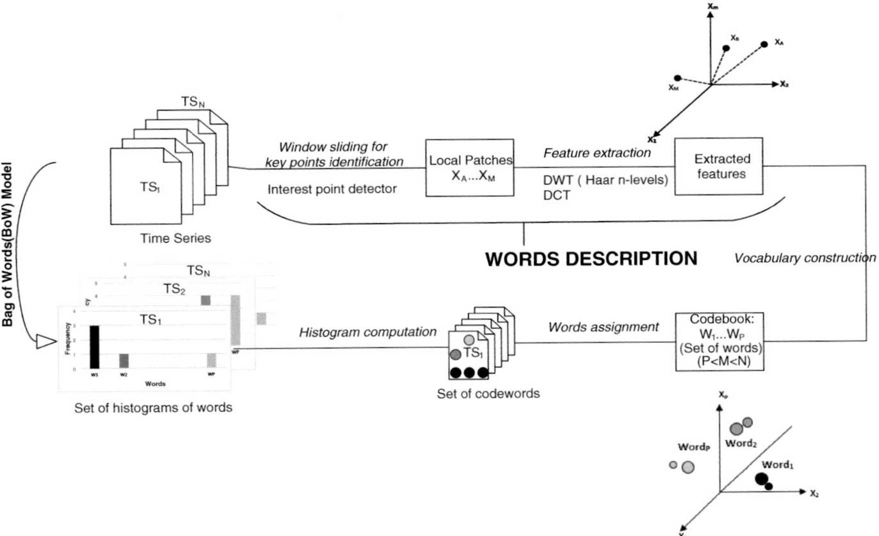

Fig. 1 Flowchart of the Bag of Words (BoW) model generation for time series

For the proposed implementation, we have performed a point detection operation with local maximums extraction from the time series waveforms (on which median filtering has been applied), followed by a window sliding at the located points. Features are then extracted from the previously identified local patches, either in the wavelet domain, through a Discrete Wavelet Transform (DWT), or in the Cosinus domain through a Discrete Cosinus Transform (DCT).

In order to detect signal anomalies, Symbolic Aggregate approXimation representations are considered of main relevance. We have chosen a DWT with a Haar wavelet. Feature vectors have been extracted to represent the local patches (X_1, X_2, …, X_M). The next step in BoW model generation, vocabulary (codebook) construction, consists of grouping into clusters the extracted features, following a K-Means clustering operation. The centroids of the clusters ($Word_1$, $Word_2$, …, $Word_P$) become the most representative values of the feature vectors [13], and are further named “words”, elements of the codebook. Other supervised and unsupervised methods can be used in codebook construction: mean-shift, supervised Gaussian mixture models [12, 15]. Despite its high computational costs [15], K-means is probably the most popular algorithm used for vector quantization. Two methods for word assignment can be applied in the next step: soft assignment, hard assignment [15]. For the current work, we have chosen a hard assignment strategy, where association is made to a single closest vocabulary word, based on the Euclidian distance function. Therefore each time series is transformed into a collection of “codewords”, based on the codebook elements, “words”. A bag of words is thus assembled. The final step is to represent the obtained set of codewords as a distribution of the codebook words’ frequencies. The BoW model for the pupillary data time series consists of converting each time series into a frequency vector of length K (number of words), and its representation as a histogram, where each position specifies the number of occurrences of the codebook word in the time series.

In classification, BoW becomes a feature set for supervised classification. Training is done through the codebook’s words and similarity measures. The most basic classifier, 1-Nearest Neighbor (1-NN), determines the word in the vocabulary located at a minimal distance. Several similarity measures have been depicted in [12]: Euclidean Distance, Chi-Squared distance, Jensen-Shannon Distance, Histogram Intersection based Distance. In the current work, we propose a classification based on SVM, proven to give better results, and more powerful in biomedical figures contexts [16].

4 Experimental Design and Results

4.1 PUPILLARY Dataset

The input dataset has been obtained from a simulation of basic driving tasks, conducted by Pedrotti et al. [1], where participants took part in a Lane Change Test (LCT) scenario (16 participants included in the experimental group), under 4 different stimuli-based stress circumstances, named trials (t0, t1, t2, t3, t4). On the experimental group 4 types of environments were applied (1 no-stress class, 3 stress classes), while no stressors were applied during the t0 trial (relaxed conditions). The experiment took place in a dedicated simulation room. For each trial (t1, t2, t3, t4), stressors were applied to the experimental group progressively: 1st trial consisted of a sample LCT, during which participants adjusted to the driving conditions; stressors were applied starting with the 2nd trial and only on the experimental group: 2nd trial —3 types of audio notifications (8.6 Hz white noise, police siren, 4 kHz tone) in case of inappropriate driving behavior every 20 s, 3rd trial—a monitoring camera and an external observer standing next to the participant, in charge with taking notes on the driving behavior, without any additional disturbances. Objective nonintrusive measures were performed on the pupil diameter, recorded at 50 Hz sampling rate with the SMI RED 4 remote video eye tracker. Samples of the normalized extracted times are shown in Fig. 2. The full description of the stimuli, experiment procedures, data acquisition and analysis are described in the experimental setup of [1]. Pupil diameter signals (PD) have been extracted in time series, following the steps: pre-processing, TEPRs extraction and normalization.

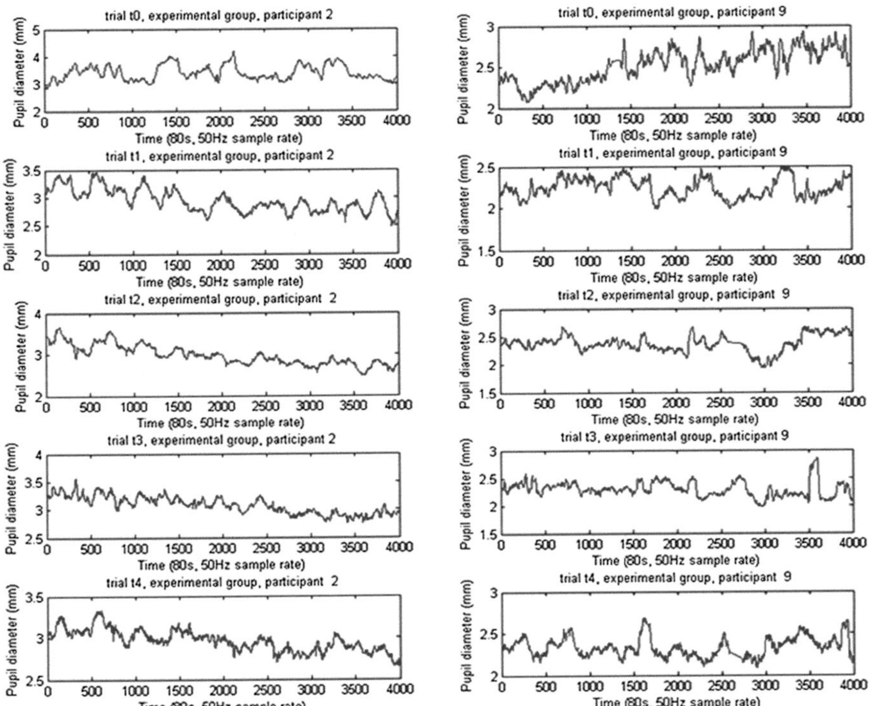

Fig. 2 Pupil diameter for all trials for 2nd and 9th participant

4.2 Experimental Results

A representation of the Pupil Diameter signal has been made using the BoW model generation. The method requires an optimal setup of processing parameters. The current section provides outcomes of each BoW stage (Fig. 1), focusing on the parameter setup.

Key points identification. For the first step, local patches identification, a window sliding step across the pupil diameter time series, is to be established. Defining a window size dimension for signal scanning is of crucial importance for a further word length definition.

The optimal value of 500 was chosen based on empiric tests, based on a pre-considered interval of 50–1000. For local patches extraction, 2 strategies were applied: (1) a standard scanning of the signal with the predefined step (2) local maximum identification and patch extraction for each identified local maximum. To identify each local maximum, a median filtering has been applied to the initial signal, although patches have been extracted from the unfiltered signal. Figure 3 illustrates 2 samples specific to 2 different trials (trial 3—te3, trial 4—te4) in the experimental group (participant 2 and 9) including: the original signal, the filtered signal, the signal on which DWT (Haar on 5 levels) has been applied and the median filtered signal with identified local maximums and minimums. Local maximums have been characterized as "points of interests", as a pupil diameter growth is related to stimuli exposure (i.e. stress stimuli for the current experiment).

Local patches and feature extraction. The second step consisted of local patches and feature extraction. For each key point previously identified (either through local maximum strategy or with predefined step scanning strategy), a local patch is extracted. A transformation has been applied to each local patch, so that signal information can be optimally described through a feature vector of lowest possible size. Each patch is initially rescaled, based on the patch average value subtraction and division by standard deviation.

For the proposed framework, signal data have been transposed from the spatial domain to frequency domain through: (1) DCT transformation followed by a quantization (2) DWT Haar on 3 levels and approximation preservation (subbands

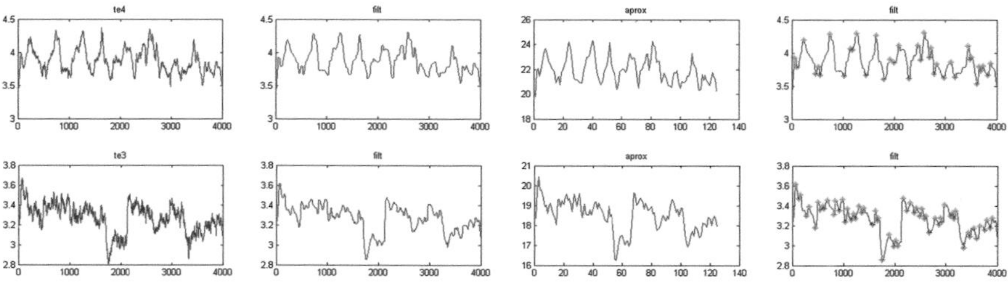

Fig. 3 Original and processed signal: filtered, approximation of DWT (Haar, 5 levels), key points detection (te3, te4)

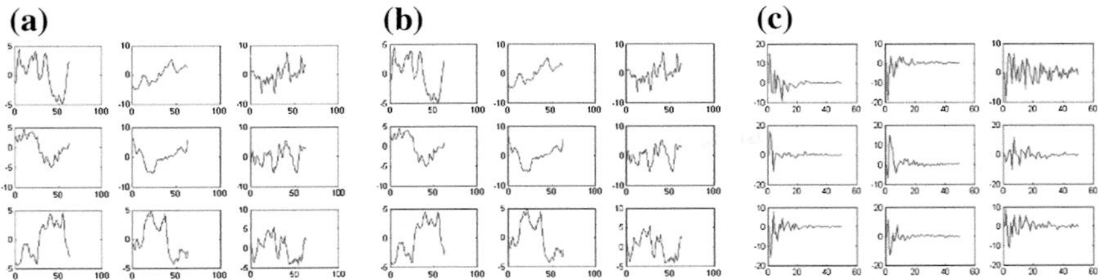

Fig. 4 Extracted local patches in the time domain (**a**), in the frequency domain: DWT (**b**), DCT (**c**)

with detail information are excluded). Following DCT transform, null values will be positioned at the end in vector of features' (as long arrays of zeros) while not null values will be positioned towards the vector's start point. Such a distribution allows a quantization in the DCT domain for each patch, by trucking each vector of features to the first 10 % values (representative enough for patch description). Therefore, through signal transposition in the frequency domain, we reduce the dimension of the representation space for each patch from 500 values to 50 (DCT), and to 63 (DWT). Figure 4a–c illustrate samples of the local patches extraction, and signal representation in the frequency domain (DWT, DCT).

Vocabulary and codebook construction. BoW vocabulary and codebook construction was performed in 2 steps: K-mean clustering and words assignment. In case of vocabulary construction, we have chosen the optimal number of words as K = 250. To each word in the codebook, a codeword is assigned. The BoW histogram is computed for pupil diameter time series (each trial), based on the frequency of each word. Samples of obtained histograms for the 3rd and 4th trials are presented in Fig. 5.

Training and test phases. For a further comparison, we have followed the classification strategy proposed by [1]. 10 participants from the experimental group were considered in the training process and 4 as test. 4 classification classes have been considered (4 trials), therefore, 40 training samples and 16 test samples have been obtained.

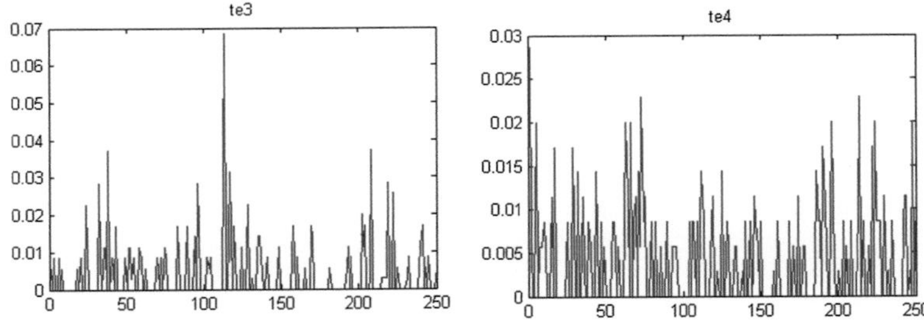

Fig. 5 BoW histograms from pupil diameter signal for trials te3, te4

Results were compared in case of a 1NN and a SVM classifier. SVM proved a slightly more accurate classification compared to the k-NN classifier. The highest recognition rate obtained, of 80 %, is close-comparable to the 79.2 % prevailed in [1], where signal classification is computed on the entire time series, after approximation in the DWT domain with Haar on 5 levels (Fig. 3—approx.). Therefore, the proposed BoW can be considered an appropriate method to be applied for stress appraisal and classification in nonintrusive experiments.

Though, a potential drawback has been identified in the insufficient collected data (4 types of stress conditions, 16 experimental participants, 13 control participants) to obtain relevant results for a further generalization used in classification of the new participants.

5 Conclusions

The current work considers an improvement method for automatic stress classification, where stress information is modeled based on pupil diameter and processed in a BoW framework with SVM classification. The major advantage of the proposed approach is given by the robust generalization, achieved thorough application on longer samples, extracted from given time series. In the related experiment, analyzed for comparison, shorter time series have been considered, so that classification accuracy is to be improved. We consider the proposed approached yet not fully validated, as the given dataset used to obtain experimental results in both: our case and the referenced case, is reduced in the amount or recorded data, therefore partially decisive. To achieve a maximal recognition rate for the classification, the input dataset needs to be consistent (number of participants, accurate experimental setup) and complex (several experiments). Based on the identified challenges a further directly-related experiment, would be to test the proposed framework on two identified larger datasets: DEAP [17], MAHNOB-HCI [18]. Even though the experiment setup is not stress-related, pupil diameter signal measurements are recorded based on given stimuli, and therefore stress is to be analyzed in an affective/emotional-context.

Acknowledgments The present work has been conducted under the 29667 Internal Grant 2014–2015, in the Technical University of Cluj-Napoca, Romania.

References

1. Pedrotti, M., Mirzaei, M.A., Tedesco, A., Chardonnet, J., Mérienne, F., Benedetto, S., Baccino, T.: Automatic stress classification with pupil diameter analysis. Int. J. Hum.-Comput. Interact. **30**, 1–17 (2014)
2. Sharma, N., Dhall, A., Gedeon, T., Goecke, R.: Modeling stress using thermal facial patterns: a spatio-temporal approach. In: Proceedings of Humaine Association Conference on Affective Computing and Intelligent Interaction, pp. 387–392 (2013)

3. Hart, S., Staveland, L.: Development of NASA-TLX (task load index): results of empirical and theoretical research. Adv. Psychol. **52**, 39–183 (1988)
4. Chen, S., Epps, J.: Automatic classification of eye activity for cognitive load measurement with emotion interference. Comput. Methods Programs Biomed. **110**(2), 111–124 (2013)
5. Giakoumis, D., Drosou, A., Cipresso, P., Tzovaras, D., Hassapis, G., Gaggioli, A., Riva, G.: Using activity-related behavioural features towards more effective automatic stress detection. PLoS One **7**(9) (2012)
6. Neerincx, M.A., Kraaij, W.: The SWELL knowledge work dataset for stress and user modeling research categories and subject descriptors. In: Proceedings of the 16th International Conference on Multimodal Interaction, pp. 291–298 (2014)
7. Bartels, M., Marshall, S.: Measuring cognitive workload across different eye tracking hardware platforms. In: Proceedings of the Symposium on Eye Tracking Research and Applications, pp. 161–164 (2012)
8. Hossain, G., Yeasin, M.: Understanding effects of cognitive load from pupillary responses using Hilbert analytic phase. In: IEEE Conferences on Computer Vision and Pattern Recognition Workshops, pp. 375–380 (2014)
9. Hussain, S., Chen, S., Calvo, A.R., Chen, F.: Classification of cognitive load from task performance & multichannel physiology during affective changes. In: Conference on Multimodal Interaction, pp. 1–4 (2011)
10. Paschero, M., Del Vescovo, G., Benucci, L., Rizzi, A., Santello, M., Fabbri, G., Mascioli, F.M.F.: A real time classifier for emotion and stress recognition in a vehicle driver. In: IEEE International Symposium on Industrial Electronics, pp. 1690–1695 (2012)
11. Ordonez, P., Armstrong, T., Oates, T., Fackler, J.: Using modified multivariate bag-of-words models to classify physiological data. In: IEEE 11th International Conference on Data Mining Workshops, pp. 534–539 (2011)
12. Wang, J., Liu, P., She, M.F.H., Nahavandi, S., Kouzani, A.: Bag-of-words representation for biomedical time series classification. Biomed. Signal Process. Control **8**(6), 634–644 (2013)
13. Merino, L.M., Meng, J., Gordon, S., Lance, B.J., Johnson, T., Paul, V., Tank, U.S.A.: A bag-of-words model for task-load prediction from EEG in complex environments. In: IEEE International Conference on Acoustics, Speech and Signal Processing, pp. 1227–1231 (2013)
14. Foggia, P., Percannella, G., Saggese, A., Vento, M.: Recognizing human actions by a bag of visual words. IEEE Int. Conference on Syst. Man Cybern. 2910–2915 (2013)
15. Zhang, Y., Jin, R., Zhou, Z.H.: Understanding bag-of-words model: a statistical framework. Int. J. Mach. Learn. Cybern. **1**(1–4), 43–52 (2010)
16. Koike, A., Takagi, T.: Classifying biomedical figures using combination of bag of keypoints and bag of words. In: Proceedings of the International Conference on Complex, Intelligent and Software Intensive Systems, pp. 848–853 (2009)
17. Koelstra, S., Muhl, C., Soleymani, M., Lee, J.S., Yazdani, A., Ebrahimi, T., Pun, T., Nijholt, A., Patras, I.: DEAP: a database for emotion analysis; using physiological signals. IEEE Trans. Affect. Comput. **3**(1), 18–31 (2012)
18. Soleymani, M., Lichtenauer, J., Pun, T., Pantic, M.: A multimodal database for affect recognition and implicit tagging. IEEE Trans. Affect. Comput. **3**(1), 42–55 (2012)

Using Kazakh Morphology Information to Improve Word Alignment for SMT

Amandyk Kartbayev

Abstract We propose an word alignment model with two core features: the ability to handle uncertainty in the morpheme matching process and in the selecting correct phrase alignments after its creation. These processes are based on the use of a morphological analysis tool and a large monolingual corpora, which is used for improving the alignment elements correspondence. The consideration of this tool is language-dependent for a special pair of the languages, although an Wikipedia data represents an adequate source of the training text that can be used in many cases, and even allows an unsupervised word segmentation. Based on these features, we propose an approach that captures the morphotactics which is common to the source text. The paper describes experiments in the general domain by using a tagset, and has been compared to a classical word alignment by the help of human judgment.

Keywords Word alignment · Kazakh morphology · Word segmentation · Machine translation · Information retrieval

1 Introduction

The growing demand of machine-translation applications that shows an witness of the creation of complex systems performing similar or even identical functions in real world. These systems, excess in the spare nature of their creation, have a limited functionality cause of mismatches in application purposes. However, integration of these systems is desirable to find information in business information systems. In recent years, a considerable research effort has been directed to evaluate the relationships between word alignment and machine translation performance, which aim at obtaining a certain degree of coordination between various kind of language pairs by automatically detecting correspondences between the elements of these alignments.

A. Kartbayev (✉)
Laboratory of Intelligent Information Systems, Al-Farabi Kazakh National University, Almaty, Kazakhstan
e-mail: a.kartbayev@gmail.com

A. Abraham et al. (eds.), *Proceedings of the Second International Afro-European Conference for Industrial Advancement AECIA 2015*, Advances in Intelligent Systems and Computing 427, DOI 10.1007/978-3-319-29504-6_34

However, there is no theoretical support from the view of providing a formulation to describe the relationship between word alignments and machine translation performance.

We examine the Kazakh language, which is the majority language in the Republic of Kazakhstan. Kazakh is part of the Kipchak branch of the Turkic language family and part of the majority Ural-Altay family, in comparison with languages like English, is very rich in morphology.

The Kazakh language which words are generated by adding affixes to the root form is called an agglutinative language. We can derive a new word by adding an affix to the root form, then make another word by adding another affix to this new word, and so on. This iteration process may continue several levels. Thus a single word in an agglutinative language may correspond to a phrase made up of several words in a non-agglutinative language [1] (Table 1).

Although the phenomena of word alignment has learned considerably after many challenges by El-Kahlout [2], Bisazza and Federico [3], these contributions are related to the use of morphology, as well as the consideration of probability distribution within the phrase pairs and the resulting alignments. These research issues in word alignment was also handling a precision of the matching process [4].

In order to demonstrate our objective, which can be used to build high quality machine translation systems [5], several applications of word alignment can be found, such as adaptation of the context-semantic disclosure. An word alignment processing is more convenient with respect to obtain m-to-n alignments, where several source words are aligned to several target words than barely segmenting the strings before the matching process. The common approaches of word alignment training are IBM Models [6] and hidden Markov model (HMM)[7], which practically use expectation-maximization (EM) algorithm [8]. The EM algorithm finds the parameters that increases the likelihood of the dependent variables. EM transfers the sentences by overlapping the actual parameters, where some rare words align to many words on the opposite sentence pair.

Since generating segments and modeling the relative features of phrases comprise a similarity measure within the parallel corpora, it makes the system too general to be applied to other kind of language pair, with the different morphotactics. However, our approach can be applied to the potential areas, which include improvements in machine translation, machine learning methods and information retrieval. Anyway,

Table 1 An example of Kazakh agglutination

Word form	English translation
stem[shol'+]	Desert
stem[shol'+]+plural[+der]	Deserts
stem[shol'+]+plural[+der]+1-st pl.[+imiz]	Our deserts
stem[shol'+]+plural[+der]+1-st pl.[+imiz]+locative[+de]	In our deserts
stem[shol'+]+plural[+der]+1-st pl.[+imiz]+locative[+de]+[+gi]	Is in our deserts

many concepts and definitions are pretty vague, which needs to be dealt within the word alignment process.

Using a Morfessor tool [9], we can find grammatical features of the word and can retrieve syntactic structure of input sentence. It clearly demonstrates the benefit of using similarly to the rule-based morphological analyzers [10], which consist in deep language expertise and a exhaustive process in system development. Unsupervised approaches use actually unlimited supplies of text and widely studied for a number of languages [11]. However, for a comprehensive survey of the rule-based morphological analyze we refer a reader to the research by Altenbek [12] and Kairakbay [13].

The article is structured as follows. Section 2 discusses the proposed model and describes the different segmentation techniques we study. And Sect. 3 presents our evaluation results.

2 Description of Our Method

Hybrid methods comprise two major groups of approaches: those that use morpheme analysis, and those that rely on probability distribution combined with techniques from machine learning in order to compare the similarity of the stems and their synonymy and ambiguity problems. An alignment process was understood as the process of establishing relations between the elements of a parallel language pair, which results in an alignment between equivalent phrases. Different alignment techniques, which enhance the quality of machine translation for Kazakh-English lasnguage pair, have been introduced in the past years in order to resolve different types of morphological segmentation of Kazakh words, relying on methods coming from areas of machine learning and linguistics. For these purposes we used Morfessor, an unsupervised analyzer and Helsinki Finite-State Toolkit (HFST) [14] for the rule-based analyze; finally we use the GIZA++ [15] tool to produce IBM Model 4 word alignment. Our morpheme analysis approach is concerned with word segmentation and as a result comparing groups of morphemes to another and detects the relations that exists between them.

Our studies try to investigate the impact of pruning technique to the overall translation quality by reduction the level of sparse phrases, which leads to higher BLEU scores [16]. We don't use a manually annotated gold standard word alignment set that the similarity measured on new sets of alignments reflects the personal opinion about a translation similarity between the instances.

2.1 Word Alignment

We suppose a phrase pair is denoted by (F, E) and with an alignment A, if any words f_j in F have a correspondence in a, with the words e_i in E. Formal definition can be described as follows: $\forall e_i \in E : (e_i, f_j) \in a \Rightarrow f_j \in F$ and $\forall f_j \in F : (e_i, f_j) \in a \Rightarrow e_i \in E$, clearly, there are $\exists e_i \in E, f_j \in F : (e_i, f_j) \in A$.

Generally, the phrase-based models are generative models that translate sequences of words in f_j into sequences of words in e_j, in difference from the word-based models that translate single words in isolation.

$$P\left(e_j \mid f_j\right) = \sum_{j=1}^{J} P\left(e_j, a_j \mid f_j\right) \tag{1}$$

Improving translation performance directly would require training the system and decoding each segmentation hypothesis, which is computationally impracticable. That we made various kind of conditional assumptions using a generative model and decomposed the posterior probability. In this notation e_j and f_i point out the two parts of a parallel corpus and a_j marked as the alignment hypothesized for f_i. If $a \mid e \sim ToUniform\,(a; I+1)$, then

$$P\left(e_j^J, a_j^J \mid f_i^I\right) = \frac{f_i}{(I+1)^J} \prod_{j=1}^{J} p\left(e_j \mid f_{a_j}\right) \tag{2}$$

We extend the alignment modeling process of Brown et al. at the following way. We assume the alignment of the target sentence e to the source sentence f is a. Let c be the tag of f for segmented morphemes. This tag is an information about the word and represents lexeme after the segmentation process. This assumption is used to link the multiple tag sequences as hidden processes, that a tagger generates a context sequence c_j for a word sequence f_j (3).

$$P\left(e_1^I, a_1^I \mid f_1^J\right) = P\left(e_1^I, a_1^I \mid c_1^J, f_1^J\right) \tag{3}$$

Then we can show Model 1 as (4):

$$P\left(e_i^I, a_i^I \mid f_j^J, c_j^J\right) = \frac{1}{(J+1)^I} \prod_{i=1}^{I} p\left(e_i \mid f_{a_i}, c_{a_i}\right) \tag{4}$$

We applied EM algorithm to estimate the phrase pairs that are consistent with the word alignments, and then assign probabilities to the obtained phrase pairs. The probability p_k of the word w to the corresponding context k is:

$$p_k(w) = \frac{p_k f_k\left(w \mid \phi_k\right)}{\sum p_i f_i\left(w \mid \phi_i\right)} \tag{5}$$

where, ϕ is the covariance matrix, and f are certain component density functions, which evaluated at each sequence. Consecutive word subsequences in the sentence pair are not longer than w words. After we use association measures to filter infrequently occurring phrase pairs by log likelihood ratio estimation [17].

Our algorithm, like a middle tier component, processes the input alignment files in a single pass. Current implementation reuses the code from https://github.com/akartbayev/clir that conducts the extraction of phrase pairs and filters out low frequency items. After the alignment processing all valid phrases have to be stored in the phrase table and should be passed further.

2.2 *Morphological Segmentation*

Kazakh is a morphologically complex language with many differences from English. We describe here the main grammar features of Kazakh that are relevant to its English translation and mostly are associated separately in English by a different order. Case suffixes are attached to the noun in Kazakh often represent the preposition in English, also the word order is pretty challenging in the context of translation to English. For Kazakh noun phrases, which correspondence to English phrases may lead to the long phrases problem that exceed the size of phrases in a phrase table.

Our job usually starts from word segmentation, which includes running morphological tools to each entry of the phrase pair. At the first step, an word segmentation process aims to get suffixes and roots from the word. Therefore, we take surface forms of the words and generate their all possible lexical forms. Also we use the vocabulary to label the initial states as the root words by parts of speech such as noun, verb, etc. The final states represent a lexeme created by affixing morphemes in each further states.

The schemes presented below are different combinations of outputs determining the removal of affixes from the analyzed words. The baseline approach is not perfect since a scheme includes several suffixes incorrectly segmented. In this case, we mainly focused on detection a few techniques for the segmentation of such word forms. In order to find an effective rule set we tested several segmentation schemes named S[1–8], some of which have described in the following Table 2.

Table 2 The segmentation schemes

Id	Schema	Examples	Translation
S1	stem	el	A nation
S2	stem+case	el + ge	To the nation
S3	stem+num+case	el + der + den	From the nation
S4	stem+poss+	el + in	His/her nation
S5	stem+poss+case	el + i +ne	To his/her nation
S6	stem+num+poss+case+case	el+der+in+de+gi	It is within their nation
S7	stem+tense	oina+dy	Played
S8	stem+suffixN+suffixN+case	oina+gan+dyk+tan	Because he/she has played

There are large amount of verbs presenting ambiguity during segmentation, which do not take personal endings, but follow conjugated main verbs. During the process, we hardly determined the border between stems and inflectional affixes, especially when the word and the suffix matches entire word in the language. In fact, there are lack of syntactic information we cannot easily distinguish among similar cases.

In order to solve the problems represented above, we have to split up Kazakh words into the morphemes and some tags which represent the morphological information expressed on the suffixation. Splitting Kazakh words in this way, we expect to reduce the sparseness produced by the agglutination being of Kazakh and the drought of training data. Anyway, the segmentation model takes into account the several segmentation options of both sides of the parallel corpus while looking for the optimal segmentation. As we discovered, words with same Part-Of-Speech (POS) tags often correspond to each other in the word alignment and may help to efficiently handle out-of-vocabulary (OOV) words by incorporating linguistic information, but it can also make the training data more sparse [18]. We also suppose that the discovery of word context relations could lead to better word alignment scores and we apply this idea using a heuristic algorithm for every single training scenarios.

To define the most convenient segmentation for our Kazakh-English system, we checked most of the segmentation options and have measured their impact on the translation quality. This application of morphological processing aims to find several best splitting options that the each Kazakh phrase ideally corresponds to one English phrase, so the deep analysis is more desirable.

3 Evaluation

For evaluation the system, three samples of text data were processed with 50k sentences each one, which were used in raw form and with special segmentation. The expert decisions about a segmentation quality were defined by our university undergraduate students. The data samples were stored randomly into a training set and a test set had one sample for each of the phrase-based Moses [19] system run. After the most of the samples were found processed correctly, which means the same interpretation of data was selected as acceptable by the experts, we decided the system was trained well, and that is a good result.

Our corpora consists of the legal documents from http://adilet.zan.kz, a content of http://akorda.kz, and Multilingual Bible texts, and the target-side language models were trained on the MultiUN [20] corpora. We conduct all experiments on a single PC, which runs the 64-bit version of Ubuntu 14.10 server edition on a 4Core Intel i7 processor with 32 GB of RAM in total. All experiment files were processed on a locally mounted hard disk. Also we expect the more significant benefits from a larger training corpora, therefore we are in the process of its construction.

We did not have a gold standard for phrase alignments, so we had to refine the obtained phrase alignments to word alignments in order to compare them with our word alignment techniques. We measure the accuracy of the alignment using

Table 3 The performance of word alignment on 50 K

Alignment	Precision	Recall	F-score
Intersection	83.0	40.8	59.0
Union	43.1	61.0	55.0
Grow-diag	51.0	57.2	54.0
Grow-diag-final	42.8	70.0	51.5

Table 4 Best performance scores

System	Precision	Recall	F-score	AER	BLEU	METEOR	TER
Baseline	57.18	28.35	38.32	36.22	30.47	47.01	49.88
Morfessor	71.12	28.31	42.49	20.19	31.90	47.34	49.37
Rule-based	89.62	29.64	45.58	09.17	33.89	49.22	48.04

precision, recall, and F-measure, as given in the equations below; here, A represents the reference alignment; T, the output alignment; A and T intersection, the correct alignments (Table 3).

$$pr = \frac{|A \cap T|}{|T|}, re = \frac{|A \cap T|}{|A|}, F - measure = \frac{2 \times pr \times re}{pr + re} \tag{6}$$

The alignment error rate (AER) values for the trained system show distinct tendencies which were consistent through the iteration of different training parameters. The values show completely the higher rates for raw lexeme than for segmented one, which seems suitable for an alignment task. Another tendency is that the differences of context receive smaller impact than the precision of segmentation. This was not clear since removing or normalization causes a change in word structure. A problem in interpreting these training results depend on the scaling of the morpheme probability, which can be of different variation, and the scale needs to be appropriate to the text domain and segmentation schemes. We assume that phrase alignment connects word classes rather than words. Consequently, the phrase translation table has to be learned directly from phrase alignment models, and an estimation of phrase distribution probability is internally part of the process (Table 4).

The system parameters were optimized with the minimum error rate training (MERT) algorithm [21], and evaluated on the out-of and in-domain test sets. All 5-gram language models were trained with the IRSTLM toolkit [22] and then were converted to binary form using KenLM for a faster execution [23]. The translation performance scores were computed using the MultEval [24]: BLEU, TER [25] and METEOR [26]; and we ran Moses several times per experiment setting, and report the best BLEU/AER combinations obtained. Our survey shows that translation quality measured by BLEU metrics is not strictly related with lower AER.

4 Conclusion and Future Work

In this paper, we learned the effect of morphological processing on SMT by making the source and target languages more similar than they usually are. The methods we use to solve most common problems are implemented as a pre-processing steps script and a middle-tier component for word alignment processing. As far as we know, dealing with nominal agglutination only does not considerable change the BLEU score of the baseline translation. However, we expected the combination of morphological analysis and phrase table refining have a positive effect on translation quality. As a result, our experiments produced not only more perfectly matching phrases, but also obtained new alignments that did not produce from the training data. Taking a closer look, we found that morphological features extracted from the source language are a valuable resource for alignment prediction. Our evaluation shows that morphological processing leads to better translations where the quality can not be measured by BLEU score. The improved model performs at slightly the same speed as the previous one, and gives an increase of about 3 BLEU over baseline translation. I think that it is a demonstration of the potential of word alignments for SMT quality, and we plan to investigate more complicated methods in the future researches, possibly adding the new alignment features to the model.

References

1. Bekbulatov, E., Kartbayev, A.: A study of certain morphological structures of Kazakh and their impact on the machine translation quality. In: IEEE 8th International Conference on Application of Information and Communication Technologies, pp. 1–5. Astana (2014)
2. Oflazer, K., El-Kahlout, D.: Exploring different representational units in English-to-Turkish statistical machine translation. In: 2nd Workshop on Statistical Machine Translation, pp. 25–32. Prague (2007)
3. Bisazza, A., Federico, M.: Morphological pre-processing for Turkish to English statistical machine translation. In: International Workshop on Spoken Language Translation 2009, pp. 129–135. Tokyo (2009)
4. Kartbayev, A.: SMT: A case study of Kazakh-English word alignment. In: Current Trends in Web Engineering, pp. 40–49. Springer, Heidelberg (2015)
5. Moore, R.: Improving IBM word alignment model 1. In: 42nd Annual Meeting on Association for Computational Linguistics, pp. 518–525. Barcelona (2004)
6. Brown, P.F., DellaPietra, V.J., DellaPietra, S.A., Mercer, R.L.: The mathematics of statistical machine translation: parameter estimation. In: Computational Linguistics, vol. 19, pp. 263–311. MIT Press Cambridge, MA (1993)
7. Vogel, S., Ney, H., Tillmann, C.: HMM-based word alignment in statistical translation. In: 16th International Conference on Computational Linguistics, pp. 836–841. Copenhagen (1996)
8. Dempster, A., Laird, N., Rubin, D.: Maximum likelihood from incomplete data via the em algorithm. J. Roy. Stat. Soc. B **39**, 1–38. Wiley-Blackwell, UK (1977)
9. Creutz, M., Lagus, K.: Unsupervised models for morpheme segmentation and morphology learning. In: ACM Transactions on Speech and Language Processing, vol. 4, article 3. Association for Computing Machinery, New York (2007)
10. Beesley, K.R., Karttunen, L.: Finite State Morphology. CSLI Publications, Palo Alto (2003)

11. Goldsmith, J.: Unsupervised learning of the morphology of a natural language. In: Computational Linguistics, vol. 27, pp. 153–98. MIT Press, Cambridge (2001)
12. Altenbek, G., Xiao-long, W.: Kazakh segmentation system of inflectional affixes. In: CIPS-SIGHAN Joint Conference on Chinese Language Processing, pp. 183–190. Beijing (2010)
13. Kairakbay, B.: A nominal paradigm of the Kazakh language. In: 11th International Conference on Finite State Methods and Natural Language Processing, pp. 108–112. St. Andrews (2013)
14. Linden, K., Silfverberg, M., Axelson, E., Hardwick, S., Pirinen, T.A.: HFST—Framework for compiling and applying morphology. In: Systems and Frameworks for Computational Morphology, pp. 67–85. Springer, Heidelberg (2011)
15. Och, F.J., Ney, H.: A systematic comparison of various statistical alignment models. In: Computational Linguistics, vol. 29, pp. 19–51. MIT Press, Cambridge (2003)
16. Papineni, K., Roukos, S., Ward, T., Zhu, W.: BLEU: A method for automatic evaluation of machine translation. In: 40th Annual Meeting of the Association for Computational Linguistics, pp. 311–318. Philadephia (2002)
17. Dunning, T.: Accurate methods for the statistics of surprise and coincidence. In: Computational Linguistics, vol. 19, pp. 61–64. MIT Press, Cambridge (1993)
18. Lee, J.-H., Lee, S.-W., Hong, G., Hwang, Y.-S., Kim, S.-B., Rim, H.-C.: A post-processing approach to statistical word alignment reflecting alignment tendency between part-of-speeches. In: 23rd International Conference on Computational Linguistics, pp. 623–629. Beijing (2010)
19. Koehn, P., Hoang, H., Birch, A., Callison-Burch, C., Federico, M., Bertoldi, N., Cowan, B., Shen, W., Moran, C., Zens, R., Dyer, C., Bojar, O., Constantin, A., Herbst, E.: Moses: Open source toolkit for statistical machine translation. In: 45th Annual Meeting of the Association for Computational Linguistics, pp. 177–180. Prague (2007)
20. Tapias, D., Rosner, M., Piperidis, S., Odjik, J., Mariani, J., Maegaard, B., Choukri, Kh., Calzolari, N.: MultiUN: a multilingual corpus from united nation documents. In: Seventh conference on International Language Resources and Evaluation, pp. 868–872. La Valletta (2010)
21. Och, F.J.: Minimum error rate training in statistical machine translation. In: 41st Annual Meeting of the Association for Computational Linguistics, pp. 160–167. Sapporo (2003)
22. Federico, M., Bertoldi, N., Cettolo, M.: IRSTLM: An open source toolkit for handling large scale language models. In: Interspeech 2008, pp. 1618–1621. Brisbane (2008)
23. Heafield, K.: Kenlm: faster and smaller language model queries. In: Sixth Workshop on Statistical Machine Translation, pp. 187–197. Edinburgh (2011)
24. Clark, J.H., Dyer, C., Lavie, A., Smith, N.A.: Better hypothesis testing for statistical machine translation: controlling for optimizer instability. In: 49th Annual Meeting of the Association for Computational Linguistics, pp. 176–181. Portland (2011)
25. Snover, M., Dorr, B., Schwartz, R., Micciulla, L., Makhoul, J.: A Study of translation edit rate with targeted human annotation. In: Association for Machine Translation in the Americas, pp. 223–231. Cambridge (2006)
26. Denkowski, M., Lavie, A.: Meteor 1.3: Automatic metric for reliable optimization and evaluation of machine translation systems. In: Workshop on Statistical Machine Translation EMNLP 2011, pp. 85–91. Edinburgh (2011)

Application of Naïve Bayes in Classification of Use Cases

Radoslav Štrba, Radim Briš, Ivo Vondrák and Svatopluk Štolfa

Abstract Our supportive method helps improve accuracy of software effort estimation, using results of classification of Use Cases. They are classified using machine-learning method called Naïve Bayes Classifier. The result of this classification helps determine the risk of underestimation of tasks in future work on the software project.

Keywords Naïve Bayes · Classification · Effort estimation · Use case

1 Introduction

"How to improve the accuracy of software development effort estimations?" It is important question for project managers in software companies. Accuracy of effort estimation depends on many parameters, e.g. we need to estimate amount of work, risk factors, testing level, remote work and other parameters. It is not possible to automatically estimate working time on some tasks with 100 % accuracy but it is possible to eliminate risk of underestimation of this task [1, 2].

R. Štrba (✉) · I. Vondrák · S. Štolfa
Department of Computer Science, VŠB - Technical University of Ostrava, 17. Listopadu 15, Ostrava-Poruba, Czech Republic
e-mail: radoslav.strba.st@vsb.cz

I. Vondrák
e-mail: ivo.vondrak@vsb.cz

S. Štolfa
e-mail: svatopluk.stolfa@vsb.cz

R. Briš
Department of Applied Mathematics, VŠB - Technical University of Ostrava, 17. Listopadu 15, Ostrava-Poruba, Czech Republic
e-mail: radim.bris@vsb.cz

A. Abraham et al. (eds.), *Proceedings of the Second International Afro-European Conference for Industrial Advancement AECIA 2015*, Advances in Intelligent Systems and Computing 427, DOI 10.1007/978-3-319-29504-6_35

Main idea of this research is to support the estimation using machine-learning method that is not so common in this area and can bring a new point of view. Our method can help to project manager to estimate the project complexity and the risk of additional work for new projects based on the analysis of use cases. The method uses knowledge base of historical use cases and provides a decision support in the form of a probability given by estimation of extra work in the project. In nutshell, the guidance for support of effort estimation based on classification of use case scenarios is provided. Use case scenarios are used in software requirements analysis for capturing and describing the functional requirements of a software system. Method that can help to project managers automatically classify use case scenarios is presented. Use case scenarios are made during the first phase of software development, which is called elaboration phase.

2 Statement of the Problem

Effort Estimation of software projects has become an important task in software engineering and project management. Old estimation methods that have been used to predict project costs developed using procedural languages are becoming inappropriate methods of estimation for the more recent projects being created with object-oriented languages. It calls for new approaches for software effort estimation supportive methods. In 2013, The Standish Group states that 43 % of software development projects were delivered late and over budget in "The Chaos manifesto 2013" [3]. These results show another increase in project success rates, with 39 % of all projects succeeding. These projects have been delivered on time, on budget, with required features and functions. Finally 18 % of projects failed because they have been cancelled prior to completion or delivered and never used. Some of reasons of project failure are lack of estimation of the staff's skill level, misunderstanding the requirements or improper software size estimation [4, 5].

3 Early Effort Estimation

Software estimation can be done at any stage within the requirements engineering process. However, performing estimation in the early stage, such as requirements elicitation means that the requirements of the software are not complete and more assumptions will need to be made in the estimation process. This could lead to poor results. There is need to find right stage within Requirements Engineering Process in which effort estimation can be done [4].

Requirements engineering process consists of several activities. Manage the scope of the system is appropriate activity for effort estimation. The functional and nonfunctional requirements are collected use cases are prioritized so the system can be delivered on expected time and budget [5, 6] (Fig. 1).

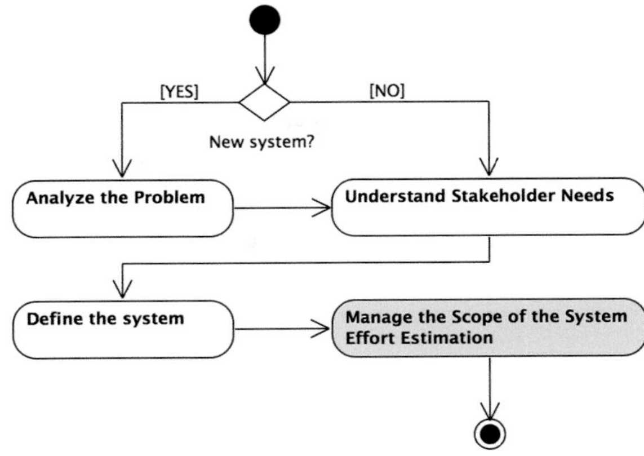

Fig. 1 Activity diagram of requirements engineering process

4 Method for Effort Estimation Supported by Classification

Classification methods can be used as supportive tool for effort estimation. Machine-learning methods help us to classify use case scenarios and predict risk factor of impropriate estimation of working time. We call working time like "extended work" parameter. Use case scenarios are used in the requirements, analysis and design stages in the software life cycle.

The use case model of the evaluated project includes a set of parameters of each use case. As parameterization method, we used existing approach detailed described in [7, 8]. Basically we use three types of parameters: descriptive, structural and really evaluated parameters.

Descriptive parameters are evaluated from the description of the use case scenario. We use following descriptive parameters: *Overall difficulty* (represents number of words, rows, etc.). *RFC* is project ID.

Structural parameters are evaluated by the structural or relational property of the use case on given project. Set of structural parameters was defined as a result from the interview with several senior project managers. We use following structural parameters and values: *NSC, which means:* New, Standard and Change functionality of the system. The parameter *Concerned activities* means how many business processes are touched by the implementation of this use case. *Use case type* can be user or sub function. Next parameters are: *remote work*, *implementation remote*, *testing level*.

Evaluated parameters are evaluated backward after project end. We use following evaluated parameter in our experiment: *Extra work* means if there was some additional work then was expected for the use case scenario. This parameter obtain binary values Yes or No. More information about parameters and parameterization, that we used is described in following papers [7–9] (Fig. 2).

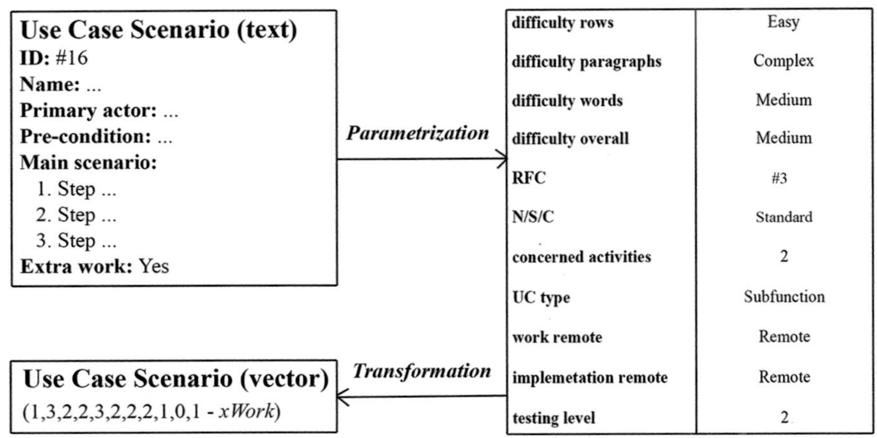

Fig. 2 Example of the parameterization and transformation of the Use Case Scenario

4.1 Data Preparation for Classification

The performance of classification algorithms in most cases depends on dataset quality. Data preprocessing techniques are essential, where the data are prepared for classification. It improves the quality of the data, and also the accuracy of the classification process. There are a number of data preprocessing techniques we used such as data cleaning. It means remove of noisy data or parameters. We also need the reducing of data size by aggregating and eliminating redundant features [10].

4.2 Classification of Use Cases

The main reason of use case classification is that we want to obtain results for effort estimation support. The extra work parameter is important for identification of underestimation. The method is focused on estimation of future parameter of the use case, based on the known parameters in the beginning of the project.

5 Naïve Bayes Classifier

The naïve Bayes classifier is a family of algorithms based on a common principle, called Bayes rule (Thomas Bayes). It is widely used for classification problems in data mining because of its simplicity and impressive classification accuracy. Classifier is a probabilistic model that assigns class labels to problem instances, represented as vectors $X = (x_1, \ldots, x_n)$ of n feature values where the class labels are drawn from some finite set. Given a set of variables x of vector X, we want to construct the posterior probability for the C among a finite set of possible classes $C = (c_1,\ldots,c_k)$ applying Bayes rule (1) [10–12].

$$P(c_k \mid x_1, \ldots, x_n) \propto P(x_1, \ldots, x_n) \times P(c_k) \tag{1}$$

$P\ (c_k \mid x_1,\ \ldots,\ x_n)$ is the posterior probability of specific class membership. If features are depended values, it assigns to instance probabilities $P\ (c_k \mid x_1,\ \ldots,\ x_n)$ for each of k possible classes. The problem is when the number of features n is too large or if a feature can take on a large number of possible values. Applying Bayes theorem, the conditional probability can be decomposed like:

$$P(c_k \mid X) = \frac{P(c_k) \times P(X \mid c_k)}{P(X)} \tag{2}$$

In other words, the Bayes theorem from formula (2) can be rewritten like formula (3), where posterior is $P(c_k \mid X)$, prior is $P(c_k)$, likehood is $P(X \mid c_k)$ and evidence is $P(X)$.

$$posterior = \frac{prior \times likehood}{evidence} \tag{3}$$

Given the intractable sample complexity for learning Bayesian classifiers, we must look for ways to reduce this complexity. The Naïve Bayes classifier does this by making a conditional independence assumption that dramatically reduces the number of parameters to be estimated when modeling $P\ (X \mid c_k)$, from our original 2 $(2n^2 - 1)$ to just $2n$.

Definition of Conditional Independence: Given random variables X, Y and Z, we say X is conditionally independent of Y given Z, if and only if the probability distribution governing X is independent of the value of Y given Z, that is in Eq. (4).

$$(\forall i,\ j,\ k)\ P\left(X = x_i \middle| \ Y = y_j,\ Z = z_k\right) = P\ (X = x_i | \ Z = z_k) \tag{4}$$

As is written above, the Naïve Bayes Classifier is based on Bayes rule that assumes the attributes of vector $X = (x_1,\ \ldots,\ x_n)$ are all conditionally independent of one another, given class c_k. The value of this assumption is that it dramatically simplifies the representation of $P\ (X \mid c_k)$, and the problem of estimating it from the training data. In case where $X = (x_1,\ x_2)$ it can be written using chain rule, where n is the number of features and k is number of classes.

$$\begin{aligned} P\ (X \mid c_k) &= P\ (x_1,\ x_2 \mid c_k) \\ &= P\ (x_1 \mid x_2,\ c_k)\ P\ (x_2 \mid c_k) \\ &= P\ (x_1 \mid c_k)\ P\ (x_2 \mid c_k) \end{aligned} \tag{5}$$

$$P(X \mid c_k) = \prod_{i=1}^{n} P(x_i \mid c_k) \tag{6}$$

Applying Bayes classification rule above (6), we label a new case X with a class level C_j that achieves the highest posterior probability. Please none that the contribution from each feature x_i can be written in several ways:

$$\frac{P(X_i \mid c_k)}{P(X_i)} = \frac{P(X_i, c_k)}{P(c_k)P(X_i)} = \frac{P(c_k \mid X_i)}{P(c_k)} \tag{7}$$

The goal is to learn a classifier that will output the probability distribution over possible values c_k of class C, for each new instance X that we ask it to classify into specific class c_k.

$$P(c_k \mid X) = \frac{P(c_k) \times P(X \mid c_k)}{\sum_j P(c_j) \times P(X \mid c_j)} \tag{8}$$

5.1 Classifier Construction

The Naïve Bayes Classifier combines probabilistic model with a decision rule. The classical way of estimating parameters in probability distributions is by maximum likelihood. This means that the estimated value of the parameter is the one that would have made the probability of the data as large as possible. If we are interested only in the most probable value of given class C, then we have the Naïve Bayes classification rule, where denominator does not depend on c_k [13]:

$$\hat{C} = \underset{k \in \{1, \ldots, K\}}{\operatorname{argmax}} \; P(c_k) \prod_{i=1}^{n} P(x_i \mid c_k) \tag{9}$$

6 Experiment—Classification Using Naïve Bayes Classifier

Naïve Bayes is a simple and very powerful classification method that you should be using on our classification problem and it is also kind of supervised training.

6.1 Process Overview

The whole process consists of five important activities. The first is Data Preprocessing, second is training, and third is testing type selection, which allows using

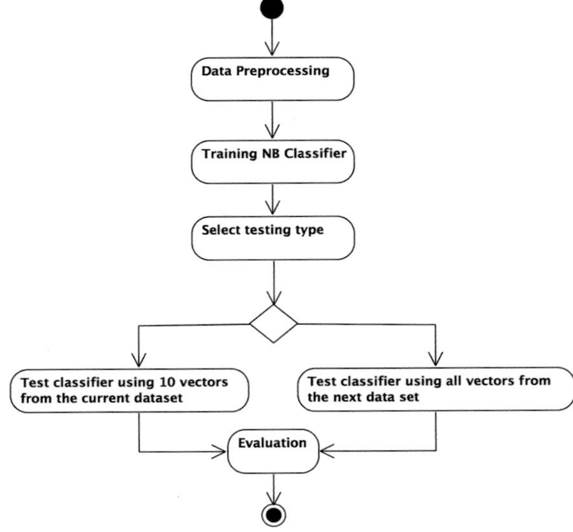

Fig. 3 Overview of classification process based on Naïve Bayes Classifier

selected vectors by project manager or all vectors from the next dataset for testing purpose. At the end of the process is activity called Evaluation (Fig. 3).

6.2 Data Preprocessing

Whole use case training set consists of 1041 parameterized use cases. The value of some parameters is the same for all items, e.g. *use case type*, *work remote* and *implementation remote*. Actually they are same for whole dataset. That is the reason, why these parameters are not important, so they can be removed.

6.3 Train Classifier

Train the classifier means that we need to tell to classifier that the given features resulted in the given category (value of extended work parameter can be *Yes* or *No*). As is written above, the whole datasets consists of 1041 Use Cases from years 2008-2013. These items are divided into 6 datasets. Each dataset includes last 10 testing items.

Data sub-sets of particular years are subsequently added into the training data set in six iterations. After the training process, we would like to determine which posterior is greater, extended work "No" or extended work "Yes". For the classification as No group the posterior example is given using formula (3):

$$posterior(\mathbf{No}) = \frac{P(\mathbf{No}) \times P(difficultyRows \mid \boldsymbol{No}) \times P(difficultyParagraphs \mid \boldsymbol{No}) \times \cdots}{evidence}$$

For the classification as Yes group the posterior example is given by similar way, using (3):

$$posterior(\mathbf{Yes}) = \frac{P(\mathbf{Yes}) \times P(difficultyRows \mid \boldsymbol{Yes}) \times P(difficultyParagraphs \mid \boldsymbol{Yes}) \times \cdots}{evidence}$$

The *evidence* example, which may be calculated like this:

$$\begin{aligned} evidence = {} & P(\mathbf{Yes}) \times P(difficultyRows \mid \boldsymbol{Yes}) \times P(difficultyParagraphs \mid \boldsymbol{Yes}) \times \cdots \\ & + P(\mathbf{No}) \times P(difficultyRows \mid \boldsymbol{No}) \times P(difficultyParagraphs \mid \boldsymbol{No}) \times \cdots \end{aligned}$$

6.4 Test Classifier Using Vectors from Current Dataset

The training accuracy is computed using training dataset, which includes all vectors from current dataset and excludes last 10 vectors. The testing accuracy is computed on testing dataset, which includes last 10 vectors from current dataset and excludes them from training dataset. You can see results in Table 1.

6.5 Test Classifier Using Vectors from the Next Dataset

Finally, we tried to test neural network using all vectors from current and all vectors from next dataset. For example, the neural network is trained using use cases from year 2008 and is tested using use cases from year 2009. Table 2 shows results of testing Naive Bayes Classifier after training process.

Table 1 Results of training using testing set of 10 use cases

Use case sets from years	Training accuracy/testing accuracy
2008	63,28/40 %
2008–2009	66,41/60 %
2008–2010	65,78/70 %
2008–2011	66,20/90 %
2008–2012	66,91/70 %
2008–2013	68,01/60 %

Table 2 Results of training using testing set of all use cases from next year

Training vectors from years	Testing accuracy (all vectors/10 vectors) from year:
2008	2009 (102 vectors) **63,76**/70 %
2008–2009	2010 (110 vectors) **66,71**/80 %
2008–2010	2011 (110 vectors) **66,43**/90 %
2008–2011	2012 (110 vectors) **66,60**/70 %
2008–2012	2013 (109 vectors) **66,94**/70 %

7 Results and Discussion

Use cases were divided into two categories. In the first category extra work on use case was needed. In this case *extra work* parameter obtains value 1. If extra work was not needed than parameter obtain value 0.

Table 1 shows results of testing process using 10 testing vectors selected by project manager from current dataset as "testing accuracy" and all training vectors also from current dataset as "training accuracy."

Table 2 shows results of testing process using 10 testing vectors selected by project manager from the next dataset as "testing accuracy" and all training vectors also from the next dataset as "training accuracy."

As the main purpose of this experiment was just to prove the possibility of Naïve Bayes usage for our purposes we have performed simple experiments. Even those simple tests showed that the accuracy for the parameters prediction could be quite high even without any special adjustments and setting of our approach.

The main advantage of Naive Bayes Classifier is simplicity and good training time. Disadvantage of this approach is lower accuracy then for example neural network based classification.

8 Conclusion and Future Work

This paper presents and discusses the usage of Naive Bayes Classifier in field of effort estimation support. Classifiers are used then for the classification of the use cases in our supportive tool for software project estimation. The results show that the method can estimate an extra work parameter with the probability for testing use cases selected by manager in most cases between 60–90 % success. Detailed results of experiments based on datasets from years 2008–2009, 2008–2010, 2008–2011 are showed in Tables 1 and 2. As we can see, the accuracy is higher when there are more data available for the training. The first dataset was trained with more than 60 % accuracy but the testing set showed 40 % accuracy using Naïve Bayes. These results seem to be promising for our purposes so far. Anyway, more experiments have to be done to demonstrate the general applicability even for our purposes.

Future work will be focused on experiments using other statistical method called logistic regression and comparison with provided Naïve Bayes based approach.

Acknowledgment Work is partially supported by Grant of SP2015/85—Knowledge modeling and its applications in software engineering, VŠB - Technical University of Ostrava, Czech Republic.

References

1. Jorgensen, M.: What we do and don't know about software development effort estimation. IEEE Softw. **31**, 37–40 (2014)
2. Jorgensen, M., Jorgensen, M.: A review of studies on expert estimation of software development effort. J. Syst. Softw. **70**, 37–60 (2004)
3. Standish Group: The CHAOS Manifesto 2013. http://www.versionone.com/assets/img/files/ChaosManifesto2013.pdf (2013)
4. Nassif, A.B., Capretz, L.F., Ho, D.: Estimating software effort using an ANN model based on use case points. In: 2012 11th International Conference on Machine Learning and Applications, pp. 42–47. IEEE (2012)
5. Gill, N.S., Sikka, S.: New complexity model for classes in object oriented system. ACM SIGSOFT Softw. Eng. Notes. **35**, 1 (2010)
6. Nassif, A.B.: Software Size and Effort Estimation from Use Case Diagrams Using Regression and Soft Computing Models (2012)
7. Stolfa, J., Kobersky, O., Kromer, P., Stolfa, S., Kopka, M., Snasel, V.: Comparison of fuzzy rules and SVM approach to the value estimation of the use case parameters. In: 2013 Joint IFSA World Congress and NAFIPS Annual Meeting (IFSA/NAFIPS), pp. 789–794. IEEE (2013)
8. Štolfa, S., Štolfa, J., Krömer, P., Koběrský, O., Kopka, M., Snášel, V.: Fuzzy rules and SVM approach to the estimation of use case parameters. In: Innovations in Bio-inspired Comoputing and Applications (IBICA), vol 237 of Advances in Intelligent Systems and Computing, pp 37–47, Springer (2014)
9. Štolfa, J., Štolfa, S., Koběrský, O., Kopka, M., Kožuszník, J., Snášel, V.: Methodology for estimating working time effort of the software project. In: CEUR Workshop Proceedings, pp. 25–37 (2012)
10. Farid, D.M., Zhang, L., Rahman, C.M., Hossain, M., Strachan, R.: Hybrid decision tree and Naïve Bayes classifiers for multi-class classification tasks. Expert Syst. Appl. **41**, 1937–1946 (2014)
11. Wu, X., Kumar, V., Ross Quinlan, J., Ghosh, J., Yang, Q., Motoda, H., McLachlan, G.J., Ng, A., Liu, B., Yu, P.S., Zhou, Z.-H., Steinbach, M., Hand, D.J., Steinberg, D.: Top 10 Algorithms in Data Mining (2008)
12. Holst, A.: The Use of a Bayesian Neural Network Model for Classification Tasks (1997)
13. Murty, M.N., Devi, V.S.: Bayes classifier. In: Pattern Recognition: An Algorithmic Approach, pp. 86–102, Springer (2011)

Part III
Nostradamus Track

Better Spectra Manipulation in SPLAT-VO

Petr Šaloun, David Andrešič, Petr Škoda and Ivan Zelinka

Abstract The current progress in Astroinformatics follows the progress in astronomical instrumentation, that produces large data sets every observing night. This amount of data is more and more processed in automated way, using multiple knowledge discovery and Big Data techniques. However, visual processing of these large spectral data sets is still crucial for many astronomical applications. We present an overview of improvements that are being prepared for SPLAT-VO, the spectral analysis tool with Virtual observatory capabilities. The overview is focused on enhancements of unconventional work with large sets of incoming data, its organization, user experience and more effective use of system resources.

Keywords SPLAT-VO · Virtual observatory · Large data sets · Spectra · Astroinformatics · Working space · Lazy loading · User-defined groups · User experience · Java

P. Šaloun (✉) · D. Andrešič · I. Zelinka
Department of Computer Science, Faculty of Electrical Engineering and Computer Science, VŠB - Technical University of Ostrava, 17. listopadu 15, 708 33 Ostrava-Poruba, Czech Republic
e-mail: petr.saloun@vsb.cz
URL: http://www.cs.vsb.cz/

D. Andrešič
e-mail: david.andresic.st@vsb.cz

I. Zelinka
e-mail: ivan.zelinka@vsb.cz

P. Škoda
Astronomical Institute of the Academy of Sciences, Fricova 298, 251 65 Ondrejov, Czech Republic
e-mail: skoda@sunstel.asu.cas.cz
URL: http://www.asu.cas.cz/

A. Abraham et al. (eds.), *Proceedings of the Second International Afro-European Conference for Industrial Advancement AECIA 2015*, Advances in Intelligent Systems and Computing 427, DOI 10.1007/978-3-319-29504-6_36

1 Introduction

The modern astronomical instrumentation is capable to produce tens of thousands of spectra of celestial objects per observing night [1]. Such amount of information is beyond human capabilities for spectral analysis and makes a very good source of homogenized, preprocessed data for application of machine learning techniques and advanced statistical processing common in Astroinformatics [1].

Visual processing is hovewer still crucial for many astronomical applications, more closely described for example in [2]. Scientists therefore needed to find a way how to be able to process all the incoming data in the most effective way. This caused a revolution in science that involved a development of new computer infrastructures across multiple fields of science.

1.1 Astroinformatics

In case of astronomy, we talk about the Astroinformatics. Using the potential of large data archives, power of supercomputer grids and abilities of newly developed protocols and software, the modern astronomer is able to easily find the resources required. Work on such complex tasks, like obtaining several spectra of multiple objects in multiple wave-lengths, or even combining them with other data and performing an extensive analysis using supercomputers, then sharing results with new communities across the world has never been easier.

A great example of this approach is the Virtual observatory (more closely described in Sect. 1.2). An infrastructure based on a (among others) server-client basis that allows to access the data and computer resources using specialized software and communication protocols. It also allows access or even participation of amateur scientists, which already proved its usefulness in several occasions.

SPLAT-VO (more closely described in Sect. 2) is an example of such VO-enabled tool. It brings a capabilities of one of the greatest tools for spectra analysis to basically everyone. Aim of this work is then to describe planned improvements of work with large incoming data sets, its organization and user experience with benefit of more efficient work with system resources in most use cases.

1.2 Virtual Observatory

Virtual observatory (VO) can be described as a collection of astronomical archives and software tools that utilizes the internet to allow international collaboration and scientific research [3]. VO offers huge data-sets with thousand fuzzy interconnected attributes for researcher, but it is necessary to select the proper one and show them.

Little more in detail, VO can be divided between client and server parts, where servers are the individual data archives, web-services or even (grids of) supercomputers performing specific tasks (generally called resources) and clients are simple software tools run and operated by astronomers. The communication among all of these components is based on a DNS-like resolving mechanism called VO resource registry that keeps the list of available resources. Clients are then accessing to these resources through specialized VO protocols specific for the required kind of data (eg. spectra or tables) or action (eg. inter-client communication or procedure call), where most of the protocols has its own set of metadata describing its subject.

For a better idea of VO architecture, please see the Fig. 1. SPLAT-VO is positioned in *Portals, User Interfaces, Tools* section, among tools like Topcat and others.

All of the specifications that the VO is based on are defined by International Virtual Observatory Alliance (IVOA)—a worldwide organisation that aims to defining the technical standards that are needed to make all the astronomical datasets and resources a seamless whole [3].

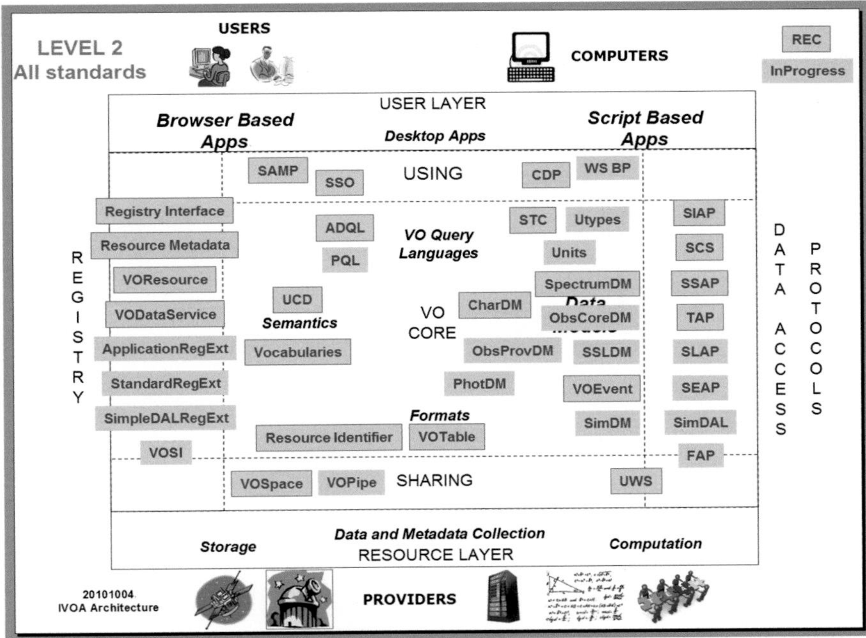

Fig. 1 Virtual observatory architecture [3]. SPLAT-VO is positioned in the *USER LAYER*

2 SPLAT-VO

The name SPLAT is based on the words SPectraL Analysis Tool [4] and it is an application for displaying, modifying and analysing astronomical spectra [6] (see Fig. 2 and Fig. 3 respectively, for example), originally developed in 2003 as a part of Starlink (and its STARJAVA package) project [6]. During its development, SPLAT was extended to include facilities that allowed an interoperability with the Virtual Observatory (see Sect. 1.2) [5].

SPLAT was also a leader project for moving future developments into the Java language, with the obvious benefits of improved portability, modern language capabilities (OOD, OOP) and core-level support for features like UIs and Internet protocols and services. This work eventually led to the formation of what became the StarJava project, containing also the well-known table processor TOPCAT [4].

Today, it is still one of the greatest tools for spectra analysis, nevertheless, as every software tool, it can be still improved in both VO-related and non-VO-related ways.

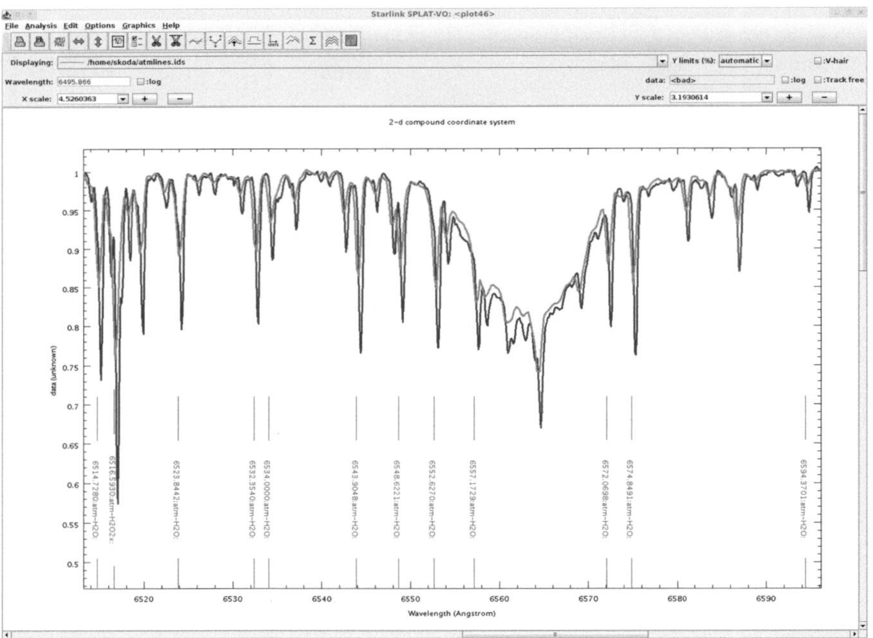

Fig. 2 SPLAT-VO at work

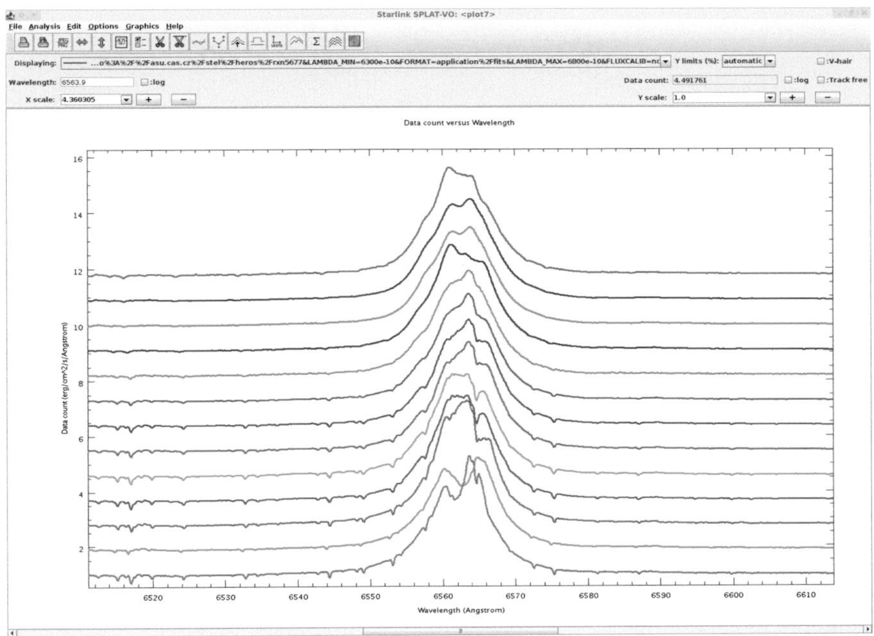

Fig. 3 Many spectra in SPLAT-VO

2.1 *Most Typical Use Cases*

For a better idea about described improvements, it is necessary to describe current capabilities of SPLAT-VO, its limits and basic internal objects first.

SPLAT-VO is currently capable of opening the spectra and light curves (represented as a spectrum with time axis instead of wavelength axis) contained in multiple formats. These data (files) can be loaded:

- localy from a file (many formats are supported);
- via SSAP[1] protocol (Virtual Observatory);
- via SAMP[2]);
- as a in-memory result of some operations performed on other spectra or light-curves.

User can also send the opened spectra and/or light-curves via SAMP protocol to other tool and/or save the spectra in a different format and/or save the entire list of spectra to a local file.

The simplified use-case model of SPLAT-VO can be seen at Fig. 4.

[1]Simple Spectra Access Protocol—www.ivoa.net/Documents/SSA/.

[2]Simple Application Messaging Protocol—www.ivoa.net/Documents/latest/SAMP.html.

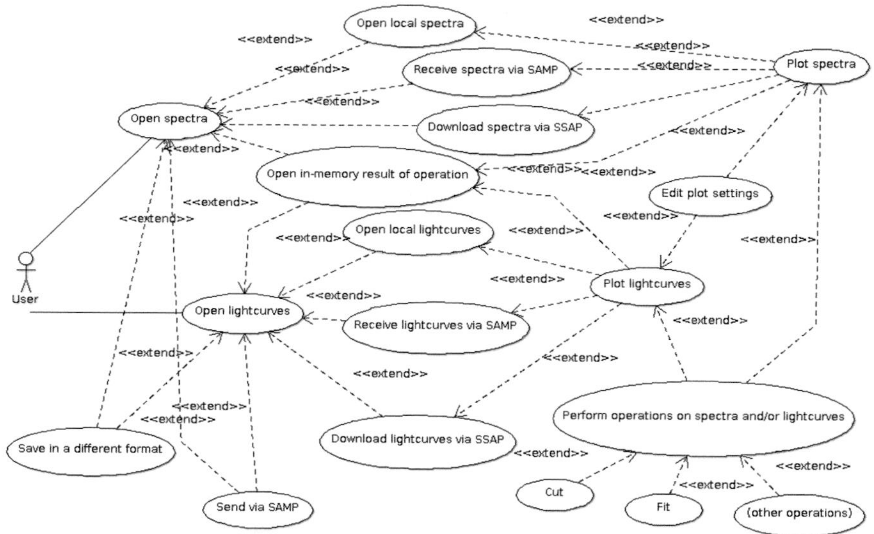

Fig. 4 Use-case model of the SPLAT-VO most typical use cases

Since the light curves are represented as a spectra, let's furthermore talk only about spectra unless said otherwise. The user can perform many operations on the spectra. Just to get an idea:

- plotting one or more spectra to a plot window;
- performing actions on a plotted spectrum/spectra (e.g. cutting, fitting);
- adjusting the visualization of plotted spectra (e.g. color or style of line etc.).

2.2 *Limits and Bottlenecks*

After integration of our previous enhancements into standard SPLAT-VO distribution, we evaluated requirements of SPLAT-VO users and its data manipulation capabilities in the current condition of VO, networking and computer performance. Based on this, we identified several bottlenecks.

For example, SPLAT-VO does not allow a direct work with a set of individual spectra in other tools, it only allows to save the entire global list as a single file or to save individual spectra one by one. Sending spectra via SAMP can be used only in cooperation with SAMP-compatible tools.

SPLAT-VO also downloads all the incoming spectra to a system temporary directory, bust most users are unaware of that and it does not include the in-memory spectra created as a result of some operations.

From the efficiency point of view, the only way how the user can restore the work is to save the global list to a file. On startup, the user thus needs to open a dialog, locate the saved global list and open it. In many cases, it would help if the user just had the previous work restored right on startup. SPLAT-VO has very limited support for organizing the opened spectra. It only has a central global list of spectra, that serves as a source of spectra for the entire application, but it lacks the clarity and organization when working with tens or even thousands of spectra.

As for the performance, when opening large sets of spectra, the user can encounter an
`java.lang.OutOfMemoryError: Java heap space` error very quickly.

As a result, a feature set of unconventional approach to work with large data sets, more closely described in Sect. 3, was suggested.

2.3 *SPLAT-VO Basic Objects*

SPLAT-VO consists of many modules but basically almost all of them use at least one of the following basic objects.

2.3.1 SpecData and RemoteSpecData

SPLAT-VO is capable of loading spectra in many data formats (e.g. FITS,[3] VOTable,[4] etc.). This is achieved by internal conversion to an abstract entity called `SpecData` (see Fig. 5), that covers all common spectra properties (data and metadata). For spectra that comes from remote sources, SPLAT-VO uses the extension of `SpecData` called `RemoteSpecData`.

This abstract representation of spectra is internally used by SPLAT-VO features.

2.3.2 SpecList

Is a singleton class that holds all spectra loaded in SPLAT-VO. In this text, the term SpecList is therefore **interchangeable with global list of spectra** (it is its implementation).

---

[3]Flexible Image Transport System—a multipurpose format for storing the scientific data, primary designed and used for astronomy [7], full specification at http://fits.gsfc.nasa.gov/.

[4]VOTable Format Definition: http://www.ivoa.net/documents/VOTable/.

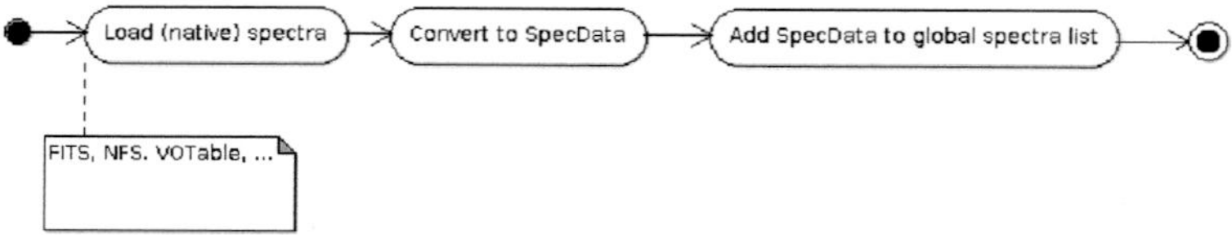

Fig. 5 Workflow of loading spectra in native format to internal SpecData format

2.3.3 GlobalSpecPlotList

Is an aggregate singleton class that provides access to the `SpecList` instance and list of all plots. It provides integrated control interfaces to both these objects and provides listeners for objects that want to be updated about changes in the lists of spectra or plots.

3 New Features and Improvements

In this section, all main planned features are described. Please note that some details may change until the final release is made.

3.1 Working Space

Simply put: this feature will save all spectra loaded in SPLAT-VO at the moment to a user-defined folder.

In current implementation, SPLAT-VO downloads all remote spectra to a system temporary directory, spectra from local files and spectra that are results of some operations are not saved anywhere. This limits the user-experience because the user has no direct access to all loaded spectra in order to use it with other tools (manual saving of the spectra or sending it via SAMP is not sufficiently general and comfortable).

With this feature, the user will be able to restore the previous work in SPLAT-VO just by starting it (SPLAT-VO will ask the user to automatically load the content of the Working space if any). The user will also be able to immediately work with all loaded spectra in other tools (the Working space directory will be editable and accessible, so the user can use other tools and access the stored spectra—the access for other tools is read-only, only SPLAT-VO will be able to write to a Working space).

More abstract view on this feature is given by UML use-case model (see Fig. 6).

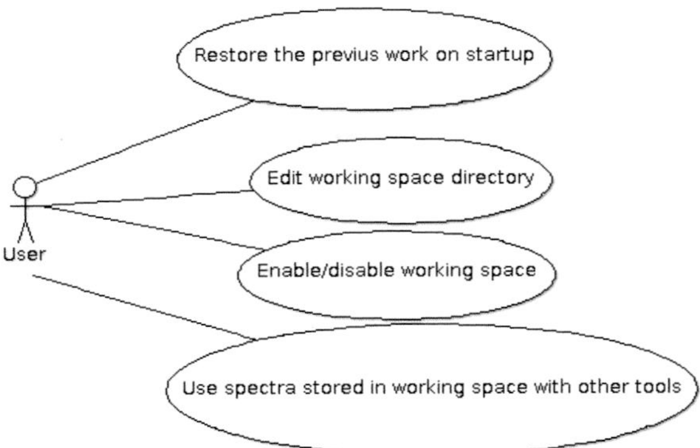

Fig. 6 Working space—use case diagram

3.2 Spectra Groups

This feature will allow the user to organize incoming spectra to groups and therefore to work with only a subset of all loaded spectra relevant to current work. In cooperation with lazy loading feature (described more closely in Sect. 3.3), this feature will also allow to significantly reduce memory consumption of SPLAT-VO because only the spectra of the currently selected group will be fully initialized.

This feature benefits from implementation of Working space (see Sect. 3.1) that ensures that all spectra are localy available (required for reference in group VOTable) and in the required format (FITS with WAVE/FLUX columns).

Each spectra group will be represented by VOTable stored in Working space. Each spectrum in the VOTable will be represented by an indiviual reference under corresponding TABLE element. All spectra will be grouped under a single RESOURCE element (please refer to Sect. 3.2 to an example of such a VOTable). The name of the VOTable in working space will be in the format of: `splatvo_{unique UUID}.sg.vot`. The user-editable name and description of the spectra group will be stored in VOTable NAME and DESCRIPTION elements.

The spectra group will serve as a source of spectra loaded to the current global spectra list (in other words: the current global spectra list will containt only the spectra of the currently selected spectra group). Each spectrum added to global list will therefore be added to the currently selected group (it can be later regrouped to one or n groups). Each spectrum will need to be assigned to at least one group and there will be a default group called `(default)`.

The chosen format has another benefit: since it complies the VOTable standard, it is readable by other VO tools.

3.2.1 Spectra Group VOTable Example

As said earlier, each spectra group will be represented by a single VOTable contained in Working space. Below can bee seen an example of such VOTable.

```
<?xml version="1.0"?>
<VOTABLE version="1.3"
         xmlns:xsi="http://www.w3.org/2001/XMLSchema-instance"
 xmlns="http://www.ivoa.net/xml/VOTable/v1.3"
 xmlns:stc="http://www.ivoa.net/xml/STC/v1.30" >
  <RESOURCE name="splatVoSpectraGroup">
    <DESCRIPTION>SPLAT-VO (ver.) Spectra Group VO-Table</DESCRIPTION>
    <TABLE name="NORMbxn0727.fits">
      <FIELD name="WAVE" datatype="double" ucd="em.wl" unit="angstrom"/>
      <FIELD name="FLUX" datatype="double" ucd="phot.flux.density"
             unit="erg/cm**2/s/angstrom"/>
      <DATA>
        <FITS>
          <STREAM href="file:///path/to/spectrum/NORMbxn0727.fits"/>
        </FITS>
      </DATA>
    </TABLE>
    <TABLE name="NORMbxn0728.fits">
      <FIELD name="WAVE" datatype="double" ucd="em.wl" unit="angstrom"/>
      <FIELD name="FLUX" datatype="double" ucd="phot.flux.density"
             unit="erg/cm**2/s/angstrom"/>
      <DATA>
        <FITS>
          <STREAM href="file:///path/to/spectrum/NORMbxn0728.fits"/>
        </FITS>
      </DATA>
    </TABLE>
  </RESOURCE>
</VOTABLE>
```

3.3 *Spectral Data Lazy Loading*

As shown at Fig. 7, spectral data consumes a considerable amount of memory. Opening a large data set can cause `java.lang.OutOfMemoryError: Java heap space` error or significantly limit the usability of the user's system.

In order to avoid this, spectral data will be loaded "lazily". This means that only spectra of one spectra group (see Sect. 3.2 or only e.g. actually plotted spectra (depends on further analysis) will be "fully loaded" (including data itslef, otherwise only metadata will be "fully loaded/initialized").

Lazy initializaton (loading of all spectral data) will be done on demand by loading the spectral data from Working space (see Sect. 3.1).

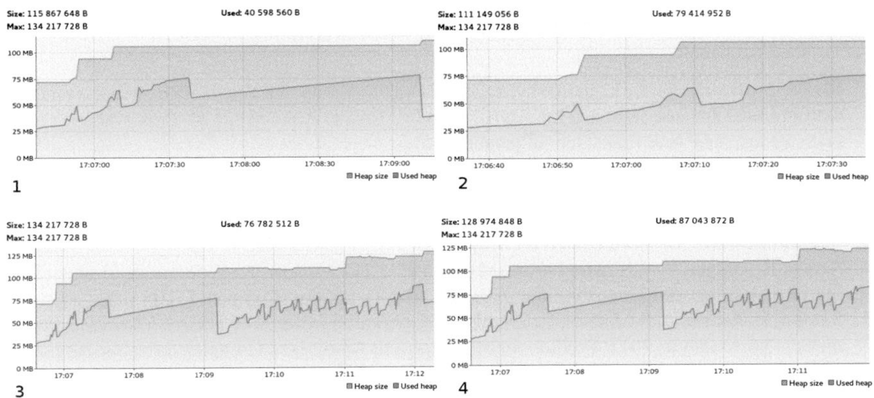

Fig. 7 Memory usage of SPLAT-VO. Figures *1* and *2* shows minimal and maximal (based on JVM garbage collection status) memory usage when no spectra are loaded. Figures *3* and *4* shows minimal and maximal memory usage after loading 60 spectra

3.4 Visual Removal of a Spectrum

When working with multiple spectra, it might be usefull to select e.g. a noisy one and conveniently remove it from Working space.

The current version of SPLAT-VO allows a visual selection of a spectrum in plot window. It highlights it and pre-selects in the local drop-down menu and global spectra list.

We added a convenient delete feature, where after pressing a Delete key or clicking on an appropriate button, the user will be asked to confirm the removal of the currently selected spectrum with the pre-checked option to remove it from the global list as well.

4 Conclusions

SPLAT-VO comes side by side with the current progress in astroinformatics and its Big Data issues, that requires appropriate end-user tools.

Implementation of all these features will bring to a regular astronomer the power of current capabilities of SPLAT-VO with more pleasant user experience, that comes with much better data and work organization, interoperability with other tools and lowered memory consumption when working with large data sets.

Acknowledgments This work was partially supported by Grant Agency of the Czech Republic—GACR P103/13/13-08195S, and by Grant of SGS No. SP2015/142, VSB-Technical University of Ostrava, Czech Republic.

References

1. Škoda, P., Bromová, P., Lopatovský, L., Palička, A., Vážný, J.: Knowledge discovery in big astronomical spectra archives. In: ADASS XXIV, Taylor, A.R., Stil, J.M. (eds.) ASP Conference Series, ASP San Francisco, in print (2015)
2. Škoda, P.: Optical spectroscopy with the technology of virtual observatory. Balt. Astron. **20**, 531–539 (2011)
3. International Virtual Observatory Alliance. http://ivoa.net/documents/Notes/IVOAArchitecture/index.html (2015). Accessed 09 May 2015
4. Škoda, P., Draper, P.W., Castro Neves, M., Andrešič, D., Jenness, T.: Spectroscopic analysis in the virtual observatory environment with SPLAT-VO. Astron. Comput. **7**, 108–120 (2014)
5. Šaloun, P., Andrešič, D., Škoda, P., Zelinka, I.: Upcoming features of SPLAT-VO in astroinformatics. Nostradamus 2013: Prediction, Modeling and Analysis of Complex Systems. Advances in Intelligent Systems and Computing, vol. 210, pp 475–486. Springer (2013). doi:10.1007/978-3-319-00542-3_47
6. Draper, P.W.: Starlink SPLAT-VO. http://star-www.dur.ac.uk/pdraper/splat/splat-vo/ (2015). Accessed 09 May 2015
7. Donahue, M., Kimball, T.: FITS file format. In: HST Data Handbook. http://www.stsci.edu/documents/dhb/web/c02_datafiles.fm2.html (1997). Accessed 28 June 2015

Optimization of Closed-Loop Poles for Limited Control Action and Robustness

Frantisek Gazdos

Abstract The presented paper exploits optimization and simulation tools of the MATLAB environment to design a robust control system in case of limitations on the controller's manipulated variable. The design is based on the polynomial approach resulting in the pole-placement problem to be solved. This is addressed numerically by means of the standard MATLAB functions for nonlinear constrained optimization to meet both robustness of the designed loop and constraints on the control input. For this purpose, convenient performance criteria are suggested together with a procedure for the optimization. Presented results of constrained robust control for the AMIRA servo-system show potential of the suggested methodology.

Keywords Optimization, constrained control • Polynomial approach • Robust control • Pole-placement problem • AMIRA servo-system

1 Introduction

Optimization is a natural part of our daily lives. Whether it is to minimize risk, time, loss or maximize profit, we are all seeking best solutions for a particular purpose under given conditions and information. Similarly, in science and industry optimization tools play a crucial role to offer best solutions respecting given conditions. In this contribution optimization techniques are fruitfully used in the field of control engineering and automation to find a suitable controller that would not only meet basic requirements, such as stability, reference tracking and disturbance rejection, but also something more, namely the limits of the controller's manipulated variable and robustness, which is important from the practical point of view. In practical

F. Gazdos (✉)
Faculty of Applied Informatics, Tomas Bata University in Zlín,
nam. T.G. Masaryka 5555, 760 01 Zlín, Czech Republic
e-mail: gazdos@fai.utb.cz

A. Abraham et al. (eds.), *Proceedings of the Second International Afro-European Conference for Industrial Advancement AECIA 2015*, Advances in Intelligent Systems and Computing 427, DOI 10.1007/978-3-319-29504-6_37

control applications there are always some limitations. The most crucial are the constraints on the control input signal—the controller's manipulated variable which is used to obtain the required course of the controlled variable. This signal is always represented by a certain physical quantity, such as a flow rate, electric current or voltage etc. which obviously have some limitations. This fact has to be carefully considered in the control system design procedure and simulation testing. Not respecting these limits can lead to serious consequences, especially when dealing with hardly controllable processes, e.g. unstable, with significant time-delay or with an inverse response [1]. In the literature there is a great number of classic methods dealing with this problem, often called anti-wind-up techniques applicable mainly to popular PI and PID controllers, e.g. [2, 3]. Among other modern control approaches the predictive control concept is also effective and popular in this field nowadays, e.g. [4, 5], although it is more computationally demanding. Although there are many sources devoted to the robust control systems design and to the constrained control separately, simultaneous solutions of both these problems are still not so common; some representative solutions can be found in e.g. [6–10]. This paper represents a contribution to this interesting and practically important topic. The methodology suggested in this paper is based on the usage of simulation and optimization tools from the MATLAB environment and the systematic algebraic control approach transforming the control system design problem to the solution of polynomial equations, e.g. [11–13]. After formulation of basic control requirements the polynomial approach enables to find both suitable structure and parameters of controllers. A natural part of the procedure for finding a suitable controller using the polynomial approach is the pole-placement problem solution, e.g. [14]. In this contribution this task is solved numerically using the standard MATLAB functions for nonlinear constrained optimization. The resultant poles (free tuning parameters) of the control loop are optimized with respect to both robustness and constraints on the controller's manipulated variable. For this purpose some suitable control quality criteria and a corresponding procedure are suggested. The whole methodology is illustrated clearly on a representative example of control system design for the AMIRA DR300 servo-system. Different control approaches for this system can be found in e.g. works [15, 16]. This paper extends previous works [17, 18] so that more parameters are optimized and the results are verified on a different nonlinear system under real-time conditions.

2 Theoretical Framework

This part recalls basics of the utilized polynomial approach for the classical control configuration and prepares the space for the methodology respecting both control input limitations and robustness of the resultant loop. Assume the classical control set-up of Fig. 1. Here, G denotes a plant to be controlled by a feedback controller Q utilizing information about the process controlled variable y and the reference (set point) signal w through the control error e and generating

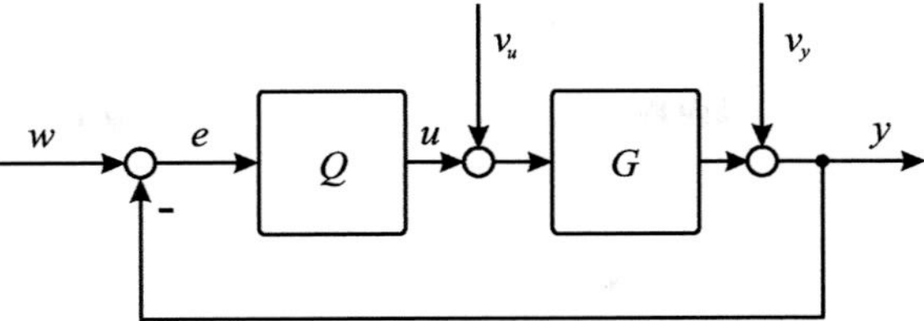

Fig. 1 Control configuration

corresponding control input (manipulated variable) u. Signals v_u and v_y represent general disturbances.

Further suppose that in the continuous-time domain both the plant and the controller can be approximated by transfer functions $G(s)$, $Q(s)$ with coprime polynomials $\{b(s), a(s)\}$, $\{q(s), p(s)\}$ according to (1) while satisfying: deg $a(s) >$ deg $b(s)$ and deg $p(s) \geq$ deg $q(s)$ (the argument s is the complex variable of the Laplace transform).

$$G(s) = \frac{b(s)}{a(s)}, \; Q(s) = \frac{q(s)}{p(s)} \tag{1}$$

The basic requirements for the control system introduced above are formulated in a common way as: *stability, asymptotic tracking of the reference signal, disturbances attenuation* and *inner properness*. Besides the above-mentioned general requirements, the control system should also be robust to cope with the real nonlinear plant (not only with an adopted simplified linear model) and possible disturbances. In addition the controller have to respect given physical limitations of the manipulated variable. All these tasks are discussed and solved further in this work.

It is straightforward that the stability condition will be fulfilled if the controller is given by a solution of the following polynomial equation with a stable polynomial $d(s)$ on its right side, e.g. [11–14]:

$$ap + bq = d. \tag{2}$$

The polynomial is called a characteristic polynomial of the closed-loop as it defines important properties, such as stability and periodic/aperiodic behavior. The polynomial equation, after a proper choice of the stable polynomial $d(s)$, is used to compute the unknown controller polynomials $q(s)$ and $p(s)$. Roots of the characteristic polynomial $d(s)$ are known as poles of the closed loop. Their proper placement influences not only stability of the loop but also the achieved control quality, i.e. a settling-time, overshoots, control input course etc. Therefore the so-called pole-placement problem is a natural part of the polynomial approach to control system design. In this work the poles are optimized numerically using the standard means of the MATLAB system to respect both limitations on the control input and robustness of the resultant loop. The procedure is described in detail further in this paper.

Further, let us assume that both the reference signal and the disturbances can be approximated by step-functions. Then it is also easy to show that to guarantee zero-control error in the steady state (despite both disturbances), the denominator polynomial of the controller $p(s)$ needs to be divisible by the s term, i.e. the controller has to include an integrator, which will be fulfilled for its denominator polynomial in the following form $p(s) = s\tilde{p}(s)$. Then the feedback controller Q(s) in (1) will be

$$Q(s) = \frac{q(s)}{s\tilde{p}(s)}, \tag{3}$$

and the polynomial Eq. (2) defining stability will be: $as\tilde{p} + bq = d$.

The inner properness of the control system is satisfied if all its parts (transfer functions) are proper. With regard to the strictly proper plant transfer function and proper controller (1), and taking into account the solvability of (2), it is possible to derive the following formulas for the degrees of the unknown polynomials $q, \tilde{p}$, and d:

$$\deg\, q(s) = \deg\, a(s), \quad \deg\, \tilde{p}(s) \geq \deg\, a(s) - 1, \quad \deg\, d(s) \geq 2\deg\, a(s). \tag{4}$$

For practical computation of the controller's polynomials $q(s)$ and $\tilde{p}(s)$ (their coefficients) it is necessary to choose a suitable stable polynomial $d(s)$ appearing on the right side of the polynomial Eq. (2). This is the so called pole-placement problem mentioned earlier, e.g. [14]. Therefore we are seeking suitable poles p_i of the designed loop to fulfill given requirements. Hence $d(s)$ can be expressed as (5) for some poles (its roots) p_i. Then the control design procedure transforms to the optimization problem of finding the right poles providing the required control quality.

$$d(s) = \prod_{i=1}^{\deg\, d} (s - p_i) \tag{5}$$

Here, it is suggested to choose the characteristic polynomial as:

$$d(s) = \prod_{i=1}^{\deg\, d} (s + \alpha_i) \tag{6}$$

for some real constants $\alpha_i > 0$. This ensures both stability of the closed loop (all poles will be negative, i.e. stable, at positions $p_i = -\alpha_i$) and aperiodic behavior as the poles are only real numbers. Now the optimization task is to find optimal values of the "tuning" parameters $\alpha_i > 0$, which is addressed in the next section.

3 Methodology

This part describes the used procedure for optimization of the closed-loop poles to meet the required control quality, i.e. loop robustness and limitation on the control input. First suitable control quality criteria are suggested and then the methods and procedure of optimization is clarified.

3.1 *Performance Criteria*

In this work the control quality is measured by two basic sub-criteria: one for assessing robustness of the designed loop—denoted as J_{rob} and the other for evaluating demands on the manipulated variable (control input) $u(t)$—indicated as J_u. As far as the loop robustness is concerned, a peak gain of the sensitivity function frequency response given by the infinity norm H_∞ is a good measure for this purpose [19]. Therefore to assess the robustness of the designed loop it is suggested to use the sensitivity function and its infinity norm H_∞. The sensitivity function is given as:

$$S = \frac{1}{1 + GQ} = \frac{ap}{ap + bq} = \frac{ap}{d}, \tag{7}$$

then, the first sub-criterion describing robustness reads:

$$J_{rob} = \|S\|_\infty = \sup_\omega |S(j\omega)|, \tag{8}$$

where ω is the frequency. The second sub-criterion J_u describes the demands on the manipulated variable $u(t)$ and is formed as follows. Let us define the achievable limits of the manipulated variable (control input) as U_{min} and U_{max} where the first one denotes the minimum allowed value of the signal and the latter one the maximum allowed value of the variable. Then, the control input has to stay within the following defined interval at all times: $u(t) \in < U_{min}; U_{max} > \forall t$. Further denote $\Delta u(t)_{\max}$ as the maximum overshoot of the manipulated variable above the given limit U_{max} and correspondingly $\Delta u(t)_{\min}$ the maximum undershoot of the manipulated variable under the given limit U_{min}. Then the sub-criterion J_u is computed simply as:

$$J_u = \Delta u(t)_{\max} + \Delta u(t)_{\min} \tag{9}$$

It is evident that the sub-criterion is equal to zero if the control input is within the desired limits and it is positive with higher values for the control input out of the required range. The situation is well illustrated in the following picture, Fig. 2, with the limits chosen as $U_{min} = -1$ and $U_{max} = 1$.

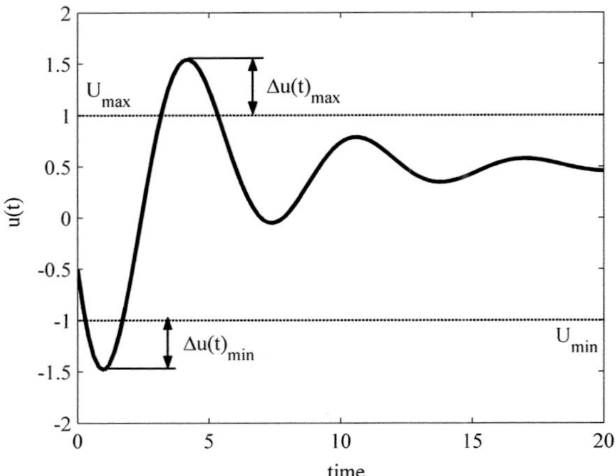

Fig. 2 Explanation of the sub-criterion J_u

Having defined the two sub-criteria to measure robustness of the loop J_{rob} and demands of the manipulated variable J_u, now it is possible to formulate the optimization problem:

$$\min_{\alpha} \ J_{rob}(\alpha) \ \text{such that}\ \alpha_i{>}0\ \text{and}\ J_u(\alpha) = 0, \tag{10}$$

where α is the vector of optimized parameters α_i. In other words, the goal is to find such stable aperiodic poles of the closed loop that provide most robust behavior and respect the given limits of the controller manipulated variable. The optimization procedure, described in the next section, is performed with the help of standard functions for optimization from the MATLAB computing system and its toolboxes.

3.2 *Optimization Process*

First, let us scale the closed-loop variables so that both the controlled output and also the reference signal are within the range from zero to one, i.e. $y(t)$, $w(t) \in <0;1>$. Then the worst-case behaviour of the control system (regarding the changes in the reference signal) can be analysed by considering the reference change of magnitude one. Therefore the designed control system is analysed (i.e. simulated) facing this condition and the control quality criteria J_{rob} and J_u from (10) are assessed for different values of the tuning parameters $\alpha_i > 0$. This is done with the help of MATLAB environment and its toolboxes for simulation and optimization. The procedure can be briefly described as:

- choose number of optimized parameters α_i;
- for every α_i choose an interval for optimization;

- find a solution of the problem specified in (10), i.e. find a minimum of the sub-criterion J_{rob} subject to the conditions such that $J_u = 0$ on the given region of α_i;
- collect the resultant parameters α_i and verify if they fulfill the given requirements.

If the algorithm for solution of the problem fails then there may not be such combination of α_i (under the given conditions) to respect the limits of the control input $u(t)$. Then the designer has several basic possibilities: try to increase the number of optimized parameters α_i (if possible), use different control structures (e.g. with a pre-filter of the reference signal), or has finally no other way than enlarging the prescribed limits of the manipulated variable $u(t)$. The optimization algorithm uses a standard MATLAB function for nonlinear constrained minimization (non-linear programming) *fmincon*. It is a gradient-based method—the trust-region-reflective algorithm based on the interior-reflective Newton method, described in detail in, e.g. [20].

4 Illustrative Example

The methodology introduced in the previous sections is now illustrated on the problem of designing a control system for the servo-system DR300 described further.

4.1 Servomechanism AMIRA DR300

The DR300 servomechanism is a product of the AMIRA company, Duisburg, Germany. It consists of two identical motors connected by a mechanical clutch. The first one is used for control of the rotation speed or the shaft angle while the other (generator) can be used for a simulation of load torque. The system is connected to a common PC by the Humusoft multifunction I/O card MF 614 and controlled in the MATLAB environment using the Real Time Toolbox. The basic control task is to regulate the angular speed of the first motor despite possible variable load generated by the second motor. The system is generally nonlinear with the static properties recorded in Fig. 3 (left) where the controlled variable (angular velocity) y is in in revolutions per minute while the manipulated variable (control voltage) u is scaled to ± 1 machine unit.

As can be seen it has a dead zone and small hysteresis. Therefore the control is performed only in the most linear part aside from the saturation limits and the dead zone. A simplified mathematical model of this nonlinear system can be derived in the state-space form (11), e.g. [16], with $i(t)$ the motor current in [A], $\omega(t)$ the

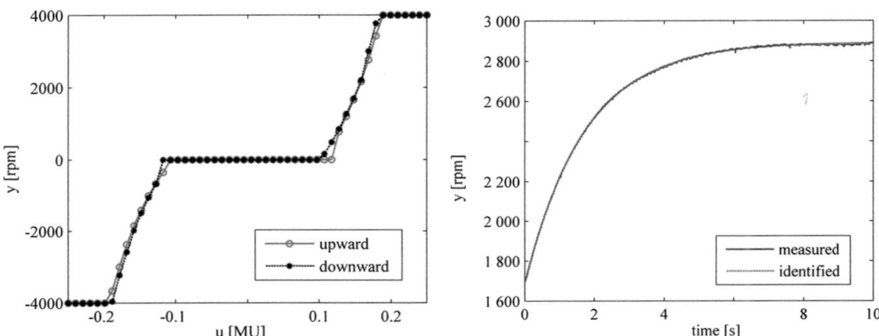

Fig. 3 Static characteristic (*left*) and measured/identified step-responses (*right*)

angular speed in [rad · s^{-1}], $u(t)$ the input voltage in [V] and $m_z(t)$ the load torque in [N · m].

$$\begin{aligned}\frac{d\,i(t)}{dt} &= -\frac{R}{L}i(t) - \frac{k_e}{L}\omega(t) + \frac{1}{L}u(t)\\ \frac{d\,\omega(t)}{dt} &= -\frac{k_m}{J}i(t) - \frac{b}{J}\omega(t) - \frac{1}{J}m_z(t)\end{aligned} \tag{11}$$

Constants R, L, k_e, k_m, b and J are parameters of the motor. The input voltage $u(t)$ is considered as the manipulated variable for the controlled variable—angular speed $\omega(t)$. The load torque $m_z(t)$ can be considered as a disturbance. An input-output model, where $U(s)$, $\Omega(s)$ are the Laplace transforms of the variables can be obtained from (11) as:

$$G(s) = \frac{\Omega(s)}{U(s)} = \frac{\frac{k_m}{LJ}}{s^2 + \left(\frac{R}{L} + \frac{b}{J}\right)s + \frac{Rb + k_e k_m}{LJ}}, \tag{12}$$

which shows that the model can be (in the linear region) identified as a 2nd order proportional system. Experimental identification provides the approximation (13), which shows that the system is stable aperiodic, relatively fast with time-constants $T_1 = 1.72$ [s], $T_2 = 0.02$ [s] and gain $k = 59766$ [rpm/MU]. The comparison of measured and identified step-responses in Fig. 3 (right) shows that the model approximates the system in the given region quite well.

$$G(s) = \frac{b(s)}{a(s)} = \frac{59766}{(1.72s+1)(0.02s+1)} = \frac{2073600}{(s+0.58)(s+59.52)} \tag{13}$$

Scaling of this model (so that the available range of $y(t)$ from 0 to 4000 [rpm], and also of $u(t)$ from 0.1 to 0.2 [MU], is tranformed to 0–1) provides the following model:

$$G(s) = \frac{1.49}{(1.72s+1)(0.02s+1)} = \frac{51.84}{(s+0.58)(s+59.52)} = \frac{51.84}{s^2+60.10s+34.70}, \tag{14}$$

which is further used for the controller design in the next section.

4.2 Control System Design and Experimental Results

Following the polynomial approach described briefly in the Sect. 2 of this paper a suitable feedback controller for this system is designed in the following general form:

$$Q(s) = \frac{q(s)}{p(s)} = \frac{q(s)}{s\tilde{p}(s)} = \frac{q_2 s^2 + q_1 s + q_0}{s(\tilde{p}_1 s + \tilde{p}_0)} \tag{15}$$

Unknown coefficients of the controller are obtained by the solution of the polynomial Eq. (2) for a given stable polynomial $d(s)$. This polynomial, in the general form (5), (6), must be according to (4) of the 4th degree. As the system has 2 different poles it is suggested to have this polynomial in the form (16) with 2 tuning parameters α_1, $\alpha_2 > 0$ (consequently there are 2 double closed-loop poles at $p_{1,2} = -\alpha_1$, $p_{3,4} = -\alpha_2$).

$$d(s) = (s+\alpha_1)^2(s+\alpha_2)^2 \tag{16}$$

Now the parameters α_1, α_2 are optimized numerically via the procedure suggested in the Sect. 3 to respect both robustness of the loop and limits of the control input signal $u(t)$, defined in this case as:

$$u(t) \in \langle 0.1; 0.2\rangle \; [MU] \quad \forall t, \;\; \text{after scaling: } u(t) \in \langle 0; 1\rangle \; [-] \quad \forall t \tag{17}$$

MATLAB computations provided the graphs on Fig. 4 of the sub-criterion J_u (9).

As can be seen, there exists a region where $J_u = 0$ (the blue-filled area inside the left graph). In other words there is a combination of parameters α_1, $\alpha_2 > 0$ for which the manipulated variable stays in the prescribed limits. The detailed contour plot of the criterion (right graph) shows the admissible area more clearly (green-filled area). Now the task is to find the most robust controller in this region, i.e. the minimum of the J_{rob} criterion (8) assessing the designed loop robustness. Complex MATLAB computations provide Fig. 5 where the robustness criterion depends on the optimized parameters, with the green-filled admissible interval for $J_u = 0$. A detailed contour plot of the criterion is also presented. As can be seen, the criterion in the admissible region for $J_u = 0$ (green area) does not change much and

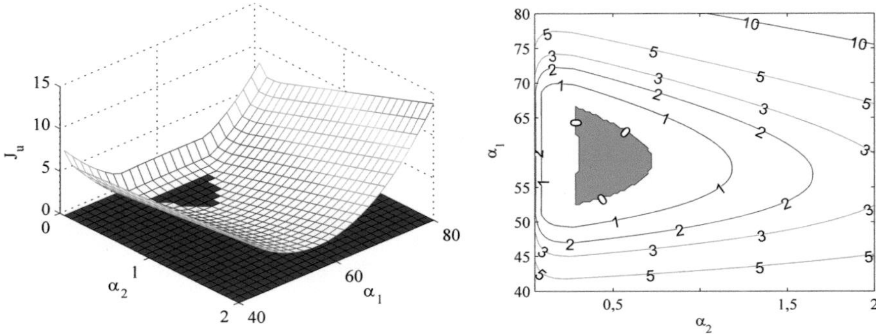

Fig. 4 Sub-criterion J_u with α_1, α_2 and its contour plot

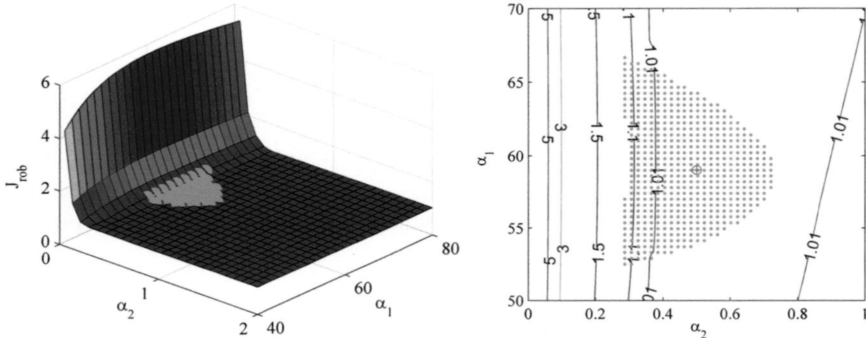

Fig. 5 Sub-criterion J_{rob} with α_1, α_2 and its contour plot

it is relatively small with is minimal values around $J_{rob} < 1.01$. Therefore it is suggested to pick the center of this area as the "optimal" setting to enable both—safe constrained control and loop robustness as well.

Consequently, the best suggested setting of the 2 tuning parameters under given conditions is for $\alpha_1 = 59$, $\alpha_2 = 0.5$. Closed-loop simulation responses for this setting are provided in Fig. 6 (left graph), together with their small variations ± 25 %.

From the recorded simulation responses above it is clear that the suggested setting provides the control input in the desired range while both the variations $\pm$ 25% give much bigger control action. Real-time measurements on the AMIRA DR300 servo-system are recorded also in Fig. 6 (right), where both, the reference tracking in different operating points and disturbance attenuation (represented by a variable load) are presented. From the recorded data it is obvious that the control system is stable, without overshoots and with good tracking not only in the identified operating point but also in different regions which shows robustness

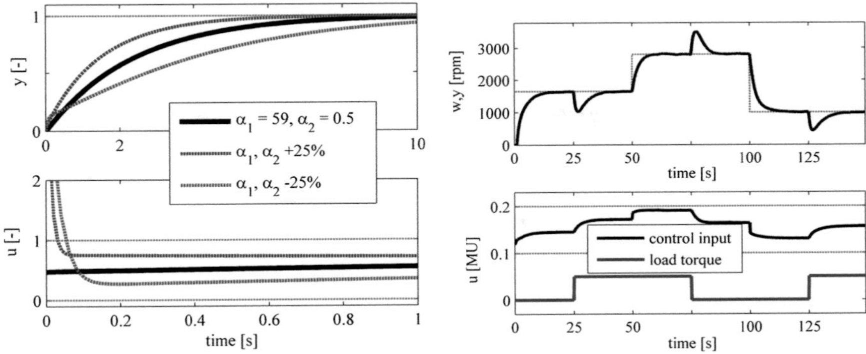

Fig. 6 Worst-case simulation responses (*left*) and real-time control for suggested setting (*right*)

of the design to model mismatch. Disturbances represented by a variable load are also compensated well, and what is important the control input signal is within the required limits at all times.

4.3 Results Discussion

From the presented information and performed experiments it is possible to summarize the following results: (1) suggested "optimal setting" of tuning parameters ($\alpha_1 = 59$, $\alpha_2 = 0.5$ resulting in closed-loop poles $p_{1,2} = -59$, $p_{3,4} = -0.5$) is very close to the poles of the identified model ($p_{1,2} = -59.52$, $p_{3,4} = -0.58$), i.e. when we prescribe closed-loop behaviour close to the dynamics of the original system, the control system is more robust and also the control input is in a reasonable range; (2) it is also expected that other control set-ups, e.g. with a pre-filter of the reference signal, can improve the presented results; (3) optimization of all the closed-loop poles may also bring improvements, however, with limited possibilities for graphical interpretation and inspection.

5 Conclusions

This contribution has presented one possible approach to robust control in case of limits on the controller's manipulated variable. The suggested methodology is based on the direct optimization of closed-loop poles using the efficient tools of the MATLAB system. For this purpose, suitable performance criteria and optimization procedure have been proposed. Experimental results on the AMIRA DR300 servo-system show the potential of the methodology under real-time conditions. Further works will be focused on the usage of different control configurations and investigate the achievable results for optimization of all the closed-loop poles. Different optimization techniques may be also studied and compared with respect to their convergence to optimal solution.

References

1. Stein, G.: Respect the unstable. IEEE Control Syst. Mag. **23**(4), 12–25 (2003)
2. Saberi, A., Stoorvogel, A.A., Sannuti, P.: Control of Linear Systems with Regulation and Input Constraints. Springer, London (2000)
3. Glattfelder, A.H., Schaufelberger, W.: Control Systems with Input and Output Constraints. Springer, London (2003)
4. Camacho, E.F., Bordons, C.: Model Predictive Control. Springer, London (2004)
5. De Doná, J.A., Goodwin, G.C., Seron, M.M.: Anti-windup and model predictive control: reflections and connections. Eur. J. Control **6**(5), 467–477 (2000)
6. Campo, P.J., Morari, M.: Robust control of processes subject to saturation nonlinearities. Comput. Chem. Eng. **14**(4–5), 343–358 (1990)
7. Miyamoto, S., Vinnicombe, G.: Robust control of plants with saturation nonlinearity based on coprime factor representations. In: 35th IEEE Conference on Decision and Control, pp. 2838–2840. Japan (1996)
8. Huba, M.: Robust constrained PID control. In: International Conference Cybernetics and Informatics, pp. 1–18. Vyšná boca, Slovak Republic (2010)
9. Vozák, D., Veselý, V.: Stable predictive control with input constraints based on variable gain approach. Int. Rev. Autom. Control **7**(2), 131–139 (2014)
10. Torchani, B., Sellami, A., Garcia, G.: Robust sliding mode control of class of linear uncertain saturated systems. Int. Rev. Autom. Control **6**(2), 134–146 (2013)
11. Kučera, V.: Diophantine equations in control—a survey. Automatica **29**, 1361–1375 (1993)
12. Hunt, K.J.: Polynomial methods in optimal control and filtering. Peter Peregrinus Ltd., London (1993)
13. Anderson, B.D.O.: From Youla-Kucera to identification, adaptive and nonlinear control. Automatica **34**, 1485–1506 (1998)
14. Kučera, V.: The pole placement equation. A survey. Kybernetika **30**(6), 578–584 (1994)
15. Bobál, V., Kubalčík, M., Chalupa, P., Dostál, P.: Self-tuning control of nonlinear servo system: comparison of LQ and predictive approach. In: 17th Mediterranaen Conference on Control and Automation, pp. 240–245. Thessaloniki, Greece (2009)
16. Roubal, J., Augusta, P., Havlena, V.: A brief introduction to control design demonstrated on laboratory model servo DR300—AMIRA. Acta Electrotechnica et Informatica **5**(4), 1–6 (2005)
17. Gazdoš, F., Marholt, J.: Robust process control with saturated control input. In: 28th European Conference on Modelling and Simulation, pp. 285–291. Brescia, Italy (2014)
18. Gazdoš, F., Marholt, J.: Simulation approach to robust constrained control. Int. Rev. Autom. Control **7**(5), 578–584 (2014)
19. Skogestad, S., Postlethwaite, I.: Multivariable Feedback Control: Analysis and Design. Wiley, Chichester (2005)
20. Coleman, T.F., Li, Y.: An interior, trust region approach for nonlinear minimization subject to bounds. SIAM J. Optim. **6**, 418–445 (1996)

New Approach to Decision Making

Magdalena Kacprzak, Bartłomiej Starosta and Katarzyna Węgrzyn-Wolska

Abstract The aim of this work is to propose a new methodology for representation and processing of imprecise ideas like *good hotel*, *good food*, etc. As an example we consider the problem of verification of a hotel's features, to choose the best place for a conference. In this approach, a vague term of *a good hotel for a conference* is modeled in paradigm of metasets, which involves splitting it into a tree-like hierarchy of related sub-concepts. Then, the techniques of Opinion Mining are used to determine the polarity and the intensity of opinions available on the Internet, that concern specific hotels. Finally, the membership grades for the sample hotels are calculated. These grades reflect each hotel's total capabilities and represent the degree to which the vague idea is satisfied by each. The result is a new system that helps in decision making.

Keywords Metaset · Partial membership · Opinion mining

1 Introduction

In many applications offering solutions to real-life problems there is a need for dealing with vague and imprecise data. Such terms often appear in natural language but are hard to represent and process in software tools and formal, mathematical models. Various approaches to this problem exist in the theory of fuzzy sets [15] or rough sets [9]. Similar, but still different approach is proposed by the theory of metasets

M. Kacprzak
Bialystok University of Technology, Bialystok, Poland
e-mail: m.kacprzak@pb.edu.pl

B. Starosta
Polish-Japanese Academy of Information Technology, Warsaw, Poland
e-mail: barstar@pjwstk.edu.pl

K. Węgrzyn-Wolska (✉)
Allianastic, ESIGETEL, Villejuif, France
e-mail: katarzyna.wegrzyn@esigetel.fr

A. Abraham et al. (eds.), *Proceedings of the Second International Afro-European Conference for Industrial Advancement AECIA 2015*, Advances in Intelligent Systems and Computing 427, DOI 10.1007/978-3-319-29504-6_38

[10]. The general idea of metaset is inspired by the method of forcing [1] in the classical set theory [5, 8]. Metasets admit partial membership, partial equality and other set-theoretic relations [10] which may be evaluated in a Boolean algebra. The certainty values for metaset relations or even compound sentences [12, 13] may also be represented as natural language terms.

The main objective of our work is to build a decision support system that will define linear order in the set of objects taking into account characteristic which is vague and unclear. Such an idea can be for example: good restaurant, interesting place to visit, etc. More specifically, having a set of certain objects, e.g. restaurants or bars, there is a need to collate them linearly from best to worst. It is a difficult and non-trivial task because: (a) criteria for evaluation may be varied, (b) it is difficult to compare some characteristics (e.g. friendly service and good food), (c) criteria often depend on the subjective preferences of customers (some prefer crowded rooms with loud music others prefer peaceful and romantic places). In this work, as an example, we consider the search for a suitable hotel for conferences and we will therefore focus on imprecise expression: *a good hotel for a conference*. On many websites there are different descriptions of hotels available in a selected village. They are often incomplete, inaccurate or simply confusing. Hotel search engines like: http://www.booking.com or http://www.trivago.com are coming to our aid by providing opinions of people who visited these hotels. In this article we introduce a basis for decision support system that will help to evaluate hotels. We achieve this goal by extracting the most important features and then representing them as metasets. Having representation of the term *a good hotel for a conference* in the form of metaset, we can evaluate specific objects. First, opinions about hotels are gathered. We use methods and techniques of Opinion Mining and on their basis we make hotel verification. The resulting degree is a numeric value which helps to verify and compare different hotels.

One of the most important tasks of our research is to analyze various imprecise concepts with the help of metasets and then to explore potential applications. The use of the theory of metasets for modeling vague terms is described in detail in [6]. In this approach, representing any imprecise concept with a metaset lies in splitting it into a tree-like hierarchy of related sub-concepts. In [6] we show how the imprecise idea of *a perfect holiday destination* is represented as a metaset of places whose membership degrees in the metaset are interpreted as their qualities. Then we use it to model and solve the problem of evaluation of the attractiveness of tourist destinations, which is helpful in solving Tourist Trip Design Problems, TTDP.

Another important aspect is the problem of modeling human attitudes towards imprecise ideas. We deal with this problem in [7], where our mechanism is applied to solve the problem of selecting the car best matching the imprecise idea of *a good car for a lady*. In this case, Opinion Mining techniques are applied to build a preference function which reflects someone's attitude towards some imprecise idea. In this approach, the outcome of the decision support tool is a numerical score assigned to a specific car, which expresses how much this car fits the needs of the client. The obtained results confirm his/her opinions shared on the Internet.

In the current paper, the focus is on verification of hotel capabilities in confrontation with the opinions available on the Internet. That's why we do not deal with the subjective preferences of one person, and we investigate whether the information placed on the official website of the hotel are confirmed in opinions of persons who have benefited from its services. For automatic analysis of posts placed on the Web, Opinion Mining techniques are applied. They help to determine polarity and intensity of a given text, i.e., whether it is positive, negative, or neutral and to what extent. To classify the intensity of opinions, we use methods introduced in [2–4]. As a result, we propose a new methodology that can be used to develop a mechanism for assigning specific numerical values to objects and on this basis to classify them.

2 Metasets

We need a tool for modeling partial satisfaction of the property *a good hotel for a conference* by different hotels. We use for this purpose a metaset—a set with partial membership relation [10, 11]. Metasets are similar to fuzzy sets [15] or rough sets [9]. A metaset may represent an imprecise notion just like a classical set represents a sharp property: it consist of elements which satisfy the property. In the case of a metaset the property may be satisfied to a degree other than full certainty. The degrees are represented by nodes of the binary tree and they may be evaluated as real numbers.

2.1 Basic Definitions

By the definition, a metaset is a relation between elements of some given set and the nodes of the binary tree $\mathbb{T}$. For simplicity, in this paper we deal with finite first-order metasets only, which we call metasets anyway.

Definition 1 A set which is either the empty set $\emptyset$ or which has the form:

$$\tau = \{ \langle \sigma, p \rangle : \sigma \text{ is a set}, p \in \mathbb{T} \}$$

is called a first-order metaset.

Informally, we interpret the pair-like structure so that the second elements of pairs are nodes of $\mathbb{T}$ which determine membership degrees of members which are first elements of pairs.

The binary tree $\mathbb{T}$ is the set of all finite binary sequences. It is ordered by the reverse prefix relation: if $p, q \in \mathbb{T}$ and p is a prefix of q, then $q \leq p$ (see Fig. 1). The root $\mathbb{1}$, which is the largest element of $\mathbb{T}$ in this ordering, is the empty sequence.

The binary sequences which are members of $\mathbb{T}$ are denoted with square brackets, e.g.: [00], [101]. If $p \in \mathbb{T}$, then we denote its direct descendants with $p.0$ and $p.1$. A

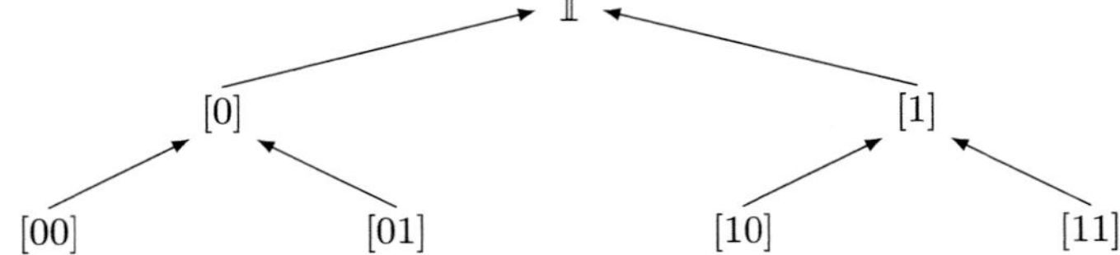

Fig. 1 The levels $\mathbb{T}_0$–$\mathbb{T}_2$ of the binary tree $\mathbb{T}$ and the ordering of nodes. Arrows point at the larger element

level $\mathbb{T}_n$ in $\mathbb{T}$ is the set of all finite binary sequences with the same length n. The level 0 consists of the empty sequence $\mathbb{1}$ only. A *branch* in $\mathbb{T}$ is an infinite binary sequence. We write $p \in C$ meaning that the binary sequence $p \in \mathbb{T}$ is a prefix of the branch C.

2.2 Interpretations

A metaset determines a collection of classical crisp sets which are called interpretations of the metaset. Each interpretation represents some particular, sharp point of view on the source metaset. For instance, there may be many particular, concrete approaches to the vague idea of *a good hotel for a conference*, shared by different experts. Interpretations of metasets are produced using branches of the binary tree.

Definition 2 Let τ be a first-order metaset and let C be a branch. The set

$$\tau_C = \{ \sigma \in \mathrm{dom}(\tau) \colon \langle \sigma, p \rangle \in \tau \wedge p \in C \}$$

is called the interpretation of the first-order metaset τ given by the branch C.

In the above definition $\mathrm{dom}(\tau) = \left\{ \sigma \colon \exists_{p \in \mathbb{T}} \ \langle \sigma, p \rangle \in \tau \right\}$ is the domain of τ.

Informally, interpreting a metaset with the given branch C involves two steps. The first step removes all the ordered pairs whose second elements are nodes which do not belong to the branch C. The second stage replaces the remaining pairs—whose second elements lie on the branch C—with their first elements.

2.3 Partial Membership

A member may belong to a metaset to a variety of degrees which are represented by nodes of $\mathbb{T}$ and which in fact make up a Boolean algebra. This property extends the classical case of two-valued membership relation and therefore enables modeling of imprecision. Membership and other set-theoretic relations for metasets are defined using interpretations.[1]

[1] For the detailed discussion of the relations or their evaluation the reader is referred to [11] or [13].

Definition 3 We say that the metaset σ belongs to the metaset τ under the condition $p \in \mathbb{T}$, whenever for each branch C containing p holds $\sigma_C \in \tau_C$. We use the notation $\sigma \epsilon_p \tau$.

Formally, we define infinitely many membership relations: each $p \in \mathbb{T}$ specifies another relation ϵ_p. Any two metasets may be simultaneously in multiple membership relations qualified by different nodes: $\sigma \epsilon_p \tau \wedge \sigma \epsilon_q \tau$. The relation $\epsilon_{\mathbb{1}}$ is equivalent to full, unconditional membership of crisp sets.

The conditional membership reflects the idea that an element σ belongs to a metaset τ whenever some conditions are fulfilled. The conditions are described by the nodes of $\mathbb{T}$.

Example 1 The ordinal number 1 is the set $\{0\}$ and 0 is just the empty set θ. Let $\tau = \{\langle 0, [0]\rangle, \langle 1, [1]\rangle\}$ and let $\sigma = \{\langle 0, [1]\rangle\}$. Let $C^0 \ni [0]$ and $C^1 \ni [1]$ be arbitrary branches containing [0] and [1], respectively. Interpretations are: $\tau_{C^0} = \{0\}$, $\tau_{C^1} = \{1\}$, $\sigma_{C^0} = 0$ and $\sigma_{C^1} = \{0\} = 1$. We see that $\sigma \epsilon_{[0]} \tau$ and $\sigma \epsilon_{[1]} \tau$. Also, $\sigma \epsilon_{\mathbb{1}} \tau$ holds.

Note, that even though interpretations of τ and σ vary depending on the branch, the metaset membership relation is preserved.

2.4 Evaluating Membership

In applications we usually need numerical evaluation of membership relation. We define it in two steps. First, we consider the smallest subset of $\mathbb{T}$ consisting of elements which determine the membership.

Definition 4 Let σ, τ be first-order metasets. The set

$$||\sigma \in \tau|| = \max\left\{ p \in \mathbb{T} \colon \sigma \epsilon_p \tau \right\}$$

is called the certainty grade for membership of σ in τ.

Here, $\max\left\{ p \in \mathbb{T} \colon \sigma \epsilon_p \tau \right\}$ denotes the set of maximum elements (in the tree ordering) of the set of nodes in $\mathbb{T}$, for which the relation $\sigma \epsilon_p \tau$ holds.

Now we define the numerical evaluation of membership taking the following assumptions. All the nodes within a level contribute equally to the membership value—none of them is distinguished. For the given $p \in \mathbb{T}$, its direct descendants $p.0$ and $p.1$ add half of the contribution of the parent p, each. Therefore, the contribution of a $p \in \mathbb{T}$ must be equal to $\frac{1}{2^{|p|}}$, where $|p|$ is the length of the sequence p.

Definition 5 Let σ, τ be first-order metasets. The following value is called the certainty value of membership of σ in τ:

$$|\sigma \in \tau| = \sum_{p \in ||\sigma \in \tau||} \frac{1}{2^{|p|}} \,.$$

One may easily see that $|\sigma \in \tau| \in [0, 1]$. If $||\sigma \in \tau|| = \{\mathbb{1}\}$, i.e., $\sigma\epsilon_{\mathbb{1}}$ holds, then $|\sigma \in \tau| = 1$. And if $||\sigma \in \tau|| = \theta$ ($\sigma\epsilon_p$ holds for no p), then $|\sigma \in \tau| = 0$.

2.5 Representing Imprecise Ideas with Metasets

Just like a set represents a collection of objects which satisfy a property given by a formula, a metaset represents a "fuzzy" collection of objects which satisfy some imprecise idea. In this paper we use a metaset to represent the imprecise term of *a good hotel for a conference*. Its members are particular hotels which match the given idea to a variety of degrees, usually different than the complete truth.

The core of the idea of representing any imprecise concept with a metaset lies in splitting it into a tree-like hierarchy of related sub-concepts. In this particular case, we require that *a good conference hotel* has proper facilities for users (*user*) and organizers (*org*). From a user point of view these facilities must assure a place for relaxation (*relax*) and professional work (*work*). Taking organizer's perspective we must assure proper rooms (*room*) and *equipment*. Participants want to relax in a *bar* or in spa and fitness locations (*spa*). For professional work during the conference they need good network access (*net*) and writing-desks in rooms for preparing notes and papers (*desk*). The organizers must provide large halls for plenary lectures (*large*) and several medium-sized places for parallel sessions (*medium*). The required *equipment* includes projectors and screens, audio sound systems, etc. (*furniture*) as well as minor office accessories like printers, photocopiers, faxes, etc. (*accessory*).

We may continue the process of splitting notions further to achieve arbitrary high precision of description, however, we stop at the 3rd level, for the sake of simplicity.

Note, that by the above description, a parent node stands for the union of two children nodes. For instance, if a hotel assures *relax*, then it means that both *net* and *desk* property are satisfied. And when *furniture* and *accessory* are present, then we say understand that *equipment* is there.

The binary tree in Fig. 2 is used throughout the paper to represent the discussed idea of *a good hotel for a conference* by means of the metaset Δ. Note, that the nodes of the tree which determine the membership degrees are natural language terms, which also describe reasons for some particular hotel to satisfy the discussed idea.

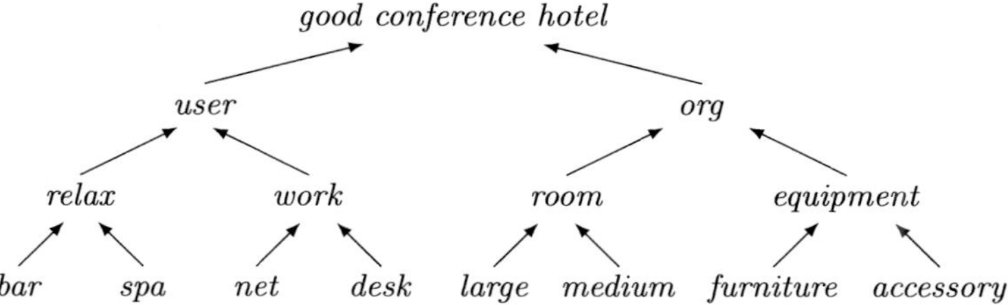

Fig. 2 The binary tree of the features describing *a good hotel for a conference*

3 Validation by Opinion Mining Techniques

More and more people share their opinions on various topics on the Internet. We can find in this way the views and experiences of a very large number of people who are not our friends, nor experts in the field, but people who may have the same tastes as ours, so their opinions can be very helpful before making our choice and to have our own ideas on a given topic. It may also consist of the evaluations of specific aspects of the product (capabilities of *a good hotel for a conference*), and feedback from readers about the official data.

Our objective is to create a system to evaluate and validate the typical choice of hotel's reservation for the conference. The main challenge is to extract from the consumers' reviews the most important opinions concerning selected hotel's features and to evaluate these opinions to validate the official data about the hotel. To perform this task we use Opinion Mining techniques.

3.1 Opinion Mining: Basic Definition

Opinion Mining consists in identifying orientation or intensity of opinion. It enables determining whether some part of document or a sentence expresses positive, negative or neutral opinion towards some object (O) or more. Also, it allows for classification of opinions according to intensity degrees.

Definition 6 An opinion is a quadruple (O, F, H, S), where O is a target object, $F = \{f_1, f_2, \dots, f_n\}$ is a set of features of the object O, H is a set of opinion's holders, S is the set of sentiment/opinion values of the opinion's holder on the feature f_i of the object O.

An object O is represented with a finite set of features, $F = \{f_1, f_2, \dots, f_n\}$. Each feature $f_i \in F$ $(i = 1, \dots, n)$ can be expressed with a finite set of words or phrases W_i, where each $W_i = \{W_{i1}, W_{i2}, \dots, W_{in}\}$ is a set of corresponding synonyms for the features.

Thus, an object O is represented as a tree or taxonomy of components F (or parts), sub-components, and so on. Each node represents a component and is associated with a set of attributes. O is the root node, which also has a set of attributes. An opinion can be expressed on any node or attribute of the node.

In general, the first step of such a process is to retrieve the information from the Web [14] (specialized websites, tweets, blogs, forums, etc.) related to the object (O: *a good hotel for a conference* in our case, presented in Example 2), to extract the opinions about the selected features (F: nodes of the binary tree) and then to classify this information according to their emotional value (polarity).

The classification of the opinion polarity consists in making decision between positive and negative status. A value called semantic orientation is created in order to demonstrate words' polarity. There are several calculation methods of the words semantic orientation (SO). The most often used method is called SO-A (Semantic Orientation from Association):

$$\text{SO-A}(word) = \Sigma_{p \in P}\, \text{A}(word, p) - \Sigma_{n \in N}\, \text{A}(word, n)$$

where

- $A(word, p)$ is the association of studied word with the positive word,
- $A(word, n)$ is equivalent negative,
- $A(word)$ is a measure of association.

If the sum is positive, the word is oriented positively, and if the sum is negative, the orientation is negative. The absolute value of the sum indicates the orientation intensity.

3.2 Validation Methodology

The purpose is to find the opinions related to the selected features, which could confirm or reject the candidate for *a good hotel for a conference*.

The general idea is the following: find on the specialized websites (booking.com, hotels.com and trivago.com) the list of potential candidates O corresponding to our needs (important features), retrieve for each hotel from candidates' list the set A of the capabilities declared in official hotel's description ($A \subset F$), extract the clients' opinions corresponding to selected features $f_i \in F$ ($i = 1, \dots, n$) and then validate the hotel's selection.

In this way the positive/neutral opinion on existing feature f_i which belongs to A validates f_i, the positive opinion on feature f_i which does not belong to A adds new feature f_i to A, the negative one cancels f_i from A set. Therefore, the polarity of opinion (negative, positive) is considered to be the significant contribution which evaluates the selection.

In order to demonstrate our methodology we use the hierarchy of conditions depicted in Fig. 2, which comprise the notion of *a good hotel for a conference*. By Definition 6, the basic components of an opinion are: object O (on which an opinion is expressed, in our case it is a hotel), feature f_i describing the object O, and sentiment/opinion (a view, attitude, or appraisal on each selected feature). According to the idea presented by the tree in Fig. 2, the set F is composed of 14 elements *{user, relax, bar, spa, work, net, desk, org, room, large, medium, equipment, furniture, accessory}*. For each element $f_i \in F$ we select manually the corresponding set W_i. For example:

$$W_{accessory} = \{microphone, printer, fax, scanner, screen, projector, \ldots\},$$
$$W_{furniture} = \{chair, seat, table, board, outlet, plug, cable, \ldots\},$$
$$W_{net} = \{wifi, internet, connection, wireless, \ldots\},$$
$$W_{spa} = \{fitness, gym, health\ facilities, sport, swimming\ pool, \ldots\}.$$

To classify intensity of opinions concerning hotel guests, we use the engine of our system [2–4]. We formalize our methodology in the following example.

Example 2 For demonstration purposes we consider here opinions on one selected hotel H. We know from the official website that its capabilities include: big conference room (*large*), desks in rooms (*desk*), Internet access (*net*), fitness locations (*spa*), bar (*bar*). So at this stage we assume for this hotel the following capabilities: $large, bar, spa, desk, net$.

To confirm these data we analyze the guest reviews about it retrieved from booking.com and trivago.com websites. The posts include opinions of this sort: "seats in conference room were not comfortable" (*–furniture*), "not able to regulate air condition" (*–furniture*), "very easy to connect my iPad to video projector" (*+accessory*), "speed color printer" (*+accessory*), "I had Wi-Fi access only near the reception" (*–net*), "fitness has always been busy" (*–spa*), "poor sport equipment" (*–spa*), "nice bar" (*+bar*).

Based on these we have validated/canceled the following capabilities for the hotel $H : large, accessory, bar, desk$.

4 Modeling with Metasets

We model the property of *a good hotel for a conference* with a metaset and its partial membership relation: if a hotel σ satisfies a property $p \in \mathbb{T}$ (see Fig. 2), then we formally write $\sigma \epsilon_p \Delta$.

By utilizing web retrieval information and opinion mining techniques we discovered that the sample hotel α has a bar, good network access, large conference rooms and the appropriate furniture, and another hotel β has proper places for relaxation and professional work, all types of rooms necessary, and is well equipped with minor office accessories:

$$\alpha \ : \ bar, net, large, furniture \ , \tag{1}$$
$$\beta \ : \ relax, work, room, accessory \ . \tag{2}$$

Formally we write:

$$\alpha\epsilon_{bar}\Delta \ \wedge \ \alpha\epsilon_{net}\Delta \ \wedge \ \alpha\epsilon_{large}\Delta \ \wedge \ \alpha\epsilon_{furniture}\Delta \ , \tag{3}$$
$$\beta\epsilon_{relax}\Delta \ \wedge \ \beta\epsilon_{work}\Delta \ \wedge \ \beta\epsilon_{room}\Delta \ \wedge \ \beta\epsilon_{accessory}\Delta \ . \tag{4}$$

By the Definition 3 we also conclude that:

$$\beta\epsilon_{user}\Delta \wedge \beta\epsilon_{room}\Delta \wedge \beta\epsilon_{accessory}\Delta . \tag{5}$$

We calculate membership grades for the sample hotels (cf. Definition 4). The grades reflect each hotel's total capabilities and represent the degree to which the vague idea is satisfied by each.

$$\begin{aligned} ||\alpha \in \Delta|| &= \{bar, net, large, furniture\} , && (6)\\ &= \{[000], [010], [100], [110]\} , && (7)\\ ||\beta \in \Delta|| &= \{user, room, accessory\} , && (8)\\ &= \{[0], [10], [111]\} . && (9) \end{aligned}$$

The numerical evaluation of the hotels' capabilities represented by conditional membership in Δ is carried out as follows (cf. Definition 5):

$$\begin{aligned} |\alpha \in \Delta| &= \frac{1}{2^{|[000]|}} + \frac{1}{2^{|[010]|}} + \frac{1}{2^{|[100]|}} + \frac{1}{2^{|[110]|}} , && (10)\\ &= \frac{1}{2^3} + \frac{1}{2^3} + \frac{1}{2^3} + \frac{1}{2^3} = \frac{4}{8} = 0.5 , && (11)\\ |\beta \in \Delta| &= \frac{1}{2^{|[0]|}} + \frac{1}{2^{|[10]|}} + \frac{1}{2^{|[111]|}} , && (12)\\ &= \frac{1}{2^2} + \frac{1}{2^1} + \frac{1}{2^3} = \frac{7}{8} = 0.875 . && (13) \end{aligned}$$

The results indicate, that β is a member of Δ to a higher degree than α. In other words, β satisfies the property *a good hotel for a conference* better than α. Therefore, the former is the better choice and it is recommended by the discussed system.

5 Conclusions

This paper shows for the first time how the paradigm of metasets and Opinion Mining techniques can be combined to obtain an excellent methodology to validate the information posted on the Internet. This is especially troublesome when the information relates to vague terms and concepts. Our approach can be applied in software decision support systems that help users to make decisions. In this paper the decision involves choosing a good hotel for a conference but can be easily applied to other similar problems.

Acknowledgments Kacprzak's contribution was supported by the Bialystok University of Technology grant S/W/1/2014.

References

1. Cohen, P.: The independence of the continuum hypothesis 1. In: Proceedings of the National Academy of Sciences of the United States of America, vol. 50, pp. 1143–1148 (1963)
2. Dziczkowski, G., Wegrzyn-Wolska, K.: Rcss—rating critics support system purpose built for movies recommendation. In: Advances in Intelligent Web Mastering. Springer (2007)
3. Dziczkowski, G., Wegrzyn-Wolska, K.: An autonomous system designed for automatic detection and rating of film. Extraction and linguistic analysis of sentiments. In: Proceedings of WIC, Sydney (2008)
4. Dziczkowski, G., Wegrzyn-Wolska, K.: Tool of the intelligence economic: recognition function of reviews critics. In: ICSOFT 2008 Proceedings (2008)
5. Jech, T.: Set Theory: The Third, Millennium edn. Revised and Expanded. Springer, Berlin (2006)
6. Kacprzak, M., Starosta, B.: An approach to making decisions with metasets. In: Trends in Contemporary Computer Science, Podlasie 2014, pp. 159–172 (2014)
7. Kacprzak, M., Starosta, B., Wegrzyn-Wolska, K.: Metasets and opinion mining in new decision support system. In: LNCS, vol. 9120, no. 2, pp. 625–636 (2015)
8. Kunen, K.: Set theory, An introduction to independence proofs. In: Studies in Logic and Foundations of Mathematics. North-Holland Publishing Company (1980)
9. Pawlak, Z.: Rough sets. Int. J. Comput. Inf. Sci. **11**, 341–356 (1982)
10. Starosta, B.: Metasets: A New Approach to Partial Membership. In: LNAI, vol. 7267, pp. 325–333. Springer (2012)
11. Starosta, B., Kosiński, W: Meta sets. Another approach to fuzziness. In: Seising, R. (ed.) Views on Fuzzy Sets and Systems from Different Perspectives. Philosophy and Logic, Criticisms and Applications, STUDFUZZ, vol. 243, pp. 509–522 (2009)
12. Starosta, B., Kosiński, W.: metasets, intuitionistic fuzzy sets and uncertainty. In: LNAI, vol. 7894, pp. 388–399. Springer (2013)
13. Starosta, B., Kosiński, W.: Metasets, certainty and uncertainty. In: Atanassov, K.T. et al. (eds.), New Trends in Fuzzy Sets, Intuitionistic Fuzzy Sets, Generalized Nets and Related Topics. Volume I: Foundations, pp. 139–165. SRI PAS (2013)
14. Wegrzyn-Wolska, K., Bougueroua, L.: Tweets mining for french presidential election. In: Proceedings of CASoN 2012. SaO Carlos, Brazil, Nov 2012
15. Zadeh, L.A.: Fuzzy sets. Inf. Control **8**, 338–353 (1965)

Modeling Lexicon Emergence as Concept Emergence in Networks

Juan Galán-Páez, Joaquín Borrego-Díaz and Gonzalo A. Aranda-Corral

Abstract A model for lexicon emergence in social networks is presented. The model is based on a modified version of classic *Naming Games*, where agents' knowledge is represented by means of formal contexts. That way it is possible to represent the effect interactions have on individual knowledge as well as the dynamics of global knowledge in the network.

1 Introduction

In Complex Systems (CS) research, the majority of recent studies on dynamics in Social Networks (SN) considers two non-exclusive perspectives. On the one hand, agent evolution is studied by considering it as part of a network, where its topology plays a relevant role. On the other hand the global perspective considers the network itself as a dynamic system on which its topology depends on local rules [13]. In the case of emergent properties based on agent interaction, modeling techniques adopt the first one, whilst the second perspective aids to understand the overall behavior of the system by means of global features and parameters.

In the field of Social Networks Analysis, human interactions lead to the formation of different kinds of (emergent) opinions. For example opinion consensus/formation (a product of collective intelligence) is very similar, in some cases, to lexicon emergence—a topic widely studied in CS field—and its modeling can be

J. Galán-Páez · J. Borrego-Díaz
Department of Computer Science and Artificial Intelligence,
University of Seville, Seville, Spain

G.A. Aranda-Corral (✉)
Department of Information Technology, Universidad de Huelva, Huelva, Spain
e-mail: garanda@us.es

A. Abraham et al. (eds.), *Proceedings of the Second International Afro-European Conference for Industrial Advancement AECIA 2015*, Advances in Intelligent Systems and Computing 427, DOI 10.1007/978-3-319-29504-6_39

exploited for developing social strategies [10]. Sentiment/opinion analysis attempts to analyze the effect and scope of these tendencies. Lexicon emergence is a research line within Language Dynamics (LD) paradigm, where agent modeling is a key tool. LD is a rapidly growing field in CS community that focuses on all processes related with emergence, evolution, change and extinction of languages [12]. The models of LD provides interesting ideas to model and predict similar social processes. The consensus (as an extension of agreement) in LD allows to study which are the key factors that drive lexicon evolution. Thus it could be interesting to adapt those to semantically enriched agent networks. Classic LD does not consider strong semantic features on agents' interactions, and Formal Concept Analysis (FCA) [8] provides a general framework in which semantic features can be added to LD (at object/attribute level) [2].

Collective consensus on concepts (opinions, new ideas, etc.) is a phenomena that is frequent in social media, and in the users own use of (weak) semantic strategies for streamlining this process (for example the use of hash-tags in Twitter or the spreading, mutation and adaptation of *memes* which has complex dynamics [1]). Moreover, sentiment/opinion analysis provides social researchers with new ideas and tools. For example, in [7] authors show how Formal Concept Analysis of twitter stream on a *Trending Topic* provides a global representation of sentiment concepts as well as a kind of new sentiment concepts based on sentiment vocabularies. This kind of study can predict viral meme evolution, for example.

The aim of this paper is to develop FCA-based Naming Games for modeling concept emergence in social networks by applying ideas from [2].

Related Work: In [4] authors design a basic FCA-based agent interaction model, intended (in that paper) for modeling consensus between bookmarking agents in order to estimate how semantic heterogeneity behaves in social bookmarking services. The particular case of NG in Social Networks is an interesting research line in CS research where the net effect on lexicon emergence is studied (see [14] for a nice introduction to the topic). The most used approaches to NG modeling are focused on lexicon emergence and semantics is not usually considered (because the own vocabulary it is not predefined). This paper is focused on the dynamics of the inherent semantic (modeled by FCA) associated to preexistent vocabularies. This basic model was enhanced to be used in NG [2], which is adapted here to social networks in turn. Several variants of NG can be studied by considering different levels of reasoning ability of agents [2]. In [2] authors show how FCA can be used to enhance models of language emergence by enriching the semantics of agent models. In this way semantics self-emergence can be modeled. For instance, an approximation to the study of self-organization and evolution of the language and its semantics could be to consider the community of users as a CS that collectively builds the semantic features of their own lexicon.

A related study is [9], where authors studied the effects of randomness on the competition between strategies in an agent-based model of tag-mediated cooperation evolving on large-scale complex networks.

2 Background

A popular approach in LD consists in modeling agents' interaction by means of the called *Naming Games* (NG) [16], which was created to explore self-organization in LD (emergence of vocabularies, in other words, the mapping between words and meanings). Naming games consist in the interaction between two agents, a speaker and a listener. The information in a language game is local to its participants: other agents are not aware of innovations or adaptations that might have come up during that interaction. Only after this innovation has been used again in other interactions, it can be spread through the population. From the basic model, a number of variants for several and specific models can be considered. The aim of the agent community is to achieve a common vocabulary. In the minimal NG each agent has its own context (object/word) and interacts according to the following steps [12]:

1. The speaker selects an object from the current context.
2. The speaker retrieves a word from its inventory associated with the chosen object, or, if its inventory is empty, invents a new word.
3. The speaker transmits the selected word to the listener.
4. If the listener has the word named by the speaker in its inventory, and that word is associated with the object chosen by the speaker, the interaction is a success. In this case, both players maintain in their inventories only the winning word, deleting all the other words that fitted the same object.
5. If the listener does not have the word named by the speaker in its inventory or the word is associated to a different object, the interaction fails: the listener updates its inventory by adding an association between the new word and the object.

NG have been considered in non-situated and situated models. In the second one, agents are placed in an artificial world, where environmental features as distance between agents or agent's neighborhood can be considered. Situated models are very interesting since the communication among agents does not obey purely random selections: communication takes place between agents that are able to do it, according to the restrictions of the model itself (for example, only neighboring agents can communicate). NG with spatially distributed agents allow modeling the emergence of different language communities by stabilization of the system [17]. This is due to the fact that the "success" of a linguistic innovation dependends on whether the group, as a whole, has adopted it or not. Likewise agent networks can be considered as in this paper.

Our interest in NG is based on their *adaptive nature*; that is to say, NG can produce changes in the lexicon of both, the speaker and listener, as side effect. Thus, agent's lexicon changes during its life within the system. In the case of this work, it is interesting to consider agents' neighbors within the network to see how their interaction model agent's lexicon. Especially in complex networks as for instance the scale-free networks, a kind of network topology very frequent in social networks.

3 Formal Concept Analysis and Implications

According R. Wille, FCA mathematizes the philosophical understanding of a concept as a unit of thoughts composed of two parts: the extent and the intent [8]. The extent covers all objects belonging to this concept, while the intent comprises all common attributes valid for all the objects under consideration. It also allows the computation of concept hierarchies from data tables. In this section, we succinctly present basic FCA elements, although it is assumed that the reader is familiar with this theory (the fundamental reference is [8]).

A formal context is represented as $M = (O, A, I)$, which consists of two sets, O (objects) and A (attributes) and a relation $I \subseteq O \times A$. Two derivation operators, both denoted by $'$, formalize the sharing of attributes for objects, and, in a dual way, the sharing of objects for attributes: Given $X \subseteq O$ and $Y \subseteq A$,

$$X' := \{a \in A \mid oIa \text{ for all } o \in X\} \text{ and } Y' := \{o \in O \mid oIa \text{ for all } a \in Y\}$$

A (formal) concept is a pair (X, Y) such that $X' = Y$ and $Y' = X$. Logical expressions used in FCA are the *implications between attributes*, pair of sets of attributes, written as $Y_1 \to Y_2$, which is true with respect to $M = (O, A, I)$ according to the following definition. A subset $T \subseteq A$ *respects* $Y_1 \to Y_2$ if $Y_1 \nsubseteq T$ or $Y_2 \subseteq T$. It is said that $Y_1 \to Y_2$ holds in M ($M \models Y_1 \to Y_2$) if for all $o \in O$, the set $\{o\}'$ respects $Y_1 \to Y_2$. See [8] for more information.

Definition 1 Let $\mathcal{L}$ be a set of implications and L an implication of M.

1. L follows from $\mathcal{L}$ ($\mathcal{L} \models L$) if each subset of A respecting $\mathcal{L}$ also respects L.
2. $\mathcal{L}$ is complete if every implication of the context follows from $\mathcal{L}$.
3. $\mathcal{L}$ is non-redundant if for each $L \in \mathcal{L}$, $\mathcal{L} \setminus \{L\} \not\models L$.
4. $\mathcal{L}$ is a basis for M if it is complete and non-redundant.

A basis called *Stem Basis* (SB) can be obtained from the *pseudo-intents* (see [8]). Although SB is used for showing the modeling, it is important to remark that SB is only an example of basis. Throughout the paper none specific property of the SB can be used, so it can be replaced by any other (implication) basis.

Roughly speaking, the goal of using FCA in NG is to analyze interactions between agents equipped with (partial) knowledge, in order to study how a collective knowledge emerges. Each interaction is a communicative act where two agents interchange new knowledge. In the experiments carried out, the creation steps have been dropped, as the aim of these models is to induce the emergence of the basis from a preexisting lexicon. For modeling communication between agents, the English lexical database *WordNet*[1] has been chosen as the global knowledge. In *WordNet* nouns, verbs, adjectives and adverbs are grouped into sets of cognitive synonyms called *Synsets*, where each of these expresses a distinct concept. *Synsets* are interlinked by means of conceptual-semantic and lexical relations.

[1] http://wordnet.princeton.edu.

A formal context associated with WordNet It is interesting to consider a real and structured language in order to exploit FCA semantic features, since the future aim of this paper is to apply the model to real world social networks (see [4]). Therefore in this paper experiments are carried out using (subsets of) WordNet. A Formal Context associated with WordNet can be obtained by considering words as objects of the context and synsets as the attributes. Synsets are sets of synonymous words, thus they can be considered as a potential definition (meaning) of each word.

Concept lattice and the basis associated to a lexical database of this size cannot be efficiently computed (in fact, conceptual descriptions of complex systems have complex structure [3]), therefore small subsets of WordNet have been considered. That way it will be possible to evaluate the soundness of the proposed models with respect to the FCA elements associated to these subsets. Likewise, only connected subsets of *WordNet* and only formed by full *synsets* (containing every word referring to that synset) have been considered. A connected subset is one such that if a network interlinking its words and synsets were built, the resulting network would be a connected graph.

Once a WordNet subset has been chosen, the same one will be used in every experiment. Two different formal contexts have been considered in the experiments presented in this work. The *global knowledge* is a formal context containing the full WordNet subset and is the knowledge that agents' community aims to achieve. Each agent has its own formal context (*local knowledge*), which is randomly initialized with some word/synset pairs drawn from the global knowledge.

3.1 Formal Contexts Within a Networked Multiagent System

Definition 2 Let $M = (O, A, I)$ be a formal context and $G = (N, E)$ be a network.

- A **linguistic distribution** d is a function $d : I \times N \to \{0, 1\}$, which induces the relation $I_d = \{(o, a) \in I \ : \ \exists n \in N \mid d((o, a), n) = 1\}$
  The **global context associated with** d is the context $M^d = (O, A, I_d)$
- Given $n \in N$, the vocabulary and the dictionary of n w.r.t d are resp.
  $V^d(n) := \{o \in O \ : \ \exists a \in A\,[\,(o, a) \in I \wedge d((o, a), n)) = 1]\}$
  $D^d(n) := \{a \in A \ : \ \exists o \in O\,[\,(o, a) \in I \wedge d((o, a), n) = 1]\}$
  and the formal context associated with n w.r.t. d is $M^d(n) := \left(V^d(n), D^d(n), I^d_n\right)$, where $I^d_n = \{(o, a) \ : \ d((o, a), n) = 1\}$
- The **context network associated with** d and G is a network where each node n is labeled by $M^d(n)$
- Given $t \in [0, 1]$, the t-collective context associated with d is $M^{d,t} := (O, A, I^{d,t})$

$$\text{where } I^{t,d} = \{(o, a) \in I \ : \ \frac{\#\{n \ : \ d((o, a), n) = 1\}}{\#N} \geq t\}$$

Given $\delta \in [0, 1]$, a random linguistic distribution according to δ is a uniform distribution d built by assigning $d((o, a), n)) = 1$ with probability δ. Note that in this case $Prob\left((o, a) \in I_d\right) = 1 - (1 - \delta)^{\#N}$.

Stability of NG can be studied with ideas from knowledge convergence in multi-agent systems (see e.g. [6]). Mainly four parameters are considered:

- #*N*, the population size (each node is an agent). The values chosen in the experiments are conditioned by the feasibility of the computation of each model.
- δ (the probability for an agent to have within its initial knowledge a pair (object-attribute)) is selected in a value range providing each pair *lemma-synset* to appear in at least one agent from the overall population, with probability $p = 0.95$.
- Given a threshold t (usually $t \geq 0.9$), the convergence (stabilization) test checks whether every existent pair *lemma-synset* is present within $M^{d,t}$.
- The size of the selected subset of WordNet is determined by both, the fact that the subsystem must contain complete synsets and by computational feasibility.

4 Modeling Communication as FCA-based Naming Games

Initial knowledge of two agents that will perform examples of communicative acts is shown in Fig. 1. The scale-free network was randomly generated following the principle of preferential attachment [5]. Each node is an agent that in each step communicates with one of its neighbors (those directly connected in the graph). The communicative process is as follows: Starting with d a linguistic random distribution according to $\delta \in [0, 1]$:

1. The world (network) is randomly built with #*N* agents (nodes). Each agent starts with an initial knowledge represented by $M^d(n)$. To obtain successful communicative games, it is necessary that the union of the initial local knowledge of each agent contains approximately all concepts within the global knowledge, hopefully $\bigcup_{n \in N} I^d_n = I$. The probability p for each pair within the selected WordNet subset to appear in the initial local knowledge, of at least one agent, is given by $p = 1 - (1 - \delta)^{\#(N)}$.
   It is recommended carrying out communication games with at least $p > 0.95$, thus the value δ to be considered depends on the number of agents N in the world (i.e. for $\#N = 200$ it is suggested that $\delta \geq 0.015$).
2. In each step, each agent (speaker) chooses randomly a listener agent between its neighbors (adjacent nodes), in order to start a communicative process (request).
3. After the simulation, the collective knowledge $M^{d,t}$ is measured.

Recall that a pair (o, a) can be considered part of the *collective knowledge* $(o, a) \in M^{d,t}$, only for a certain collective knowledge threshold $t = CK_{th}$. Thus CK_{th} denotes the minimum proportion (usually between [0.9, 1]) of agents knowing a certain pair, necessary to consider that pair as part of the collective knowledge.

There are different ways of carrying out the communicative act as well as different ways of measuring the collective knowledge (see [2]).

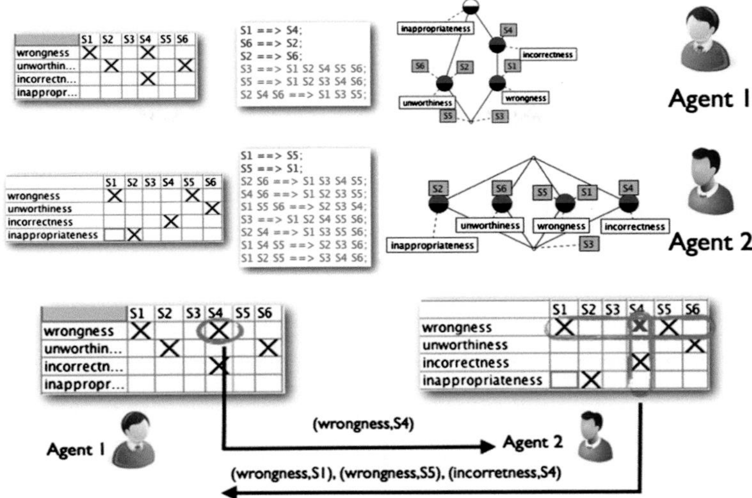

Fig. 1 Agents for the example (*top*) and Intent-extent communicative act (*bottom*)

4.1 Modeling WordNet Emergence by Intent-Extent Games

The basic model is based on the relative similarity between a *formal concept* and a *synset*. In this game, the communicative act consists in direct interchange of lemmas (objects) and synsets (attributes) between the speaker and the listener.

Communicative act: Each step, the speaker s randomly chooses a pair (o_i, a_j) from its local knowledge (formal context) and sends it (request) to the listener. The answer of the listener will consist of two sets $intent(o_i)$ and $extent(a_j)$ (relative to its own formal context). In case the listener does not have any information on o_i or a_i, it will return an empty set and will add the pair (o_i, a_j) to its local knowledge (see Fig. 1). Figure 2 shows four different states of the communicative process for this model.

Collective knowledge emergence: In order to detect and study the emerging knowledge due to agents' interactions in this communication game, the *collective knowledge* can be measured. In each time step the *error rate* between the global and the collective knowledge is measured as the difference between these.

Convergence criteria: the communication game ends when the *collective knowledge* emerged from agents' interactions is equal to the *global knowledge*. It is worthy to note that the convergence rate of the game highly depends on the collective knowledge threshold CK_{th} considered.

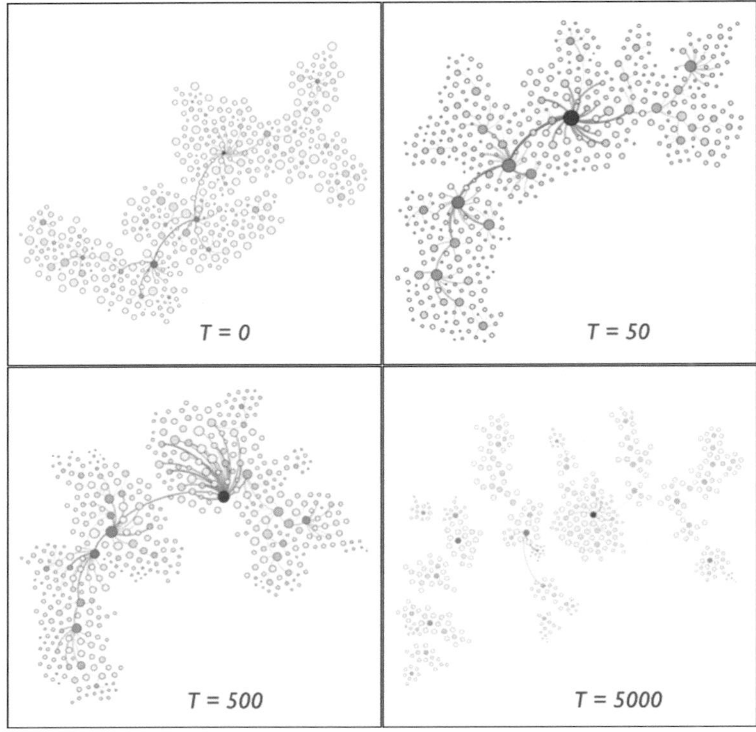

Fig. 2 Evolution of agents knowledge within the network. Node *color* denotes connectivity (darker the colour, higher the degree) and size denotes the amount of knowledge achieved. Edge *thickness* denotes knowledge difference between two agents (thicker the edge, bigger the difference)

4.2 Modeling Emergence by Implication Bases Games

This model aims to exploit the implication bases in knowledge detection tasks. In a first approach the communication process goal is the emergence of the collective knowledge by detecting and eliminating inconsistencies within agents' local knowledge. The communicative process in this case concerns to consistency issues. Each agent will contrast its knowledge with others, in order to detect and fix inconsistencies (see Fig. 3).

Communicative act: The speaker computes the SB of its local formal context $M^d(s)$, randomly chooses a rule L from it and sends it (request) to the listener l. If $M^d(l) \models L$ then it returns a positive answer and finish the communicative act. Otherwise l returns a negative answer from this context and sends to the speaker a counterexample, that is, a pair set $\{o\} \times \{o\}' \subseteq I^d_l$ such that $\{o\}' \not\models L$ in $M^d(l)$ and the speaker adds it to its local knowledge. This communicative act is similar to one step of the *attribute exploration* (cf. [8]).

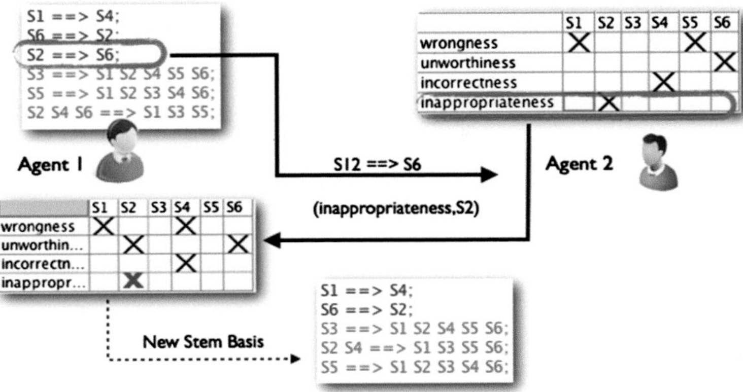

Fig. 3 Communicative act in implication basis game

Collective knowledge emergence: Since the model works with implication basis, collective knowledge has to be considered as the *collective consistent knowledge*. That is, the implication base associated with the collective knowledge has to be consistent with the base associated with the global knowledge. Since in this case, game rules have changed, the collective knowledge is measured by its consistency. In order to estimate the soundness of the emerged knowledge, it should be verified whether the true implications within the collective knowledge can be entailed from the basis of the original vocabulary. Collective implication base has to be computed from the collective knowledge. Then, its consistency is verified according the global implication base.

Convergence criteria: The game ends when it reaches the equilibrium, that is to say, when there are no more inconsistencies between agents' local knowledge. It is worthy to note that from this model, the global knowledge does not emerge (it is not the aim), but knowledge consistent with the global one.

4.3 *Modeling WordNet Emergence by Hybrid Games*

To model (in terms of language convergence and consistency) a sound language emergence, a hybrid model has been considered. Particularly, to complete the implication bases game, which stabilizes before agents' local knowledge converges to the global one. The consistency-based approach is interesting but does not provide full emergence of collective knowledge. Thus in this hybrid approach the two communicative act formerly presented will be used, one providing consistency (stem basis interactions) and the second providing direct information exchange (intent-extent interactions).

Communicative act: to combine both games, by selecting the communication type with a certain probability q. For low values of q, the agent will communicate mainly by means of the intent-extent based method (*diffusing agent*) and for high values, the agent will use mainly the implication base based method (*conciliatory agent*).

Collective knowledge emergence: In this case the both notions of collective knowledge above mentioned should be considered, in order to evaluate both, consistency and knowledge emergence.

Convergence criteria: game ends when both objectives, the emergence of consistency and the global knowledge, are achieved within agents' collective knowledge.

5 Experiments and Discussion

To study the convergence of agents' collective knowledge for each of the aforementioned NG many experiments have been carried out. The *WordNet* subset (a connected subset) selected for the experiments has a relatively small size (around 400 lemma-synset pairs) due to the high computation time of implication bases for huge formal contexts. Figure 4 shows the results of two of these experiments. Plots show the emergence of collective knowledge tending to the global knowledge (shown as % of the global knowledge). In each experiment we compare the convergence rate for the grid-based world (deeply studied in [2]) and for the network-based world. Figure 4 (top) shows knowledge convergence for the intent-extent game and Fig. 4 (bottom) for the hybrid probability-based game with $q = 0.5$. The mean collective knowledge $f(t)$ (being t the number of simulation steps) follows from $f(x) \sim a + be^{-\alpha x}$.

It should be noted that no experimental results are provided for the game based exclusively on implication bases, due to the fact that the system stabilizes before any concept exceed the collective knowledge threshold (CK_{th}) (which should be high),

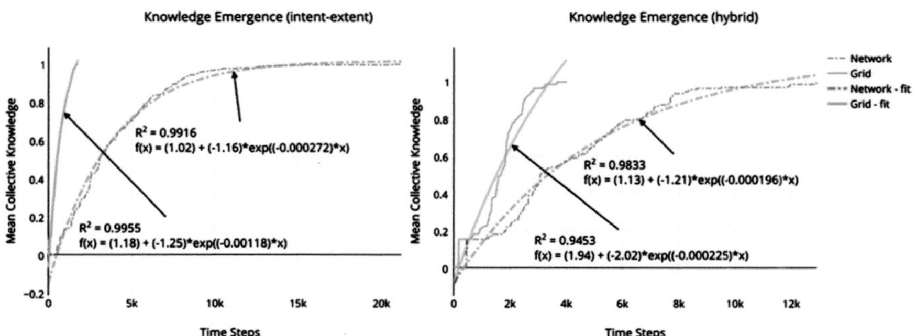

Fig. 4 Convergence of collective knowledge in communication games based on intent-extent (*left*), and hybrid games (*right*)

in order to be considered as collective knowledge. This phenomenon is similar to others in LD simulation (language competing).

Roughly speaking, intend-extend based communication produces collective knowledge fast, but will remain inconsistent (with respect to the global knowledge) until the global knowledge has been achieved. While those based on implication bases do not converge to the global knowledge, but produce partial knowledge consistent with the global one. It is worthy to note that despite the convergence rate being slower using network topology, both scenarios present the same behavior (exponential).

6 Conclusions and Future Work

In this paper a FCA-based model for lexicon emergence (and similar phenomena as sentiment emergence) in social networks is presented. Since FCA provides a solid formal semantic characterization of implicit conceptual structures of agents, the application of this model in social networks could provide some insights on the nature of information spreading and consensus in social networks. For example, the methodology presented here could be useful to model misinformation spreading in social media and networks, by enhancing other methods designed to measure the credibility of information [11]. In [2] authors show that language convergence in situated (mobile) models seems to be governed by the shared vocabulary (the mapping between words and meaning) instead of the shared language. Experiments presented in this paper show how dynamics change in models based in scale-free networks. A great amount of experiments have to be executed to deeply validate these insights.

Moreover, it could be interesting to use association rules (e.g. by Luxenburger basis [15]) instead of implications. This choice is very related with the idea of lexicon mediated by a certain confidence in the relationship.

Acknowledgments Work partially supported by TIC-6064 Excellence project (*Junta Andalucía*) and TIN2013- 41086-P (Spanish Ministry of Economy and Competitiveness), co-financed with FEDER funds.

References

1. Adamic, L.A., Lento, T.M., Adar, E., Ng, P.C.: Information evolution in social networks. CoRR(2014). arXiv:abs/1402.6792
2. Aranda-Corral, G.A., Borrego-Díaz, J., Galán-Páez, J.: Simulating language dynamics by means of concept reasoning. Bio-Inspired Models of Network, Information, and Computing Systems. Lecture Notes Institute for Computer Sciences, Social Information and Telecommunications Engineering, vol. 134, pp. 296–311. Springer (2014)
3. Aranda-Corral, G.A., Borrego-Díaz, J., Galán-Páez, J.: On the phenomenological reconstruction of complex systems-the scale-free conceptualization hypothesis. Syst. Res. Behav. Sci. **30**(6), 716–734 (2013)

4. Aranda-Corral, G.A., Borrego-Díaz, J., Giráldez-Cru, J.: Agent-mediated shared conceptualizations in tagging services. Multimed. Tools Appl. **65**(1), 5–28 (2013)
5. Barabási, A.L., Réka, A.: Emergence of scaling in random networks. Science **286**, 509–512 (1999)
6. Chli, M., de Wilde, P.: Convergence and Knowledge Processing in Multi-Agent Systems, 1st edn. Springer, London (2009)
7. Galán Páez, J., Borrego-Díaz, J.: Discovering new sentiments from the social web. CoRR (2014). arXiv:abs/1407.0374
8. Ganter, B., Wille, R.: Formal Concept Analysis: Mathematical Foundations. Springer, Berlin (1999)
9. Hadzibeganovic, T., Stauffer, D., Han, X.-P.: Randomness in the evolution of cooperation. Behav. Process. **113**, 86–93 (2015)
10. Kaur, R., Kumar, R., Bhondekar, A.P., Kapur, P.: Human opinion dynamics: an inspiration to solve complex optimization problems. Sci. Rep. **3** (2013)
11. Kumar, K.P., Geethakumari, G.: Detecting misinformation in online social networks using cognitive psychology. Hum. Centric Comput. Inf. Sci. **4**(1), 14 (2014)
12. Loreto, V., Baronchelli, A., Mukherjee, A., Puglisi, A., Tria, F.: Statistical physics of language dynamics. J. Stat. Mech. Theory Exp. **2011**(04), P04006 (2011)
13. Lozano, S.: Dynamics of social complex networks: some insights into recent research. In: Ganguly, N., Deutsch, A., Mukherjee, A. (eds.) Dynamics On and Of Complex Networks, pp. 133–143. Birkhäuser, Boston (2009)
14. Lu, Q., Korniss, G., Szymanski, B.: The naming game in social networks: community formation and consensus engineering. J. Econ. Interact. Coord. **4**(2), 221–235 (2009)
15. Luxenburger, M.: Implications partielles dans un contexte. Mathématiques, Informatique et Sciences Humaines **29** (1991)
16. Steels, L.: A self-organizing spatial vocabulary. Artif. Life **2**(3), 319–332 (1995)
17. Steels, L., McIntyre, A.: Spatially distributed naming games. Adv. Complex Syst. **1**(4), 301–323 (1999)

Dialogue Systems: Modeling and Prediction of their Dynamics

Magdalena Kacprzak, Anna Sawicka and Andrzej Zbrzezny

Abstract The aim of this paper is to propose a new model for dialogue games. This model applies interpreted systems which are commonly used to define the language semantics in epistemic and temporal logics. Such an approach allows to apply modern model checking techniques for verification properties of dialogue systems. In this way, we obtain a new method for analyzing these systems and predicting of their dynamics during the process of communication.

Keywords Formal dialogue system · Protocol · Verification

1 Introduction

Formal dialogue systems [19, 21, 24] are one of the most important trends in contemporary research on the theory of argumentation. In this approach dialogue is treated as some kind of a game played between two parties. Rules of this game define principles for the exchange of messages between parties in order to meet some assumptions. For example, in Hamblin system [4, 10] these rules prevent making argumentative mistakes (fallacies) such as begging the question (*circulus vitiosus*) [1, 22]. In contrast, Lorenzen system [13] is intended to validate formulas of some logic (e.g. classical propositional logic [8, 25] or intuitionistic logic). Different authors adopt different sets of rules or different conceptual apparatus. Generally, however, each game should have three basic categories of rules. The first one defines

M. Kacprzak (✉)
Bialystok University of Technology, Bialystok, Poland
e-mail: m.kacprzak@pb.edu.pl

A. Sawicka
Polish-Japanese Academy of Information Technology, Warsaw, Poland
e-mail: asawicka@pja.edu.pl

A. Zbrzezny
Jan Długosz University, Czestochowa, Poland
e-mail: a.zbrzezny@ajd.czest.pl

A. Abraham et al. (eds.), *Proceedings of the Second International Afro-European Conference for Industrial Advancement AECIA 2015*, Advances in Intelligent Systems and Computing 427, DOI 10.1007/978-3-319-29504-6_40

a set of moves that players can perform during the game, that is which speech acts (locutions) he/she is allowed to use. These actions express communication intentions of players. For example, the dialogue system may assume that a player can: claim something, ask, agree (concede), argue, withdraw (reject something). Rules regulating legal moves are called *locution rules*. In some games (c.f. Lorenzen game [11, 13]) players have different sets of possible moves or different restrictions on these moves. The second type of rules defines possible answers to legal moves. For example, if a player claims some sentence T, the opponent can agree with him by performing *concede* T or challenge T by performing *why* T. These rules are called *structural rules*, *possible answers rules* or *protocol*. Actually they determine a step-by-step progress of each play in accordance with rules of the game. The third group of rules applies to effects of actions. In dialogue games all actions relate to verbal expression: confirming, rejecting, questioning or arguing them. Therefore, only public declarations (commitments) of players are changed. As a result, a set containing publicly uttered statements is assigned to each player. The result of an action is a change in this set, i.e. addition of some new statement or removing the statement which is already in this set. These rules are called *commitment* or *effect rules*. In some systems, there are other categories of rules, such as *termination* and *winning rules*. Termination rule defines conditions under which the dialogue ends. Winning rule usually determines the winner. In some systems, these rules are abandoned. This is because they are designed not to indicate the winner, but to describe certain mechanisms occurring in dialogues (c.f. [14]).

The aim of our research is to develop methods of verification of protocols for dialogue games. To achieve this we are going to adopt verification techniques commonly used to test the properties of multi-agent systems (MAS). Especially, we are interested in model checking methods [15]. Main approaches in this field are based on combining bounded model checking (BMC) with symbolic verification using translations to either ordered binary decision diagrams (BDDs) [6] or propositional logic (SAT) [17]. Properties of MAS are expressed in logics which are combinations of the epistemic logic with either branching [20] or linear time temporal logic [5]. These logics are interpreted over both the subclass of interpreted systems (IS) called interleaved interpreted systems (IIS) [12] and interpreted systems themselves [3]. IIS are systems in which only one action at a time is performed in a global transition. It is the ideal semantics also to interpret the properties of dialogue games. Therefore, the first step of the research described in this paper is to give a new model for dialogue systems, i.e., to translate locution, structural and effect rules into the concept of interpreted systems. The problem of translating one terminology to another is nontrivial, also because some concepts are understood differently. For example, a protocol in dialogue systems is a set of rules of possible answers to specific actions. In MAS this term is used to the program the agent is executing and a set of possible actions is assigned to every state (not an action). Another problem is that the dynamics of a computer system is described as a change of its state. This means that after execution of an action the system moves from one state to another state. In MAS the change may affect local state of an agent, global state of the system or state of the environment. It is different in dialogue games where actions are locutions which can change players' commitments and it is the only way to affect a state of the system. In addition, the dynamics of MAS depends on knowledge or beliefs of agents. In dialogue systems we do not focus on these attributes. The choice of participant actions is based on public declarations of players, even if they do not comply with their beliefs.

This article is not the first work showing an alternative way of modelling dialogue systems. In [7] Mackenzie system was described in the terminology of game theory. The aim of this translation was to analyze dominant strategies in dialogues and test the existence of Nash equilibrium. The approach in this study will be used to examine correctness of protocols for dialogue games, to check their properties, i.e. whether Hamblin's system in fact prevents from argumentative mistakes and, consequently, to describe and analyze the dynamics of these systems. Our research is directly applicable in the analysis and evaluation of argumentative dialogues as well as protocols of communication in decision support systems like in [9].

2 Formal Framework for Natural Dialogue

There are many different dialogue systems [14, 16, 23]. To show the reconstruction of one of them in terms of interpreted systems we choose the system given in [8]. It is based on general specification for argumentation dialogue systems proposed by [18]. These dialogues are assumed to be for two parties arguing about a single dialogue *topic* T—the *proponent* P who defends T and the *opponent* O who challenges T. Both are equipped with a set of commitments that are understood as publicly incurred standpoints. Commitments are expected to be defended upon a challenge.

Definition 1 A *dialogue system* for argumentation is a pair $(\mathcal{L}, \mathcal{D})$, where $\mathcal{L}$ is a logic for argumentation and $\mathcal{D}$ is a dialogue system proper.

The elements of the above top-level definition are defined as follows.

Definition 2 A *logic for argumentation* $\mathcal{L}$ is a tuple $(L_t, R, Args, \rightarrow)$, where L_t is a logical language called the *topic language*, R is a set of *inference rules* over L_t, $Args$ is a set of arguments, and $\rightarrow$ is a binary relation of defeat defined on $Args$.

For any argument $A \in Args$, $prem(A)$ is the set of premises of A and $conc(A)$ is the conclusion of A.

Definition 3 A *dialogue system proper* is a triple $\mathcal{D} = (L_c, Pr, C)$ where L_c is a communication language, Pr is a protocol for L_c, and C is a set of effect rules of locutions in L_c.

The dialogue system from [8] assumes that in the topic language there is distinguished a sentence "The formula θ is a propositional tautology". This expression is used for questioning deductive argumentation. For convenience, we introduce the following abbreviations. Let $Taut(\theta)$ be short for "The formula θ is a propositional tautology". This sentence can be true or false. We do not state here that θ actually is a tautology. Adding this expression to the language allows for explicit questioning arguments performed by players, in which they use formal reasoning. Then, the set of arguments *Args* is a set of pairs $A = (Prem(A), Conc(A))$, where $prem(A)$ is a set of premises (a finite set of propositional formulas) and $conc(A)$ is a conclusion (a propositional formula) such that the formula $\bigwedge_{a \in prem(A)} \Rightarrow conc(A)$ is a propositional tautology.

The locution, protocol, and effect rules for the dialogue game are given below.

Locution rules The communication language assumes the following locutions:

PL1 Claim *claim* φ is performed when a player asserts that sentence φ is true and his antagonist does not have this sentence in his commitment base.

PL2 Concession *concede* φ is performed when a player asserts that sentence φ is true and his antagonist has this sentence in his commitment base.

PL3 Challenge *why* φ is performed when a player asks for proof of φ.

PL4 Argumentation φ *since* $\{\psi_1, \dots, \psi_n\}$, especially φ *since* $\{\psi_1, \dots, \psi_n, Taut(\theta)\}$, where $\theta = (\psi_1 \wedge \dots \wedge \psi_n \rightarrow \varphi)$, is performed when a player justifies statement φ with a set of premises $\{\psi_1, \dots, \psi_n\}$ and the inference rule corresponding to the formula θ, which he/she considers to be a tautology. In this locution a player may use an argumentation (e.g. inference rule) which is not correct and does not correspond to a tautology.

PL5 Retraction *retract* φ is performed when a player withdraws the statement that sentence φ is true.

Protocol rules The protocol *Pr* satisfies the following rules, where $\overline{Pr}$ is the partial function defined only for *legal dialogues*, T is the *turntaking function*, for a move m, $pl(m)$ is the *player* of the move, $s(m)$ is the *speech act* performed in m, and $t(m)$ is the *target* of m, and $s \in \{P, O\}$, d is a legal dialogue, $m \in \overline{Pr}(d)$, and $\varphi, \psi_1, \dots, \psi_n, \theta \in L_t$. For details see [8].

PP1 $pl(m) \in T(d)$.

PP2 If $d \neq d_0$ and $m \neq m_1$, then $s(m)$ is a reply to $s(t(m))$ according to L_c.

PP3 If m replies to m', then $pl(m) \neq pl(m')$.

PP4 If there is an m' in d such that $t(m) = t(m')$, then $s(m) \neq s(m')$.

PP5 For m' that surrenders to $t(m)$, m' is not an attacking counterpart of m.

PP6 If $d = \emptyset$, then $s(m)$ is of the form

(a) *claim* φ or
(b) φ *since* $\{\psi_1, \dots, \psi_n, Taut(\theta)\}$.

PP7 If m concedes the conclusion of an argument made in m', then m' does not reply to a *why* move.

PP8 If $s(m)$ is *claim* φ, then $s(m')$ for $m' \in \overline{Pr}((d, m))$ is of the form

(a) *why* φ (attack) or
(b) *concede* φ (surrender).

PP9 If $s(m)$ is *why* φ, then $s(m')$ for $m' \in \overline{Pr}((d,m))$ is of the form

(a) φ *since* $\{\psi_1, \dots, \psi_n, Taut(\psi_1 \wedge \dots \wedge \psi_n \Rightarrow \varphi)\}$ (attack) or
(b) φ *since* $\{Taut(\theta(q_1, \dots, q_n)), Taut(\theta(\alpha_1/q_1, \dots, \alpha_n/q_n)) = \beta\}$ for some formulas $\alpha_1, \dots, \alpha_n$ if $\varphi = Taut(\beta)$ (attack) or
(c) *retract* φ (surrender).

PP10 If $s(m)$ is φ *since* $\{\psi_1, \dots, \psi_n, Taut(\theta)\}$, then $s(m')$ for $m' \in \overline{Pr}((d,m))$ is of the form

(a) *why* α, where $\alpha \in \{\psi_1, \dots, \psi_n, Taut(\theta)\}$ (attack) or
(b) $\neg\varphi$ *since* $\{\beta_1, \dots, \beta_n, Taut(\beta_1 \wedge \dots \wedge \beta_n \Rightarrow \neg\varphi)\}$ or
(c) *concede* α, where $\alpha \in \{\varphi, \psi_1, \dots, \psi_n, Taut(\theta)\}$ (surrender).

PP11 If $s(m)$ is *concede* φ or *retract* φ, then $s(m')$ for $m' \in \overline{Pr}((d,m))$ is

(a) a reply (attack or surrender) to some earlier move by the other player,
(b) $\overline{Pr}((d,m)) = \emptyset$.

Rule **PP1** states that a move is legal only if made by the player-to-move. **PP2** states that a replying move must be a reply to its target according to L_c. **PP3** states that one cannot reply to one's own moves. Rule **PP4** states that if the player backtracks, the new move must be different from the original one. Finally, **PP5** states that surrenders should not be revoked. Rules **PP6** and **PP7** are inspired by liberal dialogues (see [18]). **PP6** states that each dialogue begins with either a claim or an argument. The initial claim or, if a dialogue starts with an argument, its conclusion, is the topic of the dialogue. **PP7** restricts concession of the conclusion of an argument to conclusions of counterarguments. Rules **PP8–PP11** describe possible moves after particular locution. Note that every move replies to some earlier move by the antagonist and is either an attack or a surrender.

Effect rules In our dialogue game there are two participants. Thus, we need to define commitment functions for each of them: $C_s : M^{<\infty} \times \{P, O\} \to 2^{L_t}$, where $s \in \{P, O\}$ and $M^{<\infty}$ is a set of finite sequences of moves. Locution rules, however, do not depend on the role played by the performer of the locution. The effects on commitment sets after the execution of particular moves are described below, where s denotes the speaker, $(m_0, \dots, m_n)$ is a legal dialogue, and $\varphi, \psi_1, \dots, \psi_n, \theta \in L_t$.

PE1 If $s(m_n) = claim\ \varphi$, then $C_s(m_0, m_1, \dots, m_n) = C_s(m_0, m_1, \dots, m_{n-1}) \cup \{\varphi\}$, i.e. after *claim* φ the formula φ is added to s's commitment set.
PE2 If $s(m_n) = concede\ \varphi$, then $C_s(m_0, \dots, m_n) = C_s(m_0, \dots, m_{n-1}) \cup \{\varphi\}$, i.e. after *concede* φ the formula φ is added to s's commitment set.
PE3 If $s(m_n) = why\ \varphi$, then $C_s(m_0, m_1, \dots, m_n) = C_s(m_0, m_1, \dots, m_{n-1})$, i.e. after *why* φ s's commitment set does not change.
PE4 If $s(m_n) = \varphi\ since\ \{\psi_1, \dots, \psi_n, Taut(\theta)\}$, then $C_s(m_0, m_1, \dots, m_n) = C_s(m_0, m_1, \dots, m_{n-1}) \cup \{\varphi, \psi_1, \dots, \psi_n, Taut(\theta)\}$, i.e. after this locution the formulas $\varphi, \psi_1, \dots, \psi_n, Taut(\theta)$ are added to s's commitment set.
PE5 If $s(m_n) = retract\ \varphi$, then $C_s(m_0, m_1, \dots, m_n) = C_s(m_0, m_1, \dots, m_{n-1}) \setminus \{\varphi\}$, i.e. after *retract* φ the formula φ is deleted from s's commitment set.

3 Example of Formal Natural Dialogue

The following example illustrates how formal natural dialogue can be represented in our dialogue game. This dialogue is between two friends Bob and Tom:

[1] Bob: John is unhappy.

[2] Tom: Yes, he is.

[3] Bob: I think that when John is unhappy, he drinks a lot.

[4] Tom: Why do you think so?

[5] Bob: Hmm, I'm not sure. Maybe it's not even true...

[6] Tom: I think it's not true that when John is unhappy he drinks a lot.

[7] Bob: Why do you think so?

[8] Tom: I think it's not true because we know that when John drinks a lot, he is in a good mood, and it's not true that if John is in a good mood, then he is unhappy, right? I think that these two facts make it untrue that when John is unhappy, he drinks a lot.

[9] Bob: Why do you think such reasoning is correct?

[10] Tom: It seems that I was wrong, such reasoning is not correct.

[11] Bob: So why do you think that it's not true that if John is unhappy, he drinks a lot?

[12] Tom: I don't think so anymore.

[13] Bob: So again, if John is unhappy, he drinks a lot.

[14] Tom: Why do you think so?

[15] Bob: I think this is true even though, as you have said, when John drinks a lot he is in a good mood, and it's not true that if John is in a good mood, then he is unhappy. I think that these two facts surprisingly implicate that when John is unhappy, he drinks a lot.

[16] Tom: I must admit that it seems you're right.

[17] Bob: John must be drinking a lot because when John is unhappy, he drinks a lot, and as you said, he is unhappy. I believe that these two facts implicate that he is drinking a lot.

[18] Tom: Yes, John is drinking a lot.

To construct the game based on this dialogue let us introduce some abbreviations: u – "John is unhappy", d – "John drinks a lot", g – "John is in a good mood". Our dialogue is then as follows:

[1] B: *claim* u

[2] T: *concede* u

[3] B: *claim* $u \to d$

[4] T: *why* $u \to d$

[5] B: *retract* $u \to d$

[6] T: *claim* $\neg(u \to d)$

[7] B: *why* $\neg(u \to d)$

[8] T: $\neg(u \to d)$ *since* $\{d \to g,\ \neg(g \to u),$

$$Taut(((d \to g) \wedge \neg(g \to u)) \to \neg(u \to d))\}$$

[9] B: *why* $Taut(((d \rightarrow g) \wedge \neg(g \rightarrow u)) \rightarrow \neg(u \rightarrow d))$
[10] T: *retract* $(Taut((d \rightarrow g \wedge \neg(g \rightarrow u)) \rightarrow \neg(u \rightarrow d)))$
[11] B: *why* $\neg(u \rightarrow d)$
[12] T: *retract* $\neg(u \rightarrow d)$
[13] B: *claim* $u \rightarrow d$
[14] T: *why* $u \rightarrow d$
[15] B: $u \rightarrow d$ *since* $\{d \rightarrow g, \neg(g \rightarrow u),$
$Taut(((d \rightarrow g) \wedge \neg(g \rightarrow u)) \rightarrow (u \rightarrow d))\}$
[16] T: *concede* $u \rightarrow d$
[17] B: d *since* $\{u, u \rightarrow d, Taut((u \wedge (u \rightarrow d)) \rightarrow d)\}$
[18] T: *concede* d

Bob starts a dialogue by stating that John is unhappy (move 1) and Tom agrees with him in the next move. When Bob claims in move 3 that when John is unhappy, he drinks a lot, Tom challenges him to prove it in move 4. Because Tom hasn't got anything to support his claim, he retracts it in move 5. In the next move, Tom claims the opposite—that it is not true that when John is unhappy, he drinks a lot. When Bob challenges it in move 7, Tom argues that it is true, because there are two premises and an inference rule that prove it (move 8). In move 9 Bob decides to challenge this inference rule and in move 10 Tom retracts from claiming, that this is a tautology.

We do not analyze here what happened that Tom changed his mind, but his decision was based on result of Lorenzen's game, which took place between moves 9 and 10 (we presented this process with details in [2, 8, 10]). Lorenzens dialogue games allow players to prove that a formula is a tautology of classical propositional logic.

In the next move Bob again challenges Tom's claim from move 6. Because Tom has nothing more to support his claim, he retracts it (move 12). Then Bob claims again that he was right in move 3. When asked for justification (move 14), he gives two premises and an inference rule (move 15). In move 16 Tom accepts this argument and concedes the formula from move 15. In move 17 Bob additionally argues that John is drinking a lot by giving two premises (one of which the opponent has just conceded) and another inference rule. Tom accepts this argument and concedes that John is indeed drinking a lot (move 18).

Replies in the above dialogue game derive directly from rules PP1–PP11, e.g. after move 13, according to the rule PP8 Tom has two possible moves (*why* and *concede*) and he chooses one of them in move 14.

4 Modeling Dialogue Games in Interpreted Systems

In this section we define a new model for the dialogue game presented in Sect. 2. This model uses the concept of interpreted systems and Kripke structures. In our further research we will interpret in this model formulas of a modal logic adequate to express properties describing the dynamics of dialogue systems and properties that allow prediction of players' behavior.

The set of players of a dialogue game consists of two players: W and B, $Pl = \{W, B\}$. To each player $p \in Pl$, we assign a set of actions Act_p and a set of possible local states L_p. By $\overline{p}$ we denote the opponent of p.

Every action from Act_p can influence participant's commitments. We assume that the set Act_p contains also the special empty (null) action ε_p. Players' commitments are elements of topic language and are represented by *commitment sets*. Every action (except null action) is synonymous with locution expressed by specific player. The results of locutions are determined by commitment rules and are specified afterwards. As we describe player's local state as a set of commitments, L_p denotes possible commitment sets of player p. Next, Act denotes the Cartesian product of the players actions, i.e. $Act = Act_W \times Act_B$. The global action $a \in Act$ is a pair of actions $a = (a_W, a_B)$, where $a_W \in Act_W$, $a_B \in Act_B$ and one of these actions is the empty action. This means that players can not speak at the same time.

Let G be a set of global states. A global state $g = (C_W(g), C_B(g))$, $g \in G$ is a pair of commitment sets for the players corresponding to a snapshot of the system at a given time. Given a global state $g = (C_W(g), C_B(g))$, we denote by $C_p(g)$ the commitment set of player p (by $C_{\overline{p}}(g)$ we denote the commitment set of the p's opponent). By I we denote initial global state.

Let PV be a set of propositional variables, which can be either true or false.

An *interpreted system* for a dialogue game is a tuple

$$\text{IS} = (I, \{L_p, Act_p\}_{p \in Pl}, V)$$

where $I \in G$ is the initial global state, and $V : G \to 2^{PV}$ is a valuation function.

Locutions used in players' actions are the same as described in Sect. 2, i.e., $Act_W = Act_B = \{\mathit{claim}\ \varphi, \mathit{concede}\ \varphi, \mathit{why}\ \varphi, \varphi\ \mathit{since}\ \{\psi_1, \dots, \psi_n\}, \mathit{retract}\ \varphi\}$.

The actions of the players are selected according to a *protocol function* $Pr : Act^{<\infty} \times G \to 2^{Act}$, which maps a finite sequence of actions and a global state g to the set of possible actions. This function satisfies the following rules.

(R1) A move is legal only if made by the player-to-move.

(R2) Player cannot reply to his own moves.

(R3) If the player backtracks, the new move must be different from the original one, aside from the case when player can ask *why* α after the opponent said *retract* ϕ and ϕ was previously used to justify α.

(R4) $Pr((d, (\varepsilon_W, \varepsilon_B)), I) = \{(\mathit{claim}\ \varphi, \varepsilon_B), (\varphi\ \mathit{since}\ \{\psi_1, \dots, \psi_n\}, \varepsilon_B)\}$.

(R5) $Pr((d, (\mathit{claim}\ \varphi, \varepsilon_B)), g) = \{(\varepsilon_W, \mathit{why}\ \varphi), (\varepsilon_W, \mathit{concede}\ \varphi)\}$,
$Pr((d, (\varepsilon_W, \mathit{claim}\ \varphi)), g) = \{(\mathit{why}\ \varphi, \varepsilon_B), (\mathit{concede}\ \varphi, \varepsilon_B)\}$.

(R6) $Pr((d, (\mathit{why}\ \varphi, \varepsilon_B)), g) = \{(\varepsilon_W, \varphi\ \mathit{since}\ \{\psi_1, \dots, \psi_n\}), (\varepsilon_W, \mathit{retract}\ \varphi)\}$,
$Pr((d, (\varepsilon_W, \mathit{why}\ \varphi)), g) = \{(\varphi\ \mathit{since}\ \{\psi_1, \dots, \psi_n\}, \varepsilon_B), (\mathit{retract}\ \varphi, \varepsilon_B)\}$.

(R7) $Pr((d, (\varphi\ \mathit{since}\ \{\psi_1, \dots, \psi_n\}, \varepsilon_B)), g) = \{(\varepsilon_W, \mathit{why}\ \alpha), (\varepsilon_W, \mathit{concede}\ \beta)\}$,
where $\alpha \in \{\psi_1, \dots, \psi_n\}$ and $\beta \in \{\varphi, \psi_1, \dots, \psi_n\}$,
This rule for player B is analogous.

(R8) $Pr((d,(concede\ \ \varphi,\varepsilon_B)),g) = \{(\varepsilon_W,\varepsilon_B),(\varepsilon_W,l),(\varepsilon_W,claim\ \ \alpha),(\varepsilon_W,\alpha\ \ since\ \{\psi_1,\dots,\psi_n\})\}$, where l is the response to an earlier move and α is a new thread. This rule for player B is analogous.

(R9) $Pr((d,(retract\ \ \varphi,\varepsilon_B)),g) = \{(\varepsilon_W,\varepsilon_B),(\varepsilon_W,l),(\varepsilon_W,claim\ \ \alpha),(\varepsilon_W,\alpha\ \ since\ \{\psi_1,\dots,\psi_n\}),(\varepsilon_W,why\ \beta)\}$, where l is the response to an earlier move, α is a new thread and φ was previously used to justify β.
This rule for player B is analogous.

Locution *claim* can start a dialogue or can be introduced during the dialogue, but only if it can lead to the conflict resolution and termination of discussion. Such a *claim* should refer to previous actions in the dialogue and do not start a completely new topic, which has no connection with previous ones. Therefore *claim* cannot be introduced after locutions, which demand immediate answer (locutions: *concede*, *retract* and *since*). Also, after player's *claim*, the opponent cannot utter subsequent *claim* because the answer to the first locution should reveal opponent's attitude to player's statement.

To avoid infinite dialogues, at the beginning of the game we limit the number of possible introduced atomic sentences and the number of uttered locutions *claim*. In our system, we consider only finite dialogues, that is why we need the above rules preventing infinite ones. Similar rules are constructed for locution *since*, which can also start a new thread of the discussion.

Finally, we define *global* (partial) *evolution function* $t : G \times Act \to G$, which determines results of actions. This function is symmetrical for both players.

- $t(g,(claim\ \varphi,\varepsilon_B)) = g'$ iff $\varphi \notin C_W(g)$ and $C_W(g') = C_W(g) \cup \{\varphi\}$,
- $t(g,(concede\ \varphi,\varepsilon_B)) = g'$ iff $\varphi \in C_B(g)$ and $C_W(g') = C_W(g) \cup \{\varphi\}$,
- $t(g,(why\ \varphi,\varepsilon_B)) = g'$ iff $C_W(g') = C_W(g)$,
- $t(g,(\varphi\ since\ \{\psi_1,\dots,\psi_n\},\varepsilon_B)) = g'$ iff $C_W(g') = C_W(g) \cup \{\varphi,\psi_1,\dots,\psi_n\}$,
- $t(g,(retract\ \varphi,\varepsilon_B)) = g'$ iff $C_W(g') = C_W(g) \setminus \{\varphi\}$.

With the interpreted system we associate a Kripke *model*, which is a tuple

$$M = (G, I, T, V)$$

where G is a set of the global states, I is the initial (global) state, $T \subseteq G \times G$ is a global transition relation on G defined by: $(g,g') \in T$ iff there exists an action $a \in Act$ such that $t(g,a) = g'$, and V is the valuation function.

5 Conclusions

The paper introduces a new way of modelling dialogue games. This approach uses the concept of interpreted systems. The result is a new representation for dialogue systems and their properties. In this way, we obtain a method that allows for tracking changes in dialogues and analyzing their dynamics. This is the first step to design a software tool for automatic validation of protocols for dialogues and for verification of their properties. Furthermore, our research will be used to design a platform for realization and evaluation of arguments in trust cases and for supporting standards conformity.

Acknowledgments Kacprzak's contribution was supported by the Bialystok University of Technology grant S/W/1/2014.

References

1. Budzynska, K.: Circularity in ethotic structures. Synthese **190**(15), 3185–3207 (2013)
2. Budzynska, K., Kacprzak, M., Sawicka, A., Yaskorska, O.: Dialogue Dynamics: Formal Approach, IFS PAS (2015)
3. Fagin, R., Halpern, J.Y., Moses, Y., Vardi, M.: Reasoning about Knowledge. MIT Press, Cambridge (1995)
4. Hamblin, C.: Fallacies. Methuen, London (1970)
5. Hoek, W.V., Wooldridge, M.: Model checking knowledge and time. In: Proceedings of the 9th International SPIN Workshop on Model Checking of Software (SPIN'2002). Lecture Notes in Computer Science, vol. 2318, pp. 95–111. Springer, Heidelberg (2002)
6. Jones, A.V., Lomuscio, A.: Distributed BDD-based BMC for the verification of multi-agent systems. In: Proceedings of the 9th International Conference on Autonomous Agents and Multi-Agent systems (AAMAS'2010), pp. 675–682 (2010)
7. Kacprzak, M., Dziubinski, M., Budzynska, K.: Strategies in dialogues: a game-theoretic approach. In: Proceedings of COMMA (2014)
8. Kacprzak, M., Sawicka, A.: Identification of formal fallacies in a natural dialogue. Fundamenta Informaticae **135**(4), 403–417 (2014)
9. Kacprzak, M., Starosta, B., Węgrzyn-Wolska, K.: New approach to decision making. In: Proceedings of AECIA, Nostradamus 2015, this issue (2015)
10. Kacprzak, M., Yaskorska, O.: Dialogue protocols for formal fallacies. Argumentation, 3 (2014)
11. Keiff, L.: Dialogical logic. The Stanford Encyclopedia of Philosophy (2011)
12. Lomuscio, A., Penczek, W., Qu, H.: Partial order reduction for model checking interleaved multi-agent systems. In: Proceedings of the 9th International Conference on Autonomous Agents and Multi-Agent systems (AAMAS'2010), pp. 659–666 (2010)
13. Lorenz, K., Lorenzen, P.: Dialogische logik. WBG, Darmstadt (1978)
14. Mackenzie, J.D.: Question begging in non-cumulative systems. J. Philos. Log. **8**, 117–133 (1979)
15. Meski, A., Penczek, W., Szreter, M., Wozna-Szczesniak, B., Zbrzezny, A.: BDD-versus SAT-based bounded model checking for the existential fragment of linear temporal logic with knowledge: algorithms and their performance. Auton. Agent Multi-Agent Syst. **28**, 558–604 (2014)
16. Parsons, S., Wooldridge, M., Amgoud, L.: Properties and complexity of some formal inter-agent dialogues. J. Log. Comput. **13**, 347–376 (2003)
17. Penczek, W., Lomuscio, A.: Verifying epistemic properties of multi-agent systems via bounded model checking. Fundamenta Informaticae **55**(2), 167–185 (2003)
18. Prakken, H.: Coherence and flexibility in dialogue games for argumentation. J. Log. Comput. **15**, 1009–1040 (2005)
19. Rahman, S., Tulenheimo, T.: From games to dialogues and back: towards a general frame for validity. Games Unifying Logic, Language, and Philosophy. Springer, Netherlands (2006)
20. Raimondi, F., Lomuscio, A.: Automatic verification of multi-agent systems by model checking via OBDDs. J. Appl. Log. **5**(2), 235–251 (2007)

21. Visser, J., Bex, F., Reed, C., Garssen, B.: Correspondence between the pragma-dialectical discussion model and the argument interchange format. Stud. Log. Gramm. Rhetor. **23**(36), 189–224 (2011)
22. Walton, D.: Begging the Question: Circular Reasoning as a Tactic of Argumentation. Greenwood Press, New York (1991)
23. Walton, D.N., Krabbe, E.C.W.: Commitment in Dialogue: Basic Concepts of Interpersonal Reasoning. State University of New York Press (1995)
24. Wells, S., Reed, C.: A domain specific language for describing diverse systems of dialogue. J. Appl. Log. **10**(4), 309–329 (2012)
25. Yaskorska, O., Budzynska, K., Kacprzak, M.: Proving propositional tautologies in a natural dialogue. Fundamenta Informaticae **128**(1–2), 239–253 (2013)

An Optimization Approach for the Inverse Kinematics of a Highly Redundant Robot

Paulo Costa, José Lima, Ana I. Pereira, Pedro Costa and Andry Pinto

Abstract This paper describes a robot with 12 degrees of freedom for pick-and-place operations using bricks. In addition, an optimization approach is proposed, which determines the state of each joint (that establishes the pose for the robot) based on the target position while minimizing the effort of the servomotors avoiding the inverse kinematics problem, which is a hard task for a 12 DOF robot manipulator. Therefore, it is a multi-objective optimization problem that will be solved using two optimization methods: the Stretched Simulated Annealing method and the NSGA II method. The experiments conducted in a simulation environment prove that the proposed approach is able to determine a solution for the inverse kinematics problem.

P. Costa · J. Lima · P. Costa · A. Pinto
INESC TEC - Robotics and Intelligent Systems, Rua Dr. Roberto Frias,
4200-465 Porto, Portugal
e-mail: paco@fe.up.pt

J. Lima
e-mail: jllima@ipb.pt

P. Costa
e-mail: pedrogc@fe.up.pt

A. Pinto
e-mail: andry.pinto@fe.up.pt

P. Costa · P. Costa · A. Pinto
Faculty of Engineering of University of Porto, Rua Dr. Roberto Frias,
4200-465 Porto, Portugal

J. Lima · A.I. Pereira (✉)
Polytechnic Institute of Bragança, Campus Sta Apolónia,
5301-857 Bragança, Portugal
e-mail: apereira@ipb.pt

A.I. Pereira
Algoritmi R & D Centre, University of Minho, Campus de Gualtar,
4710-057 Braga, Portugal

A. Abraham et al. (eds.), *Proceedings of the Second International Afro-European Conference for Industrial Advancement AECIA 2015*, Advances in Intelligent Systems and Computing 427, DOI 10.1007/978-3-319-29504-6_41

A real robot formed by several servomotors and a gripper is also presented in this research for validating the solutions.

Keywords Robot Manipulators · Pick-and-place · Multi-objective Optimization

1 Introduction

Architectural designs are increasing its complexity and its construction are becoming more and more difficult to implement. The awkwardness of placing components (typically bricks) fulfilling the design requirements can be solved resorting to robotics manipulation. An example of a non-standard architectural construction is presented in Fig. 1.

There are several methods already available in the market for assembling this type of structures: on the one hand, well known industrial manipulators can be used for pick-and-place operations. These manipulators have been used for several years in industrial environment and presents usually 6 degrees of freedom (6-DOF). They are an expensive solution to develop architectural arts prototypes.

On the other hand, a novelty approach based on a cable-driven robot was recently presented in [11]. The approach based on cable driven robots appears to be an interesting solution because its construction is relatively simple and inexpensive (multiple cables attached to a mobile platform) and they are easy to transport, assembly and disassemble in different construction sites due to the lightweight of the cables. Moreover, it provides large workspace, high payloads and reliable stiffness in lateral directions under external disturbances [12].

One of the main drawbacks for the cable robots applied to the architectural arts are the collision of the suspended cables within the art. A collision avoidance system is a strong possibility to avoid this type of situations however, there are situations without a proper solution. This paper presents a preliminary approach based on a 12-DOF robot manipulator that solves the collision problems. The robot is composed by an amount of servomotors installed in a consecutive and perpendicular manner

Fig. 1 Example of non-standard architecture

which makes it possible to place the end-effector in a desired (x, y, z) position with a (*pitch*, *roll*, *yaw*) orientation however, the angle for each joint must be determined. By this way, the inverse-kinematics complex model can be avoided using a physics engine (open dynamics engine—[19]).

In fact, the inverse kinematics mathematical model for the proposed robot (12 DOF manipulator) is too hard to find. Instead, an optimization approach can be used to determine the posture of robot that is required for the end-effector reach the target position while minimizing the servomotors effort. This work presents an optimization approach that uses a simulation model to find the state of each joint and to establish the pose of the robot. So, a multi-objective optimization problem was defined in order to minimize the euclidean distance to the target and the servomotors effort. To solve it, two numerical optimization methods were proposed: the first, combines the Scalarization Method with Stretched Simulated Annealing method and, the second use the well known NSGA II method for multi-objective optimization.

Moreover, a preliminary but realistic implementation of the robot is presented. This real prototype allows to validate the proposed optimization approach.

The paper is organized as follows: the related work about hyper-redundant robots and cable-driven robots are presented in Sects. 2 and 3 presents the robot model (direct and inverse kinematics). Section 4 addresses the optimization methods and algorithms used for this approach. Results are presented in Sect. 5 where the proof of the concept for this approach can be verified. Finally, major conclusions of this work are discussed in Sect. 6 where the future work direction is also pointed out.

2 Related Work

Hyper-redundant robot are used in many areas such as: Industrial inspection [7], aquatic environment [22], surgery [3], and etc. This systems offer many independent degrees of freedom and snake robots, elephant trunk robots, serpentine robots are some of representative cases. In related literature many different hyper-redundant robot are built with different principles.

One of the must approach to build a hyper-redundant robot is connecting rigid links via actuated revolute joint in a join. Another approach is parallel mechanism to connect several links together [23]. Other Hyper-redundant robot have different mechanical solutions: In [14] is used cylinder rubber pieces to construct snake robot joins, an elephant trunk robot is built in [21] features a segmented 'backbone' with a total of 32 degrees of freedom. Actuation is provided via a series of tendons routed through the structure. A slim robot were build in [13] this looks like a snake robot and composes units can stretch, shrink and bend actively is used to inspection of pipelines in plants. In this work the authors propose a new concept (bridle drive). Each unit has a bridle bellows composed of a large caliber bellows and wire lock system. A good review of the prior work had been presented in [17].

One of the major problems to solve in systems with many degrees of freedom is the resolution of inverse Kinematics. To solve inverse Kinematics there are several approaches, which can be divided into three main categories: algebraic [16] (Nonlinear optimization, Jacobian pseudo inverse,..), evolutionary computation [8] (Fuzzy logics, Artificial Neural Network, Genetics Algorithm,..) and geometric approach [9].

Cable-driven robots, also known as cable-array or cable suspended robots, control an end-effector using multiple actuated cables [1, 20]. Cables are controlled usually by a positioning system that actuates in motors for rolling and unrolling cables. These robots are capable of performing many manipulation tasks and they have several conveniences over typical robotic manipulators [5]: a smaller number of moving parts, a lower level of visual intrusion [6], a larger working area and a higher payload ratio relative to the robots weight. On the other hand, cable interferences, inaccuracies at the end-effector due to cable stretch, and the limited force in the downward directions are some the disadvantages of cable-driven robot [20]. Currently, there is a small number of cable-driven robotic systems available on the market of sports and entertainment [1], namely, the SkyCam [18] and the Cablecam [2].

3 Robot Model

This section demonstrates the kinematics model of the robot that is proposed in this research, see Fig. 3. As can be noticed in Figs. 2 and 3, the robot has 12 DOF since it is formed by 12 servomotors placed in a consecutive and perpendicular manner.

3.1 Kinematics Model

The Denavit and Hartenberg (DH) is a methodology that gives the standard kinematics characterization for robotic manipulators. It resorts to the nature of the joint

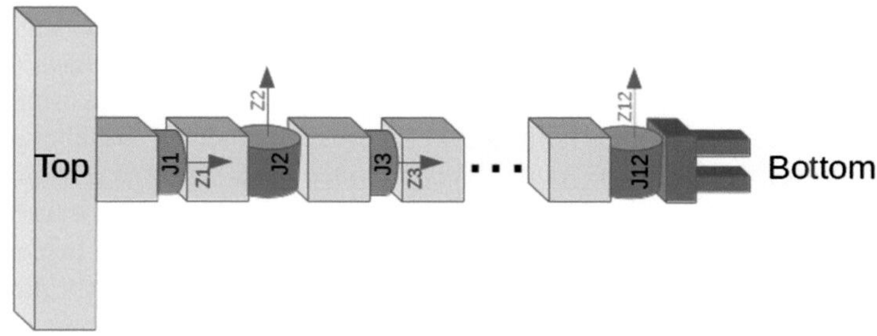

Fig. 2 Joint representation for the kinematics model of the robot

Fig. 3 The proposed robot with 12 DOF in a pick-and-place operation

(revolution or prismatic) and makes it possible to describe the final position (and orientation) of the robotic-arm based on the joint configuration (its the value).

According to the frames described in Fig. 2, the DH parameters of the proposed robot are depicted in Table 1.

The transformation matrix between the frames i and $i-1$ can be defined as:

$$^{i-1}T_i = \begin{bmatrix} c_{\theta_i} & -s_{\theta_i}.c_{\alpha_i} & s_{\theta_i}.s_{\alpha_i} & a_i.c_{\theta_i} \\ s_{\theta_i} & c_{\theta_i}.c_{\alpha_i} & -c_{\theta_i}.s_{\alpha_i} & a_i.s_{\theta_i} \\ 0 & s_{\alpha_i} & c_{\alpha_i} & d_i \\ 0 & 0 & 0 & 1 \end{bmatrix} \quad (1)$$

Table 1 Denavit-Hartenberg parameters of a 12 DOF robotic arm, where a_i is the distance between the origin of the frame o_i and z_{i-1} (is represented in blue on Fig. 2), α_i is the orientation between z_i and z_{i-1}, d_i is the distance between the x-axis of segment i and the origin o_{i-1}, and finally, θ_i is the orientation between x_i and x_{i-1}

i-link	a_i	α_i	d_i	θ_i
0	0	0°	0	0°
1	0	0°	0	θ_1^*
2	a_1	+90°	0	θ_2^*
3	a_2	−90°	0	θ_3^*
4	a_3	+90°	0	θ_4^*
...	...	...	...	...
12	a_{11}	+90°	0	θ_{12}^*

In this way, it is possible to obtain the forward kinematics model of the robotic-arm in Eq. 2, and by considering the transformations matrices of the successive links (that can be defined through Eq. 1 and the parameters of Table 1). As can be noticed, this equation gives a single solution for the final position of the robotic-arm (end-effector) given the values of θ_i of each i joint.

$$^{0}T_{12} = \prod_{i=0}^{11}(^{i}T_{i+1}) \tag{2}$$

3.2 The Problem of Inverse Kinematics

Although the solution of the kinematics model be straightforward, the same cannot be said for the inverse kinematics model: when a final position (orientation) is desired and the values of each joint (θ_i) must be determined.

The inverse kinematics can or not have a single solution (or even multiple solutions). In some cases, a unique solution can be obtained by the mathematically considerations or motion constraints (360° of joint revolution is usually not possible) however, the problem is not solvable in an efficient or close-form, which justifies the development of iterative techniques for controlling the manipulators.

For a higher number of DH parameters of a specific robotic-arm, higher is the complexity of the inverse kinematics model. In real applications, the highest number of DOFs that can be found in robotic-arms is usually 6 because the inverse kinematics problem turns it into a redundant system if more DOF is considered, see Eq. 3.

$$^{0}T_{12} = \begin{bmatrix} r_{11} & r_{12} & r_{13} & p_x \\ r_{21} & r_{22} & r_{23} & p_y \\ r_{31} & r_{32} & r_{33} & p_z \\ 0 & 0 & 0 & 1 \end{bmatrix} =^{0} T_1(\theta_1).^{1}T_2(\theta_2).^{2}T_3(\theta_3)...^{11}T_{12}(\theta_{12}) \tag{3}$$

where r_{kl} is the rotation and p_w is the position information, where $k,\ l \in \{1, 2, 3\}$, and $w \in \{x, y, z\}$. The inverse kinematic of the first and last joint (θ_1 and θ_{12}) can be found in Eqs. 4 and 5 (respectively) as a function of the known elements of Eq. 3.

$$\left[^{0}T_1(\theta_1)\right]^{-1}.^{0}T_{12} =^{1} T_2(\theta_2).^{2}T_3(\theta_3)...^{11}T_{12}(\theta_{12}), \tag{4}$$

$$\left[^{0}T_1(\theta_1).^{1}T_2(\theta_2).^{2}T_3(\theta_3)...^{10}T_{11}(\theta_{11})\right]^{-1}.^{0}T_{12} =^{11} T_{12}(\theta_{12}). \tag{5}$$

These two equations focus in only two joints (which can be extended to the others) however, they depict the complexity of the inverse kinematics problem since it requires to solve 12 simultaneous nonlinear equations, considering $r_{11}, r_{12}, \ldots, r_{33}$, p_x, p_y, p_z and the fixed link parameters.

4 Optimization Procedure

The presented optimization problem requires a multi-objective approach where it is necessary to create a path to the desired position (inverse kinematics) while minimizing the maximum effort of the servomotors. Simulation environment runs the X solution and returns the euclidean distance combined with orientation (EDO) of the end-effector (f_1). Maximum effort for servomotors (f_2) is also computed as a variable to be minimized.

In this work two global stochastic optimization methods are used. The Stretched Simulated Annealing (SSA) is a global optimization method for single-objective optimization. This method was combined with scalarization method to solve the problem. The NSGA II is a popular method based on genetic algorithm for multi-objective optimization.

4.1 *Stretched Simulated Annealing*

The multi-objective problem was formulated as a single-objective optimization problem using the scalarization method and it was used the Stretched Simulated Annealing (SSA) to solve the following resulting problem

$$\begin{aligned} \min\ & w_1 f_1 + w_2 f_2 \\ s.t.\ & w_1 + w_2 = 1 \\ & w_1, w_2 \geq 0. \end{aligned}$$

The SSA method is a well known global optimization method based on simulated annealing and on stretched technique. More details about the SSA method can be found on [15].

4.2 *NSGA II—Multi-objective Genetic Algorithm*

NSGA-II is a popular non-domination based genetic algorithm for multi-objective optimization. It is a very effective algorithm that incorporates elitism and no sharing parameter needs to be chosen a priori. The NSGA II is discussed in detail in [4].

5 Numerical Results

The numerical results were obtained using a Inter Core i7-2600 CPU 3.4 GHz with 8.0 GB of RAM.

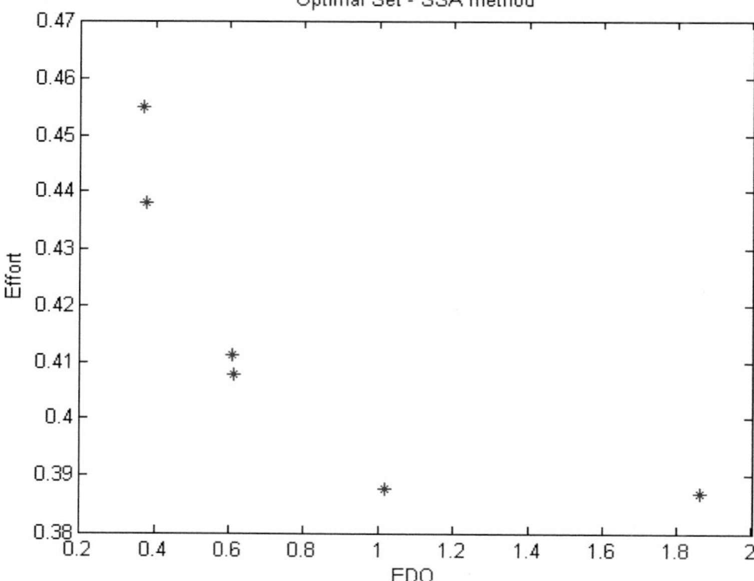

Fig. 4 Optimal set given by SSA method

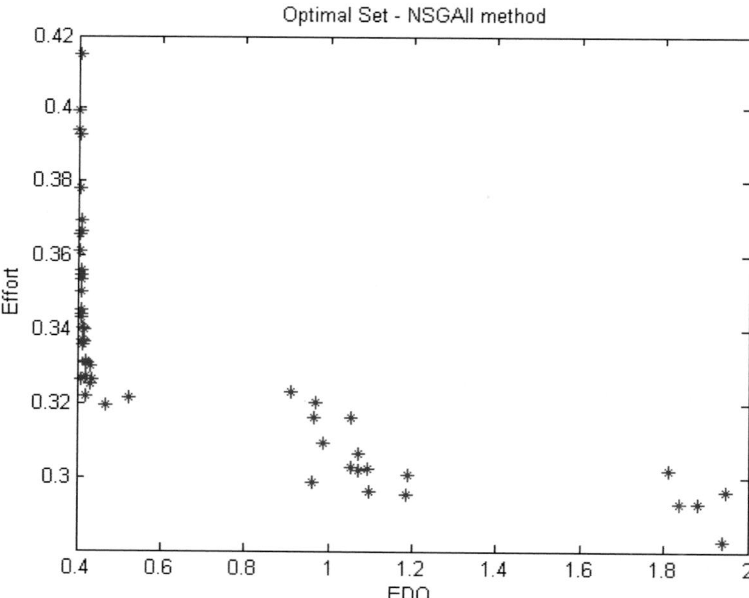

Fig. 5 Optimal set given by NSGA II method

The SSA and NSGA II methods were implemented in MatLab environment [10]. For stop criteria it was considered the maximum number of function evaluations equal to 5000. The optimal set given by SSA method can be observed in Fig. 4.

The minimum value founded by SSA method for the EDO it was 0.3694 needing 0.4549 of maximum effort from servomotors.

The optimal set given by NSGA II method is presented in Fig. 5.

The minimum value for the EDO founded by NSGA II method was 0.4020 needing 0.3993 of servomotors maximum effort.

To compare the solutions obtained by the two methods it was used the measure $M = 0.75f_1 + 0.25f_2$.

The best value was obtained using the NSGA II method with value 0.3877 (considering $f_1 = 0.4080$ and $f_2 = 0.3266$). The best solution obtained by SSA method was $M = 0.3900$ considering the solution $f_1 = 0.3694$ and $f_2 = 0.4549$.

It is possible to conclude that SSA method provides the best solution considering the EDO value and the NSGA II method provides the best solution in the terms of measure M.

6 Conclusions and Future Work

The research presented in this paper introduces the *Tentacle Robot* which is a 12 DOF robot. It can be used to pick and place operations and was developed having in mind Non-Standard Architecture and Construction parts. Robot pose can be computed resorting to optimization techniques that allows to avoid inverse kinematics complex problems. About the numerical optimization methods used, the best value was obtained using NSGA II method but the SSA method provided the best value for the EDO value.

As future work, the influence of an obstacle avoidance algorithm will be studied, with the purpose of avoid collision between robot body and objects parts. Other numerical method can be used to try to improve the obtained solutions.

Acknowledgments This work was been supported by FCT (Fundação para a Ciência e Tecnologia) in the scope of the project UID/CEC/00319/2013. This work is financed by the ERDF European Regional Development Fund through the COMPETE Programme (operational programme for competitiveness) and by National Funds through the FCT Fundação para a Ciência e Tecnologia (Portuguese Foundation for Science and Technology) within project "FCOMP-01-0124-FEDER-037281".

References

1. Borgstrom, P.H., Borgstrom, N.P., Stealey, M.J., Jordan, B., Sukhatme, G., Batalin, M.A., Kaiser, W.J.: Generation of energy efficient trajectories for nims3d, a three-dimensional cabled robot. In: IEEE International Conference on Robotics and Automation, 2008. ICRA 2008. pp. 2222–2227. IEEE (2008)

2. Cablecam. www.cablecam.com
3. Clark, J.: A novel flexible hyper-redundant surgical robot: prototype evaluation using a single incision flexible access pelvic application as a clinical exemplar. Surg. Endosc. **29**(3), 658–667 (2015)
4. Deb, K., Pratap, A., Agarwal, S., Meyarivan, T.: A fast elitist multiobjective genetic algorithm: NSGA-II. IEEE Trans. Evol. Comput. **6**(2), 182–197 (2002)
5. German, J., Jablokow, K.W., Cannon, D.J.: The cable array robot: theory and experiment. In: Proceedings 2001 ICRA. IEEE International Conference on Robotics and Automation Seoul, Korea, vol. 3, pp. 2804–2810. IEEE (2001)
6. Gouttefarde, M., Daney, D., Merlet, J.-P.: Interval-analysis-based determination of the wrench-feasible workspace of parallel cable-driven robots. IEEE Trans. Robot. **27**(1), 1–13 (2011)
7. Grzegorz, G., Hansen, M.G., Johann, B.: The OmniTread serpentine robot for industrial inspection and surveillance. Int. J. Ind. Robot. IR **32**(2), 139–148 (2005)
8. Ivanescu, M., Popescu, N., Popescu, D.: A variable length hyperredundant arm control system. In: Proceedings of the IEEE International Conference on Mechatronics and Automation, pp. 132–137 (2005)
9. Jamali, A. Khan, M.R., Rahman, M.M.: A new geometric approach to solve inverse kinematics of hyper redundant robots with variable link length. In: IEEE International Conference on Mechatronics (ICOM 2011), pp. 1–5 (2011)
10. MatLab. www.matlab.com
11. Moreira, E., Pinto, A., Costa, P., Moreira, A., Veiga, G., Lima, J., Sousa, J., Costa, P.: Cable robot for non-standard architecture and construction: a dynamic positioning system. In: IEEE International Conference on Industrial Technology, Seville, Spain (2015)
12. Oh, S.-R., Agrawal, S.: Generation of feasible set points and control of a cable robot. IEEE Trans. Robot. **22**(3), 551–558 (2006)
13. Ohno, H., Hirose, S.: Study on slime robot. In: Proceedings IEEE/RSJ International Conference Intelligent Robots and Systems, vol. 3, pp. 2218–2223 (2000)
14. Paap, K.L., Dehlwisch, M., Klassen, B.: GMD-Snake: A semi-autonomous snake-like robot. In: Proceedings 3rd International Symposium Distributed Autonomous Robotic Systems, Saitam (1996)
15. Pereira, A.I., Fernandes, E.M.G.P.: A reduction method for semi-infinite programming by means of a global stochastic approach. Optimization **58**, 713–726 (2009)
16. Perez, A., McCarthy, J.M.: Sizing a serial chain to fit a task trajectory using clifford algebra exponentials. In: IEEE International Conference on Robotics and Automation, pp. 4720–4726 (2005)
17. Shammas, E., Wolf, A., Choset, H.: Three degrees-of-freedom joint for spatial hyper-redundant robots. Mech. Mach. Theory **41**, 170–190 (2006)
18. Skycam. www.skycam.tv
19. Smith, R.: Open Dynamics Engine (2000). http://www.ode.org
20. Usher, K., Winstanley, G., Carnie, R.: Air vehicle simulator: an application for a cable array robot. In: Proceedings of the 2005 IEEE International Conference on Robotics and Automation, 2005. ICRA 2005, pp. 2241–2246 (2005)
21. Walker, I.D., Hannan, M.W.: A novel elephant trunk robot. In: Proceedings IEEE/ASME International Conference Advanced Intelligent Mechatronics, pp. 410–415. Atlanta (1999)
22. Yamada, H., Chigisaki, S., Mori, M., Takita, K., Ogami, K., Hirose, S.: Development of amphibious snake-like robot ACMR5. In: Proceedings of the International Symposium on Robotics (ISR'05), p. 133 (2005)
23. Yun, S.K., Ru, D.: Self assembly of modular manipulators with active and passive modules. In: Proceedings IEEE International Conference Robotics and Automation, pp. 1477–1482. Pasadena (2008)

Using Unsupervised Deep Learning for Human Age Estimation Problem

Klim Drobnyh

Abstract Automatic age estimation from facial images is a challenging problem upcoming in recent years. In this paper, we propose to use unsupervised deep learning features to improve accuracy of existing age estimation algorithms. The active appearance model and the bio-inspired features with a cascade of support vector machine classifiers were chosen to be the basic approaches. Experiments on the FG-NET age database demonstrated that adding unsupervised deep learning features improved accuracy for some basic models. For example, adding the features to the active appearance model gave the 5 % gain.

Keywords Machine learning · Age estimation · Unsupervised deep learning · Active appearance model · Bio-inspired feature · Support vector machine

1 Introduction

Age estimation has recently become an active research problem due to the existence of multiple interesting applications such as human-computer interaction, age-specific access control, age-based image retrieval and video retrieval systems, etc.

Using facial images might be the simplest way to estimate the age since human faces are considerably affected by the aging process. Therefore there are a lot of works that use facial images to solve this problem since 1999 [1].

This problem is a classic example of the pattern recognition task that can be stated as to classify a new unobserved sample into one of the predefined classes. In this case, classes are the ages and samples are the facial images.

One of the state-of-the-art approaches to perform a pattern recognition task is a deep learning. Deep learning is the set of algorithms that uses machine learning to

K. Drobnyh (✉)
Lobachevsky State University of Nizhny Novgorod, 23 Gagarin Avenue,
Nizhny Novgorod 603950, Russia
e-mail: klim.drobnyh@gmail.com

A. Abraham et al. (eds.), *Proceedings of the Second International Afro-European Conference for Industrial Advancement AECIA 2015*, Advances in Intelligent Systems and Computing 427, DOI 10.1007/978-3-319-29504-6_42

find the best feature representation of input data. A. Coats and A.Y. Ng proposed new unsupervised deep learning algorithm that uses K-means clustering [2].

In this paper, we propose to use unsupervised deep learning features to improve the accuracy of existing age estimation algorithms.

2 Basic Approaches

The majority of approaches that give the best accuracy are based on the following representations:

1. Active appearance model [3]
2. Bio-inspired features [4]

2.1 Active Appearance Model

An active appearance model is a statistical model that describes both the shape and grey-level appearance of the object of interest. This model can be generalized to almost any valid case. Brief description of the algorithm is presented below.

First of all, an active shape model is built. Input data is a training set of labelled images where each object is marked with key points.

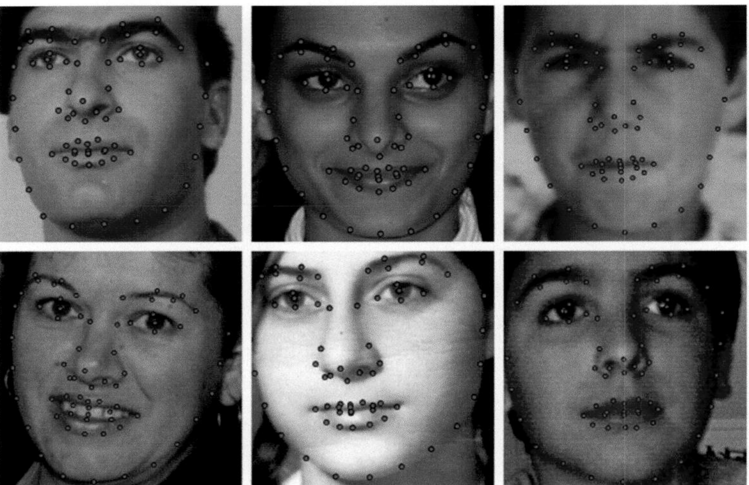

Fig. 1 Labelled face example from FG-NET database

Every set $\{(x_1, y_1), (x_2, y_2), \ldots, (x_n, y_n)\}$ can be combined to a single shape vector for each image (sample):

$$\boldsymbol{x} = (x_1, \ldots, x_n, y_1, \ldots, y_n)^{\mathsf{T}} \quad . \tag{1}$$

Then the shape model can be built. The shape of an object is normally assumed to be independent of the position, orientation and scale. So a set of training shapes needs to be aligned to minimize the sum of squared distances between each shape and the mean:

$$D = \sum |\boldsymbol{x}_i - \overline{\boldsymbol{x}}|^2 \quad . \tag{2}$$

Next, principle component analysis (PCA) is applied to the set of shapes to obtain the entire shape model:

$$\boldsymbol{x} \approx \overline{\boldsymbol{x}} + \boldsymbol{P}_s \boldsymbol{b}_s \quad , \tag{3}$$

where $\overline{\boldsymbol{x}}$ is a mean shape, $\boldsymbol{P}_s$ is a matrix containing eigenvectors of covariation matrix and $\boldsymbol{b}_s$ is a set of the shape parameters.

To build a statistical model of the grey-level appearance each sample is warped to the mean shape (using a triangulation algorithm and bilinear approximation). An example of aligned textures is shown on the Fig. 2. Gray intensity values are collected from warped images over the main shape region to obtain texture vectors $\boldsymbol{t}_i$. After that the texture vectors are aligned to the mean in the same way as the shapes.

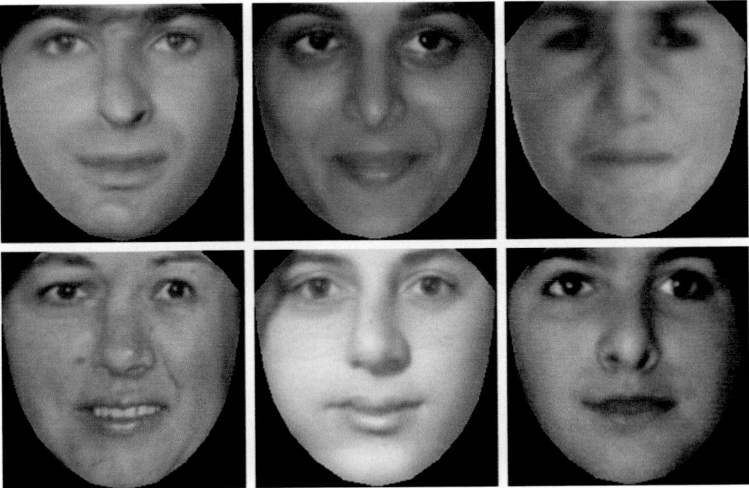

Fig. 2 Aligned textures obtained from the AAM implementation

As the next step, PCA is applied to get the entire texture model:

$$\boldsymbol{g} \approx \overline{\boldsymbol{g}} + \boldsymbol{P}_g \boldsymbol{b}_g \ . \tag{4}$$

Finally, PCA is performed once again to obtain appearance parameters.

2.2 Bio-inspired Features

Bio-inspired features (BIF) were created for approximate modelling of object recognition process in the cortex. The model contains alternated simple and complex layers. BIF model designed for the age estimation problem contains only one simple and one complex layers. The input is a gray-scale facial image. The steps of the algorithm are briefly described below:

1. Convolve the image with the appropriate Gabor filter for every predefined angle θ_i:
$$G_i(x,y) = \exp\left(-\frac{(X^2 + \gamma^2 Y^2)}{2\sigma^2}\right) \times \cos\left(\frac{2\pi}{\lambda}X\right) \ , \tag{5}$$
where $X = x\cos\theta_i + y\sin\theta_i$ and $Y = -x\sin\theta_i + y\cos\theta_i$
2. Post-process filter results (MAX pooling)
3. Calculate statistical features: standard deviation over small areas
4. Gather all features in a vector
5. Perform PCA to get a smaller dimensional representation of the features

3 Proposed Approach

We propose to use unsupervised deep learning features to improve prediction accuracy. First of all, a large set of small patches is obtained (for example, k patches from every input image). Let n be the size of the patches, every patch can be presented as an n^2 vector. Before applying K-means clusterization, patches should be normalized:

$$\tilde{\boldsymbol{x}}_j^{(i)} = \frac{\boldsymbol{x}_j^{(i)} - mean(\boldsymbol{x}_j^{(i)})}{\sqrt{var(\boldsymbol{x}_j^{(i)}) + 10}} \ , \tag{6}$$

where $\boldsymbol{x}_j^{(i)}$ are the unnormalized patches obtained from jth image.

After normalization, K-means clusterization algorithm is applied to obtain m clusters: $\left\{c_i \ : \ i = \overline{1,m}\right\}$.

For every input image:

1. Collect all possible $n \times n$ patches
2. Calculate the nearest cluster for every collected patch:

$$h_j^i = |\{\tilde{x}_j^{(l)},\ i = \arg\min_{a=1}^{m} \|\tilde{x}_j^{(l)} - c_a\|\}|\ . \tag{7}$$

3. Calculate histogram:

$$H_j = \{\tilde{h}_j^i,\ i = \overline{1,m}\};\ \ \tilde{h}_j^i = \frac{h_j^i}{\sum_{l=1}^{m} h_j^l}\ . \tag{8}$$

The calculated histograms are the new feature representation.

4 Experiments

4.1 Database

FG-NET [5] database was used for estimating the accuracy. The database contains 1002 images with wide variations of pose, expression, and lighting. There are 82 subjects who contributed from 6 to 18 images taken at different ages (0–69).

FG-NET database also contains 68-point annotations for the images (Fig. 1 shows an example).

4.2 Bio-inspired Features

Enhanced bio-inspired features [6] were chosen to be implemented. The difference is in the input data: aligned textures (100×100 pixels) obtained from the AAM implementation are used as the input data in the EBIF algorithm (Fig. 2 shows an example).

4.3 Unsupervised Deep Learning

Aligned textures (200×200 pixels) that were obtained from the AAM implementation and smoothed by Gaussian filter (diameter $= 9, \sigma = 3.2$) were used as the input images. 1500 11×11 patches from each image were extracted and the K-means algorithm was applied to get 1000 clusters. Then all possible patches were collected to build the histograms. Obtained histograms were added to other features and PCA was called to reduce feature space dimension (Figs. 3 and 4).

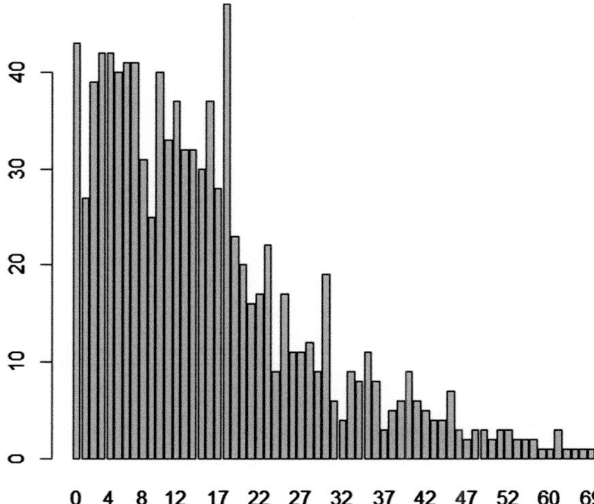

Fig. 3 Age distribution in the FG-NET database

Fig. 4 An example of clusters obtained by K-means algorithm

4.4 Classifier

To solve classification problem, a cascade of binary classifiers was built. Classifier $f_i(\boldsymbol{x})$ returns 1 if age of the subject is greater than i on the corresponding image and 0 otherwise. Then the final cascade classifier is defined as follows:

$$f(\boldsymbol{x}) = \sum_{i=0}^{68} f_i(\boldsymbol{x}) \ . \tag{9}$$

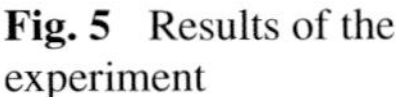

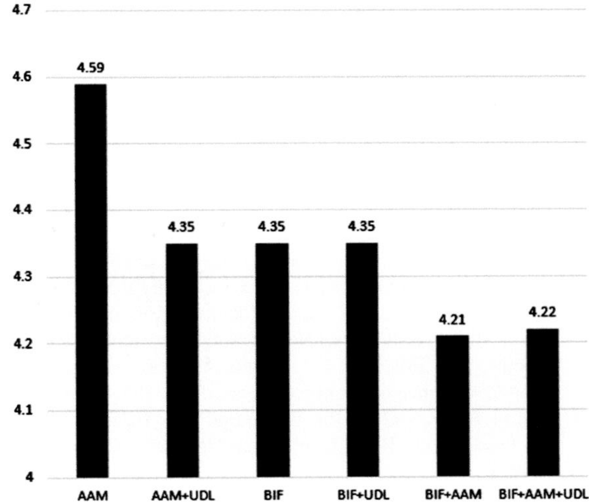

Fig. 5 Results of the experiment

Each classifier is an RBF SVM classifier. Cross-validation algorithm was used to determine optimal parameters for each classifier.

4.5 *Error Calculation*

The algorithms were tested through the leave-one-person-out approach: for each person, his or her images comprised the test set and the images of others made up the training set. Mean Absolute Error (MAE, in ages) was used to measure a model accuracy.

4.6 *Results*

Results on the Fig. 5 show that using unsupervised deep learning features changed (reduced) only the active appearance model's classification error. The results illustrate that the proposed approach can be used to improve the estimation accuracy.

5 Conclusion

In this paper, using unsupervised deep learning for human age estimation problem was proposed and investigated. Two basic approaches were implemented:

active appearance model and bio-inspired features. The experiments shown that simple adding an unsupervised deep learning features to existing model can improve prediction accuracy.

References

1. Kwon, Y.H., da Vitoria Lobo, N.: Age classification from facial images. In: Proceedings IEEE Conference Computer Vision and Pattern Recognition, pp. 762–767 (1999)
2. Coates, A., Ng, A.Y.: Learning feature representations with K-means. In: Montavon, G., Orr, G.B., Müller, K-R. (eds.) Neural Networks: Tricks of the Trade, ch. 22, Springer (2012)
3. Cootes, T.F., Edwards, G.J., Taylor, C.J.: Active Appearance Models. In: Proceedings European Conference on Computer Vision, pp. 484–498 (1998)
4. Mu, G.W., Guo, G.D., Fu, Y., Huang, T.S.: Human age estimation using bio-inspired features. In: Proceedings IEEE Conference Computer Vision and Pattern Recognition, pp. 112–119 (2009)
5. Panis, G., Lanitis, A., Tsapatsoulis, N., Cootes, T.F.: An overview of research on facial aging using the FG-NET aging database, IET Biom. J. (in press, 2015)
6. El Dib, M.Y., El-Saban, M.: Human age estimation using enhanced bio-inspired features (EBIF). In: Proceedings 17th IEEE International Conference Image Processing, pp. 1589–1592 (2010)

Two Methods of Hybrid Adaptive Control Applied on Nonlinear Plant

Jiri Vojtesek and Petr Dostal

Abstract The hybrid adaptive control with a delta External Linear Model is presented in this work. The adaptive approach is based on the choice of the External Linear Model of the originally nonlinear system parameters of which are identified recursively during the control and parameters of the controller are recomputed according to results of this estimation too. Two methods, Pole-placement and LQ approach, are compared with similar results. Advantage of these methods is that they are easily programmable and the result can be affected by tuning parameters in both methods. The polynomial approach with 2DOF control configuration satisfies basic control requirements and moreover, it could suppress overshoots of the output variable. Proposed methods were tested by the simulation on a very complex nonlinear system represented by the continuous stirred tank reactor with cooling in the jacket.

Keywords Adaptive control · Polynomial approach · Recursive identification · Nonlinear system · Simulation · 2DOF · Pole-placement method · LQ approach

1 Introduction

The control of the nonlinear technological system is always very challenging [1]. The most of the processes especially in the chemical and biochemical industry are very complex with numbers of variables and relations [2, 3].

J. Vojtesek (✉) · P. Dostal
Faculty of Applied Informatics, Tomas Bata University in Zlin,
Nam. T.G. Masaryka 5555, 760 01 Zlin, Czech Republic
e-mail: vojtesek@fai.utb.cz
URL: http://www.utb.cz/fai

P. Dostal
e-mail: dostalp@fai.utb.cz

A. Abraham et al. (eds.), *Proceedings of the Second International Afro-European Conference for Industrial Advancement AECIA 2015*, Advances in Intelligent Systems and Computing 427, DOI 10.1007/978-3-319-29504-6_43

The modelling and simulation help us not only with the understanding of the system's behaviour but it is also important preliminary step for the choice of the optimal control strategy and initial settings of the controller.

The adaptive approach [4] used in the work belongs to the rank of so called "modern control methods" with various benefits compared to the classical controllers with fixed parameters. Moreover, the adaptive control has great theoretical background with various modifications and improvements which makes this method more efficient and powerful. Adaptation from the classical control point of view could be done for example by the change of the structure, parameters of the controller, adaptation with the reference signal etc. [5]. Adaptive approach is also combined with other modern control techniques such as predictive control [6], nonlinear control [7] etc.

The "adaptivity" here is satisfied by the choice of the External Linear Model (ELM) of the originally nonlinear system, parameters of which are identified recursive and so they can copy changes that could occur. The choice of the ELM comes from the dynamic analysis which shows the dynamic behaviour of the controlled output. There are various types of ELM, such as a continuous-time (CT) and a discrete-time (DT) models. CT models are more accurate but online identification is not so easy as the estimation of DT models. Somewhere between these two types are so called Delta models (δ-models) [8] which belong to the class of DT models but as the input and output variables are recomputed to the sampling period, parameters of the δ-model approaches to the CT model [9].

The polynomial approach [10] satisfies basic control requirements such as a stability, a reference signal tracking and a disturbance attenuation. The simplest configuration is with the one degree-of-freedom (1DOF) which has controller only in the feedback part. On the other hand, two degrees-of-freedom (2DOF) control configuration and two-feedback-controllers (TFC) has controller separated into two parts—a feedback and a feedforward part. These configuration can increase the quality of the control.

Two methods compared here are both used for designing of the stable optional polynomial on the right side of Diophantine equations used in polynomial synthesis. The Pole-placement method [11] is simple tool which can be disadvantage in some cases. It is good to combine this method somehow with the estimation of ELM. This is done by the spectral factorization of the polynomial in the denominator of the ELM's transfer function. The second, more sophisticated, Linear-Quadratic (LQ) approach [12] is based on the minimization of the cost function and it provides also very good control results. Both methods offers also tuning parameters which can affect the course of the controlled output variable.

The Continuous Stirred-Tank Reactor (CSTR) with so called Van der Vusse reaction inside [13] is considered as a controlled plant in this work. The mathematical model is derived with the use of material and balances inside and it is described by the set of four nonlinear ordinary differential equations which are solved numerically. As the system offers more input and output variables, the temperature of the reactant is be controlled by the change of the input heat removal of the cooling liquid.

All experiments in this contribution are simulations made in the mathematical software Matlab. The next step is the verification of these methods on the real system.

2 Hybrid Adaptive Control

The adaptive control is one of the modern techniques which can be used for various nonlinear systems as it was shown for example in [14]. The origin of this method can be found in the nature where plants, animals or even human beings adopt their behaviour to the actual conditions and environment they are living in. You can find adapting system in everyday living for example in washing machines, dishwasher, air-condition etc.

The adaptive approach used in this work can be described in following steps:

1. Describe the system and the most important variables.
2. Measure or simulate steady-state and dynamic characteristics.
3. Choose External Linear Model from the measured/simulated dynamic characteristic.
4. Derive the controller with the use of control synthesis.
5. Do experiments on the real system or abstract model with proposed controller.

The preliminary step is the knowledge about the steady-state and dynamic behaviour. The static analysis helps us with the choice of the optional working point which could be for example combination of input variables which produces best outputs etc. On the contrary, the dynamic behaviour helps with the choice of the *External Linear Model (ELM)* of the controller, usually nonlinear, system.

The ELM could be described mathematical by the continuous-time, $G(s)$, or the discrete-time, $G(z)$ linear transfer function in the polynomial form:

$$G(s) = \frac{b(s)}{a(s)}; \qquad G(z) = \frac{b(z)}{a(z)} \tag{1}$$

2.1 Control Synthesis

The polynomial synthesis is then employed for the design of the controller. The *one degree-of-freedom (1DOF) scheme* displayed in Fig. 1 is the simplest control configuration with the controller in the feedback part.

The block $G(s)$ in Fig. 1 and similarly in Fig. 2 represents the ELM (1), $Q(s)$ and $R(s)$ are transfer functions of the controller, w is a reference signal (wanted value), $e = w - y$ denotes a control error, u is used for the action value, v is disturbance and y is controlled output variable.

It was proofed for example in [15], that the *two degrees-of-freedom (2DOF) scheme* with the feedback, $Q(s)$, and feedforward, $R(s)$, has better efficiency especially

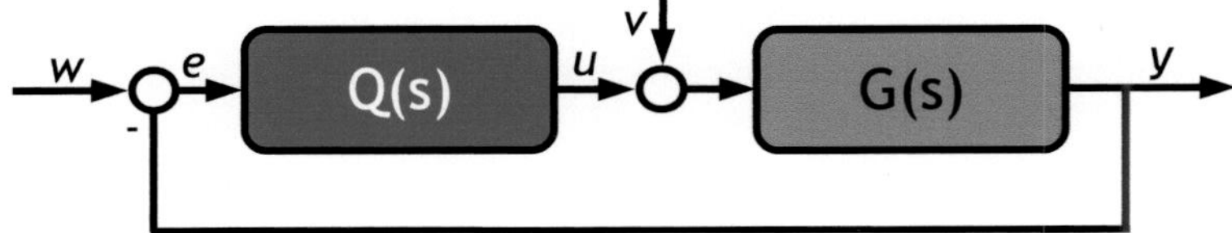

Fig. 1 The one degree-of-freedom (1DOF) control configuration

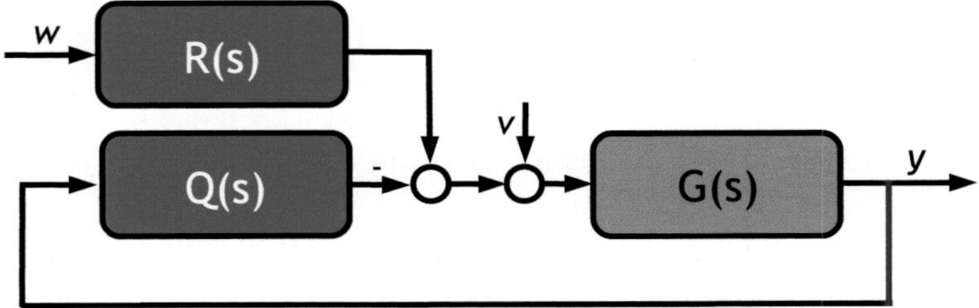

Fig. 2 The two degrees-of-freedom (2DOF) control configuration

in the overshoot compensation. This control scheme is shown in Fig. 2 and it will be used in this work.

The general form of the controlled system's transfer function $G(s)$ is designed with the help of the dynamic analysis. The adaptation is then satisfied by the identification of polynomials $a(s)$ and $b(s)$ in this transfer function $G(s)$. The recursive least-squares method can be used successfully here because it has good theoretical background [16] and it is easily programmable.

The last step of the controller design is to define relations for computation of the polynomials $q(s)$, $p(s)$ and $r(s)$ in transfer functions $Q(s)$ and $R(s)$ of the controller. As it was allready mentioned, used polynomial approach [10] satisfies basic control requirements such as a stability of the control system, a disturbance attenuation and a reference signal tracking. Moreover, it could be easily programmable and implemented in the industrial computers.

General forms of the controller's transfer functions are

$$Q(s) = \frac{q(s)}{s \cdot p(s)}; \quad R(s) = \frac{r(s)}{s \cdot r(s)}; \tag{2}$$

In 2DOF control configuration, parameters of polynomials $q(s)$, $p(s)$ and $r(s)$ are computed from the set of Diophantine equations:

$$\begin{aligned} a(s) \cdot s \cdot p(s) + b(s) \cdot q(s) = d(s) \\ t(s) \cdot s + b(s) \cdot r(s) = d(s) \end{aligned} \tag{3}$$

2.2 *Recursive Identification of ELM*

As it can be seen, the ELM (1) in a continuous-time (CT) form is expected in the previous part. This model is more accurate than the discrete-time(DT) model. On the other hand, the online identification of the CT model is much more complex then those in DT.

Compromise between CT and DT models can be found in δ-identification models [8] which belong to the class of DT models but their parameters are close to the CT ones for enough small sampling period. That is why call this approach "*Hybrid adaptive control*"—the polynomial synthesis is derived in CT and the identification uses δ-models which are in fact DT models

The delta model introduces a new complex variable γ as an alternative to complex variables s in continuous-time and z in discrete-time and [8]

$$\gamma = \frac{z-1}{\beta \cdot T_v \cdot z + (1-\beta) \cdot T_v} \tag{4}$$

where T_v as a sampling period and there are several types of δ-models for various values of β. So called *forward δ -model* for $\beta = 0$ was used here and operator γ is then $\gamma = (z-1)/T_v$.

The recursive identification then estimates the vector of parameters, θ_δ, with known values in the data vector, φ_δ, from the ARX model in vector form:

$$y_\delta(k) = \theta_\delta^T(k) \cdot \phi_\delta(k-1) + e(k) \tag{5}$$

where $e(k)$ is a general random immeasurable component. The simple *Recursive Least-Squares (RLS) method* was used in this work because it can be easily programmable and modified with the exponential and the directional forgetting modifications which could increase the efficiency and the accuracy of the computing.

3 Choice of the Polynomial $d(s)$

The polynomial $d(s)$ on the right hand of the polynomial equation (3) is stable optional polynomial roots of which affects the result of the control.

3.1 *Pole-Placement Method*

This stable polynomial could be chosen in several ways. The most simplest way is to use the *Pole-placement method* which splits the polynomial $d(s)$ in various roots, generally

$$d(s) = \prod_{i=1}^{\deg d(s)} \left(s - s_i\right) \tag{6}$$

where s_i denote roots and $s_i = \alpha_i + \omega_i \cdot j$. The value of $\omega_i = 0$ for the aperiodical response and $\alpha_i < 0$ for stable polynomial $d(s)$.

It is clear that the number of roots depends on the degree of the polynomial $d(s)$. For example, the fourth degree $d(s)$ produces four different roots, two double roots, one triple and one simple root or one quadruple root.

Disadvantage of this method is in its simplicity there is no general rule, how we can choose the roots whether is better to take multiple or simple roots, which rank is better etc. In this case is good to do simulations on the mathematical which are quicker, cheaper and safer and can help us with the choice of these roots.

The second way is to connect the choice of the polynomial $d(s)$ somehow with the controlled system. This could be done for example by the *Spectral factorization* of the polynomial $a(s)$ in the denominator of the system's transfer function (1):

$$n^*(s) \cdot n(s) = a^*(s) \cdot a(s) \tag{7}$$

and the polynomial $d(s)$ is then split into two polynomials:

$$d(s) = n(s) \cdot (s + \alpha)^{\deg d - \deg n} \tag{8}$$

where $n(s)$ comes from (7) and the second part is simple root choice from the *Pole-placement method.*

The Spectral factorization of the polynomial $a(s)$ satisfies that the polynomial $n(s)$ is always stable even if the recursive identification estimates parameters of $a(s)$ as an unstable which could happened for example at the very beginning of the identification, when the estimator do not have enough information about the system and needs some time for the stabilization of the identification.

3.2 Linear-Quadratic (LQ) Approach

The second method for the choice of the polynomial $d(s)$ is the use of *Linear-Quadratic (LQ) theory* which is based on the minimizing of the cost function J_{LQ}:

$$J_{LQ} = \int_0^\infty \left\{\mu_{LQ} \cdot e^2(t) + \varphi_{LQ} \cdot \dot{u}^2(t)\right\} dt \tag{9}$$

where $e(t)$ is control error, $\dot{u}(t)$ denotes increment of the input variable and $\varphi_{LQ} > 0$ a $\mu_{LQ}0$ are weight coefficients, the choice of which affects the speed and quality of the control.

The stable polynomial $d(s)$ is then separated into two parts (polynomials) $n(s)$ a $g(s)$:

$$d(s) = n(s) \cdot g(s) \tag{10}$$

where polynomial $n(s)$ is obtained by the spectral factorization of the system's polynomial $a(s)$ as it is described above in (7) and polynomial $g(s)$ comes from the minimization of the cost function J_{LQ} from (9), which mathematically means:

$$(a(s) \cdot f(s))^* \cdot \varphi_{LQ} \cdot a(s) \cdot f(s) + b^*(s) \cdot \mu_{LQ} \cdot b(s) = g^*(s) \cdot g(s) \tag{11}$$

4 Simulation Experiment

Two approaches in hybrid adaptive control with Pole-placement and LQ methods described above were tested on a mathematical model of the Continuous Stirred-Tank Reactor (CSTR) [13] with cooling in the jacket, scheme of which is shown in Fig. 3. This chemical reactor is a typical nonlinear system with lumped parameters. The reaction inside is so called *Van der Vusse reaction* $A \rightarrow B \rightarrow C, 2A \rightarrow D$.

The mathematical model is constructed with the knowledge of material and heat balances inside the system and results in the set of Ordinary Differential Equations (ODEs) [13] 1

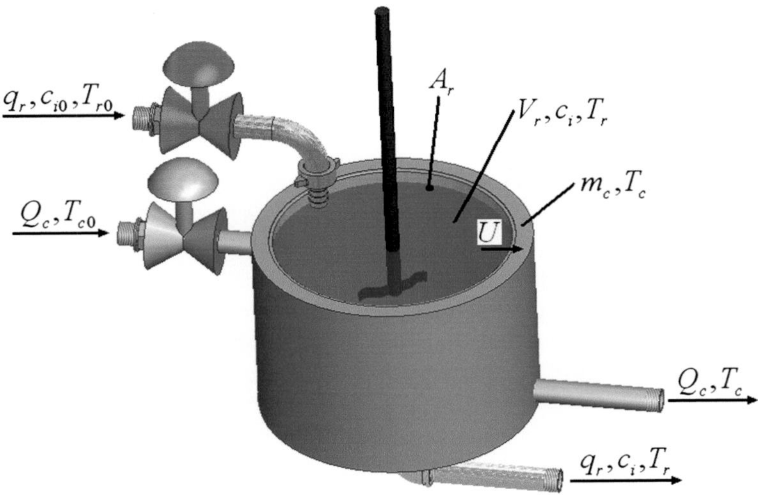

Fig. 3 The continuous stirred-tank reactor (CSTR)

$$
\begin{aligned}
\frac{dc_A}{dt} &= \frac{q_r}{V_r}\left(c_{A0} - c_A\right) - k_1 c_A - k_3 c_A^2 \\
\frac{dc_B}{dt} &= -\frac{q_r}{V_r} c_B + k_1 c_A - k_2 c_B \\
\frac{dT_r}{dt} &= \frac{q_r}{V_r}\left(T_{r0} - T_r\right) - \frac{h_r}{\rho_r c_{pr}} + \frac{A_r U}{V_r \rho_r c_{pr}}\left(T_c - T_r\right) \\
\frac{dT_c}{dt} &= \frac{1}{m_c c_{pc}}\left(Q_c + A_r U\left(T_r - T_c\right)\right)
\end{aligned}
\tag{12}
$$

where t in is the time, c are concentrations, T represents temperatures, c_p is used for specific heat capacities, q_r means the volumetric flow rate of the reactant, Q_c is the heat removal of the cooling liquid, V_r is volume of the reactant, ρ stands for densities, A_r is the heat exchange surface and U is the heat transfer coefficient. Indexes $(\cdot)_A$ and $(\cdot)_B$ belong to compounds A and B, respectively, $(\cdot)_r$ denotes the reactant mixture, $(\cdot)_c$ cooling liquid and $(\cdot)_0$ are feed (inlet) values.

The variable h_r is a reaction heat and k_{1-3} in (1) denotes reaction rates which are computed from

$$
\begin{aligned}
h_r &= h_1 \cdot k_1 \cdot c_A + h_2 \cdot k_2 \cdot c_B + h_3 \cdot k_3 \cdot c_A^2 \\
k_j\left(T_r\right) &= k_{0j} \cdot \exp\left(\frac{-E_j}{RT_r}\right), \ \text{for}\, j = 1, 2, 3
\end{aligned}
\tag{13}
$$

where h_i stands for reaction enthalpies. Reaction rates k_{1-3} in the second equation are nonlinear functions of the reactants temperature computed via *Arrhenius law* with k_{0j} as rate constants, E_j are activation energies and R means gas constant.

Fixed parameters of the system are shown in Table 1 [13].

The steady-state characteristic presented for example in [14] shows high nonlinearity of the system and results in optimal working point which is in this case combination of the input volumetric flow rate of the reactant, $q_r^s = 2.365 \cdot 10^{-3}\,\text{m}^3.\text{ss}^{-1}$, and the heat removal of the cooling, $Q_c^s = -18.56\,\text{kJ.min}^{-1}$. Steady-state values of state variables for this working point are

$$
\begin{aligned}
c_A^s &= 2.14\,\text{kmol.m}^{-3} \qquad & c_B^s &= 1.09\,\text{kmol.m}^{-3} \\
T_r^s &= 387.34\,\text{K} & T_c^s &= 386.06\,\text{K}
\end{aligned}
\tag{14}
$$

There are four theoretical input variables—the volumetric flow srate of the reactant, q_r, the heat removal, Q_c, the input concentration of compound A, c_{A0} and the input temperature of the reactant, T_{r0}. The output variables are both concentrations of compounds A and B c_A, c_B and temperatures of the reactant and the cooling T_r and T_c.

The input variable in this case is the change of the heat removal Q_c from its working point Q_c^s and the output, controlled, variable is the change of the reactant's temperature, T_r, from its steady-state value T_r^s:

$$
u(t) = \frac{Q_c(t) - Q_c^s}{Q_c^s}\ [\%] \qquad y(t) = T_r(t) - T_r^s\ [K]
\tag{15}
$$

The dynamic characteristics was done for three negative and three positive step changes of $u(t)$ and the results are shown in Fig. 4.

Table 1 Fixed parameters of the CSTR

Name of the parameter	Symbol and value of the parameter
Volume of the reactant	$V_r = 0.01\ \mathrm{m}^{-3}$
Density of the reactant	$\rho_r = 934.2\ \mathrm{kg.m}^{-3}$
Heat capacity of the reactant	$c_{pr} = 3.01\ \mathrm{kJ.kg}^{-1}.\mathrm{K}^{-1}$
Weight of the coolant	$m_c = 5\ \mathrm{kg}$
Heat capacity of the coolant	$c_{pc} = 2.0\ \mathrm{kJ.kg}^{-1}.\mathrm{K}^{-1}$
Surface of the cooling jacket	$A_r = 0.215\ \mathrm{m}^2$
Heat transfer coefficient	$U = 67.2\ \mathrm{kJ.min}^{-1}.\mathrm{m}^{-2}.\mathrm{K}^{-1}$
Pre-exponential factor for reaction 1	$k_{01} = 2.145 \cdot 10^{10}\ \mathrm{min}^{-1}$
Pre-exponential factor for reaction 2	$k_{02} = 2.145 \cdot 10^{10}\ \mathrm{min}^{-1}$
Pre-exponential factor for reaction 3	$k_{03} = 1.5072 \cdot 10^{8}\ \mathrm{min}^{-1}.\mathrm{kmol}^{-1}$
Activation energy of reaction 1 to R	$E_1/R = 9758.3\ \mathrm{K}$
Activation energy of reaction 2 to R	$E_2/R = 9758.3\ \mathrm{K}$
Activation energy of reaction 3 to R	$E_3/R = 8560\ \mathrm{K}$
Enthalpy of reaction 1	$h_1 = -4200\ \mathrm{kJ.kmol}^{-1}$
Enthalpy of reaction 2	$h_2 = 11000\ \mathrm{kJ.kmol}^{-1}$
Enthalpy of reaction 3	$h_3 = 41850\ \mathrm{kJ.kmol}^{-1}$
Input concentration of compound A	$c_{A0} = 5.1\ \mathrm{kmol.m}^{-3}$
Input temperature of the reactant	$T_{r0} = 387.05\ \mathrm{K}$

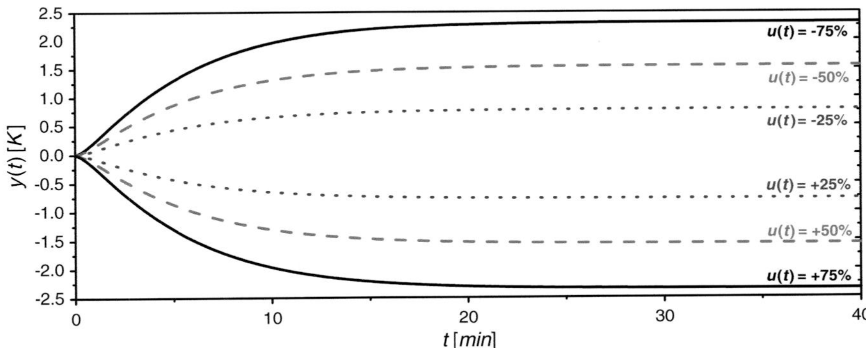

Fig. 4 Results of dynamic analysis for various steps of $u(t)$

Presented results of the dynamic analysis show that this output could be expressed by the second order transfer function with relative order one:

$$G(s) = \frac{b(s)}{a(s)} = \frac{b_1 s + b_0}{s^2 + a_1 s + a_0} \tag{16}$$

which is also the ELM of the controlled output used later in the adaptive control.

4.1 Adaptive Control with Pole-Placement Method

The first study was done for the hybrid adaptive control with the Pole-placement method and Spectral factorization.

The degrees of polynomials $q(s)$, $p(s)$, $r(s)$ and $d(s)$ in the Diophantine equation (3) are for the step changes of the reference signal and the disturbance and the ELM (16):

$$\begin{array}{ll} \deg q(s) = \deg a(s) = 2 & \deg p(s) = \deg a(s) - 1 = 1 \\ \deg r(s) = 0 & \deg d(s) = 2 \cdot \deg a(s) = 4 \end{array} \tag{17}$$

which means that transfer functions of the controller (2) are

$$Q(s) = \frac{q_2 s^2 + q_1 s + q_0}{s \cdot (s + p_0)}; \quad R(s) = \frac{r_0}{s \cdot (s + p_0)} \tag{18}$$

and the stable optional polynomial $d(s)$ on the right side of (3) is

$$d(s) = n(s) \cdot (s + \alpha)^2 \tag{19}$$

for $n(s) = s^2 + n_1 s + n_0$ computed from the Spectral factorization of $a(s)$.

The vector of parameters in δ identification, θ_δ, has four parts and similarly the data vector, φ_δ, has four parts:

$$\begin{aligned} \theta_\delta &= \left[a'_1, a'_0, b'_1, b'_0\right]^T \\ \phi_\delta(k-1) &= \left[-y_\delta(k-1), -y_\delta(k-2), u_\delta(k-1), u_\delta(k-2)\right]^T \end{aligned} \tag{20}$$

where

$$\begin{array}{ll} y_\delta(k) = \frac{y(k) - 2y(k-1) + y(k-2)}{T_v^2} & \\ y_\delta(k-1) = \frac{y(k-1) - y(k-2)}{T_v} & u_\delta(k-1) = \frac{u(k-1) - u(k-2)}{T_v} \\ y_\delta(k-2) = y(k-2) & u_\delta(k-2) = u(k-2) \end{array} \tag{21}$$

The simulation was done for the sampling period $T_v = 0.3$ min and various step changes of the reference temperature of the reactant, $w(t)$ and three values of the parameter $\alpha = 0.05, 0.08$ and 0.12 as a tuning parameter of the control. Results are displayed in Figs. 5 and 6.

Obtained results show that proposed 2DOF controller produces good control responses even the controlled system has nonlinear behaviour. Increasing value of the tuning parameter α reflects in quicker course of the output variable. As it was already mentioned, 2DOF control configuration reduce overshoots. You can see, that the output responses have minimal overshoots. The course of the input variable $u(t)$ is very similar to the output variable—bigger value of α affects mainly the speed of changes of the input variable.

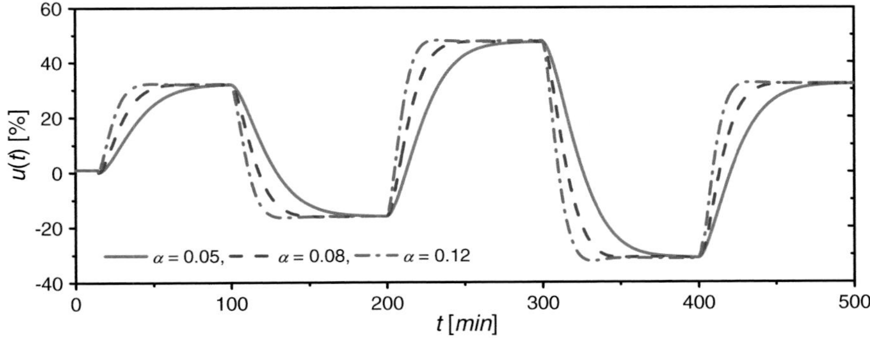

Fig. 5 Course of the input variable $u(t)$ for various values of α, Pole-placement method

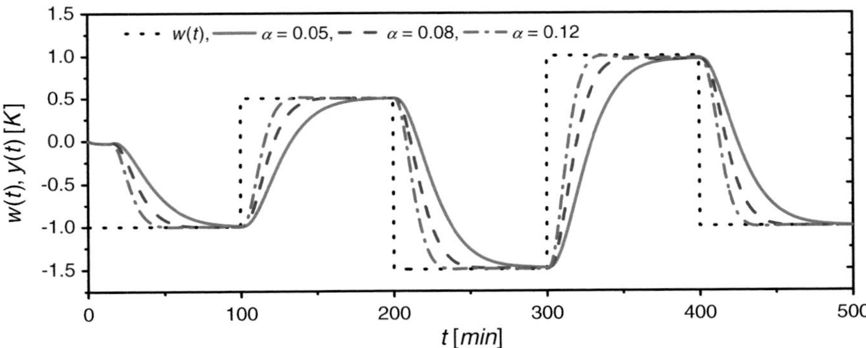

Fig. 6 Courses of the reference signal $w(t)$ and the output variable $y(t)$ for various α, Pole-placement method

4.2 Adaptive Control with LQ Approach

The second method employs LQ approach which have two weighting variables that acts as tuning parameters of the controller. The parameter μ_{LQ} gives weight to the control error $e(t)$ and the parameter φ_{LQ} is important if we care about the increment of the input variable $\dot{u}(t)$.

Degrees of the controller's polynomials and the polynomial $d(s)$ are in this case

$$\begin{array}{cc} \deg q(s) = \deg a(s) = 2 & \deg p(s) \geq \deg a(s) - 1 = 2 \\ \deg r(s) = 0 & \deg d(s) = 2 \cdot \deg a(s) + 1 = 5 \end{array} \tag{22}$$

and the transfer functions $Q(s)$ and $R(s)$ are then

$$Q(s) = \frac{q_2 s^2 + q_1 s + q_0}{s \cdot \left(s^2 + p_1 s + p_0\right)}; \quad R(s) = \frac{r_0}{s \cdot \left(s^2 + p_1 s + p_0\right)} \tag{23}$$

with the polynomial $d(s)$

$$d(s) = n(s) \cdot g(s) \tag{24}$$

where $n(s) = s^2 + n_1 s + n_0$ and $g(s) = g_3 s^3 + g_2 s^2 + g_1 s + g_0$ are products of spectral factorizations (7) and (11). The data vector and the vector of parameters are the same as in previous case because we have the same second order ELM with relative order one (16).

It was shown for example in [17], that there is important only the ratio $\mu_{LQ} : \varphi_{LQ}$. That is why we expect $\mu_{LQ} = 1$ and only changes of $\varphi_{LQ} = 0.03, 0.15$ and 0.5 were done. The simulation was done for the same step changes and sampling period as in the previous case and you can find simulation results in Figs. 7 and 8.

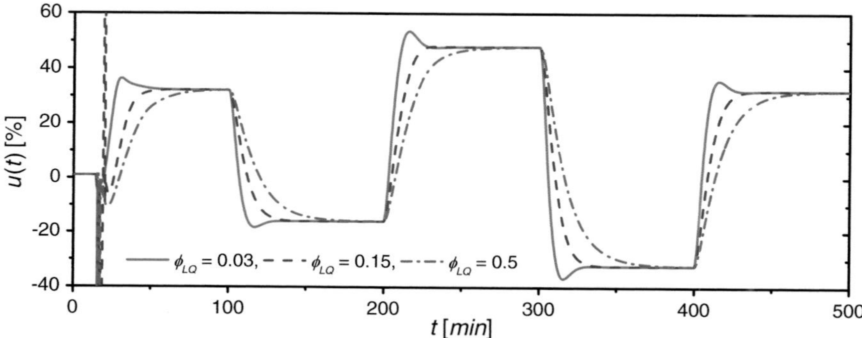

Fig. 7 Course of the input variable $u(t)$ for various values of φ_{LQ}, LQ approach

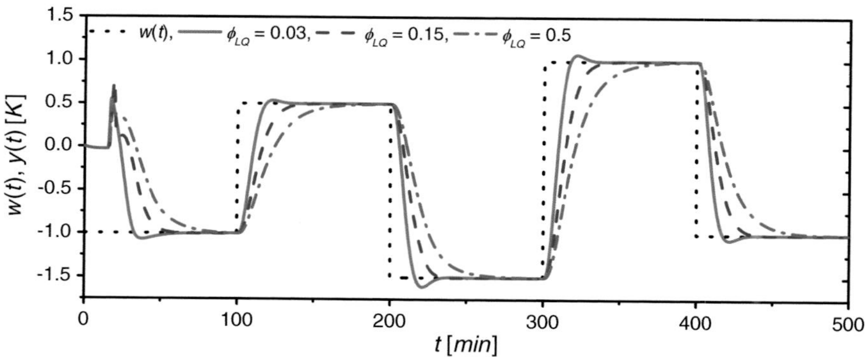

Fig. 8 Courses of the reference signal $w(t)$ and the output variable $y(t)$ for various φ_{LQ}, LQ approach

The dynamics of the output variable $y(t)$ and the input variable $u(t)$ are quicker with decreasing value of φ_{LQ}. You can find also small overshoot of the output variable for the lowest value φ_{LQ}, but we can say that this method also cope with the control very good.

5 Conclusion

The paper presents two approaches in hybrid adaptive control of the nonlinear system. An adaptivity is satisfied by the recursive identification of parameters of the ELM as a linear representation of the originally nonlinear system. The structure and computation relations comes from the polynomial synthesis and as parameters of the controller are computed from polynomials in nominator and denominator of the transfer function representing ELM, they can vary during the control according to results of the identification. The control loop satisfies basic control requirements such as a stability, a reference signal tracking and a disturbance attenuation. Moreover, both methods offers tuning parameter which can affect the result of control—the Pole-placement method has a position of the root α and increasing value of this pole results in quicker output response. On the other hand, the second, LQ, approach has similar results with two weighting parameters as a tuning variables. The change of one of them was tested and it was proofed, that increasing value of φ_{LQ} provides slower control responses. Both methods provides good control results which is supported by the use of 2DOF control configuration which minimizes mainly the overshoots of the output variable. These methods could be used for control of the real system which will be the nest step after this simulation study.

References

1. Corriou, J.-P.: Process Control. Theory and Applications. Springer, London (2004)
2. Ingham, J., Dunn, I.J., Heinzle, E., Prenosil, J.E.: Chemical Engineering Dynamics. An Introduction to Modeling and Computer Simulation. Second. Completely Revised Edition. VCH Verlagsgesellshaft. Weinheim (2000). ISBN: 3-527-29776-6
3. Dusek, F, Honc, D.: Control of a simple constrained MIMO system with steady state optimization—Case study. In: Proceedings of the IASTED International Conference on Modelling, Identification, and Control, MIC (2009)
4. Astrom, K.J., Wittenmark, B.: Adaptive Control. Addison Wesley, Reading (1989)
5. Bobal, V., Bhm, J., Fessl, J., Machacek, J.: Digital Self-tuning Controllers: Algorithms Implementation and Applications. Advanced Textbooks in Control and Signal Processing. Springer, London (2005). ISBN: 1-85233-980-2
6. Honc, D., Dusek, F.: State-space constrained model predictive control. In: Proceedings—27th European Conference on Modelling and Simulation, ECMS (2013)
7. Vojtesek, J., Dostal, P.: Control of concentration inside CSTR using nonlinear adaptive controller. In: Nostradamus 2014: Prediction, Modeling and Analysis of Complex Systems. Advances in Intelligent Systems and Computing, vol. 289, 2014, pp. 195–204 (2014)

8. Mukhopadhyay, S., Patra, A.G., Rao, G.P.: New class of discrete-time models for continuos-time systeme. Int. J. Control **1992**(55), 1161–1187 (1992)
9. Stericker, D.L., Sinha, N.K.: Identification of continuous-time systems from samples of input-output data using the δ-operator. Control-Theory Adv. Technol **9**(1993), 113–125 (1993)
10. Kucera, V.: Diophantine equations in control—A survey. Automatica **29**(1993), 1361–1375 (1993)
11. Kucera, V.: Analysis and Design of Discrete Linear Control Systems. Prentice-Hall, London (1991). ISBN: 0-13-033085-X
12. Dostal, P., Bobal, V., Blaha, M.: One approach to adaptive control of nonlinear processes. In: Proceedings of IFAC Workshop on Adaptation and Learning in Control and Signal Processing ALCOSP 2001, pp. 407–412. Cernobbio-Como, Italy (2001)
13. Chen, H., Kremling, A., Allgwer, F.: Nonlinear predictive control of a benchmark CSTR. In: Proceedings of 3rd European Control Conference. Rome, Italy (1995)
14. Vojtesek, J., Dostal, P.: From steady-state and dynamic analysis to adaptive control of the CSTR reactor. In: Proceedings of 19th European Conference on Modelling and Simulation ESM 2005, pp. 591–598. Riga, Latvia (2005)
15. Vojtesek, J., Dostal, P.: Adaptive LQ approach used in conductivity control inside continuous-stirred tank reactor, international federation of automatic control. In: Proceedings of the 17th IFAC World Congress, Soul, pp. 12929–12934 (2008). ISBN-ISSN 978-1-1234-7890-2
16. Mikles, J., Fikar, M.: Process Modelling, Identification, and Control. Springer, Berlin (2007)
17. Vojtesek, J., Dostal, P.: Effect of Weighting Factors in Adaptive LQ Control. Nostradamus 2013: Prediction, Modeling and Analysis of Complex Systems. Advances in Intelligent Systems and Computing, vol. 210, pp 265–274 (2013)

Prediction of NO_X Concentration Time Series Using the Chaos Theory

Radko Kříž and Petra Lešáková

Abstract This paper is aimed at analysis of NOx concentration time series. At first we estimated the time delay and the embedding dimension, which is needed for the Lyapunov exponent estimation and for the phase space reconstruction. Subsequently we computed the largest Lyapunov exponent, which is one of the important indicators of chaos. Then we estimated the correlation dimension and Kolmogorov entropy. The results indicated that chaotic behaviors obviously exist in NOx concentration time series. Finally we computed predictions using a radial basis function and polynomials to fit global nonlinear functions to the data.

Keywords Prediction · NO_x · Chaos theory · Time series analysis · Gaussian radial basis function

1 Introduction

NO_x is term for nitrogen dioxide (NO_2) and nitric oxide (NO). They are usually produced from the reaction of nitrogen (N_2) and oxygen (O_2) in the air during combustion. In fact, the atmosphere itself is the source of much of the nitrogen leading to the formation of NO_x. Air is mainly composed of N_2 (78 %) and O_2 (21 %), which together are the major gases of the atmosphere [1]. Such high temperature and high pressure conditions exist in internal combustion engines, like those in automobiles (known as "mobile sources"). Thus, NO_x is one of the major mobile source air pollutants. These conditions of high temperature and pressure can also exist in boilers such as those in power plants, so NO_x is also commonly found in high concentrations leaving fossil fuel power generating stations [1]. Limiting

R. Kříž (✉) · P. Lešáková
University of Pardubice, Pardubice, Czech Republic
e-mail: radko.kriz@upce.cz

P. Lešáková
e-mail: petra.lk@seznam.cz

A. Abraham et al. (eds.), *Proceedings of the Second International Afro-European Conference for Industrial Advancement AECIA 2015*, Advances in Intelligent Systems and Computing 427, DOI 10.1007/978-3-319-29504-6_44

NO_x production demands the precise control of the amount of air used in combustion. The burning of biomass creates additional pollutants including NO_x [2].

Both NO and NO_2 are harmful and toxic to humans, although atmospheric concentrations of nitrogen oxides are usually well below the concentrations expected to lead to adverse health effects. There is some evidence that long-term exposure to NO_2 at concentrations above 40–100 $\mu g \cdot m^{-3}$ may decrease lung function and increase the risk of respiratory symptoms [3]. Nitrogen oxides play key roles in important reactants in O_3 formation [1]. Environmental legislation creates continual pressure on finding new solutions that are both economically advantageous and environmentally friendly [4].

Recent advances in the modeling of air pollutants have led to the application of concepts including nonlinear modeling techniques. Numerous new methods of time series analysis have been developed for dealing with nonlinear data e.g. [5–7]. The approaches based on chaos theory are widely acceptable due to the assumption that, it is possible to predict the future state of the system based on the single scalar time series assuming that, all the information regarding the external forcing factors is contained in that single time series [8]. The prediction of NO_x concentration has been studied in several works based on different unconventional approaches such as neural networks [8–11], Petri nets [12], hierarchical fuzzy inference systems [13], radial basis function [14, 15], ship plume parameterization [16].

2 Input Data

This study was based on hourly NO_x concentration data from Prague—Libuš from 1.1.2011 to 31.12.2013. Several missing values were ignored. The data were provided by the Czech Hydrometeorological Institute for a related diploma thesis that is aimed at nonlinear and chaotic behavior of air pollutants time series (Fig. 1).

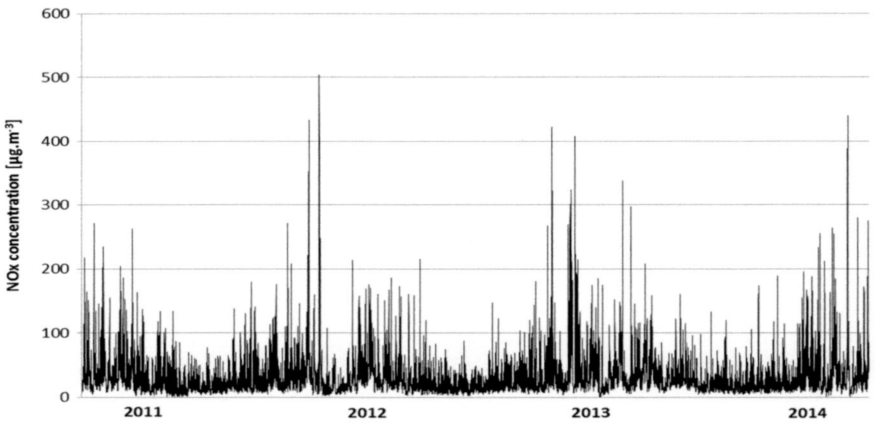

Fig. 1 Hourly averaged NOx concentration data from Prague—Libuš

3 Analysis

Chaos theory allows for the reconstruction of phase space from time series, which can be used for specifying the system states [17]. This analysis is based on Takens embedding theorems [18]. Takens' theorem transforms the prediction problem from time extrapolation to phase space interpolation [19].

Let there be given a time series x_1, x_2, ..., x_N which is embedded into the m-dimensional phase space by the time delay vectors. A point in the phase space is given as:τ

$$Y_n = x_n, x_{n-\tau}, \ldots, x_{n-(m-1)\tau}\, n = 1, 2, \ldots, N-(m-1)\tau, \tag{1}$$

where τ is the time delay and m is the embedding dimension. Different choices of τ and m yield different reconstructed trajectories. Kodba et al. [20] discuss How we can determine optimal τ and m. Fraser and Swinney [21] introduced the mutual information between x_n and $x_{n+\tau}$ as a suitable quantity for determining τ. The mutual information between x_n and $x_{n+\tau}$ quantifies the amount of information we have about the state $x_{n+\tau}$ presuming we know the state x_n. Now we can define mutual information function:

$$I(\tau) = -\sum_{h=1}^{j}\sum_{k=1}^{j} P_{h,k}(\tau)\ln\frac{P_{h,k}(\tau)}{P_h P_k}, \tag{2}$$

where P_h and P_k denote the probabilities that the variable assumes a value inside the hth and kth bins, respectively, and $P_{h,k}(\tau)$ is the joint probability that x_n is in bin h and $x_{n+\tau}$ is in bin k. The first minimum of $I(\tau)$ then marks the optimal choice for the time delay.

The embedding dimension m can be chosen using the "false nearest neighbors'" method. This method measures the percentage of close neighboring points in a given dimension that remain so in the next highest dimension. The minimum embedding dimension capable of containing the reconstructed attractor is that for which the percentage of false nearest neighbors drops to zero for a given tolerance level μ.

In order to calculate the fraction of false nearest neighbors the following algorithm is used according to Kennel et al. [22]. Given a point $p(i)$ in the m-dimensional embedding space, one first has to find a neighbor $p(j)$, so that

$$\|p(i) - p(j)\| \leq \mu. \tag{3}$$

We then calculate the normalized distance R_i between the $(m + 1)$th embedding coordinate of points $p(i)$ and $p(j)$ according to the equation:

$$R_i = \frac{|x_{i+m\tau} - x_{j+m\tau}|}{\|p(i) - p(j)\|}. \tag{4}$$

If R_i is larger than a given threshold R_{tr}, then $p(i)$ is marked as having a false nearest neighbor. Equation (4) has to be applied for the whole time series and for various $m = 1, 2, \ldots$ until the fraction of points for which $R_i > R_{tr}$ is negligible [20].

Lyapunov exponent λ of a dynamical system is a quantity that characterizes the rate of separation of infinitesimally close trajectories. The largest Lyapunov exponent can be defined as follows:

$$\lambda = \lim_{\substack{\delta Z_0 \to 0 \\ t \to \infty}} \frac{1}{t} \ln \frac{|\delta Z(t)|}{|\delta Z_0|}. \tag{5}$$

A positive largest Lyapunov exponent is usually taken as an indication that the system is chaotic. We have used the Rosenstein algorithm [23], which counts the largest Lyapunov exponent as follows:

$$\lambda_1(i) = \frac{1}{i\Delta t} \cdot \frac{1}{(M-i)} \sum_{j=1}^{M-i} \ln \frac{d_j(i)}{d_j(0)}, \tag{6}$$

where $d_j(i)$ is distance from the j point to its nearest neighbor after i time steps and M is the number of reconstructed points. For more information see [23].

One of the principal characteristics which is used for description of strange attractors is a dimension. The dimension reflects the complexity and strangeness of an attractor [24]. Notice that fractal dimension is a real number. A non-integer dimension does not imply chaotic dynamic, but all strange attractors must have non-integer fractal dimensions. For more information see [25]. In practice, fractal dimension cannot be computed easily. The method is based on the concept of correlation dimension D_C suggested by Grassberger et Procaccia [26]. D_C describes the dimensionality of the underlying process in relation to its geometrical reconstruction in phase space. D_C quantifies the “strangeness” of an attractor [27]. Define the correlation integral $C(\varepsilon)$ for set of M data:

$$C(\varepsilon) = \frac{1}{M(M-1)} \sum_{\substack{i,j=1 \\ i \neq j}}^{M} \Theta(\varepsilon - \|x_i - x_j\|), \tag{7}$$

where Θ is the Heaviside step function. Euclidean metric is used for all calculations in this paper. When a lower limit exists, the correlation dimension is then defined as

$$D_C = \lim_{\substack{\varepsilon \to 0 \\ M \to \infty}} \frac{\partial \ln(C(\varepsilon))}{\partial \ln(\varepsilon)}. \tag{8}$$

In real cases, the limitation imposed in the above equation is not possible due to the limited amount of available data. Hence, D_C can be obtained by plotting $lnC(\varepsilon)$

against $ln(\varepsilon)$. The slopes of the curves for different embedding dimensions m give the values of D_C [26]. The saturation of D_C at a certain values of m indicates that the process generating time series is not random but rather deterministic [28].

The $C(\varepsilon)$ can be regarded as an average density of points where the local density is obtained by a kernel estimator with a step kernel. Very practical is the replacement of the step kernel by the Gaussian kernel [28]. Gaussian kernel correlation integral $C_G(\varepsilon)$ can be obtained from usual $C(\varepsilon)$ via

$$C_G(\varepsilon) = \frac{1}{2\varepsilon^2} \int_0^\infty d\tilde{\varepsilon} e^{-\frac{\tilde{\varepsilon}^2}{4\varepsilon^2}} \tilde{\varepsilon} C(\tilde{\varepsilon}). \tag{9}$$

If $C(\varepsilon)$ is given at discrete values of ε, the Gaussian kernel correlation integral $C_G(\varepsilon)$ can be carried out numerically by interpolating $C(\varepsilon)$ with pure power laws [27].

Entropies are an information theoretical concept to characterize the amount of information needed to predict the next measurement with a certain precision. The most popular one is the Kolmogorov-Sinai entropy. When analyzing time series we are usually dealing with distributions of delay vectors with delay τ in an m-dimensional reconstructed phase space. The m dependence of correlation integral C_q in the limit of large m can then be expressed as

$$\begin{matrix} C_q(m, \varepsilon) \\ \varepsilon \to 0 \\ m \to \infty \end{matrix} = \alpha m e^{-(q-1)h_q \tau m} \varepsilon^{(q-1)D_q}, \tag{10}$$

which defines the order q entropy h_q. [27] Second order entropy is called Kolmogorov entropy. An algorithm for the determination of the Kolmogorov entropy is given in Cohen et Procaccia [29].

4 Prediction

Predictability is one way how correlations between data express themselves [28]. Most properties of chaotic systems are much more easily determined from the governing equations than from a time series. Unfortunately, the governing equations are usually not known, except for well controlled laboratory experiments. Analyzing an empirical model, and maybe synthetic time series data generated from it, can provide a valuable consistency test for the results of time series analysis. Chaotic dynamical systems generically show the phenomenon of structural instability [6].

According [8], using the reconstructed phase space for m and τ, a functional relationship f between the current state $X(t)$ and future state $X(t + T)$ can be given as

$$X(t+T)=f(X(t)), \tag{11}$$

where T represents the number of time steps ahead that one wishes to perform the prediction. Function f represents the approximation to unknown dynamical system. It is shown that for sufficiently large values of the embedding dimension and if some additional conditions are satisfied, the reconstructed trajectory has the same topological and geometrical properties as the system's phase space trajectory [18]. The predictive mapping can be expressed as

$$X(t+T)=f_p(X(t)). \tag{12}$$

The aim is to find the predictor f_p, so that $x(t + T)$ can be predicted based on the reconstructed time series. If the time series is chaotic, then f_p is necessarily non-linear. Several local and global approaches are available in the literature to find the function f_p [30].

The idea of locally linear predictions is following. If there is a good reason to assume that the relation

$$s_{n+1}=f(s_n) \tag{13}$$

is fulfilled by the experimental data in good approximation (say, within 5 %) for some unknown f and that f is smooth, predictions can be improved by fitting local linear models. They can be considered as the local Taylor expansion of the unknown f, and are easily determined by minimizing

$$\sigma^2 = \sum_{s_j \in U_n} (s_{n+1} - a_n s_j - b_n)^2, \tag{14}$$

with respect to a_n and b_n, where U_n is the ε-neighborhood of s_n, excluding s_n, as before. Then, the prediction is

$$\hat{s}_{n+1}=a_n s_n + b_n \tag{15}$$

The minimization problem can be solved through a set of coupled linear equations, a standard linear algebra problem [28].

The local linear fits are very flexible, but can go wrong on parts of the phase space where the points do not span the available space dimensions and where the inverse of the matrix involved in the solution of the minimization does not exist. Moreover, very often a large set of different linear maps is unsatisfying. Therefore many authors suggested fitting global nonlinear functions to the data, i.e. to solve

$$\sigma^2 = \sum_n (s_{n+1} - f_p(s_n))^2, \tag{16}$$

where f_p is now a nonlinear function in closed form with parameters p, with respect to which the minimization is done. The results depend on how far the chosen ansatz f_p is suited to model the unknown nonlinear function, and on how well the data are deterministic at all [28]. A radial basis function (RBF) is a real-valued function whose value depends only on the distance from the origin, or alternatively on the distance from some other point x_i, called a center, or alternatively on the distance from some other point c, called a center, so that

$$\Phi(x, c) = \Phi(\|x - c\|) \tag{17}$$

Gaussian kernel is used in this analysis.

$$\Phi(x) = e^{-c^2x^2} \tag{18}$$

Hence the prediction made is:

$$f_P = a_0 + \sum_{i=1}^{n} a_i\Phi(\|x - x_i\|) \tag{19}$$

respectively

$$x_{t+1} = a_0 + \sum_{i=1}^{n} a_i e^{-(c\|x - x_i\|)^2} \tag{20}$$

Polynomial model where f_p is modeled by polynomial

$$P(x) = \sum_{i=0}^{n} a_i x^i \tag{21}$$

where n is a natural number, the coefficients $a_0, \ldots, a_n$ are elements of R, and X is a formal symbol, whose powers X_i are just placeholders for the corresponding coefficients ai, so that the given formal expression is just a way to encode the sequence (a_0, a_1, ...), where there is an n such that $a_i = 0$ for *all* $i > n$. n is called order of polynomial.

5 Results and Discussion

We will use the mutual information approach to determine the time delay τ and the false nearest neighbor method to determine the minimal sufficient embedding dimension m. τ is estimated from the graph in Fig. 2a. The first minimum of the mutual information function $I(\tau)$ (2) marks the optimal choice for the time delay. Thus, the time delay τ is 7. The embedding dimension m is chosen using the "false nearest neighbors" method, estimated from the graph in Fig. 2b. The minimum

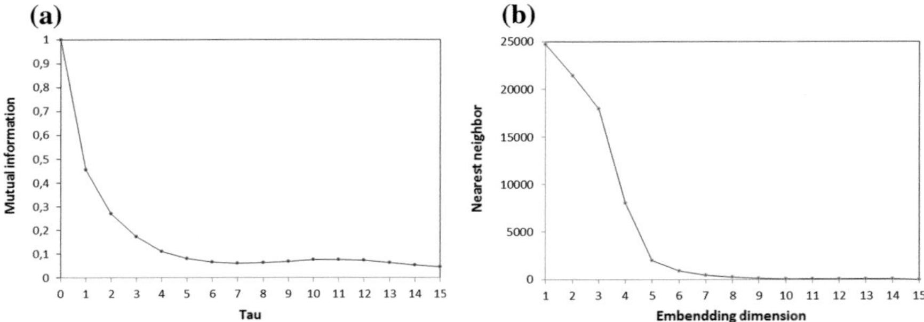

Fig. 2 Mutual average information and fraction of false nearest neighbors for the NO_x concentration time series

embedding dimension capable of containing the reconstructed attractor is that for which the percentage of false nearest neighbors drops to zero for a given tolerance level μ. Thus, the embedding dimension m is 10.

We calculate the largest Lyapunov exponent as was shown above. We used the Rosenstein algorithm. The calculation of the largest Lyapunov exponent depends on the estimation of the embedding dimension. The value of the largest Lyapunov exponent was estimated at 0,034 for embedding dimension 10. A positive largest Lyapunov exponent is one of the necessary conditions for chaotic behavior.

Next, we estimate the correlation dimension. Relationship between the Gaussian kernel correlation integral $C_G(\varepsilon)$ and radius ε on ln-ln scale with embedding dimensions m from 1 to 15 is shown in Fig. 3. The saturation value of the correlation dimension for NO_x concentration time series is 2,20. The value of the Kolmogorov entropy NO_x concentration time series was calculated at 10,11 for embedding dimension 5 to 10 (Fig. 4).

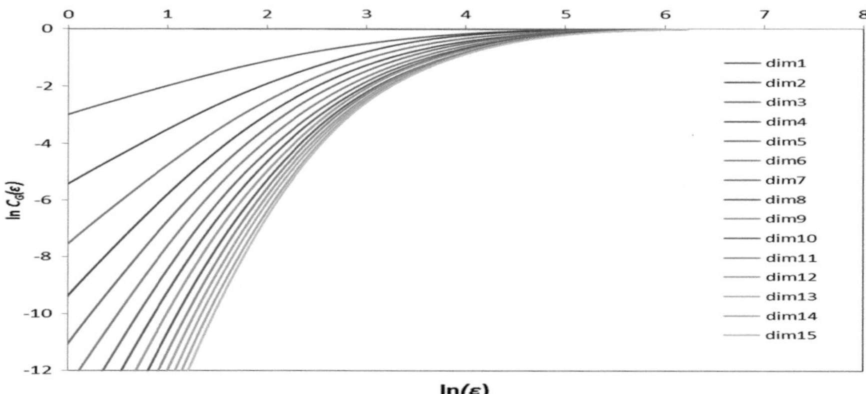

Fig. 3 Ln $C_G(\varepsilon)$ versus ln(ε) plots for the NO_x concentration time series

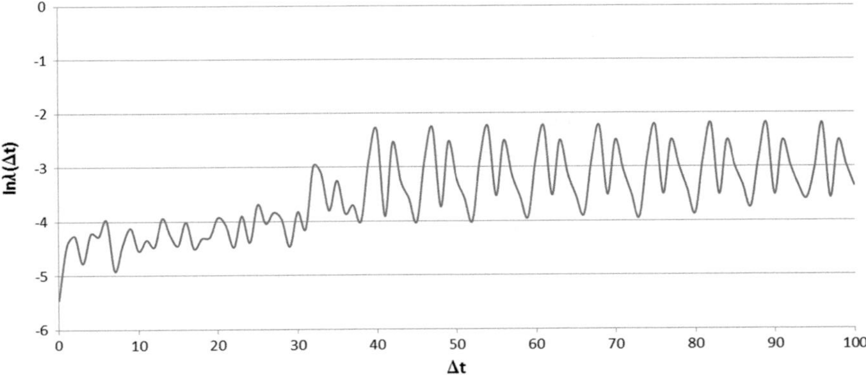

Fig. 4 Ln $\lambda(\Delta t)$ versus Δt plots for the NO_x concentration time series

Finally we computed predictions using a Gaussian radial basis function and polynomials to fit global non-linear functions to the data. The 3-order polynomial approach proved to be the best, however it can only used to predict a maximum of 100 h. 1000 h forecasts are displayed in Fig. 5. This figure clearly displays that RBF is much better than the polynomial fit.

This prediction using RBF for the first few hours is very good even though this is an area with a high volatility. In many cases the prediction correctly estimated the peaks and trends of the NO_x concentration time series. Predictions using RBF can be used in long term.

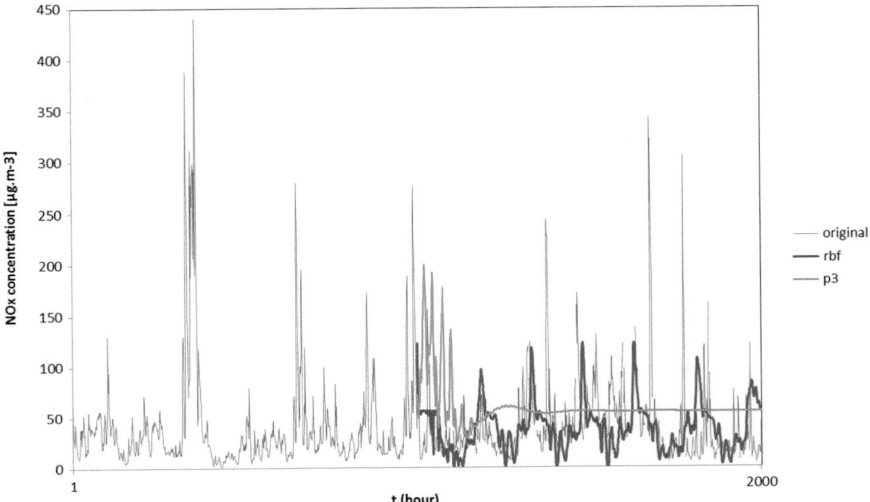

Fig. 5 Prediction of the NOx concentration time series using RBF and polynomials

6 Conclusion

We shown in this paper that the NO_x concentration time series is chaotic. First, we computed the values of the time delay $\tau = 7$ and the embedding dimension $m = 10$. The estimated largest Lyapunov exponent is 0,034. If the correlation dimension is low, the largest Lyapunov exponent is positive and the Kolmogorov entropy has a finite positive value, chaos is probably present. From these estimations it can be concluded that NO_x concentration time series is chaotic. Finally we computed predictions using a Gaussian RBF to fit global nonlinear functions to the data. Considering all these findings, we recommend the Gaussian RBF to fit global non-linear functions as one of the methods used for prediction. As it may not be reliable under certain circumstances, it should be used in combination with other prediction methods.

Acknowledgments This work was supported by SGS 40520/20SG/450009 UPCE.

References

1. Vallero, D.A.: Fundamentals of Air Pollution, 4th edn. Academic Press, Amsterdam and Boston (2007)
2. Baťa, R., Půlkrábková, P.: The importance of modelling the environmental impacts of a biomass based electric power generation for public safety. WSEAS Trans. Environ. Dev. **9**(4) (2013)
3. World Health Organization: Health aspects of air pollution with particulate matter, ozone and nitrogen dioxide, p. 48. 13–15 2003. Accessed 19 Nov 2011
4. Baťa, R., Kadlecová, P.: Modeling of economic and environmental impacts of energy generation using of waste paper. E + M Ekonomie a Management **2** (2011)
5. Abarbanel, H.D.I.: Analysis of Observed Chaotic Data, 272 pp. Springer, New York (1996)
6. Kantz, H., Schreiber, T.: Nonlinear Time Series Analysis, 304 pp. Cambridge University Press, Cambridge (1997)
7. Miksovsky, J., Raidl, A.: Testing the performance of three nonlinear methods of time series analysis for prediction and downscaling of European daily temperatures. Nonlinear Process. Geophys. **12**(6), 979–991 (2005)
8. Chelani, A.B., Singh, R.N., Devotta, S.: Nonlinear dynamical characterization and prediction of ambient nitrogen dioxide concentration. Water, Air Soil Pollut. **166**(1–4), 121–138 (2005)
9. Zhou, H., Cen, K., Fan, J.: Modeling and optimization of the NOx emission characteristics of a tangentially fired boiler with artificial neural networks. Energy **29**(1), 167–183 (2004)
10. Hájek, P., Olej, V.: Air quality modelling by Kohonen's self-organizing feature maps and LVQ neural networks. WSEAS Trans. environment and Dev. **4**(1), 45–55 (2008)
11. Kolehmainen, M., Martikainen, H., Ruuskanen, J.: Neural networks and periodic components used in air quality forecasting. Atmos. Environ. **35**(5), 815–825 (2001)
12. Baťa, R.: Modeling of LCA-chain segment for Biofuels as an instrument for the protection of the population. WSEAS Trans. Environ. Dev. **9**(2) (2013)
13. Hájek, P., Olej, V.: Air quality indices and their modelling by hierarchical fuzzy inference systems. WSEAS Trans. Environ. Dev. **10**, 661–672 (2009)

14. Kříž, R.: Chaos in nitrogen dioxide concentration time series and its prediction. In: Nostradamus 2014: Prediction, Modeling and Analysis of Complex Systems Advances in Intelligent Systems and Computing, vol. 289, pp 365–376 (2014)
15. Peng, H., Ozaki, T., Haggan-Ozaki, V., Toyoda, Y.: A parameter optimization method for radial basis function type models. IEEE Trans. Neural Netw. **14**(2), 432–438 (2003)
16. Huszar, P., et al.: Modeling the regional impact of ship emissions on NOx and ozone levels over the Eastern Atlantic and Western Europe using ship plume parameterization. Atmos. Chem. Phys. **10**(14), 6645–6660 (2010)
17. Abarbanel, H.D., Brown, R., Sidorowich, J.J., Tsimring, L.S.: The analysis of observed chaotic data in physical systems. Rev. Mod. Phys. **65**(4), 1331 (1993)
18. Takens, F.: Detecting strange attractors in turbulence. In: Dynamical Systems and Turbulence, Warwick 1980, pp. 366–381. Springer, Berlin (1981)
19. Kříž, R.: Chaotic Analysis of the GDP Time Series. In: Nostradamus 2013: Prediction, Modeling and Analysis of Complex Systems, pp. 353–362. Springer (2013)
20. Kodba, S., Perc, M., Marhl, M.: Detecting chaos from a time series. Eur. J. Phys. **26**(1), 205 (2005)
21. Fraser, A.M., Swinney, H.L.: Independent coordinates for strange attractors from mutual information. Phys. Rev. A **33**(2), 1134 (1986)
22. Kennel, M.B., Brown, R., Abarbanel, H.D.: Determining embedding dimension for phase-space reconstruction using a geometrical construction. Phys. Rev. A **45**(6), 3403 (1992)
23. Rosenstein, M.T., Collins, J.J., De Luca, C.J.: A practical method for calculating largest Lyapunov exponents from small data sets. Physica D: Nonlinear Phenomena **65**(1), 117–134 (1993)
24. Raidl, A.: Estimating the fractal dimension, K2-entropy, and the predictability of the atmosphere. Cz. J. Phys. **46**(4), 293–328 (1996)
25. Kříž, R., Kratochvíl, Š.: Analyses of the Chaotic behavior of the electricity price series. In: ISCS 2013: Interdisciplinary Symposium on Complex Systems, p. 215–226. Springer, Berlin (2014)
26. Grassberger, P., Procaccia, I.: Characterization of strange attractors. Phys. Rev. Lett. **50**(5), 346–349 (1983)
27. Schreiber, T.: Interdisciplinary application of nonlinear time series methods. Phys. Rep. **308** (1), 1–64 (1999)
28. Hegger, R., Kantz, H., Schreiber, T.: Practical implementation of nonlinear time series methods: The TISEAN package. Chaos **9**(2), 413–435 (1999)
29. Cohen, A., Procaccia, I.: Computing the Kolmogorov entropy from time signals of dissipative and conservative dynamical systems. Phys. Rev. A **31**, 1872 (1985)
30. Farmer, D.J.: Sidorowich, J.J.: Predicting chaotic time series. Phys. Rev. Lett. **59**, 85–848 (1987)

Predicting the Direction of Movement for Stock Price Index Using Machine Learning Methods

Pınar Tüfekci

Abstract Stock price prediction with high accuracy may offer significant opportunities for the investors who make decisions on making profit or having high gains over the stocks in stock markets. The aim of this study is to predict the movement directions (UP/DOWN) of the Istanbul Stock Exchange National 100 (ISE National 100) Index accurately for short-term futures by using three machine learning methods, which are Logistic Regression (LR), Support Vector Machines (SVMs), and Multilayer Perceptron (MLP). Two datasets used in this study are composed of sessional and daily points of data over a 5-year period from November 2007 to November 2012. During the prediction of the movement directions, the following factors were taken into account; data of macroeconomic indicators, gold prices, oil prices, foreign exchange prices, stock price indexes in various countries, and the data of ISE National 100 index for past sessions and prior days, which are used as input variables in the datasets. In connection with that, the most effective features of these input variables were determined by using some feature selection methods. As a result, the movement directions of ISE National 100 were predicted with higher accuracies by using reduced datasets than original datasets and the best performances were found by LR classifier.

Keywords Data mining • Prediction of stock market prices • Istanbul stock exchange National 100 Index

---

P. Tüfekci (✉)
Çorlu Engineering Faculty, Computer Engineering Department,
Namık Kemal University, 59860 Çorlu, Tekirdağ, Turkey
e-mail: ptufekci@nku.edu.tr

A. Abraham et al. (eds.), *Proceedings of the Second International Afro-European Conference for Industrial Advancement AECIA 2015*, Advances in Intelligent Systems and Computing 427, DOI 10.1007/978-3-319-29504-6_45

1 Introduction

The stock market is an environment where the investors trade for the company stocks and derivatives and also a huge system where millions of different traders manipulate thousands of instruments every day. Obviously, predicting stock market index is considered highly complicated and often a difficult task because it is essentially; dynamic, non-linear, nonparametric, and chaotic in nature. In addition, stock market's movements might be influenced by so many factors, such as; macro-economic indicators, political events, firms' policies, general economic conditions, commodity price index, bank rate, bank exchange rate, investors' expectations, institutional investors' expectations, institutional investors' choices, movements of other stock markets, psychology of traders, etc [1, 2].

Financial time series prediction intends to find underlying patterns, trends, cycles, and forecasts the future using historical and currently observable data [3, 4]. Recently, some of the machine learning approaches, such as the LR [5], SVMs [6–10], and artificial neural networks (ANNs) [1, 11–15] have been extensively applied to the prediction process of financial time series, including financial stock market prediction, based on its past and present data [16].

The purpose of this study is to predict the movement directions of the ISE National 100 Index, which is composed of National Market companies except investment trust in Turkey. In order to predict the movement directions, which are "UP" and "DOWN", the classification methods, which are the LR, SVMs, and MLP classifiers, are applied to two datasets. The datasets are composed of sessional and daily points of data over a 5-year period from November 2007 to November 2012.

During the prediction of the movement directions, the following factors were taken into account; data of macroeconomic indicators, gold prices, oil prices, foreign exchange prices, stock price indexes in various countries, and the data of the ISE National 100 Index for past sessions and prior days which are used as input variables in the datasets. To determine which factors in the datasets are more effective on the prediction of the movement directions, some feature selection methods, which are filter-based and wrapper-based methods, are applied to these datasets. Thus, the movement directions of the ISE National 100 Index were predicted for the closing of the current session with 60.79 % accuracy by using reduced dataset D1 with 29 features and for the closing of the current day with 58.49 % accuracy by using reduced dataset D2 with 22 features. The best performances were found for both datasets by the LR classifier.

The rest of the paper is constructed as follows: Sect. 2 introduces feature selection process. Section 3 introduces the classifiers used in this study. Section 4 mentions comparative experiments and results. The conclusion and future works are summarized in the final section.

2 Feature Selection Process

Data pre-processing is a significant process that contains the processes of cleaning, integration, transformation, and reduction of data for using quality data in machine learning (ML) algorithms. Data sets may variety in dimension from two to hundred features, and many of these features may be irrelevant or redundant. Feature subset selection decreases the data set dimension by removing irrelevant and redundant features from an original feature set. The objective of feature subset selection is to procure a minimum set of original features. Using the decreased set of original features enables ML algorithms to operate faster and more effectively. However, it helps to predicate more correctly by increasing learning accuracy of ML algorithms and improving result comprehensibility [17].

The feature selection process has three basic stages such as generation, evaluation and stopping criteria as shown in Fig. 1. The validation stage, which checks the validity of the selected subset and compares the results to find the best feature subset, may not be a stage of the feature selection process [18].

The feature selection process begins by inputting an original feature set, which includes n number of features or input variables. At the first stage of feature selection, which is called subset generation, a search strategy is used for producing possible feature subsets of the original feature set for evaluation. There are some several search procedures to find the optimal subset of the original feature set [18]. In this study, as indicated in Fig. 1, four search procedures are chosen for this task such as Best First (BF), Genetic Search (GS), and Greedy Stepwise Backward (GSB) and Forward (GSF) Direction procedures in WEKA [19]. After the candidate subsets are generated, an evaluation algorithm determines a current best subset. The evaluation algorithms are divided into two broad groups such as Filter and Wrapper methods based on their dependence on the inductive algorithm that will finally use the selected subset. In this study, a filter-based feature selection method, CFS Subset Evaluator [20] and two wrapper-based feature selection methods, Wrapper Subset Evaluators with LR and SVMs are adopted to the datasets [21].

A stopping criterion is needed for stopping the search and preventing an exhaustive search of subsets. The feature selection process halts by outputting the selected subset of features, which is then validated [22].

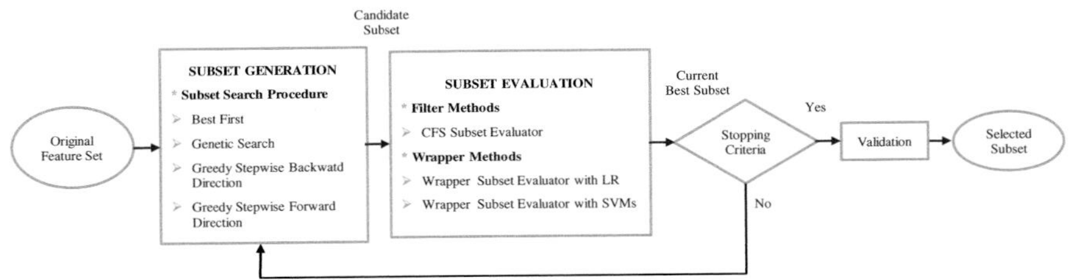

Fig. 1 The feature selection process applied in this study

3 Machine Learning Methods for Classification

A machine learning algorithm estimates an unknown dependency between the inputs, which are independent variables, and output, which is a dependent variable, from a dataset. In most problem domains, there is no functional relationship $y = f(x)$ between y and x where x is input variable and y is output. In this case, the relationship between the inputs and output has to be described more generally by a probability distribution $P(x,y)$ [23].

There are two different approaches for a data classification task. The first considers only a dichotomous distinction between the two classes, assigns the class labels 0 or 1 to an unknown data item [24]. SVMs are an example of this type approach. The second attempts to model $P(y \mid x)$; this yields not only a class label for a data item, but also a probability of class membership. LR and ANNs (such as MLP) are members of second approach [25]. Because of that in this study, LR, SVMs and MLP methods, which are stated by the WEKA statistical analysis package [19], are applied for the task of data classification. These classification methods are generated as learning algorithms to predict the direction of ISEN 100 such as "UP" or "DOWN" movement. A brief summary of the classifiers is explained by following subsections.

3.1 Logistic Regression (LR)

The LR is a nonlinear regression technique for prediction of dichotomous (binary) dependent variables in terms of the independent variables (covariates). In producing the LR equation, the maximum-likelihood ratio was used to determine the statistical significance of the variables. In logistic regression models, dependent variable is always in categorical form and has two or more levels. Independent variables may be in a numerical or categorical form. We consider the situation where we observe a binary outcome variable y and a vector x of covariates for each of N [26]. The dependent variable can represent the status of the target value (e.g., UP, Y = 1 or DOWN, Y = 0).

3.2 Support Vector Machines (SVMs)

Support vector machines are kernel based learning algorithm for solving classification and regression problems. The SVMs build optimal separating boundaries between data sets by solving a constrained quadratic optimization problem [24]. The algorithm of SVMs defines the best hyper-plan that separates the set of points associated with different class labels with a maximum-margin. The unlabeled examples are then classified by deciding on which side of the hyper-surface they reside by using different kernel functions, varying degrees of nonlinearity and flexibility can be included in the model. Because they can be derived from advanced statistical ideas, and bounds on the generalization error can be calculated for them [27].

3.3 Multilayer Perceptron (MLP)

A perceptron that has a single layer of weights can only approximate linear functions of the input. Feed-forward networks with intermediate or hidden layers between the input and output layers which are called MLP; do not have limitations like single layer perceptron. The MLP is a feed forward artificial neural network model that consists of neurons with massively weighted interconnections, where signals always travel in the direction of the output layer. The input signals are sent by the input layer to the hidden layer without performing any operations. Then the hidden and output layers multiply the input signals by a set of weights, and either linearly or nonlinearly transforms results into output values. These weights are optimized to obtain reasonable prediction's accuracy [28].

4 Comparative Experiments and Results

The purpose of this study is to choose a predictive model which accurately predicts the movement directions of the ISE National 100 Index such as "UP" or "DOWN" movement using three classifiers. The layout of the proposed prediction model, which is used in our experiments, is illustrated in Fig. 2 and the parts of the structure are explained in the following subsections:

4.1 Original Data

The original data used in this study was presented by a total of 2501 sessional data points from November 2007 to November 2012 that was obtained from the Central Bank of Republic of Turkey (CBRT) Electronic Data Delivery System and a web-site-related stock market (http://www.finnet.com.tr). The factors that employed 55 input variables were divided mainly through six categories: the major macroeconomic indicators, gold prices, oil prices, foreign exchange prices, stock price indexes in other countries, and some data related on the ISE National 100 for previous session as shown in Fig. 3.

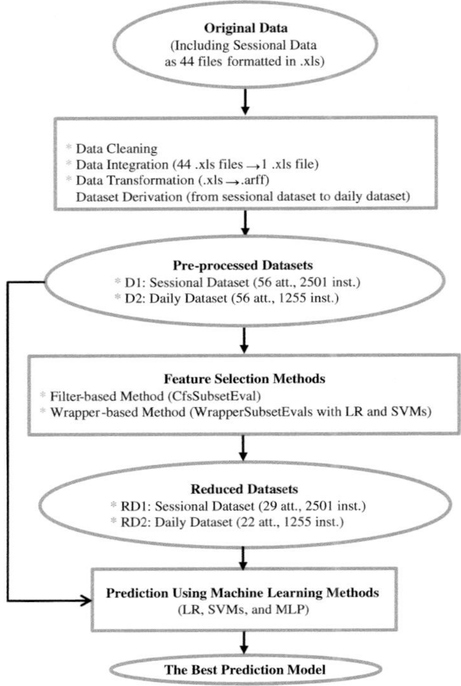

Fig. 2 The flow diagram of the prediction process

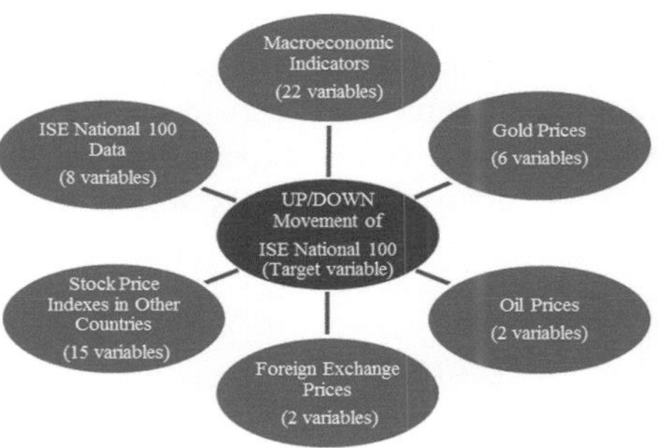

Fig. 3 Contents of the original data

These input variables, which are indicated in Table 1, influence the target variable which is the UP/DOWN movement of ISE National 100 in the current session.

$$\text{Difference} = \text{C} - \text{P} \tag{1}$$

In (1), C represents the closing price of the ISE National 100 Index in current session, and P represents the closing price of the ISE National 100 Index in

Table 1 Features of the datasets

Data type		Feature abb.	Feature names	Time of taken value	
				For D1 (sessional)	For D2 (daily)
Input variables	Macro-economic indicators	F1	Probability of buying durable goods	Previous month	Previous month
		F2	Probability of buying durable goods next 6 months	Previous month	Previous month
		F3	Expectation about the direction of price changes	Previous month	Previous month
		F4	The general economic situation previous 3 months	Previous month	Previous month
		F5	The general economic situation next 3 months	Previous month	Previous month
		F6	Capacity Utilization Rate of manufacturing industry (CUR)	Previous month	Previous month
		F7	Industrial production index-manufacturing industry	Previous month	Previous month
		F8	Job opportunities next 6 months	Previous month	Previous month
		F9	Probability of buying or building a home	Previous month	Previous month
		F10	Exchange rates-unit labor cost based real effective exchange rate	Previous month	Previous month
		F11	Exchange rates-PPI based real effective exchange rate	Previous month	Previous month
		F12	Probability of buying a car	Previous month	Previous month

(continued)

Table 1 (continued)

Data type		Feature abb.	Feature names	Time of taken value	
				For D1 (sessional)	For D2 (daily)
		F13	Industrial Production Index (IPI)	Previous month	Previous month
		F14	Purchasing Power Parity (PPP) previous 6 month	Previous month	Previous month
		F15	Purchasing Power Parity (PPP) next 6 month	Previous month	Previous month
		F16	Consumer Price Index (CPI)	Previous month	Previous month
		F17	Consumer Confidence Index	Previous month	Previous month
		F18	Consumption of debt financing next 3 month	Previous month	Previous month
		F19	Producer Price Index (PPI)	Previous month	Previous month
		F20	KYD B-type fund index	Previous day	Previous day
		F21	KYD O/N repo index	Previous day	Previous day
		F22	All bond index KYD BONDS	Previous day	Previous day
	Gold prices	F23	1 Ounce gold (XAU) London selling price (USD/XAU)	Previous month	Previous month
		F24	Cumhuriyet gold selling price	Previous month	Previous month
		F25	Istanbul gold exchange gold market index closing price	Previous day	Previous day
		F26	Istanbul gold exchange gold (USD/XAU) closing price	Previous day	Previous day
		F27	Istanbul gold exchange gold market index opening price	Current day	Current day
		F28	Istanbul gold exchange gold (USD/XAU) opening price	Current day	Current day

(continued)

Table 1 (continued)

Data type		Feature abb.	Feature names	Time of taken value	
				For D1 (sessional)	For D2 (daily)
	Oil prices	F29	Brent oil future closing price	Previous day	Previous day
		F30	Brent oil future opening price	Current day	Current day
	Foreign exchange prices	F31	EURO/USD closing price	Previous day	Previous day
		F32	EURO/USD opening price	Current day	Current day
	Stock price indexes in other countries	F33	BRSP Bovespa closing price	Previous day	Previous day
		F34	BUSE Merval closing price	Previous day	Previous day
		F35	CAC 40 closing price	Previous day	Previous day
		F36	DAX closing price	Previous day	Previous day
		F37	Dow Jones average 30 industrial closing price	Previous day	Previous day
		F38	EURONEXT 100 ID closing price	Previous day	Previous day
		F39	Hang Seng closing price	Previous day	Previous day
		F40	FTSE 100 closing price	Previous day	Previous day
		F41	MIB 30 index closing price	Previous day	Previous day
		F42	Nasdaq composite closing price	Previous day	Previous day
		F43	NIKKEI 225 closing price	Previous day	Previous day
		F44	NYSE composite closing price	Previous day	Previous day
		F45	Russian RTS index USD closing price	Previous day	Previous day
		F46	Shanghai composite closing price	Previous day	Previous day
		F47	S&P 500 closing price	Previous day	Previous day

(continued)

Table 1 (continued)

Data type		Feature abb.	Feature names	Time of taken value	
				For D1 (sessional)	For D2 (daily)
	ISE National 100 data	F48	ISE National 100 opening price	Previous session	Previous day
		F49	ISE National 100 low	Previous session	Previous day
		F50	ISE National 100 high	Previous session	Previous day
		F51	ISE National 100 moving average	Previous session	Previous day
		F52	ISE National 100 traded value	Previous session	Previous day
		F53	ISE National 100 stocks traded	Previous session	Previous day
		F54	ISE National 100 closing price	Previous session	Previous day
		F55	ISE National 100 opening price	Current session	Current session
	Target variable	F56	UP/DOWN movement of ISE National 100	Current session	Current day

previous session or day. If the difference is equal or bigger than zero, the target value is assigned as "UP". On the other hand, if the difference is less than zero, the target value is assigned as "DOWN" in the datasets.

4.2 Pre-processing

Data pre-processing is a significant stage for complicated data. It has a very large impact on the success of predictions. Therefore, the original dataset including sessional data consists of 44 files formatted in .xls. At the beginning of the pre-processing stage, each data employs the 55 variables, which belong to the factors, and then is cleaned from any noises and incompatible data. Afterwards, each data is merged to eliminate any duplicated data and integrated into the data set. After that, the data set is integrated as a .xls file, which has been transformed into an .arff format that is necessary for processing in WEKA tool. At the end of that, the first dataset including sessional data has obtained 2501 instances in total. To derive another dataset including daily data from the dataset including sessional, some instances are removed. Consequently, second dataset has also been obtained with totaling 1255 instances. Each dataset is divided into 2 as a train set, which is 75 % of the dataset, and a test set which 25 % of the dataset is as indicated in Table 2.

Table 2 Used dataset in this study

Dataset	Content of dataset	# att.	# inst. of train set	# inst. of test set	# total inst.
D1	Sessional data	56	1871	630	2501
D2	Daily data	56	940	315	1255

4.3 Feature Selection Process

After the data pre-processing, the datasets are applied to the process of feature selection to eliminate irrelevant and redundant features in the datasets and to reduce the sizes of the datasets. In this study, the WEKA tool has been used to get the results of measurements and to compare the performances of each feature selection algorithm. To select the feature subsets for two datasets, CFS SubSetEval Evaluator and Wrapper SubsetEval Evaluators with several search procedures, which are the BF, GS, GSB reedy and GSF, are chosen with their default threshold settings. The best evaluators and search procedures for each dataset, which have the highest performance for the prediction of the movement directions of the ISE National 100 Index, are explored with some experiments. For this purpose, the classifiers have applied to the reduced datasets RD1 and RD2. The results are evaluated using the accuracy measure and compared with each other, shown in Table 3.

As a result of this comparison, for both datasets the most successful accuracies of the predictions are obtained by the classifier LR. For RD1 including 29 attributes, the best performance is found by using GS search and Wrapper Subset Evaluator with SVMs with a 60.79 % accuracy. For RD2 including 15 attributes, the best performance is found by using GS search and Wrapper Subset Evaluator with LR with a 58.49 % accuracy. That indicates that, the GS is suitable as a search procedure in the feature selection method for both datasets with a 58.07 % mean accuracy. Thus, the best feature subsets of RD1 and RD2 are determined in this process.

4.4 Prediction Process and Results

In this study, three classification methods are applied to the selected subsets to choose the best feature subset of each dataset. After choosing the best feature subset, for determining the most successful model for each dataset, the results for the original datasets and the reduced datasets are compared with each other as shown in Table 4. According to the accuracies in this table, the feature selection process has increased the prediction performance of each dataset. For dataset D1, the prediction accuracy has been increased from 54.76 to 60.79 % by the LR classifier. For dataset D2, the prediction accuracy has been increased from 56.51 to

Table 3 The results of the feature selection process

Feature selection methods	Evals.	Search Proc.	RD1				RD2				Mean		
			# red. att.	LR	SMO	MLP	# red. att.	LR	SMO	MLP	LR	SMO	MLP
Filter-based method	CFS subset eval	BF	2	52.86	52.86	52.86	5	55.35	55.35	55.35	54.11	54.11	54.11
		GS	3	52.86	52.86	52.86	13	48.11	55.35	53.46	50.49	54.11	53.16
		GSB	2	52.86	52.86	52.86	5	55.35	55.35	55.35	54.11	54.11	54.11
		GSF	2	52.86	52.86	52.86	5	55.35	55.35	55.35	54.11	54.11	54.11
Wrapper-based method	Wrapper subset eval with LR	BF	19	60.63	45.56	48.09	15	48.43	44.03	58.18	54.53	44.80	53.14
		GS	29	56.98	50.00	51.27	15	**58.49**	55.35	55.03	57.74	52.68	53.15
		GSB	39	54.92	51.90	53.97	43	57.55	47.48	48.43	56.24	49.69	51.20
		GSF	3	52.86	51.43	52.86	11	47.80	47.48	46.86	50.33	49.46	49.86
	Wrapper subset eval with SVMs	BF	12	47.94	46.03	53.49	4	55.35	55.35	55.35	51.65	50.69	54.42
		GS	29	**60.79**	45.71	53.65	22	55.35	55.35	55.66	**58.07**	50.53	54.66
		GSB	55	56.03	54.60	53.02	55	55.97	47.80	47.48	56.00	51.20	50.25
		GSF	8	49.21	46.03	50.79	3	55.35	55.35	55.35	52.28	50.69	53.07

Table 4 The results of the prediction process

Original datasets								Reduced datasets							
Dataset	# att.	LR		SVMs		MLP		Dataset	# att.	LR		SVMs		MLP	
		Train acc.	Test acc.	Train acc.	Test acc.	Train acc.	Test acc.			Train acc.	Test acc.	Train acc.	Test acc.	Train acc.	Test acc.
D1	56	71.51	**54.76**	53.23	52.86	70.55	50.32	RD1	29	70.71	**60.79**	55.96	45.71	66.27	53.65
D2	56	69.15	**56.51**	57.77	46.03	83.72	49.52	RD2	15	68.76	**58.49**	56.43	55.35	65.25	55.03
Mean		70.33	**55.64**	55.50	49.45	77.14	49.92			69.74	**59.64**	56.20	50.53	65.76	54.34

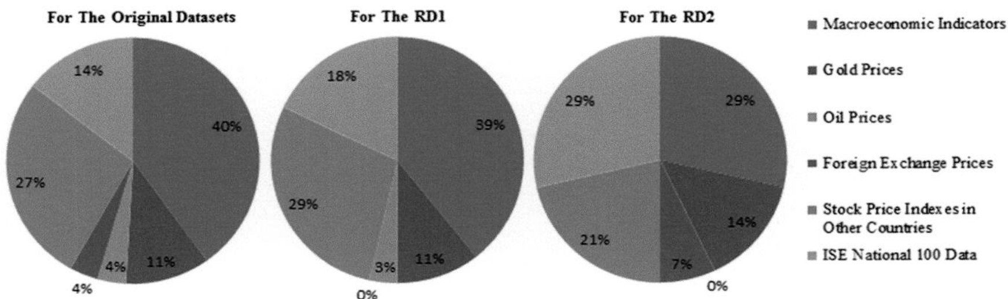

Fig. 4 Distribution of the features for the original and reduced datasets RD1 and RD2

58.49 % by the LR classifier. The LR classifier for both datasets gives the highest accuracy values.

The distribution as percentage of the features for the original datasets and the reduced datasets RD1 and RD2 with the best feature subsets is shown in Fig. 4. According to the feature distribution of RD1, it has been seen that the macroeconomic indicators with 39 %, stock price indexes in other countries with 29 %, ISE National 100 data with 18 %, gold prices with 11 % and oil prices with 3 % have influence on the prediction of the closing of the sessional ISE National 100 index, respectively. Foreign exchange prices are not located in the best feature subset of RD1. Correspondingly, according to the feature distribution of RD2, it also has been seen that the macroeconomic indicators and ISE National 100 data with the same percentage of 29 %, stock price indexes in other countries with 21 %, ISE National 100 data with 21 %, gold prices with 14 % and foreign exchange prices with 7 % have influence on the prediction of the closing of the daily ISE National 100 index, respectively. Oil prices are not located in the best feature subset of RD2.

5 Conclusion

During this study, the movement directions of the ISE National 100 Index for short-term futures were predicted for two datasets by using three machine learning methods, which are LR, SVMs, and MLP. In addition, the impacts of macroeconomic indicators, gold prices, oil prices, foreign exchange prices, stock price indexes in various countries, and the ISE National 100 data for past sessions and prior days on the prediction of the movement directions of the ISE National 100 Index were explored by using feature selection methods, which are filter-based and wrapper-based methods. Consequently, the most important factors were determined for two datasets which consisted of sessional and daily data for a period of 5 years from 2007 to 2012.

As a result, the prediction of the movement directions of the ISE National 100 Index was obtained with higher accuracies by using reduced datasets compared to the original datasets. Thus, the movement directions of the ISE National 100 Index were predicted for the closing of the current session with 60.79 % accuracy by using reduced dataset RD1 with 29 features and for the closing of the current day with 58.49 % accuracy by using reduced dataset RD2 with 22 features. However, the best performance of each dataset was found by the LR classifier. Afterwards, the MLP and SVMs classifiers respectively followed the LR classifier according to the mean result values.

The future works may be planned to predict the movement directions of the ISE National 100 Index for the datasets which consist of weekly and monthly data.

References

1. Yudong, Z., Lenan, W.: Stock market prediction of S&P 500 via combination of improved BCO approach and BP neural network. Expert Syst. Appl. (2009)
2. Tan, T.Z., Quek, C., Ng, G.S.: Brain inspired genetic complimentary learning for stock market prediction. IEEE congr. Evol. Comput. **3**, 2653–2660 (2005)
3. Yumlu, M.S., Gurgen, F.S., Okay, N.: Financial time series prediction using mixture of experts. In: Computer and Information Sciences, Lecture Notes in Computer Science, vol. 2869, pp. 553–560
4. Yümlü, S., Gürgen, F.S., Okay, N.: A comparison of global, recurrent and smoothed-piecewise neural models for Istanbul stock exchange (ISE) prediction. Pattern Recogn. Lett. (2005)
5. Alrasheedi, M,: Predicting up/down direction using linear discriminant analysis and logit model: the case of SABIC price index. Res. J. Bus. Manag. **6**(4), 121–133 (2012)
6. Huang, W., Nakamori, Y., Wang, S.Y.: Forecasting stock market movement direction with support vector machine. Comput. Oper. Res. **32**, 2513–2522 (2005)
7. Manish, K., Thenmozhi, M.: Support vector machines approach to predict the S&P CNX NIFTY index returns. In: Proceedings of 10th Indian Institute of Capital Markets Conference (2006)
8. Pai, P.F., Lin, C.S.: A hybrid ARIMA and support vector machines model in stock price forecasting. Omega **33**, 497–505 (2005)
9. Tay, F.E.H., Cao, L.J.: Application of support vector machines in financial time series forecasting. Omega **29**, 309–317 (2001)
10. Pai, P.F., Wei, W.R.: Predicting movement directions of stock index futures by support vector models with data preprocessing. In: Ind. Eng. Eng. Manag. (2007)
11. Armano, G., Marchesi, M., Murru, A.: A hybrid genetic-neural architecture for stock indexes forecasting. Inf. Sci. **170**, 3–33 (2005)
12. Avcı, E.: Forecasting daily and sessional returns of the ISE-100 index with neural network models. J. Dogus Univ. **8**(2), 128–142 (2007)
13. Chen, A.S., Leung, M.T., Daouk, H.: Application of neural networks to an emerging financial market: forecasting and trading the Taiwan stock index. Comput. Oper. Res. **30**(6) (2003)
14. Egeli, B., Ozturan, M., Badur, B.: Stock market prediction using artificial neural networks. In: Proceedings of the 3rd Hawaii International Conference on Business, Honolulu, Hawaii (2003)
15. Karaatli, M., Gungor, I., Demir, Y., Kalayci, S.: Estimating stock market movements with neural network approach. J. Balikesir Univ. **2**(1), 22–48 (2005)
16. Boyacıoğlu, M.A., Avcı, D.: An adaptive network-based fuzzy inference system for the prediction of stock market return: the case of the Istanbul stock exchange. Expert Syst. Appl. (2010)
17. Han, J., Kamber, M.: Data Mining: Concepts and Techniques. Morgan Kauffmann Publishers (2001)

18. Dash, M., Liu, H.: Consistency-based search in feature selection. Artif. Intell. **151**, 155–176 (2003)
19. Machine Learning Group at University of Waikato. http://www.cs.waikato.ac.nz/ml/weka/ (2015). Accessed 01 July 2015
20. Hall, M.A.: Correlation-based feature selection for machine learning. Ph.D. thesis, Department of Computer Science, University of Waikato, Hamilton, New Zealand (1998)
21. Kohavi, R., John, G.: Wrappers for feature subset selection. Artif. Intell. **97**, 273–324 (1997)
22. Hall, M.A., Holmes, G.: Benchmarking attribute selection techniques for discrete class data mining. IEEE Trans. Knowl. Data Eng. **15**, 1437–1447 (2003)
23. El-Sebakhy, E.: Constrained estimation functional networks for statistical pattern recognition problems: theory and methodology. Int. J. Pattern Recognit. Artif. Intell. (2007)
24. Dreiseitl, S., Ohno-Machado, L.: Logistic regression and artificial neural network classification models: a methodology review. J. Biomed. Inform. **35**(5–6) (2002)
25. Gutierrez, P.A., Hervas, C., Fernandez, J.C., Jurado-Exposito, M., Pena-Barragan, J.M., Lopez-Granados, F.: Structural simplification of hybrid neuro-logistic regression models in multispectral analysis of remote sensed data. In: Neural Netw. World, **19** (2009)
26. Chen, M.Y.: Predicting corporate financial distress based on integration of decision tree classification and logistic regression, Expert Syst. Appl. **38(9)** (2011)
27. Hassan, S., Mihalcea, R., Banea, C.: Random walk term weighting for improved text classification. Int. J. Seman. Comput. **1**(4) (2007)
28. Haykin, S.: Neural Networks, A Comprehensive Foundation, 2nd edn. Prentice Hall (1999)

The Directed Inference for the Kosinski's Fuzzy Number Model

Piotr Prokopowicz

Abstract The Kosinski's Fuzzy Number (KFN) model (former name the Ordered Fuzzy Number) is a tool for processing an imprecise information interpreted in a similar way as the classical convex fuzzy numbers. The specificity of KFNs is an additional property for the fuzzy number—the direction. Thanks to that, the calculations can be done as flexibly as with real numbers. Especially, we are not doomed to get the more fuzzy results after many arithmetical operations. Apart good calculations, the direction also has additional potential in the interpretation of fuzzy data. It can be treated as a direction of process, not only the value. For example "an income is high and process is growing" is a different situation than "an income is high, but process is lowering". The direction of KFN can be used to represent difference between these sentences. Since we deal with an additional information, there is need for the new methods which let benefit a full potential of KFNs in the modeling of linguistic data. This paper introduces algorithm of the inference operation for fuzzy rule constructed with the KFNs. Presented proposal consider the direction of values in processing. It bases on the ideas presented in previous studies on this subject—the Direction Determinant. It was proposed as the general basic tool for defining methods, where we need sensitivity for the direction.

Keywords Kosinski's Fuzzy Numbers · Ordered Fuzzy Numbers · Directed fuzzy · Fuzzy inference · Directed fuzzy inference

1 Introduction

Fuzzy sets are useful tool for processing an information described linguistically. They allow to create a formal model for the situation described by common language. In case where the knowledge is incomplete or available data is inaccurate, linguistic description may be the only possibility.

P. Prokopowicz (✉)
Institute of Mechanics and Applied Computer Science, Kazimierz Wielki University, Bydgoszcz, Poland
e-mail: piotrekp@ukw.edu.pl

A. Abraham et al. (eds.), *Proceedings of the Second International Afro-European Conference for Industrial Advancement AECIA 2015*, Advances in Intelligent Systems and Computing 427, DOI 10.1007/978-3-319-29504-6_46

When we deal with quantitative imprecise values, the special kind of fuzzy sets—the fuzzy numbers are used (usually convex fuzzy numbers). Unfortunately, calculation mechanisms used commonly cause that the small number of arithmetic operations can easily give in the result a drastic increase of imprecision. So usefulness of outcome can be lost.

There is some propositions which eliminate main calculations defects. The Kosinski's Fuzzy Number (KFN) model is one of them. The specificity of this model is the new feature of fuzzy value—the direction. It is represented as an independent order of the characteristic parts of fuzzy number. Some of properties of KFNs model, were presented in the publications [1–4].

One of main applications of fuzzy sets is a fuzzy system. Its core is a base of rules. There is many publications which consist a preview of basic conceptions of fuzzy sets and modeling fuzzy systems [5–8]. The basis for the processing of fuzzy rules are the methods of inference. They describes algorithms for transferring given fuzzy input into fuzzy answer. Generally these methods are based on implications. However, there are also popular solutions like the MIN or PROD, which formally are not the implications, but their practical usefulness is proved. If we deal with quantitative imprecise data, we can use the KFNs instead of classical fuzzy numbers. We can ignore the direction and use the same methods. However, if we want process an additional information contained in the new model, we need the methods sensitive for the direction.

Remark 1 As the **sensitivity for the direction** we understand the situation where is possible to get a different result if we change only the direction of the KFNs used in processing.

It is worth to underline, that is general conception, not a definition. There can be many different situations, many significantly different methods of processing and configurations of direction of the KFNs. Not every direction's change must generate a difference in the results. This is postulate only for possibility of the changes in some situations. If so, then we can call such method/operation has property of *sensitivity for the direction*.

In this paper the inference method for the KFN model is presented. It is a tool for modeling fuzzy systems where we want to preserve and process the idea behind the direction. It uses the concepts introduced in the previous studies on the subject. Especially, it bases on the "Direction Determinant" and is compatible with other direction sensitive propositions as compatibility and aggregation already presented in [9, 10]. Before short introduction to the Kosinski's Fuzzy Numbers its name need some explanation. The model previously was called "Ordered Fuzzy Numbers". This name underlined an individual order of parts of fuzzy number, which is independent from the domain of membership function. To honor the late Professor Witold Kosiński, his work, the contribution and commitment to the development, analysis and popularization of the "Ordered Fuzzy Numbers" model from the paper [10] author uses new name the "Kosinski's Fuzzy Numbers".

2 Kosinski's Fuzzy Number (KFN)

In the series of papers [1–3, 11–14] were introduced and developed main concepts of the idea of Kosinski's Fuzzy Numbers. As they are unusual model of fuzzy information, before introducing a formal definition, it can be useful to clarify a background concept. It is not new observation [15] that the each convex membership function of fuzzy number could be split into two parts: a first is non-decreasing and a second non-increasing (see Fig. 1). The KFNs idea is generally based on such point of view. However, the new model defines an order between non-decreasing and non-increasing part of fuzzy number as independent from the domain values.

Such assumptions leads to the new possibilities in calculations and processing of imprecise data. Some of them presents this paper. Following the papers [1–3, 11, 12, 16] fuzzy number will be identified with the pair of functions defined on the interval [0, 1].

Definition 1 The Kosinski's Fuzzy Number (KFN in short) A is an ordered pair of two continuous functions

$$A = (f_A, g_A) \tag{1}$$

called the up-part and the down-part, respectively, both defined on the closed interval [0, 1] with values in **R**.

If the functions f and g are monotonic (Fig. 2a), they are also invertible and possess the corresponding inverse functions defined on a real axis with the values in [0,1]. If these two inversed functions are not connected, we linking them with constant function with the value 1. In such way we receive an object which directly represents the classical fuzzy number. For the finalization of transformation, we need to mark an order of f and g with an arrow on the graph (see Fig. 2b). Notice that pairs (f, g) and (g,f) are the two different Kosinski's Fuzzy Numbers, unless $f = g$. They differ by their orientation or direction. The interpretations for this orientation and its relations with the real world problems are explained in the [12] and [14].

For the later use it will be more convenient to adopt the following indications of KFN boundaries:

$$UP = (s, 1^-),\ CONST = [1^-, 1^+],\ DOWN = (1^+, e). \tag{2}$$

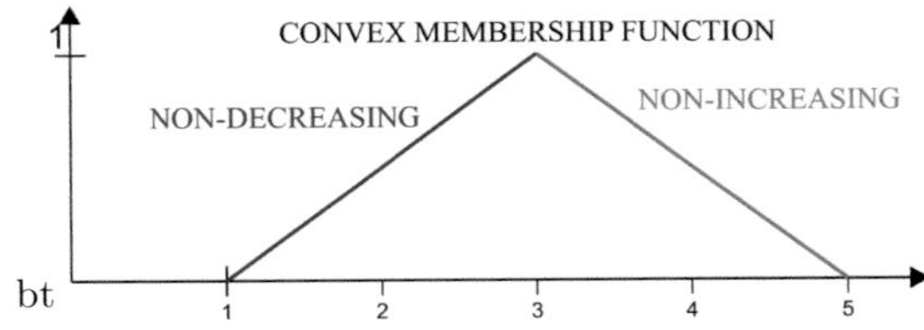

Fig. 1 Parts of the convex membership function

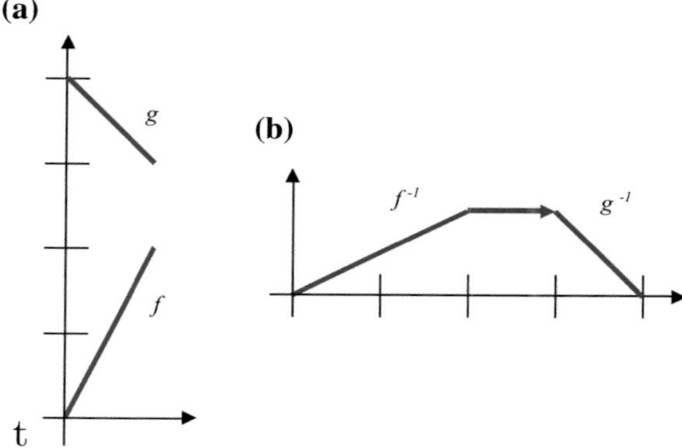

Fig. 2 **a** Kosinski's Fuzzy Number. **b** The Kosinski's Fuzzy Number as convex fuzzy number with an *arrow*

Here it is need to be noted, these intervals can be improper intervals, which have been already discussed in the framework of the extended interval arithmetic by Kaucher in [17] and called by him directed intervals, i.e. such $[n, m]$ where n may be greater than m.

For monotonous f and g we may point the membership function:

$\mu(x) = f^{-1}(x)$, if $x \in [f(0), f(1)] = [s, 1^-]$, and
$\mu(x) = g^{-1}(x)$, if $x \in [g(1), g(0)] = [1^+, e]$ and
$\mu(x) = 1$ when $x \in [1^-, 1^+]$.

It is worth to point out that a class of Kosinski's Fuzzy Numbers represents the whole class of convex fuzzy numbers with continuous membership functions (regarding "classical fuzzy numbers" [5, 6, 8, 15]).

Calculations on Kosinski's Fuzzy Numbers were analyzed and discussed among others in the papers [3, 4, 18]. The paper [4] presents also some extended examples of calculations. Operations on KFNs we define as operations with the UP and DOWN parts as follows:

Definition 2 Let $A = (f_A, g_A)$, $B = (f_B, g_B)$ and $C = (f_C, g_C)$ are mathematical objects called Kosinski's Fuzzy Numbers. The sum $C = A + B$, subtraction $C = A - B$, product $C = A \cdot B$, and division $C = A \div B$ are defined by formula

$$f_C(y) = f_A(y) \star f_B(y) \qquad \wedge \qquad g_C(y) = g_A(y) \star g_B(y) \tag{3}$$

"$\star$" replaces "+", "−", "·", and "/". The A/B is determined only if B does not contain zero. The $y \in [0, 1]$ is the domain of functions f and g.

2.1 Modeling Potential of the KFNs

Key element of the KFN model is an order of the parts of fuzzy number. It is independent from the real numbers universe. This feature is named a 'Direction'. Proposition of real world interpretation for the new property was presented in [12, 14]. KFNs are considered in these papers as the values representing an observation, which passes in time, regardless of the values of the analyzed data. So, the time dependence can be a natural interpretation of the Direction.

However, introducing the Direction, has also other consequences. There is some inconsistency with the convex fuzzy numbers. This is understandable since the KFNs are a specific extension of classic model and thus unprecedented elements appear—the improper OFNs. This aspect was commented in [2, 4, 11]. However, thanks to the new property, also new potential for the practical use of OFN appears. We get a new quality associated with the Direction. The work [19] present the practical use in modeling financial data, and [20] diversity of opinions in social networks. In [21] the application of KFNs for ant colony optimization algorithm is presented.

3 Tools for Considering 'Direction' in the Processing of Imprecise Data

The new proposition in this paper consider a direction of KFNs. It bases on the ideas presented earlier in [9, 10]. As these ideas are specific for the KFN model, here will be presented short introduction to them.

We start with supporting structure, which will simplify further description.

3.1 PART Function

The PART function as the result presents the information about part of KFN, which contains the given argument.

Definition 3 For the KFN A defined on X the PART function $X \to Y$ is determined as follows:

$$PART_A(x) = y = \begin{cases} CONST_A : x \in CONST_A, \\ UP_A : x \in UP_A, \\ DOWN_A : x \in DOWN_A, \\ NONE_A : x \in NONE_A. \end{cases} \tag{4}$$

where: $x \in X$, $Y = \{CONST_A, UP_A, DOWN_A, NONE_A\}$, $CONST_A$—a subset of X for which the membership function of A number is equal to 1; UP_A—a subset of X for which the inverse of the up-part has values; $DOWN_A$—a subset of X for which the

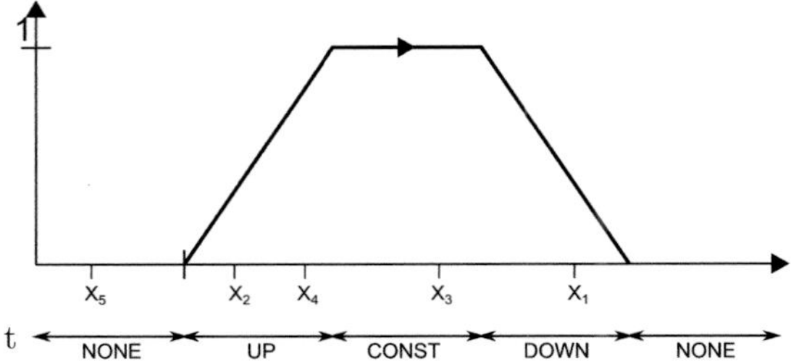

Fig. 3 Parts of the support of KFN

inverse of the down-part has values; $NONE_A$—a subset of X for which the membership function of A number is 0. Figure 3 illustrate the effect of *PART* function. For this example we have the following results: $PART(x_1) = DOWN$, $PART(x_2) = UP$, $PART(x_3) = CONST$, $PART(x_4) = UP$, $PART(x_5) = NONE$ Fuzzy numbers are fuzzy sets defined over the space (or subspace) of real numbers. Thus the sets UP, CONST and DOWN can be treated as numerical intervals.

3.2 *The Direction Determinant*

The direction of the KFN is an additional property which has a different meaning than the membership. So, for processing an full information contained in the KFN, we need additional parameter which will represent this property.

Direction determinant (see [9, 10]) represents a kind of direction 'intensity' of an analyzed argument.

Definition 4 Let A denote the KFN, and x is an element of the support. **Proportional direction determinant** of x in relation to A marked as dir_x^A is calculated as a the result of directional function $D : suppA \rightarrow (-1; 1)$ for the argument x in the following way:

$$dir_x^A = D_A(x) = \begin{cases} 0 & : for \;\; PART(x) \; = \; CONST \\ \frac{(x-1^-)}{(1^- - s)} & : for \;\; PART(x) \; = \; UP \\ \frac{(x-1^+)}{(e-1^+)} & : for \;\; PART(x) \; = \; DOWN \end{cases} \tag{5}$$

The above-mentioned determinant is called proportional, because it is calculated from the ratio of the position of support of the considered argument in relation to the whole fuzzy boundary of OFN, to which this argument belongs. It is illustrated on the Fig. 4.

If the degree of membership is equal to zero, the direction determinant is undefined, because the argument is not a part of function domain D (the value is outside the KFN support). However, if arguments are in the *CONST* interval, we have the direction determinant that is equal to zero, what is justified, as these are the values

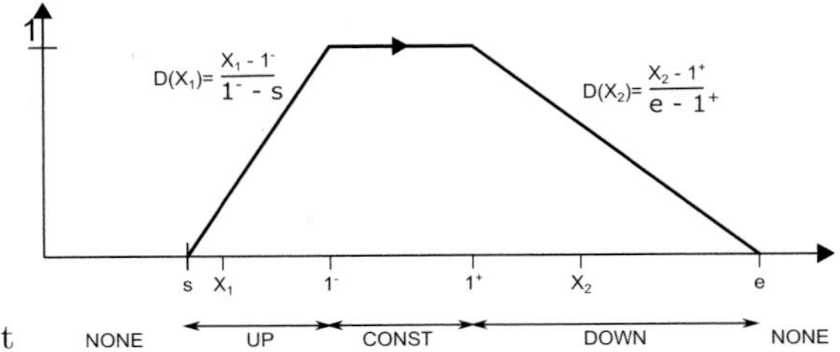

Fig. 4 Proportional direction determinant calculations

about which we have no doubt—their membership is full (equal to 1). According to this intuition we have that, the closer the arguments are to the kernel of fuzzy number, their direction 'intensity' (that is the direction determinant) is smaller. Presented parameter is a key tool for defining methods which are **sensitive to the direction**.

3.3 *Directed Fuzzy Compatibility*

The fuzzy statement 'A is B' where A and B are fuzzy sets is a base for the analysis where we want to apply the fuzzy sets and their imprecise mechanisms. Calculation result of this statement can be called a similarity or compatibility of A with B. In this section the idea for calculating compatibility between two KFNs is presented. Originally it was proposed and analyzed in [10].

Definition 5 For Kosinski's Fuzzy Numbers A and B the result of statement 'A is B' called **directed fuzzy compatibility** and labeled $COMP_{AB}$ is composed of two values: truth value T_{AB} and Direction Determinant D_{AB} calculated as follows:

$$COMP_{AB} = (T_{AB}, D_{AB}) \tag{6}$$

$$T_{AB} = max(min(\mu_A(x), \mu_B(x))) \, : \, x \in \mathbf{X} \tag{7}$$

If T_{AB} is zero, then D_{AB} is unspecified, else

$$D_{AB} = D_B(x_0), \, x_0 = x \, : \, \mu_B(x) = T_{AB} \tag{8}$$

where: $X \subset \mathbf{R}$—a domain of given KFNs, $\mu_A(x)$, $\mu_B(x)$—membership functions of A and B, D_B—Direction Determinant of B for given x.

The D_{AB} can be interpreted as an indication of shifts of A to B. The negative values means the shift in the direction of *up-part* of B, the positive—shift in the direction of *down-part* of B. Such behavior can be also understand as a kind of directed relative dependence between values. Next part of this paper presents a proposition of new inference method, which uses Direction Determinant.

4 Directed Inference Operation

Presented here an inference mechanism is based on the 'generalized modus ponens', where main role play a rule type:

$$\textbf{IF}\ X\ is\ A\ \textbf{THEN}\ Y\ is\ B \tag{9}$$

where A, B—are fuzzy values which are modeling a rule, X, Y—input and output variable. In the generalized modus ponens, where the data are represented by fuzzy numbers (or sets), the whole mechanisms of inference is closed in the mathematical rule. This rule describes an algorithm for calculation the answer—Y value. Sometimes it is also called an inference operator. New proposition is dedicated for KFN model, so in the rule (9) values are presented as such objects. The statement 'X is A' is calculated as compatibility between KFNs. The method was presented in [10] and shortly introduced in previous section.

Definition 6 For the rule in formula (9) let A and B are the KFNs. Let X be the input value also represented by a KFN. The result of 'X is A' is calculated as directed fuzzy compatibility—$COMP_{XA} = (T_{XA}, D_{XA})$, where T_{XA} is the truth value and D_{XA} is the Direction Determinant part of $COMP_{XA}$. The '**Directed Inference by the Multiplication with a Shift**' we call the calculations of answer Y of the rule like follows: if $T_{XA} = 0$ the inference is not calculated, in other case

$$\begin{aligned} & Y = B + |D_{XA}| \cdot c \\ where: & \\ c = & \begin{cases} s - B & : D_{XA} < 0 \\ e - B & : D_{XA} > 0 \end{cases} \end{aligned} \tag{10}$$

4.1 Examples

It is worth of noting that this is not the classical logical inference. The truth value of the premise part of rule is used to check whether the rule can be implemented at all. The specificity of the proposition is that the inference is made through arithmetic operations. We do processing of the quantitative data with calculations. It is in general new aspect of processing fuzzy data. Often in the classical fuzzy systems the input data is quantitative and is represented by the fuzzy numbers. However, after an inference we usually loss the quantitative context and in next steps we deal with the fuzzy sets which are not fuzzy numbers. Characteristic aspect of the presented method is that for activated rule, we always get as outcome another KFN.

For better understanding of the proposed method, an example will be useful. Lets assume for the rule from formula (9) we have KFNs A like on the Fig. 5a and

B—on Fig. 5b. On the Fig. 5a we also can find the input value X. According to the Definition 5 "X is A" is $COMP_{XA} = (T_{XA} = 0.66; D_{XA} = -0.33)$. Using new inference we get the result shown on the Fig. 5c. On the next Fig. 6a we have a situation where the X KFN value changed only a direction (but does not change the shape). This time the result of "X is A" is $COMP_{XA} = (T_{XA} = 0.66; D_{XA} = 0.33)$. As we can see on the Fig. 6b the result of inference was changed. This is related to the change of Direction Determinant.

We can note that if the D_{XA} will be closer to -1, the result of inference will be the narrow fuzzy number situated at *UP* part side of support of KFN B. On the other hand, when the D_{XA} will tend to 1, the result of inference is aimed at extreme values of support but on the *DOWN* side. Finally, when $D_{XA} = 0$ and the $T_{XA} = 1$ it means that the X is fully compatible with A. So, the result of inference is exactly the number B—the value from conclusion. Practically, in the applications (a fuzzy system like—for example), as a result of inference should be considered a pair of values. The first is the truth value of the premise part of rule. The second is the KFN calculated in accordance with the definition 6.

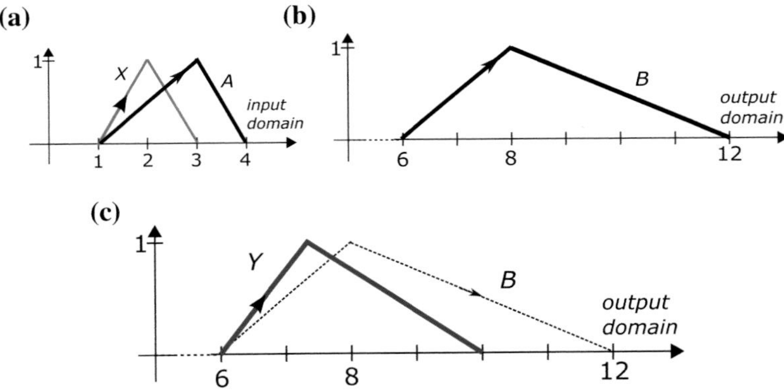

Fig. 5 **a** Example KFNs X and A. **b** KFN B from rule conclusion. **c** Y the result of inference operation

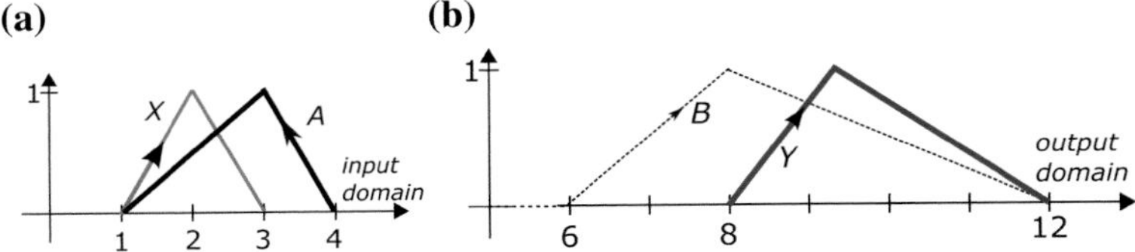

Fig. 6 **a** KFN A with opposite direction while X is the same like before. **b** The new Y result of inference operation

5 Summary and Conclusions

In the paper the new inference method for imprecise values represented by fuzzy numbers was presented. It is named '**Directed Inference by the Multiplication with a Shift**'. Method is dedicated for the model of Kosinski's Fuzzy Numbers. The intention of the new proposal, was to use it as a part of the fuzzy system which process the direction of information.

The KFN model allow for flexible calculations between real numbers and fuzzy numbers, without growing fuzziness. As the benefit of such flexibility, proposed inference is using at the basis arithmetical operations, where the whole KFN is calculated with the real numbers which represent characteristic bounds of the support. Thanks to that we have a method for processing a fuzzy numbers without loss of their quantitative context.

Important aspect of the proposition is using additional feature of KFN—the direction. New inference operation is a method *sensitive for the direction* as the change of the direction of processed values can change also the result. In fact, proposition uses lately introduced parameter called Direction Determinant. So, it is more likely sensitive for the changes of that determinant. It is significant, because if a change of direction only, will not change a result of compatibility of KFNs from premise part of a rule, then also proposed inference operation will not generate change in the result.

References

1. Kosiński W., Prokopowicz P., Ślęzak D.: On algebraic operations on fuzzy reals. In: Rutkowski, L., Kacprzyk, J. (eds.), Advances in Soft Computing. Proceedings of the Sixth International Conference on Neutral Networks and Soft Computing, Zakopane, Poland, 11–15 June 2002, pp. 54–61. Physica-Verlag, Heidelberg (2003)
2. Kosiński W., Prokopowicz P., Ślęzak D.: Ordered fuzzy numbers. Bull. Polish Acad. Sci. Ser. Sci. Math. **51**(3), 327–338 (2003)
3. Kosiński, W.: On fuzzy number calculus. Int. J. Appl. Math. Comput. Sci. **16**(1), 51–57 (2006)
4. Prokopowicz, P.: Flexible and Simple Methods of Calculations on Fuzzy Numbers with the Ordered Fuzzy Numbers Model, LNCS (LNAI), vol. 7894, pp. 365–375. Springer, Heidelberg (2013)
5. Pedrycz, W., Gominde, F.: An Introduction to Fuzzy Sets. The MIT Press, Cambridge (MA), London (1998)
6. Wagenknecht, M., Hampel, R., Schneider, V.: Computational aspects of fuzzy arithmetic based on archimedean *t*-norms. Fuzzy Sets Syst. **123**(1), 49–62 (2001)
7. Piegat, A.: Fuzzy Modeling and Control. Studies in Fuzziness and Soft Computing. Springer, Berlin, Heidelberg (2001)
8. Buckley James, J., Eslami, E.: An Introduction to Fuzzy Logic and Fuzzy Sets, Physica-Verlag, A Springer-Verlag Company, Heidelberg (2005)
9. Prokopowicz, P., Malek, S.: Aggregation Operator for Ordered Fuzzy Numbers Concerning the Direction, LNCS (LNAI), vol. 8467, pp. 267–278. Springer International Publishing, Switzerland (2014)
10. Prokopowicz P., Pedrycz W.: The directed compatibility between ordered fuzzy numbers—a base tool for a direction sensitive fuzzy information processing. In: Proceedings of ICAISC

2015, Part I LNAI, vol. 9119 pp. 249–259. Springer International Publishing, Switzerland (2015)

11. Kosiński W., P. Prokopowicz P., Ślęzak D.: On algebraic operations on fuzzy numbers. In: Intelligent Information Processing and Web Mining. ASC, vol. 22, pp. 353–362. Springer (2003)
12. Kosiński, W., Prokopowicz, P.:Fuzziness—representation of dynamic changes, using ordered fuzzy numbers arithmetic, new dimensions in fuzzy logic and related technologies. In: Proceedings of the 5th EUSFLAT, Ostrava, Czech Republic, vol. I, pp. 449–456. University of Ostrava (2007)
13. Prokopowicz, P.: Adaptation of Rules in the Fuzzy Control System Using the Arithmetic of Ordered Fuzzy Numbers. LNCS (LNAI), vol. 5097, pp. 306–316. Springer, Heidelberg (2008)
14. Kosiński, W., Prokopowicz, P., Kacprzak, D.: Fuzziness representation of dynamic changes by ordered fuzzy numbers. In: Seising, R. (ed.) Views on Fuzzy Sets and Systems from Different Perspectives. STUDFUZZ, vol. 243, pp. 485–508. Springer, Heidelberg (2009)
15. Nguyen, H.T.: A note on the extension principle for fuzzy sets. J. Math. Anal. Appl. **64**, 369–380 (1978)
16. Kosiński, W., Prokopowicz, P., Rosa, A.: Defuzzification functionals of ordered fuzzy numbers. IEEE Trans. Fuzzy Syst. **21**(6), 1163–1169 (2013). doi:10.1109/TFUZZ.2013.2243456
17. Kaucher, E.: Interval analysis in the extended interval space IR. Comput. Suppl. **2**, 33–49 (1980)
18. Koleśnik, R., Prokopowicz, P., Kosiński, W.: Fuzzy calculator—usefull tool for programming with fuzzy algebra. In: ICAISC 2004. LNCS (LNAI), vol. 3070, pp. 320–325. Springer, Heidelberg (2004)
19. Kacprzak, D., Kosiński, W., Kosiński, K.: Financial Stock Data and Ordered Fuzzy Numbers, LNCS (LNAI), vol. 7894, pp. 259–270. Springer, Heidelberg (2013)
20. Kacprzak, M., Kosiński, W., Wegrzyn-Wolska, K.: Diversity of opinion evaluated by ordered fuzzy numbers, LNCS (LNAI), vol. 7894, pp. 271–281. Springer, Heidelberg (2013)
21. Czerniak, J., Apiecionek, L., Zarzycki, H.: Application of ordered fuzzy numbers in a new OFNAnt algorithm based on ant colony optimization. Beyond databases, architectures, and structures. Commun. Comput. Inf. Sci. **424**, 259–270 (2014)

Application of a Double-Impact-Element for Vibrations Attenuation of a Machine Mounted on Nonlinear Support

Marek Lampart and Jaroslav Zapoměl

Abstract The goal of this paper is to concentrate on dynamics and vibration attenuation of a machine coupled with a baseplate by nonlinear disc springs and damped by a double-impact-element. The system motion is governed by a set of three mutually coupled second-order ordinary differential equations. Analysis of behavior of the linearized system shows that multiple motion states can occur in the resonance area which corresponds to multiple solutions of the linearized equation of motion of the machine. To reduce maximum amplitude of the machine vibrations a double-impact-element damper has been chosen. Then the most significant system nonlinearities are stiffness of the support spring elements and internal impacts. As the main results it is shown that the double impact damping device massively attenuates vibrations of the machine frame for a suitable choice of the clearances between the impact elements in dependence on the excitation frequency of the baseplate.

1 Introduction

The impacts of solids are characterized by short duration of the bodies collisions, very large impact forces and by near sudden changes of the system state parameters. The experience and theoretical analyses show that the behavior of such systems is highly nonlinear, highly sensitive to initial conditions and instantaneous excitation

M. Lampart (✉)
Department of Applied Mathematics & IT4 Innovations,
VŠB - Technical University of Ostrava, Ostrava, Czech Republic
e-mail: marek.lampart@vsb.cz
URL: http://www.vsb.cz

J. Zapoměl
Department of Applied Mechanics, VŠB - Technical University of Ostrava,
Ostrava, Czech Republic
e-mail: jaroslav.zapomel@vsb.cz

J. Zapoměl
Department of Dynamics and Vibrations, Institute of Thermomechanics,
Prague, Czech Republic

A. Abraham et al. (eds.), *Proceedings of the Second International Afro-European Conference for Industrial Advancement AECIA 2015*, Advances in Intelligent Systems and Computing 427, DOI 10.1007/978-3-319-29504-6_47

effects, leading frequently to irregular vibrations and nearly unpredictable movements. The behavior of each system where the body collisions take place is different, and therefore, each of them must be investigated individually.

Because of the practical importance, a good deal of attention is focused on analysis of vibro-impact systems, where the vibrations are governed by the momentum transfer and mechanical energy dissipation through the body collisions. This is utilized for impact dampers applied to attenuate high-amplitude oscillations, such as those appearing in subharmonic, self-excited and chaotic vibrations.

Even though the problem of impacts is very old, the new possibilities of its investigation enabled by efficient computational simulations appeared at the end of the 20th century. Chaterjee et al. [1] studied the dynamic behavior of an impact damper for vibration attenuation of an externally loaded and self-excited cart moving in one direction. The damping was produced by impacts of a point body colliding with the cart walls.

A new mechanical model with clearances for a gear transmission was reported in [6]. The model has time varying boundaries and impacts between two gears occur at different locations. The horizontal movement of a cart excited by a rotating particle and damped by an impact damper formed by a point body bouncing on the cart walls was investigated by de Souza et al. [9].

Wagg [10] studied periodic sticking motion of a two degree-of-freedom impact oscillator for several parameter ranges. He concentrated on influence of the forcing frequency and the coefficient of restitution on the system vibrations and described three types of post-bifurcation behavior which occur if the coefficient of restitution takes a zero value. Metallidis and Natsiavas [8] generalized their previous works on the dynamics of discrete oscillators and investigated the response of a continuous system with clearance and motion-limiting constraints. The approach used by the authors is based on the exact solution form obtained within time intervals where the system parameters remain constant. Results of the parametric studies demonstrated a strong influence of super- and subharmonic modes on the system vibrations.

Luo et al. [7] analyzed oscillation of a two degrees-of-freedom system with a clearance subjected to harmonic oscillations concentrating on influence of the exciting frequency and the clearance width. Based on the sampling ranges of dynamic parameters, investigation of their influence on impact velocities, existence regions and correlative distribution of different types of periodic-impact motions of the system was the main objective of the analysis. The computational simulations confirmed a chaotic character of the oscillations. Consequently, the influence of the impact body to the cart mass ratio on its suppression was examined.

In this paper, a system formed by a machine mounted on a frame flexibly coupled with a baseplate and of an impact damper consisting of two impact elements is analyzed. A new contribution of the presented work consists of investigating the system oscillations as a result of the baseplate vibrations. The emphasis is put on observing the influence of the inner impacts on the character and reduction of the system vibration in dependence on the width of the clearances between the impact bodies and the damper housing.

The investigated system is of a great practical importance as it represents a simplified model of a machine excited by a ground vibration and damped by an impact damper. Results of the performed simulations contribute to better understanding of the dynamic behavior of such technological devices and of impact systems in general.

2 The Vibrating System

The investigated system consists of a machine mounted on a frame (body 1, Fig. 1) and of a baseplate (body 2, Fig. 1), with which the machine casing is coupled by disc springs and a viscous damping element. The machine and the baseplate can move in a vertical direction. Vibration of the baseplate is the main source of the system excitation. The values of parameters are summarized in Table 3.

The main goal was to evaluate amplitude of the machine vibrations and if excessive to propose a damping device for their reduction.

In the computational model all bodies are considered as absolutely rigid. The disc springs coupling the machine frame and the baseplate have a nonlinear cubic characteristic

$$F_K = k_1 \varDelta + k_3 \varDelta^3 \tag{1}$$

where F_K is the spring force, k_1, k_3 are the stiffness parameters and $\varDelta$ is the spring deformation (compression or extension).

The investigated system has one degree of freedom. The equation of motion takes the form:

$$m\ddot{y} = F_{ZK} + F_{ZB} - mg + F_{p1}, \tag{2}$$

where y is vertical displacement of the machine frame, $(\dot{\ })$, $(\ddot{\ })$ denote the first and second derivative with respect to time respectively.

Here, $F_{\bullet}$ stand for the forces:

$$\begin{aligned} F_{ZK} &= -k_3(y - y_z)^3 - k_1(y - y_z), \\ F_{ZB} &= -b(\dot{y} - \dot{y}_z), \end{aligned}$$

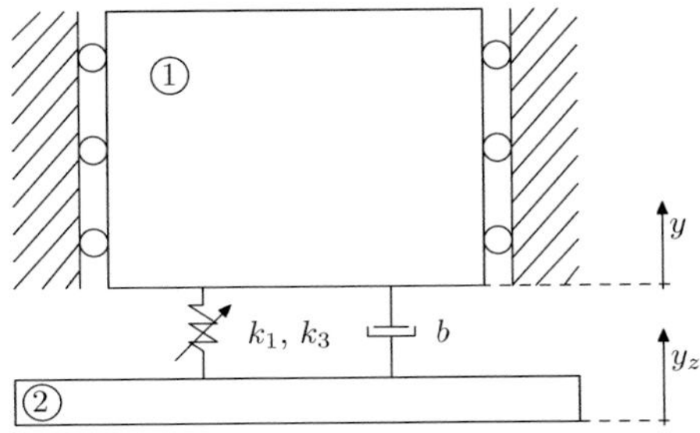

Fig. 1 Model of vibrating system without damping device

$$F_{p1} = mg. \tag{3}$$

Excitation of the baseplate is governed by

$$y_z(t) = A\,(1 - \mathrm{e}^{-\alpha t})\sin(\omega t) \tag{4}$$

where A is the amplitude, α is the constant determining how fast the vibration of the baseplate becomes steady state and ω stands for the excitation frequency, so $y_z(0) = 0$ and $\dot{y}_z(0) = 0$.

After introducing the substitutions:

$$\Delta y = y - z,$$

$$\Delta \dot{y} = \dot{y} - \dot{z},$$

$$\Delta \ddot{y} = \ddot{y} - \ddot{z},$$

and taking into account (3) the equation of motion (2) describing the steady state vibrations of the system (for $t \gg 0$) is transformed to

$$m\Delta\ddot{y} + b\Delta\dot{y} + k_1\Delta y + k_3\Delta y^3 = mA\omega^2\sin(\omega t). \tag{5}$$

The equivalent stiffness of the nonlinear disc spring element obtained by application of a direct linearization method takes the form

$$k_e = \frac{5}{7}k_3\Delta Y^2. \tag{6}$$

After substitution of (6) into (5) the linearized equation of motion reads

$$m\Delta\ddot{y} + b\Delta\dot{y} + (k_1 + \frac{5}{7}k_3\Delta Y^2)\Delta y = mA\omega^2\sin(\omega t). \tag{7}$$

Calculation of the relative amplitude of its steady state solution arrives at solving a cubic algebraic equation

$$a_2^2B^3 + 2a_2(a_1 - \omega^2)B^2 + [a_3 + (a_1 - \omega^2)^2]B - A^2\omega^4 = 0 \tag{8}$$

where $a_1 = k_1/m$, $a_2 = 5k_3/(7m)$ and $a_3 = (b\omega/m)^2$.

The frequency response of the relative motion of the machine with respect to the baseplate is drawn in Fig. 2.

The results show the stiffening of the system with rising excitation frequency and possibility of occurrence of multiple operating states.

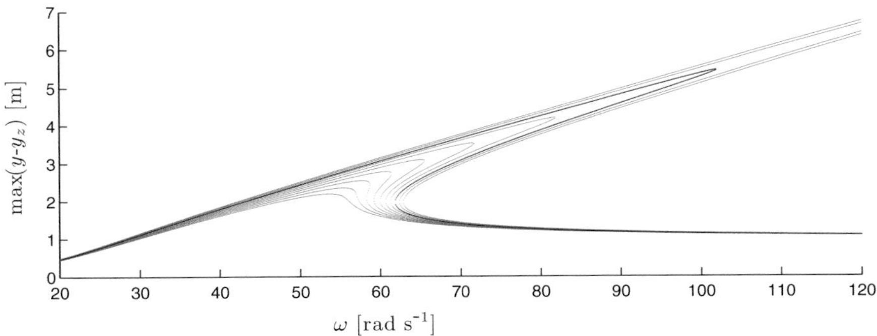

Fig. 2 Dependence of max $(y - y_z)$ on the excitation frequency ω with dependence on damping coefficient of the suspension b form 2400 N s m^{-1} (line with the highest maximal value) to 3300 N s m^{-1} (line with the lowest maximal value) by 100 N s m^{-1}, *black* one stands for 2600 N s m^{-1}

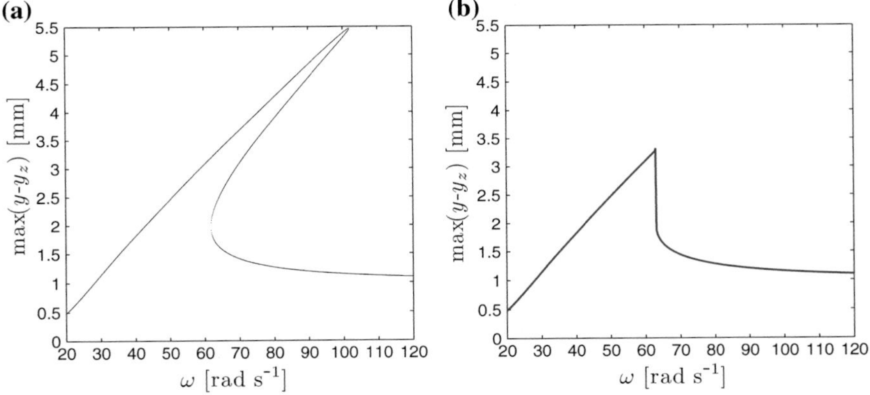

Fig. 3 Dependence of max $(y - y_z)$ on the excitation frequency ω with the suspension $b = 2600$ N s m $^{-1}$, here **a** corresponds to the linearized and **b** to nonlinear system

The same results obtained by solving the linearized (7) and nonlinear (5) equations of motions are depicted in Fig. 3 for the damping coefficient $b = 2600$ N s $^{-1}$.

To reduce the vibration amplitude in the resonance area the impact damping elements were proposed.

3 The Vibrating System with an Impact Damping Device

The proposed damping device consist of two impact elements (body 3 and 4) elastically coupled with the damper housing as shown in Fig. 4. The impact bodies can move only in a vertical direction and are separated mutually and from the housing by the middle, upper and lower clearances that limits their vibration amplitude.

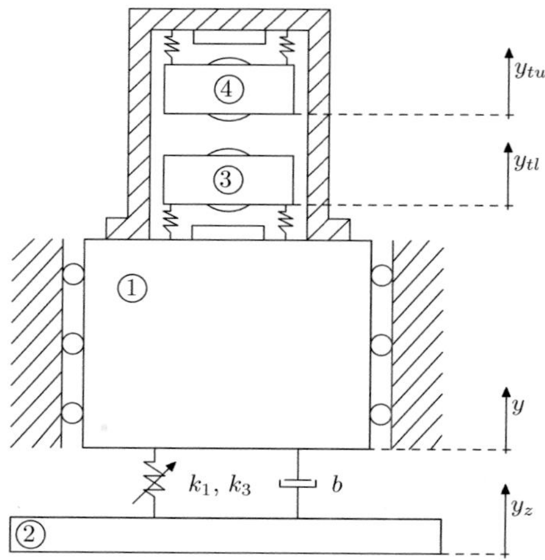

Fig. 4 Model of vibrating system with damping device

The next investigations are focused on analyzing the influence of the upper and lower clearances and the mass of the impact bodies respectively on attenuation of the machine oscillation and character of its motion.

In the computational model all bodies are considered as absolutely rigid except the contact areas between the impact elements and the damper housing.

The investigated system has three degrees of freedom. The equations of motion have been derived by means of the Lagrange equations of the second kind. An alternative approach utilizing the liberation method is discussed in [3]. Its instantaneous position is defined by three generalized coordinates: y—vertical displacement of the machine casing, y_{tu}—vertical displacement of the upper impact body, y_{tl}—vertical displacement of the lower impact body:

$$\begin{aligned} m\ddot{y} &= -(k_{tu}+k_{tl})y + k_{tu}y_{tu} + k_{tl}y_{tl} + \\ &\quad +F_{ZK} + F_{ZB} + F_{Ru} - F_{Rl} - mg + F_{p2}, \\ m_{tl}\ddot{y}_{tl} &= -F_{Rs} + F_{Rl} - k_{tl}(y_{tl} - y) - m_{tl}g, \\ m_{tu}\ddot{y}_{tu} &= F_{Rs} - F_{Ru} - k_{tu}(y_{tu} - y) - m_{tu}g. \end{aligned} \tag{9}$$

Here, $F_{\bullet}$ stand for the forces:

$$F_{p2} = (m + m_{tl} + m_{tu})g,$$

$$F_{Ru} = \begin{cases} k(y_{tu} - c_u - y) & \text{if } y_{tu} - c_u - y > 0, \\ 0 & \text{if } y_{tu} - c_u - y \leq 0, \end{cases}$$

$$F_{Rs} = \begin{cases} k(y_{tl} - c_s - y_{tu}) & \text{if } y_{tl} - c_s - y_{tu} > 0, \\ 0 & \text{if } y_{tl} - c_s - y_{tu} \leq 0, \end{cases}$$

$$F_{Rl} = \begin{cases} k(y - c_l - y_{tl}) & \text{if } y - c_l - y_{tl} > 0, \\ 0 & \text{if } y - c_l - y_{tl} \leq 0, \end{cases}$$

where c_u, c_s and c_l denote the clearances between the frame and the upper impact element, upper and lower impact elements and the frame and the lower impact element, respectively, referred to the case when the system takes the equilibrium position.

It can be assumed, without loss of generality, that in the beginning, the system is at rest and takes the equilibrium position with no contacts between the impact bodies and the machine frame. Then the initial conditions are given as follows

$$\begin{aligned} \dot{y}(0) = 0,\ \dot{y}_{tu}(0) = 0,\ \dot{y}_{tl}(0) = 0, \\ k_3 y(0)^3 + k_1 y(0) + (m + m_{tu} + m_{tl})g - F_{p2} = 0, \\ k_{tu}(y_{tu}(0) - y(0)) + m_{tu}g = 0, \\ k_{tl}(y_{tl}(0) - y(0)) + m_{tl}g = 0. \end{aligned} \tag{10}$$

The last three conditions of (10) can be simplified and solved in algebraic form using Cardano's formulas, analogous proof was provided in [2].

The research of the paper naturally continues the topic started in [4] and [5] (for more details see references therein).

4 Evaluation of the Efficiency of the Damping Elements

The following simulations are for the system parameters summarized in Table 3 and the mass of the impact elements $m_{tu} = m_{tl} = 15$ kg. Next, parameters c_u, c_s and c_l are assumed to be equal for simplicity. The baseplate excitation frequencies ω and the clearances c_u, c_s, c_l were taken as variables in the performed analysis. In the following, behavior of the system in dependence on the baseplate excitation and clearances is discussed.

It is evident from the simulations that for the clearances 2, 3, 4 and 5 mm the resonance peak is shifted to the lower excitation frequencies, and the width of the frequency range in which amplitude of vibrations reaches high values is going down. These observations are summarized in Table 1 and are shown in Figs. 5, 6 and 7a.

Table 1 Frequency ranges referred to high vibrations amplitude ($c_u = c_s = c_l = 5$ mm)

Amplitude (mm)	Without damping device (rad s^{-1})	With damping device (rad s^{-1})
1.25	47.85	47.85
1.50	31.89	31.89
1.75	23.92	23.92
2.00	20.51	12.53
2.25	17.09	9.12
2.50	12.53	7.97
2.75	9.12	5.69
3.00	4.56	3.42
3.25	1.14	0

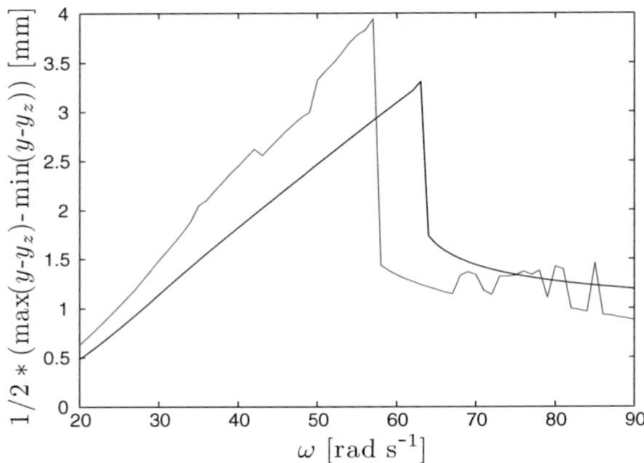

Fig. 5 Dependence of the maximal amplitude of the machine relative motion on the excitation frequency for $c_u = c_s = c_l = 3$ mm

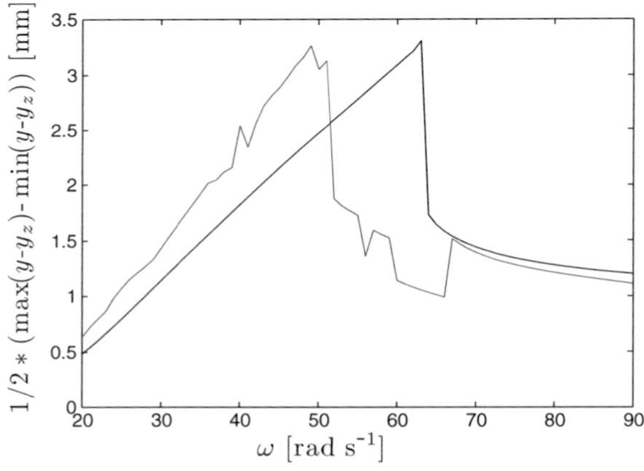

Fig. 6 Dependence of the maximal amplitude of the machine relative motion on the excitation frequency for $c_u = c_s = c_l = 4$ mm

For clearances from about 8 mm the original resonance peak is split into two ones, see Fig. 7b. One peak is shifted to lower and the other to higher frequencies. The value of the peak corresponding to the lower frequency is higher than that of the original resonance peak, value of the second peak is a bit lower. For the clearance of 10 mm the damper behaves as inertial in the region close to the excitation frequency of 58 rad s^{-1} and the vibration amplitude approaches to zero there.

Amplitude of the vibration for low excitation frequencies approaches to the same value for all studied widths of the clearances and the same holds for the high frequencies.

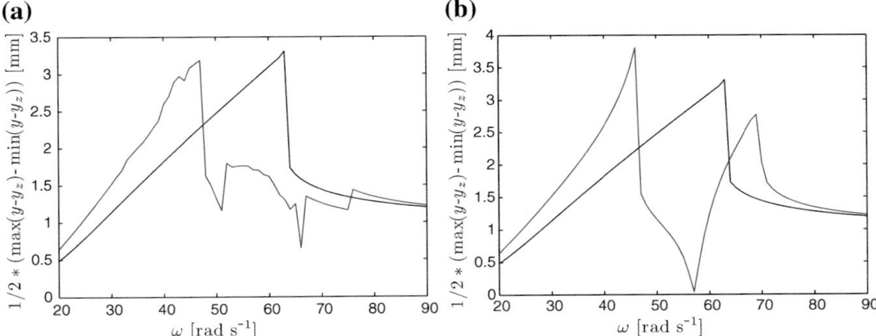

Fig. 7 Dependence of the maximal amplitude of the machine relative motion on the excitation frequency for: **a** $c_u = c_s = c_l = 5$ mm and **b** $c_u = c_s = c_l = 20$ mm

Table 2 Optimum settings of the system parameters

Excitation frequency (rad s^{-1})	Amplitude (mm)
20–46	2
46–60	10
60–80	4

If there is a possibility to control the width of the clearance in dependence on the angular frequency excitation the results of simulations show to set the optimal parameters, as given in Table 2. Such settings minimize the maximum value of the resonant peak, so that it is lower then the original one (Table 3).

Table 3 Parameters of the system (9)

Value	Quantity	Format	Description
m	140	kg	Mass of the damping body
m_{tu}	15	kg	Mass of the upper impact element
m_{tl}	15	kg	mass of the lower impact element
k_1	1.5×10^5	N m^{-1}	Linear stiffness coefficient
k_3	6×10^{10}	N m^{-3}	Cubic stiffness coefficient
b	2.6×10^3	N s m^{-1}	Damping coefficient of the suspension
k	4×10^7	N m^{-1}	Contact stiffness parameter
k_{tu}	8×10^4	N m^{-1}	Coupling stiffness of the upper impact element
k_{tl}	8×10^4	N m^{-1}	Coupling stiffness of the lower impact element
α	1	s^{-1}	Parameter of the baseplate excitation
ω		rad s^{-1}	Baseplate excitation frequency
A	1	mm	Amplitude of y_z
c_u	2	mm	Upper clearance
c_s	2	mm	Middle clearance
c_l	2	mm	Lower clearance

5 Conclusions

In this paper, it was analyzed behavior of a newly designed double impact element with soft stops intended to attenuation of vibration of a machine supported by nonlinear spring elements in dependence on the excitation frequency and namely the width of the clearances. This model was inspired by real frequently occurring technological problems when machines are excited by the ground vibrations. The equations of motions were solved numerically by the explicit Runge-Kutta method using Matlab. The computational simulations showed that the excitation frequency of vibrations of the baseplate played a key role here. These simulations also proved that a suitable application of the double-element-impact damper enables to achieve a significant decrease of the machine vibration amplitude.

Acknowledgments This work was supported by the European Regional Development Fund in the IT4Innovations Centre of Excellence Project (CZ.1.05/1.1.00/02.0070). The work was also supported by the Grant Agency of the Czech Republic, Grant No. 15-06621S.

References

1. Chaterjee, A.K., Mallik, A., Ghosh, A.: On impact dampers for non-linear vibration systems. J. Sound Vib. **187**, 403–420 (1995)
2. Lampart, M., Zapoměl, J.: Dynamics of the electromechanical system with impact element. J. Sound Vib. **332**, 701–713 (2013)
3. Lampart, M., Zapoměl, J.: Dynamical properties of the electromechanical system damped by impact element with soft stops. Int. J. Appl. Mech. **6**, 1450016 (2014)
4. Lampart, M., Zapoměl, J.: Vibration attenuation of an electromechanical system coupled with plate springs damped by an impact element. To appear in Int. J. Appl. Mech. (2015)
5. Lampart, M., Zapoměl, J.: Vibration attenuation of the electromechanical system by a double impact element. To appear in Proceedings of The Second Euro-China Conference on Intelligent Data Analysis and Applications (2015)
6. Luo, A.C.J., O'Connor, D.: Periodic motions with impacting chatter and stick in a gear transmission system. ASME J. Vib. Acoust. **131** (2009)
7. Luo G.W., Lv, X.H., Shi, Y.Q.: Vibro-impact dynamics of a two-degree-of freedom periodically-forced system with a clearance: diversity and parameter matching of periodic-impact motions. Int. J. Non-Linear Mech. **65**, 1–286 (2014)
8. Metallidis, P., Natsiavas, S.: Vibration of a continuous system with clearance and motion constraints. Int. J. Non-Linear Mech. **35**, 675–690 (2000)
9. de Souza, S.L.T., Caldas, I.L., Viana, R.L., Balthazar, J.M., Brasil, R.M.L.R.F.: Impact dampers for controlling chaos in systems with limited power supply. J. Sound Vib. **279**, 955–967 (2005)
10. Wagg, D.J.: Periodic sticking motion in a two-degree-of-freedom impact oscillator. Int. J. Non-Linear Mech. **40**, 1076–1087 (2005)

Prediction of Football Match Results in Turkish Super League Games

Pınar Tüfekci

Abstract This paper presents a model that predicts the results (Home Win/Draw/Away Win) of the football matches played in the Turkish Super League, by using three machine learning classification methods, which are Support Vector Machines (SVMs), Bagging with REP Tree (BREP), and Random Forest (RF). The dataset used in this study includes the data of 70 features, which are composed of 69 input variables relating to statistical data of home and away teams, and a target variable in 1222 total football games in a 4-year period between 2009 and 2013. In connection, the most effective features of these input variables were determined as a reduced dataset by using some feature selection methods. The results showed that the match outcomes were predicted using the reduced dataset better than using the original dataset and the RF classifier produced the best result.

Keywords Data mining · Prediction competitions · Prediction sports

1 Introduction

Prediction the outcome of any sport game is an attractive and difficult problem [1]. Large quantities of historical data from different media outlets regarding the structure and outcomes of sporting events, in the form of a variety of numerically or symbolically represented factors are assumed to contribute to solve this problem [2]. Moreover, forecasting sports is comprised of win/draw/loss results, and exact scores, and both perspectives are used by the bookmakers on the bets [3, 4].

Recently, some of the machine learning approaches have proposed for the prediction of the sporting events such as football (soccer), tennis, bowl games, and horseracing. In this paper we address the issues related to the match result (home

P. Tüfekci (✉)
Çorlu Engineering Faculty, Computer Engineering Department,
Namık Kemal University, 59860 Çorlu, Tekirdağ, Turkey
e-mail: ptufekci@nku.edu.tr

A. Abraham et al. (eds.), *Proceedings of the Second International Afro-European Conference for Industrial Advancement AECIA 2015*, Advances in Intelligent Systems and Computing 427, DOI 10.1007/978-3-319-29504-6_48

win, draw or away win) forecasting in association football. In recent studies relating to football games, when the goal is to predict the result of a football match, the classification techniques from the field of machine learning such as Bayesian network models [5–7], Artificial Neural Networks (ANNs) [2, 8], Support Vector Machines (SVMs), and Decision Trees (DT) [2] are applied to a dataset, which covers a set of past data based on the statistical data of both competitor teams. In addition, when the goal is to predict the exact goal score of a match (0–0, 2–1, 1–3, etc.) [9], the regression techniques of the field of the machine learning are applied to the dataset [2, 10].

The purpose of this study is to predict the results of the football matches played in the Turkish Super League. In order to predict the match results, which are "Home Win", "Draw" and "Away Win", the classification methods, which are the SVMs, BREP, and RF classifiers, are applied to a dataset. The dataset includes the data of 70 features, which are composed of 69 input variables relating to statistical data of home and away teams, and a target variable in 1222 total football games between 2009 and 2013.

During the prediction of the match results, the following features were taken into account; general features of the match, features of a Home Team (HT) and an Away Team (AT) at home, away, and overall. To determine which features in the dataset are more effective in predicting the match result, some feature selection methods, which are filter-based and wrapper-based methods, are applied to the dataset and then a reduced dataset including 38 features has been found. As a result, the prediction of the results of the football matches played in the Turkish Super League was obtained with higher accuracies by using the reduced dataset compared to the original dataset. Thus, the match results were predicted with a 70.87 % prediction accuracy by using the reduced dataset. Furthermore, the best performance was found by the RF classifier.

The rest of the paper is organized as follows: Sect. 2 introduces feature selection process. Section 3 introduces the classifiers used in this study. Section 4 mentions comparative experiments and results. The conclusion and future works are summarized in the final section.

2 Feature Selection Process

Data pre-processing is a significant stage that contains the processes of cleaning, integration, transformation, and reduction of data for using quality data in machine learning (ML) algorithms. Data sets may variety in dimension from two to hundreds of features, and many of these features may be irrelevant or redundant. Feature subset selection decreases the data set dimension by removing irrelevant and redundant features from an original feature set. The objective of feature subset selection is to procure a minimum set of original features. Using the decreased set of original features enables ML algorithms to operate faster and more effectively.

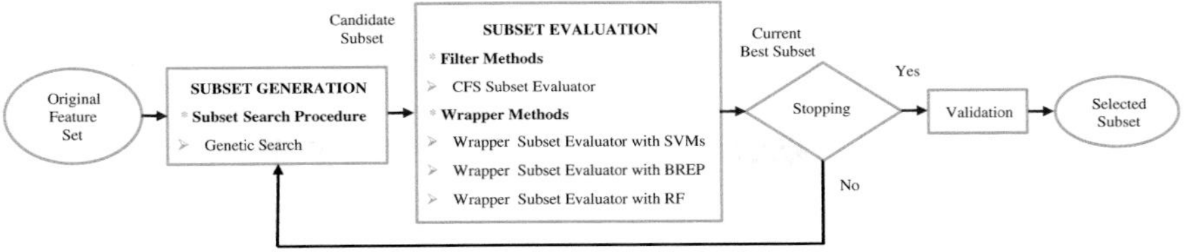

Fig. 1 The feature selection process applied in this study

However, it helps to predicate more correctly by increasing learning accuracy of ML algorithms and improving result comprehensibility [11].

The feature selection process has three basic stages such as generation, evaluation and stopping criteria as shown in Fig. 1. The validation stage, which checks the validity of the selected subset and compares the results to find the best feature subset, may not be a stage of the feature selection process [12].

The feature selection process begins by inputting an original feature set, which includes n number of features or input variables. At the first stage of feature selection, which is called subset generation, a search strategy is used for producing possible feature subsets of the original feature set for evaluation. There are several search procedures to find the optimal subset of the original feature set [12]. In this study, as indicated in Fig. 1, Genetic Search is chosen for this task in WEKA [13]. After the candidate subsets are generated, an evaluation algorithm determines a current best subset. The evaluation algorithms are divided into two broad groups such as Filter and Wrapper methods based on their dependence on the inductive algorithm that will finally use the selected subset. In this study, a filter-based feature selection method, CFS Subset Evaluator [14] and three wrapper-based feature selection methods, which are the Wrapper Subset Evaluators with SVMs, BREP, and RF, are adopted to the dataset [15].

A stopping criterion is needed for stopping the search and preventing an exhaustive search of subsets. The feature selection process halts by outputting the selected subset of features, which is then validated [16].

3 Machine Learning Methods for Classification

A machine learning algorithm estimates an unknown dependency between the inputs, which are independent variables, and output, which is a dependent variable, from a dataset. In most problem domains, there is no functional relationship $y = f(x)$ between y and x where x is input variable and y is output. In this case, the relationship between the inputs and output has to be described more generally by a probability distribution $P(x, y)$ [17].

In this study, SVMs, BREP, and RF methods, which are stated by the WEKA statistical analysis package [13], are applied for the task of data classification. These classification methods are generated as learning algorithms to predict the match results such as "Home Win", "Draw" or "Away Win" results. In order to objectively compare performance of these classification methods, tenfold cross-validation (CV) was applied as a validation scheme where the dataset is partitioned into equal sized subsets. In the general case of K-fold CV, at each machine learning experiment one subset is used for validation (i.e. to test the predictive model) and the rest is for training [18]. For the generalization of results, each cell in the following tables (Tables 3 and 4) corresponds to tenfold CV. A brief summary of the classifiers is explained by following subsections.

3.1 Support Vector Machines (SVMs)

Support vector machines are kernel based learning algorithm for solving classification and regression problems. The SVMs build optimal separating boundaries between data sets by solving a constrained quadratic optimization problem [19]. The algorithm of SVMs defines the best hyper-plan that separates the set of points associated with different class labels with a maximum-margin. The unlabeled examples are then classified by deciding on which side of the hyper-surface they reside by using different kernel functions, varying degrees of nonlinearity and flexibility can be included in the model. Because they can be derived from advanced statistical ideas, and bounds on the generalization error can be calculated for them [20].

3.2 Bagging REP Tree (BREP)

Bagging or bootstrap aggregating is general technique for improving prediction rules by creating various sets of the training sets. Bagging algorithm is applied to tree-based methods such as REP Trees to reduce the variance associated with prediction, and therefore, increase the accuracy of the resulting predictions. Bagging can be formalized as follows

$$\hat{y}_{BAG} = \frac{1}{B} \sum_{b=1}^{B} \varnothing(x; T_b) \tag{1}$$

where B is the number of bootstrap samples of training set T and x is the input. $\hat{y}_{BAG}$ is the average of the different estimated trees. A bootstrap sample is randomly drawn from the training set, but with replacement. The purpose is to create numerous similar training sets using, sampling and train a new function for each of these sets. The functions learned from these sets are then used collectively to predict the test set [21, 22].

3.3 *Random Forest (RF)*

The RF, which is an advanced method of machine learning, has been proposed as one of the most successful classifiers based on assemble learning algorithm. It achieves the classification task by constructing an ensemble of randomized classification and regression trees (CART) [23]. For a given training dataset, $A = \{(x_1, y_1), (x_2, y_2), \ldots, (x_n, y_n)\}$, where $x_i = 1, 2, \ldots, n$, is a variable or vector and y_i is its corresponding property or class label. The basic RF algorithm is presented as follows:

1. Bootstrap sample: Each training set is drawn with replacement from the original dataset *A*. Bootstrapping allows replacement, so that some of the samples will be repeated in the sample, while others will be “left out” of the sample. The “left out” samples constitute the “Out-ofbag (OOB)”.
2. Growing trees: For each bootstrap sample, a tree is grown. *m* variables (m_{try}) are selected at random from all *n* variables ($m_{try} \leq n$) and the best split of all m_{try} is used at each node. Each tree is grown to the largest extent (until no further splitting is possible) and no pruning of the trees occurs.
3. OOB error estimate: Each tree is constructed on the bootstrap sample. The OOB samples are not used and therefore regarded as a test set to provide an unbiased estimate of the prediction accuracy. Each OOB sample is put down the constructed trees to get a classification.
4. Variable importance: RF has the ability to rank the variable importance. For each tree grown in the forest, put down the OOB and count the number of votes cast for the correct class. Permute the value of variable *m* in OOB randomly and put these samples down the tree. Count the number of votes for the correct class in the variable-m-permuted OOB data. Again count the number of votes for the correct class in the untouched OOB data. Subtracting the two counts and averaging this number over all trees in the forest is the raw importance score for variable *m*. Finally importance score will be computed depending on the correlations between trees [24].

4 Comparative Experiments and Results

The purpose of this study is to choose a predictive model which accurately predicts the results of the football matches played in the Turkish Super League. The prediction results, which are “Home Win”, “Draw”, or “Away Win”, are obtained by using three classifiers. The layout of the proposed prediction model, which is used in our experiments, is illustrated in Fig. 2 and the parts of the structure are explained in the following subsections:

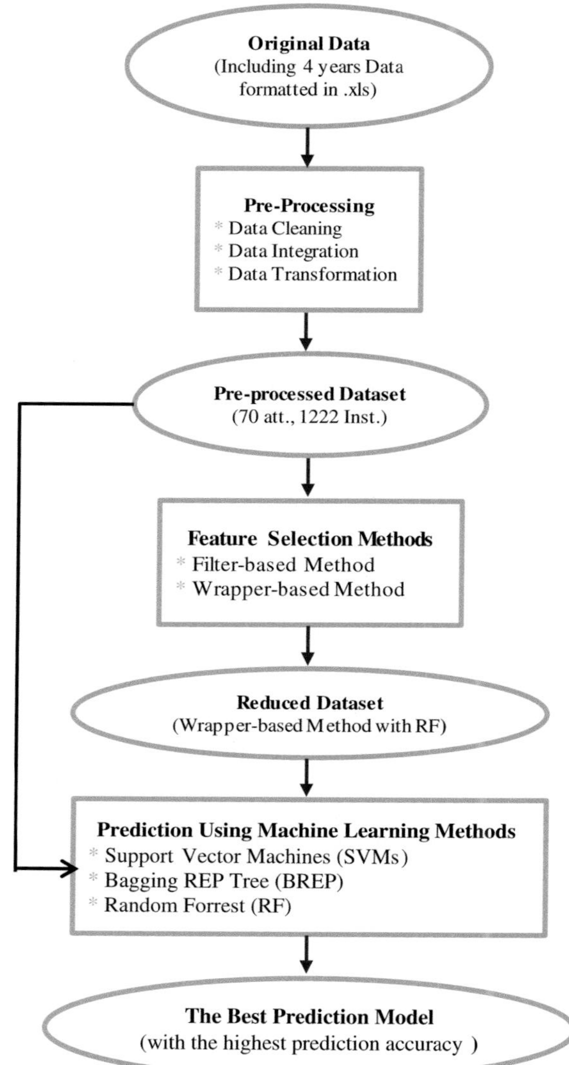

Fig. 2 The flow diagram of the prediction process

4.1 Original Data

The original data used in this study was collected from a variety of sports databases available on the web, including www.mackolik.com, www.fenersk.com, www.sahadan.com and www.tff.org. First, the data of each season were collected from

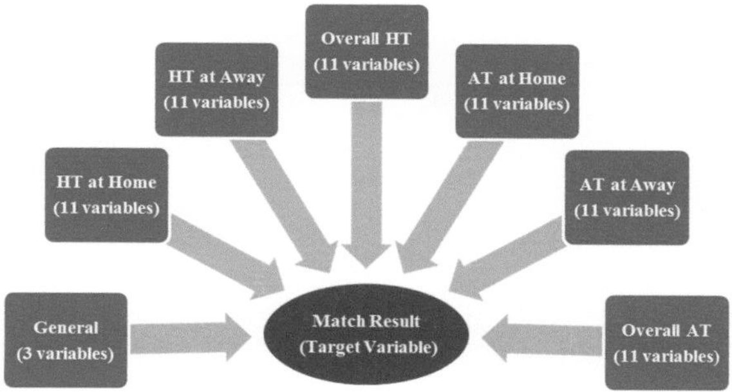

Fig. 3 Contents of the original data

these web sites as 4 different datasets and when they were merged to have a complete dataset, which included 1222 football games, representing a complete set of four seasons of Turkish Super League played between 2009 and 2013. The features that employed 69 input variables were divided mainly through seven categories: general features of a match, features of a HT at home, features of a HT at away, features of overall HT, features of an AT at home, features of an AT at away, features of overall AT, as shown in Fig. 3.

4.2 *Pre-processing*

Data pre-processing is a significant stage for complicated data. It has a very large impact on the success of predictions. Therefore, the original dataset includes four datasets for four seasons between 2009 and 2013. Each dataset of a season consists of 34 files formatted in.xls, which are composed of 70 features for the football games played, totaly 34 weeks of a season indicated in Table 1.

At the beginning of the pre-processing stage, each data employs the 70 variables, which are indicated in Table 2, and then is cleaned from all noises and incompatible data. Afterwards, each data is merged to eliminate any duplicated data and integrated into the data set. Then, the dataset is integrated as a .xls file, which has been transformed into an .arff format that is necessary for processing into WEKA tool.

Table 1 Content of the dataset used in this study

Content of dataset	# att.	# inst.
2009–2010 season	70	305
2010–2011 season	70	306
2011–2012 season	70	306
2012–2013 season	70	305
2009–2013 seasons	70	1222

Table 2 Features of the dataset

General features			Away Team (AT) features		
General	F1	Derby game	AT at home	F37	AT played games at home
	F2	Penalty without spectators		F38	AT games won at home
	F3	Stadium ban		F39	AT games drawn at home
Home Team (HT) features				F40	AT games lost at home
HT at home	F4	HT games played at home		F41	AT total goals for at home
	F5	HT games won at home		F42	AT total goals against at home
	F6	HT games drawn at home		F43	AT goal difference at home
	F7	HT games lost at home		F44	AT points at home
	F8	HT total goals for at home		F45	AT goals for per game at home
	F9	HT total goals against at home		F46	AT goals against per game at home
	F10	HT goal difference at home		F47	AT percent of under-over 2.5 goals at home
	F11	HT points at home	AT at away	F48	AT games played at away
	F12	HT goals for per game at home		F49	AT games won at away
	F13	HT goals against per game at home		F50	AT games drawn at away
	F14	HT percent of under-over 2.5 goals at home		F51	AT games lost at away
HT at home	F15	HT games played at away		F52	AT goals for at away
	F16	HT games won at away		F53	AT goals against at away
	F17	HT games drawn at away		F54	AT goal difference at away
	F18	HT games lost at away		F55	AT points at away
	F19	HT total goals for at away		F56	AT goals for per game at away
	F20	HT total goals against at away		F57	AT goals against per game at away
	F21	HT goal difference at away		F58	AT percent of under-over 2.5 goals at away

(continued)

Table 2 (continued)

General features			Away Team (AT) features		
	F22	HT points at away	Overall AT	F59	AT games played
	F23	HT goals for per game at away		F60	AT games won
	F24	HT goals against per game at away		F61	AT games drawn
	F25	HT percent of under-over 2.5 goals at away		F62	AT games lost
Overall HT	F26	HT games played		F63	AT total goals for
	F27	HT games won		F64	AT total goals against
	F28	HT games drawn		F65	AT total goal difference
	F29	HT games lost		F66	AT goals for per game
	F30	HT total goals for		F67	AT goals against per game
	F31	HT total goals against		F68	AT points
	F32	HT total goal difference		F69	AT percent of under-over 2.5 goals
	F33	HT goals for per game	Target variable	F70	Match result (fulltime)
	F34	HT goals against per game			
	F35	HT points			
	F36	HT percent of under-over 2.5 goals			

4.3 Feature Selection Process

After the data pre-processing, the dataset is applied to the process of feature selection to eliminate irrelevant and redundant features in the dataset and to reduce its size. In this study, the WEKA tool has been used to get the results of measurements and to compare the performances of each feature selection algorithm. To select the feature subsets for the dataset, CFS SubSetEval Evaluator and Wrapper SubsetEval Evaluators with Genetic Search procedure, are chosen with their default threshold settings. The best evaluator with the search procedure, which has the highest performance for the predictions of the match results, is explored with some experiments.

For this purpose, the classifiers have been applied to the reduced dataset and the results, which are evaluated using the accuracy measure, are compared with each other shown in Table 3. Moreover, the Genetic Search procedure is used as the search procedure with all the evaluators for determining the reduced dataset.

As a result of this comparison, the best performance for the reduced dataset, which is produced by applying Wrapper Subset Evaluator with RF classifier, is found with a 70.87% prediction accuracy by the RF classifier. The number of

Table 3 The results of the feature selection process

Feature selection method (with genetic search)	# att.	SVMs	BREP	RF
CFS SubSetEval	18	67.18	66.28	70.79
Wrapper SubSetEval with SVMs	34	68.66	68.00	70.38
Wrapper SubSetEval with BREP	35	68.25	67.68	70.38
Wrapper SubSetEval with RF	38	67.68	69.48	**70.87**
Mean		67.94	67.86	**70.61**

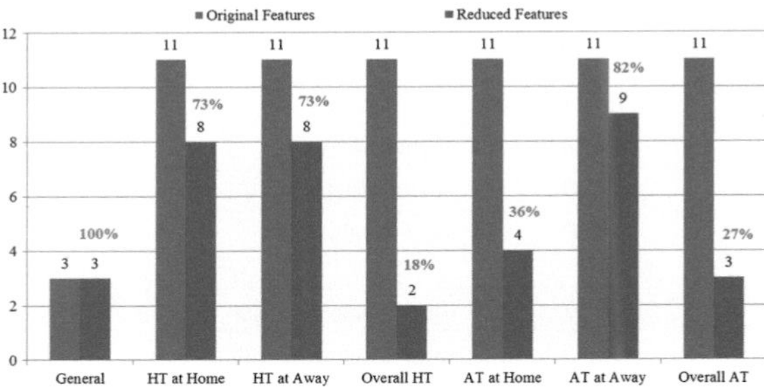

Fig. 4 Distribution of the features for the original and reduced datasets

attributes of the original dataset is reduced from 70 attributes to 38 attributes to obtain the most successful prediction accuracy. According to mean values in Table 3, the best classifier is found as the RF classifier with a 70.61 % prediction mean accuracy. The distribution of the features for the original dataset and the reduced dataset including 38 attributes, which are determined by Wrapper SubSetEval Evaluator with RF, is shown in Fig. 4.

Figure 4 indicates that the features used in the reduced dataset comprise all of the general features of a match, and 16 features of the 22 total HT features at home and away, and 2 features of the 11 features of the overall HT. In addition, 4 features of the 11 features of AT at home, and 9 features of the 11 features of AT at away, and 3 features of the 11 features of the overall AT are used in the reduced dataset for predicting the target variable, which is the match result. The most effective features with the usage rate at 100 % are related on the general features of a match, which specify if the match is a derby match, and there are any penalties of without spectators, or stadium bans. Afterwards, the features of AT at away with 82 %, the features of HT at home and away with 73 %, the features of AT at home with 36 %, overall with 27 %, and the features of HT overall with 18 % are used in the prediction of the match result respectively as indicated in Fig. 4.

4.4 Prediction Process and Results

In the prediction process, three classification methods are applied to the original and reduced datasets, for determining the most successful model. Thus, the results for the original and reduced datasets are compared with each other as shown in Table 4.

Table 4 The results of the prediction process

Dataset	# att.	SVMs	BREP	RF
The original dataset	70	67.10	67.43	69.97
The reduced dataset	38	67.68	69.48	70.87
Mean		67.39	68.46	70.42

According to the accuracies in Table 4, the feature selection process has increased the prediction performance. The prediction accuracies have been increased from 67.10 to 67.68 % by the SVMs classifier, from 67.43 to 69.48 % by the BREP classifier, and from 69.97 to 70.87 by the RF classifier. The RF classifier for both datasets, which are the original and reduced datasets, gives the highest accuracy values with a 70.42 % prediction mean accuracy.

5 Conclusion

During this study, the results of the football matches played in the Turkish Super League were predicted by using three machine learning classification methods, which are Support Vector Machines, Bagging with REP Tree, and Random Forest. In addition, the most effective features of the dataset consisting of 70 features, which are composed of 69 input variables relating to statistical data of home and away teams, and a target variable of 1222 total football games during a period between 2009 and 2013, were explored by using feature selection methods, which are filter-based and wrapper-based methods. Consequently, the most important factors were determined as the reduced dataset including 38 features.

According to the reduced dataset, the most effective features with the usage rate at 100 % are related on the general features of a match, which specify if the match is a derby match, and if there are any penalties without spectators, or stadium bans. Afterwards, the features of AT at away with 82 %, the features of HT at home and away with 73 %, the features of AT at home with 36 %, overall with 27 %, and the features of HT overall with 18 % are used in the prediction of the match result respectively. Hence, the most effective features of the dataset were determined as a reduced dataset and then applied to the classifiers for the prediction process.

As a result, the prediction of the results of the football matches played in the Turkish Super League was obtained with higher accuracies by using the reduced dataset compared to the original dataset. Thus, the match results were predicted with a 70.87 % prediction accuracy by using the reduced dataset including 38 features. However, the best performance was found by the RF classifier. Afterwards, the BREP and SVMs classifiers respectively followed the RF classifier according to the mean result values.

Future work plan is to predict the match results by artificial neural networks (ANNs).

References

1. Stekler, H.O., Sendor, D., Verlander, R.: Issues in sports forecasting. Int. J. Forecast. **26**, 606–621 (2010)
2. Delen, D., Cogdell, D., Kasap, N.: A comparative analysis of data mining methods in predicting NCAA bowl outcomes. Int. J. Forecast. **28**, 543–552 (2012)
3. Baker, R.D., McHale, I.G.: Forecasting exact scores in National Football League games. Int. J. Forecast. **29**, 122–130 (2013)
4. Hvattum, L.M., Arntzen, H.: Using ELO ratings for match result prediction in association football. Int. J. Forecast. **26**, 460–470 (2010)
5. Constantinou, A.C., Fenton, N.E., Neil, M.: pi-footbal: A Bayesian network model for forecasting Association Football match outcomes. Knowl.-Based Syst. **36**, 322–339 (2012)
6. Baio, G., Blangiardo, M.: Bayesian hierarchical model for the prediction of football results. J. Appl. Stat. **37**(2), 253–264 (2010)
7. Joseph, A., Fenton, N., Neil, M.: Predicting football results using Bayesian nets and other machine learning techniques. Knowl.-Based Syst. **7**, 544–553 (2006)
8. Tsakonas, A., Dounias, G., Shtovba, S., Vivdyuk, V.: Soft computing-based result prediction of football games. In: The First International Conference on Inductive Modelling (ICIM'2002). Lviv, Ukraine (2002)
9. Dixon, M., Coles, S.: Modelling association football scores and inefficienies in the football betting market. Appl. Stat. **46**, 265–280 (1997)
10. Goddard, J.: Regression models for forecasting goals and match results in association football. Int. J. Forecast. **21**, 331–340 (2005)
11. Han, J., Kamber, M.: Data mining: concepts and techniques. Morgan Kauffmann Publishers, San Francisco (2001)
12. Dash, M., Liu, H.: Consistency-based search in feature selection. Artif. Intell. **151**, 155–176 (2003)
13. Machine Learning Group at University of Waikato. www.cs.waikato.ac.nz/ml/weka/ (2015). Accessed 09 June 2015
14. Hall, M.A.: Correlation-based feature selection for machine learning. Ph.D. thesis, Department of Computer Science, University of Waikato, Hamilton, New Zeland (1998)
15. Kohavi, R., John, G.: Wrappers for feature subset selection. Artif. Intell. **97**, 273–324 (1997)
16. Hall, M.A., Holmes, G.: Benchmarking attribute selection techniques for discrete class data mining. IEEE Trans. Knowl. Data Eng. **15**, 1437–1447 (2003)
17. El-Sebakhy, E.: Constrained estimation functional networks for statistical pattern recognition problems: theory and methodolgy. Int. J. Pattern Recogn. Artif. Intell. (2007)
18. Alpaydın, E.: Introduction to Machine Learning, 2nd ed. MIT Press (2010)
19. Dreiseitl, S., Ohno-Machado, L.: Logistic regression and artificial neural network classification models: a methodology review. J. Biomed. Inf. **35**(5–6) (2002)
20. Hassan, S., Mihalcea, R., Banea, C.: Random walk term weighting for improved text classification. Int. Seman. Comput. **1**(4) (2007)
21. Breiman, L.: Bagging predictors. Mach. Learn. **24**(2), 123–140 (1996)
22. D'Haen, J., Van den Poel, D.: Temporary staffing services: a data mining perspective. In: IEEE 12th International Conference on Data Mining Workshops (2012)
23. Breiman, L.: Random forests. Mach. Learn. **45**, 5–32 (2001)
24. Zeng, L., Watson, D.G., Johnston, B.F., Clark, R.L., Edrada-Ebel, R., Elseheri, W.: A chemometric study of chromatograms of tea extracts by correlation optimization warping in conjunction with PCA, support vector machines and random forest data modeling. Analytica Chimica Acta **642** (2009)

Chaos PSO with Super-Sized Swarm—Initial Study

Michal Pluhacek, Roman Senkerik and Ivan Zelinka

Abstract In this paper it is investigated the possibility of improving the performance of PSO algorithm with super-sized population. The performance fo canonical PSO with super-sized swarm has been investigated previously and showed promising results. In this study four different chaotic systems are used as pseudo-random number generators for the PSO algorithm. The IEEE CEC'13 benchmark set is used to evaluate the performance of the method.

Keywords Particle swarm optimization · Chaos · PSO · Evolutionary algorithm · Optimization

1 Introduction

The Particle Swarm Optimization algorithm (PSO) [1, 2] is one of the most popular and widely used Evolutionary Computation Techniques (ECTs). The PSO is also one of the most prominent representatives of the "Swarm intelligence" [2]. The main area of application is in global optimization. In recent years, the demand for fast and effective optimizers is increasing and as a reaction for this, the PSO is intensively studied and regularly modified [3–9]. Recently the performance of PSO

M. Pluhacek (✉) · R. Senkerik
Faculty of Applied Informatics, Tomas Bata University in Zlin,
T.G. Masaryka 5555, 760 01 Zlín, Czech Republic
e-mail: pluhacek@fai.utb.cz

R. Senkerik
e-mail: senkerik@fai.utb.cz

I. Zelinka
Faculty of Electrical Engineering and Computer Science,
VŠB-Technical University of Ostrava, 17. listopadu 15, 708 33 Ostrava-Poruba,
Czech Republic
e-mail: ivan.zelinka@vsb.cz

A. Abraham et al. (eds.), *Proceedings of the Second International Afro-European Conference for Industrial Advancement AECIA 2015*, Advances in Intelligent Systems and Computing 427, DOI 10.1007/978-3-319-29504-6_49

algorithm with super-sized population was investigated [10]. It has been shown that with swarm size of 2000 it is possible to achieve better results than with typically used swarm size of 50 and within the same limit of cost function (CF) evaluations (FEs). In this study it is investigated whether it is possible to further improve the performance of PSO algorithm with super-sized swarm by implementation of chaotic sequences [11]. The implementation of chaotic sequences into ECTs is very effective way of performance improving [12, 13].

2 Particle Swarm Optimization

A brief description of PSO algorithm follows in this section. The PSO algorithm takes inspiration from the natural swarm behavior of birds and fish. It was firstly introduced by Eberhart and Kennedy in 1995 [1]. Each particle in the population represents a candidate solution for the optimization problem that is defined by the cost function. In each iteration of the algorithm, a new location (combination of CF parameters) for the particle is calculated based on its previous location and velocity vector (velocity vector contains particle velocity for each dimension of the problem).

According to the method of selection of the swarm or subswarm for best solution information spreading, the PSO algorithms are noted as global PSO (GPSO) or local PSO (LPSO). Within this research the PSO algorithm with global topology (GPSO) [6] was utilized. The chaotic PRNG is used in the main GPSO formula (1), which determines a new "velocity", thus directly affects the position of each particle in the next iteration.

$$v_{ij}^{t+1} = w \cdot v_{ij}^{t} + c_1 \cdot Rand \cdot (pBest_{ij} - x_{ij}^{t}) + c_2 \cdot Rand \cdot (gBest_j - x_{ij}^{t}) \tag{1}$$

where

v_i^{t+1}	New velocity of the *i*th particle in iteration $t + 1$
W	Inertia weight value
v_i^t	Current velocity of the *i*th particle in iteration t
c_1, c_2	Priority factors
$pBest_i$	Local (personal) best solution found by the *i*th particle
$gBest$	Best solution found in a population
x_{ij}^t	Current position of the *i*th particle (component j of dimension D) in iteration t
$Rand$	Pseudo random number, interval (0, 1). CPRNG is applied only here

The maximum velocity was limited to 0.2 times the range as it is usual. The new position of each particle is then given by (2), where x_i^{t+1} is the new particle position:

$$x_i^{t+1} = x_i^t + v_i^{t+1} \tag{2}$$

Finally the linear decreasing inertia weight [5, 7] is used in the typically referred GPSO design that was used in this study. The dynamic inertia weight is meant to slow the particles over time thus to improve the local search capability in the later phase of the optimization. The inertia weight has two control parameters w_{start} and w_{end}. A new w for each iteration is given by (3), where t stands for current iteration number and n stands for the total number of iterations. The values used in this study were $w_{start} = 0.9$ and $w_{end} = 0.4$.

$$w = w_{start} - \frac{((w_{start} - w_{end}) \cdot t)}{n} \quad (3)$$

3 Chaotic Maps

In this section four discrete dissipative chaotic systems (maps) are described. These four chaotic maps were used as CPRNG's for the velocity calculation in PSO (See (1)). The choice was based on previous research [13, 14].

3.1 Lozi Chaotic Map

The Lozi map is a simple discrete two-dimensional chaotic map. The map equations are given in (4). The typical parameter values are: a = 1.7 and b = 0.5 with respect to [11]. For these values, the system exhibits typical chaotic behavior and with this parameter setting it is used in the most research papers and other literature sources. The x, y plot of Lozi map with the typical setting is depicted in Fig. 1.

$$\begin{aligned} X_{n+1} &= 1 - a|X_n| + bY_n \\ Y_{n+1} &= X_n \end{aligned} \quad (4)$$

3.2 Dissipative Standard Map

The Dissipative standard map is a two-dimensional chaotic map [11]. The parameters used in this work are $b = 0.6$ and $k = 8.8$ based on previous experiments [13, 14] and suggestions in literature [11]. The x, y plot of Dissipative standard map is given in Fig. 2. The map equations are given in (5).

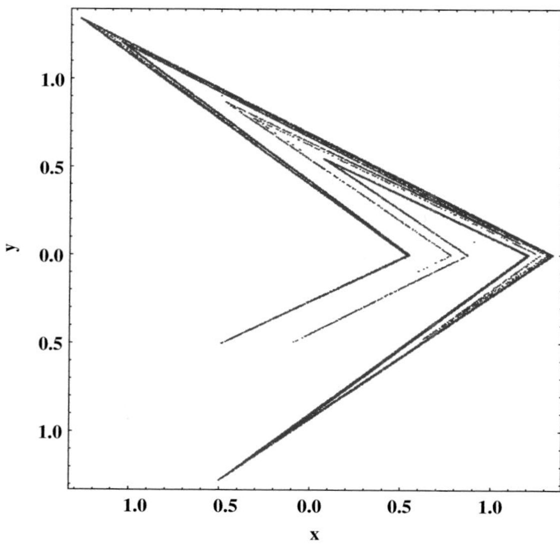

Fig. 1 x, y plot of Lozi map

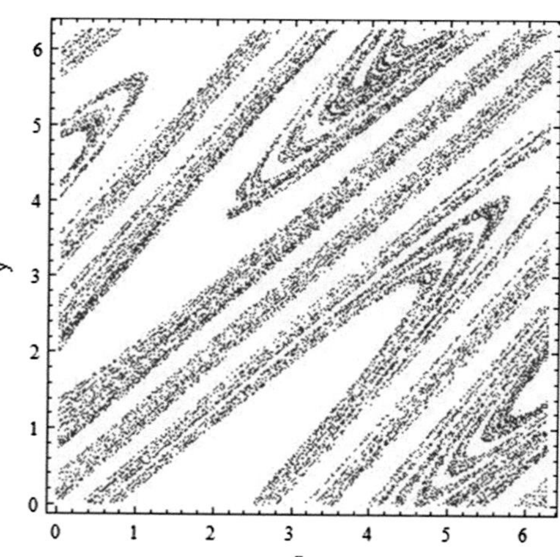

Fig. 2 x, y plot of Dissipative standard map

$$
\begin{aligned}
X_{n+1} &= X_n + Y_{n+1} (\mathrm{mod} 2\pi) \\
Y_{n+1} &= bY_n + k \sin X_n (\mathrm{mod} 2\pi)
\end{aligned}
\tag{5}
$$

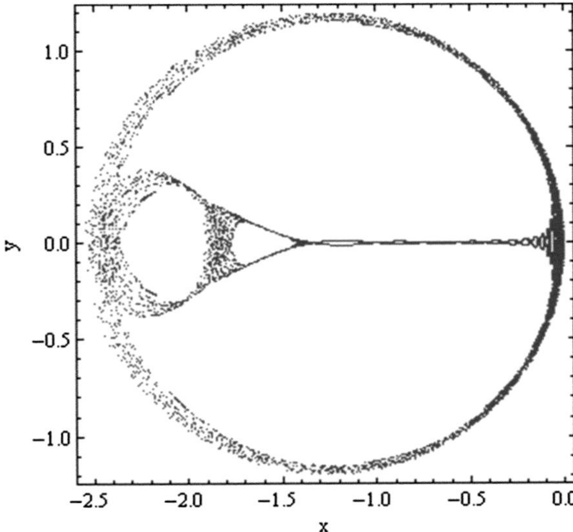

Fig. 3 x, y plot of Burgers map

3.3 *Burgers Chaotic Map*

The Burgers map (See Fig. 3) is a discretization of a pair of coupled differential equations The map equations are given in (6) with control parameters $a = 0.75$ and $b = 1.75$ as suggested in [11].

$$\begin{aligned} X_{n+1} &= aX_n - Y_n^2 \\ Y_{n+1} &= bY_n + X_n Y_n \end{aligned} \tag{6}$$

3.4 *Tinkerbell Map*

The Tinkerbell map is a two-dimensional complex discrete-time dynamical system given by (7) with following control parameters: $a = 0.9$, $b = -0.6$, $c = 2$ and $d = 0.5$ [11]. The x, y plot of the Tinkerbell map is given in Fig. 4.

$$\begin{aligned} X_{n+1} &= X_n^2 - Y_n^2 + aX_n + bY_n \\ Y_{n+1} &= 2X_n Y_n + cX_n + dY_n \end{aligned} \tag{7}$$

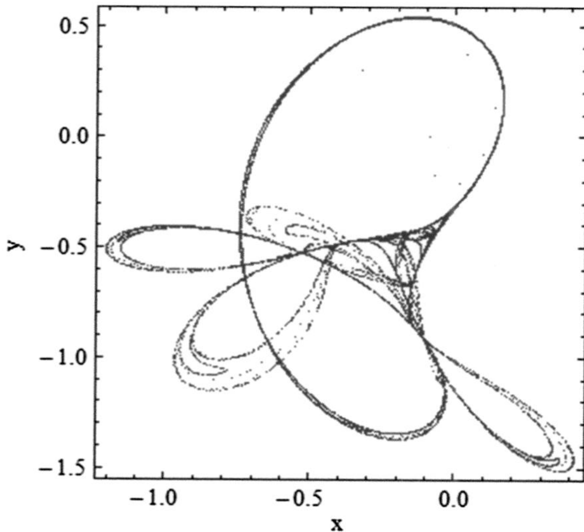

Fig. 4 x, y plot of Tinkerbell map

4 Experiment Setup

In this study, several different PSO variants were compared:

- PSO with canonical pseudo random number generator—noted PSO
- PSO with CPRNG based on Lozi map—noted PSO Lozi
- PSO with CPRNG based on Dissipative standard map—noted PSO Disi
- PSO with CPRNG based on Burgers map—noted PSO Burger
- PSO with CPRNG based on Tinkerbell map—noted PSO Tinker

The CEC'13 benchmark set [15] was evaluated for the dimension setting 10 with the population size 2000 and number of iterations 50. In accordance with previous research [10] the acceleration constants c_1 and c_2 were set to 1.

Other controlling parameters of the PSO were set to typical values as follows:
$w_{start} = 0.9$;
$w_{end} = 0.4$;
$v_{max} = 0.2$;

5 Results

In this section the results are summarized. Table 1 contains the mean results for all PSO variants described above. Furthermore the performance of chaos enhanced PSO variants is compared with the best performing (on CEC'13 Benchmark) PSO-based algorithm [16] in Table 2. The best results are given in bold numbers.

Table 1 Mean results comparison

$f(x)$	PSO	PSO Lozi	PSO Disi	PSO Burger	PSO Tinker
$f(1)$	−1.400E + 03	−1.400E + 03	−1.400E + 03	−1.400E + 03	−1.400E + 03
$f(2)$	**1.534E + 05**	2.340E + 05	2.589E + 05	4.066E + 05	4.022E + 05
$f(3)$	1.764E + 06	**1.688E + 06**	1.809E + 06	2.544E + 06	4.320E + 06
$f(4)$	−1.940E + 02	−3.148E + 02	−4.088E + 02	−2.198E + 02	**−4.142E + 02**
$f(5)$	−1.000E + 03	−1.000E + 03	−1.000E + 03	−1.000E + 03	−1.000E + 03
$f(6)$	−8.924E + 02	−8.902E + 02	−8.895E + 02	−8.896E + 02	**−8.946E + 02**
$f(7)$	−7.938E + 02	−7.951E + 02	−7.952E + 02	**−7.962E + 02**	−7.957E + 02
$f(8)$	−6.796E + 02	**−6.797E + 02**	−6.796E + 02	−6.796E + 02	−6.796E + 02
$f(9)$	−5.967E + 02	**−5.970E + 02**	−5.967E + 02	−5.969E + 02	**−5.970E + 02**
$f(10)$	**−4.998E + 02**	−4.997E + 02	**−4.998E + 02**	−4.995E + 02	−4.996E + 02
$f(11)$	−3.941E + 02	−3.939E + 02	−3.942E + 02	**−3.952E + 02**	−3.946E + 02
$f(12)$	−2.880E + 02	−2.881E + 02	−2.869E + 02	−2.869E + 02	**−2.886E + 02**
$f(13)$	**−1.829E + 02**	−1.810E + 02	−1.815E + 02	−1.823E + 02	−1.811E + 02
$f(14)$	2.226E + 02	2.201E + 02	2.008E + 02	2.034E + 02	**1.992E + 02**
$f(15)$	7.580E + 02	7.635E + 02	**7.141E + 02**	7.197E + 02	7.813E + 02
$f(16)$	2.010E + 02	2.010E + 02	2.010E + 02	2.010E + 02	2.010E + 02
$f(17)$	3.199E + 02	3.205E + 02	**3.192E + 02**	3.224E + 02	3.208E + 02
$f(18)$	4.226E + 02	**4.221E + 02**	4.250E + 02	4.295E + 02	4.279E + 02
$f(19)$	5.009E + 02	5.008E + 02	5.008E + 02	5.008E + 02	5.008E + 02
$f(20)$	**6.025E + 02**	6.026E + 02	**6.025E + 02**	6.028E + 02	6.026E + 02
$f(21)$	1.096E + 03	1.083E + 03	1.090E + 03	**1.081E + 03**	1.090E + 03
$f(22)$	1.186E + 03	1.162E + 03	**1.155E + 03**	1.162E + 03	1.189E + 03
$f(23)$	1.786E + 03	1.785E + 03	1.806E + 03	**1.769E + 03**	1.800E + 03
$f(24)$	1.196E + 03	1.204E + 03	1.202E + 03	1.201E + 03	**1.190E + 03**
$f(25)$	1.292E + 03	1.297E + 03	1.298E + 03	1.299E + 03	**1.289E + 03**
$f(26)$	1.385E + 03	1.385E + 03	1.390E + 03	**1.381E + 03**	1.389E + 03
$f(27)$	1.632E + 03	1.638E + 03	1.636E + 03	1.628E + 03	**1.626E + 03**
$f(28)$	**1.664E + 03**	1.711E + 03	1.698E + 03	1.680E + 03	1.684E + 03

Table 2 Mean results comparison

$f(x)$	fk–PSO	PSO Lozi	PSO Disi	PSO Burger	PSO Tinker
$f(1)$	−1.400E + 03	−1.400E + 03	−1.400E + 03	−1.400E + 03	−1.400E + 03
$f(2)$	**1.430E + 05**	2.340E + 05	2.589E + 05	4.066E + 05	4.022E + 05
$f(3)$	**6.740E + 05**	1.688E + 06	1.809E + 06	2.544E + 06	4.320E + 06
$f(4)$	**−6.840E + 02**	−3.148E + 02	−4.088E + 02	−2.198E + 02	−4.142E + 02
$f(5)$	−1.000E + 03	−1.000E + 03	−1.000E + 03	−1.000E + 03	−1.000E + 03
$f(6)$	**−8.970E + 02**	−8.902E + 02	−8.895E + 02	−8.896E + 02	−8.946E + 02
$f(7)$	**−7.980E + 02**	−7.951E + 02	−7.952E + 02	−7.962E + 02	−7.957E + 02

(continued)

Table 2 (continued)

f(*x*)	fk–PSO	PSO Lozi	PSO Disi	PSO Burger	PSO Tinker
f(8)	**−6.800E + 02**	−6.797E + 02	−6.796E + 02	−6.796E + 02	−6.796E + 02
f(9)	**−5.970E + 02**	**−5.970E + 02**	−5.967E + 02	−5.969E + 02	**−5.970E + 02**
f(10)	−4.990E + 02	−4.997E + 02	**−4.998E + 02**	−4.995E + 02	−4.996E + 02
f(11)	**−4.000E + 02**	−3.939E + 02	−3.942E + 02	−3.952E + 02	−3.946E + 02
f(12)	**−2.930E + 02**	−2.881E + 02	−2.869E + 02	−2.869E + 02	−2.886E + 02
f(13)	**−1.890E + 02**	−1.810E + 02	−1.815E + 02	−1.823E + 02	−1.811E + 02
f(14)	**−6.220E + 01**	2.201E + 02	2.008E + 02	2.034E + 02	1.992E + 02
f(15)	**5.540E + 02**	7.635E + 02	7.141E + 02	7.197E + 02	7.813E + 02
f(16)	**2.000E + 02**	2.010E + 02	2.010E + 02	2.010E + 02	2.010E + 02
f(17)	**3.110E + 02**	3.205E + 02	3.192E + 02	3.224E + 02	3.208E + 02
f(18)	**4.160E + 02**	4.221E + 02	4.250E + 02	4.295E + 02	4.279E + 02
f(19)	5.010E + 02	5.008E + 02	5.008E + 02	5.008E + 02	5.008E + 02
f(20)	6.030E + 02	6.026E + 02	**6.025E + 02**	6.028E + 02	6.026E + 02
f(21)	**1.080E + 03**	1.083E + 03	1.090E + 03	1.081E + 03	1.090E + 03
f(22)	**9.220E + 02**	1.162E + 03	1.155E + 03	1.162E + 03	1.189E + 03
f(23)	**1.420E + 03**	1.785E + 03	1.806E + 03	1.769E + 03	1.800E + 03
f(24)	1.200E + 03	1.204E + 03	1.202E + 03	1.201E + 03	**1.190E + 03**
f(25)	1.310E + 03	1.297E + 03	1.298E + 03	1.299E + 03	**1.289E + 03**
f(26)	1.390E + 03	1.385E + 03	1.390E + 03	**1.381E + 03**	1.389E + 03
f(27)	1.670E + 03	1.638E + 03	1.636E + 03	1.628E + 03	**1.626E + 03**
f(28)	1.730E + 03	1.711E + 03	1.698E + 03	**1.680E + 03**	1.684E + 03

6 Conclusion

In this initial research study it was investigated the possibility of improving the performance of PSO algorithm with super-sized population by implementation of chaotic pseudo-random number generator. Four different CPRNGs were tested. From Table 1 it follows that in general it is possible to improve the performance of the PSO by implementation of CPRNG however the selection of particular chaotic systems seems to be very problem-dependant.

Furthermore from Table 2 it follows that in the case of very complex problems (f(24)–f(28)) it is possible to obtain better results than by the state-of-art method fk-PSO. Future research will focus on tuning experiments regarding the population size and acceleration constants.

Acknowledgments This work was supported by Grant Agency of the Czech Republic—GACR P103/15/06700S, further by financial support of research project NPU I No. MSMT-7778/2014 by the Ministry of Education of the Czech Republic and also by the European Regional Development Fund under the Project CEBIA-Tech No. CZ.1.05/2.1.00/03.0089, partially supported by Grant of SGS No. SP2015/142 of VSB-Technical University of Ostrava, Czech Republic and by Internal Grant Agency of Tomas Bata University under the projects No. IGA/FAI/2015/057.

References

1. Kennedy, J., Eberhart, R.: Particle swarm optimization. In: IEEE International Conference on Neural Networks, 1995, pp. 1942-1948
2. Kennedy, J., Eberhart, R.C., Shi, Y.: Swarm Intelligence. Morgan Kaufmann Publishers (2001)
3. Liang, J., Suganthan, P.N.: Dynamic multi-swarm particle swarm optimizer. In: Swarm Intelligence Symposium, SIS 2005, pp. 124–129 (2005)
4. Liang, J.J., Qin, A.K., Suganthan, P.N., Baskar, S.: Comprehensive learning particle swarm optimizer for global optimization of multimodal functions. IEEE Trans. Evol. Comput. **10**(3), 281–295 (2006)
5. Nickabadi, A., Ebadzadeh, M.M., Safabakhsh, R.: A novel particle swarm optimization algorithm with adaptive inertia weight. Appl. Soft Comput. **11**(4), 3658–3670 (2011)
6. Zhi-Hui, Z., Jun, Z., Yun, L., Yu-hui, S.: Orthogonal learning particle swarm optimization. IEEE Trans. Evol. Comput. **15**(6), 832–847 (2011)
7. Yuhui, S., Eberhart, R.: A modified particle swarm optimizer. In: IEEE World Congress on Computational Intelligence, pp. 69–73, 4–9 May 1998
8. Kennedy, J., Mendes, R.: Population structure and particle swarm performance. In: Proceedings of the 2002 Congress on Evolutionary Computation, CEC '02, pp. 1671–1676 (2002)
9. van den Bergh, F., Engelbrecht, A.P.: A study of particle swarm optimization particle trajectories. Inf. Sci. **176**(8), 937–971 (2006)
10. Pluhacek, M., Senkerik, R., Zelinka, I.: The initial study on the potential of super-sized swarm in PSO. Adv. Intell. Syst. Comput. Mendel **378**(2015), 127–135 (2015)
11. Sprott, J.C.: Chaos and Time-Series Analysis. Oxford University Press (2003)
12. Caponetto, R., Fortuna, L., Fazzino, S., Xibilia, M.G.: Chaotic sequences to improve the performance of evolutionary algorithms. IEEE Trans. Evol. Comput. **7**(3), 289–304 (2003)
13. Pluhacek, M., Senkerik, R., Zelinka, I.: Particle swarm optimization algorithm driven by multichaotic number generator. Soft. Comput. **18**(4), 631–639 (2014). doi:10.1007/s00500-014-1222-z
14. Pluhacek, M., Senkerik, R., Davendra, D., Kominkova Oplatkova Z., Zelinka I.: On the behavior and performance of chaos driven PSO algorithm with inertia weight. Comput. Math. Appl. **66,** 122–134 (2013)
15. Liang J. J., Qu B-Y., Suganthan P.N., Hernández-Díaz Alfredo G.: Problem Definitions and Evaluation Criteria for the CEC 2013 Special Session and Competition on Real-Parameter Optimization. Technical report 201212, Computational Intelligence Laboratory, Zhengzhou University, Zhengzhou China and Technical Report, Nanyang Technological University, Singapore, Jan 2013
16. Nepomuceno, F.V., Engelbrecht, A.P.: A self-adaptive heterogeneous pso for real-parameter optimization. In: Proceedings of 2013 IEEE Congress on Evolutionary Computation (CEC), pp. 361–368 (2013)

New Approach to H_∞ Filtering for Fuzzy Systems in Finite Frequency Domain

D. El Hellani, A. El Hajjaji and R. Ceschi

Abstract This paper is concerned with the problem of finite frequency (FF) H_∞ full-order filtering design for discrete-time nonlinear systems, in the Takagi-Sugeno form. We choose basis dependent Lyapunov function and using the finite frequency l_2 gain definition, and the generalized S-procedure lemma, we propose sufficient conditions, ensuring that the filtering error system is stable and has a minimized $H\infty$ attenuation level in the low-, middle-, and high-frequency domains. In order to linearize and relax the obtained conditions, we apply the Finsler's lemma twice, and we obtain a set of new sufficient conditions for the existence of the H_∞ filter in terms of linear matrix inequalities (LMIs). Then, we show how we can calculate the filter gain matrices by using these conditions. Numerical examples are given to illustrate the effectiveness and the less conservatism of the proposed approach in comparison with the existing methods.

1 Introduction

It is well known that the state estimation problem is a very important and challenging part of automatic control because of increasing needs for efficiency and robustness in many aspects of sciences and engineering applications. Due to its advantages, the H_∞ filtering theory has undergone growing research efforts in various practical situations, see [1–9] and the references therein.

D. El Hellani (✉) · A. El Hajjaji
Modeling, Information, System (MIS) Laboratory, University of Picardie Jules Verne, UFR Sciences, Street Saint Leu, 80000 Amiens, France
e-mail: doha.elhellani@u-picardie.fr; doha-el-hellani@hotmail.com

D. El Hellani · R. Ceschi
ESIGETEL, EFREI Group, 30-32 Avenue of the Republic, 94800 Villejuif, France
e-mail: roger.ceschi@groupe-efrei.fr

A. Abraham et al. (eds.), *Proceedings of the Second International Afro-European Conference for Industrial Advancement AECIA 2015*, Advances in Intelligent Systems and Computing 427, DOI 10.1007/978-3-319-29504-6_50

On the other hand, the well-known Takagi-Sugeno model has been widely used in the last decade because it is a powerful technic that can be used to approximate a large class of complex nonlinear systems by some local linear subsystems. There are many results in the literature reported on filtering for T-S nonlinear systems, among them [3–6].

Recently, the filter synthesis in a finite frequency (FF) domain (that is, low-, middle- and high-frequency domain) has been addressed, and there have appeared many results in this domain. Note that when the external disturbances belong to a certain frequency range which is known beforehand, it is not favorable to minimize the H_∞ norm in the full frequency domain, because this may introduce some conservatism and poor system performance. Among the results in this domain [2, 5, 10, 11].

Motivated by these results mentioned above, we develop in this paper a new approach concerning the FF H_∞ filtering for T-S nonlinear discrete-time systems. We propose a fuzzy Lyapunov function, and based on the FF l_2 gain definition [5], and the generalized S-procedure, we develop a set of sufficient conditions that guarantee the stability and the H_∞ performance for the filtering error system in the frequency domain of the noise input. Then, we use the Finsler's lemma and some matrix propriety to linearize and relax the obtained condition, so that we can use to determine the filter parameters. It will be shown that, this condition can be used to find a suitable filter even in the full frequency domain, and it is still less conservative that some latest ones. Noting that the advantage of our proposed approach lies in the application of the Finsler's lemma twice, which means that additional slack variables are generated, providing less-conservative design procedure compared with the existing results.

The rest of this paper is organized as follows. In Sect. 2, we formulate the problem under consideration and we give some linear matrix inequality (LMI) technique and matrix property lemmas, which are helpful to derive our results. Section 3 presents our proposed FF H_∞ filtering approach in details. In Sect. 4, we provide illustrative examples showing that a lower H_∞ performance index is attained compared with the existing methods. Conclusions are given in Sect. 5.

Notation: A^T denotes the transpose of a matrix A. R^{nxm} and H^{nxm} stand for the nxm-dimensional real and Hermitian matrices respectively. $He(A)$ denotes $A+A^*$. $l_2[0,\infty)$ is the space of square-summable vector functions over $[0,\infty)$.

2 Problem Statement and Preliminaries

2.1 Problem Statement

Consider the discrete-time system represented by T-S fuzzy dynamic model as follows:

Plant rule i: If $\sigma_1(k)$ is N_{i1} and … and $\sigma_p(k)$ is N_{ip}, then:

$$\begin{cases} X(k+1) = A_iX(k) + B_iW(k), & A_i \in R^{n\times n}, B_i \in R^n \\ Y(k) = C_iX(k) + D_iW(k), & C_i \in R^{y\times n}, D_i \in R^{y\times m} \\ Z(k) = L_iX(k), & L_i \in R^{q\times n} \end{cases} \tag{1}$$
$$i = 1, 2, \ldots, \ell, \text{ we denote } \wp = \{1, 2, \ldots, \ell\}$$

where $X \in R^n$ is the state vector, $Y \in R^y$ is the measurement output, $Z \in R^q$ is the signal to be estimated, $W \in R^m$ is the noise input that belongs to $l_2[0, \infty)$ with known frequency ranges, the matrices $(A_i, B_i, C_i, D_i, L_i)$ are constant. The fuzzy set is N_{ij}, ℓ is the number of If-Then rules, and $[\sigma_1(k), \sigma_2(k), \ldots, \sigma_p(k)]$ is the premise variables vector.

The fuzzy basis functions are given by:

$$h_i(\sigma(k)) = \frac{\prod_{j=1}^{p} N_{ij}(\sigma_j(k))}{\sum\limits_{i=1}^{\ell} \prod_{j=1}^{p} N_{ij}(\sigma_j(k))} \tag{2}$$

We have:

$$h_i(\sigma(k)) \geq 0, i = 1, 2, \ldots \ell, \text{ and } \sum_{i=1}^{\ell} h_i(\sigma(k)) = 1 \tag{3}$$

where $N_{ij}(\sigma_j(k))$ is the grade of membership of $\sigma_j(k)$ in N_{ij}.

In the rest, h_i and h_i^+ denote $h_i(\sigma(k))$ and $h_i(\sigma(k+1))$ respectively.

A more compact presentation of the T-S fuzzy model is given by:

$$\begin{cases} X(k+1) = A(h)X(k) + B(h)W(k) \\ Y(k) = C(h)X(k) + D(h)W(k) \\ Z(k) = L(h)X(k) \end{cases} \tag{4}$$

$$A(h) = \sum_{i=1}^{\ell} h_iA_i,\ B(h) = \sum_{i=1}^{\ell} h_iB_i,\ C(h) = \sum_{i=1}^{\ell} h_iC_i,\ L(h) = \sum_{i=1}^{\ell} h_iL_i,\ D(h) = \sum_{i=1}^{\ell} h_iD_i \tag{5}$$

We consider a full-order filter for system (4) of the following form:

$$\begin{cases} X_f(k+1) = A_f(h)X_f(k) + B_f(h)Y(k) & A_f(h) \in R^{n\times n}, B_f(h) \in R^{n\times y} \\ Z_f(k) = C_f(h)X_f(k) + D_f(h)Y(k) & C_f(h) \in R^{q\times n}, D_f(h) \in R^{q\times y} \end{cases} \tag{6}$$

where X_f, Z_f and Y are the state, the output and the input of the filter, respectively. The fuzzy-basis dependent matrices $A_f(h), B_f(h)$, $C_f(h)$, and $D_f(h)$ are to be determined.

We combine systems (4) and (6), we obtain the following filtering error system:

$$\begin{cases} \hat{X}(k+1) = \hat{A}(h)\hat{X}(k) + \hat{B}(h)W(k) \\ E(k) = \hat{C}(h)\hat{X}(k) + \hat{D}(h)W(k) \end{cases} \tag{7}$$

where $E(k) = Z(k) - Z_f(k)$ is the filtering error, $\hat{X}(k) = \begin{bmatrix} X^T(k) & X_f^T(k) \end{bmatrix}^T$ and:

$$\begin{aligned} \hat{A}(h) &= \begin{bmatrix} A(h) & 0 \\ B_f(h)D(h) & A_f(h) \end{bmatrix}, \ \hat{B}(h) = \begin{bmatrix} B(h) \\ B_f(h)D(h) \end{bmatrix}, \\ \hat{C}(h) &= [\, L(h) - D_f(h)C(h) \quad -C_f(h)\,], \ \hat{D}(h) = -D_f(h)D(h) \end{aligned} \tag{8}$$

Consider γ a positive scalar, our main objective is to find a filter of form (6) such that the filtering error system is asymptotically stable when $W(k) = 0$, and the l_2 gain from the noise signal $W(k)$ to the estimation error $E(k)$ of (7) is smaller or equal to γ in the FF domain of the external disturbances $W(k)$, under zeros initial conditions $(\hat{X}(0) = 0)$.

Definition 1 [5] The filtering error system in (7) is said to have a FF l_2 gain γ if the inequality:

$$\sum_{k=0}^{\infty} E^T(k)E(k) \leq \gamma^2 W^T(k)W(k) \tag{9}$$

holds for any nonzero $W(k) \in l_2[0, \infty)$ and under zero initial conditions, the following hold:

$$\sum_{k=0}^{\infty} (X_{k+1} - X_k)(X_{k+1} - X_k)^T \leq (2\sin\frac{\theta_l}{2})^2 \sum_{k=0}^{\infty} X_k X_k^T \text{ for low frequency range.} \tag{10}$$

$$\sum_{k=0}^{\infty} (X_{k+1} - X_k)(X_{k+1} - X_k)^T \geq (2\sin\frac{\theta_h}{2})^2 \sum_{k=0}^{\infty} X_k X_k^T \text{ for high frequency range.} \tag{11}$$

$$e^{j\theta_w} \sum_{k=0}^{\infty} (X_{k+1} - e^{j\theta_1}X_k)(X_{k+1} - e^{j\theta_2}X_k)^T \leq 0 \text{ for middle frequency range.} \tag{12}$$

where $\theta_w = (\theta_2 - \theta_1)/2$.

θ denotes the frequency of noise signal $W(k)$, and its range is assumed to be known, see the table (Table 1).

Table 1 Different frequency ranges

	Low freq.	Middle freq.	High freq.
θ	$\lvert\theta\rvert \leq \theta_l$	$\theta_1 < \theta < \theta_2$	$\lvert\theta\rvert \geq \theta_h$

2.2 Preliminaries

The following lemmas are essential to prove the subsequent theorems.

Lemma 1 [12] (Finsler's lemma) *Let* $\mathrm{E}\in R^{n}, \mathrm{T}\in R^{m\times n}$ *and* $\Sigma=\Sigma^{T}\in R^{n\times n}$ *be given such that* $rank(\mathrm{T})<n$. *The following statements are equivalent:*

i. $\mathrm{E}^{T}\Sigma\mathrm{E}<0\,for\,all\,\mathrm{TE}=0, \mathrm{T}\neq 0$
ii. $N_{\mathrm{T}}^{T}\Sigma N_{\mathrm{T}}<0$
iii. $\exists\mathrm{M}\in R^{nxm}$ *such that* $\Sigma+M\mathrm{T}+\mathrm{T}^{T}M^{T}<0$

Lemma 2 [13] (the generalized S-procedure) *Let* $\mathrm{P}\in H_{n}$ *and a one-vector-lossless set* $\mathrm{K}\subset H_{n}$ *be given such that* $\mathrm{P}=\mathrm{P}^{*}\geq 0$, *then the following two statements are equivalents:*

i. $\rho^{*}\mathrm{P}\rho>0, \forall\rho\in\Xi, \ \Xi=\left\{\rho\in C^{n}:\rho\neq 0, \rho^{*}M\rho\geq 0, \forall M\in\mathrm{K}\right\}$
ii. *There exists* $M\in\mathrm{K}$ *such that* $\mathrm{P}>M$.

3 Main Results

3.1 Finite Frequency H Filtering Analysis

This subsection is devoted to H$_\infty$ finite frequency filtering examination. The following theorem provides sufficient conditions under which the filtering error system in (7) is asymptotically stable with a prescribed H$_\infty$ level γ, and it will be used later to find the filter parameters.

Theorem 1 (middle frequency range) *Consider a discrete time system in (1) with noise input W which belongs to a known middle frequency interval, let scalar* $\gamma>0$ *and a filter of form (6) be given. The filtering error system in (7) is asymptotically stable with H performance level* γ *if there exist matrices* $y_2\in R^{n\times n}$, $y_7\in R^{q\times q}, Q>0(Q=Q^{T})\in R^{2n\times 2n}$, *fuzzy basis dependent matrices* $P(h)>0(P(h)=P^{T}(h)), P_s(h)>0(P_s(h)=P_s^{T}(h)), S(h)(S(h)=S^{T}(h)),\ F(h), G(h), F_s(h), G_s(h)\in R^{2n\times 2n}, M(h)\in R^{q\times 2n}, N(h)\in R^{m\times 2n},\ R(h)\in R^{q\times q},\ y_1(h), y_3(h),\ y_4(h), y_5(h), y_{s1}(h), y_{s2}(h)\in R^{n\times n},\ y_{s3}(h)\in R^{4n\times n}, y_6(h)R^{q\times n},\ y_8(h)\in R^{(m+2n+q)\times n}$, *such that*:

$$\Xi_{M}(h)+He\{Y(h)X(h)\}<0 \tag{13}$$

$$\Xi_{s}(h)+He\{Y_s(h)X_s(h)\}<0 \tag{14}$$

where

$$X(h)=\begin{bmatrix} -I_n & 0_{n\times n} & A(h) & 0_{n\times n} & B(h) & -I_n & 0_{n\times n} & 0_{n\times q} \\ 0_{n\times n} & -I_n & B_f(h)C(h) & A_f(h) & B_f(h)D(h) & 0_{n\times n} & -I_n & 0_{n\times q} \\ 0_{q\times n} & 0_{q\times n} & L(h)-D_f(h)C(h) & -C_f(h) & -D_f(h)D(h) & 0_{q\times n} & 0_{q\times n} & -I_q \end{bmatrix}$$

$$\Xi_M(h)=\begin{bmatrix} P(h^+)-S(h^+) & e^{j\theta_c}Q & 0_{2n,q} & 0_{2n,m} & F(h) & 0_{2n,q} \\ e^{-j\theta_c}Q & -P(h)+S(h)-2\cos(\theta_w)Q & 0_{2n,q} & 0_{2n,m} & G(h) & 0_{2n,q} \\ 0_{q,2n} & 0_{q,2n} & I_q & 0_{q,m} & M(h) & R(h) \\ 0_{m,2n} & 0_{m,2n} & 0_{m,q} & -\gamma^2 I_m & N(h) & 0_{m,q} \\ F^T(h) & G^T(h) & M^T(h) & N^T(h) & 0_{2n,2n} & 0_{2n,q} \\ 0_{q,2n} & 0_{q,2n} & R^T(h) & 0_{q,m} & 0_{q,2n} & 0_{q,q} \end{bmatrix}$$

$$Y(h)=\begin{bmatrix} y_1(h) & y_2 & 0_{n\times q} \\ y_3(h) & y_2 & 0_{n\times q} \\ y_4(h) & 0_{n\times n} & 0_{n\times q} \\ y_5(h) & 0_{n\times n} & 0_{n\times q} \\ y_6(h) & 0_{q\times n} & y_7 \\ y_8(h) & 0_{(q+m+2n)\times n} & 0_{(q+m+2n)\times q} \end{bmatrix},\ E_s(h)=\begin{bmatrix} P_s(h^+) & 0_{2n\times 2n} & F_s(h) \\ 0_{2n\times 2n} & -P_s(h) & G_s(h) \\ F_s^T(h) & G_s^T(h) & 0_{2n\times 2n} \end{bmatrix}$$

$$Y_s(h)=\begin{bmatrix} y_{s1}(h) & y_2 \\ y_{s2}(h) & y_2 \\ y_{s3}(h) & 0_{4n\times n} \end{bmatrix},\ X_s(h)=\begin{bmatrix} -I_n & 0_{n\times n} & A(h) & 0_{n\times n} & -I_n & 0_{n\times n} \\ 0_{n\times n} & -I_n & B_f(h)C(h) & A_f(h) & 0_{n\times n} & -I_n \end{bmatrix}$$

Proof It is omitted due to the limited space. □

3.2 Finite Frequency H Filter Design

In this subsection, we will present a finite frequency H_∞ filter design approach.

Theorem 2 (middle frequency range) *Consider a discrete time system in (1) with noise input W which belongs to a known middle frequency interval, and let a constant $\gamma>0$ be given. If there exist matrices $y_2\in R^{n\times n}, y_7\in R^{q\times q}, Q>0\ (Q=Q^T)$ $\in R^{2n\times 2n}, P_i>0(P_i=P_i^T),\ P_{si}>0(P_{si}=P_{si}^T),\ S_i(S_i=S_i^T), F_i, G_i,\ F_{si}, G_{si}\in R^{2n\times 2n}$, $M_i\in R^{q\times 2n},\ N_i\in R^{m\times 2n},\ R_i\in R^{q\times q}, y_{1i}, y_{3i},\ y_{4i}, y_{5i}, y_{s1i}, y_{s2i}\in R^{n\times n},\ y_{6i}\in R^{q\times n}$, $y_{s3i}\in R^{4n\times n},\ y_{8i}\in R^{(m+2n+q)\times n},\ a_{fi}\in R^{n\times n},\ b_{fi}\in R^{n\times y},\ c_{fi}\in R^{q\times n},\ d_{fi}\in R^{q\times y}$ $X_{iil}(X_{iil}=X_{iil}^T), X_{ijl}(X_{ijl}=X_{ijl}^T)_{i<j}\ \in R^{(6n+2q+m)\times(6n+2q+m)},\ X_{siil}\ (X_{siil}=X_{siil}^T), X_{sijl}$ $(X_{sijl}=X_{sijl}^T)_{i<j}\in R^{6n\times 6n}$, such that:*

$$\Xi_{Mil}+He\{\Omega_{ii}\}+\mathrm{X}_{iil}<0,\ (i,l\in\wp) \tag{15}$$

$$\Xi_{Mil}+He\{\Omega_{ij}\}+\Xi_{il}+He\{\Omega_{ji}\}+\mathrm{X}_{ijl}+\mathrm{X}_{ijl}^T<0,\ (i<j,i,j,l\in\wp) \tag{16}$$

$$\Xi_{sil}+He\{\Omega_{sii}\}+X_{siil}<0,\ \ (i,l\in\wp) \tag{17}$$

$$\Xi_{sil}+He\{\Omega_{sij}\}+\Xi_{sil}+He\{\Omega_{sji}\}+X_{sijl}+X_{sijl}^{T}<0,\ \ (i<j\,i,j,l\in\wp) \tag{18}$$

$$X_l=\begin{bmatrix} X_{11l} & X_{12l} & \dots & X_{1\ell l}\\ X_{12l}^T & X_{22l} & \dots & X_{2\ell l}\\ \vdots & \vdots & \ddots & \vdots\\ X_{1\ell l}^T & X_{2\ell l}^T & \dots & X_{\ell\ell l}\end{bmatrix}>0,\ \ l\in\wp \tag{19}$$

$$X_{sl}=\begin{bmatrix} X_{s11l} & X_{s12l} & \dots & X_{s1\ell l}\\ X_{s12l}^T & X_{s22l} & \dots & X_{s2\ell l}\\ \vdots & \vdots & \ddots & \vdots\\ X_{s1\ell l}^T & X_{s2\ell l}^T & \dots & X_{s\ell\ell l}\end{bmatrix}>0,\ l\in\wp \tag{20}$$

$$\Xi_{Mil}=\begin{bmatrix} P_l-S_l & e^{j\theta_c}Q & 0_{2n,q} & 0_{2n,m} & F_i & 0_{2n,q}\\ e^{-j\theta_c}Q & -P_i+S_i-2\cos(\theta_w)Q & 0_{2n,q} & 0_{2n,m} & G_i & 0_{2n,q}\\ 0_{q,2n} & 0_{q,2n} & I_q & 0_{q,m} & M_i & R_i\\ 0_{m,2n} & 0_{m,2n} & 0_{m,q} & -\gamma^2 I_m & N_i & 0_{m,q}\\ F_i^T & G_i^T & M_i^T & N_i^T & 0_{2n,2n} & 0_{2n,q}\\ 0_{q,2n} & 0_{q,2n} & R_i^T & 0_{q,m} & 0_{q,2n} & 0_{q,q}\end{bmatrix}$$

$$\Omega_{ij}=\left[\begin{matrix} -y_{1i} & -y_2 & y_{1i}A_j+b_{fi}C_j & a_{fi}\\ -y_{3i} & -y_2 & y_{3i}A_j+b_{fi}C_j & a_{fi}\\ -y_{4i} & 0_{n\times n} & y_{4i}A_j & 0_{n\times n}\\ -y_{5i} & 0_{n\times n} & y_{5i}A_j & 0_{n\times n}\\ -y_{6i} & 0_{q\times n} & y_{6i}A_j+y_7L_i-d_{fi}C_j & -c_{fi}\\ -y_{8i} & 0_{(2n+m+q)\times n} & y_{8i}A_j & 0_{(2n+m+q)\times n}\end{matrix}\right.$$
$$\left.\begin{matrix} 0_{n\times q} & y_{1i}B_j+b_{fi}D_j & -y_{1i} & -y_2 & 0_{n\times q}\\ 0_{n\times q} & y_{3i}B_j+b_{fi}D_j & -y_{3i} & -y_2 & 0_{n\times q}\\ 0_{n\times q} & y_{4i}B_j & -y_{4i} & 0_{n\times n} & 0_{n\times q}\\ 0_{n\times q} & y_{5i}B_j & -y_{5i} & 0_{n\times n} & 0_{n\times q}\\ -y_7 & y_{6i}B_j-d_{fi}D_j & -y_{6i} & 0_{q\times n} & -y_7\\ 0_{(2n+m+q)\times q} & y_{8i}B_j & -y_{8i} & 0_{(2n+m+q)\times n} & 0_{(2n+m+q)\times q}\end{matrix}\right]$$

$$E_{sil}=\begin{bmatrix} P_{sl} & 0_{2n\times 2n} & F_{si}\\ 0_{2n\times 2n} & -P_{si} & G_{si}\\ F_{si}^T & G_{si}^T & 0_{2n\times 2n}\end{bmatrix}$$

$$\Omega_{sij}=\begin{bmatrix} -y_{s1i} & -y_2 & y_{s1i}A_j+b_{fi}C_j & a_{fi} & -y_{s1i} & -y_2\\ -y_{s2i} & -y_2 & y_{s2i}A_j+b_{fi}C_j & a_{fi} & -y_{s2i} & -y_2\\ -y_{s3i} & 0_{4n\times n} & y_{s3i}A_j & 0_{4n\times n} & -y_{s3i} & 0_{4n\times n}\end{bmatrix}$$

Then there exists a fuzzy filter (6) such that the filtering error system in (7) is asymptotically stable and has a finite frequency l_2 gain γ. In this case, a suitable full-order H filter realization is given by:

$$A_f(h) = y_2^{-1} a_f(h) = y_2^{-1} \sum_{i=1}^{\ell} h_i a_{fi};\ B_f(h) = y_2^{-1} b_f(h) = y_2^{-1} \sum_{i=1}^{\ell} h_i b_{fi}$$
$$C_f(h) = y_7^{-1} c_f(h) = y_7^{-1} \sum_{i=1}^{\ell} h_i c_{fi};\ D_f(h) = y_7^{-1} d_f(h) = y_7^{-1} \sum_{i=1}^{\ell} h_i d_{fi} \tag{21}$$

Proof Suppose that inequalities (14)–(20) has feasible solutions with the decision variables defined in theorem (2). Now, taking (13)–(14) and introducing the following change of variables:

$$a_f(h) = y_2 A_f(h) \quad b_f(h) = y_2 B_f(h) \quad c_f(h) = y_7 C_f(h) \quad d_f(h) = y_7 D_f(h) \tag{22}$$

Then we have:

$$\begin{aligned}
\Xi(h) + He\{Y(h)X(h)\} &= \sum_{i=1}^{\ell}\sum_{j=1}^{\ell}\sum_{l=1}^{\ell} h_i h_j h_l^+ \left(\Xi_{il} + He\{\Omega_{ij}\}\right) = \sum_{i=1}^{\ell}\sum_{l=1}^{\ell} h_i^2 h_l^+ \left(\Xi_{il} + He\{\Omega_{ii}\}\right) \\
&\quad + \sum_{i=1}^{\ell}\sum_{i\le j}^{\ell}\sum_{l=1}^{\ell} h_i h_j h_l^+ \left(\Xi_{il} + He\{\Omega_{ij}\}\right) \sum_{i=1}^{\ell}\sum_{i\le j}^{\ell}\sum_{l=1}^{\ell} h_i h_j h_l^+ \left(\Xi_{il} + He\{\Omega_{ji}\}\right) \\
&\le -\sum_{i=1}^{\ell}\sum_{l=1}^{\ell} h_i^2 h_l^+ X_{iil} - \sum_{i=1}^{\ell}\sum_{i\le j}^{\ell}\sum_{l=1}^{\ell} h_i h_j h_l^+ \left(X_{ijl} + X_{ijl}^T\right) \\
&= \sum_{l=1}^{\ell} h_l^+ \left(\begin{bmatrix} h_1 \\ h_2 \\ \vdots \\ h_\ell \end{bmatrix}^T \begin{bmatrix} X_{11l} & X_{12l} & \dots & X_1 \\ X_{12l}^T & X_{22l} & \dots & X_{2\ell l} \\ \vdots & \vdots & \ddots & \vdots \\ X_{1\ell l}^T & X_{2\ell l}^T & \dots & X_{\ell\ell l} \end{bmatrix} \begin{bmatrix} h_1 \\ h_2 \\ \vdots \\ h_\ell \end{bmatrix} \right) \le 0.
\end{aligned}$$

Which implies that all the conditions of Theorem 1 hold, and this completes the proof.

Similarly we can establish the condition for the filtering error system stability. □

Theorem 3 (middle frequency range) *the problem of finding the suitable filter gains that gives the smallest achievable H performance and guarantees the asymptotic stability of the filtering error system (7) in the FF range (middle frequency) of the noise input known beforehand, can be solved via the following optimization problem:*

$$\min_{\substack{y_2, y_7, Q, P_i, P_{si}, S_i, F_i, G_i, F_{si}, G_{si}, M_i, N_i, R_i, \\ y_{1i}, y_{3i}, y_{4i}, y_{5i}, y_{s1i}, y_{s2i}, y_{6i}, y_{s3i}, y_{8i}, a_{fi}, b_{fi}, \\ c_{fi}, d_{fi}, X_{iil}, X_{ijl}, X_{siil}, X_{sijl}}} \gamma^2$$

subject to $Q > 0$, $P_i > 0$, $P_{si} > 0$, *and (15)–(20).*
Then a suitable filter is given by (21).

Theorem 4 (low frequency range) *the problem of finding the suitable filter gains that gives the smallest achievable H performance and guarantees the asymptotic stability of the filtering error system (7) in the FF range (low frequency) of the noise input known beforehand, can be solved via the following optimization problem:*

$$\min_{\substack{y_2, y_7, Q, P_i, P_{si}, S_i, F_i, G_i, F_{si}, G_{si}, M_i, N_i, R_i, \\ y_{1i}, y_{3i}, y_{4i}, y_{5i}, y_{s1i}, y_{s2i}, y_{6i}, y_{s3i}, y_{8i}, a_{fi}, b_{fi}, \\ c_{fi}, d_{fi}, X_{iil}, X_{ijl}, X_{siil}, X_{sijl}}} \gamma^2$$

subject to $Q > 0$, $P_i > 0$, $P_{si} > 0$, *(23), (24) and (17)–(20).*
Then a suitable filter is given by (21).

$$\Xi_{Lil} + He\{\Omega_{ii}\} + X_{iil} < 0, \ (i, l \in \wp) \tag{23}$$

$$\Xi_{Lil} + He\{\Omega_{ij}\} + \Xi_{il} + He\{\Omega_{ji}\} + X_{ijl} + X_{ijl}^T < 0, \ \ (i < j, i, j, l \in \wp) \tag{24}$$

where $\Xi_{Lil} = \begin{bmatrix} P_l - S_l & Q & 0_{2n,q} & 0_{2n,m} & F_i & 0_{2n,q} \\ Q & -P_i + S_i - 2\cos(\theta_l)Q & 0_{2n,q} & 0_{2n,m} & G_i & 0_{2n,q} \\ 0_{q,2n} & 0_{q,2n} & I_q & 0_{q,m} & M_i & R_i \\ 0_{m,2n} & 0_{m,2n} & 0_{m,q} & -\gamma^2 I_m & N_i & 0_{m,q} \\ F_i^T & G_i^T & M_i^T & N_i^T & 0_{2n,2n} & 0_{2n,q} \\ 0_{q,2n} & 0_{q,2n} & R_i^T & 0_{q,m} & 0_{q,2n} & 0_{q,q} \end{bmatrix}$

Theorem 5 (high frequency range) *The problem of finding the suitable filter gains that gives the smallest achievable H performance and guarantees the asymptotic stability of the filtering error system (7) in the FF range (high frequency) of the noise input known beforehand, can be solved via the following optimization problem:*

$$\min_{\substack{y_2, y_7, Q, P_i, P_{si}, S_i, F_i, G_i, F_{si}, G_{si}, M_i, N_i, R_i, \\ y_{1i}, y_{3i}, y_{4i}, y_{5i}, y_{s1i}, y_{s2i}, y_{6i}, y_{s3i}, y_{8i}, a_{fi}, b_{fi}, \\ c_{fi}, d_{fi}, X_{iil}, X_{ijl}, X_{siil}, X_{sijl}}} \gamma^2$$

subject to $Q > 0$, $P_i > 0$, $P_{si} > 0$, *(25),(26) and (17)–(20).*

Then a suitable filter is given by (21).

$$\Xi_{Hil} + He\{\Omega_{ii}\} + \mathrm{X}_{iil} < 0, \quad (i, l \in \wp) \tag{25}$$

$$\Xi_{Hil} + He\{\Omega_{ij}\} + \Xi_{il} + He\{\Omega_{ji}\} + \mathrm{X}_{ijl} + \mathrm{X}_{ijl}^T < 0, \quad (i < j, i, j, l \in \wp) \tag{26}$$

where $\Xi_{Hil} = \begin{bmatrix} P_l - S_l & -Q & 0_{2n,q} & 0_{2n,m} & F_i & 0_{2n,q} \\ -Q & -P_i + S_i + 2\cos(\theta_h)Q & 0_{2n,q} & 0_{2n,m} & G_i & 0_{2n,q} \\ 0_{q,2n} & 0_{q,2n} & I_q & 0_{q,m} & M_i & R_i \\ 0_{m,2n} & 0_{m,2n} & 0_{m,q} & -\gamma^2 I_m & N_i & 0_{m,q} \\ F_i^T & G_i^T & M_i^T & N_i^T & 0_{2n,2n} & 0_{2n,q} \\ 0_{q,2n} & 0_{q,2n} & R_i^T & 0_{q,m} & 0_{q,2n} & 0_{q,q} \end{bmatrix}$

Proof Following the same lines for that of Theorem 2, we can immediately conclude the LMI conditions corresponding to l_2 gain and the stability of the estimation error system (7), in the low- and high-frequency ranges.

Remark 1 In this result we apply the Finsler's lemma twice to linearize and relax the matrix inequalities of Theorem 1, however, there is no condition on the additional matrix variables introduced, and they are different from each other for $i = 1, 2, \ldots, \ell$. Furthermore, it is expected that the results will be less conservatives due to the freedom degrees added by those slack variables.

Remark 2 If we take $Q = 0$, we can use Theorem 2 to solve the H_∞ filtering problem in the entire frequency domain of the T-S nonlinear discrete-time systems.

4 Illustrative Examples

Example 1: Consider the following discrete-time fuzzy system that was studied in [5]:

Plant rule 1: If $X_1(k)$ is $N_1(X_1(k))$, then:

$$X(k+1) = \begin{bmatrix} 0.998 & 1 \\ -0.02 & 0.98 \end{bmatrix} X(k) + \begin{bmatrix} 0 \\ 0.02 \end{bmatrix} W(k)$$

$$Y(k) = [1 \quad 0] X(k) + W(k)$$

$$Z(k) = [0 \quad 1] X(k)$$

Plant rule 2: If $X_1(k)$ is $N_2(X_1(k))$, then:

$$X(k+1) = \begin{bmatrix} 0.908 & 1 \\ -0.02 & 0.98 \end{bmatrix} X(k) + \begin{bmatrix} 0 \\ 0.02 \end{bmatrix} W(k)$$

$$Y(k) = [1 \quad 0]X(k) + W(k)$$

$$Z(k) = [0 \quad 1]X(k)$$

The membership functions are given by:

$$N_1(X_1(k)) = \begin{cases} \frac{X_1(k)+3}{3}, & for -3 \le X_1(k) \le 0 \\ 0, & for\ X_1(k) < -3 \\ \frac{3-X_1(k)}{3}, & for\ 0 \le X_1(k) \le 3 \\ 0, & for\ X_1(k) > 3 \end{cases} \quad \text{And} \quad N_2(X_1(k)) = 1 - N_1(X_1(k)).$$

In Table 2, we evaluate the minimum noise attenuation level γ in different frequency ranges and we compare our result with [5].

We consider that $|\theta| \le 0.1$, i.e., the frequency range of $W(k)$ is [0 5], and we choose $W(k) = \sin(2kT)$. Figures 1 and 2 show the estimation error $E(k)$ obtained by Theorem 2 given in [5] and Theorem 4, respectively. Figure 3 shows the simulated value of γ. It can be easily seen that our proposed approach leads to better results than the method developed in [5].

Table 2 Minimum noise attenuation level in different frequency ranges

	Theorem 4	Theorem 2 [5]	Theorem 5	Theorem 3 [5]
$|\theta| \le 0.1$	0.0252	0.1695	–	–
$|\theta| \le \pi$	0.2140	Infeasible	–	–
$|\theta| \ge 0.24$	–	–	0.0684	0.0898

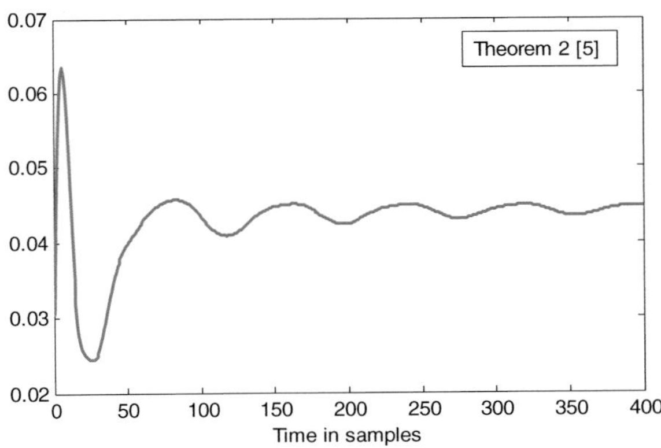

Fig. 1 Estimation error subject to sinusoidal disturbance $W(k) = \sin(2kT)$

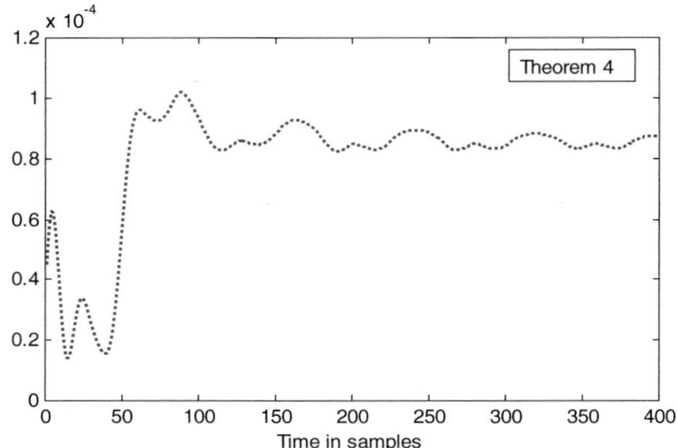

Fig. 2 Estimation error subject to sinusoidal disturbance $W(k) = \sin(2kT)$

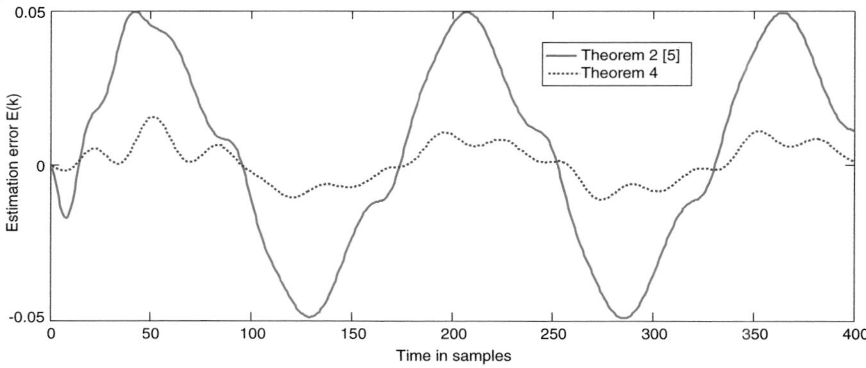

Fig. 3 Value of $\sqrt{\sum_{i=0}^{k} E^T(i)E(i)/\sum_{i=0}^{k} W^T(i)W(i)}$, with $(W(k) = \sin(2kT))$

5 Conclusion

A new approach has been developed to study the finite frequency H_∞ filtering problem for T-S nonlinear discrete-time systems. Using a basis dependent Lyapunov functional and by the aid of the S-procedure lemma, a set of LMIs conditions has been proposed, under which the filtering error system is stable and satisfies a minimum l_2 gain. The potential of our new structure lies in the additional slack variables introduced by applying the Finsler's lemma twice to provide extra free dimensions in the solution space of the H_∞ optimization. The effectiveness of the developed results has been demonstrated by an illustrative example.

References

1. Su, Y., Chen, B., Lin, C., Wang, Z., Lin, Y.: H_∞ filter design for TS nonlinear discrete-time with interval time delays. In: 2010 Chinese Control and Decision Conference, pp. 3595–3600 (2010)
2. Gaoand, H.,. LI, X.: H_∞filtering for discrete-time state-delayed systems with finite frequency specifications. IEEE Trans. Autom. Control **56**(12), 2935–2941 (2011)
3. Su, X., Shi, P., Wu, L., Song, Y.D.: A novel approach to filter design for t-s fuzzy discrete-time systems with time-varying delay. In: IEEE Trans. Fuzzy Syst. **20**(6), 1114–1129 (2012)
4. Su, Y., Zhou, Q., Gao, F.: A new delay-range-dependent H∞ filter design for T-S nonlinear systems. J. Franklin Inst. **351**(6), 3305–3321 (2014)
5. Ding, D.W., Yang, G.H.: Fuzzy filter design for nonlinear systems in finite-frequency domain. IEEE Trans. Fuzzy Syst. **18**(5), 935–945 (2010)
6. Zhang, H., Shi, Y., Mehr, A.S.: On H_∞ filtering for discrete-time Takagi-Sugeno fuzzy systems. IEEE Trans. Fuzzy Syst. **20**(2), 396–401 (2012)
7. Zhang, H., Wang, J.: Robust H_∞ filtering of discrete-time networked nonlinear systems approximated by Takagi-Sugeno fuzzy technique. In: American Control Conference (ACC), pp. 5373–5378. IEEE (2014)
8. Assawinchaichote, W., Nguang, S.K.: Synthesis of a robust H_∞ fuzzy filter for uncertain nonlinear dynamical systems. 2010 2nd ICETC, vol. 5, pp. 61–65 (2010)
9. Chang, X.H., Park, J.H., Tang, Z.: New approach to H_∞ filtering for discrete-time systems with polytopic uncertainties. Signal Process. **113**, 147–158 (2015)
10. Wang, H., Yang, G.H.: A finite frequency approach to filter design for uncertain discrete-time systems. Int. J. Adapt. Control Signal Process. **22**(6), 533–550 (2008)
11. Ding, D., Yang, G.: Finite frequency H_∞ filtering for uncertain discrete-time switched linear systems. Prog. Nat. Sci. **19**(11), 1625–1633 (2009)
12. Oliveira, M.C.: Stability test for constrained linear systems. Lect. Notes Control Inf. Sci. **268**, 241–257 (2001)
13. Ebihara, Y., Maedaand, K., Hagiwara, T.: Generalized S-procedure for inequality conditions on one-vector-lossless sets and linear system analysis. In: 43rd IEEE Conference on Decision and Control, CDC, pp. 1272–1277. IEEE (2004)

PSO as Complex Network—Capturing the Inner Dynamics—Initial Study

Michal Pluhacek, Jakub Janostik, Roman Senkerik, Ivan Zelinka and Donald Davendra

Abstract This paper presents an initial proposal of methodology for converting the inner dynamics of PSO algorithm into complex network. The motivation is in the recent trend of adaptive methods for improving the performance of evolutionary computational techniques. It seems very likely that the complex network and its statistical characteristics can be used within those adaptive approaches. The methodology described in this paper manages to put significant amount of information about the inner dynamics of PSO algorithm into a complex network.

Keywords Particle Swarm Optimization · PSO · Evolutionary algorithm · Complex network

1 Introduction

The Evolutionary Computation Techniques (ECTs) are a very successful group of heuristic algorithms based on evolutionary principles. Among them the Particle Swarm Optimization algorithm (PSO) [1–4] holds a prominent position as one of

M. Pluhacek (✉) · J. Janostik · R. Senkerik
Faculty of Applied Informatics, Tomas Bata University in Zlin,
T.G. Masaryka 5555, 760 01 Zlin, Czech Republic
e-mail: pluhacek@fai.utb.cz

J. Janostik
e-mail: janostik@fai.utb.cz

R. Senkerik
e-mail: senkerik@fai.utb.cz

I. Zelinka · D. Davendra
Faculty of Electrical Engineering and Computer Science, VŠB-Technical
University of Ostrava, 17. listopadu 15, 708 33 Ostrava-Poruba, Czech Republic
e-mail: Ivan.zelinka@vsb.cz

D. Davendra
e-mail: Donald.davendra@vsb.cz

A. Abraham et al. (eds.), *Proceedings of the Second International Afro-European Conference for Industrial Advancement AECIA 2015*, Advances in Intelligent Systems and Computing 427, DOI 10.1007/978-3-319-29504-6_51

the most popular and widely used. Recently it has been proposed that the evolutionary computational techniques may be analyzed using a transformation of the inner dynamic of the algorithm into a complex network [5–9]. Given that the previously presented results and studies for other ECTs seem very promising it is investigated in this paper the possibility of converting the inner dynamic of PSO algorithm into a complex network.

The PSO algorithm is one of the Swarm Intelligence based methods and as such present a particular challenge for the complex network creation. There is no simple identification of parents and offspring as it is in many other ECTs.

The motivation for this research is that it seems like that if the inner dynamic of PSO algorithm can be successfully transformed into a complex network of reasonable structure and density than advanced statistical methods for complex network analysis can be used to derive otherwise hidden information. This information can then be used for various adaptive or learning approaches or for a prediction of the behavior of the swarm and algorithm as a whole (e.g. prediction of premature convergence).

2 Particle Swarm Optimization

The PSO algorithm takes inspiration from the natural swarm behavior of birds and fish. It was firstly introduced by Eberhart and Kennedy in 1995 [1]. Each particle in the population represents a candidate solution for the optimization problem that is defined by the cost function. In each iteration of the algorithm, a new location (combination of CF parameters) for the particle is calculated based on its previous location and velocity vector (velocity vector contains particle velocity for each dimension of the problem).

According to the method of selection of the swarm or sub-swarm for best solution information spreading, the PSO algorithms are noted as global PSO (GPSO) or local PSO (LPSO). Within this research the PSO algorithm with global topology (GPSO) [6] was utilized. The chaotic PRNG is used in the main GPSO formula (1), which determines a new "velocity", thus directly affects the position of each particle in the next iteration.

$$v_{ij}^{t+1} = w \cdot v_{ij}^{t} + c_1 \cdot Rand \cdot (pBest_{ij} - x_{ij}^{t}) + c_2 \cdot Rand \cdot (gBest_j - x_{ij}^{t}) \tag{1}$$

where

v_i^{t+1} New velocity of the *i*th particle in iteration $t + 1$.

w Inertia weight value.

v_i^t Current velocity of the *i*th particle in iteration t.

c_1, c_2 Priority factors.

pBest$_i$ Local (personal) best solution found by the *i*th particle.
gBest Best solution found in a population.
x_{ij}^t Current position of the *i*th particle (component *j* of dimension *D*) in iteration *t*.
Rand Pseudo random number, interval (0, 1). CPRNG is applied only here.

The maximum velocity was limited to 0.2 times the range as it is usual. The new position of each particle is then given by (2), where x_i^{t+1} is the new particle position:

$$x_i^{t+1} = x_i^t + v_i^{t+1} \tag{2}$$

Finally the linear decreasing inertia weight [3, 4] is used in the typically referred GPSO design that was used in this study. The dynamic inertia weight is meant to slow the particles over time thus to improve the local search capability in the later phase of the optimization. The inertia weight has two control parameters w_{start} and w_{end}. A new w for each iteration is given by (3), where t stands for current iteration number and n stands for the total number of iterations. The values used in this study were $w_{start} = 0.9$ and $w_{end} = 0.4$.

$$w = w_{start} - \frac{((w_{start} - w_{end}) \cdot t)}{n} \tag{3}$$

3 Experiment

In the experiment a PSO with typical defaults setting was used to optimize the commonly used Schwefel's benchmark function for 100 iterations with population size set to 30.

In order to create a network with relation to the inner dynamic of PSO it is necessary to identify the communication within the swarm. From (1) it is clear that the only communication channel within the swarm is the shared knowledge o the location of global best solution (gBest).

Based on this fact it was decided to track the updating particle on every *gBest* update. Therefore the authorship of the actual *gBest* is known and it may be stated that the particle that triggered the last *gBest* update is affecting all other particles and communicating with all other members of the swarm simultaneously.

In this initial study the main interest was in the communications that leads to CF improvement. Therefore only communication leading to improvement of the particles personal best (*pBest*) was tracked. The link was created between the particle that has improved and particle that triggered the current *gBest*'s update.

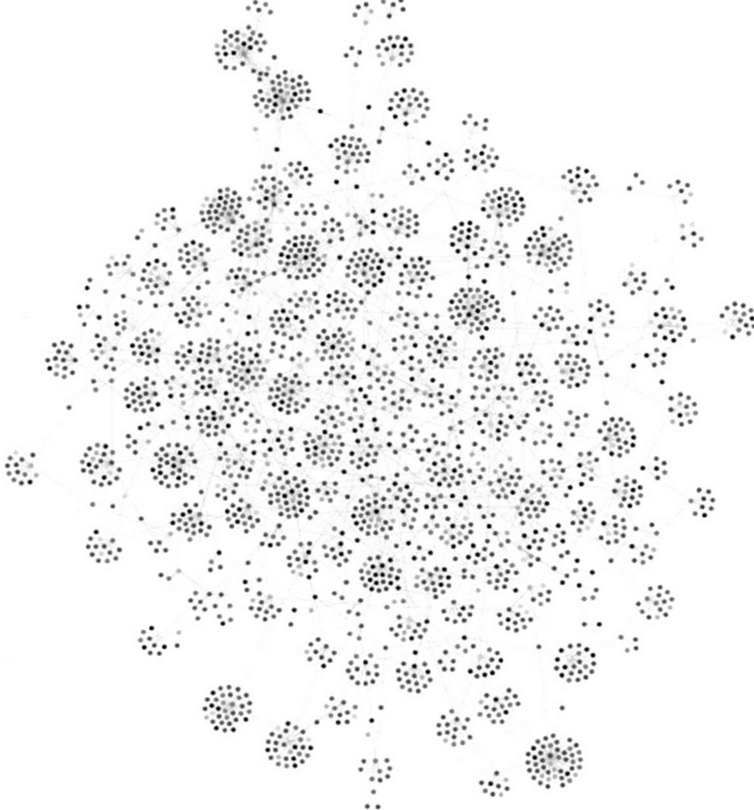

Fig. 1 PSO dynamic as complex network—complete view

However this approach leads to a network with low number of nodes and very high number of links. That is not favorable for statistical analysis of the network.

The time component needed to be also tracked. The information of particle and iteration was combined to create a record of evolution of particles during the iterations of the algorithm.

In Fig. 1 the created complex network is visualized. Nodes of similar color represent particles with same ID during different iterations. All links are from particle that triggered *gBest* update to particle that has improved based on that *gBest*.

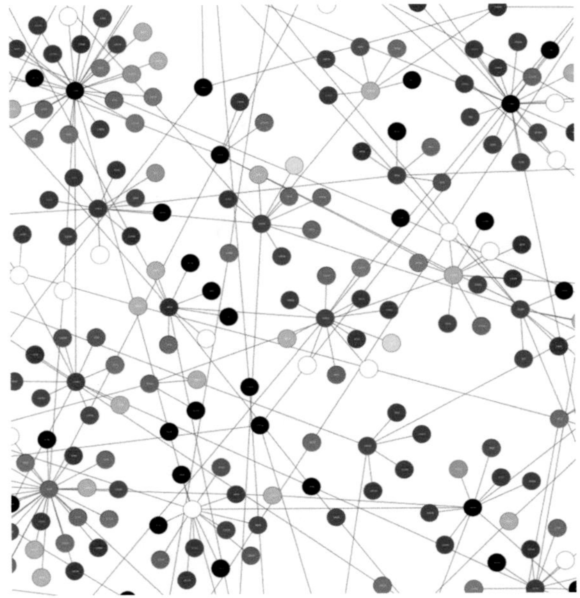

Fig. 2 PSO dynamic as complex network—partial view

In Fig. 2 a zoomed partial view of the network is presented. It is possible to clearly see the density of the network and links of various lengths.

Close look on a single cluster in the network is presented in Fig. 3. The numbers in nodes represent a code for a particle ID and current iteration. That way it is possible to track exactly the development of the network and the communication that happens within the swarm. On this example cluster it can be observed a single *gBest* update led to improvement of multiple particles in different iterations.

Two large interconnected clusters are depicted in Fig. 4 and finally in Fig. 5 it is depicted the dead-end where the particle that triggered the gBest update did not improve further until the end of the optimization.

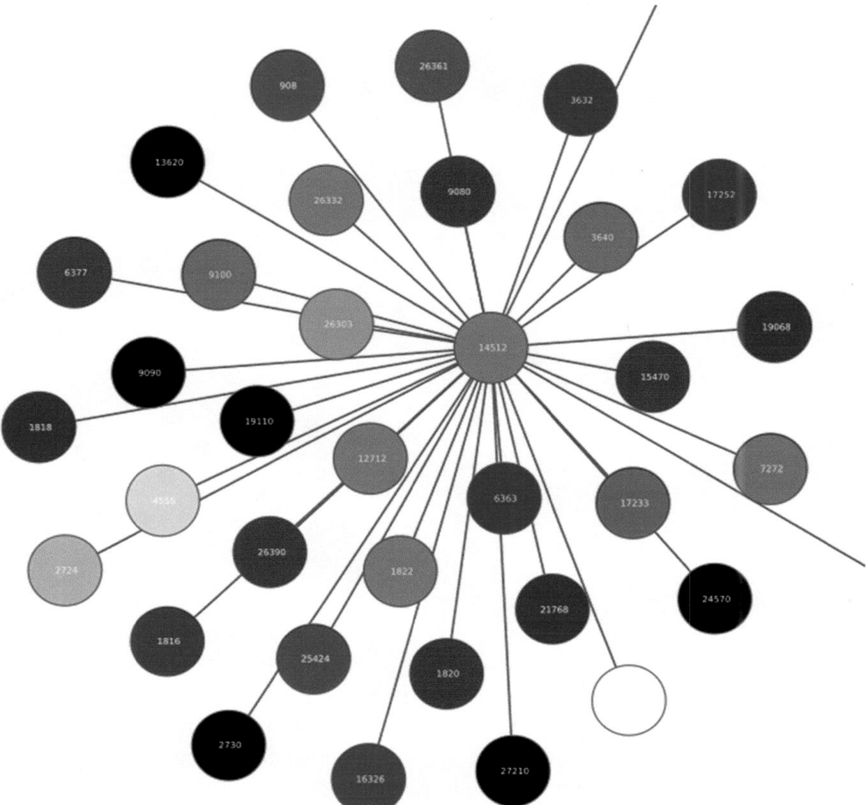

Fig. 3 PSO dynamic as complex network—close view

4 Conclusion

As was presented in the above described visualization it is possible to contain a large amount of information about the inner dynamic of PSO algorithm in a form of complex network. Even though that many useful data can be gathered from the network and it visualization by simple methods the main aim for application is to

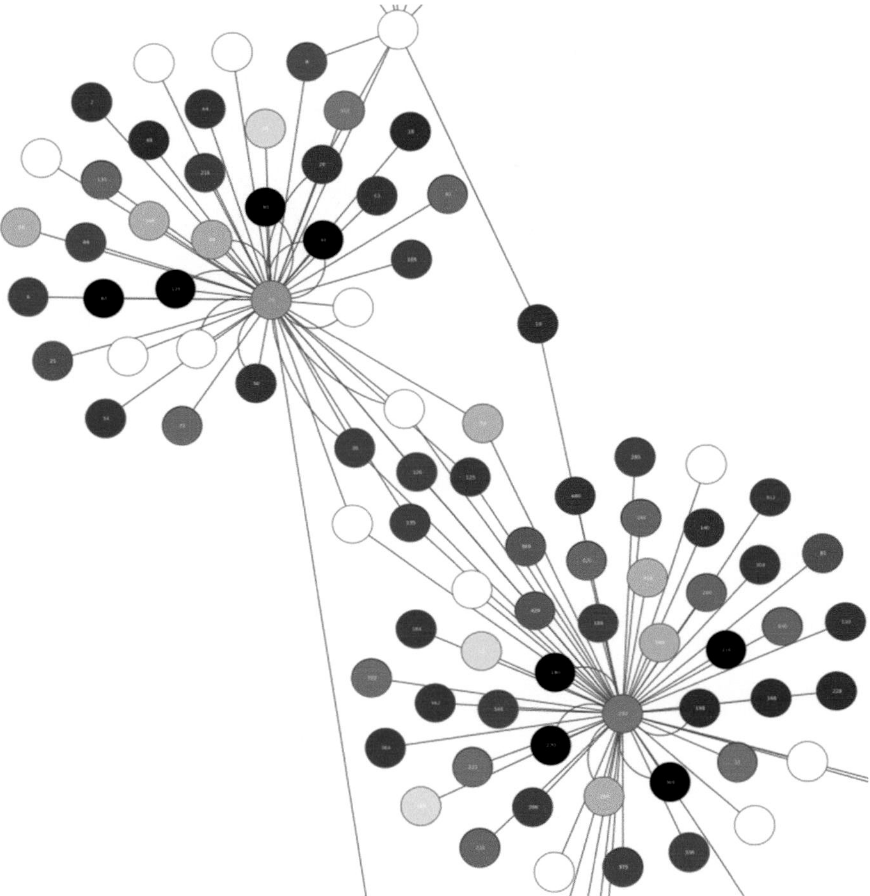

Fig. 4 PSO dynamic as complex network—notable part

use advanced statistical methods to analyze the complex network a use the results as a input for adaptive or learning mechanism. Furthermore a possibility of prediction of pathological states (such as premature convergence or stagnation) with the help of complex network analysis will be investigated in the future research. The methodology for complex network creation presented in this work will be utilized.

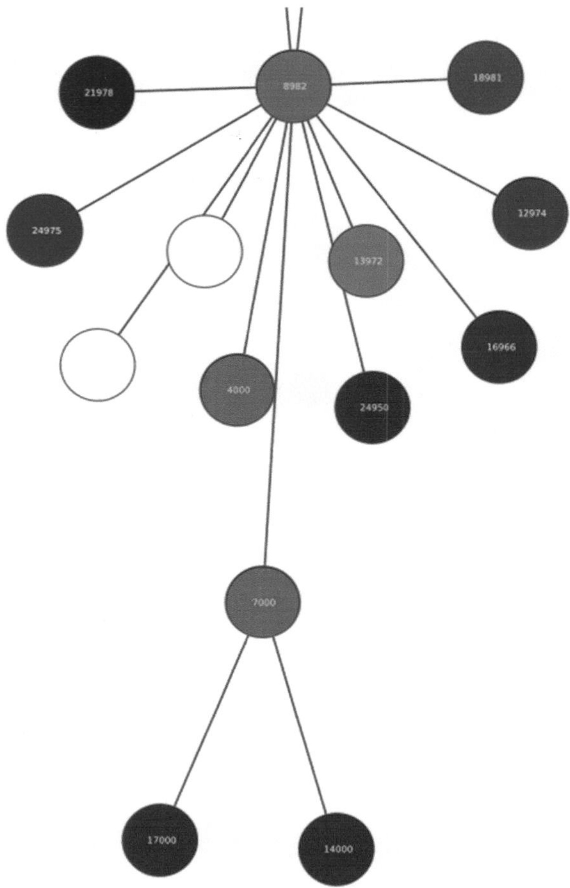

Fig. 5 PSO dynamic as complex network—notable part

Acknowledgments This work was supported by Grant Agency of the Czech Republic - GACR P103/15/06700S, further by financial support of research project NPU I No. MSMT-7778/2014 by the Ministry of Education of the Czech Republic and also by the European Regional Development Fund under the Project CEBIA-Tech No. CZ.1.05/2.1.00/03.0089, partially supported by Grant of SGS No. SP2015/142 and SP2015/141 of VSB - Technical University of Ostrava, Czech Republic and by Internal Grant Agency of Tomas Bata University under the projects No. IGA/FAI/2015/057 and IGA/FAI/2015/061.

References

1. Kennedy, J., Eberhart, R.: Particle swarm optimization. In: IEEE International Conference on Neural Networks, pp. 1942–1948 (1995)
2. Kennedy, J., Eberhart, R.C., Shi, Y.: Swarm Intelligence. Morgan Kaufmann Publishers (2001)

3. Nickabadi, A., Ebadzadeh, M.M., Safabakhsh, R.: A novel particle swarm optimization algorithm with adaptive inertia weight. Appl. Soft Comput. **11**(4), 3658–3670 (2011)
4. Yuhui, S., Eberhart, R.: A modified particle swarm optimizer. In: IEEE World Congress on Computational Intelligence, 4–9 May 1998, pp. 69–73
5. Zelinka, I., Davendra, D., Enkek, R., Jaek, R.: Do evolutionary algorithm dynamics create complex network structures? Complex Syst. **2**, 0891–2513, **20**, 127–140
6. Zelinka, I., Davendra, D.D., Chadli, M., Senkerik, R., Dao, T.T., Skanderova, L.: Evolutionary dynamics as the structure of complex networks. In: Zelinka, I., Snasel, V., Abraham, A. (eds.) Handbook of Optimization. ISRL, vol. 38, pp. 215–243. Springer, Heidelberg (2013)
7. Zelinka, I.: Investigation on relationship between complex network and evolutionary algorithms dynamics. AIP Conf. Proc. **1389**(1), 1011–1014 (2011)
8. Davendra, D., Zelinka, I., Senkerik, R., Pluhacek, M.: Complex network analysis of discrete self-organising migrating algorithm. In: Zelinka, I., Suganthan, P., Chen, G., Snasel, V., Abraham, A., Rossler, O. (eds.) Nostradamus 2014: Prediction, Modeling and Analysis of Complex Systems, Advances in Intelligent Systems and Computing, pp. 161–174. Springer, Berlin, Heidelberg (2014)
9. Davendra, D., Zelinka, I, Metlicka, M., Senkerik, R., Pluhacek, M.: Complex network analysis of differential evolution algorithm applied to flowshop with no-wait problem. In: 2014 IEEE Symposium on Differential Evolution (SDE), pp. 1, 8, 9–12 Dec 2014

Particle Swarm Optimizer with Diversity Measure Based on Swarm Representation in Complex Network

Jakub Janostik, Michal Pluhacek, Roman Senkerik and Ivan Zelinka

Abstract In this paper a alternative approach to the diversity guided particle swarm optimization (PSO) is investigated. The PSO shows acceptable performance on well-known test problems, however tends to suffer from premature convergence on multi-modal test problems. This premature convergence can be avoided by increasing diversity in search space. In this paper we introduce diversity measure based on graph representation of swam evolution and we discuss possibilities of graph representation of swarm population in adaptive control of PSO algorithm. Based on our findings we concluded, that network representation of evolution population and its subsequent analysis can be used in adaptive control, in various degrees of success.

Keywords Particle swarm optimization · PSO · Graph · Complex network · Evolutionary algorithm · Adaptive control

---

J. Janostik (✉) · M. Pluhacek · R. Senkerik
Faculty of Applied Informatics, Tomas Bata University in Zlin, T.G. Masaryka 5555, 760 01 Zlin, Czech Republic
e-mail: janostik@fai.utb.cz

M. Pluhacek
e-mail: pluhacek@fai.utb.cz

R. Senkerik
e-mail: senkerik@fai.utb.cz

I. Zelinka
Faculty of Electrical Engineering and Computer Science, VŠB-Technical University of Ostrava, 17. listopadu 15, 708 33 Ostrava-Poruba, Czech Republic
e-mail: ivan.zelinka@vsb.cz

A. Abraham et al. (eds.), *Proceedings of the Second International Afro-European Conference for Industrial Advancement AECIA 2015*, Advances in Intelligent Systems and Computing 427, DOI 10.1007/978-3-319-29504-6_52

1 Introduction

The particle swarm optimization algorithm (PSO) is one of the most widely used Evolutionary Computation Techniques (ECT's). It was first introduce by Kennedy et al. in 1995 [1, 2]. Even thought it was already shown that this model is comparable in performance with traditional algorithms such as the genetic algorithm or simulated annealing [3–6], PSO suffers from premature convergence during multi-modal optimization. This results in performance loss and sub-optimal solutions. Reason for this seems to be information flow limited to only communication between swarm individual and globally best individual. During evolution process we can observe decline in diversity which leads to clustering of particles, leaving the PSO unable to escape local minima. Consequently, the swarm exhibits low diversity and fitness stagnation [7].

Several paper have been written on model improving optimization algorithm and trying to avoid sub-optimal solutions. Many solutions have been suggested, mostly in regards to genetic algorithms (GA) [5, 8, 9]. PSO model updates include the addition of queen particle, the introduction of subpopulations or the alternation of the neighborhood topology [5, 10–12].

Among more recent proposals, the Diversity-Guided Evolutionary Algorithm (DGEA), by Ursem can be mentioned [13]. In his paper Ursem proposes redefining of the Gaussian mutation to be directed mutation instead. This idea is then implemented by Riget and Vesterstrom, who use directed mutation to adaptively control swarm evolution [7]. In their paper they propose attractive and repulsive PSO (ARPSO). This modified model uses attraction and repulsion phase. By measuring diversity they alternate between these phases.

We have adopted this idea and modified repulsion phase to be guided by the information gathered by the means of graph analysis and the complex network analysis over graph constructed from swarm evolution.

2 Particle Swarm Optimization

A brief description of PSO algorithm follows in this section. The PSO was firstly introduced by Eberhart and Kennedy in 1995 [1]. Each particle in the population represents a candidate solution for the optimization problem that is defined by the cost function (CF). In each iteration of the algorithm, a new location (combination of CF parameters) for the particle is calculated based on its previous location and velocity vector (velocity vector contains particle velocity for each dimension of the problem).

According to the method of selection of the swarm for best solution information spreading, the PSO algorithms are noted as global PSO (GPSO) or local PSO (LPSO). Within this research the PSO algorithm with global topology (GPSO) [14] was utilized.

The new "velocity" is determined by following formula (1).

$$v_{ij}^{t+1} = w \cdot v_{ij}^{t} + c_1 \cdot Rand \cdot (pBest_{ij} - x_{ij}^{t}) + c_2 \cdot Rand \cdot (gBest_j - x_{ij}^{t}) \quad (1)$$

where:

v_i^{t+1}	New velocity of the *i*th particle in iteration $t + 1$.
w	Inertia weight value.
v_i^t	Current velocity of the *i*th particle in iteration t.
c_1, c_2	Priority factors.
$pBest_i$	Local (personal) best solution found by the *i*th particle.
$gBest$	Best solution found in a population.
x_{ij}^t	Current position of the *i*th particle (component j of dimension D) in iteration t.
$Rand$	Pseudo random number, interval (0, 1). CPRNG is applied only here.

The maximum velocity was limited to 0.2 times the range as it is usual. The new position of each particle is then given by (2), where x_i^{t+1} is the new particle position:

$$x_i^{t+1} = x_i^t + v_i^{t+1} \quad (2)$$

Finally the linear decreasing inertia weight [15, 16] is used in the typically referred GPSO design that was used in this study. The dynamic inertia weight is meant to slow the particles over time thus to improve the local search capability in the later phase of the optimization. The inertia weight has two control parameters w_{start} and w_{end}. A new w for each iteration is given by (3), where t stands for current iteration number and n stands for the total number of iterations. The values used in this study were $w_{start} = 0.9$ and $w_{end} = 0.4$.

$$w = w_{start} - \frac{((w_{start} - w_{end}) \cdot t)}{n} \quad (3)$$

3 Network Representation

Our network representation of PSO algorithm builds on findings of Davendra and Zelinka in their paper Complex Network Analysis of Differential Evolution Algorithm applied to Flowshop with No-Wait problem [17]. In their paper they were able to prove that interaction between individuals during differential evolution (DE) represented as graph exhibits non-trivial features of complex networks (CN). Their network was constructed from individuals as nodes and their interaction as edges between them.

Because during PSO only interaction in the swarm is between individual and globally best individual, graphs constructed in the same way were not sufficient for use in adaptive control of evolution process. New ways of representing the swarm in graph had to be implemented. Several possible ways of handling this problem were inspected.

In this article we introduced new approach. Nodes of our network consists of swarm individuals in each iteration. Therefore individual can be represented in graph up to N times, where N equals number of iterations inside the measured window. Individuals are included in the graph only if they took part in interaction with other individual. Interactions are defined at the end of each iteration. After the population is sorted, edge is created for every two individuals who changed places in the sorting process. This way active individuals create binding with individuals

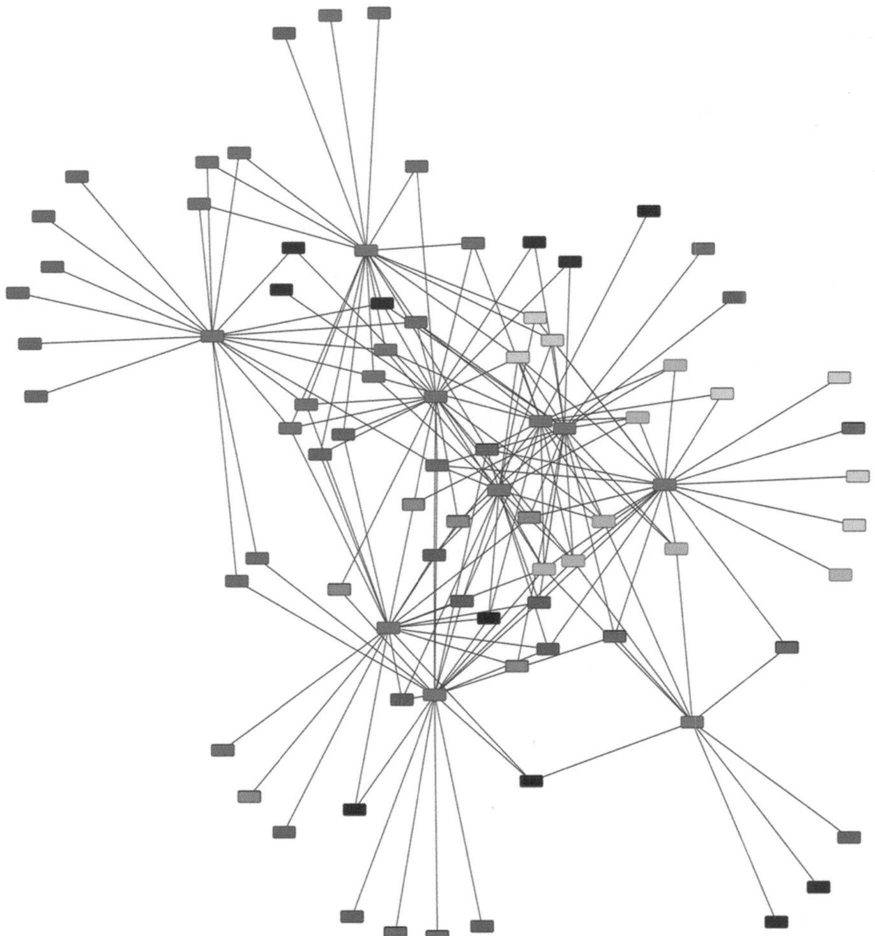

Fig. 1 Generated network for swarm of size 10 running for 10 iterations. Nodes with same color belong to same iteration

which are stagnating and not part of optimization process. This network is being constructed after every iteration until limiting number of iterations is reached. Then network is analyzed, action is taken and network is discarded. Evolution process then continues as before (Fig. 1).

4 Proposed Diversity Guided PSO Model

Same as ARPSO proposed by Riget and Vesterstrom [7], we define two phases. Attractive and repulsive phase. During attractive phase the model behaves in the same way as basic PSO, with the exception graph construction in the background.

In the ARPSO model, the repulsive phase is triggered by measuring diversity and comparing it to the limiting value. We chose fixed number of iteration before repulsion phase is triggered because long periods of high diversity would lead to graph bloating and decrease of performance of algorithm. During repulsive phase we divide swarm into up to K subpopulations. We create division by finding shortest paths between first and last occurrence of K best individuals. Such shortest path contains best individual and every individual he interacted with. After division is complete we trigger repulsion for each subpopulation, where each individual is repulsed to the best individual in each population *iBest*. This repulsion phase last for fixed period of iterations. Repulsion is achieved using formula:

$$v_{ij}^{t+1} = w \cdot v_{ij}^{t} - c_1 \cdot Rand \cdot (pBest_{ij} - x_{ij}^{t}) + c_2 \cdot Rand \cdot (iBest_j - x_{ij}^{t}) \tag{4}$$

Repulsive phase can be describes in following pseudocode:

```
PROGRAM REPULSE
    WHILE size(paths) < LIMIT_PATHS && hasMoreNodes(graph)
        paths += extract_next_path()
        remove_node_cccurences_from_graph(paths)
    paths += construct_path_from_remaining_nodes()
    FOR path IN paths
        FOR REPULSION_DURATION
            repulse_in_path(path)
```

From pseudo code we can observe several properties. Number of extracted paths can be fewer than defined limiting number of paths. This can occur when whole population stagnates. In which case single path will be constructed from entire swarm and repulsion will occur for whole population. Also, because we construct path from every node which wasn't part of any interaction, during highly active phase of evolution, limit set for number of extracted paths can be exceeded by one.

As Riget and Vesterstrom deducted in their paper [7], during repulsion phase, close to none improvement occurs. Therefore fitness computations during repulsion phase are skipped. This is done to improve computation times of the algorithm.

5 Experiment Setup

Our model was tested against basic PSO implementation. We tested both models on following functions.

Griewank,

$$f_1(\vec{x}) = \frac{1}{4000}\sum_{i=1}^{n} x_i^2 - \prod_{i=1}^{n} \cos\left(\frac{x_i}{\sqrt{i}}\right) + 1 \quad , where \ -600 \le x_i \le 600 \tag{5}$$

Ackey,

$$f_2(\vec{x}) = e + 20 - 20 \cdot \exp\left(-0.2 \cdot \sqrt{\frac{1}{n}\sum_{i=1}^{n} x_i^2}\right) - \exp\left(\frac{1}{n}\sum_{i=1}^{n} \cos(2\pi \cdot x_i)\right) \quad , where \ -30 \le x_i \le 30 \tag{6}$$

Rosenbrock

$$f_3(\vec{x}) = \sum_{i=1}^{n-1} 100 \cdot \left(x_{i+1} - x_i^2\right)^2 + (x_i - 1)^2 \quad , where \ -100 \le x_i \le 100 \tag{7}$$

Rastrigin

$$f_4(\vec{x}) = \sum_{i=1}^{n} x_i^2 + 10 - 10 \cdot \cos(2\pi x_i) \quad , where \ -5.12 \le x_i \le 5.12 \tag{8}$$

These function have been selected as widely used benchmark multimodal functions. All test functions have global minimum at (0, 0 … 0) with a fitness value of 0.

In the experiment basic PSO and our proposed model were compared. The experiment was conducted between implementations over first 1000 iterations with varying population size of 50 and 70 individuals. In this experiments final fitness value and computation duration were measured.

6 Results

In this section, the results for each test function are summarized to a statistical overview (Tables 1, 2, 3 and 4).

Table 1 Performance comparison for Rastrigin test function between basic PSO and proposed model (PSO*) for varying population size of 50 and 70

	PSO 50	PSO* 50	PSO 70	PSO* 70
Mean CF value	72.4524	43.1546	56.1157	44.1581
Std. dev.	33.0305	18.0353	17.6823	14.9404
CF value median	70.6781	36.9852	57.4907	42.6537
Max. CF value	95.3461	79.9244	102.6788	67.8557
Min. CF value	50.6028	19.8487	29.4514	17.2041

Table 2 Performance comparison for Rosenbrock test function between basic PSO and proposed model (PSO*) for varying population size of 50 and 70

	PSO 50	PSO* 50	PSO 70	PSO* 70
Mean CF value	50.3647	27.4340	39.3269	24.4512
Std. dev.	26.5502	28.3511	30.0355	20.0357
CF value median	33.3364	18.8838	25.6310	18.8470
Max. CF value	84.4643	135.4004	112.6849	84.1558
Min. CF value	22.4506	17.8376	15.1080	12.7633

Table 3 Performance comparison for Ackley test function between basic PSO and proposed model (PSO*) for varying population size of 50 and 70

	PSO 50	PSO* 50	PSO 70	PSO* 70
Mean CF value	6.4842	0.5930	5.0961	0.4176
Std. dev.	3.0506	0.9238	0.9946	0.8846
CF value median	5.4848	0.2499	5.2262	0.0347
Max. CF value	10.0301	3.0755	7.2524	3.0639
Min. CF value	4.4110	9.1978E-05	3.4468	4.0863E-05

Table 4 Performance comparison for Griewank test function between basic PSO and proposed model (PSO*) for varying population size of 50 and 70

	PSO 50	PSO* 50	PSO 70	PSO* 70
Mean CF value	3.0895	0.3168	1.5426	0.2561
Std. dev.	1.4716	0.3422	0.4751	0.3310
CF value median	2.7051	0.1668	1.4166	0.0626
Max. CF value	4.5084	1.0352	2.9306	0.9590
Min. CF value	1.8247	1.0013E-09	0.9926	1.6934E-05

Table 5 Computational times for proposed model

	Rastrigin	Rosenbrock	Ackley	Griewank
Computational time	13.36T	10.8T	16.5T	17.18T

T is mean time the basic PSO implementation takes to complete 1000 cycles

From the results presented in Tables 1, 2, 3 and 4 we can conclude that proposed model offers improvements in evolution process. Presented results show significant improvement in all test function. As can be seen in Table 5, proposed model suffers from increased computational time. This is mainly caused by demanding shortest path algorithm. This value can be lowered by tuning size of the window during which the network is constructed and shortening the repulsive phase. Also during measurements the graph was constructed during entire evolution. This can be changed and graph can be constructed only if diversity is decreased below limiting value or we observe stagnation in population.

7 Conclusion

We can conclude that using graph representation of swarm evolution and following adaptive control using graph and complex network analysis can be done. Our model improved process of evolution. We found solution faster and were able to locate more precise results. As a negative side we see increase in computational time. Because of this we do not propose this model as a direct competitor against models such as ARPSO [7], but it can be seen as proof of concept for graph and complex network analysis as a tool in adaptive control of evolutionary computations.

8 Future Work

This or similar methods can be implemented in different, more suitable algorithms, such as differential evolution or firefly algorithm, where graph can be constructed from direct interaction between individuals. In this paper we chose shortest path as a tool for partitioning swarm and subsequent diversification of population. For this other tools of network analysis can be used, such as finding clusters or computing network degree per population individual. For future work we are considering investigating mentioned methods and implementing adaptive control for more algorithms with different properties.

Acknowledgments This work was supported by Grant Agency of the Czech Republic - GACR P103/15/06700S, further by financial support of research project NPU I No. MSMT-7778/2014 by the Ministry of Education of the Czech Republic and also by the European Regional Development Fund under the Project CEBIA-Tech No. CZ.1.05/2.1.00/03.0089, partially supported by Grant of

SGS No. SP2015/142, VŠB - Technical University of Ostrava, Czech Republic and by Internal Grant Agency of Tomas Bata University under the project No. IGA/FAI/2015/057 and IGA/FAI/2015/061.

References

1. Kennedy, J., Eberhart, R.: Particle swarm optimization. In: IEEE International Conference on Neural Networks, pp. 1942–1948 (1995)
2. Kennedy, J., Eberhart, R.C., Shi, Y.: Swarm Intelligence. Morgan Kaufmann Publishers (2001)
3. Angeline, P.J.: Evolutionary optimization versus particle swarm optimization: philosophy and performance differences. In: Evolutionary Programming VII, Lecture Notes in Computer Science, vol. 1447, pp. 601–610. Springer (1998)
4. Eberhart, R.C., Shi, Y.: Comparison between genetic algorithms and particle swarm optimization. In: Evolutionary Programming VII, Lecture Notes in Computer Science, vol. 1447, pp. 611–616. Springer (1998)
5. Krink, T., Vesterstrøm, J., Riget, J.: Particle swarm optimization with spatial particle extension. To appear in: Proceedings of the Congress on Evolutionary Computation 2002 (CEC-2002)
6. Vesterstrøm, J., Riget, J., Krink, T.: Division of labor in particle swarm optimization. To appear in: Proceedings of the Congress on Evolutionary Computation 2002 (CEC-2002)
7. Riget, J., Vestterstrom, J.S.: A diversity-guided particle swarm optimizer the ARPSO. Technical report, EVAlife, Department of Computer Science, University of Aarhus, Denmark (2002)
8. Back, et al.: Handbook on Evolutionary Computation. IOP Publishing Ltd. and Oxford University Press. Chapter 6.3 and 6.4
9. DeJong, K.A.: An analysis of the behavior of a class of genetic adaptive systems. Ph.D. thesis, University of Michigan (1975)
10. Clerc, M.: The Swarm and the Queen: Towards a Deterministic and Adaptive Particle Swarm Optimization
11. Kennedy, J.: Small worlds and mega-minds: effects of neighborhood topology on particle swarm performance. In: Proceedings of the 1999 Congress of Evolutionary Computation, vol. 3, 1931–1938. IEEE Press
12. Løvbjerg, M., Rasmussen, T.K., Krink, T.: Hybrid particle swarm optimiser with breeding and subpopulations. In: Proceedings of the third Genetic and Evolutionary Computation Conference (GECCO-2001)
13. Ursem, R.K.: Diversity-guided evolutionary algorithms. In submission for the Parallel Problem Solving form Nature Conference (PPSN VII)
14. Zhan, Z.-H., Zhang, J., Li, Y., Shi, Y.-H.: Orthogonal learning particle swarm optimization. IEEE Trans. Evol. Comput. **15**(6), 832–847 (2011)
15. Nickabadi, A., Ebadzadeh, M.M., Safabakhsh, R.: A novel particle swarm optimization algorithm with adaptive inertia weight. Appl. Soft Comput. **11**(4), 3658–3670 (2011)
16. Yuhui, S., Eberhart, R.: A modified particle swarm optimizer. In: IEEE World Congress on Computational Intelligence, 4–9 May 1998, pp. 69–73
17. Davendra, D., Zelinka, I., Metlicka, M., Senkerik, R., Pluhacek, M.: Complex network analysis of differential evolution algorithm applied to flowshop with no-wait problem. In: 2014 IEEE Symposium on Differential Evolution (SDE), pp. 1, 8, 9–12 Dec 2014

Capturing Inner Dynamics of Firefly Algorithm in Complex Network—Initial Study

Jakub Janostik, Michal Pluhacek, Roman Senkerik, Ivan Zelinka and Frantisek Spacek

Abstract In this paper the idea of capturing inner dynamics of evolutionary process in complex network is discussed. Complex network, in itself, can contain large number of information about running evolutionary process. Consequently this information can be used in subsequent adaptive control of algorithm. We present initial study of methodology for capturing the information about algorithm and present possible uses for created networks. For this study the Firefly algorithm was selected.

Keywords Firefly algorithm · Complex network · Evolutionary algorithm · Adaptive control

1 Introduction

As was recently proposed, evolutionary computational techniques (ECTs) can by analyzed by the means of transformation of evolution process into complex network. Already presented results show promising areas for further studies in

J. Janostik (✉) · M. Pluhacek · R. Senkerik · F. Spacek
Faculty of Applied Informatics, Tomas Bata University in Zlin, T.G. Masaryka 5555, 760 01 Zlin, Czech Republic
e-mail: janostik@fai.utb.cz

M. Pluhacek
e-mail: pluhacek@fai.utb.cz

R. Senkerik
e-mail: senkerik@fai.utb.cz

F. Spacek
e-mail: spacek@fai.utb.cz

I. Zelinka
Faculty of Electrical Engineering and Computer Science, VŠB-Technical University of Ostrava, 17. listopadu 15, 708 33 Ostrava-Poruba, Czech Republic
e-mail: ivan.zelinka@vsb.cz

A. Abraham et al. (eds.), *Proceedings of the Second International Afro-European Conference for Industrial Advancement AECIA 2015*, Advances in Intelligent Systems and Computing 427, DOI 10.1007/978-3-319-29504-6_53

analyzing created networks [1–4]. In this article we wanted to explore the idea of applying this knowledge and create network describing swarm intelligence based evolutionary technique. For this research Firefly algorithm (FA) was selected as a suitable candidate for its high information flow between large numbers of particles in each iteration.

The motivation for this research is high likelihood of FA creating large density and highly complex network structures from which important information can be derived. Subsequently this information can be later used for predicting swarm behavior or creating new adaptive or learning approaches.

2 Firefly Algorithm

Firefly algorithm was first presented by Yang in at Cambridge University [5, 6]. FA is based on simplified behavior of fireflies in night. Following rules were established to describe mentioned behavior [5–7]:

1. All fireflies are unisex so that fireflies will attract each other regardless of their sex.
2. The attractiveness is proportional to the brightness, and they both decrease as their distance increases. This means that for any two flashing fireflies, the less bright one will move towards the brighter one. Firefly will move randomly if there is no brighter one.
3. The brightness of a firefly is determined by the landscape of the objective function.

Firefly's attractiveness is determined by its light intensity, which is proportional to the encoded objective function. The brightness $I(r)$ varies with the distance r monotonically and exponentially. That is,

$$I(r) = \frac{I_0}{1 + \lambda r^2}, \tag{1}$$

where I_0 is the initial brightness and λ is the light absorption coefficient. Similarly, the attractiveness of a firefly can be defined using following formula:

$$A(r) = \frac{A_0}{1 + \lambda r^2}, \tag{2}$$

where A_0 is the initial attractiveness.

If a firefly located at $\acute{x} = (\acute{x}_1, \acute{x}_2, \ldots, \acute{x}_n)$ is brighter than firefly located at $x = (x_1, x_2, \ldots x_n)$, the firefly located at x will move towards one located at $\acute{x}$.

The algorithm can be summarized as follows [8].

1. Generate a random solution set $\{x_1, x_2, \ldots x_n\}$.
2. Compute intensity $\{I_1, I_2, \ldots I_n\}$ for each member of solution set.
3. Move firefly towards other brighter firefly if possible, move it randomly if not.
4. Update solution set.
5. If a termination criterion is fulfilled terminate algorithm. Go to step 2 otherwise.

3 Experiment

For the experiment we used FA with default setting. Algorithm optimized Schwefel's benchmark function for 100 iterations with population of size 30. Rather low iteration count and population size were chosen to keep generated networks clear and small, so the visualization is possible.

Before the networks were constructed, methodology for capturing information flow during evolution process, had to be implemented. We came up with two processes that resulted in the basic weighted oriented network and the time capturing oriented network.

3.1 Basic Weighted Oriented Network

In order to create network we decided to visualize every firefly as a node. Connection between nodes is plotted for every successful interaction between fireflies. Successful interaction is defined as such interaction where one of the individuals gets improved. In the case of FA it is when firefly flies towards another and improves own brightness. This leads to network presented in Fig. 1. Duplicate connections were omitted from the network in the sake of clarity.

Because across multiple iterations of algorithm there may multiple connections between nodes we decided to improve upon the design by weighting the connections. If there is connection between the firefly A and B it starts with weight 1. If in another iteration there is another successful interaction between the firefly A and B, new connection is not created but the weight of the existing connection is incremented by 1. At the end of evolution the weight is normalized. If the firefly gets improved by another in every iteration, at the end of the evolution their connection will have weight 1. If it never gets improved their connection will have weight 0.

In Fig. 2 we can see network where connections have their weights visually distinguished. In the top left corner we can see one dominant firefly which improved entire population more than 70 % of iterations (blue lines). On bottom right side we can observe few fireflies which took part in improvement of population only less than 30 % of iterations (red lines). Also from the network we can see that most of the fireflies improved one another only in between 30 and 70 % of iterations.

3.2 Time Capturing Oriented Network

Even though previous network yields interesting information, its size is limited by the size of population. This may not be favorable for all tools of network analysis. We decided to tackle this issue by adding time component to the constructed network. In the new model nodes are not created by fireflies alone but by fireflies and the iteration in which they were created. This way every firefly will get represented up to number of times that is equal to the number of iterations. Because of this there can't be same connection between nodes, so the weights become unnecessary. Created network can be seen in Fig. 3.

For easier readability every connection received color for the iteration in which it was created. In Fig. 4 we can see connections in iteration 1 being colored red and in every subsequent iteration color gradually changes into blue (last iteration). For the sake of clarity only first 10 iterations are presented.

In Fig. 5 if we compare red and blue part we can see that blue part seems more clustered, in other words it has more connections. From this we can deduce that

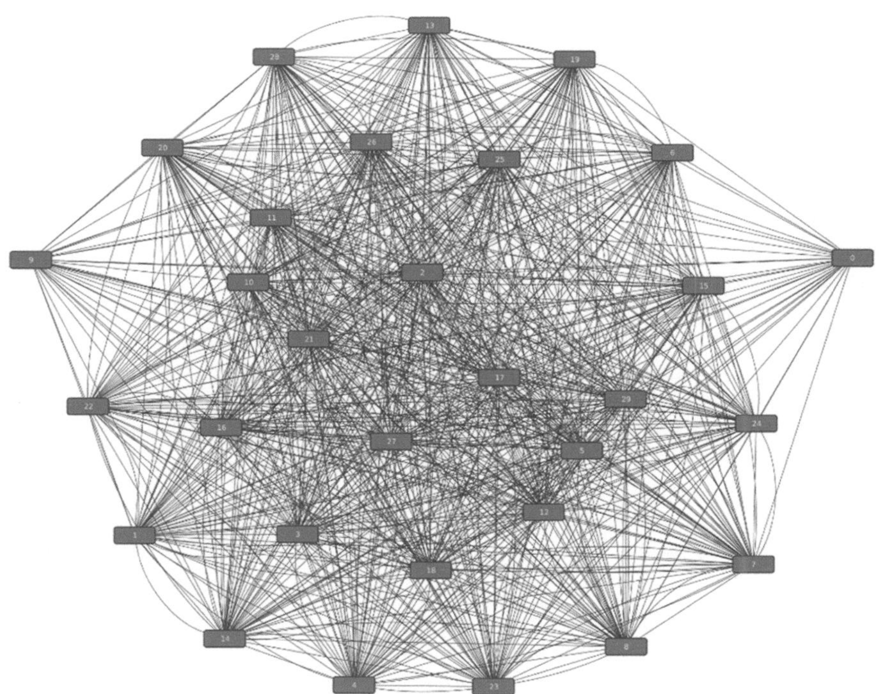

Fig. 1 Basic weighted oriented network for population of size 30 after 100 iterations

firefly population became more successful in searching optima after fifth iteration. Difference in densities was highlighted in Fig. 5, where iteration 0 was highlighted in red and iteration 10 in blue.

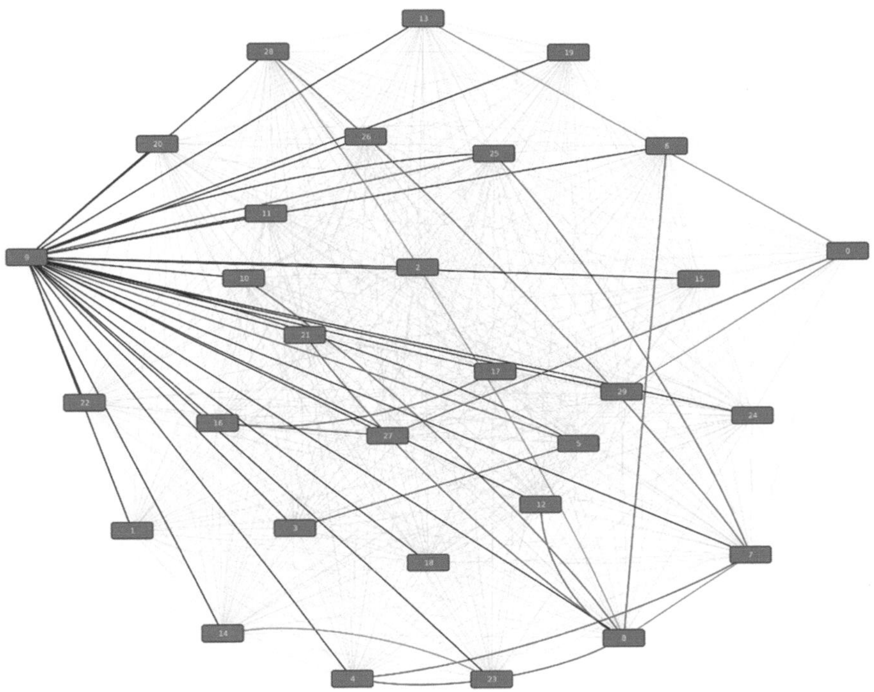

Fig. 2 Basic weighted oriented network for population of size 30 after 100 iterations with visually highlighted weights (*Blue* 0.7 < weight, *Grey* 0.3 < weight < 0.7, *Red* weight < 0.3)

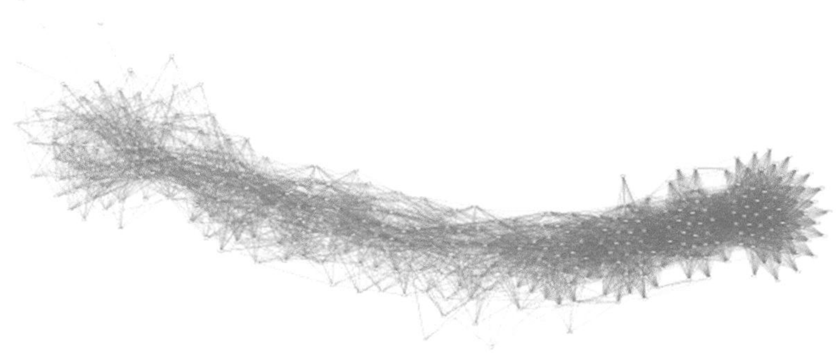

Fig. 3 Time capturing oriented network

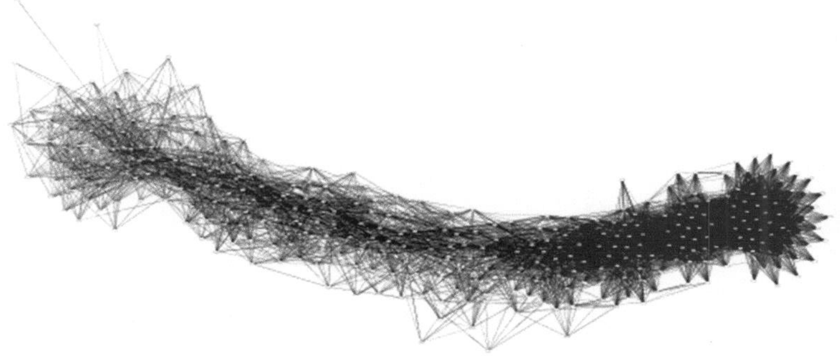

Fig. 4 Time capturing oriented network with visually encoded iteration order (*Red* iteration 1, *Blue* iteration 10)

Fig. 5 Time capturing oriented network with highlighted iteration 0 and 10

4 Conclusion

From the presented results we concluded that it is possible to construct complex network from swarm intelligence based evolutionary technique and for this network to have information about evolution process. These and similar network structures can be possibly used to find potentially dominant or dormant individuals and from statistical analysis of such networks pathological states as stagnation or premature convergence can possibly be predicted. Such information could be used for creation of tools in adaptive and learning mechanisms.

Acknowledgments This work was supported by Grant Agency of the Czech Republic - GACR P103/15/06700S, further by financial support of research project NPU I No. MSMT-7778/2014 by the Ministry of Education of the Czech Republic and also by the European Regional Development Fund under the Project CEBIA-Tech No. CZ.1.05/2.1.00/03.0089, partially supported by Grant of SGS No. SP2015/142, VŠB - Technical University of Ostrava, Czech Republic and by Internal Grant Agency of Tomas Bata University under the project No. IGA/FAI/2015/057, IGA/FAI/2015/061 and IGA/FAI/2015/031.

References

1. Zelinka, I., Davendra, D., Enkek, R., Jaek, R.: Do evolutionary algorithm dynamics create complex network structures? Complex Syst. **2**, 0891–2513, **20**, 127–140
2. Zelinka, I., Davendra, D.D., Chadli, M., Senkerik, R., Dao, T.T., Skanderova, L.: Evolutionary dynamics as the structure of complex networks. In: Zelinka, I., Snasel, V., Abraham, A. (eds.) Handbook of Optimization. ISRL, vol. 38, pp. 215–243. Springer, Heidelberg (2013)
3. Zelinka, I.: Investigation on relationship between complex network and evolutionary algorithms dynamics. AIP Conf. Proc. **1389**(1), 1011–1014 (2011)
4. Davendra, D., Zelinka, I., Senkerik, R., Pluhacek, M.: Complex network analysis of discrete self-organising migrating algorithm. In: Zelinka, I., Suganthan, P., Chen, G., Snasel, V., Abraham, A., Rossler, O. (eds.) Nostradamus 2014: Prediction, Modeling and Analysis of Complex Systems, Advances in Intelligent Systems and Computing, pp. 161–174. Springer, Berlin, Heidelberg (2014)
5. Yang, X.S.: Nature-Inspired Metaheuristic Algorithms. Luniver Press, UK (2008)
6. Yang, X.S.: Firefly algorithms for multimodal optimization. In: Watanabe, O., Zeugmann, T. (eds.) Proceedings of 5th Symposium on Stochastic Algorithms, Foundations and Applications, Lecture Notes in Computer Science, vol. 5792, pp. 169–178 (2009)
7. Yang, X.S., He, X.: Firefly Algorithm: Recent Advances and Applications (2013)
8. Tilahun, S.L., Ong, H.C.: Modified firefly algorithm. J. Appl. Math. **2012** (2012). ISSN:1110-757X

Nature-Inspired Phenomenons Modelling Method for Managing Complex Waste Disposal Systems

Malgorzata Zabinska, Wojciech Turek, Krzysztof Cetnarowicz and Pawel Pawlowicz

Abstract People produce vast quantities of garbage, what has become the real problem for the world. Despite multi-use recycled packages or existing incinerators, garbage volume still increases and it is necessary to remove it from cities and finally dispose on garbage dumps. The problem is how to do it optimally at the minimal outlay of work and cost. To solve such a problem we could use mobile robots. However the task to manage a group of robots emptying garbage containers located in streets is very complex. Since it seems that a centralized system for such management would be very compound, not flexible and inefficient we propose decentralized approach in the paper. The presented decentralized system is realized as an agent system where agents perform nature-inspired algorithms. We present our proposal of the system and results of simulation examination.

Keywords Mobile-robot management · Agent system · Nature inspired

1 Introduction

People produce vast quantities of garbage, what has become the real problem for the world. Despite multi-use recycled packages or existing incinerators, garbage volume still increases and it is necessary to remove it from cities and finally dispose on garbage dumps. The problem is how to do it optimally at the minimal outlay of work and cost, when the distance between a huge city and the garbage dump is meaningful. To solve such a problem we could use mobile robots; fast development of

M. Zabinska (✉) · W. Turek · K. Cetnarowicz · P. Pawlowicz
AGH University of Science and Technology, Al. Mickiewicza 30,
30-059 Krakow, Poland
e-mail: zabinska@agh.edu.pl

W. Turek
e-mail: wojciech.turek@agh.edu.pl

K. Cetnarowicz
e-mail: cetnar@agh.edu.pl

A. Abraham et al. (eds.), *Proceedings of the Second International Afro-European Conference for Industrial Advancement AECIA 2015*, Advances in Intelligent Systems and Computing 427, DOI 10.1007/978-3-319-29504-6_54

robotics allows assuming that in the near future such solutions may become common. Even nowadays automated warehouse management systems are introduces, which has some resemblance to garbage disposal problem. It is necessary to build the appropriate control systems, in more detail—management systems functioning in such a way that they could interact with their environment, other robots and are aware of their existence in the environment. Fulfilment of these requirements leads directly to the concept of creation the autonomous agents control systems. In the paper we presented the system for management of mobile robots which take care of waste/garbage disposal.

2 Problem Definition

The problem of garbage removal management may be considered basing on the real urban environment which comprises streets and crossroads [1]. Dustbins are located in the given places (streets, crossroads) with known locations (Fig. 1). The containers (dustbins) are filled in with unknown intensity. The intensity (frequency) of filling in is variable, unknown and impossible to predict. Each dustbin has given capacity, level of current filling l, and defined two levels of filling in: l_{min}—defining minimal filling and l_{max} defining maximal filling. The minimal level is the signal for emptying the container. A container should be emptied when its level l is below l_{max}. In parallel we consider the cyberspace where the computer system functions according to the real environment model as a graph with the marked containers (Fig. 1). The containers are equipped with sensors measuring the level of filling in, as an attribute of the container model which is transferred on-line to the system in the cyberspace.

The containers are emptied continuously by vehicles (garbage trucks or mobile robots) circulating in the streets. We assume that the container (a dustbin) is emptied by a garbage truck when the level of filling in is greater than zero. The number of vehicles (garbage trucks, robots) circulating simultaneously in the city is defined, but it is possible to activate the additional vehicle (or deactivate) at any moment of system functioning.

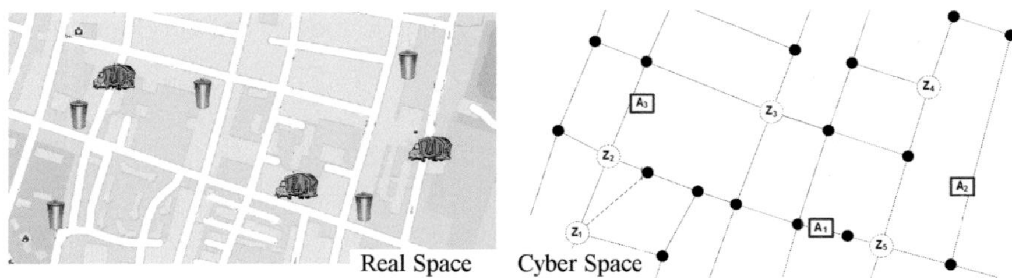

Fig. 1 Fragment of a city road system (Real Space) and the corresponding model in the Cyber Space

The problem is to continuously plan the routes of vehicles (garbage trucks, robots) in such a way that containers are emptied with the regard to the defined constraints (to avoid the overflow) and in the optimal way according to some assumed criteria.

The solution of the problem may be formulated as the choice made by each vehicle (a robot, a garbage truck) arriving at the given graph node. The inspiration for the algorithms presented in this paper is hunting by a predator for a victim in the forest where the predator does not see the victim and to choose the further way it navigates with the use of the smell emitted by the victim. Similarly the garbage containers (models of dustbins) emit smell which disperses in the graph by edges. The agents representing robots have the sense of smell, they recognize it and at nodes they choose edges with the most intensive smell. In the paper we presented a proposal of agent system for control of mobile robots group (or garbage trucks) where the real world is mapped to the model in the cyberspace. The mobile robots making the group of collaborating (in a way) robots will be associated in the virtual world with agents which can control them. Such a model enables simulation of the real environment without a need to build a prototype. The results of experiments will serve to indicate an optimal or at least suboptimal way of mobile robots management when they perform their tasks with regard to the described principles and constraints and enable identification of parameters which have decisive influence on the quality of the real system functioning.

3 Planning for Multiple Entities

The presented problem belongs in a way to the class of the vehicle routing problems (VRP). It also resembles the Travelling Salesman Problem (TSP) [5] or the Multiple Travelling Salesman Problem (mTSP). We can solve these problems in several ways including heuristic algorithms (like ant algorithm [3]). The essential difference of the presented problem is that locations which are to be visited are not known at the very beginning, but they appear and disappear as a result of filling in and emptying containers and finally the given node (a container i.e. a dustbin) according to needs should be visited by the vehicle again.

Applying robots for solving different tasks, e.g. waste collection and disposal it is necessary to coordinate their actions. Problem of mobile robots coordination as planning, tasks assignment and mobility check is described in [4]. Planning algorithms are usually complex and time consuming, because the process of plan realisation is frequently changed by unexpected situations. The algorithms are based on knowledge about environment, robots and tasks what require intensive communication between elements of the system.

As for tasks assignment in robot systems (MRTA) there are two approaches mentioned in the sources: centralised and decentralised approach. Centralised solutions apply one algorithm on one computer that plans all the decisions concerning allocation of tasks to robots. In most cases it is related to the choice of robots group leader [2]. Such a centralised approach may give theoretically the best solution optimal at the given criteria. However it has got many disadvantages as vulnerability to failures, slowing down the system at intensive communication, lack of reaction on frequent parameter changes, and others resulting from the centralised approach.

It seems that decentralised approach could be much more appropriate to solve the mentioned problem. Especially application of predefined negotiation protocols which define what influences the information exchange and what reaction of robots should be in the given situations [8] could be less vulnerable to errors, quicker and enable adjustment of an algorithm to changing conditions.

There is a group of solutions which solve the allocation tasks without obvious communication between robots when decision taking is based on observations of other robots and environment (e.g. assignment of low-level tasks [7]). Such solutions based on a swarm concept are more reliable because system functioning does not depend on failures of the particular robots. However creation of fully autonomous algorithm for the complex task of MRTA class is not simple. A solution presented in our previous paper [6] uses a concept of signal propagation in virtual grid, where only one type of signal was used. The concept is used in this work for solving different problem.

4 Nature-Inspired Model of Robots Management

The solution of the shown garbage disposal problem was based on the concept of two spaces, viz. Real space and Cyber space [1]. In the real space, there are active mobile robots who should realize the appearing tasks. Thus physical process of garbage disposal by robots (garbage trucks) circulating in streets takes place in the real space. In the cyber space the model of environment is created as a graph mapping street network with models of containers (dustbins). The dustbins (containers) are in nodes and have attributes describing the level of their filling. There are agents circulating in the graph who represent robots. Each robot from the real space is associated with an agent acting in the cyber space. An agent can transfer in the graph (in cyber space) between nodes by edges. The agent moving in the graph in the cyber space directs the associated robot through the streets of the environment in the real space. The task of the agent is such controlling of the robot that the latter one visits places with the filled in containers (dust bins) to be emptied. The agent realizes this task moving appropriately in the graph in such a way to reach these models of containers which filling in level is satisfactory (filling in attribute).

In the graph the containers, agents and edges can emit signals that are received by agents who use the received signals. Four algorithms of emitting and sensing signals are considered: the algorithm using repulsive pheromones, the algorithm based on attracting smell of garbage, the algorithm using the repulsive smell of agents, and the algorithm of random choice of the path.

Repulsive Pheromones

The algorithm applied a modification of ants moving when looking for nutrition. Ants choose such a way which has been chosen before by the other one and was marked by leaving there its pheromone. The quality of the path depends on the intensity of the signal. The high intensity means that many ants have used it so it is recommended. Such a script is applied by ants delivering food to the anthill. The other behavior is shown by species looking for the new sources of food. The ants-searchers do not choose such often used paths [3]. The analogy was used in the algorithm for the considered agents. The agent moving within the graph leaves its individual pheromone which evaporates with time. After reaching the node, the agents choose such an edge which has the least intensity (or its lack), thus the pheromone acts repulsively on the agent. As a result the agent directs towards areas do not visited by others (or not visited for a relatively long time). It assumes that there may be containers which are not emptied. If two or more edges were qualified for the choice, then the edge is chosen at random.

Attractive Smell of Waste

Models of garbage containers (dustbins) emit smells in the cyberspace that disperse on edges directly outgoing from the node and dependently on the level of signal propagation are distributed on the subsequent edges according to the assumed schema. The intensity of smell depends on the level of filling in the container. The signal as smell of garbage is attractive for agents and as a result the agent chooses the further way along the edge which is the source of the highest intensity of smell coming in. If two or more edges are qualified for the choice, then the appropriate edge is chosen at random.

Repulsive Smell of Robots

Algorithm of waste signal has one essential disadvantage. In case of simultaneous action of two or more agents, there is the high risk that some agents could decide to move towards the same container basing on the smell. It may cause origination of unnecessary aggregation of robots (agents) at the given dustbin that can be emptied by one robot only. We can prevent it by emission of additional smell by agents. The way is analogous as for the waste signal and the substantial difference is that such smell is repulsive for the other agents (it does not affect the agent which emits it). As a result, the agent representing the garbage truck, choosing the edge it was to move along, takes into consideration both signals, and more precisely for each edge it calculates the edge k, where the intensity of smell SkZ is maximal and the intensity of other agents SkA is minimal.

Random Choice of the Paths

The algorithm consists in random choice of an edge from a set of edges K. The drawing takes place with the equal probability for each edge. This method has been used to compare the results obtained this way with the results obtained as a result of application of the algorithms like the nature-inspired ones.

5 Simulation and Results

The agent system for management of mobile robots group was implemented in C++. The constructed prototype system, beside the realized management of mobile agents group (robots), makes also a simulation tool which allows examination of influence of many parameters and algorithms of control upon the obtained results.

The built prototype simulator of agent system for management of mobile robots group enables work in two modes. The first mode is the detail analysis and tracing of agents and the simulation course the mode enables tracing of agents movements and observation of changes and events in the system. The second mode is examination of influence of the chosen parameter on the obtained results—the examination is done within the given number of cycles for the given range of the parameter, and on the basis of the partial results the average rates are computed.

The simulation system enables verification of agent environment behaviour in different structure of the environment (graph making the simulation environment) and number of agents (robots, dustbins) that take part in the simulation. It allows testing strategy of moving agents/garbage trucks in the given environment with different range and the way of dispersion of smell signals, frequency of filling in the containers, time for emptying containers, speed of transferring between nodes and intensity of agents pheromone evaporation.

5.1 Results of Experiments

The presented results have been obtained as a result of the series of simulations (examination of parameters) performed on the test graph that was experimental environment for the created system. The graph used for tests was a regular one in the form of a grid if size 20×20 (400 nodes and 760 edges). The garbage containers (dustbins) were allocated at random in 40 nodes (10 % of the total number). At the beginning of the simulation the containers were filled in with the random amount of waste (selected at random, i.e. the drawn level of filling in). In the chosen node (the node with index 0) there was the given number of garbage trucks with the same starting parameters.

To evaluate the quality of the simulated management system, the indicators were assumed which analysis serves to evaluate the properties of the simulated system of garbage disposal according to various criteria.

The defined indicators are the following:

- Number of emptying,
- Average penalty max,
- Average filling in of garbage containers (dustbins),
- Total route (cost),
- Volume of the collected garbage,

- Average cost of emptying the container expressed as the ratio of the Total route (cost) to the Number of emptying,
- Efficiency of the garbage truck expressed as the ratio of the Volume of collected garbage to the Number of emptying,

5.1.1 Influence of Frequency of Filling the Containers

The important indicator of the system quality evaluation is the value of penalty max increased by 1 when the level of filling of emptied container is greater than l_{max}. Frequency of filling in is defined as average time *dt* between two addition of the given amount of garbage. It informs about percentage of emptying the garbage containers which were overflowed. The value of the indicator should be as low as possible. As we can see in Fig. 2 the value is very high for very frequent filling in the garbage containers. It is the evidence that examined garbage trucks are not able to keep up with emptying the containers. Application of algorithms with garbage signals and garbage truck signals causes that we quickly obtain satisfactory results. The lowest efficiency is shown for the algorithm with pheromone path of agents.

Further essential quality indicator (Fig. 3) is the average filling of the containers in time (level l_{max} was assumed as 100 %). Additional important information is the level to which the containers were have been overflowed. In this case the most effective is the algorithm of garbage with garbage trucks signal and signal of garbage themselves.

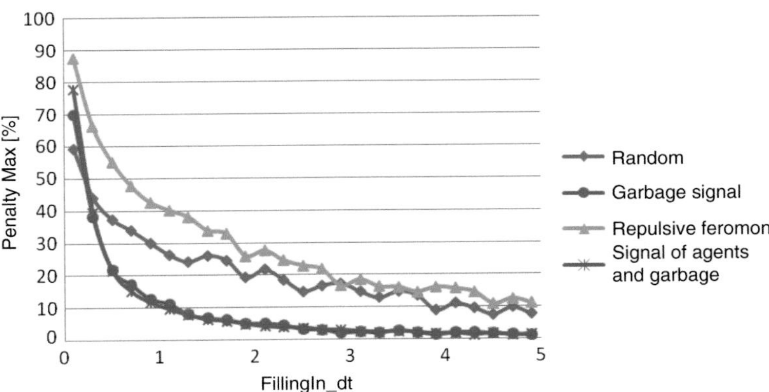

Fig. 2 Dependency between frequency of filling in the containers and the containers overflow

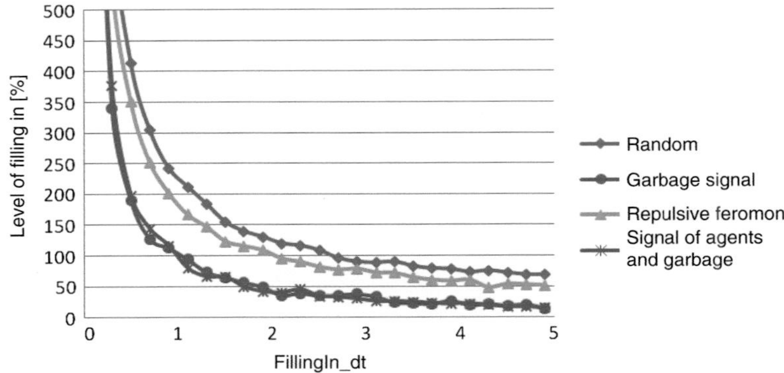

Fig. 3 Influence of containers filing in frequency on average level of containers filling

5.1.2 Influence of the Number of Agents (Robots, Garbage Trucks) on the Quality of Emptying Process

Results of conducted simulation depending to the number of simultaneously working trucks for all four algorithms shown in the following figures. For each number of robots the simulation was performed ten times and presented data expresses the average values.

The number of simultaneously acting agents (robots, trucks) has substantial influence on the penalty indicator max (Fig. 4. For algorithms where the garbage signal or garbage and truck signal matters, there is a vast improvement of the value of indicator up to the moment of execution of four robots at the same time. Further increasing of robots number has only slight influence on the indicator improvement. This means that the four trucks are enough for the examined environment to provide the small ratio of overflowed containers. This conclusion concerns algorithms using waste

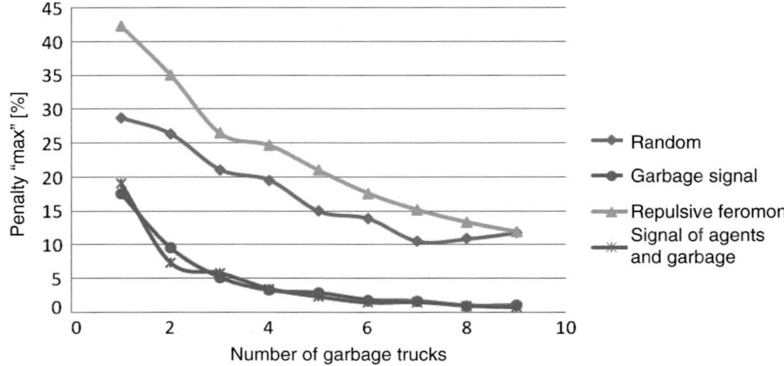

Fig. 4 Dependence between the garbage trucks number in the system and containers overflow

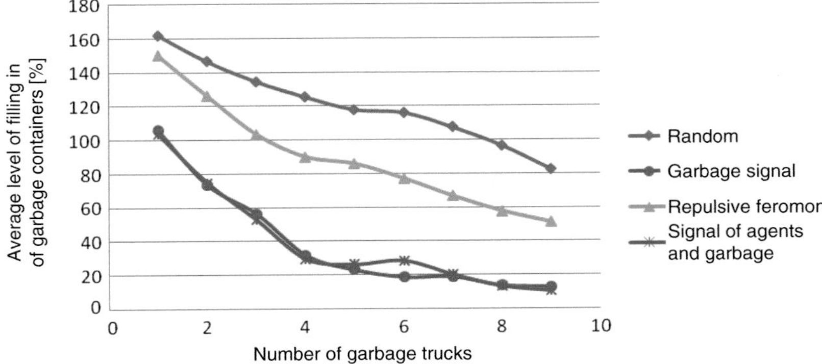

Fig. 5 Influence of the number of garbage trucks/robots on the value of average filling in of the containers

signal or the signal of garbage and trucks together. In case of the other two algorithms addition of each garbage truck decreases ratio of overflowed containers. It is worth noticing that the use of two garbage trucks working simultaneously moving according to the algorithms with garbage signal or garbage and truck signal together gives better results than nine simultaneously working trucks working according to random algorithms or algorithm using repulsive pheromones.

In the Fig. 5 we presented the dependency of average filling in of containers from the number of garbage trucks (robots). Let us assume that at the correct emptying, the average filling in does not be greater than 60 % (l_{max} 100 %). Such a result we obtain already for three garbage trucks moving according to the algorithm waste signal or waste and trucks signal together whereas for eight garbage trucks moving according to the algorithm repulsive pheromones. For the random algorithm the result is not satisfactory.

5.2 *Summary of Simulation Results*

The study of three major factors suggest that the best results are obtained while using the garbage signal and agents algorithm simultaneously. Slightly worse results are obtained while conducting experiments with the use of selective garbage signal algorithm. Such conclusion can be reach while taking into consideration the optimal value of examined parameters. While designating above mentioned value, it is not enough to take into account only one indicator, because this usually will not provide the full view of the solution. The indicators such as penalty max and average filling should be analysed simultaneously, since they complement each other. Calculation of optimal number of trucks basing on two indicators depend on such estimation so that none of them do not deviate significantly from the norm. Assuming that the

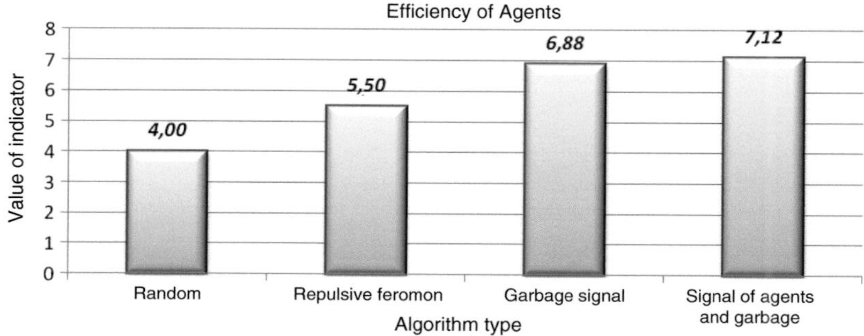

Fig. 6 Efficiency of agents/robots for the applied algorithms/srtategies. The resuls of simulation on a graph 20×20 with five robots and the level of signal propagation equal to 4

penalty max value should be less than 10 % and the average filling should be in range 40–60 %, then from Fig. 5 we can conclude that the results that fulfill this criteria are reached for three garbage trucks. Similarly the other optimal parameters value can be designated.

Similar sequence of algorithms (Fig. 6) can be observed while analysing strategies regarding obtained effectiveness of agents/robots. This indicator is the ratio of the amount of collected garbage to cost, that is the sum of the length of travelled distance (the number of edges) by all robots (trucks). Also in this case the best result is obtained by experiment with the use of garbage and agents signal algorithm.

6 Conclusions

The simulation model of the system for control of mobile robots applying agent architecture was created. The area of system application is the problem of efficient disposal of waste in the urban environment with the use of autonomous mobile robots. The problem has more and more important meaning together with the development of civilisation. Very important feature is the reflection of the real environment in the virtual space and introducing software agents related to the real robots (garbage trucks). Agents reside in the cyberspace and have ability to recognize the surrounded world and they control robots according to the chosen strategy. It enables independence of control system from hardware platform (types of robots), thus the system may be applied with almost every type of robots (by selection of the appropriate communication protocols). The shown model and the system for simulation enable further works on development of the presented algorithms as well as new approaches inspired by nature.

Acknowledgments The research presented in the paper has received founding from the Polish National Science Centre under grant DEC-2011/01/D/ST6/06146.

References

1. Cetnarowicz, K.: A Perspective on Agent Systems: Paradigm, Formalism, Examples. Studies in Computational Intelligence
2. Dias, M., Stentz, A.: A market approach to multirobot coordination. Technical report (2001)
3. Dorigo, M., Maniezzo, V.: The ant system: optimization by a colony of cooperating agents. IEEE Trans. Syst. Man Cybern. Part B **26**, 1–13 (1996)
4. Gerkey, B.P., Matarii, M.J.: A formal analysis and taxonomy of task allocation in multi-robot systems. Int. J. Robot. Res. **23**(9), 939–954 (2004)
5. Reinert, G.: The Traveling Salesman: Computational Solutions for TSP Applications. Lecture Notes in Computer Science
6. Zabinska, M., Sosnicki, T., Turek, W., Cetnarowicz, K.: Robot task allocation using signal propagation model. Procedia Comput. Sci. **18**, 1505–1514 (2013). 2013 International Conference on Computational Science
7. Zhang, D., Xie, G., Yu, J., Wang, L.: An adaptive task assignment method for multiple mobile robots via swarm intelligence approach. In: Computational Intelligence in Robotics and Automation, pp. 415–420, June 2005
8. Zheng, X., Koenig, S.: K-swaps: cooperative negotiation for solving task-allocation problems. In: IJCAI'09, pp. 373–378, San Francisco, CA, USA (2009)

Preliminary Study on the Randomization and Sequencing for the Chaos Embedded Heuristic

Roman Senkerik, Michal Pluhacek, Ivan Zelinka, Donald Davendra and Jakub Janostik

Abstract This research deals with the hybridization of the two softcomputing fields, which are chaos theory and evolutionary computation. This paper investigates the utilization of the time-continuous chaotic system, which is UEDA oscillator, as the chaotic pseudo random number generator. (CPRNG). Repeated simulations were performed investigating the influence of the oscillator sampling time to the selected heuristic, which is differential evolution algorithm (DE). Through the utilization of time-continuous systems and with different sampling times from very small to bigger, it is possible to fully keep, suppress or remove the hidden complex chaotic dynamics from the generated data series. Experiments are focused on the preliminary investigation, whether the different randomization given by particular CPRNG or hidden complex chaotic dynamics providing the unique sequencing are beneficial to the heuristic performance. Initial experiments were performed on the selected test function in several dimension settings.

Keywords Differential evolution · Complex dynamics · Deterministic chaos · Randomization · UEDA oscillator

R. Senkerik (✉) · M. Pluhacek · J. Janostik
Faculty of Applied Informatics, Tomas Bata University in Zlin,
Nam T.G. Masaryka 5555, 760 01 Zlin, Czech Republic
e-mail: senkerik@fai.utb.cz

M. Pluhacek
e-mail: pluhacek@fai.utb.cz

J. Janostik
e-mail: janostik@fai.utb.cz

I. Zelinka · D. Davendra
Faculty of Electrical Engineering and Computer Science,
Technical University of Ostrava, 17. Listopadu 15,
708 33 Ostrava-Poruba, Czech Republic
e-mail: ivan.zelinka@vsb.cz

D. Davendra
e-mail: donald.davendra@vsb.cz

A. Abraham et al. (eds.), *Proceedings of the Second International Afro-European Conference for Industrial Advancement AECIA 2015*, Advances in Intelligent Systems and Computing 427, DOI 10.1007/978-3-319-29504-6_55

1 Introduction

This research deals with the hybridization of the two softcomputing fields, which are the complex dynamics given by chaos theory and evolutionary computation techniques (ECT's). Currently the Differential Evolution (DE) [1] is known as powerful heuristic for many difficult and complex optimization problems.

A number of DE variants have been recently developed [2] with the emphasis on adaptivity/selfadaptivity, ensemble approach or utilization for discrete domain problems. The importance of randomization as a compensation of limited amount of search moves is stated in the survey paper [3]. This idea has been carried out in subsequent studies describing various techniques to modify the randomization process [4, 5] and especially in [6], where the sampling of the points is tested from modified distribution. Together with this persistent development in such mainstream research topics, the basic concept of chaos driven DE have been introduced.

A chaotic approach for heuristic generally uses the chaotic map or sampled time-continuous chaotic system in the place of a pseudo random number generator [7].

Initial research studies presenting DE with chaotic searching algorithm [8], self-adaptive chaos differential evolution (SACDE) [9] were followed by the direct embedding of chaotic systems in the form of chaos pseudo random number generator (CPRNG) into the DE (ChaosDE) [10].

Also the PSO (Particle Swarm Optimization) algorithm with elements of chaos was introduced as CPSO [11]. Later on the chaos embedded PSO with inertia weigh strategy was closely investigated [12], followed by the introduction of a PSO strategy driven alternately by two chaotic systems [13]. Recently the chaos driven firefly algorithm [14] and chaos embedded artificial bee colony algorithm have been introduced [15].

The organization of this paper is following: Firstly, the motivation for this research is proposed. The next sections are focused on the description of the concept of chaos driven DE and the experiment background. Results and conclusion follow afterwards.

2 Motivation

This research is an extension and continuation of the previous successful initial experiment with chaos driven DE (ChaosDE) and PSO, where the positive influence of hidden complex dynamics for the heuristic performance has been experimentally shown. Nevertheless, the questions, as to why it works, why the correlated chaotic time series may be beneficial, they remain. Many research studies have been carried out focusing on the utilization of different randomization type, different PRNGs, distributions etc. Thus we decided to design simple experiment, to show, whether the chaos embedded concept belongs to the group of "utilization of different PRNG" or the unique chaos dynamics and sequencing is the key of performance improvements. Since the sequencing in chaotic series generated by discrete chaotic systems is given directly by the discrete nature and mathematical notations of the used chaotic map, a different type of experiment was performed and presented here. It is focused on the time-continuous chaotic systems, to be more precise, on the investigating the influence of the oscillator sampling time to the selected heuristic, which is DE. Through the utilization of time-continuous systems and with different sampling times from very small to bigger, it is possible to fully keep, suppress or remove the hidden complex chaotic dynamics from the generated data series.

In this experiment, the canonical DE strategy DE/Rand/1/Bin and the ChaosDE strategy driven by discretized UEDA oscillators (ChaosDE) were used.

3 Differential Evolution

DE is a population-based optimization method that works on real-number-coded individuals [1]. DE is quite robust, fast, and effective, with global optimization ability. It does not require the objective function to be differentiable, and it works well even with noisy and time-dependent objective functions. Due to a limited space and the aims of this paper, the detailed description of well known differential evolution algorithm basic principles is insignificant and hence omitted. Please refer to [1, 16] for the detailed description of the used DERand1Bin strategy (both for ChaosDE and Canonical DE) as well as for the complete description of all other strategies.

4 The Concept of ChaosDE with Time-Continuous Systems

The general idea of CPRNG is to replace the default PRNG with the chaotic system. As the chaotic system is a set of equations with a static start position, we created a random start position of the system, in order to have different start position for different experiments. Thus we are utilizing the typical feature of chaotic systems, which is extreme sensitivity to the initial conditions, popularly known as "butterfly effect". This random position is initialized with the default PRNG, as a one-off randomizer. Once the start position of the chaotic system has been obtained, the system generates the next sequence using its current position.

Generally there exist many other approaches as to how to deal with the negative numbers as well as with the scaling of the wide range of the numbers given by the chaotic systems into the typical range 0–1:

- Finding of the maximum value of the pre-generated long discrete sequence and dividing of all the values in the sequence with such a maxval number.
- Shifting of all values to the positive numbers (avoiding of ABS command) and scaling.

5 UEDA Oscillator

UEDA oscillator is the simple example of driven pendulums, which represent some of the most significant examples of chaos and regularity.

The UEDA system can be simply considered as a special case of intensively studied Duffing oscillator that has both a linear and cubic restoring force. Ueda oscillator represents the both biologically and physically important dynamical model exhibiting chaotic motion. It can be used to explore much physical behavior in biological systems [17].

The UEDA chaotic system equations are given in (1). The parameters are: $a = 1.0$ $b = 0.05$, $c = 7.5$ and $\omega = 1.0$ as suggested in [18]. The graphical outputs are organized in following way: the left part of the graphic grid Fig. 1 shows x, y parametric plots of the chaotic system, whereas the right part depicts the typical chaotic behavior of the utilized chaotic system, represented by the examples of direct output for the variable x.

$$\begin{aligned} \frac{dx}{dt} &= y \\ \frac{dy}{dt} &= -ax^3 - by + c\sin\omega t \end{aligned} \tag{1}$$

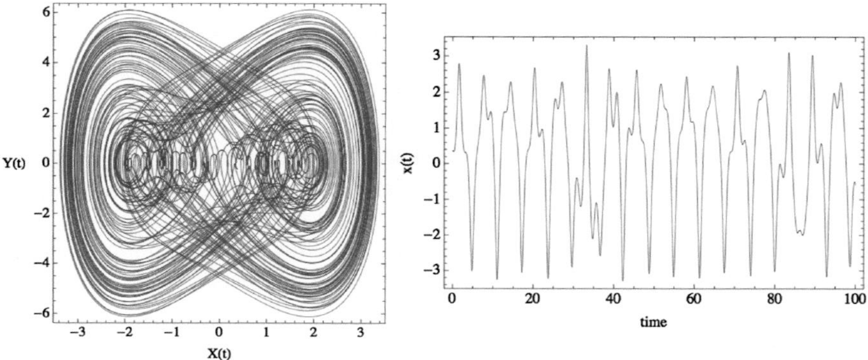

Fig. 1 x, y parametric plot of the UEDA oscillator (*left*); Simulation of the UEDA oscillator (variable x) (*right*)

Table 1 Parameter set up for ChaosDE and Canonical DE

DE parameter	Value
Popsize	{50, 50, 75}
F (for ChaosDE)	0.4
CR (for ChaosDE)	0.4
F (for Canonical DE)	0.5
CR (for Canonical DE)	0.9
Dim	{10, 20, 30}
Max. Generations	{1000, 1000, 1500}
Max Cost Function Evaluations (CFE)	{50000, 50000, 75000}

6 Experiment Design

For the purpose of ChaosDE performance comparison within this initial research, the multimodal Schwefel's test function (2) was selected:

$$f(x) = \sum_{i=1}^{\dim} -x_i \sin\left(\sqrt{|x_i|}\right) \tag{2}$$

Function minimum:
Position for E_n: $(x_1, x_2 \ldots x_n) = (420.969, 420.969, \ldots, 420.969)$
Value for E_n: $y = -418.983 \cdot dim$

The parameter settings for both canonical DE and ChaosDE were obtained analytically based on numerous experiments and simulations (see Table 1). Experiments were performed in the environment of *Wolfram Mathematica*; canonical DE therefore has used the built-in *Wolfram Mathematica* pseudo random number generator *Wolfram Cellular Automata* representing traditional pseudorandom number generator in comparisons. All experiments used different initialization, i.e. different initial population was generated within the each run of Canonical or Chaos driven DE. The maximum number of generations was fixed at 1000 generations for 10*D* and 20*D* problems; and 1500 generations for 30*D* problem. This allowed the possibility to analyze the progress of DE within a limited number of generations and cost function evaluations.

7 Results

Statistical results of the selected experiments are shown in comprehensive Tables 2, 4 and 6 which represent the simple statistics for cost function (CF) values, e.g. average, median, maximum values, standard deviations and minimum values

Table 2 Simple results statistics for the Canonical DE and adaptive ChaosDE Schwefel's function, *dim* = 10, max. Generations = 1000

DE Version	Avg CF	Median CF	Max CF	Min CF	StdDev
Canonical DE	**−4189.62**	−4189.82	**−4188.7**	**−4189.83**	**0.425073**
ChaosDE—Sampling 0.1 s	−4098.64	**−4189.83**	−3559.49	**−4189.83**	209.0547
ChaosDE—Sampling 0.5 s	−3766.93	−3710.45	−3307.99	−4189.67	377.1445
ChaosDE—Sampling 1.0 s	−3480.44	−3411.2	−2881.27	−4189.69	436.9247

Table 3 Comparison of progress towards the minimum for the Schwefel's function, *dim* = 10, max. Generations = 1000

DE Version	Generation No. 250	Generation No. 500	Generation No. 750	Generation No. 1000
Canonical DE	**−3024.34**	−3502.34	**−4017.97**	**−4189.62**
ChaosDE—Sampling 0.1 s	−2879.33	**−3574.31**	−3979.95	−4098.64
ChaosDE—Sampling 0.5 s	−2609.5	−2909.7	−3366.64	−3766.93
ChaosDE—Sampling 1.0 s	−2574.59	−2800.19	−3132.67	−3480.44

representing the best individual solution for all 50 repeated runs of DE/ChaosDE and different *Dimension* settings (10*D*, 20*D* and 30*D*).

Tables 3, 5 and 7 compare the progress of two versions of adaptive ChaosDE and Canonical DE. This table contains the average CF values for the particular generation No. from all 50 runs. The bold values within the all Tables 2, 3, 4, 5, 6 and 7 depict the best obtained results.

The graphical comparisons of the time evolution of average CF values for all 50 runs of three versions of ChaosDE with different UEDA oscillator sampling rates and canonical DERand1Bin strategy are depicted in Figs. 2, 3 and 4. Finally Fig. 5a–d show the influence of sampling rate to the distribution of pseudo random numbers given by particular CPRNG (left figures) and to the dynamics inside the generated data series (right figures).

Table 4 Simple results statistics for the Canonical DE and adaptive ChaosDE Schwefel's function, *dim* = 20, max. Generations = 1000

DE Version	Avg CF	Median CF	Max CF	Min CF	StdDev
Canonical DE	−4583.05	−4531.29	−4265.43	−5238.75	**268.4226**
ChaosDE—Sampling 0.1 s	**−6250.01**	**−6346.65**	**−4947.92**	**−7535.51**	952.9218
ChaosDE—Sampling 0.5 s	−4246.65	−4004.84	−3668.64	−5504.73	562.4262
ChaosDE—Sampling 1.0 s	−4508.7	−4161.67	−3912.97	−5840.88	699.4066

Table 5 Comparison of progress towards the minimum for the Schwefel's function, *dim* = 20, max. Generations = 1000

DE Version	Generation No. 250	Generation No. 500	Generation No. 750	Generation No. 1000
Canonical DE	**−3955.27**	−4287.11	−4422.72	−4583.05
ChaosDE—Sampling 0.1 s	−3781.28	**−4618.32**	**−5431.47**	**−6250.01**
ChaosDE—Sampling 0.5 s	−3536.42	−3740.76	−3958.08	−4246.65
ChaosDE—Sampling 1.0 s	−3443.07	−3758.27	−4081.61	−4508.7

Table 6 Simple results statistics for the Canonical DE and adaptive ChaosDE Schwefel's function, *dim* = 30, max. Generations = 1500

DE Version	Avg CF	Median CF	Max CF	Min CF	StdDev
Canonical DE	−5686.39	−5577.68	−5260.29	−6265.26	**347.9753**
ChaosDE—Sampling 0.1 s	**−7222.92**	**−7284.94**	**−5284.72**	**−8235.16**	835.3056
ChaosDE—Sampling 0.5 s	−5539.82	−5433.4	−4794.82	−6544.66	609.7107
ChaosDE—Sampling 1.0 s	−5179.78	−4966.64	−4639.47	−5983.05	444.3675

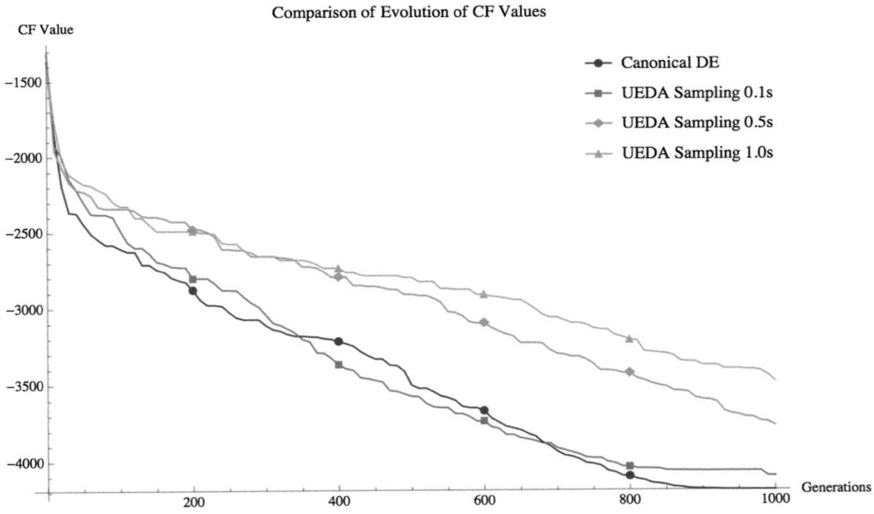

Fig. 2 Comparison of the time evolution of avg. CF values for the all 50 runs of Canonical DE, and three versions of ChaosDE with different samling rates of UEDA oscillator as CPRNG. Schwefel's function, *Dim* = 10

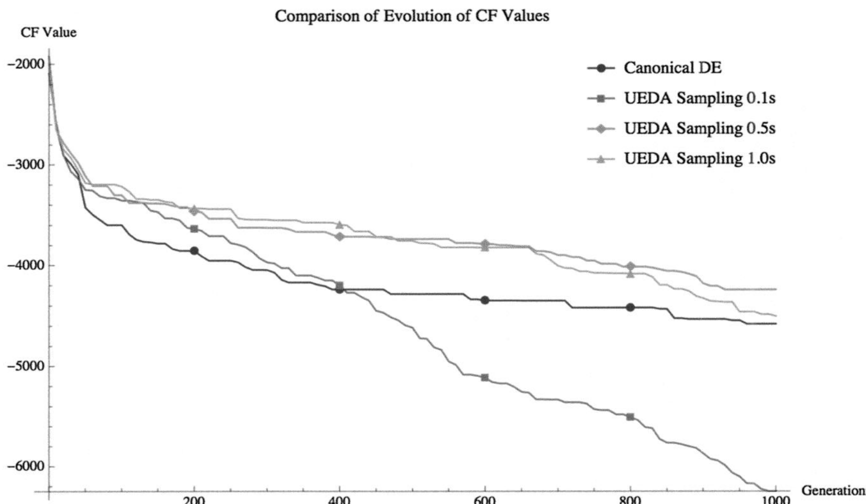

Fig. 3 Comparison of the time evolution of avg. CF values for the all 50 runs of Canonical DE, and three versions of ChaosDE with different samling rates of UEDA oscillator as CPRNG. Schwefel's function, $Dim = 20$

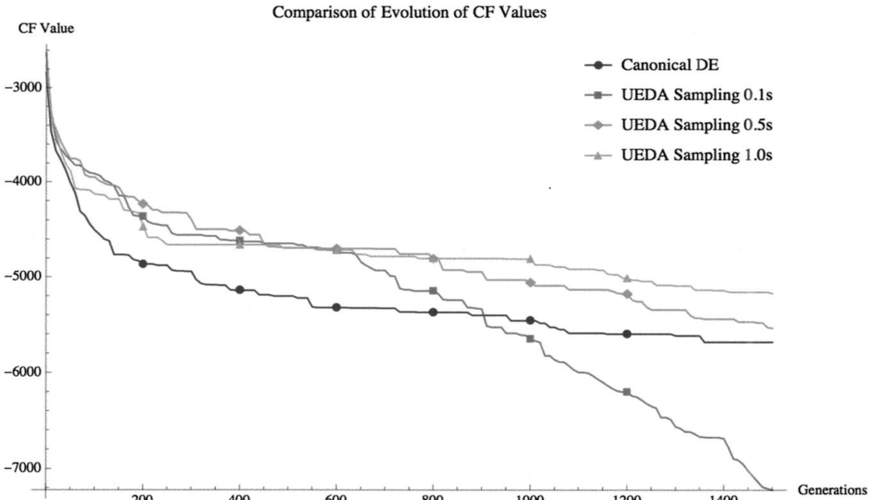

Fig. 4 Comparison of the time evolution of avg. CF values for the all 50 runs of Canonical DE, and three versions of ChaosDE with different samling rates of UEDA oscillator as CPRNG. Schwefel's function, $Dim = 30$

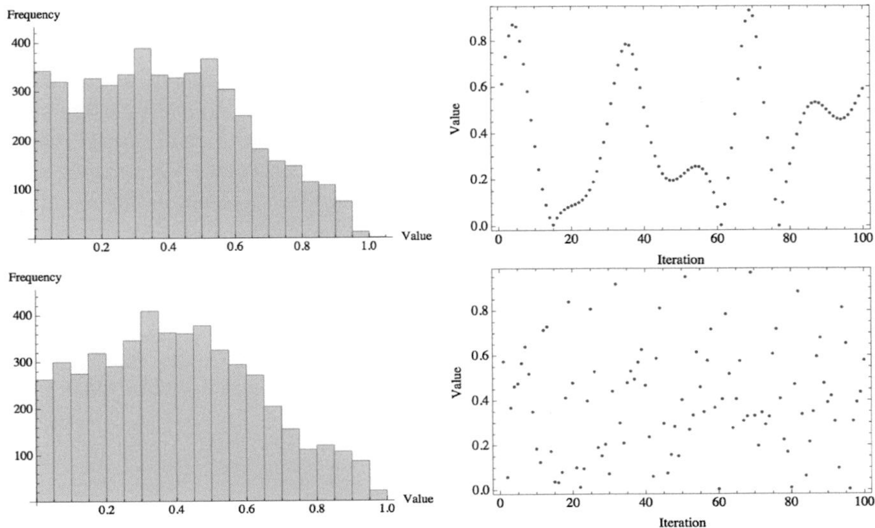

Fig. 5 Comparison of the influence of sampling rate to the distribution of numbers given by UEDA CPRNG; *Left* Histogram of the distribution of real numbers transferred into the range <0 – 1>; *Right* Example of the chaotic dynamics: range <0 – 1> generated by means of UEDA oscillator sampled with the particular sampling rate—variable x; Sampling rates from up to down: 0.1 s, 0.5 s

Table 7 Comparison of progress towards the minimum for the Schwefel's function, dim = 30, max. Generations = 1500

DE Version	Generation No. 250	Generation No. 500	Generation No. 750	Generation No. 1500
Canonical DE	**–4878.15**	**–5205.08**	**–5370.8**	–5686.39
ChaosDE—Sampling 0.1 s	–4459.74	–4649.26	–5137.75	**–7222.92**
ChaosDE—Sampling 0.5 s	–4324.86	–4693.77	–4758.6	–5539.82
ChaosDE—Sampling 1.0 s	–4662.34	–4698.82	–4785.59	–5179.78

8 Conclusion

Obtained graphical comparisons and data in Tables 2, 3, 4, 5, 6 and 7 and Figs. 2, 3 and 4 support the claim that chaos driven heuristic is very sensitive to the hidden chaotic dynamics driving the CPRNG and not sensitive to the distribution of CPRNG. The sensitivity is growing together with the increase of the dimensionality of the problems to be solved. Such a chaotic dynamics can be significantly changed by the selection of sampling time in the case of the time-continuous systems. In case of discrete systems, the simplest way for changing the influence to the heuristic is to swap currently used chaotic system for different one, or to change the internal parameters of discrete chaotic systems.

Another important phenomenon was discovered—Only sampling rate of 0.5 s keeps the information about the chaotic dynamics (as in Fig. 5b) and by using such chaotic dynamics driving the heuristic, its performance is significantly better. Other settings of sampling rates have given comparable (or even worse) performance with canonical version of heuristic. Distributions of CPRNGs were almost identical.

Acknowledgments This work was supported by Grant Agency of the Czech Republic—GACR P103/15/06700S, further by financial support of research project NPU I No. MSMT-7778/2014 by the Ministry of Education of the Czech Republic and also by the European Regional Development Fund under the Project CEBIA-Tech No. CZ.1.05/2.1.00/03.0089, partially supported by Grant of SGS No. SP2015/142 and SP2015/141 of VSB—Technical University of Ostrava, Czech Republic and by Internal Grant Agency of Tomas Bata University under the projects No. IGA/FAI/2015/057 and IGA/FAI/2015/061.

References

1. Price, K.V.: An introduction to differential evolution. In: Corne, D., Dorigo, M., Glover, F. (eds.) New Ideas in Optimization, pp. 79–108. McGraw-Hill Ltd. (1999)
2. Das, S., Suganthan, P.N.: Differential evolution: a survey of the state-of-the-art. IEEE Trans. Evol. Comput. **15**(1), 4–31 (2011). doi:10.1109/TEVC.2010.2059031
3. Neri, F., Tirronen, V.: Recent advances in differential evolution: a survey and experimental analysis. Artif. Intell. Rev. **33**(1–2), 61–106 (2010)
4. Weber, M., Neri, F., Tirronen, V.: A study on scale factor in distributed differential evolution. Inf. Sci. **181**(12), 2488–2511 (2011)
5. Neri, F., Iacca, G., Mininno, E.: Disturbed exploitation compact differential evolution for limited memory optimization problems. Inf. Sci. **181**(12), 2469–2487 (2011)
6. Iacca, G., Caraffini, F., Neri, F.: Compact differential evolution light: high performance despite limited memory requirement and modest computational overhead. J. Comput. Sci. Technol. **27**(5), 1056–1076 (2012)
7. Aydin, I., Karakose, M., Akin, E.: Chaotic-based hybrid negative selection algorithm and its applications in fault and anomaly detection. Expert Syst. Appl. **37**(7), 5285–5294 (2010)
8. Liang, W., Zhang, L., Wang, M.: The chaos differential evolution optimization algorithm and its application to support vector regression machine. J. Softw. **6**(7), 1297–1304 (2011)
9. Zhenyu, G., Bo, C., Min, Y., Binggang, C.: Self-adaptive chaos differential evolution. In: Jiao, L., Wang, L., Gao, X.-B., Liu, J., Wu, F. (eds.) Advances in Natural Computation. Lecture Notes in Computer Science, vol. 4221, pp. 972–975. Springer, Berlin Heidelberg (2006)
10. Davendra, D., Zelinka, I., Senkerik, R.: Chaos driven evolutionary algorithms for the task of PID control. Comput. Math. Appl. **60**(4), 1088–1104 (2010)
11. Coelho, L.D.S., Mariani, V.C.: A novel chaotic particle swarm optimization approach using Hénon map and implicit filtering local search for economic load dispatch. Chaos, Solitons Fractals **39**(2), 510–518 (2009)
12. Pluhacek, M., Senkerik, R., Davendra, D., Kominkova Oplatkova, Z., Zelinka, I.: On the behavior and performance of chaos driven PSO algorithm with inertia weight. Comput. Math. Appl. **66**(2), 122–134 (2013)

13. Pluhacek, M., Senkerik, R., Zelinka, I., Davendra, D.: Chaos PSO algorithm driven alternately by two different chaotic maps—an initial study. In: Proceedings of the 2013 IEEE Congress on Evolutionary Computation (CEC), pp. 2444–2449, 20–23 June 2013
14. Metlicka, M., Davendra, D.: Chaos-driven discrete artificial bee colony. In: Proceedings of the 2014 IEEE Congress on Evolutionary Computation (CEC), pp. 2947–2954 (2014)
15. Gandomi, A.H., Yang, X.S., Talatahari, S., Alavi, A.H.: Firefly algorithm with chaos. Commun. Nonlinear Sci. Numer. Simul. **18**(1), 89–98 (2013)
16. Price, K.V., Storn, R.M., Lampinen, J.A.: Differential Evolution—A Practical Approach to Global Optimization. Natural Computing Series. Springer, Berlin Heidelberg (2005)
17. Bharti, L., Yuasa, M.: Energy variability and chaos in ueda oscillator. http://www.rist.kindai.ac.jp/no.23/yuasa-EVCUO.pdf
19. Sprott, J.C.: Chaos and Time-Series Analysis. Oxford University Press (2003)

Part IV
Special Session on Advances in Image Processing and Colorization

Visual Quality Improvement of Color-Embedded Gray Images for a Reversible Color-to-Gray Algorithm

Takahiko Horiuchi, Jiqiang Lin and Keita Hirai

Abstract In this paper, we propose a method to improve the visual quality of color-embedded gray images for a reversible color-to-gray algorithm. The reversible color-to-gray algorithm is an effective technique that considers several perspectives such as running cost, data quantity, and security in practical applications. Many reversible color-to-gray algorithms have been proposed, but most of these studies have only evaluated already-recovered color images. Although the visual quality of color-embedded gray images is also important, there has not been much assessment in this area. In this paper, we present a Fourier-based reversible color-to-gray algorithm to realize an improved visual quality of color-embedded gray images while maintaining the quality of the recovered color image. We also construct a color image recovery system from a printed gray image captured by a monochrome camera. Experimental results show that the proposed algorithm can improve the visual quality of both the color-embedded gray image and the recovered image in the closed-computer world, but the algorithm does not perform well for a printed gray images.

Keywords Color-to-gray · Visual quality · Color encoding · Information hiding · Color restoration

T. Horiuchi (✉) · J. Lin · K. Hirai
Graduate School of Advanced Integration Science, Chiba University, Yayoi-Cho 1-33, Inage-ku, Chiba 263-8522, Japan
e-mail: horiuchi@faculty.chiba-u.jp

K. Hirai
e-mail: hirai@faculty.chiba-u.jp

A. Abraham et al. (eds.), *Proceedings of the Second International Afro-European Conference for Industrial Advancement AECIA 2015*, Advances in Intelligent Systems and Computing 427, DOI 10.1007/978-3-319-29504-6_56

1 Introduction

The technological development of various types of imaging devices has facilitated the processing of color images. However, there remains a considerable gap for processing color images in the same manner as gray images with regard to the running cost, data quantity, security, etc. In recent years, many algorithms have been proposed for the conversion of color images to gray images that can subsequently be decoded and reconverted to color [1–10]. However, most of the conventional algorithms have only been evaluated using recovered color images. Although the visual quality of color-embedded gray images is also important, their assessment has not been reported. Figure 1 shows examples of color-embedded gray images that use a few conventional wavelet-based algorithms. Figure 1a shows an original color image, while Fig. 1b–d represent close-up images of color-embedded gray images generated by Refs. [2, 3, 6], respectively. As shown in the figure, the visual quality differs significantly depending on the method employed.

In this paper, we propose a new algorithm that can improve the visual quality of both color-embedded gray images and recovered images compared to previous studies. This is done by effectively distributing color information effectively in the Fourier domain. Then in order to verify the feasibility of the proposed algorithm, we perform experiments using digital image data and printed images.

Fig. 1 Color-embedded *gray images* obtained by previous algorithms. **a** Original color image. **b** Ref. [2]. **c** Ref. [3]. **d** Ref. [6]

2 Reversible Color-to-Gray Algorithm

As shown in Fig. 1, conventional wavelet-based algorithms result in a visible checker pattern artifact. We propose a new color-to-gray mapping algorithm and color-recovery method by removing the checker-pattern artifact using Fourier transform.

2.1 Color-to-Gray Step

(E1) Convert images from RGB into Y, Cb, Cr.
(E2) Use the Fourier transform on the luminance Y, and chrominance Cb and Cr:

$$Y \rightarrow F_Y(u,v),\ \ Cr \rightarrow F_{Cr}(u,v),\ \ Cb \rightarrow F_{Cb}(u,v) \tag{1}$$

(E3) Decompose the power spectra into low-pass and high-pass power spectra for luna-chroma components:

$$\begin{aligned} &F_Y(u,v) \rightarrow \left\{F_Y^{(H)}(u,v), F_Y^{(L)}(u,v)\right\},\ F_{Cr}(u,v) \rightarrow \left\{F_{Cr}^{(H)}(u,v),\ F_{Cr}^{(L)}(u,v)\right\}, \\ &F_{Cb}(u,v) \rightarrow \left\{F_{Cb}^{(H)}(u,v), F_{Cb}^{(L)}(u,v)\right\} \end{aligned} \tag{2}$$

(E4) Replace the high-frequency component of the luminance component using two kinds of low frequency chrominance components:

$$F_Y^{(H_1)}(u,v) \leftarrow F_{Cr}^{(L)}(u,v),\ \ F_Y^{(H_2)}(u,v) \leftarrow F_{Cb}^{(L)}(u,v) \tag{3}$$

(E5) Take the inverse Fourier transform to obtain the color-embedded gray image, i.e.,

$$\left\{F_Y^{(H_1)}(u,v),\ F_Y^{(H_2)}(u,v),\ F_Y^{(L)}(u,v)\right\} \rightarrow Y' \tag{4}$$

(E6) Image Y′ is the resulting gray image.

Figure 2 shows the procedure by the proposed color-to-gray conversion method.

There are various ways of embedding chrominance components in (E4). We selected the arrangement empirically, as shown in Fig. 3.

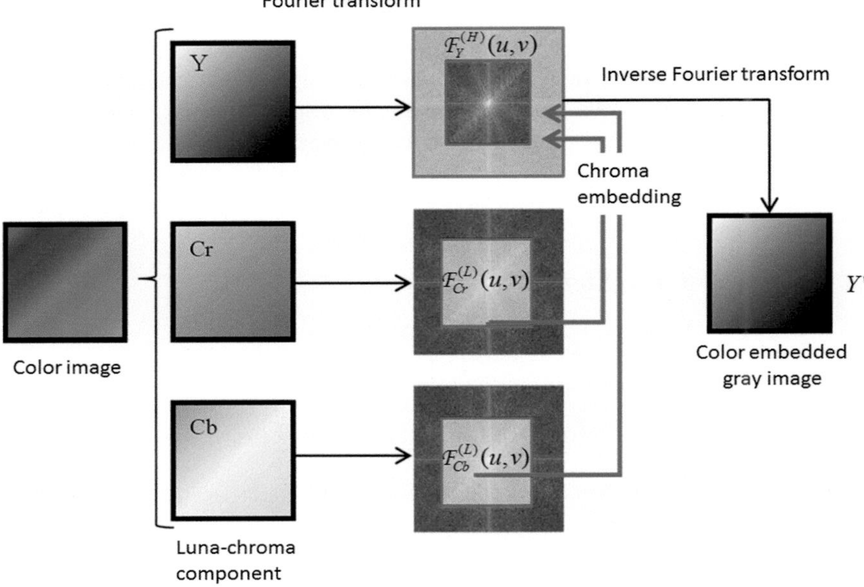

Fig. 2 The proposed recovery of color from *gray image*

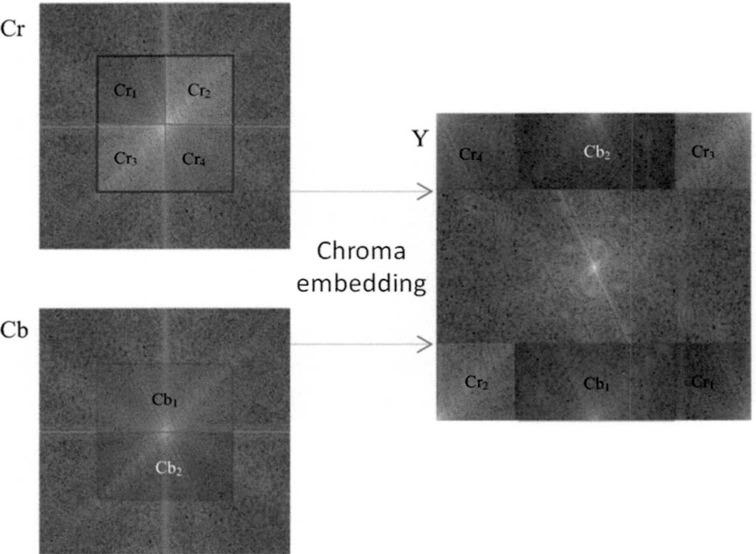

Fig. 3 Arrangement of chroma-embedding in (E4) used in our algorithm

2.2 Recovery Step

(R1) Read the gray textured image.
(R2) Use the Fourier transform to convert the gray image into power spectra:

$$Y' \rightarrow \left\{ F_Y^{(H_1)}(u,v),\ F_Y^{(H_2)}(u,v),\ F_Y^{(L)}(u,v) \right\} \tag{5}$$

(R3) Decompose the power spectra of the gray image into low-pass power spectra for chrominance components:

$$F_{Cr}^{(L)}(u,v) \leftarrow F_Y^{(H_1)}(u,v),\quad F_{Cb}^{(L)}(u,v) \leftarrow F_Y^{(H_2)}(u,v) \tag{6}$$

(R4) Set 0 as the high-pass spectra of the luna-chroma components:

$$F_Y^{(H)}(u,v) \leftarrow 0,\quad F_{Cr}^{(H)}(u,v) \leftarrow 0,\quad F_{Cb}^{(H)}(u,v) \leftarrow 0 \tag{7}$$

(R5) Use the inverse Fourier transform to recovery Y'', Cr'', Cb''.
(R6) Convert the Y'', Cr'', Cb'' planes back to RGB.

Figure 4 shows the procedures employed for the proposed color-recovery method.

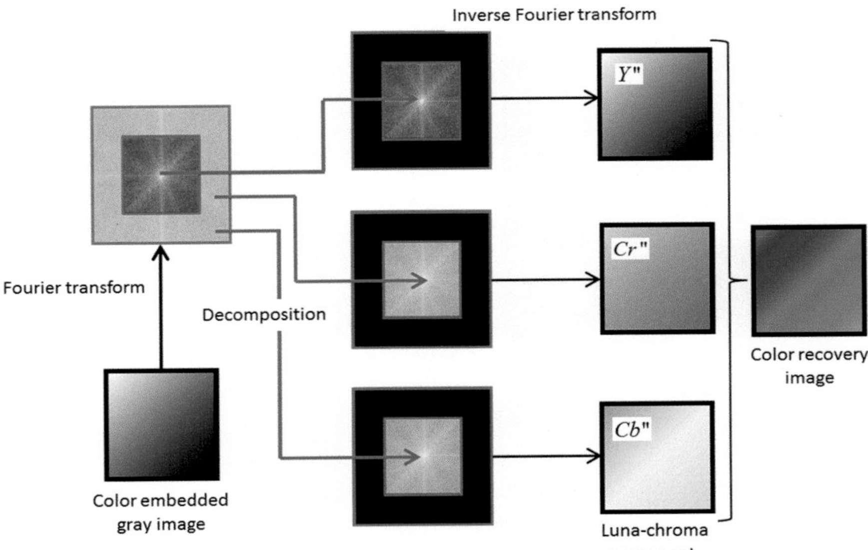

Fig. 4 The proposed recovery of color from *gray image*

3 Color Image Recovery System

3.1 System Construction

Figure 5 shows the system that we constructed. Our system consists of a high-speed monochrome CCD camera (Baumer TXG50), an sRGB color display, an LED ring light, and a PC (CPU: Intel Core i7-3770 with 3.40 GHz; Memory: 8.00 GB; OS: Microsoft Windows 7). The camera can capture the 8-bit gray image at 15 fps with an image size of 2448 × 2050. We calibrated in advance vignetting and an uneven illuminance.

3.2 Calibration Algorithm

We used the calibration algorithms developed in Ref. [11].

Calibration Pattern
We need to correct the gray level, position, and resolution, which are distorted by the printing and capturing processes. We used the calibration pattern is to correct

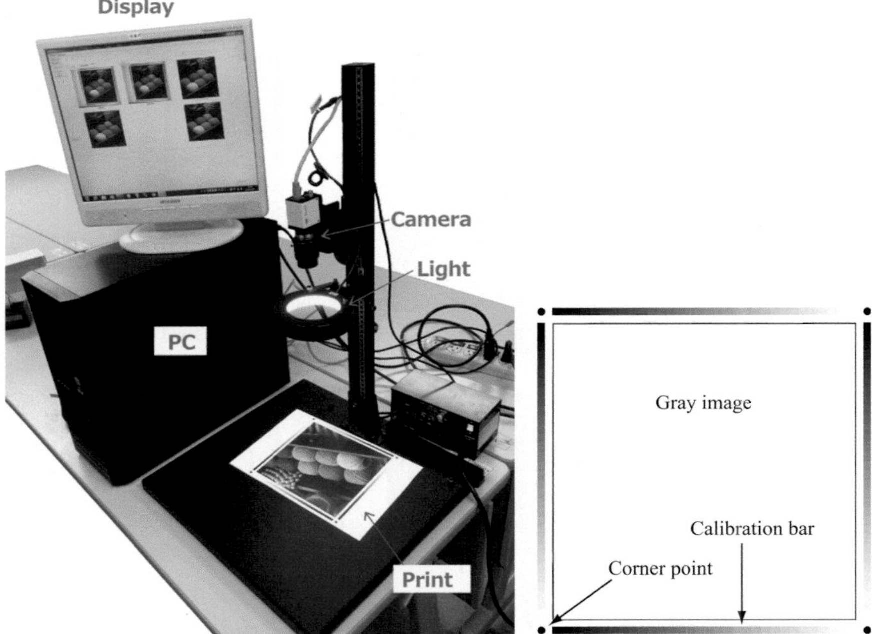

Fig. 5 Constructed system and calibration pattern

those image characteristics. The pattern has four corner points to detect the image size and position, and it also has four gray bars to detect the gray-level transition, as shown in Fig. 5. In order to detect the spatial variation of the illuminant, four bars are required.

Geometric Calibration
Our system assumes that a gray image is placed at an arbitrary position. The distance between the camera and the image is fixed. Therefore, the rotation and shift parameters can be detected from the captured gray image. We realize the geometric correction by detecting the four corner points.

Gray Level Calibration
Because of the printer and camera device characteristics, the captured gray level of the 8-bit intensity at each pixel is generally different from the original gray level. Further, the illuminant distribution causes a spatial variation in intensities. We can correct the gray level using the help of the four calibration bars, which can be identified by connecting the corner points detected during the geometric calibration step.

Resolution Adjustment
Finally, we adjusted the resolution of the captured image is adjusted. In general, the size of the captured image is higher than the size of the original image. Therefore, printed halftones often appear in the high-resolution captured image. Our system assumes that the image size of the original color image is known. We performed a down-sampling process is performed to adjust the image size.

4 Experiments

We tested the performance of the constructed color-image-recovery algorithm using several standard images as shown in Fig. 6. Figure 7a, b show color-embedded gray images obtained using a conventional algorithm [6] and the proposed algorithm, respectively, while Fig. 7c shows their close-up images. As shown in the figure and

Fig. 6 Sample color images used in the experiment: Bride, Silver, and Wool

the table, the low-frequency texture pattern was reduced, and we obtained an accurate gray reproduction was obtained by using the proposed algorithm. In order to verify the feasibility of the results, we also performed an objective assessment. Because the color-embedded gray images have texturized patterns, pixel-to-pixel assessments such as the peak-to-noise ratio (PSNR) are not appropriate. We used the S-CIELAB metric, which can evaluate the "perceptual fidelity." Table 1 shows the S-CIELAB metric for the results in Fig. 7. We assumed that the viewing distance of the S-CIELAB was 50 cm. With the exception of the image "silver," the S-CIELAB values of our results were also improved.

Fig. 7 Color embedded *gray images*. **a** Conventional results. **b** Our results. **c** Close-up image: *left* conventional result, *right* our result

Figure 8 shows the color recovery results from Fig. 7. The objective evaluations by the color difference CIEDE2000 are shown in Table 1. As shown in the figure and table, our results also improved the color-recovery accuracy.

We also evaluated the printed images using our system described in Sect. 3. Each test image was down-sampled to 200 × 200 pixels, and color information was embedded using the algorithm described in Sect. 2. The color images were printed on a sheet of white paper WPHO (A4 size; thickness: 0.23 mm) using a laser printer Canon LBP 9500C with a resolution of 1600 dpi. The printed image was placed under the camera in Fig. 5.

Fig. 8 Color recovery images. **a** Conventional results. **b** Our results. **c** Close-up image: *left* conventional result, *right* our result

Table 1 Color difference between the original gray image and the color-embedded image

		Digital image		Printed image
		Gray (S-CIELAB)	Color (CIEDE2000)	Color (CIEDE2000)
Bride	Ref. [6]	0.005	2.02	3.89
	Proposed	0.002	1.45	5.17
Silver	Ref. [6]	0.034	3.62	6.41
	Proposed	0.057	3.01	9.26
Wool	Ref. [6]	0.019	3.68	6.16
	Proposed	0.001	2.89	9.99

All algorithms were implemented on a PC using C# and C++ programming languages. The processing speed was 670 ms for a captured frame using a PC with the specifications described in Sect. 3. Therefore, our system can recover a color image at a speed of 1.5 fps. Objective evaluations by the color difference CIEDE2000 are shown in Table 1. By comparing these results with those of the digital image, the color-recovery accuracy of the printed images decreased. Although the visual quality of gray images was improved by the proposed algorithm, it became difficult to distinguish the printed gray. Therefore, the color-recovery error increased.

5 Conclusions

In this paper, we improved the visual quality of a color-embedded gray image for a reversible color-to-gray algorithm, while maintaining the color-recovery accuracy. We developed the proposed algorithm based on a Fourier spectra replacement. Experimental results showed that the proposed system can improve the visual quality of the color-embedded gray image and accurately recover the color image in the digital world. However, in the real world, the color recovery accuracy of printed images decreased due to the discriminate difficulty of printed gray.

References

1. Campisi, P., Kundur, D., Hatzinakos, D., Neri, A.: Compressive data hiding: an unconventional approach for improved color image coding. EURASIP J. Appl. Sig. Process. **2002**(2), 152–163 (2002)
2. de Queiroz, R.L., Braun, K.: Color to gray and back: color embedding into textured gray images. IEEE Trans. Image Process. **15**(6), 1464–1470 (2006)
3. Ko, K.-W., Kwon, O.-S., Son, C.-H., Ha, Y.-H.: Color embedding and recovery based on wavelet packet transform. J. Imaging Sci. Technol. **52**(1), 010501-1–010501-10 (2007)
4. Chaumont, M., Puech, W.: A grey-level image embedding its color palette. In: Proceedings of ICIP, vol. I, pp. 389–392 (2007)

5. de Queiroz, R.L.: Reversible color to gray mapping using subband domain texturization. Pattern Recogn. Lett. (2008). doi:10.1016/j.patrec.2008.11.010
6. Horiuchi, T., Nohara, F., Tominaga, S.: Accurate reversible color-to-gray mapping algorithm without distortion conditions. Pattern Recogn. Lett. **31**, 2405–2414 (2010)
7. Semary, N.A., Hadhoud, M.M., El-Sayed, H.M.: Morphological decolorization system for colors hiding. In: Proceedings of INFOS2012, pp. 130–137 (2012)
8. Chaumonta, M., Puechb, W., Lahanier, C.: Securing color information of an image by concealing the color palette. J. Syst. Softw. **86**(3), 809–825 (2013)
9. Son, C., Choo, H.: Color recovery of black-and-white halftoned images via categorized color-embedding look-up tables. Digit. Signal Process. **28**, 93–105 (2014)
10. Kekre, H.B., Thepade, S.D., Chaturvedi, R.N.: "Color to Gray and Back" using DST-DCT, Haar-DCT, Walsh-DCT, Hartley-DCT, Slant-DCT, Kekre-DCT hybrid wavelet transforms. In: Proceedings of 3rd International Conference on Soft Computing for Problem Solving Advances in Intelligent and Soft Computing, vol. 259, pp. 613–623 (2014)
11. Lin, J., Horiuchi, T., Hirai, K., Tominaga, S.: Color image recovery system from printed gray image. In: Proceedings of Southwest Symposium on Image Analysis and Interpretation, pp. 41–44 (2014)

Watermarking 3D Triangular Mesh Models Using Intelligent Vertex Selection

Mona M. Soliman, Aboul Ella Hassanien and Hoda M. Onsi

Abstract Watermarking provides a mechanism for copyright protection or ownership assertion of digital media by embedding information in the data. In this paper, a blind robust 3D triangular mesh watermarking approach is presented using one of Computational Intelligent (CI) techniques named neural network; the role of this neural network is to select the best vertices that can be used as watermark carrier. This watermark position has to guarantee minimum distortion of 3D model and maximum robustness for watermark bits extraction. First, we select the best position of watermark carrier vertices using smoothing feature clustering. This clustering stage is performed using one of unsupervised neural network types which is a Self Organizing Maps (SOM). Then, watermark bits stream are embedded in the selected marked vertices using local statistical measures such as; mean and standard deviation. Experimental results show that our watermarking algorithm is robust since watermarks can be extracted without mesh alignment or re-meshing under a variety of attacks, including noise addition, cropping, smoothing filtering, rotation, translation, and scale. Our work had been compared with other work of blind 3D watermarking models and proves its efficiency in terms of both robustness and imperceptibility.

Keywords 3D watermarking · Smoothing curvature · Self Organizing Map · Unsupervised learning

M.M. Soliman (✉) · A.E. Hassanien · H.M. Onsi
Faculty of Computers and Information, Cairo University, Cairo, Egypt
e-mail: mona.solyman@fci-cu.edu.eg
URL: http://www.egyptscience.net

M.M. Soliman · A.E. Hassanien · H.M. Onsi
Scientific Research Group in Egypt (SRGE), Cairo, Egypt
e-mail: h.onsi@fci-cu.edu.eg

A. Abraham et al. (eds.), *Proceedings of the Second International Afro-European Conference for Industrial Advancement AECIA 2015*, Advances in Intelligent Systems and Computing 427, DOI 10.1007/978-3-319-29504-6_57

1 Introduction

Unfortunately, like digital images and audio/video clips, 3D graphic models can be easily duplicated and redistributed without any loss of quality by a pirate. Therefore, under this background, a critical demand exists on the intellectual property protection of 3D models. Digital watermarking [1] is considered as an efficient solution to solve this emerging problem. Watermarking has variety of applications like: Copyright Protection, Owner Identification, Copy Protection, Data Hiding, and Digital Rights Management (DRM) [2–4].

This paper focuses on introducing a robust and blind mesh watermarking approach. The proposed approach use one of computational intelligence techniques in selecting best vertex positions to be watermark carriers. The watermark bits stream are only embedded on these specified set of vertices. These vertices are selected by utilizing Self Organization Maps (SOM). SOM is a kind of competitive neural network in which the networks learn to form their own classifications of the training data without external help. Watermark bits sequence are inserted based on the mean and standard deviation of selected vertices' neighbors in the original mesh.

The remainder of this paper is organized as follow. Section 2 reviews related work of watermark 3D model using Computational Intelligent techniques. Section 3 discusses in details the proposed watermarking approach. Section 4 shows the experimental results. Conclusions are discussed in Sect. 5.

2 State of the Art on 3D Watermarking Using CI

Computational intelligence (CI) is the study of adaptive mechanisms to enable or facilitate intelligent behavior in complex and changing environments [5]. CI is considered as one of the most important increasing fields, which attract a large number of researchers and scientists working in different areas.

There still exist few watermarking methods for 3D watermarking using Computational Intelligent techniques compared with the amount of algorithms available for traditional media such as audio, image and video. In this section we explore some of this published work. In [6] Matwani et al. proposed a novel approach to overcome the challenge of selecting best vertices to be watermark carriers by using Artificial Neural Networks (ANN). In his proposed approach a Feature vectors representing the geometry of the vertex and its surrounding vertices are extracted and used to train and simulate ANN. ANN is used as a classifier to determine which vertices should be selected for watermarking. Another propose approach by Matwani et al. is introduced in [7] where author built a non blind 3D watermarking system using support vector machines (SVM). This proposed approach utilizes SVM as a binary classifier for the selection of vertices for watermark embedding. The performance of this algorithm is heavily dependent on the quality of training feature vectors and the size of the training set. In [8] Hu et al. used quadratic programming (QP) for constrained optimization of 3D meshes. It proposed a histogram-based method for watermarking 3D polygonal meshes by using quadratic programming to minimize the mean square error between the original mesh and the watermarked mesh.

However, this method has difficulties in dealing with large meshes because of the complexity limitations of existing QP solvers. Soliman et al. in [9] proposed a novel approach of building a non blind 3D watermarking system. This system based on clustering 3D vertices into appropriate or inappropriate candidates for watermark insertion using Self Organization Map (SOM). In [10] introduced a novel robust and blind mesh watermarking schema by converting the original triangle mesh into a multi-resolution format, consisting of a coarse base mesh and a sequence of refinement, the watermark bits are inserted through progressive mesh level of details, and extracted at refinement stage without any need for the original model. The watermark insertion is performed only on set of marked vertices come out from Self Organization Maps (SOM) clustering neural network.

Recent work utilizes CI in embedding 3D watermark is proposed by Bors and Luo [11]. This work proposes a new approach to 3D watermarking by ensuring the optimal preservation of mesh surfaces using Levenberg–Marquardt optimization method.

3 The Proposed 3D Watermarking Approach

The objective of this proposed work is to utilize an intelligent algorithm such as unsupervised learning using SOM. The main goal of using SOM is to watermark a 3D model in those locations which will produce imperceptible distortions in the final watermarked model. This approach for 3D mesh watermarking is shown in the following figure. The proposed approach can be divided into four basic phases (Fig. 1).

3.1 Phase 1: Smoothness Feature Extraction

The curvature of a curve is the measure of its deviation from a straight line in a neighborhood of a given point, and the curvature becomes greater as this deviation becomes greater [12]. In this work we used angle variation between surface normals and the average normal corresponding to a vertex to determine the vertex smoothness measure [13]. Such a smoothness measure reflects the local geometry of a surface or region that is if the region is flat, the angles will be small in magnitude. On the contrary in peak regions the angles will be high in magnitude.

The feature vector is a set of angles derived by computing the orientation of the surface normal to the average normal of the triangular faces that form a 1-ring neighborhood for a vertex.

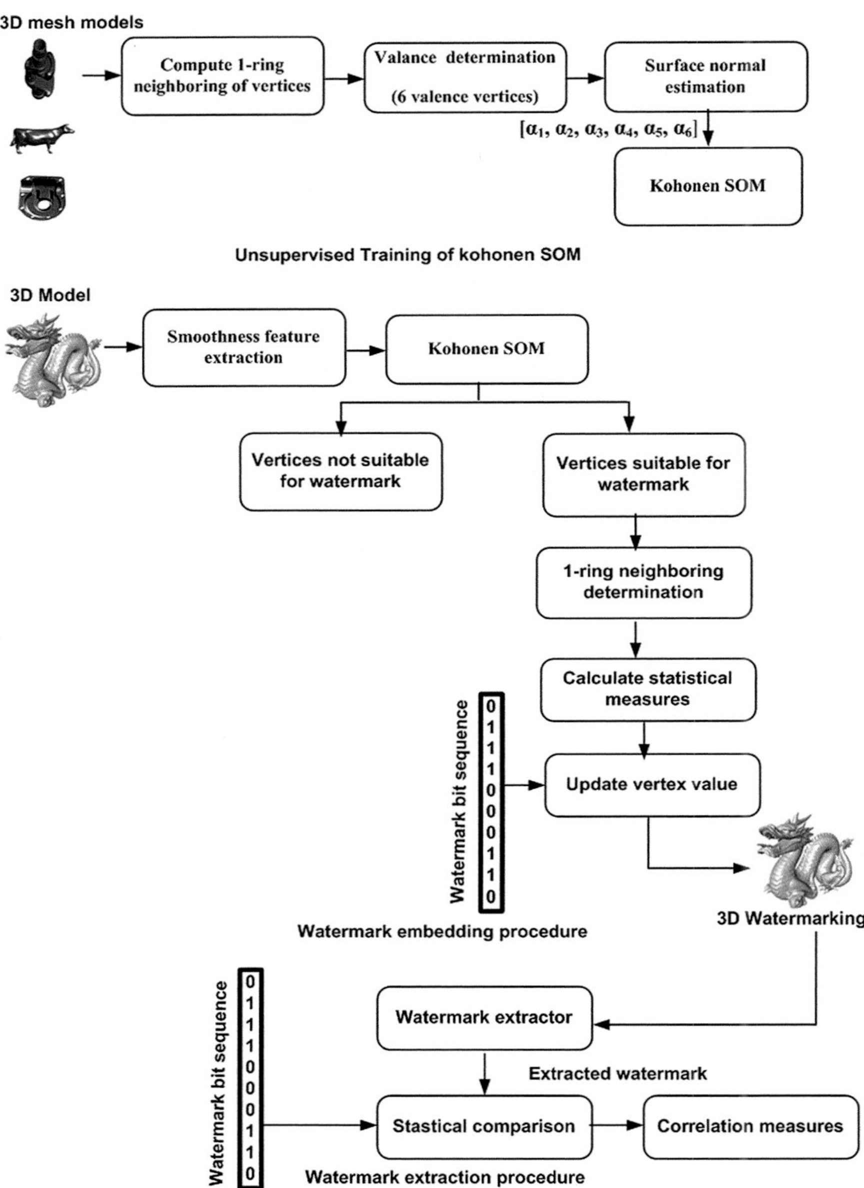

Fig. 1 Proposed approach of 3D mesh watermarking with intelligent vertex selection

3.2 Phase 2: Intelligent Vertex Selection

In this phase of intelligent vertex selection we use Kohonen SOM in clustering whole mesh vertices into three clusters (High magnitude, medium magnitude, and low magnitude). The set of feature vectors used in the unsupervised training of Kohonen SOM are set of smoothness feature vectors extracted from phase 1. If we aim to achieve high imperceptibility both flat and peak regions cannot be used to hide watermark bits. We have to neglect these regions and mark their vertices as unsuitable watermark carrier [10]. We only consider angles of medium magnitude value to be watermark carriers, so vertices with smoothness feature of medium magnitude will be marked as suitable watermark carriers.

3.3 Phase 3: Watermark Embedding Procedure

Phase-2 using Kohonen SOM result in set of marked vertices v^* which are most suitable carriers for watermark bit stream w. The embedding of watermark bits stream w is performed by searching for the immediate neighbors Vn_x of marked vertices v^*, estimating their mean μ and standard deviation σ. The embedding of the watermark stream is performed according to equations: "Eq. (1)" in case insertion of 1 from watermark sequence, and "Eq. (2)" in case insertion of 0 from watermark sequence.

$$v^*_{x'} = \mu(v^*\{Vn_x\}) + 2 * \sigma(v^*\{Vn_x\}) \tag{1}$$

$$v^*_{x'} = \mu(v^*\{Vn_x\}) - 2 * \sigma(v^*\{Vn_x\}) \tag{2}$$

3.4 Phase 4: Watermark Extraction Procedure

Here we implement a blind watermarking procedure thus there is no need to the original model during the extraction process. To extract the watermark, we only need the position of marked vertices v^*. Therefore, we need to store these data for the correct watermark extraction. Once these locations are detected watermark bits w are extracted by comparing the value of marked vertices with their immediate neighbors' mean and standard deviation.

4 Experimental Results and Discussion

The performance of the proposed 3D watermarking scheme is tested on several mesh models. Here the results are presented for five of them where these models are described as: Venus (100,759 vertices), Bunny (34,835 vertices), Rabbit (70,658 vertices), Elephant (24,955 vertices), and Dragon (50,000 vertices). We use 128 binary bit streams as watermark bits embedded on the original mesh. The experimental results show different results for the proposed approach in terms of both imperceptibility and robustness. Our proposed model will be compared to a well-known blind 3D watermarking method proposed by Cho et al. [14]. The basic idea proposed in [14] is to split the distribution of vertex norm into distinct sets called bins and embed one bit per bin. The distribution of vertex norms is modified by either shifting the mean value of the distribution according to the watermark bit to be embedded and or changing its variance.

4.1 Baseline Evaluation

The quality of the watermarked objects can be measured by two measures: Vertex Signal-to-Noise Ratio (VSNR) that quantifies the visual differences between the original and watermarked models. The other measure we consider in our baseline evaluation is the symmetric Haussdorff distance. The most popular implementation that has been widely used in the literature to compare results is the Metro tool [15]. Table 1 shows the values of VSNR, Haussdorff distance for the tested five models after inserting 128 binary bits sequence as watermark data. It also show the time per second of embedding procedure and extraction procedure.

4.2 Robustness Evaluation

The resistance of the embedded watermark has been tested under different types of attacks. These set of attacks including: Similarity Transformations, random noise attack, smoothing attack, and Cropping attack.

The robustness is evaluated in terms of correlation coefficient between the extracted watermark bit sequence w_n' and the originally inserted one w_n. Tables 2 and 3 show the correlation results for different types of attacks with different parameters for the Cho method [14] and our proposed approach.

4.3 Discussion and Comparison with the Methods of Cho et al.

As Cho approach cannot stand for cropping attacks [14], so we only report our results for cropping attacks. Cropping attack is defined by number of cropped vertices where their z coordinates values exceeds certain value. As shown in these tables our proposed approach depends on intelligent vertex selection in watermark insertion

Table 1 Baseline evaluation of proposed watermarking approach

Model	Embedd time	Extract time	VSNR	Haussdroff
Bunny	2.97	0.084	105.12	1.95
Dragon	4.18	0.966	112.56	2.04
Venus	8.94	0.110	129.76	0.066
Rabbit	6.23	0.965	128.24	0.460
Elephant	2.17	0.0971	106.31	0.185

Table 2 Robustness against uniform noise attack

Model	Amplitude	Cho approach [14]		Proposed approach	
	%	Corr	$Haussdorff_{10^{-3}}$	Corr	$Haussdorff_{10^{-3}}$
Bunny	0.1	0.654	0.32	0.908	0.31
	0.3	0.166	0.89	0.917	0.90
	0.5	0.055	1.47	0.876	1.48
	0.7	0.034	2.07	0.770	2.06
	0.9	0.031	2.62	0.724	2.64
Dragon	0.1	0.740	0.25	0.915	0.43
	0.3	0.33	0.80	0.92	0.85
	0.5	0.157	1.33	0.831	1.33
	0.7	0.126	1.80	0.719	1.81
	0.9	0.100	2.21	0.604	2.20
Venus	0.1	0.510	0.20	0.949	0.24
	0.3	0.130	0.70	0.936	0.70
	0.5	0.067	1.17	0.811	1.17
	0.7	0.036	1.63	0.700	1.64
	0.9	0.051	2.10	0.580	2.11
Rabbit	0.1	0.619	0.30	0.927	0.30
	0.3	0.151	1.08	0.783	1.08
	0.5	0.120	1.80	0.573	1.90
	0.7	0.122	2.51	0.45	2.41
	0.9	0.059	3.18	0.385	3.18
Elephant	0.1	0.57	0.45	0.98	0.48
	0.3	0.174	1.34	0.94	1.36
	0.5	0.086	2.23	0.83	2.24
	0.7	0.038	3.11	0.718	3.11
	0.9	0.023	3.93	0.597	3.90

Table 3 Robustness against cropping attack

Mesh model	No. of cropped vertices	Proposed approach corr
Bunny	5916	0.454
	3616	0.609
	746	0.833
Dragon	9666	0.838
	5604	0.851
	3156	0.833
	681	0.884
	21	0.918
Rabbit	15374	0.755
	11326	0.838
	6511	0.902
	3389	0.934
	151	0.952
Venus	17678	0.807
	10936	0.835
	4927	0.870

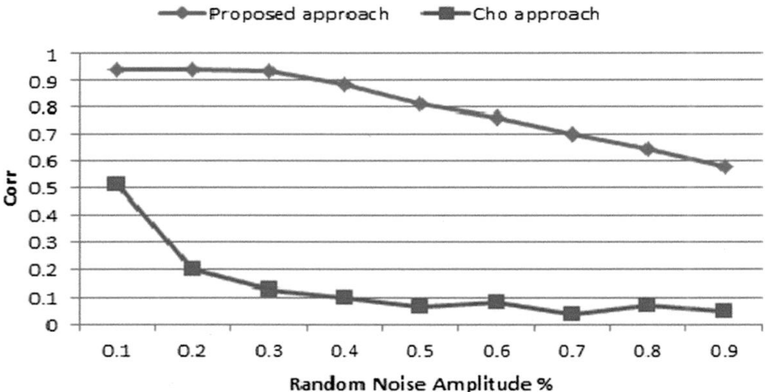

Fig. 2 Robustness response of random noise attack for Venus model

provide high robustness correlation value compared with Cho approach. In case of random noise attack shown in Table 2 the correlation value of proposed approach is higher than Cho approach for all models and at all different amplitude values. On average, we can still successfully extract up to 57 % of the watermark bits under 0.90 % noise addition. Figure 2 show such a difference in correlation value for Venus model as an example. Regarding the smoothing noise attack, although our proposed approach does not give high correlation value for some models for different values of iteration number but it prove its efficiency of high correlation values for all models

Table 4 Robustness against smoothness attack

Model	Iterations	Cho approach [14]		Proposed approach	
	No.	Corr	$Haussdorff_{10^{-3}}$	Corr	$Haussdorff_{10^{-3}}$
Bunny	5	0.900	0.08	0.88	0.18
	15	0.578	0.15	0.845	0.19
	25	0.370	0.23	0.526	0.25
	35	0.132	0.26	0.288	0.32
	50	0.07	0.40	0.165	0.41
Dragon	5	0.936	0.07	0.902	0.17
	15	0.675	0.15	0.683	0.20
	25	0.476	0.24	0.468	0.26
	35	0.331	0.35	0.334	0.34
	50	0.238	0.45	0.244	0.46
Rabbit	5	0.983	0.69	0.918	0.07
	15	0.903	0.07	0.886	0.07
	25	0.724	0.11	0.637	0.10
	35	0.611	0.13	0.604	0.12
	50	0.496	0.17	0.454	0.17
Venus	5	0.96	0.03	0.91	0.05
	15	0.827	0.06	0.90	0.06
	25	0.553	0.09	0.815	0.09
	35	0.45	0.12	0.573	0.12
	50	0.195	0.15	0.377	0.16
Elephant	0.1	0.90	0.07	0.98	0.17
	0.3	0.722	0.16	0.93	0.19
	0.5	0.533	0.25	0.883	0.27
	0.7	0.431	0.33	0.866	0.35
	0.9	0.298	0.45	0.669	0.45

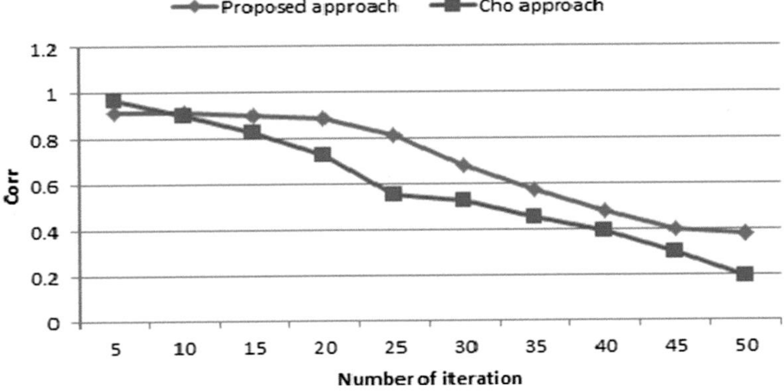

Fig. 3 Robustness response of smooth noise attack for Venus model ($\lambda = 0.03$)

when number of iterations increased as shown in Table 4. Figure 3 show different correlation values in case of smoothing attacks at different iteration numbers for Venus model. For cropping attacks as shown in Table 3 our approach report a high correlation value and superior upon Cho approach that cannot stand for cropping attacks at all.

5 Conclusions

In this paper, a new robust and blind 3D mesh watermarking approach is proposed. This approach utilize one of Computational intelligent (CI) methods in building 3D Blind and robust watermarking system. This work aims to insert layer of intelligent in the selection of watermark carrier vertices. This is performed by using kohonen SOM as a clustering method aims to classify 3D mesh vertices as suitable and non-suitable candidates for being watermark carriers. The embedding is accomplished by utilizing some statistical values from selected vertices immediate neighbors, this result into a blind and robust 3D watermarking mesh model with high correlation value against different types of attacks.

References

1. Wanga, K., Lavoue, G., Denis, F., Baskurt, A.: Robust and blind mesh watermarking based on volume moments. Comput. Graph. **35**(1), 1–19 (2011)
2. Gaber, T., Zhang, N.: A novel method for supporting fairness in digital license reselling. In: The Fifth International Conference on Internet Monitoring and Protection (ICIMP), IEEE, pp. 89–98 (2010)
3. Gaber, T., Zhang, N.: A license revocation protocol supporting digital license reselling in a consumer-to-consumer model. Int. J. Online Mark. (IJOM) **2**(1), 38–49 (2012)
4. Gaber, T., Zhang, N., Hassanien, A.: A novel approach to allow multiple resales of DRM-protected contents. In: The 8th IEEE International Conference on Computer Engineering and Systems (ICCES), Cairo, Egypt, pp. 86–91 (2013)
5. Darwish, A., Abraham, A.: The use of computational intelligence in digital watermarking: review challenges, and new trends. Neural Netw. World **01**, 277–297 (2011)
6. Motwani, M.C., Bryant, B.D., Dascalu, S.M., Harris, F.C.: 3D multimedia protection using artificial neural network. In: 7th IEEE Conference on Consumer Communications and Networking (CCNC), pp. 1–5 (2010)
7. Motwani, R.C., Motwani, M.C., Bryant, B.D., Harris, F.C., Akshata, J.R., Agarwal, S.: Watermark embedder optimization for 3D Mesh objects using classification based approach. ICSAP 125–129 (2010)
8. Hu, R., Alface, P., Macq, B.: Constrained optimisation of 3D polygonal mesh watermarking by quadratic programming. In: Acoustics, IEEE International Conference on Speech and Signal Processing, ICASSP, pp. 1501–1504 (2009)
9. Soliman, M.M., Hassanien, A., Onsi, H.M.: Robust watermarking approach for 3D triangular mesh using self organization map. In: 8th International Conference on Computer Engineering and Systems (ICCES), pp. 99–104 (2013)

10. Soliman, M.M., Hassanien, A.E., Onsi, H.M.: A blind robust 3D-watermarking scheme based on progressive mesh and self organization maps. advances in security of information and communication networks communications in computer and information. Science **381**, 131–142 (2013)
11. Bors, A., Luo, M.: Optimized 3D watermarking for minimal surface distortion. IEEE Trans. Image Process. **22**(5), 1822–1835 (2013)
12. Nigam, S., Agrawal, V.: Review. Curvature approximation on triangular meshes. Int. J. Eng. Sci. Innov. Technol. (IJESIT) **2**(3), 330–339 (2013)
13. Motwani, R.C., Motwani, M.C., Harris, F.C., Bryant, B.D.: Watermark embedder optimization for 3D mesh objects using classification based approach. In: International Conference on Signal Acquisition and Processing (2010). doi:10.1109/ICSAP.2010.83
14. Cho, J.W., Prost, R., Jung, H.Y.: An Oblivious watermarking for 3-D polygonal meshes using distribution of vertex norms. IEEE Trans. Signal Process. **55**(1), 144–152 (2005)
15. Alface, P., Craene, M., Macq, B.: Three-dimensional image quality measurement for the benchmarking of 3D watermarking schemes. In: Proceedings of SPIE (5681) (2005)

Generic Obstacle Detection for Mobile Devices Using a Dynamic Intermediate Representation

Radu Danescu, Andra Petrovai, Razvan Itu and Sergiu Nedevschi

Abstract Obstacles are environment elements that need to be detected reliably and accurately. While active sensors and stereovision have the advantage of obtaining 3D data directly, monocular vision is easy to set up, and can benefit from the increasing computational power of smart mobile devices. Several driving assistance application are available for mobile devices, but they are mostly targeted for simple scenarios and a limited range of obstacle shapes and poses. This paper presents a technique for generic, shape independent real-time obstacle detection for mobile devices, based on a dynamic, free form 3D representation of the environment: the particle based occupancy grid. Monocular vision is used to update the occupancy grid, and the occupancy grid tracked cells are used for extracting the obstacles as cuboids having position, size, orientation and speed. The easy to set up system is able to reliably detect most obstacles in urban traffic, and its measurement accuracy is comparable to a stereovision system.

Keywords Obstacle detection · Occupancy grid · Monocular vision · Advanced driving assistance system · Mobile device

R. Danescu (✉) · A. Petrovai · R. Itu · S. Nedevschi
Computer Science Department, Technical University of Cluj-Napoca, Cluj-Napoca, Romania
e-mail: radu.danescu@cs.utcluj.ro

A. Petrovai
e-mail: andrapetrovai@mail.utcluj.ro

R. Itu
e-mail: itu.razvan@gmail.com

S. Nedevschi
e-mail: sergiu.nedevschi@cs.utcluj.ro

A. Abraham et al. (eds.), *Proceedings of the Second International Afro-European Conference for Industrial Advancement AECIA 2015*, Advances in Intelligent Systems and Computing 427, DOI 10.1007/978-3-319-29504-6_58

1 Introduction

Driving is a process of continuous sensing, processing the sensory information, making decisions and putting these decisions into action. Some parts of this loop can be done by humans, and some by machines. All decisions related to driving are based on sensorial information, and this is why, for decades, researchers have tried to devise better sensors, and better and faster algorithms, to help perceive the environment, help the driver and increase traffic safety.

While the most research and development effort has been dedicated to adding intelligent sensing capabilities to the vehicles themselves, especially to the higher end ones, driving assistance applications on mobile devices are a less costly alternative. Facilitated by the rapid evolution of the processing power of the smart mobile devices, computer vision for driving assistance on mobile devices may quickly become a reliable tool for traffic safety.

The set of driving assistance applications available includes iOnRoad, Drivea, and Movon FCW. iOnRoad [1] is one of the first augmented driving application available for iOS and Android, using the smartphone's camera and GPS receiver to provide time to collision to the detected vehicle in front of the ego-vehicle, and for lane departure warning. A similar application is Drivea [2], which provides obstacle collision warning, lane departure warning, and warning about speeding. A newer application, Movon FCW [3] is one of the few applications that can detect incoming cars, or cars that are not in full view, and can provide actual distance information.

In the research community, most of the recent obstacle detection efforts have been focused on sensors or sensorial solutions that are able to deliver 3D data directly: radar, laser, or stereovision [4]. While some mobile devices are equipped with stereo cameras, which can be used for short distance accurate detection of obstacles [5], most devices are only capable of monocular vision.

Monocular obstacle detection faces multiple challenges, due to the lack of direct 3D information. For the identification of the obstacle regions in the image, solutions include color based segmentation [6, 7], motion based analysis [8, 9], or appearance analysis, either through heuristics such as symmetry [10, 11], or through the use of machine learning [12, 13]. For extracting the 3D position of the obstacle, monocular vision uses constraints imposed on the structure of the environment, such as the condition that the road is flat, which leads to the measurement of the points on the road by Inverse Perspective Mapping (IPM) [14–16].

The method proposed in this paper relies on segmenting obstacle areas from an IPM image, and then, instead of using these areas to find the obstacles, these areas are used to update a probabilistic occupancy grid [17], which is able to filter out some detection errors, and extract dynamic properties for each cell. The estimated occupied grid cells, having also speed and orientation, are grouped into obstacles. In this way, no assumption about the nature of the obstacle is made, and the detected obstacle has position, size, speed and orientation.

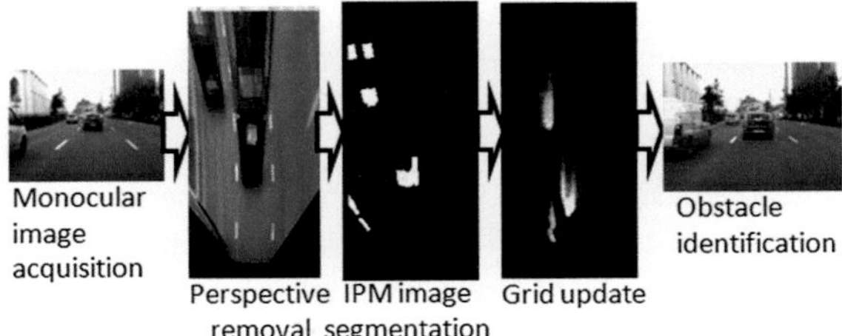

Fig. 1 Overview of the obstacle detection process

2 Method Overview

The process starts with the acquisition of an image from the mobile device's camera. The image is then transformed into a bird eye view of the road by Inverse Perspective Mapping, using the camera calibration parameters. The IPM image is segmented along rays starting from the camera position, and a hypothetic obstacle map is generated. The obstacle map is used to update the particle based occupancy grid. In the update process, the vehicle motion information (speed and yaw rate) is obtained from the available sensors of the mobile device (GPS for speed, gyroscope or accelerometer for yaw rate). The occupancy grid attaches to each cell of the top view world an occupancy probability and a speed probability density.

The estimated grid cells occupancy probabilities, and the speed vectors, are used in the process of obstacle extraction. The process is depicted in Fig. 1.

3 Algorithm Description

3.1 Removing the Perspective Effect

Assuming that the road is flat, the image can be remapped in such a way that the road markings become parallel, as seen from above, and the lateral and longitudinal coordinates in the real world of a road feature are proportional to the pixel coordinates (Fig. 2). In the IPM image the obstacles are deformed, since they do not

Fig. 2 Removal of the perspective effect. The obstacle areas will expand radially from the camera position point

belong to the road surface. Only the point of contact between the obstacle and the road is relevant. The camera is calibrated, therefore the correspondence between any point in the 3D world and its corresponding projected pixel in the image plane is known. The 3D coordinate system is assumed to have the origin on the ground, below the camera, the x axis pointing to the driver's right, the z axis indicating the direction of travel, and the y axis the height above the road (xOz) plane.

The removal of the perspective effect is performed using the following algorithm:

Algorithm: IPM Transformation
Input: Source image I
Output: IPM image I_T
For each pixel of coordinates (u_T, v_T) of I_T
$x = k\, u_T + x_0$
$z = j\, v_T + z_0$
$y = 0$
$(u, v) = \text{Projection}(x_W, y_W, z_W)$
$I_T(u_T, v_T) = I(u, v)$
End For

The constants k, j, x_0 and z_0 are chosen in such a way that the most relevant portion of the road plane is displayed in the remapped image. The projection function uses the camera intrinsic and extrinsic parameters for mapping a 3D point, expressed in world coordinates, to the image plane.

The completely flat road assumption is restrictive, and the world will often not comply with it, due to the road geometry and, most often, to the balancing of the observer vehicle. This error is taken into account in the grid estimation process.

3.2 Processing the IPM Image

From the top view IPM image we have to produce a binary one, indicating the areas corresponding to the contact between the obstacle and the road. Since we want our system to be as form-independent as possible, we will not use shape-based clues, such as symmetry.

An upright obstacle has the tendency, in the IPM image, to expand radially from the camera. We'll use this property, and scan the IPM image with rays for each angular degree of the field of view, each ray converging in the camera position point. If an obstacle is present on such a ray, its starting position must be found, as it is the contact point between the obstacle and the road.

In order to find the obstacle distance along the scan ray, the grayscale values of the IPM image corresponding to the ray are recorded in a vector $g_\alpha(d)$, α denoting the angle of the ray, and d being the distance from the camera. For each candidate obstacle distance d, the following values are computed:

$$\mu_{\alpha,P}(d) = \frac{1}{d - 3 - d\_\min_{\alpha} + 1} \sum_{k=d\_min_{\alpha}}^{d-3} g_{\alpha}(k) \tag{1}$$

$$\mu_{\alpha,D}(d) = \frac{1}{d\_\max_{\alpha} - d - 3 + 1} \sum_{k=d+3}^{d\_\max_{\alpha}} g_{\alpha}(k) \tag{2}$$

$$\mu_{\alpha,M}(d) = \frac{1}{7} \sum_{k=d-3}^{d+3} g_{\alpha}(k) \tag{3}$$

These quantities are the mean intensity values for the proximal region ($\mu_{\alpha,P}$), middle region ($\mu_{\alpha,M}$) and distal region ($\mu_{\alpha,D}$) of a ray of angle α with respect to the camera axis, at the hypothetical obstacle distance d. Based on these mean values, we have to decide if a distance d may in fact be a distance to an obstacle touching the road. First, a binary function of d is designed, indicating whether the intensity profile of the ray of angle α supports the hypothesis of an obstacle being present at the distance d.

$$\omega_{\alpha}(d) = \begin{cases} 1, \ \textit{if} \ (\mu_{\alpha,P}(d) - \mu_{\alpha,M}(d)) > \sigma \,\& \\ (\mu_{\alpha,D}(d) - \mu_{\alpha,M}(d) > \sigma)| \\ ((\mu_{\alpha,P}(d) - \mu_{\alpha,D}(d) > \sigma) \\ 0, \ \textit{otherwise} \end{cases} \tag{4}$$

In Eq. (4), σ is the graylevel standard deviation of the IPM image, a measure of the image contrast. Intuitively, Eq. (4) states that an obstacle point of contact with the road should be darker than the road leading to it, and also darker than the rest of the obstacle along the ray. The shadow area below a car passes this test. Darker areas that continue along the ray are also taken into consideration.

If multiple candidate distances for a ray have $\omega_{\alpha}(d) > 0$, the best candidate has to be selected. A continuous score function is defined:

$$\varsigma_{\alpha}(d) = \left(\left|\mu_{\alpha,P}(d) - \mu_{\alpha,M}(d)\right| \cdot \left|\mu_{\alpha,D}(d) - \mu_{\alpha,M}(d)\right| \cdot \left|\mu_{\alpha,P}(d) - \mu_{\alpha,D}(d)\right|\right)^{\frac{1}{3}} \tag{5}$$

Out of all the values d that have $\omega_{\alpha}(d) > 0$, the one having a maximum $\varsigma_{\alpha}(d)$ is kept as the obstacle position. The detected obstacle positions along the rays are used to build a binary measurement image (see Fig. 1, third image).

3.3 Using the Particle-Based Occupancy Grid as an Intermediate Dynamic Representation of the Scene

The particle based occupancy grid is a freeform dynamic bird eye view world representation, able to estimate an occupancy probability and a speed vector for each cell. Each cell of the grid, corresponding to a 20 × 20 cm area of the road surface, contains a population of particles, each having a speed vector. The occupancy probability is proportional to the number of particles in a cell, and the speed vector of the cell is the resultant of the speed vectors of the cell's particles. The tracking mechanism has two main phases: prediction and update. The prediction is achieved by moving particles from one cell to another, using the laws of motion and the particles' speed vectors.

The update process multiplies or destroys the particles in a cell, based on the measurement data, the binary obstacle/no obstacle map generated by processing the IPM image. For each cell of the grid, the measurement probability density under the assumption of the cell being occupied is computed using the bivariate Gaussian function, using the column distance d_{column} and the row distance d_{row} between the cell position and the nearest binary obstacle cell.

$$p_{occupied} = \frac{1}{2\pi\sigma_{row}\sigma_{column}} e^{-\frac{1}{2}\left(\left(\frac{d_{row}}{\sigma_{row}}\right)^2 + \left(\frac{d_{column}}{\sigma_{column}}\right)^2\right)} \quad (6)$$

In Eq. (6), σ_{row} and σ_{column} stand for the expected measurement error standard deviations expressed in grid coordinates row and column, which are in fact the measurement errors on the Z and X coordinates, scaled down by the cell size $D = 20$ cm.

The expected measurement error depends on the characteristics of the detection process (the sensor model). In our case, the relation between an image pixel and its real world counterpart is done using the perspective projection. To simplify, we can assume that knowing the image line coordinate of a point, and the camera extrinsic parameters, we know the angle θ between the optical ray passing through this point and the vertical. Knowing the camera height h above ground, this ray crosses the road at a distance:

$$z = h \tan\theta. \quad (7)$$

Any error in detecting the image line position of the obstacle translates linearly in an error of the angle θ, σ_θ, which we'll assume to be constant at 0.25°.

The relation between the uncertainty of z and the uncertainty of θ is:

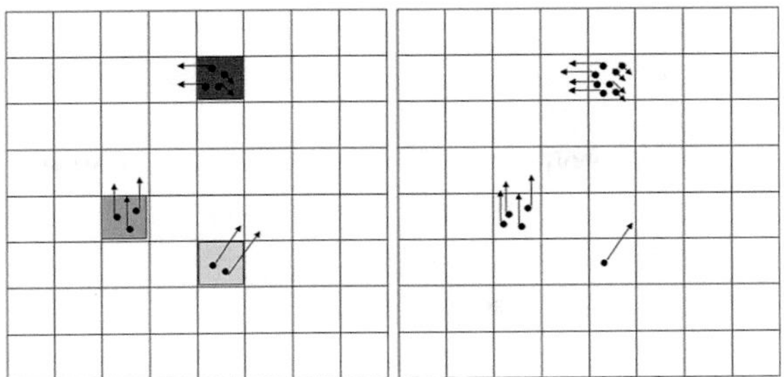

Fig. 3 Particle population control using the measurement. *Left* predicted particles and the occupied measurement probability, encoded as a *gray* level (darker = higher). *Right* the new particle population, after measurement

$$\sigma_z = \frac{dz}{d\theta}\sigma_\theta. \tag{8}$$

This can be written as:

$$\sigma_z = h(1 + \tan^2\theta)\sigma_\theta. \tag{9}$$

Using (7) we can replace the tangent, and obtain the final equation for the uncertainty of z. Other sources of errors are modeled by a constant offset σ_{z0}:

$$\sigma_z = h\left(1 + z^2/h^2\right)\sigma_\theta + \sigma_{z0}. \tag{10}$$

From σ_z we can compute the uncertainty of the lateral position x:

$$\sigma_x = x\sigma_z/z + \sigma_{x0}. \tag{11}$$

Dividing σ_z and σ_x by the grid cell size D, the uncertainties in grid row and column coordinates, σ_{row} and σ_{column}, are found.

Based on $p_{occupied}$, the particles in a cell are either multiplied or destroyed. Thus, the particle population is attuned with the measurement (Fig. 3).

More details about the particle world model representation, the particle motion mode, the compensation of the ego motion, and the population control mechanism, can be found in [17].

3.4 Detection of Individual Obstacles

The dynamic grid's cells that hold a significant number of particles are considered occupied, and therefore parts of obstacles. A grouping algorithm, which takes into account the proximity of occupied cells, but also the agreement of their speed

Fig. 4 Obstacle detection results. *Left* the dynamic occupancy grid; *right* the extracted cuboids

vectors, extracts connected regions from the grid. An oriented rectangle shape is matched to these connected regions, and the 3D object is extracted as a cuboid (obviously, the height of the object above the road cannot be computed, and a fixed value is set). The resulted objects have length, width, orientation and speed, and can be therefore classified into static and dynamic. An object is declared to be dynamic if its longitudinal speed exceeds 2 m/s, or if the lateral speed exceeds 1 m/s. The process is seen in Fig. 4: the stationary vehicles are shown in green, and the moving vehicle is shown in red.

4 Implementation for Android Mobile Devices

The solution was implemented as an Android application, using Java for the user interface, and native C++ for the algorithm processing workload (Fig. 5).

The calibration process is simplified by the automatic retrieval of the camera's horizontal and vertical field of view angles using the Android API. The fixed image resolution of 640 × 480 pixels, combined with the known field of view, give us the focal distance in pixels. The position of the principal point is assumed to be in the

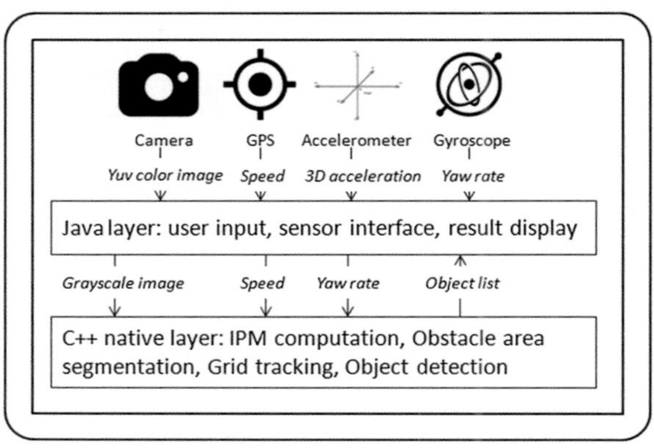

Fig. 5 The architecture of the solution

center of the image. The user is asked to tune the pitch angle and the camera height, while the IPM image is shown with a distance grid superimposed, so that the user can adjust these parameters until top view scene is in accordance to the grid.

The grid tracking algorithm requires the motion parameters of the host vehicle, the speed and the yaw rate. The speed is provided by the mobile device's GPS receiver. For the yaw rate, the onboard gyroscope is used when available; otherwise, the lateral acceleration combined with the speed is used to infer the yaw rate.

5 Tests and Results

The algorithm was tested on recorded sequences from mostly urban scenarios, under a various range of lighting conditions. Obstacles from a wide range of shapes, speeds and orientations have been successfully detected, and their static and dynamic properties successfully measured. Some examples are shown in Fig. 6: the top row of each panel shows the IPM image, the binary obstacle grid, and the estimated occupancy grid with color coding of speed and orientation for each cell, and the bottom row shows the perspective image and the projected cuboids.

The qualitative analysis shows that most of the obstacles are correctly detected, but there are still problems with objects of low contrast (Fig. 6, third panel), and with strong shadows in bright daylight (Fig. 6, fourth panel). A video showing results for an 11 min drive sequence (recorded and processed offline) can be seen at http://vimeo.com/107588262.

The system was tested on multiple mobile devices: the Samsung Galaxy S5 phone, and the LG Gpad 8.3 and the Samsung Galaxy Tab Pro tablets (Fig. 7). Depending on the complexity of the scene, the system's processing time ranged from 50 to 80 ms. The resulted framerate ranged between 9 and 15 fps, as the whole frame cycle includes the acquisition process and the result display.

In order to assess the accuracy of the obstacle measurement process, the obstacle range obtained from the proposed solution is compared with the range provided by a stereovision-based obstacle tracking system. The monocular solution was tested

Fig. 6 Successful detection for cars of multiple orientations, and pedestrians (*panels 1*, *2* and *3*); a detection failure (*panel 4*), and false positives (*panel 5*)

Fig. 7 Obstacle detection using a smartphone (*left*) and a tablet (*right*), mounted inside the vehicle

on the left frame of the stereo pair, and the stereo system used both images. While noisier, the estimations of the monocular system follow closely the stereovision results. For the whole sequence, the Root Mean Square Error was 1.33 m, and the Mean Absolute Error 1.02 m.

6 Conclusion and Future Work

The proposed technique shows promising results, detecting a large variety of obstacles, and measuring their characteristics with good accuracy. The concept of combining the occupancy grid with IPM and ray-based segmentation has proved to be a working choice, producing results comparable to stereovision. The solution was deployed and is able to work in real time on mobile platforms.

There is significant room for improvement. Due to monocular vision limitations, the false positives or the false negatives are always a threat, especially for a generic obstacle detection system. The reduction of the error rate can be achieved by using color data (already available) or optical flow.

Acknowledgments Research supported by grants of the Romanian Authority for Scientific Research, projects PN-II-PCE-2011-3-1086 (MultiSens) and PN-II-PCCA-2011-3.2-0742 (SmartCoDrive).

References

1. iOnRoad Augmented Driving Pro. http://www.ionroad.com/
2. Drivea—Driving Assistant App. http://blog.ezemobileapps.com/drivea-the-driving-assistant-app/
3. Movon Corporation: Movon FCW. https://play.google.com/store/apps/details?id=com.movon.fcw
4. Sivaraman, S., Trivedi, M.M.: Looking at vehicles on the road: a survey of vision-based vehicle detection, tracking, and behavior analysis. IEEE Trans. Intell. Transp. **14**, 1773–1794 (2013)

5. Petrovai, A., Costea, A., Oniga, F., Nedevschi, S.: Obstacle detection using stereovision for Android-based mobile devices. In: IEEE 10th International Conference on Intelligent Computer Communication and Processing, pp. 141–147. IEEE Press (2014)
6. Haltakov, V., Belzner, H., Ilic, S.: Scene understanding from a moving camera for object detection and free space estimation. In: IEEE Intelligent Vehicles Symposium, pp. 105–110. IEEE Press (2012)
7. Wybo, S., Tsishkou, D., Vestri, C., Abad, F., Bougnoux, S., Bendahan, R.: Monocular vision obstacles detection for autonomous navigation. In: IEEE/RSJ International Conference on Intelligent Robots and Systems (IROS), pp. 4190–4196. IEEE Press (2008)
8. Reisman, P., Mano, O., Avidan, S., Shashua, A.: Crowd detection in video sequences. In: IEEE Intelligent Vehicles Symposium, pp. 66–71. IEEE Press (2004)
9. Zhang, X., Jiang, P., Wang, F.: Overtaking vehicle detection using a spatio-temporal CRF. In: IEEE Intelligent Vehicles Symposium, pp. 338–343. IEEE Press (2014)
10. Graefe, V., Efenberg, W.: A novel approach for the detection of vehicles on freeways by real-time vision. In: IEEE Intelligent Vehicles Symposium, pp. 363–368 (1996)
11. Lim, K.H., Ang, L.M., Seng, K.P., Chin, S.W.: Lane-vehicle detection and tracking. In: International Multi Conference of Engineers and Computer Scientists, pp. 1124–1129 (2009)
12. Sivaraman, S., Trivedi, M.M.: A general active-learning framework for on-road vehicle recognition and tracking. IEEE Trans. Intell. Transp. **11**, 267–276 (2010)
13. Arrospide, J., Salgado, L., Marinas, J.: HOG-like gradient-based descriptor for visual vehicle detection. In: IEEE Intelligent Vehicles Symposium, pp. 223–228. IEEE Press (2012)
14. Bertozzi, M., Broggi, A.: GOLD: a parallel real-time stereo vision system for generic obstacle and lane detection. IEEE Trans. Image Process. **7**, 62–81 (1998)
15. Tuohy, S., O'Cualain, D., Jones, E., Glavin, M.: Distance determination for an automobile environment using inverse perspective mapping in OpenCV. In: IET Signal and Systems Conference, pp. 100–105. IET (2010)
16. Itu, R., Danescu, R.: An efficient obstacle awareness application for android mobile devices. In: IEEE 10th International Conference on Intelligent Computer Communication and Processing, pp. 157–163. IEEE Press (2014)
17. Danescu, R., Oniga, F., Nedevschi, S.: Modeling and tracking the driving environment with a particle-based occupancy grid. IEEE Trans. Intell. Transp. **12**, 1331–1342 (2011)

Wolf Local Thresholding Approach for Liver Image Segmentation in CT Images

Abdalla Mostafa, Mohamed Abd Elfattah, Ahmed Fouad, Aboul Ella Hassanien and Hesham Hefny

Abstract This paper enhances the usage of level set method to get a reliable liver image segmentation in CT images. The approach depends on a preprocessing phase to enhance the liver's edges. This phase is performed in two ways using the morphological operations and wolf local thresholding. The first way starts with applying the morphological operations on the image to clean image annotation and bed lines. Then, it applies contrast stretching and texture filters. The other way applies the wolf local threshold to each point in the image. It uses a window or a mask to calculate the average and standard deviation to apply an iterative threshold. Each way is followed by a step of connecting ribs to separate the flesh and skin from liver's region. The last step is to use level set method to segment the whole liver. A set of 47 images taken in pre-contrast phase, were used to test the approach. Validating the approach is done using similarity index measure. The obtained experimental results showed that the overall accuracy presented by the proposed approach results in 93.19 % accuracy for using morphological operations, and 93.30 % accuracy for using Wolf local thresholding.

A. Mostafa · H. Hefny
Institute of Statistical Studies and Researches, Cairo University, Giza, Egypt
e-mail: abdalla_mosta75@yahoo.com

M.A. Elfattah
Faculty of Computers and Information, Mansoura University, Mansoura, Egypt
e-mail: m_abdelfatah@ymail.com

A. Fouad
Faculty of Computers and Information, Suez Canal University, Ismaïlia, Egypt
e-mail: ahmed_fouad@ci.suez.edu.eg

A.E. Hassanien (✉)
Faculty of Computers and Information, Cairo University, Giza, Egypt
e-mail: aboitcairo@gmail.com
URL: http://www.egyptscience.net

A. Mostafa · M.A. Elfattah · A. Fouad · A.E. Hassanien
Scientific Research Group in Egypt (SRGE), Giza, Egypt

A. Abraham et al. (eds.), *Proceedings of the Second International Afro-European Conference for Industrial Advancement AECIA 2015*, Advances in Intelligent Systems and Computing 427, DOI 10.1007/978-3-319-29504-6_59

Keywords Region growing · Morphological operations · Local thresholding · Filtering · Segmentation · Level-set

1 Introduction

Every CAD (Computer-aided diagnosis) system should include a phase for segmenting the required organ. CAD system includes the phases of preprocessing, segmentation and classification. Liver is situated under the diaphragm in the upper abdominal cavity, and is held in place by several ligaments. It is a reddish-brown colour organ , but it may have different colours according to different lesions. Colours could be dark blue for cyst, dark brown for cirrhosis, yellow for fatty, green for biliary cirrhosis. These colours have different intensity values in CT grey images. This is one of the difficulties in liver segmentation. Also, falciform ligament separates the liver lobes as a boundary, which could be a kind of difficulty in segmenting the liver from other regions. Other difficulty, could be blood vessels as portal vein and hepatic artery [9].

Filtering is a preprocessing phase that stresses features, gets rid of noise and distortion. Another step that comes after filtering is stressing the boundaries of liver. Stressing of boundaries is performed in two ways in this paper, the morphological operations and Wolf local thresholding. Then, the image is passed to segmentation phase where liver is separated from other organs, using Caselles' level-set segmentation method. Liver segmentation requires a smoothing filter to hide the unnecessary liver details such as arteries, veins and lesions. Also, morphological operations and local thresholding can be used to stress the boundaries of liver. The anatomical structure of liver is separated from the abdomen through the segmentation process. The main key used for segmentation, relies on the intensity values of cells in the image. Images with JPG format are implemented in this paper to segment liver and extract regions of interest (ROIs). It has intensity values range between 0 and 512 [1]. In CT images, there is a similarity between liver intensity and some other organs, such as spleen, heart, skin and muscles [10]. So, extracting liver region is suitable for the following steps of ROIs' segmentation and features extraction. The proposed approach in this paper smooths image and uses Wolf local thresholding and some morphological operations to separate the whole liver using level set method. Finally, testing the result of segmentation is done using similarity index measure.

Reinhard Beicher et al., used graph-cut to segment liver and applied a user manual enhancement for the segmented liver [2]. Sheng Wu used a hierarchical contextual active contour (HCAC) to segment liver [12]. Xuechen Li et al. used fuzzu clustering combined with level set to extract liver from the abdomen in CT images [7]. Huang W. et al. used feature extraction in the process of segmentation [5]. They extracted 4 features from liver CT image including neighbourhood mean, neighbourhood variance, Laws texture, and Unsers sum-and-difference histograms.

The remainder of this paper is ordered as follows. Section 2 describes the used Wolf local thresholding method. Details of the proposed approach is shown in Sect. 3. Section 4 shows the experimental results and analysis. while, conclusions and future work are discussed in Sect. 5.

2 Wolf Local Thresholding

Thresholding is the process of partitioning the regions of an image using intensity or properties. It separates the image into two classes, depending on the threshold value. Thresholding could be global or local. While global threshold is a general value that applies to all pixels in the image, local threshold computes a threshold for every pixel. It extracts some properties of the pixel depending on the neighbour pixels. To segment an image using general thresholding, the histogram is used to get the peaks' values. Using gradient and laplacian operators can help to deepen the valley between histogram peaks [6]. This can determine the boundaries of the objects. But, for local thresholding, standard deviation and mean are calculated for every point. For a pixel, the value is true if the intensity value is greater than the standard deviation and mean of this pixel. This results in a binary image [4]. Thresholding can be applied using different techniques as, mean, histogram and edge maximization techniques [3].

Wolf local thresholding, which is applied in this paper, depends on **calculating means and standard deviation for each pixel**. It calculates threshold as follows.

$$T = \sqrt{(1-k) * mean + \frac{k * dev}{R * (mean - M)} + k * M} \tag{1}$$

where T is the threshold, k is an initial threshold provided by user, dev is the standard deviation, R is the maximum deviation, and M is the minimum intensity value in the image [11]. The used window is a mask where the center of the mask is the pixel which we calculate its threshold. This method is used **iteratively** in this paper, by increasing the used mask with a small k.

3 The Proposed CT Liver Segmentation Approach

The proposed approach of liver segmentation is comprised of the following fundamental building phases: (1) **Pre-processing phase** In the proposed approach, a pre-processing algorithm is performed before segmentation, performing some main functions to remove noise, enhance contrast, stress boundaries of liver. It also connects ribs around the liver, and removes annotations and bed lines from the image, (2) **Region growing phase:** In the second phase, region growing segmentation method is used to separate the whole liver and tested against the morphological operations, and (3) **Level set phase:** In the second phase, Level set segmentation method is used to separate the whole liver, and tested against morphological operations and Wolf local thresholding.

Figure 1 illustrates the architecture of the proposed liver image segmentation approach. The following will describe these phases in detail.

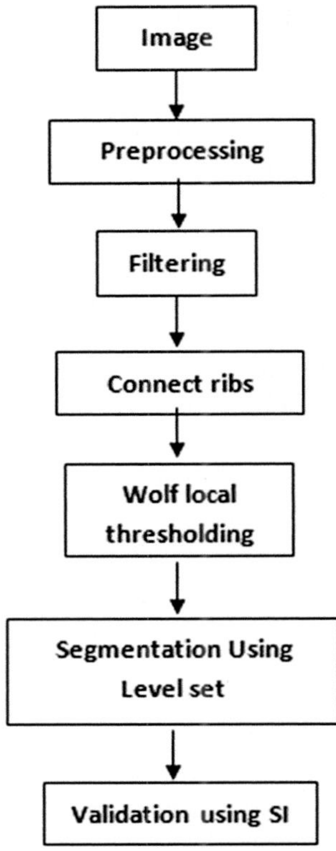

Fig. 1 Liver image segmentation approach: phases

3.1 Preprocessing Phase: Applying Filters and Ribs Connection

In this paper, filters play a main role in segmentation process. Filters are used to enhance the accuracy of CAD systems for lesions detection. The very little details in liver as veins, arteries and ligaments, are concealed using smoothing concept. It enables the segmentation method to grow the initial seed points and move the contour outward till extracting the whole liver. On the other hand, combining contrast concept with thresholding or morphological operations, could be used to emphasize the boundaries of the extracted liver.

According to this, the most suitable filters could be contrast stretching, average, gaussian and Gabor. For the purpose of connecting ribs, a step of cleaning image is performed. It removes the annotation of patient's data and the machine's bed, to facilitate the next operation of connecting ribs. Cleaning image step starts by using open operation to remove annotations. This operation removes all little objects that have white pixels fewer than 5500 pixels, which represents the characters of image annotation and machine's bed. Then, it fills all holes inside the abdomen. Finally, the edges of abdomen are eroded to conceal a part of the skin and flesh around ribs.

In brief, The steps of opening, closing, fill and erosion clean image by removing the annotation of the image. It erodes the skin of the abdomen and removes the bed of the machine. Ribs boundary algorithm is used, to handle the problem of muscles and flesh which are adjacent and connected to liver [8]. It uses **contrast stretching to stress the bones intensity values** in the image. A global threshold close to bones white colour, aids to extract the pixels of ribs boundaries. A thick black line is drawn to connect the pixels of the ribs bones. The ribs colour is changed into black colour. This way, a better connection between the ribs is guaranteed. Contrast stretching and structure filters are used to manipulate the problem of similarity between liver intensity values and other organs. Then average filter is used to smooth the little details of the liver like veins and ligaments to ease the next segmentation process.

3.2 *Region Growing Phase*

Region growing is one of the well known segmentation methods. Its simplicity and speedy performance make it worthy to improve. It starts with seed points representing the initial contour, and moves outward in the direction of the boundary of object. Region growing is implemented in this paper to extract the whole liver from the abdomen.

Algorithm (1) demonstrates the steps of liver segmentation proposed approach using region growing.

Algorithm 1 Liver segmentation using region growing

1: Clean the image by removing the annotations, machine bed and belly flesh. This is done using the morphological operations.
2: Apply **contrast stretching filter** to the image to deepen the edges of liver.
3: Connect ribs by tracing the white pixels of bones, drawing a line between them , and changing their white colour into black.
4: Apply **texture filter** to increase stressing liver edges. Combining texture filter with contrast stretching filter acts as edge segmentation.
5:
6: Segment the liver using region growing segmentation method, starting with seed points and moving towards the liver boundary.
7: Fill all holes inside the extracted liver, which may result from using texture filter.

3.3 *Level Set Phase*

Level set is one of region-base segmentation techniques. As we used it in the form of Caselles, which improved the geodesic active contour. Level set method is improved using the two proposed preprocessing steps of morphological operations and Wolf local thresholding. These two steps deepen the boundaries of the liver before the image is passed to the segmentation process. Both act as an edge-base segmentation method, rather than a filtering technique. Algorithm (2) gives the steps of liver segmentation using level set.

Algorithm 2 Liver segmentation using level set

1: Pick up a set of seed points that acts as an initial contour. It can be only one seed point. But, the less the number of mask's seed points, the more the expensive calculations.
2: Use gradient and gaussian filter to compute feature image.
3: Determine the curve's narrow band which is a thick line or space around all curve's pixels.
4: Get curvature and use gradient descent to minimize energy.
5: Find a signed distance function according to equation (3). It is on the boundary of object when it equals to zero.
6: Evolve the curve outwards.
7: Repeat the second step until obtaining the segmented liver.

4 Experimental Results and Discussion

The experiments on the proposed approach have been applied using 47 real CT images of patients in the first phase of CT scan before patient injection of contrast materials. Region growing and Caselles level set methods are compared to the proposed approach. The resultant image is compared to a manual segmented image of liver. Figure 2 shows the result of the algorithm. Figure 3 shows the cleaned image

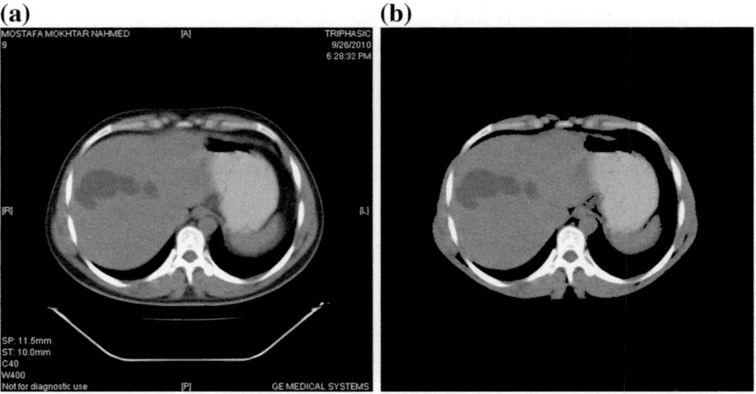

Fig. 2 Cleaning image annotations **a** original image **b** cleaned image

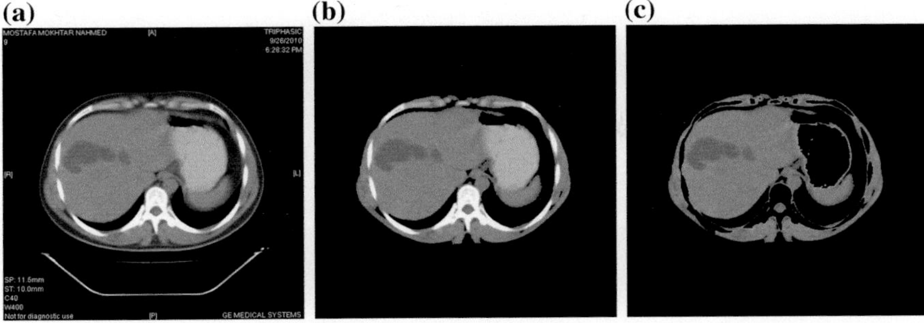

Fig. 3 Cleaning image and connecting ribs in *black* **a** original image **b** cleaned image **c** connected ribs

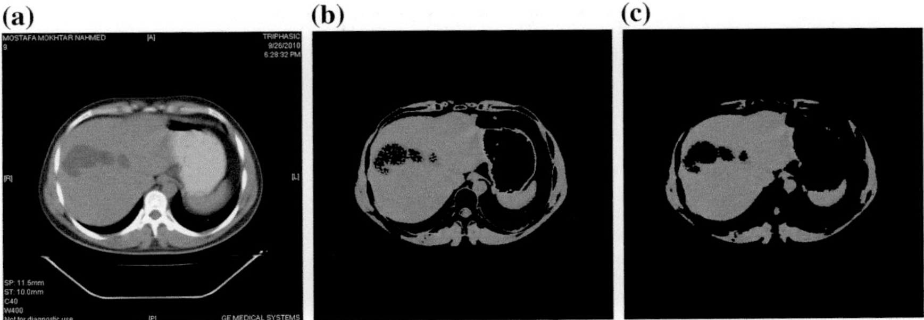

Fig. 4 Adjust contrast and texture filters to stress the boundaries **a** Original image, **b** Contrast stretching filter, **c** Texture filter

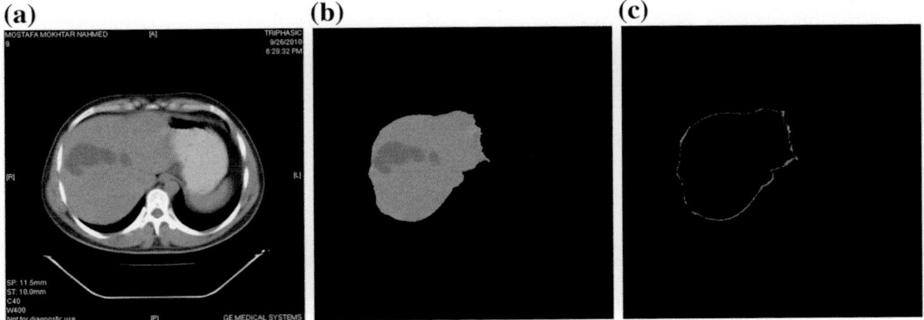

Fig. 5 Segmented image compared to the annotated image **a** Original image, **b** Segmented image, **c** Difference between segmented and annotated image

and the result of connecting ribs, changing its ribs' white colour into black. Figure 4 shows the result of the process of connecting ribs using adjust contrast and texture filters.

Figure 5 shows the difference between the segmented image and annotated one. Figure 6 demonstrates some results of segmented liver.

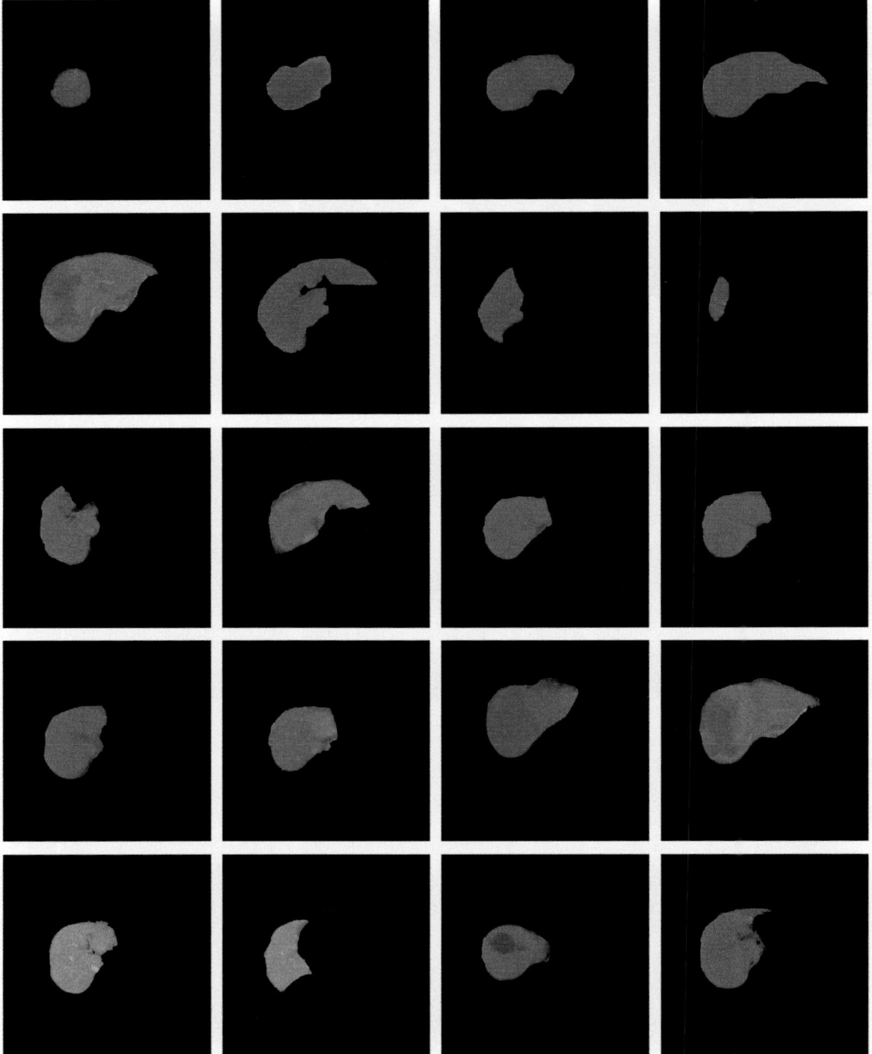

Fig. 6 Different segmented liver in different slices in CT images

Evaluation is performed using SI where SI is the similarity index defined using the following equations:

$$SI(I_{auto}, I_{man}) = \frac{I_{auto} \bigcap I_{man}}{I_{auto} \bigcup I_{man}} \tag{2}$$

where I_{auto} is the binary automated segmented image, resulting from the phase of segmentation of the whole liver in the used approach and I_{man} is the binary image, segmented manually by a radiology specialist.

The columns in Table 1 are used as follows: (M1) represents the results of region growing method. (M2) for enhanced region growing using morphological operations. (M3) represents the level set. (M4) enhanced level set using morphological operations. (M5) represents the enhanced level set using Wolf local thresholding. Table 1 shows the detailed result of different approaches for every image of the tested dataset.

According to Table 2, the average performance of liver images segmentation is improved using the proposed approach. Segmentation using region growing has an average result of SI = 84 %, which is improved using morphological operations to

Table 1 Results of tested approaches

Image	M1	M2	M3	M4	M5	Image	M1	M2	M3	M4	M5
1	0.881	0.942	0.961	0.980	0.968	2	0.927	0.936	0.904	0.920	0.928
3	0.831	0.876	0.928	0.960	0.922	4	0.740	0.782	0.880	0.950	0.904
5	0.868	0.903	0.910	0.960	0.922	6	0.849	0.938	0.890	0.930	0.930
7	0.956	0.945	0.890	0.940	0.922	8	0.934	0.936	0.908	0.940	0.880
9	0.953	0.942	0.948	0.970	0.962	10	0.942	0.928	0.950	0.950	0.938
11	0.923	0.905	0.970	0.950	0.973	12	0.939	0.924	0.930	0.930	0.963
13	0.890	0.866	0.920	0.930	0.945	14	0.893	0.880	0.930	0.930	0.936
15	0.757	0.772	0.960	0.930	0.949	16	0.917	0.959	0.890	0.930	0.932
17	0.902	0.946	0.960	0.900	0.957	18	0.919	0.943	0.950	0.920	0.957
19	0.938	0.950	0.920	0.950	0.947	20	0.910	0.937	0.900	0.920	0.900
21	0.898	0.933	0.915	0.940	0.940	22	0.622	0.891	0.880	0.940	0.912
23	0.536	0.897	0.870	0.950	0.880	24	0.880	0.872	0.940	0.930	0.941
25	0.708	0.939	0.918	0.950	0.953	26	0.660	0.843	0.930	0.900	0.892
27	0.848	0.950	0.888	0.928	0.945	28	0.616	0.807	0.920	0.932	0.950
29	0.696	0.911	0.870	0.984	0.890	30	0.682	0.927	0.920	0.935	0.928
31	0.820	0.904	0.910	0.954	0.906	32	0.913	0.954	0.970	0.920	0.941
33	0.930	0.941	0.940	0.960	0.960	34	0.921	0.950	0.960	0.930	0.946
35	0.908	0.923	0.950	0.950	0.954	36	0.920	0.948	0.946	0.933	0.920
37	0.947	0.957	0.930	0.930	0.926	38	0.941	0.935	0.910	0.910	0.930
39	0.942	0.936	0.919	0.860	0.890	40	0.866	0.915	0.934	0.888	0.896
41	0.909	0.943	0.915	0.920	0.920	42	0.888	0.952	0.862	0.873	0.951
43	0.856	0.902	0.870	0.883	0.949	44	0.855	0.901	0.866	0.890	0.946
45	0.913	0.901	0.901	0.910	0.955	46	0.671	0.881	0.909	0.950	0.951
47	0.661	0.855	0.905	0.960	0.942						
Result	0.849	0.912	0.918	0.932	0.933						

Table 2 Comparison of results of different tested approaches

ID	Method	Result
1	Region growing	0.849
2	Enhanced region growing using morphological operations	0.912
3	Level set	0.918
4	Enhanced level set using morphological operations	0.932
5	Enhanced level set using Wolf local thresholding	0.933

91.23 %. And level set segmentation has an average result of 91.8 %. This result is improved using the morphological operations to 93.19 and 93.30 % using Wolf local threshold.

5 Conclusion and Future Work

Liver segmentation is a complicated process, built on many steps for preprocessing and segmentation. The preprocessing is involved in separating and distinguishing liver region from the regions of similar intensities of other organs like spleen, flesh and muscles. The algorithm of combining ribs connecting combined contrast stretching, morphological operations and Wolf local thresholding, can help to handle this problem. Liver segmentation using region growing has an average result of SI = 84 %, which is improved using morphological operations to 91.23 %. And level set segmentation has an average result of 91.8 %. This result is improved using the morphological operations to 93.19 and 93.30 % using Wolf local threshold. Optimization techniques are a good trend that includes genetic algorithm, firefly algorithm, Wolf swarm, fish swarm, simulated annealing, and bacterial foraging. Our future work is to use these optimization techniques in segmentation and classification of the anomalies of liver.

References

1. Bandkman, I.N.: Handbook of Medical Imaaging Processing and Analysis, 2nd edn. Academic Press (2008)
2. Beichel, R., Bauer, C., Bornik, A., Sorantin, E., Bischof, H.: Liver segmentation in CT data: a segmentation refinement approach. In: Proceedings of 3D Segmentation in The Clinic: A Grand Challenge, pp. 235–245 (2007)
3. Dalitz, C., Niederrhein, H.: Soft thresholding for visual image enhancement, Fachbereich Elektrotechnik & Informatik (2014)
4. Gonzales, R.C., Woods, R.E.: Image Segmentation, in Digital Image Processing, 3rd edn. Pearson Printice Hall, New jersy (2008)

5. Huang, W., et al.: A semi-automatic approach to the segmentation of liver parenchyma from 3D CT images with extreme learning machine. In: 34th Annual International Conference of the IEEE EMBS (2012)
6. Jun, Z., Jinglu, H.: Image segmentation based on 2D Otsu method with histogram analysis. In: International Conference on Computer Science and Software Engineering (2008)
7. Li, X., Luo, S., Li, J.: Liver segmentation from CT image using fuzzy clustering and level set. J. Signal Inf. Process. **4**, 36–42 (2013)
8. Mostafa, A., Hefny, H., Ghali, N., Hassanien, A., Schaefer, G.: Evaluating the effects of image filters in CT liver CAD system. In: International Conference on Biomedical and Health Informatics. The Chinese University of Hong Kong, Hong Kong SAR, on 5–7 January 2012
9. Sherlock, S.: Sherlock's Diseases of the Liver and Biliary System, 12th ed. (2011)
10. Sherlock, S., Summerfield, J.: A Colour Atlas of Liver Disease. Wolfe, London (2013)
11. Wolf, C., Jolion, J., Chassaing, F.: Text localization, enhancement and binarization in multimedia documents. Pattern Recognition International Conference **2** (2002)
12. Wu Yi, S., Cheng, X.G., Za Zhi, X.:Three-dimensional CT liver image segmentation based on hierarchical contextual active contour, J. Biomed. Eng. (2014)

Semi-automated System for Cup to Disc Measurement for Diagnosing Glaucoma Using Classification Paradigm

Taras Kotyk, Sayan Chakraborty, Nilanjan Dey, Tarek Gaber, Aboul Ella Hassanien and Vaclav Snasel

Abstract Recently, Glaucoma has become one of the major retinal diseases. In order to detect such retinal diseases, cup to disc ratio measurement is a vital index of Glaucoma, as the Glaucomatous neuropathy increases the *cup to disc ratio* when the excavation of the optic cup is increased. In this paper, a semi-automated system to detect both of optic cup and optic disc and to measure cup to disc ratio has been proposed. The proposed system firstly, uses an object detection function from red channel of the retinal images. Then further using threshold values, the optic cup and optic disc are detected. Although, for several images manual tuning is needed as the object detection function as well as the threshold value fail to detect the optic cup and optic disc correctly. The manually tuned images and the automatically detected images are further used to determine the error in the system which leads to the categorizing of the images. These images are later post-processed using Haralick texture features. Haralick texture features' obtained values are trained using back

T. Kotyk
Department of Human Anatomy, Ivano-Frankivsk National Medical University, Ivano-Frankivs'k, Ukraine

S. Chakraborty · N. Dey
Department of CSE, Bengal College of Engineering, Durgapur, India

T. Gaber (✉)
IT4Innovation, VŠB-TU of Ostrava, Ostrava, Czech Republic
e-mail: tmgaber@gmail.com

A.E. Hassanien
Faculty of Computers and Information, Cairo University, Cairo, Egypt

T. Gaber · V. Snasel
Faculty of Computers and Informatics, Suez Canal University, Ismaïlia, Egypt

V. Snasel
Department of Computer Science and IT4Innovations, VŠB-TU of Ostrava, Ostrava, Czech Republic

A. Abraham et al. (eds.), *Proceedings of the Second International Afro-European Conference for Industrial Advancement AECIA 2015*, Advances in Intelligent Systems and Computing 427, DOI 10.1007/978-3-319-29504-6_60

propagation neural network to determine the system's accuracy. The proposed system was evaluated using RIM-ONE database. By increasing the absolute error, system's accuracy is evaluated. The proposed system's accuracy is 86.43 % at 0.5 error value.

Keywords Cup to disc ratio • Optic cup • Optic disc • Haralick features • Glaucoma detection • Back-propagation neural network

1 Introduction

Ophthalmology has become an integral part of tele-medicine recently. Ophthalmology refers to the medicine that is mainly used to deal with the diseases of the retina or the eye. Glaucoma is considered one of the major causes of blindness around the world. It causes blindness for approximately 13 % of visually impaired people. Physiologically, Glaucoma is recognized by the deterioration of optic nerve cells. It is one of the well-known diseases that have given a lot of trouble to the researchers working in ophthalmology. Glaucoma [1] can be detected using cup-to-disc ratio measurement. The eye's blind spot is often called as the optic disc. Optic disc [2] can also be said as the entrance path of the optic nerve and blood vessels. Optic disc is generally flat or cupped. Glaucoma [3] causes extra intraocular pressure on optic disc which leads to additional pathological cupping inside the optic disc. No nerve fibers are located inside the white cup. Glaucoma [4, 5] causes the cup to enlarge until disc area is completely occupied. Hence, cup to disc ratio measurement can lead to glaucoma detection. The cup to disc ratio (CDR) refers to the proportion of the disc [5] contained by the cup. It's an index to measure the glaucoma. 0.3 to 0.5 should be the CDR for normal eye. Researchers [3, 4] have found that that increment of CDR refers to the retina containing glaucoma. If the CDR [6] is 0.8 then vision is completely lost. It's quite clear that CDR is one of the most efficient techniques for glaucoma detection and hence an automated system is heavily required to measure CDR [7] which will ultimately lead to assessment of glaucoma.

In this current paper, a semi-automated framework has been proposed which will lead to measurement of cup to disc ratio using optic cup and optic disc [8] detection. The proposed system has two main parts, the first part would detect the optic disc and cup automatically, whereas the other part will require medical expertise to manually tune the optic disc [9] and optic cup detection. The results of these are then assessed to find out the error which helps to categorize the image. In order to check system's accuracy haralick [10, 11] texture features are applied to those categorized images. The results of haralick texture features are trained using back-propagation neural network.

The rest of this paper is organized as follows: Sect. 2 gives an overview about the related work and Sect. 3 presents the proposed system. The results and the discussion are reported in Sect. 4. The paper is concluded in Sect. 5.

2 Related Works

Extensive research has been done in order to detect glaucoma as well as to measure cup to disc ratio in various ways. Earlier in 1998, Garway-Heath et al. studied the relation between optic disc [8] and vertical cup to disc ratio. In this work, total 200 images were assessed to determine vertical cup to disc ratio and optic disc and relation between them were established using linear regression. In 2002, Beaumont and Kang discussed several topics on glaucoma, such as cup to disc ratio [7] (CDR), intraocular pressure, etc. In this work, they proposed various techniques to detect glaucoma from retinal images and presented a comparative study among them. Crowston et al. in 2004 carried out a blue mountain eye study [9] to determine optic disc's diameter on vertical cup to disc ratio. A large data-pool of 6678 was used in this study to observe the changes of optic disc diameter. In 2009, Wong et al., proposed an ARGALI [1] based cup to disc ratio measurement for glaucoma detection. In this paper, ARGALI that is an automated system for cup to disc ratio measurement was used and also SVM [12, 13] was used as an alternative of ARGALI to determine CDR. Muramatsu et al. proposed an automated system for cup to disc ratio [10] measurement using image segmentation. This work also required help from medical experts to segment the images manually. In 2012, Murthi and Madheswaran [4] proposed a novel cup to disc ratio enhancement detection technique. In their proposal, they mainly focused on enhancing the detection technique of CDR using core components of ARGALI. In the same year, Dey et al. proposed another CDR technique [5] using Harris corner detection. This work included Harris corner detection technique to assess glaucoma using CDR from fundus retinal images. Later, in 2012 also, Fondon et al. introduced an automated [5, 6] system using Active Contours and Color Clustering to detect cup to disc ratio. In this study, ground truth images were provided by an expert ophthalmologist to compare the results of the automated system. In 2014, Salazar-Gonzalez et al. proposed a new technique of blood vessel segmentation and optic disk from the fundus retinal images. This method was applicable on modern ophthalmology as the morphology of the blood vessel and optic disk is an important indicator for several retinal diseases. In 2012, Malek et al., introduced a novel [2, 11] method automatically locate and boundary detect of the optic disk in two phases.

From the discussion above, it is quite clear that most of the CDR measurement techniques required medical expertise in order to determine the system's robustness. In this paper, cup to disc ratio has been measured in a novel way that is presented in Sect. 3.

3 Semi-automated System for Optic Cup and Disc Detection

The proposed system has two stages: automatic detection and semi-automatic detection. As it will be explained below, the automatic detection is performed first and if it is failed, the semi-automatic (with help of the ophthalmologist) will be executed.

3.1 Automated Detection

In this current work, RIM-ONE database [14] is used. In the proposed framework, as shown in Fig. 1, the RGB color images are firstly loaded. Red, green and blue channels are then separated from the RGB images afterwards. Green channel is doubled to produce double green channel images (DGCI). An exponential transformation is performed on DGCI images. After erosion, dilation and using a median filter (with radius 1), the DGCI images are converted into binary images.

Afterwards, red channel of the images are enhanced using histogram equalization. Enhanced red channel of those images are converted into binary and further optimized by smoothening of detected boundaries (erode and dilate, use median filter (with radius 1). The proposed framework uses roundness >0.7 of largest size particles as the object detection function. Using object detection function, optical discs are detected. If the largest particle has circularity is less than 0.10, then the image is inverted, and the object detection function starts one more time on inverted images. If function can't find particle with roundness greater or equal to 0.7 or having circularity less than 0.15 then on enhanced red channels an exponential transformation is performed. Further, those images are binarized, eroded, dilated along with applying hole-filling function, median filter (with radius 1). The object detection function is applied one more time on these optimized inverted enhanced red channels.

3.2 Semi-automated Detection

If the optic disc can't be detected even after inverting the previous image, then manual tuning is needed. Manual tuning [15] refers to human interacting to the proposed system which uses the draggable points located on the fundus image [16] to tune the obtained optic disc. This makes the system a semi-automated framework. If the optic disc is detected from those images then also the boundaries are not proper ellipse. Using fit-ellipse function this problem is solved. Coordinates of these ellipses are stored in arrays (x_1, y_1). Then, boundaries of optical disc on DGCI are restored. Outside of the optic disc is cleared to prevent gaining additional

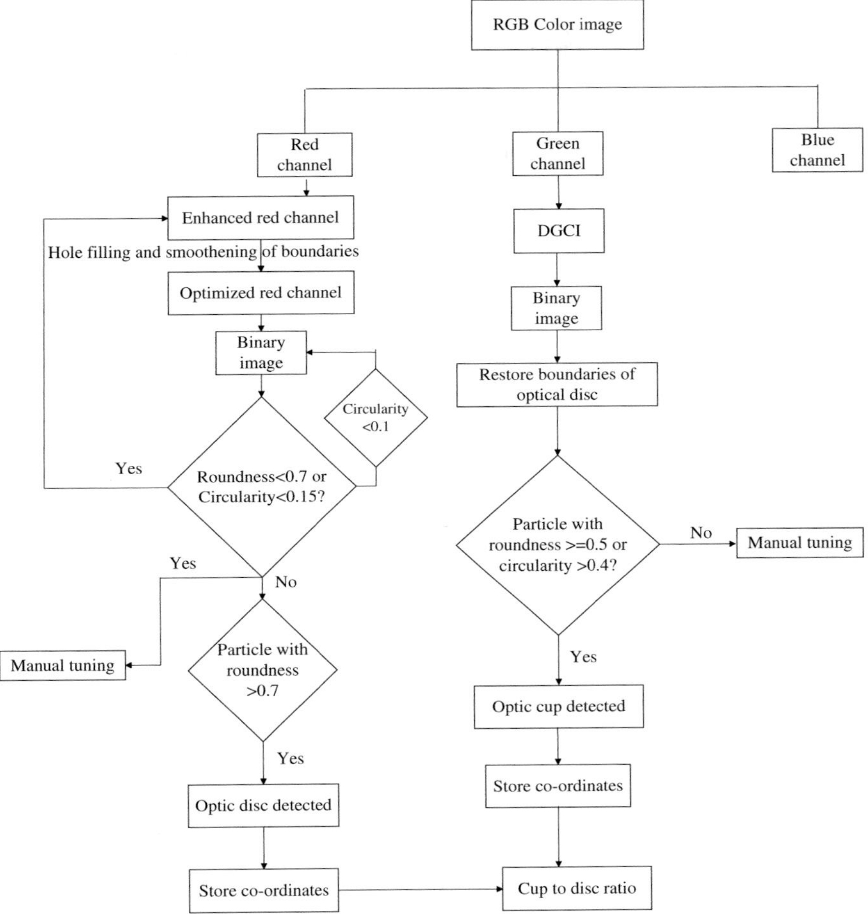

Fig. 1 Block diagram of the automated system

boundaries of *optic cup* that are located outside of optical disc [4] boundaries. If any object having the largest particle with roundness greater than 0.5 and circularity greater than 0.4 then optic cup is detected in those images. If it can't detect optic cup, then manual tuning is needed. If the framework succeeds, then the coordinates are stored in arrays (x_2, y_2). Centroid of cup is measured by the coordinates of point (x_0, y_0), which are mean values of x_2 and y_2, respectively. Distance between centroid to all points of cup [7] boundaries is further measured. Cup diameter is calculated as double radius of optic cup. The same steps are used to measure diameter of the optic disc. Cup to disc ratio is calculated as the diameter of cup to diameter. The framework is shown in Fig. 1.

3.3 Performance Evaluation Using Classification Paradigm

After the calculation of cup to disc ratio, detection of optic disc and optic cup, the system uses three medical experts' manually tuned resultant images as ground truth. Mean values of the ground truth results are calculated and the mean values are subtracted from the automated system's results. The obtained values are termed as absolute error in this work. According to the error values the retinal images are categorized as "good quality images (having error less than 0.1) for automatic detection" and "degraded images (having error greater than or equals to 0.1)". For different absolute error values (varies from 0.1 to 0.5) the images [8] are categorized. Haralick texture features are detected from those images. The obtained results of Haralick texture features are further post-processed using back propagation neural network, for different error values to determine the accuracy of the proposed system.

4 Results and Discussion

The proposed semi-automated framework is implemented using ImageJ version 1.49von a CPU Intel core i5th generation 2.7 GHz processor based system with Mac OS X 10.10.3 operating system. The performance evaluation is done using MATLAB R2012a with 2.20 GHz Intel i3rd generation computer with Microsoft Windows 7 operating system. As previously discussed, at first using the automated system the retinal images were processed and optic disc as well as the optic cup were found. Figure 2a, b show the original image and optic cup respectively. Figure 2c, d shows the optic disc and *optic cup* along with the disc. This detection was done using the proposed algorithm.

Figure 3 shows the effect of the proposed algorithm on good images. Figure 3a, d are the original images. Figure 3b, e shows the *optic cup* and optical disc detected. The green circle is the *optic cup* and the blue circle is the optical disc. In Fig. 3c, f, red, green and blue circles indicate *optic cup* and optical disc detected by three experts.

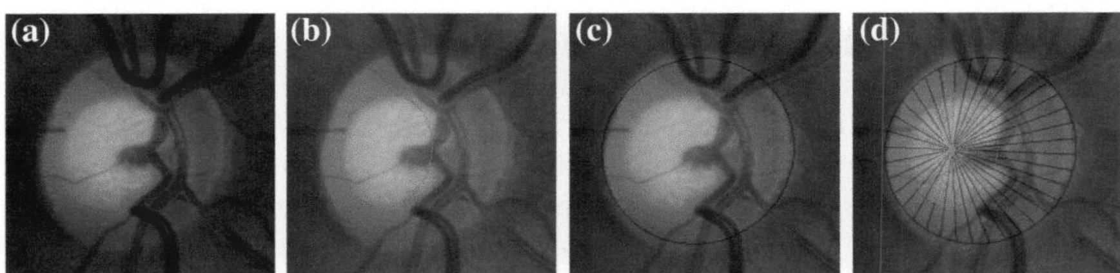

Fig. 2 Detection of *optic cup* and optical disc by algorithm, **a** original image, **b** *optic cup*, **c** optical disc, **d** *optic cup* and disc

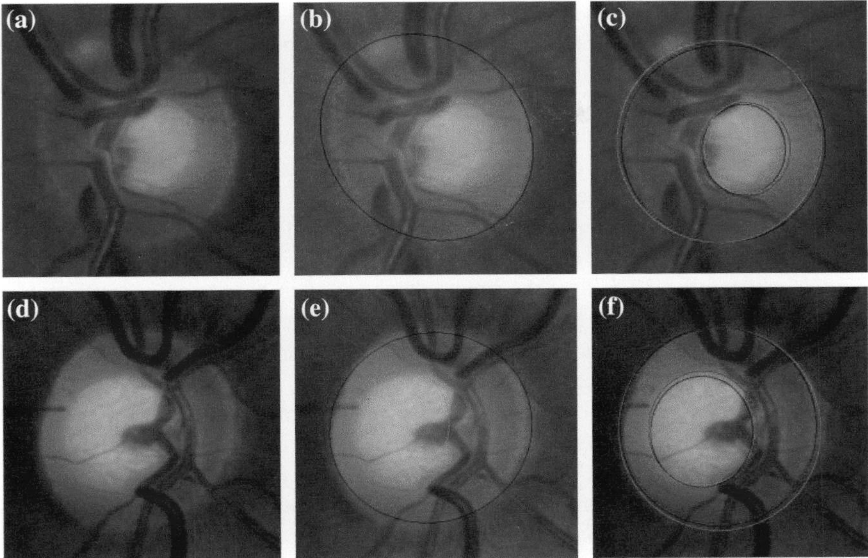

Fig. 3 Results of algorithm on a "good images", **a**, **d** original images; **b**, **e** *optic cup* (*green*) and optical disc (*blue*) detected by algorithm; **c**, **f** *optic cup* and optical disc detected by three experts

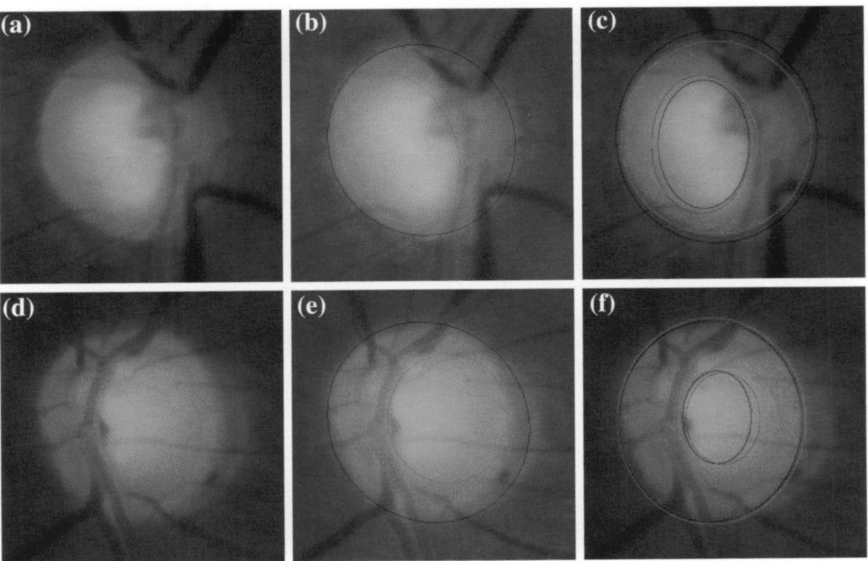

Fig. 4 Results of algorithm on a "degraded images", where boundaries of cup and disc are not clear **a**, **d** original images; **b**, **e** *optic cup* (*green*) and optical disc (*blue*) detected by algorithm; **c**, **f** optic cup and disc detected by three experts

The effect of the proposed algorithm on bad images is shown in Fig. 4. Green and blue circles in Fig. 4b, e indicates optic cup and optical disc respectively. The red, green and blue circles in Fig. 4c, f refers to the optic cup and disc detected by experts.

Figure 5 shows the visual representation of optic cup and disc detected by the proposed system. In this work, the error was calculated using the obtained result and the mean result of three experts.

Figure 6a is the visual representation of the values of absolute error between the results obtained by proposed algorithm and the mean results of three experts. Figure 6 shows that 37.5 % of the image database were processed correctly when the error (= result obtained by proposed algorithm − mean result obtained by three experts). 24.5, 13.5, 7, 4 and 12.5 % were processed correctly for errors with 0.2,0 0.3, 0.4, 0.5 and 0.6 respectively.

Figure 6b shows cup to disc ratio for each image that was used in the proposed framework. The dataset were sorted according to the error which was shown in Fig. 6a. In Fig. 6b, some of the images are found to be outside the graph are along the x-axis. Those images are the cases where the proposed system was unable to

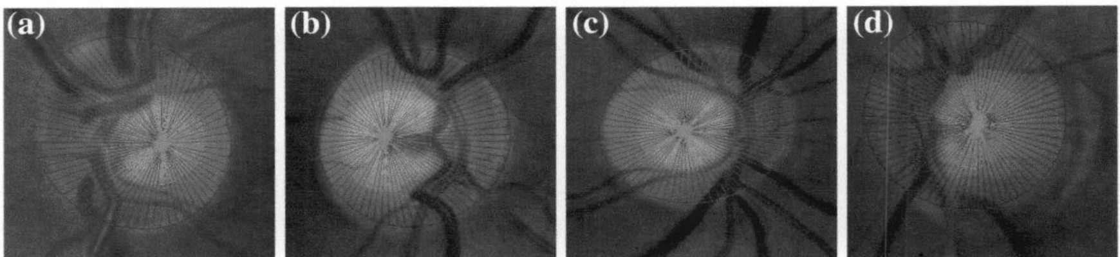

Fig. 5 Visual representation of *optic cup* and disc

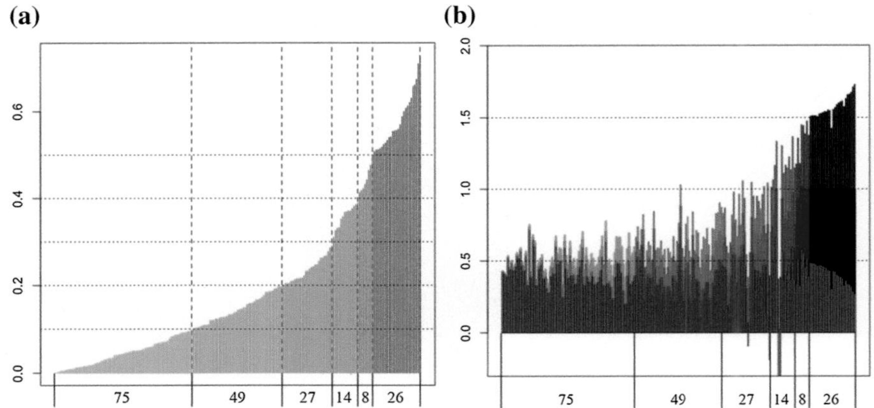

Fig. 6 No. of images versus **a** absolute error, **b** cup to disc ratio with error

detect optic cup or optic disc correctly. In those cases, the absolute error (cup to disc ratio calculated by automated system—cup to disc ratio calculated by experts) was below 0. Hence, in graph the plots of those images are below the zero value of x-axis. These cases are shown in Fig. 7. The four images shown in Fig. 7 had CDR −0.237, −0.29, −0.3531 and −0.357.

In order to make the system more robust the proposed framework was further modified by a classification system. In this system back propagation neural network with scaled conjugate algorithm was used. As previously noted the images were categorized as good quality images and degraded images with error value taken as 0.1. Upon classification of the good quality images and degraded images using Haralick features and back propagation network (80 % training, 10 % testing and 10 % validation) it was observed that the accuracy of the system was 62.81 %, which is pretty low. In order to validate further the error was increased to 0.5. The accuracy results that are shown in Table 1 indicate that system accuracy increase proportionally with error value. Table 1 shows that highest accuracy was gained for error 0.5 value, whether the system was least accurate when the error value was 0.1. From Table 1 it can be said that when the error is high then the system is more accurate. Higher error indicates the set of good quality images and degraded images being more distinct from each other.

Hence, it will be safe to conclude that if the dataset contain distinct images of the retina then the system accuracy will be higher and the semi-automated system will be more robust.

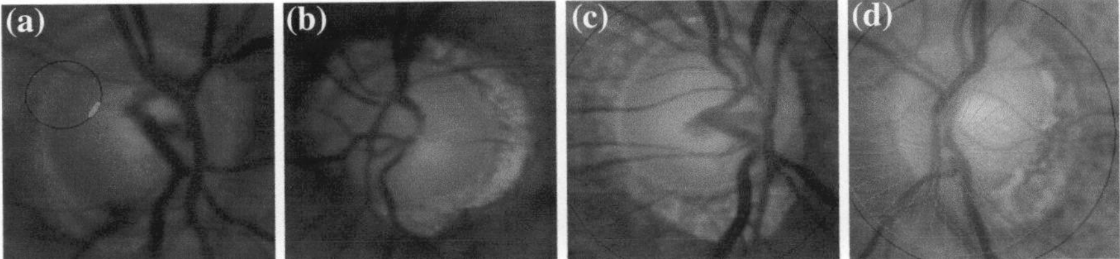

Fig. 7 Images having error less than zero

Table 1 Accuracy measurement using BPNN

Error	Accuracy (ANN-SCG-80-10-10) (%)
0.5	86.43
0.4	82.91
0.3	75.87
0.2	70.35
0.1	62.81

5 Conclusion

This paper proposed a framework (or a system) to detect the optic cup and optic disc. Firstly, it is automatically detected the optic cup and optic disc and if failed based on a help from three medical expert's, the images were manually tuned to draw the optic cup and optic disc. The results of automated and semi-automated system were subtracted and the obtained values were used to categorize the images. Further, Haralick features were applied on those categorized images. The obtained results were post-processed using a neural network to assess the system's accuracy. Results showed that having higher errors made the system more accurate. Hence, it can be said that the more distinct the images, the better accuracy is provided by the system. In future, making the system fully automated with more robustness can be a good topic of research. Future work may also include using different texture features to determine system's accuracy. Using other glaucoma detection technique or changing the threshold in various steps of this algorithm can open a wide area of research in this domain.

Acknowledgments This paper has been elaborated in the framework of the project "New creative teams in priorities of scientific research", reg. no. CZ.1.07/2.3.00/30.0055, supported by Operational Programme Education for Competitiveness and co-financed by the European Social Fund and the state budget of the Czech Republic and supported by the IT4Innovations Centre of Excellence project (CZ.1.05/1.1.00/02.0070), funded by the European Regional Development Fund and the national budget of the Czech Republic via the Research and Development for Innovations Operational Programme and by Project SP2015/146 Parallel processing of Big data 2 of the Student Grant System, VSB Technical University of Ostrava.

References

1. Wong, D.W., Liu, J., Lim, J.H., Tan, N.M., Zhang, Z., Lu, S., Li, H., Teo, M.H., Chan, K.L., Wong, T.Y.: Intelligent fusion of cup-to-disc ratio determination methods for glaucoma detection in ARGALI. In: Conference Proceedings IEEE Engineering in Medicine and Biology Society, vol. 2009, pp. 5777–5780 (2009)
2. Salazar-Gonzalez, A., Kaba, D., Yongmin, L., Xiaohui, L.: Segmentation of the blood vessels and optic disk in retinal images. IEEE J. Biomed. Health Inf. **18**(6), 1874–1886 (2014)
3. Malek, J., Ben Abdallah, M., Mansour, A., Tourki, R.: Automated optic disc detection in retinal images by applying region-based active contour model in a variational level set formulation. In: 2012 International Conference on Computer Vision in Remote Sensing (CVRS), 16–18 Dec 2012, Xiamen, pp. 39–44 (2012)
4. Murthi, A., Madheswaran, M.: Enhancement of optic cup to disc ratio detection in glaucoma diagnosis. In: International Conference on Computer Communication and Informatics (ICCCI), pp. 1–5 (2012)
5. Dey, N., Roy, A.B., Das, A., Chaudhuri, S.S.: *Optic cup* to disc ratio measurement for glaucoma diagnosis using harris corner. In: 2012 Third International Conference on Computing Communication & Networking Technologies (ICCCNT), 26–28 July 2012, pp. 1–5 (2012)
6. Fondón, I., Núñez, F., Tirado, M., Jiménez, S., Alemany, P., Abbas, Q., Serrano, C., Acha, B.: Automatic cup-to-disc ratio estimation using active contours and color clustering in fundus images for glaucoma diagnosis. Image Anal. Recogn. **7325**, 390–399 (2012)

7. Beaumont, P.E., Kang, H.K.: Cup-to-disc ratio, intraocular pressure, and primary open-angle glaucoma in retinal venous occlusion. Ophthalmology **109**(2), 282–286 (2002)
8. Garway-Heath, D., Ruben, S., Viswanathan, A., Hitchings, R.: Vertical cup/disc ratio in relation to optic disc size: its value in the assessment of the glaucoma suspect. Br. J. Ophthalmol. **82**(10), 1118–1124 (1998)
9. Crowston, J.G., Hopley, C.R., Healey, P.R., Lee, A., Mitchell, P.: The effect of optic disc diameter on vertical cup to disc ratio percentiles in a population based cohort: the Blue Mountains Eye Study. Br. J. Ophthalmol. **88**(6), 766–770 (2004)
10. Muramatsu, C., Nakagawa, T., Sawada, A., Hatanaka, Y., Yamamoto, T., Fujita, H.: Automated determination of cup-to-disc ratio for classification of glaucomatous and normal eyes on stereo retinal fundus images. J. Biomed. Opt. **7260**, 72603L1–72603L5 (2011)
11. Samanta, S., Ahmed, S.S., Salem, M-A., Nath, M.M., Dey, S.S., Chowdhury, N.: Haralick features based automated glaucoma classification using back propagation neural network. In: International Conf on Frontiers of Intelligent Computing: Theory and Applications (FICTA 2014), vol. 327, pp 351–358 (2015)
12. Tharwat, A., Gaber, T. Hassanien, A.E.: Cattle identification based on muzzle images using Gabor features and SVM classifier. In: Advanced Machine Learning Technologies and Applications. Springer International Publishing, vol. 488, pp. 236–247 (2014)
13. Tharwat, A., Gaber, T. Hassanien, A.E., Hassanien, H.A., Tolba, M.F.: Cattle Identification using muzzle print images based on texture features approach. In: Proceedings of the Fifth International Conference on Innovations in Bio-Inspired Computing and Applications IBICA 2014. Springer International Publishing, vol. 303, pp. 217–227 (2014)
14. http://medimrg.webs.ull.es/download/602/
15. Chakraborty, S., Acharjee, S., Bose, S., Mukherjee, A., Dey, N.: A semi-automated system for optic nerve head segmentation in digital retinal images. In: 2014 International Conference on Information Technology (2014)
16. Dey, N., Pal, M., Das, A.: A session based watermarking technique within the NROI of retinal fundus images for authentication using DWT, spread spectrum and Harris corner detection. Int. J. Modern Eng. Res. **2**(3), 749–757 (2012)

Part V
Special Session on Data Processing, Protocols, and Applications in Wireless Sensor Networks

Big Data Pre-processing Techniques Within the Wireless Sensors Networks

Mohamed Mostafa Fouad, Tarek Gaber, Maamoun Ahmed, Nour E. Oweis and Vaclav Snasel

Abstract The recent advances in sensors and communications technologies have emerged the interaction between physical resources and the need for sufficient storage volumes for keeping the continuously generated data. These storage volumes are one of the components of the Big Data to be used in future prediction processes in a broad range of fields. Usually, these data are not ready for analysis as they are incomplete or redundant. Therefore one of the current challenge related to the Big Data is how to save relevant data and discard noisy and redundant data. On the other hand, Wireless Sensor Networks (WSNs) (as a source of Big Data) use a number of techniques that significantly reduce the required data transmissions ratio. These techniques not only improve the operational lifetime of these networks but also raise the level of the refinement at the Big Data side. This article gives an overview and classifications of the data reduction and compression techniques proposed to do data pre-processing in-networks (i.e. in-WSNs). It compares and discusses which of these techniques would be adopted or modified to enhance the functionality of the WSNs while minimizing any further pre-processing at the Big Data side, thus reducing the computational and storage cost at the Big Data side.

M.M. Fouad (✉)
Arab Academy for Science, Technology, and Maritime Transport, Cairo, Egypt
e-mail: mohamed_mostafa@aast.edu
URL: http://www.egyptscience.net

M.M. Fouad · T. Gaber · N.E. Oweis · V. Snasel
IT4Innovations, VŠB-Technical University of Ostrava, Ostrava, Czech Republic

T. Gaber
Faculty of Computers and Informatics, Suez Canal University, Ismailia, Egypt

N.E. Oweis · V. Snasel
FEECS, Department of Computer Science, VŠB-TUO, Ostrava, Czech Republic

M. Ahmed
Aqaba University of Technology (AUT), Aqaba, Jordan

M.M. Fouad · T. Gaber
Scientific Research Group in Egypt (SRGE), Cairo, Egypt

A. Abraham et al. (eds.), *Proceedings of the Second International Afro-European Conference for Industrial Advancement AECIA 2015*, Advances in Intelligent Systems and Computing 427, DOI 10.1007/978-3-319-29504-6_61

Keywords Big Data · Wireless Sensor Networks · Data compression · Data reduction

1 Introduction

Wireless Sensor Networks (WSNs) are characterized by its low-cost, and auto-configuration capabilities. Therefore, they applied for ambient monitoring in a number of modern applications such as: agricultural management, smart building, health care monitoring, and smart homes [1]. These applications produce a huge amount of a non-stop streaming data which poses different challenges due to the WSNs storage restriction and their limitations in power and computational capabilities [2]. Other challenges related to the WSNs application include: time-sensitive applications. For example, in industrial control, automotive or aerospace systems, a WSNs system not only depends on capturing data but also on the time at which results can be produced [3]. The expected generated data from these WSNs applications make the WSNs a very crucial source of the Big Data. The previously mentioned WSNs limitations along with the potential benefits of using such huge amount of generated data, pose challenges for researchers and developers to efficiently design specific protocols to minimize the size and the number of messages required to be exchanged among the sensor nodes [4]. In particular, data reduction and compression techniques could have possible solutions for these limitations. However, their traditional techniques could not directly applied to the WSNs applications. Hence, a modified or a new data reduction and compression techniques for WSNs applications should be proposed.

This article gives an overview and classifies the data reduction and compression techniques proposed to achieve data pre-processing in-networks. It then discusses which of these techniques would enhance the functionality of the WSNs while minimizing any further pre-processing at the Big Data side, thus reducing the computational and storage cost at the Big Data side.

The remainder of this article is organized as follows. A brief introduction of the Big Data is introduced in Sect. 2 while the WSNs-based data compression techniques are given in Sect. 3. Section 4 presents the WSNs-based data reduction techniques, and finally Sect. 5 highlights the main concludes of the article.

2 The Big Data

The big amount of the data and the variety of the input data source, e.g. WSNs, social networks, cloud servers, web server logs, produce a different level of heterogeneity in data formats, structure, and meaning. The traditional database architecture and its processing approaches become inapplicable to deal with this too fast streaming and huge unstructured amount of data. Therefore, there is a need for alternative cost-effective approaches to extract values from such huge amount of data. Manipulating, managing, and analyzing this volume, variety and different level of heterogeneity have emerged the paradigm of "Big Data". This new paradigm has changed the perspective of a number of fields including military, business, medical, etc. [5, 6].

Although the classification of Big Data techniques are laying under the artificial intelligent and specially under the machine learning techniques, it is not intended to learn the machine how to think like human. These techniques are used to predict and extract hidden information from a huge amount of structured or unstructured data [5]. In other words, any prediction approach should characterize different aspects of the Big Data. Such aspects are identified through the *6V's* characteristics (Volume, Velocity, Variety, Veracity, Viability, and Value) [6]. The growing aspects of the Big Data, as given in Table 1, has also prompted a series of input sources, storage platforms and management software. To be able to utilize these infrastructures and achieve the features of the Big Data, proper pre-processing approaches should be applied to clean redundancy, biases, noisy and abnormality within the data.

The Big Data analytics are vital processes to get new values and relations from this Big Data. However, this analytics faces many challenges. One of these challenges is how to effectively store this amount of data and remove the redundant one. For the big data collected from WSNs, this paper aims to give an overview and discuss a number of the state-of-art of the data processing techniques (data compression and reduction) designed for WSNs to enhance the performance of these networks. However, as source of the Big Data in many applications, these techniques could also help the Big Data side by doing the pre-processing of the data in-network (i.e. within the WSNs), thus reducing the storage space and improving the analysis performance of the Big Data.

Table 1 Different aspects of the Big Data Paradigm

Applications aspect, (notable uses)	Medical and health care, telecommunication, financial, industrial, web, social network, search engine, marketing, science and engineering
Management aspect, (Processing)	+ **Platforms**: IBM, SAP, Microsoft, Intel, INFOBRIGHT, Hortonwork, Kognitio, ORACLE, and Amazon. + **Extract, Analysis and Visualize**: NoSQL, MapReduce, Hadoop, and R Language, + **Storage**: DAS, NAS, SAN, Cloud, Data Center
Pre-processing aspect	Mining, fusion, compression, reduction, etc.
Data aspect, (data types)	Structure, semi-structure, unstructured, or multi-structure
Input aspect, (Data sources)	WSN, IoT, social network, search engine, biomedical, mobile, RFID, NFC, machine log data, cloud computing, smart phone, smart city, and smart poster

3 Data Compression over WSNs

The main operations consuming energy in the WSNs are: the sensing, processing, and communication. Therefore, any technique, without trading off the performance of the whole wireless network, can directly or indirectly reduces the complexity of one or more of these operations, is highly required for the WSNs applications. Data compression techniques are of these techniques which could improve the WSNs performance. However, most of the traditional compression techniques are not suitable for the WSNs due to the resources limitations of such networks [7].

Recently, there has been an increase in the number of data compression algorithms to deal with reducing data exchanged between sensor nodes and thus reducing the power consumption of the WSN nodes. Srisooksai et al. [8] expressed the hardware constrained related to the sensor node and then based on the applications, classified the data compression algorithms into two main categories: distributed and local data compression approaches. While the distributed data compression approaches are appropriate for dense networks, the local approaches are most suitable for sparse networks where data compression is performed at the node side without any collaboration with other sensor nodes.

Usually the distributed data compression approaches suit well for the dynamic data change over time occurred within the monitored environment. The survey in [8] had generally classified the distributed compression approaches into four main techniques: Distributed Source Modeling (DSM), Distributed Transform Coding (DTC), Distributed Source Coding (DSC), and Compressed Sensing (CS) techniques.

The DSM technique searches for a model that suit the required physical measurements acquired by a specific group of sensor nodes. An example of the DSM is work done by Zhang et al. [9], where they proposed a cloud computing-based architecture for WSNs. The architecture choses a number of nodes (virtual sinks) to form a cloud computing platform managed by Hadoop. Therefore all sensed data are compressed and processed in distribution manner. Although this architecture increases the ratio of packets successfully reaching the sink, reduces the number of end-to-end hops, and minimizes the energy consumption, it may not work probably if the selection of the cloud nodes was not chosen carefully.

The DTC techniques convert the readings by the sensors into coefficients based on their individual characteristics. Usually, these approaches are suitable for multimedia WSNs. The work done by Gaeta et al. [10] is the best example of these approaches, where they proposed a robust local transform-based scheme that based on the fuzzy transform (F-transform) with a wavelet based scheme. The F-transform is known for its suitability for video and image based WSNs [11, 12]. As these approaches are suitable for large sized data types (e.g. videos and images), they would be efficient to dealing with the diversity data formats of the Big Data paradigm.

Distributed Source Coding (DSC): It is about transmitting a single sensor signal where the signals of other closely related sensors are compressed according to the spatial correlation to that main signal. The performance of this type of compression is relevant to the stability of the sensor readings therefore it is more suitable for correlation patterns [13]. The Compressive Distributed Sensing (CDS) scheme [14] is an example of the DSC approach where the data gathering and projection generating processes are integrated. The authors had also extended their scheme to support lossy compression for error fault tolerance.

For local data compression approaches, usually they are categorized as lossless and lossy compression schemes because of the low memory and processing capabilities. Aquanet project [15] developed for monitoring water resources in Chania city in Greece, is example of lossless compression on WSNs, where a sensor nodes were deployed to measure the water level at a number of water tanks. The water level within a tank is represented through a binary encoding string through applying statistical compression models such as Huffman Encoding, GolombRice and Exponential Golomb-Rice codes. Although, the results of some coding models (such as Exponential Golomb-Rice) was not achieving a high compression ratio, they may be ideal for the minimum memory requirements and to their simple implementations.

Compressed sensing (CS): It is an emerging field based on single-signal sensing and compression. The main idea is that linear projections of a sparse signal contains enough details for information recovery [16]. The application of using the CS approach in [17] eliminates the need for digital signal processing (DSP) through integrating it with an electrocardiogram (ECG) sensor and it achieved high accuracy and low energy consumption. The main advantage of the integrated compressed sensing analog-domain front-end (CS-AFE) with the sensor board, is the reduction of hardware complexity through the usage of pipelined multiplication scheme.

One of the important bridges between the in-network wireless sensor compression and the Big Data is the compression approaches of multiple data types. Recently, a sensor node is often equipped with various sensors to observe different environmental phenomena. The distributed data compression approaches [18] was tested using multiple data types. However, the performances were slightly worse for compressing multiple data types. On the other hand, local compression approaches were tested using multiple data types within the approach proposed by Gana et al. [19] which showed a better performance than the schemes using the distributed approaches. However, as mentioned early, the local algorithms exploited only temporal correlation in WSNs, which means that the approach performs compression locally on each sensor with no distributed collaboration among other sensors and regardless the WSN topology. Hence, such algorithms are not suitable for dense WSNs.

3.1 Compression Techniques Comparison and Discussion

Table 2 shows a summary of a comparison from which few remarks can be noticed. Firstly, the Distributed Source Modeling is a suitable compression technique for the distributed WSNs and its applications are promising for the Big Data analytics. Secondly, although the Distributed Transform Coding technique could be a suitable for

Table 2 Comparison between different data compression techniques

Reference	Application	Data compresssion rate	Energy saving	Network topology	Type of data compression
[9]	Undef	Hadoop data compression mechanism	High	Distributed	DSM
[10]	Multimedia applications	Undef	Fair	Sparse	DTC, local and lossy compression
[15]	Water resources montoring	Wyner-Ziv rate [20]	High	Sparse	DSC, local, lossless, and lossy compression
[17]	Bio-signal acquisition system	Competitive compression ratio over a wavelet-based compression	High	Sparse	CS and lossy compression

large data compression such as videos, and images, the energy saving is still a matter to be considered. Finally, the Distributed Source Coding and Compressed Sensing techniques could achieve competitive compression rates for sparse networks.

4 Data Reduction over WSNs

One of the characteristics of the Big Data is the high and the large sample size. The former brings spurious correlations between unrelated covariates and responses which could lead to wrong scientific conclusions, noise accumulation, and incidental homogeneity whereas the latter combined with the high dimensionality could cause heavy computational cost and algorithmic instability [21]. Data reduction is a data pre-processing technique which can increase storage capacity, thus reducing costs. The main aim of the data reduction techniques is to remove unnecessary data. This is a very useful way for WSNs as it leads to saving energy of these networks through minimizing the required communication cost [22]. As the WSNs is a source of Big Data, so it is very important to address the data reduction problem which would help in two main points: minimizing the energy consumption and increasing the overall network lifetime as well as supporting the efficiency of the storage capacity at the big data side.

The main idea of the data reduction techniques in WSNs is that the sensor nodes are forced to stop data transmission when they are confident about the possibility of using the existing, past and proximity observations to regenerate the future sensed data at the sensor/sink. In this way, the used energy resources for the data transmission can be conserved [23]. The data reduction techniques for WSNs could be classified as Filter based, Cluster Head based, and Data Prediction Based Reduction.

Filter-based Data Reduction Techniques Such techniques depend on the idea of transmitting data with some prediction values which are compared with data counter values. If mismatching is happened, then the sink node sends an alert message and then both ends switched to the initialization mode [22, 24]. Utilizing this concept, Stojkoska et al. [25], have proposed a data reduction technique using a variable step size LMS algorithm. Also McCorrie et al. [26] proposed data reduction in WSNs by using selective filtering for the engine monitoring problem: exhaust gas temperature (EGT) of gas turbine engine. They did this by addressing the problem through suggesting a solution to the accurate state predictions using a skewed double exponentially weighted moving average filter. Low transmission rates are achieved even when significant temperature step changes occur.

Cluster Head-based Data Reduction Techniques Misbahuddin et al. [27] suggest a data reduction algorithm which applied to clustered WSNs and allowing sensor nodes and cluster heads to preserve history of transmitted data and received data samples, respectively. This enables the developed algorithm to reduce payload size, thus energy consumption that is very important for the lifetime Big Data source, i.e. WSNs. The approach in [28] is another example using the clustering. This approach utilizes the idea of the similarity among cluster sensor nodes. In each cluster, the cluster head reports its data. When a monitoring query is requested by a user, the answer to this query can be returned as soon as the cluster head has the required information, thus the reaction time can significantly be saved. This is because few sensor nodes are participate in the data communication, which will lead to data reduction.

Data Prediction Based Reduction The main idea behind the data prediction is the possibility of predicting some value using some techniques and then choosing the data accordingly. Generally, there are three types of data prediction which are used to achieve data reduction [29]: Stochastic approach, Time-Series forecasting, and algorithmic approach. Singh et al. [30] proposed a data reduction method applying prediction mechanism to support prolonging the WSNs lifetime through developing an energy efficient routing scheme. Based on simulation experiments, the author reported that the proposed method can increasing the overall lifetime of WSNs. Ashouri et al. [29] also introduced an approach for data reduction within the WSNs. It is based on the fact that monitoring environmental conditions by WSN system shows a stability in continuous data sequences whereas the changes in physical phenomena are relatively slow comparing with a constant and predictable rate. This is noticeable within a time interval for few hours. Utilizing this fact, Ashouri et al. proposed an efficient data collection approach integrating data prediction with data-aware clustering mechanisms which achieved a good data communication reduction and low computation cost, thus support long lifetime for WSNs.

4.1 Reduction Techniques Comparison and Discussion

Table 3 summarize a comparison between the technique given above. From this table, the following remarks can be noticed. Firstly, the cluster-based topology is the most suitable network topology for the data reduction techniques. The cluster head-based data reduction technique achieves the highest data reduction rate. Thirdly, the higher data reduction rate is achieved, the more energy can be saved, thus prolonging the WSNs lifetime. Fourthly, the Filter-based data reduction techniques, namely the work in [26], has achieved the highest reduction, but the network topology was overlooked. This solution could be further evaluated under different topologies and then could be adopted to reduce the data transmission in-network which will save the WSN's resources (energy, memory, etc.). In conclusion, the data reduction techniques for WSNs data cannot only lead to saving time and space while

Table 3 Comparison between different data reduction techniques

Reference	Application	Data reduction rate	Energy saving	Network topology	Type of data reduction
[25]	Temperature measurements	95 %	High	Star network	Filter-based data reduction techniques
[26]	Aircraft (gas turbine engine exhaust gas temperature (EGT) monitoring application)	95, 99.8, and 91 % in the take-off, cruise, and landing phases, respectively	High	Undef	Filter-based data reduction techniques
[25]	Temperature measurements	97 %	High	Cluster-based topology	Cluster head-based
[27]	Undef	80 % (when payload is 64 bit)	Good	Cluster-based topology	Cluster head-based
[28]	Real sound and light data sets collected	60 %	Fair	Tree network	Cluster head-based
[29]	Simulations (temperature data)	70 %	Good	Cluster-based topology	Data prediction based reduction
[30]	Simulations	Undef	One method improvise WSN lifetime by a factor of 1.65 and the other by 2.1	Cluster-based topology	Data prediction based reduction

applying the Big Data processing techniques but also improve the performance of the WSNs themselves, i.e., saving communication resources, prolong its lifetime. Also this could lead to the reduction in the Big Data side and the assurance of up-to-dated data delivery.

5 Conclusions

This article has discussed the relation between the WSNs and the Big Data. It showed that the WSNs are considered one of the biggest sources of the Big Data in many applications. As a source of generating Big Data, the article discussed that it is important to apply pre-processing techniques (data compression and reduction) within this source to maximize the benefit of both the WSNs and the Big Data. Then it showed that one of the current challenge related to the Big Data is how to save relevant data and discard noisy and redundant data. Hence, the article presented a number of existed solutions for the data compression and reduction techniques to be applied in-network such that significantly reduces the number of required data transmissions, thus not only improving the lifetime of the WSNs but also improve the level of refinement at the Big Data side. The article then conducted a comparison between these techniques and it concluded that there are some techniques would be adopted for some Big Data application and other need a modification to enhance the functionality of the WSNs while minimizing any further pre-processing at the Big Data side.

Acknowledgments This paper has been elaborated in the framework of the project New creative teams in priorities of scientific research, reg. no. CZ.1.07/2.3.00/30.0055, supported by Operational Programme Education for Competitiveness and co-financed by the European Social Fund and the state budget of the Czech Republic and supported by the IT4Innovations Centre of Excellence project (CZ.1.05/1.1.00/02.0070), funded by the European Regional Development Fund and the national budget of the Czech Republic via the Research and Development for Innovations Operational Programme, and by Project SP2015/146 "Parallel processing of Big data 2" of the Student Grand System, VSB-Technical University of Ostrava.

References

1. Prasad, P.: Recent trend in wireless sensor network and its applications: a survey. Sens. Rev. **35**(2), 229–236 (2015)
2. Fouad, M.M., Aboul, E.H.: Key pre-distribution techniques for WSN security services. In: Bio-inspiring Cyber Security and Cloud Services: Trends and Innovations, pp. 265–283. Springer, Berlin (2014)
3. Tardioli, D., Sicignano, D.: A wireless multi-hop protocol for real-time applications. Comput. Commun. **55**, 4–21 (2015)
4. Hammoudeh, M., Robert, N.: Adaptive routing in wireless sensor networks: QoS optimisation for enhanced application performance. Inf. Fusion **22**, 3–15 (2015)
5. Mayer-Schonberger, V., Kenneth, C.: Big data: a revolution that will transform how we live, work, and think. Houghton Mifflin Harcourt (2013)

6. Yin, S., Kaynak, O.: Big data for modern industry: challenges and trends. Proc. IEEE (2015)
7. Razzaque, M.A., Chris, B., Simon, D.: Compression in wireless sensor networks: a survey and comparative evaluation. ACM Trans. Sens. Netw. (TOSN) **10**(1), 5 (2013)
8. Srisooksai, T., Keamarungsi, K., Lamsrichan, P., Araki, K.: Practical data compression in wireless sensor networks: a survey. J. Netw. Comput. Appl. **35**, 37–59 (2012)
9. Sartipi, M.: On the rate-distortion performance of compressive sensing in wireless sensor networks. In: International Conference on, IEEE Computing, Networking and Communications (ICNC), pp. 168–172 (2013)
10. Zhang, P., Zheng, Y., Hamlin, S.: A novel architecture based on cloud computing for wireless sensor network. In: Proceedings of the 2nd International Conference on Computer Science and Electronics Engineering. Atlantis Press (2013)
11. Gaeta, M., Loia, V., Tomasiello, S.: Multisignal 1-D compression by F-transform for wireless sensor networks applications. Appl. Soft Comput. **30**, 329–340 (2015)
12. Hurtik, P., Perfilieva, I.: Advances in intelligent. Syst. Res. **32**(2013), 521–526 (2013)
13. Alhilal, M.S., Adel, S., Abdullah, Al.-D.: Image-based object identification for efficient event-driven sensing in wireless multimedia sensor networks. Int. J. Distrib. Sens. Netw. (2015)
14. Qaisar, S., Rana, M.B., Wafa, I., Muqaddas, N., Sungyoung, L.: Compressive sensing: from theory to applications, a survey. J. Commun. Netw. **15**(5), 443–456 (2013)
15. Duarte, M.F., Shriram, S., Dror, B., Michael, B.W., Richard, G.B.: Distributed compressed sensing of jointly sparse signals. In: Asilomar Conference on Signals, Systems and Computers, pp. 1537–1541 (2005)
16. Gangopadhyay, D., Emily, G.A., Anna, M.R.D., Karthik, N., Subhanshu, G., David, J.A.: Compressed sensing analog front-end for bio-sensor applications. IEEE J. Solid-State Circuits **49**(2), 426–438 (2014)
17. Thanh, D., Nirupama, B., Wu-chi, Feng.: Robust data compression for irregular wireless sensor networks using logical mapping, ISRN Sens. Netw. Vol. (2013)
18. Gana, J., Li-Minn Ang, K., Seng, K.P.: Performance comparison of data compression algorithms for environmental monitoring wireless sensor networks. Int. J. Comput. Appl. Technol. **46**(1), 65–75 (2013)
19. Holzinger, A.: Biomedical Informatics: Discovering Knowledge in Big Data, 1st edn. Springer Publishing Company, Incorporated (2014)
20. Burdakis, S., Antonios, I., Antonios, D.: Compressed data acquisition from water tanks. In: Proceedings of the 1st ACM International Workshop on Cyber-Physical Systems for Smart Water Networks, p. 2. ACM (2015)
21. Misbahuddin, S., Mahjabeen, T., Samia, S.: An efficient lossless data reduction algorithm for cluster based wireless sensor network. In: International Conference on Collaboration Technologies and Systems (CTS), pp. 287–290. IEEE (2014)
22. McCorrie, D.J., Elena, G., Keith, B., Nigel, P., Roger, H.: Predictive Data Reduction in Wireless Sensor Networks Using Selective Filtering for Engine Monitoring. Wireless Sensor and Mobile Ad-Hoc Networks, pp. 129–148. Springer, New York (2015)
23. Anjan, D.: An enhanced data reduction mechanism to gather data for mining sensor association rules. In: 2nd National Conference on Emerging Trends and Applications in Computer Science (NCETACS), pp. 1–4, IEEE (2011)
24. Bayer, I.K., and Surendar, S.: Least square approximation technique for energy conservation in wireless sensor networks. In: International Conference on Control, Instrumentation, Communication and Computational Technologies (ICCICCT). IEEE (2014)
25. Wyner, A.D., Ziv, J.: The rate-distortion function for source coding with side information at the decoder. IEEE Trans. Inf. Theory **22**(1), 1–10 (1976)
26. Yang, Z., Ren, K., Liu, C.: Efficient data collection with spatial clustering in time constraint WSN applications. Pervasive Computing and the Networked World. Springer, Berlin (2013)
27. Stojkoska, B., Mahoski, K.: Comparison of Different Data Prediction Methods for Wireless Sensor Networks. CIIT, Bitola (2013)
28. Singh, D.P., Vikrant, B., Surender, K.S.: Prolonging the lifetime of wireless sensor networks using prediction based data reduction scheme. In: 2014 International Conference on Signal Processing and Integrated Networks (SPIN). IEEE (2014)

29. Stojkoska, B., Dimitar, S., Danco, D.: Data prediction in WSN using variable step size LMS algorithm. In: SENSORCOMM 2011, The Fifth International Conference on Sensor Technologies and Applications (2011)
30. Ashouri, M., et al.: PDC: Prediction-based data-aware clustering in wireless sensor networks. J. Parallel Distrib. Comput. (2015)

An Adaptive PSO-Based Sink Node Localization Approach for Wireless Sensor Networks

Mohamed Mostafa Fouad, Vaclav Snasel and Aboul Ella Hassanien

Abstract Localization in Wireless Sensor Networks (WSNs) is the process of estimating sensors' positions. Localization have been used for identifying: geographic routing, network coverage, network connectivity, object tracking, and other WSNs-based techniques. Many protocols have been proposed as solutions for the localization problem. However, still some protocols need more efforts such as; the Topology Control protocols (TC). These protocols seeking to reduce the number of active nodes/links without any prior knowledge about the nodes' physical positions. Therefore, this paper proposes an optimized approach to determine the best location for a sink node within a topology control protocol using the Adaptive Particle Swarm Optimization (APSO). The evaluation matrix of the proposed approach is based on the number of active nodes, the time intervals for constructing topologies, and the operational lifetime of the network. The proposed approach was tested against the performance of the standard PSO (SPSO). The simulation results show that the APSO technique surpasses the SPSO and it provides the best lifetime for a substantially longer network's operation time.

Keywords Adaptive particle swarm optmization · Topology control protocol · Localization · Sink node · Optimization · Wireless sensor networks

M.M. Fouad (✉)
Arab Academy for Science, Technology, and Maritime Transport, Cairo, Egypt
e-mail: mohamed_mostafa@aast.edu
URL: http://www.egyptscience.net

M.M. Fouad · V. Snasel
IT4Innovations, VŠB-Technical University of Ostrava, Ostrava, Czech Republic

V. Snasel
Faculty of Electrical Engineering and Computer Science, VŠB-TUO, Ostrava, Czech Republic

A.E. Hassanien
Faculty of Computers and Information, Cairo University, Cairo, Egypt

M.M. Fouad · A.E. Hassanien
Scientific Research Group in Egypt (SRGE), Cairo, Egypt

A. Abraham et al. (eds.), *Proceedings of the Second International Afro-European Conference for Industrial Advancement AECIA 2015*, Advances in Intelligent Systems and Computing 427, DOI 10.1007/978-3-319-29504-6_62

1 Introduction

Wireless Sensor Networks have been ubiquitous and are being in several domains, as for examples: environmental monitoring [1], structure health monitoring [2], healthcare monitoring [3], and surveillance [4]. These WSNs consists of a number of low-cost wireless sensor devices that have the ability to sense their environments and to collaborate locally to deliver all acquired data to a centralized and powerful node; that tagged as a base-station (a sink node). Although there are recent powerful sensors that come over the regular constrains attached with early sensor nodes such limitations in memory, power supply, and processing capabilities, still finding the suitable nodes' locations and establishing an efficient routing between them is a rich area for researchers.

Usually, there are three deployment strategies for sensor nodes; the predefined (planned), the random (unplanned), and the hybrid (a switch between the predefined and the random strategies) deployments. Despite the fact that the network structure and the routing performance of the predefined deployment is guarantee, the realization and the lifetime of the other two strategies are unpredictable. Since their performance rely on the efficiency of the in-routing establishment process. Normally, this process is based on the self-configuration behavior which performed between the network's nodes.

The communication Topology control (TC) approaches are of the important protocols for designing sensor networks. These protocols selecting a minimum subset of nodes for routing coverage within a wireless sensor network. The reduction of the fully connected network into a subset of nodes provides a number of advantages, such cutbacks in: communication overhead, number of exchanged messages, collision, and energy consumption. Figure 1 illustrates the reduced topology of an example of a fully connected mesh network.

Generally, designing a topology control protocol in WSNs relies on three complementary protocols: Neighbor-discovery Protocol, Topology Construction and Topology Maintenance. Within the neighbor-discovery (the initialization), each node uses its maximum transmission power to discover its neighbors. Due to its design objectives to reduce the collision and energy consumption ratio, the topology construction

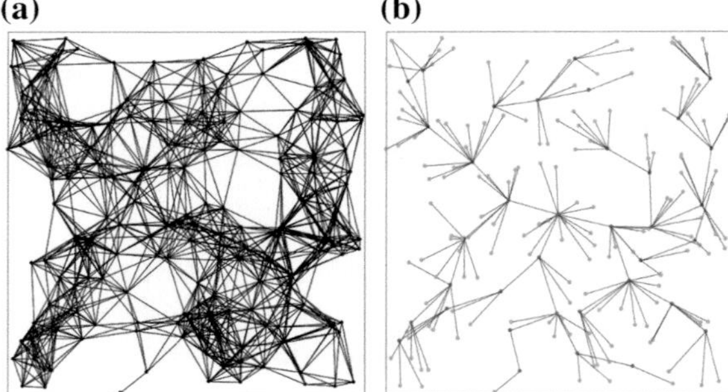

Fig. 1 Topology reduction output. **a** Fully connnected mesh topology. **b** Reduced topology

protocols activating a subset of nodes those will improve the network connectivity and coverage. In a further step and based on the necessity to extend the lifetime of a wireless sensor network topology maintenance protocols are iteratively applied.

However, the TC protocols enhance the performance and the stability of the wireless senor network, a number of them worked with the assumption that the sink node is centralized within the deployment area, such as A3 [5], A3Cov [7]. Finding the best location of the sink node in a random deployment manner could add more improvement to the WSNs; e.g. for the scenario of dropping nodes away from the center of deployment area, the nodes will increase their radio transmission signal to broadcast to the in-center sink node. This scenario increases the transmission power consumption per node, thus as a consequence it reduces their lifetime. Accordingly, this article adapt the A3 topology control protocol to optimize the sink node position within the deployment area.

The remainder of this article is organized as follows: a brief background of main concepts is introduced in Sect. 2. The proposed optimization approach is described in Sect. 3. Section 4 shows performance evaluations of the proposed approach, and finally Sect. 5 highlights the main concludes of the paper and future directions.

2 Background

2.1 Topology Control Protocols

All of the topology control protocol are based on neighbor's discovery process. At most of them the nodes know their own positions and through an adapted transmission range they detect their neighbors. Some of the techniques use GPS and antennas to determine the location and the orientation of the nodes in the deployment area [8]. However, adding an energy consuming chip like the GPS to the limited resources sensors could not be considered as an efficient strategy. Another strategy is to reduce the transmission power of sensors to remove unnecessary interconnected links [6]. Although this reduction of the number of links extends the network's lifetime, still there is no need to leave all nodes active and operating simultaneously within the network. In order to remove unnecessary nodes, the work of [7] gives a suggestion to turn off redundant nodes for a future use.

Selecting a subset of nodes to build a reduces topology, is done through establishing a Connected Dominating Set (CDS) of active nodes those providing complete communication coverage [9, 10]. The A3 (a tree) algorithm [5] provide a solution for establishing a minimum CDS through based on a weighted distance-energy-based metric. However, this may not be enough to establish a close to optimal CDS since it works with a restriction parameter. The assumption of the centrality of the sink node is intended parameter.

2.2 Node Localization

Localization has been used in a number of WSNs applications, but it is still a problem to determining the best locations (positions) of the wireless sensor nodes. Although, localization is critical for WSN-based dynamic systems such as AGVs (automated guided vehicles), and robot navigation [11], it is even more critical for static wireless sensor networks. Usually, static wireless sensors in many applications are deployed randomly without any prior knowledge about their locations. In these situations the nodes' location information and their current neighbors are important to establish the wireless network and to start the intended tasks [12]. Despite the fact that node localization problem has been addressed using many population based optimization approaches, still the Particle Swarm Optimization approach (PSO) is most feasible solution for such problem due to its low computation requirements and its rapid convergence. For example, Dan, and Xian [13] propose a distributed localization algorithm based on two-phase PSO algorithm to solve the flip ambiguity phenomena. They improve the efficiency and the precision that their algorithm reduces the initial search space through using a bounding box method. Moreover, the PSO could be applied to assist other approaches towards node localization in WSNs such as the work of [14] where the authors purpose a particle swarm optimization assisted Extended Kalman Filter (PSO-EKF) for Localization in WSNs. The main observed result that the accuracy of node localization by using PSO-EKF was higher than the application of the EKF alone.

2.3 Adaptive Particle Swarm Optimization (APSO)

The standard PSO (SPSO) applying a collection of particles those trying to change their positions (current local solutions) towards a single promising position (global best solution). The movements are based on a change in their velocity throughout the search space. But, the particles may become inactive due to their trapping in local optimum or mimic flying but within quite a small space by velocity close to zero. Therefore, to overcome this problem, the APSO provides a replacement procedure for the inactive particles to maintain the social behavior of the swarm.

Therefore, the application of the APSO is initiated as the SPSO optimization technique, where a there is a population of random solutions. Each particle (solution) denoted by p_i has a position pos_i in a D-dimensional search space. This position is dynamically changed through a calculated velocity v_i according to both; the particle and the population previous experiences, best local particle (pBest), best global particle (gBest), respectively.

The Adaptive PSO only different by its Eq. 1 which check the satisfactory of the particle to be replaced by a new one, where F_i is the relative error function, and ε is a predefined critical constant (a very small number, very close to zero). The relative error function investigate the fitness of particle i to the fitness of the global best particle; as it is illustrated in Eq. 2 [15].

$$|\Delta F_i| < \varepsilon \tag{1}$$

$$|\Delta F_i| = f(F_i, F_g Best) \tag{2}$$

Algorithm 1 rely on these equations for the application of the APSO optimization technique toward solving the sink node localization problem within WSNs.

3 The Proposed Optimization Approach

It is possible to view the node localization in WSNs as an optimization problem that locates the optimal position of each node. The objective of the applied APSO localization algorithm is to help TC protocols to discover the proper location of a sink node that will minimize the number of active nodes as well as maximizing the network's operational lifetime. Even though the static sink node may constrain the operational lifetime of the network, if its neighboring sensors has been exhausted [16], it is still a case to be considered in some applications. Therefore, may be this process seems easy for mobile-based wireless sensor devices which can move smoothly to the best sink node position, with the supervision of a fitness function. However, the self-mobility feature of the stationary deployment wireless nodes is constrained. Therefore, the proposed approach assumes the random deployment of N particles $(P = \{p_1, p_2, \cdots, p_N\})$ within the deployment area. Each p_i is donated through a set of basic features: the particle position $(pos_i(t) = x_i(t), y_i(t))$, its velocity $(v_i(t))$, its best experience (the best local), and the best ever found particle by all particles (the best global). The basic use of particle is to allow them to clone the nearest stationary node (step 6 in Algorithm 1). The cloning mechanism includes: the node's position, number of its neighbors, and its residual energy. The algorithm proceed as our previous presented work in [17] through the SPSO technique until step number 12, where each particle (except the gBest) goes through an evaluation process. This process inspect the relative error ratio between the fitness of particle j to the fitness of the global best particle. According to this inspection, a replacement process is performed to any inactive particles. After T_{max} iterations, the final delivery of the algorithm is the fittest node position that could act as a sink node within the current deployment scenario.

Algorithm 1 APSO-based Sink Node Localization Algorithm

1. Initialize v_i, $pBest_{p_i}$ and the $gBest$ as close to zero as possible per each particles.
2. Initialize $\omega, c_1, and\ c_2$
3-Initialize the max numbers of iteration T_{max}
4-Initialize the fitness function $f(p_i)$ per each particle

5- While($i \neq N$) for each particle do

6- Clone the Nearest Sensor to p_i,
7- Compute the fitness function $f(p_i)$ using Eq. 3

$$f(p_i) = \alpha_1 \left|N(p_i)\right| + \alpha_2 \sum_{p \in N(p_i)} p.e + \alpha_3 d_p \tag{3}$$

Where α_1, α_2, and α_3 are random numbers ranged in $[0, 1]$. While $N(p_i)$ refers to the sensors neighbors for the particle p_i, the $p.e$ refers to the residual energy within a neighbor node $p \in N(p_i)$ and d_p is the Euclidean distance between the position of the particle p and the center of the deployment area.

8- Get the Local Best Particle for each Particle $pBest_{p_i}$
9- Apply Eq. 4 to choose the best global particle $gBest$, in the current iteration, from the list of the detected local best particles $pBest_{p_i}$.

$$gBest = Max\{pBest_k | k \in P\} \tag{4}$$

10- Calculate the velocity $v_i(t+1)$ per each particle with Eq. 5:

$$v_i(t+1) = \omega v_i(t) + c_1 r_1 (pBest_i - v_i(t)) + c_2 r_2 (gBest_j - v_i(t)) \tag{5}$$

While t denotes the iteration counter and v_i represents particle velocity, the ω parameter is a constant inertia-weight that controls velocity of the exploration within the search space. Also, r_1 andr_2 are random numbers in the range $[0, 1]$. Whereas c_1 represents the cognitive coefficient, c_2 represents the social coefficient towards the best solution [18].
11- The particle use the new velocity to shift its position towards the best optimal solution (Eq. 6):

$$pos_i(t+1) = pos_i(t) + v_i(t+1) \tag{6}$$

The APSO Algorithm Replacement Process

12- While($j \neq N$) for each particle do

13- IF ($p_j \neq gBest$ && $\left|F(p_j)\right| < \varepsilon$))

14- THEN $ReplaceParticle(p_j)$ AND Goto line 6

15- ELSE *ContinuewithNextIteration*
16- *SinkID*= Nearest node to the *gBest* of the final iteration

4 Performance Evaluation

The proposed approach was coded and tested using Atarraya Simulator [5, 6]; which is a Java-based simulator for topology control algorithms. In each of the simulation scenarios, the nodes have been assigned randomly. Table 1 shows the simulation parameters adapted for testing scenarios.

The first set of experimental scenarios were employed to evaluate the topology construction performance without any optimization approaches in comparison with the application of both the APSO and the SPSO. The testing scenarios without any optimization will be denotes as "*OrgTC*" as an abbreviation for Original Topology Control protocol. While the application of the SPSO applies the Gaussian Jump [17, 19], as an equivalent to replace inactive particles feature of the APSO, to void trapping in the local minima.

Figure 2 shows that there is a positive correlation between the number of active nodes in the topology construction and the application of the APSO algorithm. That the ratio of active nodes, used to establishing the CDS, reached 9 % with the APSO-based localization compared to 10 % with both the *OrgTC* and the *SPSO*. Further statistical tests revealed that the time required to reduce the fully connected graph into a subset of active nodes has been reduced within a range of [12 % : 16 %] by the proposed approach (as illustrated by Fig. 3).

The previous experimental results had supported the network availability, which was proven through the significant extension appeared within the network's operational lifetime through the application of the dynamic global time-based topology

Table 1 Simulation parameters

Parameter	Value
Distribution area	600 m * 600 m
Sensor nodes distributed	100, 300, 500, 700, 900
Sensor node module	MICA2 Mote
Node communication range	100 m
Node sensing range	20 m
Node distribution	Uniform
Initial energy distribution	Uniform
Max energy	2000 milliampere-hour, (mA-h)
Fitness function probability α_1	0.4
Fitness function probability α_2	0.1
Fitness function probability α_3	0.5
The inertia weight ω	0.8 [20]
The acceleration constants (c_1, c_2)	2 [18]
The critical constants ε	$le - 4$ [15]
Random numbers (r_1, r_2)	0.5

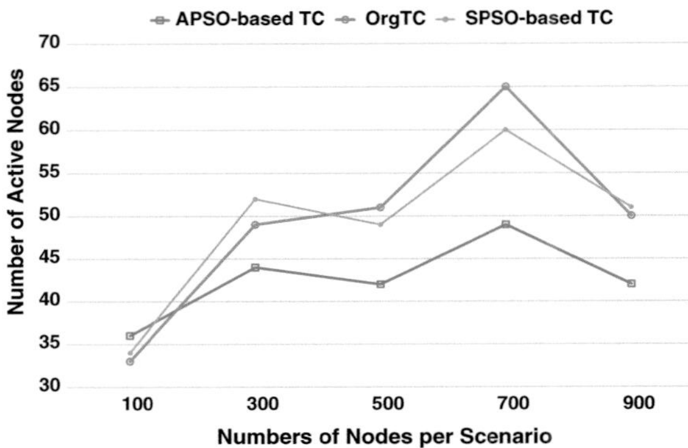

Fig. 2 Number of active nodes in topology construction phase

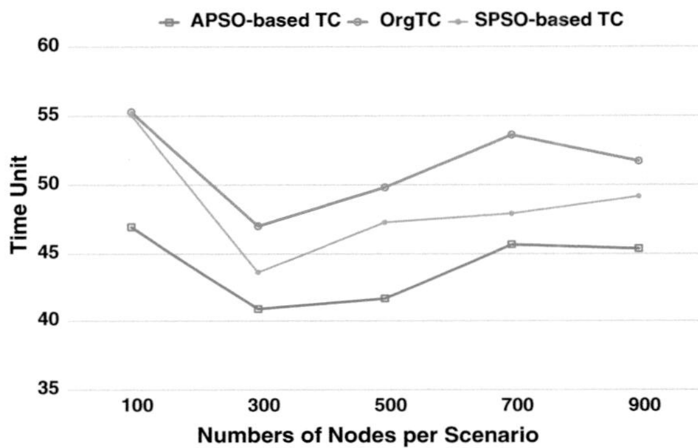

Fig. 3 Initial topology reduction time

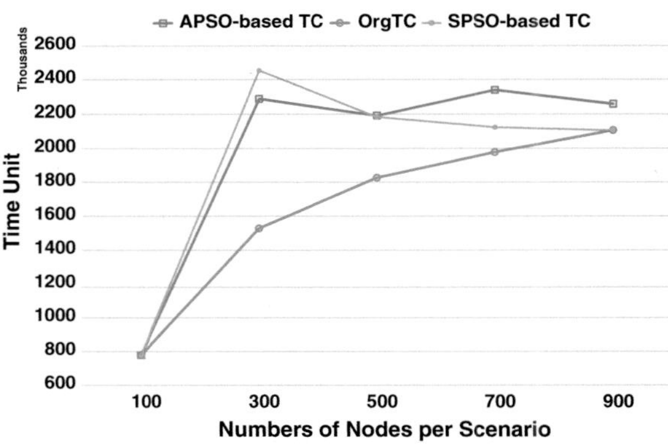

Fig. 4 The network's operational lifetime

recreation (DGTTRec) maintenance protocol. It was used to dynamically reconstruct on-the-fly new topology of active nodes [7]. Figure 4 illustrates the network's lifetime obtained through a number of test scenarios. It shows that the improvement ratio increases with the network size for the APSO-based topology. Another finding, that the SPSO approach may be suitable for low network's size ranged from

100 to 500 nodes with increase lifetime ratio from 20 to 60 % compared with the *OrgTC*. However, the *APSO* showed its scalability with different network's sizes, as for the operational lifetime, the increase ratio is ranged from 7 to 50 % over the *OrgTC* protocol.

5 Conclusions and Future Directions

Enhancing energy efficiency and reducing radio interference are among the most important issues in WSNs. Topology control protocols provide a suitable solution; through the management of the transmission power and the use of the local information for establishing a reduced topology that preserve the connectivity and the coverage characteristics of the sensor network. Although the ability of these protocols to augment the network's lifetime, still there are a number of improvements can be applied to enhance their performance.

Since the Adaptive PSO optimization had provided an improvement over the Standard PSO, the paper proposes the use of this optimization based approach for the sink node localization for WSNs topology control protocols. In order to evaluate the proposed approach, it was benchmarked against both a topology without any optimization and the application of a Standard PSO-based optimization approach. The simulation results showed that in addition to the significant improvement for topology reduction and construction time, the primary contribution of APSO-based sink node localization approach is the maximization of the network's lifetime. As a future work, the authors intends to use other optimization techniques to gain more efficient energy conservation within the WSNs.

Acknowledgments This paper has been elaborated in the framework of the project New creative teams in priorities of scientific research, reg. no. CZ.1.07/2.3.00/30.0055, supported by Operational Programme Education for Competitiveness and co-financed by the European Social Fund and the state budget of the Czech Republic and supported by the IT4Innovations Centre of Excellence project (CZ.1.05/1.1.00/ 02.0070), funded by the European Regional Development Fund and the national budget of the Czech Republic via the Research and Development for Innovations Operational Programme, and by Project SP2015/146 "Parallel processing of Big data 2" of the Student Grand System, VSB - Technical University of Ostrava.

References

1. El-Bendary, N., Fouad, M.M.M., Ramadan, R.A., Banerjee, S., Hassanien, A.E.: Smart environmental monitoring using wireless sensor networks. Wireless Sensor Networks: From Theory to Applications Book, USA (2013). ISBN: 9781466518100
2. Mohamed, A., Fouad, M.M.M., Elhariri, E., El-Bendary, N., Zawbaa, H.M., Tahoun, M., Hassanien, A.E.: RoadMonitor: An intelligent road surface condition monitoring system. Intelligent Systems' 2014, pp. 377–387. Springer International Publishing (2015)

3. Fouad, M.M.M., El-Bendary, N., Ramadan, R.A., Hassanien, A.E.: Wireless sensor networks: a medical perspective. Wireless Sensor Networks: From Theory to Applications Book, CRC Press, USA (2013). ISBN: 9781466518100
4. Irgan, K., Unsalan, C., Baydere, S.: Low-cost prioritization of image blocks in wireless sensor networks for border surveillance. J. Netw. Comput. Appl. **38**, 54–64 (2014)
5. Wightman, P.M., Labrador, M.A.: A3: a topology construction algorithm for wireless sensor networks. In: Global Telecommunications Conference, IEEE GLOBECOM, pp. 1–6 (2008)
6. Labrador, M.A., Wightman, P.M.: Topology Control in Wireless Sensor Networks: with a companion simulation tool for teaching and research. Springer Science & Business Media (2009)
7. Wightman, P.M., Labrador, M.A.: A3Cov: a new topology construction protocol for connected area coverage in WSN. In: Wireless Communications and Networking Conference (WCNC), pp. 522–527. IEEE (2011)
8. Jurdak, R., Sommer, P., Kusy, B., Kottege, N., Crossman, C., McKeown, A., Westcott, D.: Camazotz: multimodal activity-based GPS sampling. In: International Conference on Information Processing in Sensor Networks (IPSN), ACM/IEEE, pp. 67–78. IEEE (2013)
9. Guha, S., Khuller, S.: Approximation algorithms for connected dominating sets. In: Proceedings of European Symposium on Algorithms, pp. 179–193 (1996)
10. Thilagavathi, S., Gnanasambandan Geetha, B.: Energy aware swarm optimization with intercluster search for wireless sensor network. Sci. World J. (2015)
11. Pinto, A.M., Moreira, A.P., Costa, P.G.: A localization method based on map-matching and particle swarm optimization. J. Intell. Robot. Syst. **77**(2), 313–326 (2015)
12. Han, G., Xu, H., Duong, T.Q., Jiang, J., Hara, T.: Localization algorithms of wireless sensor networks: a survey. Telecommun. Syst. **52**(4), 2419–2436 (2013)
13. Li, D., Wen, X.B.: An improved PSO algorithm for distributed localization in wireless sensor networks. Int. J. Distrib. Sens. Netw. (2015)
14. Janapati, R.C., Balaswamy, C., Soundararajan, K.: Enhanced mechanism for localization in Wireless Sensor Networks using PSO assisted Extended Kalman Filter Algorithm (PSO-EKF). In: International Conference on Communication, Information & Computing Technology (ICCICT), 2015, pp. 1–6. IEEE (2015)
15. Xie, X.-F., Zhang, W.-J., Yang, Z.-L.: Adaptive particle swarm optimizationon individual level. In: International Conference on Signal Processing, pp. 1215–1218 (2002)
16. Hu, Yi-Fan, Ding, Yong-Sheng, Ren, Li-Hong, Hao, Kuang-Rong, Han, Hua: An endocrine cooperative particle swarm optimization algorithm for routing recovery problem of wireless sensor networks with multiple mobile sinks. Inf. Sci. **300**, 100–113 (2015)
17. Fouad, M.M., Snasel, V., Hassanien, A.E.: Energy-Aware sink node localization algorithm for wireless sensor networks. Int. J. Distrib. Sens. Netw. 2015, Article ID 810356, (2015). doi:10.1155/2015/810356
18. Shi, Y., Eberhart, R.C.: Empirical study of particle swarm optimization. In: Proceedings of the Congress on Evolutionary Computation (CEC'99), vol. 3. IEEE (1999)
19. Krohling, R.: Gaussian particle swarm with jumps. In: The IEEE Congress on Evolutionary Computation, vol. 2, pp. 1226–1231. IEEE (2005)
20. Charalampakis, A.E., Dimou, C.K.: Identification of BoucWen hysteretic systems using particle swarm optimization. Comput. Struct. **88**(21), 1197–1205 (2010)

Investigating the Impact of Traffic Type on the Performance of on Demand Routing Protocols and Power Consumption in MANET

Uchenna Odih, Panos Bakalis and Predrag Rapajic

Abstract Design of a power-aware protocol in mobile Ad hoc network (MANETs) and its evaluation requires good knowledge of mobile nodes power consumption behavior. Some studies have been reported in the literature to evaluate the performance of MANETs On-demand routing protocols under Transmission control protocol (TCP) and Constant bit rate (CBR) traffic type. However, little work has been done to evaluate this performance in terms of their power consumption behavior under different traffic types. This paper investigates and quantifies the impact of traffic type (TCP and CBR) on the power consumption, network throughput and end to end delay in MANET using OPNET Modeler 17.1 network simulator. Simulation results show that AODV/CBR recorded the lowest power consumption after 900 s of continuous packet transmission as compare to AODV/TCP, DSR/CBR and DSR/TCP. Also, AODV/CBR deliver the optimum performance in all other chosen metrics, thus a better choice for an efficient and power aware routing protocol in MANET.

Keywords MANET · DSR · AODV · TCP · CBR · OPNET

U. Odih (✉) · P. Bakalis · P. Rapajic
Department of Electrical, Electronic and Computer Engineering, School of Engineering, University of Greenwich, Medway Campus, Kent, UK
e-mail: U.A.Odih@greenwich.ac.uk; ou05@gre.ac.uk

P. Bakalis
e-mail: P.Bakalis@gre.ac.uk

P. Rapajic
e-mail: P.Rapajic@gre.ac.uk

A. Abraham et al. (eds.), *Proceedings of the Second International Afro-European Conference for Industrial Advancement AECIA 2015*, Advances in Intelligent Systems and Computing 427, DOI 10.1007/978-3-319-29504-6_63

1 Introduction

In recent time, Mobile Ad Hoc Networks (MANETs) has been receiving increasing attention as the available mobile wireless network architecture capable of supporting the promise of Internet of Things (IOT) technology; where users can send, receive, share and monitor information anywhere, anytime without any fixed or centralized management system [1].

Mobile ad hoc networks (MANETs) represent a network systems that comprise wireless mobile nodes that can freely and dynamically self-organize into temporary network topologies; which allows devices to seamlessly internetwork among each other in a specify area with no pre-existing communication infrastructure [2, 3]. Examples of MANET areas of application are emergency search-and-rescue operations, conference meetings, acquisition of medical data in hospitals, underwater and underground network where infrastructure based network is not possible.

In order to establish communication between mobile nodes in a particular network, a routing protocol is used to discover routes and forward packet from the source node to the destination node. The performance of mobile nodes in MANET depends on the routing protocol employed [4]. It has been stated in [5] that route discovery and packet routing are the major cause of power consumption in MANET.

Therefore, as mobile nodes operating in MANET depends on battery power, efficient utilization of node battery power becomes more important in the design and choice of routing protocol to be employed in any particular mobile ad hoc network setup [4]. The design of routing protocols for mobile ad hoc networks is challenging, thus, requires good knowledge of the mobile nodes power consumption behavior together with other factors affecting its performance [6, 7].

The choice of traffic type (TCP and CBR) incorporated in the routing protocols play an important role in the performance of these routing protocols and the general performance of the MANET network. The scope of this paper focuses on investigating and evaluating the impact of these traffic types on the performance of DSR and AODV routing protocol using power consumption as the major performance metric.

The organization of the paper is as follows. Section 2 presents related work. Section 3 describes MANET on-demand routing protocols considered and Sect. 4 describes the traffic types considered for the comparative analysis in this study. Section 5, presents the system model adopted for the network modelling. In Sect. 6, performance metrics used in this study were discussed. Section 7, presents simulation tools and parameters used in the analysis. Section 8, we present our simulation results and the performance comparison of the two routing protocols. Finally, we conclude in Sect. 9.

2 Related Work

Over the years, several performance evaluations of MANET on-demand and proactive routing protocols have been carried out in the literature focusing on the impact of various factors such as mobility model, network size, traffic type, node speed and pause time on the overall performance of the MANET.

In the literature studied [7–11], authors, examined and presented results on the impact of these factors on both reactive and proactive routing protocols, but focusing only on the following performance metrics; throughput, average routing overhead, end to end delay, packet delivery ratio, traffic sent and traffic received. Also, authors in [12] carried out comparative study of On-demand and Proactive routing protocol, using CBR as traffic type.

In their work, the authors investigate the effect of network size, mobility and number of nodes on the performance of these routing protocols using throughput, good put, routing load and end-to-end delay as their performance metrics.

Again these studies, failed to evaluate or quantified the effects of these factors on the power consumption and life time of the individual nodes; which is the most influential performance metrics to be used when considering the performance of MANET, due to its 100 % reliability on battery power.

However, authors in [13, 14] evaluates the impact of these factors on both reactive and proactive routing protocols, their work is slightly related to this work as they include energy consumption as one of their performance metric. They use ns-2 network simulator and in their energy model, they considered the network to be free from overhearing power, thereby did not take into consideration of the overhearing power in their proposed energy model.

Therefore, evaluating the effects of these factors based on power consumption will help guide the design choice for a more efficient power aware on-demand routing protocols. This work is similar to work done in [15], where the author examine and evaluate the impact of traffic type on the performance of MAC layer transport protocol in wireless multi-hop networks using NS2 simulation tool. In the simulation result presented by the author, it show that CBR performed better than TCP. They attributed this performance as a result of CBR not setting up handshaking process before commencing packet forwarding and its non-congestion technique.

In this work, OPNET network simulator was used and the overhearing power was incorporated to the power model used in this work to fully express the actual power consumed for a successful transmission of packet from source node to destination node.

3 Overview of MANET On-demand Routing Protocols

On demand routing protocol also known as reactive routing protocols obtain routing information on request, by using a connection establishment on-demand process. Mobile nodes employed with reactive routing protocols, do not maintain the network topology information and also they do not exchange routing information periodically [16]. The two main on-demand routing protocol used in MANET are; Dynamic source routing (DSR) and Ad hoc On-demand distance vector (AODV) routing protocol.

3.1 Dynamic Source Routing

Dynamic Source Routing (DSR) protocol presented in [3, 8, 9, 17, 18] is an on-demand routing protocol that is based on the source routing technique, where each source node contain a source route information of other nodes in its packet's header. This source route, lists all the addresses of all the intermediate nodes between the source node and destination node responsible for relaying the packet to the destination node.

When a source node wants to communicate with destination node, it first checks its route cache to see if there is available route to destination. If there is no available route information in its route cache, then the source node will initiate a route request process by broadcasting a route request message 'RREQ' to all the intermediate nodes. If the route request is successful, the source node receives a route reply packet listing a sequence of hops through which it may reach to the destination. Also DSR, consists of two major mechanisms: route discovery and route maintenance.

3.2 Ad Hoc On-demand Distance Vector Routing Protocol

Ad Hoc on-demand Distance Vector protocol (AODV) [8, 12, 18, 17] is a combination of both DSR and Destination sequence distant vector routing protocol (DSDV). It uses the basic principle of on-demand mechanism of Route Discovery and Route Maintenance from DSR, in addition to the use of hop-by-hop routing, sequence numbers, route table and periodic beacons from DSDV.

In AODV, route request is initiated whenever a source node wishes to send packet to another node (destination node), if there is no routing information from source to the destination node found in its routing table. The source node thereby initiate route request by broadcasting a route request control message "RREQ" to all its intermediate nodes, demanding route to the destination. If any intermediate node, knows the route to the destination, it replies with a route reply control message "RREP" that propagates through the reverse route to the source node.

4 Traffic Models

In MANET, different traffic models are used for packet routing. In this work two major traffic models are considered for the analysis.

4.1 Constant Bit Rate (CBR)

CBR is non-connection oriented traffic model that sends traffic at a constant bit rate. In CBR, establishment phase of connection between source node and destination node is not required. It is a one way direction, thus, it is tailored for any type of data for which the end systems require predictable response time and a static amount of bandwidth to be continuously available for the life-time of the connection. Examples of these applications include; video conferencing and voice services [12].

4.2 Transmission Control Protocol (TCP)

TCP is a connection oriented and reliable transport protocol used in both fixed and mobile wireless network. TCP uses acknowledgement to ensure reliable data transfer between source and destination. If a source fails to receive acknowledgement from the receiver during a certain period of time, the packet is assumed lost and then TCP is required to retransmit the packet again [12].

5 System Propagation Model

In this work, free space propagation model used in [13] was adopted. The model was used to calculate the received power by the destination node in wireless mobile network. The free space model assumes that in signal propagation between source and destination nodes, there is clear line of slight (LOS) path between the source and destination node. Equation (1) calculates the received power [13].

$$\mathbf{Prx} = \frac{PtxGtxGrx\lambda^2}{(\mathbf{4}\pi)^2 d^2} \tag{1}$$

where Ptx is the power transmission (in watts), Gtx and Grx are the antenna gains of the source node and destination node respectively, λ is the wave length and d is the distance between the source and the destination node.

6 Performance Metrics

In order to evaluate the impact CBR and TCP has on power consumption in MANET On-demand routing protocols, the following performance metrics were selected for the evaluation and analysis.

6.1 *Total Transmission Power Consumed (POWTx)*

This is the actual node power consumed for a successful transmission of packet from source node to destination node. The total power consumption is highly dependent on the receiving process. The receiving process includes packet reception and overhearing. The overhearing power depends highly on the radio range, as all nodes within this radio share the same radio channel.

In this work, overhearing power is included in the power model, to represent the total actual power consumed during packet transmission. In wireless communication, overhearing power is indirect power consumption by nodes that are within same communication range. Therefore, to calculate the total power consumed during packet transmission, overhearing power should be taking into consideration.

In this work, the overhearing power is calculated as

$$\mathbf{HC} * (\mathbf{Prx}) \tag{2}$$

where *HC*, and *Prx* are number of intermediate nodes between source node and destination node and overhearing node reception power respectively. Therefore, the total power consumed for any successful packet transmission is calculated as;

$$\mathbf{POWTx} = \mathbf{Ptx} * \mathbf{Rtxn} + (\mathbf{HC} * (\mathbf{Prx})) \tag{3}$$

where *Ptx*, *Rtnx*, (*HC***Prx*) are transmission power, No of retransmission attempts, overhearing power respectively.

6.1.1 Network Throughput

This is the ratio of the total number of successful packets received (in bits/sec) at destination node. It can be calculated as

$$Throughput = (Pkt/Ttx) \tag{4}$$

where *Pkt*, and *Ttx* is the number of successful packet received by destination node and *Ttx* is the total time spent for the whole routing process respectively.

6.1.2 End to End Delay (ETE)

This is the maximum time (in seconds) taken for a successful transmission of a packet in bits from source node to destination node.

The end to end delay can be calculated as;

$$ETE = Ttx - Ttr \tag{5}$$

where Ttx and Ttr are time at which first packet was successfully transmitted by source and time at which first packet successfully arrived at the destination node respectively.

7 Simulation Parameters

Simulations of all the scenarios were performed using OPNET Modeler 17.1 for 900 simulated seconds in a network topology of 1000 m × 1000 m; with 50 mobile nodes constantly in motion with a uniform speed. The nodes are spread randomly over the network. The random waypoint model was used as the mobility model for the simulation. Random CBR and TCP traffic were generated using OPNET application profile from the OPNET object pallet.

In this network model, each node starts its journey from a random location to a random destination with a randomly chosen speed (uniformly distributed between 0–10 m/s), at zero seconds pause time, which reflect the real life mobile ad hoc network scenario. The summary of the simulation parameters are shown in Table 1.

Table 1 Simulation parameter

Parameter	Value
Simulation software	OPNET 17.1
Network topology	1000 × 1000 m
Routing protocol	AODV, DSR
Simulation time	900 s
Number of nodes	50
Traffic type model	TCP, CBR
Pause time	0 s
Max node speed	10 m/s
Mobility model	Random waypoint
Mac layer	IEEE 802.11g
Transmission power	0.1 W
Propagation model	Free space model

8 Simulation Results and Performance Comparison

In this section, results to compare and evaluate the effect of CBR over TCP on the general performance of MANET On-demand routing protocol with respect to power consumption, network throughput and end-to end delay were presented.

8.1 Result 1a: Average Throughput Versus Simulation Time for AODV and DSR Under TCP Traffic Type

In this section, the performance of AODV, and DSR under TCP traffic type is been evaluated. Figure 1a as shown, present the response of Network throughput against simulation time for AODV/DSR under TCP traffic scenarios. The result shows that AODV out performed DSR under TCP traffic type because of it low overhead routing information carried by AODV enabled nodes. TCP setting up handshaking process before commencing packet forwarding and its congestion control technique also contributed immensely in the performance gained by AODV/TCP, as congested route is been avoided.

With TCP ability to provide reliable connection-oriented packet delivery, AODV/TCP is suitable for real time applications that required reliable communication. The DSR poor performance is attributed as a result of high processing time spent processing heavy overhead routing information associated with DSR enabled node, the frequent change in network topology and continuous learning of new route as a result of network partition caused by link break.

8.2 Result 1b: Average Throughput Versus Simulation Time for AODV and DSR Under CBR Traffic Type

In this section, simulation result shows that AODV still out performed DSR in terms of network throughput. This high performance as shown in Fig. 1b is attributed as a combined result of CBR non connection oriented technique in packet forwarding and also, AODV low overheard routing information.

With AODV low overhead routing information, packet routing get more attention as less time is being spent in processing the overhead information as against DSR, where all the nodes in the network carries routing information of all other nodes within the network.

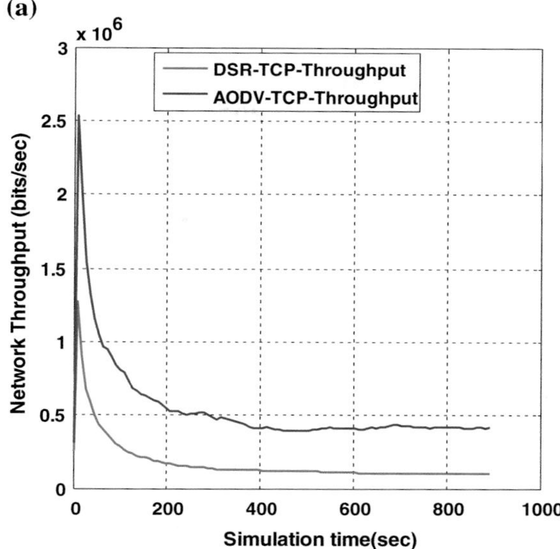

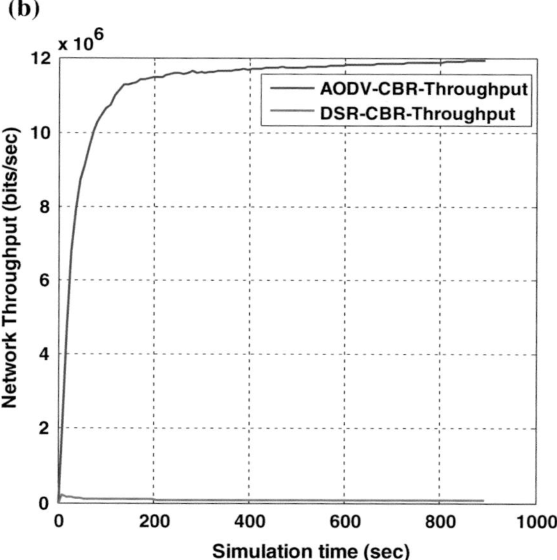

Fig. 1 **a** Network throughput versus simulation time (AODV/DSR under TCP traffic type). **b** Network throughput versus simulation time (AODV/DSR under CBR traffic type)

8.3 Result 2a: Average End-to-End Delay Versus Simulation Time for AODV and DSR Under CBR Traffic Type

Figure 2a as shown; present the average end-to-end delay against simulation time for AODV and DSR under CBR traffic scenarios. The result shows that AODV out

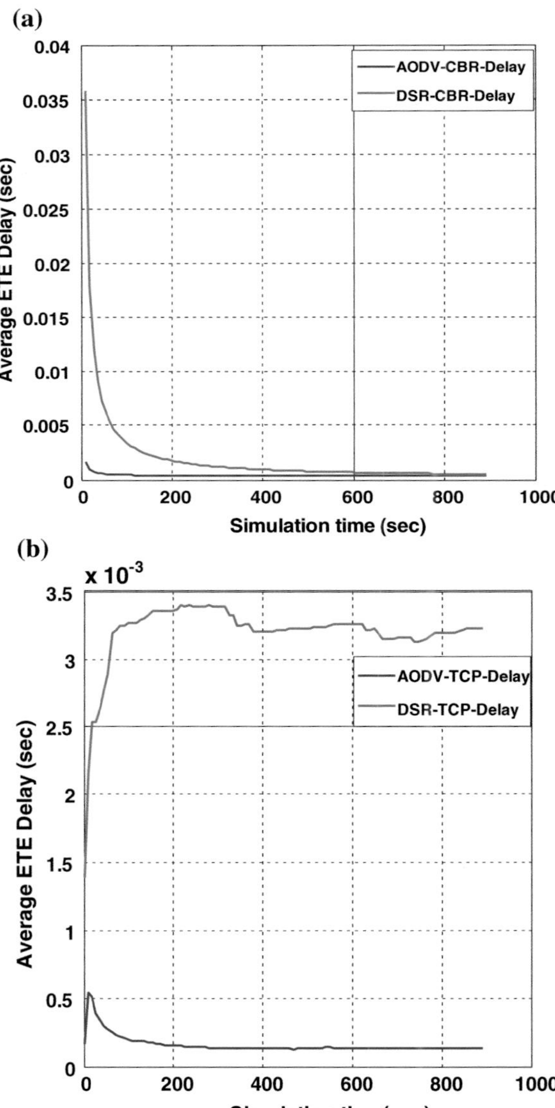

Fig. 2 **a** Average end-to-end delay versus simulation time (AODV/DSR under CBR traffic type). **b** Average end-to-end delay versus simulation time (AODV/DSR under TCP traffic type)

performed DSR under CBR traffic type. This result is attributed to the low overhead routing information been carried by AODV enabled nodes.

The DSR poor performance is attributed as a result of high processing time spent processing heavy overhead routing information being carried by DSR enabled node, and continuous learning of new route caused by link break due the frequent change in network topology.

8.4 *Result 2b: Average End-to-End Delay Versus Simulation Time for AODV and DSR Under TCP Traffic Type*

In this section, simulation result as shown in Fig. 2a shows that AODV out performed DSR. The recorded high end-to-end delay by DSR throughout the simulation is attributed to the heavy overhead routing information being processed by individual node. With TCP connection oriented nature, more time is been spent in packet routing as source node need to establish a connection link with the receiving node before packet can be sent.

Thus, more time been spent by DSR enabled nodes in both processing the overhead information and packet forwarding.

8.5 *Result 3a: Total Power Consumed Versus Simulation Time for AODV and DSR Under TCP Traffic Type*

In this section, the simulation result shows that AODV out performed DSR. At the beginning of the simulation, both AODV and DSR recorded high power consumption, which is as a result of route discovery process. After 10 min of the simulation time, AODV shows better performance than DSR.

This performance is attributed to AODV low overhead routing information. However, DSR high power consumption behavior is due to large overhead routing information and the general TCP transport mechanism, whereby a node that have packet to send, first establish contact with its destination before commencing on the packet forwarding, thus, slightly increasing the level of power consumed by mobile nodes, as some amount of power will be spent in the handshaking process before the actual packet forwarding process.

8.6 Result 3b: Total Power Consumed Versus Simulation Time for AODV and DSR Under CBR Traffic Type

In this section, simulation result presented for the total amount of power consumed for a successful routing of packet by AODV and DSR enabled source node to destination node, under CBR traffic type. The result as shown in Fig. 3b shows that AODV out performed DSR.

However, both routing protocols recorded high power consumption rate under CBR traffic type, which is as a result CBR transport mechanism, whereby a node

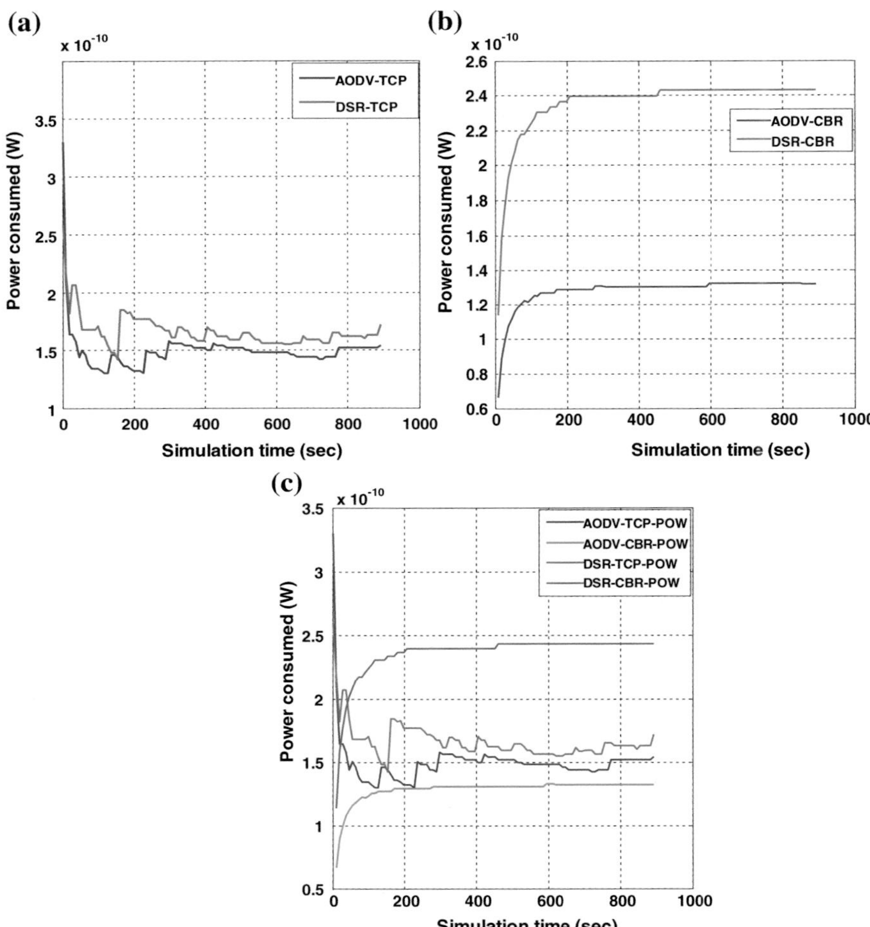

Fig. 3 **a** Total power consumed versus simulation time (AODV/DSR under TCP traffic type). **b** Total power consumed versus simulation time (AODV/DSR under CBR traffic type). **c** Comparison of total power consumed by AODV and DSR under TCP and CBR traffic type

that have packet to send, don't have to first establish contact with its destination before packet forwarding, thus increasing the number of packet lost and number of retransmission attempt required.

As a result of these, DSR/CBR enabled node consumed very high amount of power during packet routing process.

8.7 Result 3c: Comparison of Total Power Consumed by AODV and DSR Under TCP and CBR Traffic Type

With power consumption behavior of the MANET routing protocols the main focus of this work, it's important to bring together the power consumption behavior of the AODV and DSR under TCP and CBR traffic type to clearly show how each protocol perform under different traffic type.

In this section, the simulation result presented clearly shows that AODV under CBR had the best performance in terms of the total power spent for successful packet transmission from a source node to the destination node. Generally, AODV routing protocol as shown in Fig. 3c is said to be the best performing routing protocol to be use in a small network as simulated in this work.

Furthermore, the massive gain achieved in the power consumption behavior, good throughput and very low end-to-end delay demonstrated by AODV under CBR traffic outweighed that of AODV under TCP traffic type. Thus, AODV under CBR traffic type is the best choice, as it will prolong network life time, increased throughput and promote general network performance.

9 Conclusion

Packet Routing in MANET has been identified as the major cause of power consumption. With this in mind, there is need to include power consumption metric when evaluating the performance of these routing protocols under different network conditions.

This paper evaluate and present results of the impact TCP and CBR traffic type has on power consumption of MANET DSR and AODV routing protocols; using power consumed, network throughput and end-to-end delay as performance metrics. Simulated results show that AODV under CBR traffic deliver the optimum performance with low power consumption, high network throughput and low end-to-end delay. This performance is attributed to CBR non congestion and non-self-checking technique.

Also, the results obtained show that power conservation was achieved using AODV under both CBR and TCP, resulting in better network performance, and efficient resource utilization in MANET. Thus, AODV/CBR in this work is the best routing protocol and traffic type for non-real time application that required low latency packet delivery while AODV/TCP is the best routing protocol for connection oriented packet delivery for real time applications.

To conclude, it is very crucial to examine and select suitable routing protocol and traffic type for any proposed MANET setup, in order to perform optimization if needed before the actual implementation in real life environment, as this will help to achieve effective network performance, network resource utilization and general service delivery.

References

1. Conti, M., Giordano, S.: Mobile ad hoc networking: milestones, challenges, and new research directions. IEEE Commun. Mag. **52**(1), 85–96 (2014)
2. Chlamtac, I., Conti, M., Liu, J.: Mobile ad hoc networking: imperatives and challenges. Ad Hoc Netw. **1**(1), 13–64 (2003)
3. Royer, E., Toh, C.-K.: A review of current routing protocols for ad hoc mobile wireless networks. IEEE Pers. Commun. **6**(2), 46–55 (1999)
4. Toh, C.: Maximum battery life routing to support ubiquitous mobile computing in wireless ad hoc networks. IEEE Commun. Mag. **39**(6), 138–147 (2001)
5. Mohsin, A., Abu Bakar, K., Adekiigbe, A., Ghafoor, K.: A Survey of energy-aware routing protocols in mobile ad-hoc networks: trends and challenges. Netw. Protoc. Algorithms **4**(2) 2012
6. Gomez, J., Campbell, A.T., Naghshineh, M., Bisdikian, C.: Conserving transmission power in wireless ad hoc networks. In: Proceedings of Ninth International Conference on Network Protocols. ICNP 2001, 11–14 Nov 2001
7. Saini, T.K., Kumar, S., Dhaka, M.K.: Analysis of routing protocols using UDP traffic under dynamic network topology. In: IEEE International Advance Computing Conference (IACC), 21–22 Feb 2014
8. Broch, J., et al.: A performance comparison of multi-hopwireless ad hoc network routing protocols. In: Mobicom 98 Proceedings of the 4th Annual ACM/IEEE International Computing and Networking. pp. 85–97. ACM, New York (1998)
9. Lawal, B., Bakalis, P.: Performance evaluation of CBR and TCP traffic models on manet using dsr routing protocol. In: CMC'10 Proceedings of 2010 International Conference on Communication and Mobile Computing, vol. 3, pp. 318–322. IEEE Computer Society Washington, DC (2010)
10. Jain, R., Khairnar, N.B., Shrivastava, L.: Comparative study of three mobile ad-hoc network routing protocols under different traffic source. In: 2011 International Conference on Communication Systems and Network Technologies (CSNT), 3–5 June 2011
11. Perkins, D.D., Hughes, H.D., Owen, C.B.: Factors affecting the performance of ad hoc networks. In: IEEE International Conference on Communications, vol. 4, pp. 2048–2052 (2002)
12. Mbarushimana, C., Shahrabi, A.: Comparative study of reactive and proactive routing protocols performance in mobile ad hoc networks. In: 21st International Conference on Advanced Information Networking and Applications Workshops, vol. 2, 21–23 May 2007
13. Rhattoy, A., Zatni, A.: Physical propagation and traffic load impact on the performance of routing protocols and energy consumption in Manet. In: 2012 International Conference on Multimedia Computing and Systems (ICMCS), 10–12 May 2012

14. Cano, J.C., Manzoni, P.A.: Performance comparison of energy consumption for mobile ad hoc networks routing protocols. In: 8th International Symposium on Modeling, Analysis and Simulation of Computer and Telecommunication Systems, 29 Aug–1 Sept 2000
15. Giannoulis, S., et al.: TCP versus UDP performance evaluation for CBR traffic on wireless multihop network. http://www.researchgate.net/publication/228700971 (2015). Accessed 09 Aug 2015
16. Abusalah, L., Khokhar, A., Guizani, M.: A survey of secure mobile Ad Hoc routing protocols. IEEE Commun. Surv. Tutorials **10**(4), 78–93 (2008)
17. Liu, C., Kaiser, J.: A survey of mobile ad hoc network routing protocols. University of Ulm Technical Report Series, Nr. 2003–08, 21 Oct 2005
18. Al-Maashri, A., Ould-Khaoua, M.: Performance analysis of MANET routing protocols in the presence of self-similar traffic. In: 31st IEEE Conference on Local Computer Networks (2013)

Author Index

A. Abraham et al. (eds.), *Proceedings of the Second International Afro-European Conference for Industrial Advancement AECIA 2015*, Advances in Intelligent Systems and Computing 427, DOI 10.1007/978-3-319-29504-6

Z

Printed in the United States
By Bookmasters